Contents

Preface to second edition

The first edition of the **Collins Advanced Science** series represented a breakthrough in comprehensive support and appeal to A-level pupils and teachers that received wide critical acclaim. With the advent of AS and A2 level specifications, the second edition has been fully revised and updated to build on the strengths of the first edition.

The text has been augmented to increase the breadth of examinations board coverage. Each of the major examination boards' specifications are now represented, including the International Baccalaureate. With that in mind, examination questions from across the boards have been included for student practice at the ends of chapters.

For the teacher

Each chapter in the second edition of **Collins Advanced Science: Human Biology** includes an opening page that identifies the AS and A2 nature of each section in the chapter, relating to the majority of examination specifications. The section headings can be checked for the detail against individual specifications.

To support the textbook, a comprehensive **Teacher Support CD-ROM** is available. This CD-ROM includes answers to the examination questions contained within the text, illustrations for use with OHPs and white boards, further questions for in-class and homework use, together with suggestions for practical exercises.

Collins Advanced Science: Biology and Human Biology Teacher Support CD-ROM
ISBN 000 713939 4

For the student

Maintaining the book's strong support for self study, the design of the second edition updates its visual appeal. The content is written in an accessible style and there are plenty of interesting real-life applications of the subject, frequent self-test questions and 'Remember This' boxes to aid understanding. The Assignments at the end of each chapter give practice and portfolio evidence for Key Skills requirements.

Dedication

For the Boyle family: Lynda, Charlotte, James and Alex. And anyone who loves the subject. *Mike Boyle*
To my dear mother, Anne Senior, for her unfailing love, support and encouragement. With much love. *Kathryn Senior*

Examination questions

The publisher thanks examination boards for their permission to reproduce examination questions. Questions are acknowledged as follows:

AQA	The Assessment and Qualifications Alliance
Edexcel	Edexcel Foundation
IBO	International Baccalaureate Organization
OCR	Oxford Cambridge and RSA Examinations
WJEC	Welsh Joint Education Committee

01

What's in a cell?

Escherichia coli O157

In 1996, meat infected with a bacterium called _Escherichia coli_ O157 (often abbreviated to _E. coli_ O157) was sold and served in Wishaw in Scotland with devastating results. In the food poisoning outbreak that followed, 501 people became ill and 21 of them died. Most of the dead were pensioners who had eaten the contaminated meat at a church lunch.

This tragic incident illustrates the delicate balance that exists between humans and the bacteria that live in and around us. Other strains of _E. coli_ live quite happily in our guts, doing a useful job, but _E. coli_ O157 is more unpredictable. Some people become infected and just get a bad bout of diarrhoea, but an unlucky 5% or so, can develop severe or even fatal illness after ingesting as few as 10 bacterial cells. This strain of _E. coli_ produces toxins that cause cells in the lining of the intestine and in the kidney to self-destruct. The damage is particularly bad in the very old and the very young, which means that pensioners and children are particularly at risk.

The first case of _E. coli_ O157 causing food poisoning wasn't reported until 1980, so where did it come from? No-one can really be sure, but experts think that the genes that code for its deadly toxins were transferred to _E. coli_ from other bacteria, on short loops of DNA called plasmids. These are swapped from one cell to another during conjugation, a sort of prokaryotic sex encounter. But, whatever the mechanism, there is some hopeful news. A class of antibiotic is being developed that may be able to kill all strains of _E. coli_ O157. It is not yet ready to test in people, but if all trials are completed successfully, we could have an effective treatment for _E. coli_ O157 by about 2010.

Fig 1.1 This painting illustrates something of the complexity of an animal cell. It was commissioned by the Science Museum on behalf of the Biochemical Society to provide the centrepiece of a permanent exhibition opened in 1987

1 THE COMPLEXITY OF CELLS

When you first looked down a microscope at human or animal cells, and saw something like Fig 1.9, you may have been disappointed. But appearances are deceptive. These little bags with a dot in the middle are extremely complex, and infinitely more sophisticated than you could imagine. Fig 1.1 attempts to show some of the detail that is impossible to show with a light microscope.

As complex as cells are, in your body right now cells are dying and being replaced at a staggering rate. A human body contains about 50 million million cells. Every day we make new skin cells to replace those that wear away, new red blood cells to replace those that die and new cells to replenish the lining of our digestive system. In this chapter we explore the basic types of cells, how they work and the techniques that have revealed their structure and function.

2 THE CELL THEORY

Following the invention and refinement of the microscope came the **cell theory**, a general acceptance that all living things are made of cells. Modern cell theory has three central ideas:

- The cell is the smallest independent unit of life.
- The cell is the basic living unit of all organisms: all organisms are made up of one or more cells.
- Cells arise from other cells by cell division. They cannot arise spontaneously.

As you study biology in more detail, you will discover that there are exceptions to every rule. In the case of cell theory, that exception is the **virus**. Viruses do not have a cellular structure or organisation, and whether they are actually living organisms is a subject of debate.

3 USING CELL STRUCTURE TO CLASSIFY ORGANISMS

Organisms can be classified on the basis of the internal organisation of their individual cells. With the exception of viruses, all organisms are either **prokaryotic** or **eukaryotic**. Prokaryotic cells are relatively simple, they have no separate nucleus and show little organisation. Bacteria are prokaryotes. In contrast, eukaryotic cells are larger and show much more internal organisation. Animals, plants, fungi and protoctists are all eukaryotes. The main differences between prokaryotes and eukaryotes are shown in Table 1.1.

Table 1.1 Differences between prokaryotes and eukaryotes

	Prokaryotes	Eukaryotes
Organisms	bacteria	plants, animals, protoctists, fungi
Diameter of cells	0.1–10 µm	10–100 µm
Site of genetic material	DNA in cytoplasm	DNA inside distinct nucleus
Organisation of genetic material	DNA is circular; no histone proteins; DNA does not condense at cell division	DNA is linear; attached to histone proteins; condenses into visible chromosomes before cell division
Internal structure	few organelles	many organelles with complex membrane systems
Cell walls	always present	present in plants and fungi and some protoctists; never in animals
Flagella	have simple flagella	have modified cilia that consist of microtubules in a distinctive '9 + 2' arrangement (see Fig 1.29)

4 PROKARYOTES – THE FIRST ORGANISMS?

A few billion years ago, the first living organisms to evolve on Earth were probably prokaryotes. The term literally means 'before the nucleus' because the genetic material (DNA) of these organisms is not enclosed by a membrane, and therefore they do not have a true nucleus.

It is tempting to think of prokaryotes as inferior to eukaryotes, but in some ways they have achieved greater success. They have been on Earth more than twice as long as eukaryotes, they are present in greater numbers (there are more bacteria living on your skin than there are people on Earth) and they occupy an enormous number of different habitats. Some bacteria, for example, are able to live in volcanic springs at temperatures as high as 90 °C.

 REMEMBER THIS

No matter how far you take the study of biology, all you need to know about microscopic distances is these two simple facts:

1 mm = 1000 µm

1 µm = 1000 nm

1 millimetre = 1000 micrometres

1 micrometre = 1000 nanometres

! SCIENCE FEATURE How small is small?

Cells and the molecules they contain are very small. When we study them, we must think about measurements that are minute beyond our imagination. It is easy, for example, to develop a mental block when confronted by the statement 'The nanometre is 10^{-9} m.'

The two units commonly used to describe microscopic objects are the **micrometre** (μm) (commonly called the micron) and the **nanometre** (nm). Starting with a familiar unit, the millimetre (mm), one thousandth of one millimetre is known as a micrometre (μm). The micrometre is used to describe cells and organelles. An average animal cell is 30 to 50 μm across; the

nucleus has a diameter of about 10 μm. Plant cells can reach 150 μm or more in length.

When describing small cellular components and molecules, the useful unit is the nanometre (nm). A nanometre is one thousandth of a micrometre. As a rough guide, the light microscope reveals structures that can be measured in micrometres, but you need an electron microscope to see objects measured in nanometres.

Fig 1.2 Units of measurement; how they are used and how they relate to each other

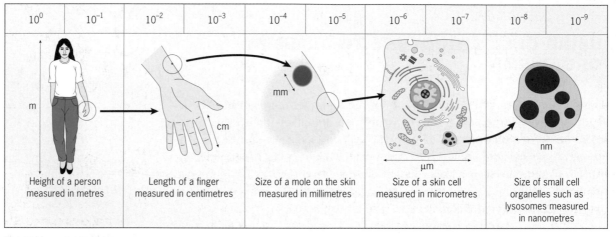

10^0	10^{-1}	10^{-2}	10^{-3}	10^{-4}	10^{-5}	10^{-6}	10^{-7}	10^{-8}	10^{-9}

Height of a person measured in metres

Length of a finger measured in centimetres

Size of a mole on the skin measured in millimetres

Size of a skin cell measured in micrometres

Size of small cell organelles such as lysosomes measured in nanometres

? TEST YOURSELF

A　**a)** Estimate the diameter (in micrometres) of a full stop on this page.
b) How many nanometres are there in 1 millimetre?

BACTERIA

Most bacteria are spherical or rod-shaped cells, several micrometres long (see Figs 1.3 and 1.4) . Their rigid protective **cell wall** is made of **peptidoglycan**, a substance unique to bacteria. Beneath the cell wall is the **cell surface membrane**, which is similar in structure to the membrane of eukaryotic cells. This completely encloses the contents of the cell. Some

(left) **Fig 1.3** Scanning electron micrograph showing *Staphylococcus aureus* cells. These spherical cells are called cocci (singular coccus), and are one of the two main types of bacterial cell (the others are bacilli (singular bacillus); see Fig 1.4). *S. aureus* is commonly found on human skin where it usually does no harm. However, it can cause serious infection if it enters the body through a cut or wound.

(right) **Fig 1.4** Scanning electron micrograph showing *Salmonella enteritidis* cells. These sausage-shaped cells are called bacilli. This species of *Salmonella* can cause severe food poisoning.

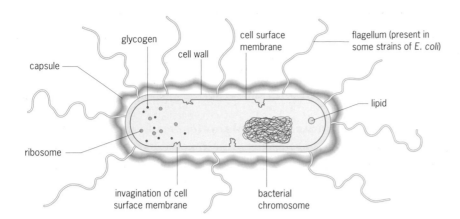

Fig 1.5 The structure of a bacterial cell illustrated by *Escherichia coli*

SEE QUESTIONS 1 AND 2

types of bacteria, such as *Neisseria meningitidis*, which can cause meningitis, also have a **capsule**. This is a sticky coat outside the cell wall that prevents the bacterium from drying out, or being digested by host intestinal enzymes or being attacked by the host's immune system.

Fig 1.5 shows the internal structure of a rod-shaped bacterium, *Escherichia coli*.

Inside the cell surface membrane, the bacterial cell is a single cytoplasmic compartment that contains DNA, RNA, proteins and small molecules. Bacteria have a circular piece of DNA in a region of the cytoplasm known as the **nucleoid**. There are smaller rings of DNA, known as **plasmids**, elsewhere in the cytoplasm. Plasmids are of great interest to biologists because they often contain genes that code for antibiotic resistance, and can be used to carry genes between cells in genetic engineering (see Chapter 27).

Bacteria feed by **extracellular digestion**. They release enzymes into the surrounding medium and absorb the soluble products. **Glycogen granules** and **lipid droplets** often occur in the bacterial cytoplasm and provide a limited store of these polymers. Bacteria can synthesise a wide variety of enzymes, and some species are able to digest unlikely substances such as oil and plastic. Proteins are synthesised on **ribosomes** (page 13), and cell respiration occurs on **mesosomes**, inner extensions of the cell surface membrane.

Some bacteria are **motile**: they can swim. They have thin fibres called **flagella** (singular flagellum) that are corkscrew-shaped and that rotate, propelling the bacteria in different directions. Other fibres, called **pili**, enable the bacteria to perform a primitive form of sex, as shown in Fig 1.6. During **conjugation**, the pili become joined and form a channel to allow plasmids to be swapped from one cell to another.

The roles of bacteria in disease and in the environment are discussed in Chapters 29 and 35, respectively.

? TEST YOURSELF

B Some bacteria can digest, or break down, oil and plastic. Suggest how this could be useful to people.

✔ REMEMBER THIS

Until the invention of the microscope, people had no idea that cells existed. They knew nothing of bacteria and had no idea about how infectious diseases spread.

Fig 1.6 Conjugation in two bacterial cells

5 EUKARYOTES: ANIMAL CELLS

All species of animals, plants, **fungi** and **protoctists** are made of eukaryotic cells. The term **eukaryote** means 'true nucleus', because the DNA of eukaryotic cells is confined to a definite area inside the cell enclosed by a **nuclear envelope**.

Eukaryotic cells also have other **organelles** that form compartments. By being in a compartment the chemicals involved in a particular process, such as respiration or photosynthesis, are kept separate from the rest of the cytoplasm. This allows the chemical reactions of the process to take place quickly and efficiently. This high degree of internal organisation is one of the reasons why eukaryotic cells are larger than prokaryotic cells. The fluid that occupies the space between organelles is the **cytosol**, a solution containing a complex mixture of enzymes, the products of digestion (amino acids, sugars, etc) and waste materials.

INSIDE AN ANIMAL CELL

If you look at an animal cell under a **light microscope**, the right staining and illumination techniques, a good quality microscope and a bit of luck will allow you to see the **nucleus, nucleolus**, the **chromosomes** in a dividing cell, and even the **Golgi complex, mitochondria** and food storage particles. But to be sure of seeing the inside of a cell in more intricate detail, you really need an **electron microscope**.

The **electron micrograph** in Fig 1.7 shows the internal structure of an animal cell. You can see far more of the cell's components – the **endoplasmic reticulum**, the internal features of the mitochondria, the cell surface membrane, **lysosomes, ribosomes** and **cytoskeleton** are all now visible.

In this chapter, we look at the detailed structure of each individual organelle and find out how each type contributes to the function of the cell as a whole.

Table 1.2 summarises the functions of some of the major organelles and structures in the eukaryotic cell.

REMEMBER THIS

1 mm = 1000 µm
1 µm = 1000 nm

Table 1.2 Summary of the functions of major eukaryotic cell organelles and structures

Organelle	Occurrence	Size	Function
Nucleus	usually one per cell	10 µm	site of the nuclear material – the DNA
Nucleolus	inside nucleus	1–2 µm	manufacture of ribosomes
Mitochondrion	numerous in cytoplasm; up to 1000 per cell	1–10 µm	aerobic respiration
Rough endoplasmic reticulum	continuous throughout cytoplasm	extensive membrane network	isolation and transport of newly synthesised proteins
Smooth endoplasmic reticulum	usually small patches in cytoplasm	variable	synthesis of some lipids and steroids
Ribosome	free in cytoplasm or attached to rough ER	20 nm	site of protein synthesis
Golgi body	free in cytoplasm	variable	modification and synthesis of chemicals
Lysosome	free in cytoplasm	100 nm	digestion of unwanted material
Chloroplast	cytoplasm of some plant cells, eg mesophyll	4–10 µm	site of photosynthesis
Vacuole	usually large, single fluid filled space in plant; smaller and more numerous in animals	up to 90 per cent of volume of whole plant cell	storage of salts, sugars and pigments; creates turgor pressure by interaction with cell wall
Cell surface membrane	encloses the cytoplasm of all cells	7–10 µm	exchange and transport of materials into and out of the cell
Cell wall	surrounds all plant cells	thickness varies	provides rigidity and strength

Fig 1.7 A micrograph of a human cell with an interpretive diagram. It shows that the cytoplasm of animal cells contains a complex system of membrane-bound organelles

Using microscopes

THE LIGHT MICROSCOPE

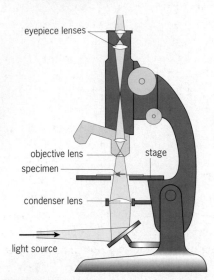

Fig 1.8 The standard compound microscope

Fig 1.8 shows the structure of a **light** or **optical microscope**. This instrument is also known as a **compound microscope** because two lenses, the **eyepiece lens** and the **objective lens**, are combined to produce a much greater **magnification** than is possible with a single lens. The total magnification is calculated by multiplying the magnification of the two lenses together. For example, if the eyepiece lens has a magnification of ×10 and the objective lens is ×50, the total magnification is ×500.

The light microscope has powers of magnification of up to ×1500, good enough to see cells, larger organelles and individual bacteria, but not powerful enough to reveal smaller structures such as cell surface membranes, viruses or individual molecules. Table 1.3 shows organelles in animal and plant cells that are visible with a light microscope. Fig 1.9 shows a light micrograph of an animal cell.

Fig 1.9 The light microscope can reveal the basic features of animal cells; this is what cheek cells look like

Table 1.3 Structures normally visible in animal and plant cells with the light microscope

Organelle	Function	Animal cells	Plant cells
Nucleus	control of cell activities	✓	✓
Vacuole	storage and support	✗	✓
Chloroplasts	photosynthesis	✗	✓
Cell wall	support	✗	✓

An important feature of a microscope is its **resolving power**, which should not be confused with the magnification. Two objects close together may appear as one single image when viewed under the light microscope. Increasing the magnification does not allow you to **resolve** the two objects into separate images; the objects just appear to be a larger single image. Resolving power is as important as magnification when investigating structural details.

The limitation of the light microscope is due to the nature of light itself. The wavelength of light determines the maximum effective magnification and the resolving power. The wavelength of visible light is around 500–650 nm and the resolving power – the resolution – of the light microscope is 200 nm (0.2 μm), so two objects separated by less than 200 nm appear as one object.

THE ELECTRON MICROSCOPE (EM)

Fig 1.10 shows the essential features of an electron microscope (EM). Invented in the 1930s, the present-day version of the EM can magnify up to 500 000 times and has a resolution of 1 nm. In other words, the EM can resolve two objects that are only 1 nm apart. As many biological molecules are larger than 1 nm, the EM can be used to study the arrangement of individual molecules that make up structures in a cell.

The development of the EM has had a huge impact on biology. A magnification of half a million means that an object the size of a full stop becomes over 200 m in diameter – the length of two football pitches. Organelles that are only blurred images when viewed with a light microscope can now be studied in great detail. Many new cell structures have been discovered using electron microscopy.

While the light microscope uses lenses to focus a beam of light, the electron microscope uses electromagnets to focus a beam of electrons. The wavelength of the electrons is much smaller than the wavelength of light, so the resolving power is much greater.

The main disadvantage of the EM is that the electron beam must travel in a vacuum because, being so small,

cathode
(electron gun)

anode

electron beam

condenser lens
(electromagnetic
'lens')

vacuum

specimen

objective lens

projector lens

black and white
image

fluorescent
screen

camera
chamber

Fig 1.10 The basic features of the transmission electron microscope. The beam of electrons is focused by electromagnets before passing through the specimen, and the image appears on a fluorescent screen

electrons are scattered when they hit air molecules. Specimens for the EM must therefore be prepared (killed, dehydrated and fixed) so that they retain their structure inside a vacuum. Such harsh preparation methods can damage cells, and cause **artefacts** – features that do not exist in the living cell – to appear. For example, microsomes, tiny vesicles surrounded by ribosomes, were seen when animal cells were examined using the electron microscope. At first, cell biologists thought that these were organelles they had not noticed before, but they later

realised that microsomes were fragments of endoplasmic reticulum (see page 11) produced by the fixing process. The main differences between electron microscopy and light microscopy are shown in Table 1.4.

SCANNING AND TRANSMISSION ELECTRON MICROSCOPES

There are two main types of electron microscope – the **transmission electron microscope** and the **scanning electron microscope**.

In a transmission electron microscope (TEM), a beam of electrons is transmitted through the specimen. The specimen must be thin and it is stained using electron-dense substances such as heavy metal salts. These substances deflect electrons in the beam and the pattern that the remaining electrons produce as they pass through the specimen is converted into an image. The electron micrograph in Fig 1.7 was produced using the TEM.

We can use scanning electron microscopes (SEMs) to study relatively large three-dimensional objects. Thin sections are not necessary because the SEM records the electrons that bounce off the surface of the object rather than passing through it. Fig 1.11 shows an example of the detailed three-dimensional image that an SEM produces. Although the SEM does not have the resolving power of the TEM, it is more versatile and can be used to observe many kinds of intact structures. The person operating the SEM can move the specimen about, to look at the surface of the structure from a variety of angles – a bit like taking aerial views of the countryside.

Fig 1.11 Scanning electron micrograph showing the surface of the human tongue

Table 1.4 Comparing light and electron microscopes

	Light microscope	Electron microscope
Illumination	light	electrons
Focused by	lenses	magnets
Maximum magnification	×1500	×500 000
Resolving power	200 nm (0.2 µm)	1 nm
Specimens	living or dead	dead
Preparation of specimens	often simple	more complex
Cost of equipment	relatively cheap	very expensive
Images in colour	yes	no (colour is added by computer)

? TEST YOURSELF

C If a light microscope had an eyepiece lens of ×25 and an objective lens of ×40, what would the total magnification be?

SEE QUESTION 3

THE CELL SURFACE MEMBRANE

The **cell surface membrane** is the boundary between the cell and its environment. It has little mechanical strength but plays a vital role in controlling which materials pass in and out of the cell.

Although basically a double layer of **phospholipid molecules**, arranged tail to tail, the cell surface membrane is a complex structure, studded with proteins. These can be embedded in the membrane or they can penetrate the **bilayer** forming **pores** (holes or channels – see Fig 1.12 and the section on the cell surface membrane in Chapter 4) through which molecules can pass. Chapter 4 covers the structure and function of the cell surface membrane in detail.

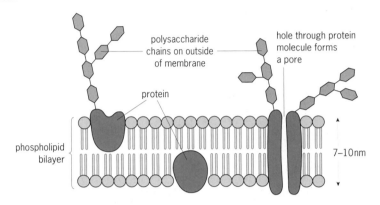

Fig 1.12 The cell surface membrane is a complex organelle but its structure can be represented by this simplified diagram

THE NUCLEUS

The **nucleus** is the largest and most prominent organelle in the cell. Almost all eukaryote cells have a nucleus – red blood cells in mammals and phloem cells in plants are an exception. Every nucleus is surrounded by a **nuclear envelope**. As Fig 1.13 shows, this consists of two membranes that are separated by a gap of 20 to 40 nm.

The nucleus is usually spherical and about 10 μm in diameter. It contains the cell's DNA, which carries information that allows the cell to divide and carry out all its cellular processes When the cell is not actively dividing (as in Fig 1.13), the DNA is spread throughout the nucleus as chromatin. Close examination of the chromatin reveals two different levels of density. Dark-staining chromatin, consisting of tightly packed DNA, is known as **heterochromatin**; the lighter, more loosely packed material is called

Fig 1.13 As the electron micrograph and the diagram show, the nucleus is bounded by the nuclear envelope, a double membrane which contains many pores, each one about 100 nm in diameter. The nuclear pores represent about 15 per cent of the total surface area of the nuclear envelope, indicating the heavy traffic of materials in and out of the nucleus. The nucleus contains the cell's genetic material that exists as chromatin, loosely packed DNA attached to proteins called histones. The nucleolus is clearly visible

euchromatin. Euchromatin contains the DNA that is being actively read to produce proteins; in heterochromatin, the DNA is packed together, and is not being read.

Individual segments of DNA called genes contain the information necessary to make individual proteins, including the enzymes that control most of the cell's activities. In fact, a central concept in biology that is true for all cells, prokaryotes and eukaryotes, is that:

Many genes code for making enzymes that, in turn, control the activities of the cell.

When a cell is dividing, its chromosomes become visible, as Fig 1.14 shows.

You can study enzymes in Chapter 5 and we look at protein synthesis in Chapter 24. We cover cell division in detail in Chapter 23.

Nuclei also have one or more nucleoli (Fig 1.15). These dark-staining, spherical structures are ribosome-producing centres: they synthesise ribosomal RNA and package it with ribosomal proteins to make ribosomes.

Fig 1.14 Before a cell divides, its DNA condenses into visible chromosomes. For further information about cell division, see Chapter 23

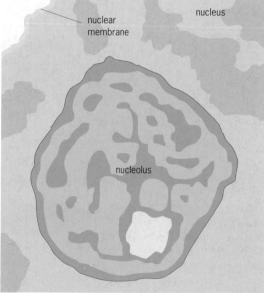

Fig 1.15 The nucleolus has a highly organised structure; its DNA codes for the RNA and proteins that are used to make ribosomes

ENDOPLASMIC RETICULUM (ER)

The nuclear envelope joins with the membrane of the **endoplasmic reticulum** (ER), a system of complex tunnels that are spread throughout the cell.

On much of the outside surface of the ER in a eukaryotic cell are the sites of attachment for ribosomes. This gives it a grainy appearance (Fig 1.16) and its name, **rough ER**.

The main function of rough ER is to keep together and transport the proteins made on the ribosomes. Instead of simply diffusing away into the cytoplasm, newly made proteins are threaded through pores in the membrane and accumulate in the space called the **ER lumen** (Fig 1.17). Here, they are free to fold into their normal three-dimensional shape. Not surprisingly, a mature cell that makes and secretes large amounts of protein – such as one

Fig 1.16 Electron micrograph (right) and interpretive diagram (left) showing rough ER. The ER is a large sheet of membrane that is folded over on itself many times, forming stacked layers called cisternae. The space inside the cisternae, the ER lumen, forms an extensive transport system throughout the cytoplasm

that makes digestive enzymes – has rough ER that occupies as much as 90 per cent of the total volume of the cytoplasm. The rough ER is also a storage unit for enzymes and other proteins.

Small vesicles containing newly synthesised proteins pinch off from the ends of the rough ER and either fuse with the Golgi complex or pass directly to the cell surface membrane.

ER with no ribosomes attached is known as **smooth ER** (Fig 1.18). Smooth ER tends to occur in small areas that are not continuous with the nuclear membrane. Smooth ER is not involved in protein synthesis but is the site of steroid (lipid hormone) production. It also contains enzymes that **detoxify**, or make harmless, a wide variety of organic molecules, and it acts as a storage site for calcium in skeletal muscle cells.

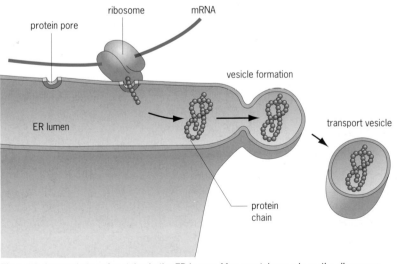

Fig 1.17 Accumulation of proteins in the ER lumen. Many proteins made on the ribosomes are threaded through pores in the ER membrane

RIBOSOMES

Ribosomes are small, dense organelles, about 20 nm in diameter, present in great numbers in the cell. Most are attached to the surface of rough ER but they can occur free in the cytoplasm, as in Fig 1.19. This artist's impression of protein synthesis shows the ribosome's distinctive shape. Ribosomes are made from a combination of ribosomal RNA and protein (65 per cent RNA : 35 per cent protein).

Fig 1.18 Smooth ER has no attached ribosomes. It is usually not as abundant as rough ER, although it is common in the cells of the liver, gut and some glands

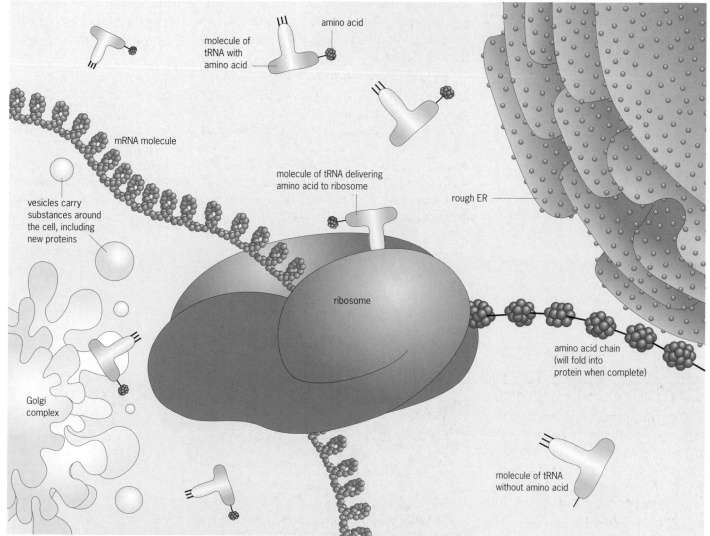

Fig 1.19 An artist's impression of protein synthesis. A ribosome can be thought of as a giant enzyme on which a protein is assembled. Transfer RNA molecules (tRNA) bring specific amino acids to the ribosome. Each one is added to the growing amino acid chain according to the code on the messenger RNA molecule

Fig 1.20 The Golgi complex with a very large number of vesicles. The forming face is at the foot and secretory vesicles are at the top

Ribosomes are involved in protein synthesis. They assemble amino acids in the right order to produce new proteins. The ribosome uses the code on messenger RNA (mRNA) to put amino acids together in chains to form specific proteins. This is another central concept in biology that you will learn more about later on:

> **A gene is a piece of DNA that codes for a particular protein. A copy of the gene in the form of messenger RNA passes out of the nucleus and travels to the ribosome where it controls protein synthesis.**

Protein synthesis is covered in detail in Chapter 24.

Generally, proteins that are to be used inside the cell are made on free ribosomes, while those that are to be secreted out of the cell are made on ribosomes that are bound to ER membranes.

THE GOLGI COMPLEX

The **Golgi complex**, also called the **Golgi apparatus**, is a tightly-packed group of flattened cavities or vesicles (Fig 1.20). The whole organelle is a shifting, flexible structure; vesicles are constantly being added at one side and lost from the other. Generally, vesicles fuse with the forming face (the one nearest to the nucleus) and leave from the maturing face (the one nearest to the cell surface membrane).

Fig 1.21 summarises the relationship between the rough ER and the Golgi complex.

✔ REMEMBER THIS

A vesicle is a small spherical organelle, bounded by a single membrane, which is used to store and transport material around the cell.

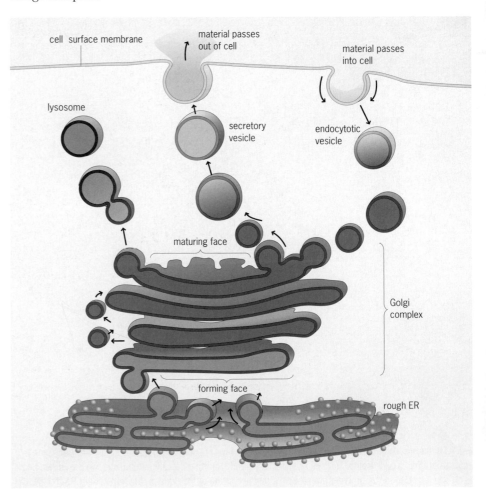

Fig 1.21 The various functions of the Golgi complex are summarised in this diagram. Vesicles from the ER or from outside the cell bring material to the Golgi complex. After processing, material passes out of the cell or enters lysosomes

The Golgi complex appears to be involved with the synthesis and modification of proteins, lipids and carbohydrates. Studies have shown that proteins made on the ribosomes attached to ER are packaged into vesicles by the ER. Some of the vesicles join with the Golgi complex, and the proteins they contain are modified before they are secreted out of the cell.

CELL FRACTIONATION

In order to study the function of a particular organelle it is often helpful to isolate it from the rest of the cell. This can be done by cell fractionation, as outlined in Fig 1.22.

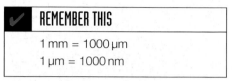

REMEMBER THIS

1 mm = 1000 µm

1 µm = 1000 nm

1 Liver tissue is placed in a fluid that is an **ice-cold isotonic buffer**. When the cell is broken up, membranes are ruptured and many compounds that do not normally mix are brought together. The low temperature minimises unwanted reactions, including self-digestion by digestive enzymes. The isotonic solution prevents organelles gaining or losing water by osmosis, and the buffer prevents any changes in pH.

2 Tissue cut into small pieces   **3** Tissue put into blender/homogeniser to break up whole cell   **4** Mixture filtered to remove debris   **5** Filtrate spun in centrifuge

Principle: The organelles will separate out in a particular order, according to their density and shape. After a time, the sediment containing a particular type of organelle can be separated from the supernatant (the liquid that contains the remaining organelles). The exact times vary from tissue to tissue.

	Organelle	Centrifuge setting (g)	Time (minutes)
First to separate out	Nuclei	800–1000	5–10
	Mitochondria Lysosomes	10 000–20 000	15–20
	Rough ER	50 000–80 000	30–50
	Plasma membranes Smooth ER	80 000–100 000	60
Last to separate out	Free ribosomes	150 000–300 000	> 60

Fig 1.22 Cell fractionation: how to isolate particular types of organelle from liver tissue

Fig 1.23 Lysosomes are simply bags of digestive enzymes. They can be distinguished from other vesicles in the cell only by using a stain specific for the chemicals inside the lysosomes

LYSOSOMES

Lysosomes are small **vesicles** 0.2–0.5 μm in diameter (Fig 1.23) that contain a mixture of digestive enzymes called **lytic enzymes**. It is important that the membrane of lysosomes remains intact because if the enzymes leak out they could digest vital molecules in the cell.

Why do cells need such potentially lethal structures? They have several uses:

● To supply the enzymes which destroy old or surplus organelles.

● To digest material taken into the cell. After a white cell has engulfed a bacterium, for example, lysosomes discharge enzymes into the vacuole (Fig 1.24) and digest the organism. This process is called **phagocytosis**.

● To destroy whole cells and tissues. Parts of tissues and organs often need to be removed after they have performed their function. The muscle of the uterus is reduced after giving birth and milk-producing tissue is destroyed after weaning. Bone is also constantly made and reabsorbed throughout life.

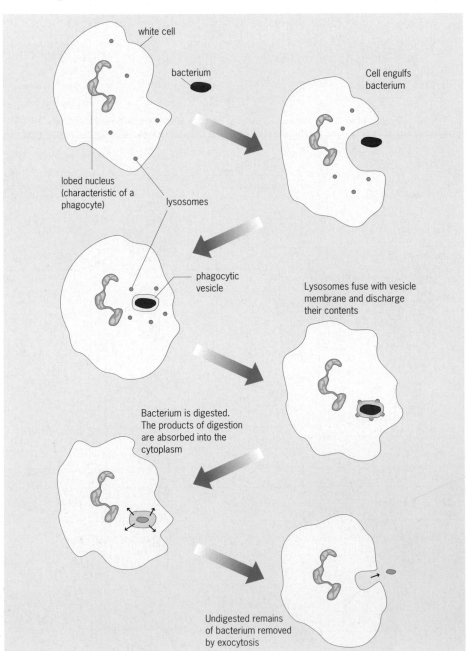

Fig 1.24 One of the functions of lysosomes is to digest material taken into the cell by the process of phagocytosis (see also Chapter 4). Here, a white cell, known as a phagocyte, is ingesting a bacterium. Lysosomes discharge their enzymes into the temporary vacuole and the bacterium is digested

Lysosomes are similar in structure to many other vesicles in the cell and are thought to be made in the same way: inactive digestive enzymes from the rough ER pass through the Golgi complex where they are activated and packaged into vesicles.

MITOCHONDRIA

Mitochondria are relatively large, individual organelles that occur in large numbers in most cells. They are usually spherical or elongated (sausage-shaped) and are 0.5–1.5 μm wide and 3–10 μm long. Their function is to make **ATP** via the process of **aerobic respiration**. ATP is a molecule that diffuses around the cell and provides instant chemical energy to the processes that require it (see Chapter 34).

As Fig 1.25 shows, mitochondria have a double membrane; the outer membrane is smooth while the inner one is folded. This arrangement gives a large internal surface area on which the complex reactions of aerobic respiration can take place. Mitochondria are particularly abundant in metabolically active cells, tissues such as muscle and tissues involved in active transport (see Chapter 4).

The function of mitochondria is covered in more detail in Chapter 34.

? TEST YOURSELF

H Why are digestive enzymes synthesised in an inactive form?

SEE QUESTIONS 4 AND 5

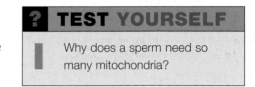

? TEST YOURSELF

I Why does a sperm need so many mitochondria?

Fig 1.25 Mitochondria are the sites of aerobic respiration in the cell. The inner membrane is folded into cristae, which provide a large surface area that is occupied by many of the enzymes involved in cell respiration. Generally, the more metabolically active the cell, the more cristae there are in the mitochondria

outer membrane

matrix (fluid)

cristae (folds)

ribosome

inner membrane

! SCIENCE FEATURE Autoradiography

Electron microscopes are useful tools for investigating cell structure, but they don't reveal much about the processes going on inside a living cell. After World War II, work on the atomic bomb produced chemicals called **radioactive isotopes** as a by-product. These are very useful for studying metabolism because they behave in the same way as non-radioactive atoms and their movements through the cell can be traced (Fig 1.26).

After cells are exposed to a radioactive isotope such as **carbon-14,** they are washed to remove excess label and then fixed, embedded, sectioned and mounted on microscope slides. The slides are then taken to a dark room and coated with photographic emulsion that has been heated to make it melt. After coating, the slides are cooled and kept in the dark for various time periods from a few days to several weeks. During this time, radioactive decay causes a chemical change to occur in the film of photographic emulsion. When this is developed using standard photographic techniques and the slide is examined under the light microscope, silver grains can be seen in the areas exposed by the radioactivity. The pattern of grains shows the structures of the cell that have incorporated the radioactive label.

1 Living cells are exposed to the radioactive isotope ●

2 Cells are washed (to remove any excess isotope) and sectioned

3 Section put onto microscope slide and dipped into photographic emulsion

4 Slides are incubated in the dark and the isotope reacts with the light-sensitive chemicals. The emulsion is then developed like a normal photograph and, when examined under the microscope, the position of the isotope in the cell can be seen

Fig 1.26 The basic method of autoradiography

CENTRIOLES AND THE CYTOSKELETON

ribosome

cell surface membrane

endoplasmic reticulum

microfilament

microtubule

mitochondrion

Fig 1.27 Inside the cell is a network of tubules which forms the cytoskeleton

Between the organelles of all eukaryotic cells is a complex network of fine threads called **filaments**. These vary in size and chemical composition. As Fig 1.27 shows, they help to form a supportive scaffolding known as the **cytoskeleton** ('cell skeleton'). This:

● supports the whole cell,
● maintains cell shape, as in red blood cells,
● organises and moves organelles,
● forms the spindle during cell division,
● moves the whole cell.

This last function indicates that the cytoskeleton is not a rigid structure; it can be assembled or dismantled in seconds – as happens in a moving white blood cell.

Centrioles are short bundles of filaments, set at right angles to each other (Fig 1.28). They are found in a clear area of cytoplasm known as the **centrosome**. Centrioles occur in all animal cells but are absent from the cells of plants. For a long time their function was thought to be just the formation of the **spindle** – the cradle of threads that guide chromosomes during cell division. In addition to spindle formation, the centrioles act as the centre of formation for the whole cytoskeleton and they have therefore become known as **microtubule organising centres**.

Microtubules make up more complex external cell structures called **cilia**. These tiny hair-like projections are used for cell movement. The tail of a sperm is, for example, a modified cilium. In human cells, cilia contain a bundle of microtubules in the '9 + 2' arrangement shown in Fig 1.29.

Fig 1.28 The centrioles are two bundles of fibres that form the cytoskeleton of the cell and also give rise to the spindle during cell division

? TEST YOURSELF

J How many of the following cells would fit into a line 1 cm long?
a) Animal cells of 20 μm
b) Plant cells of 50 μm
c) Bacterial cells of 2 μm

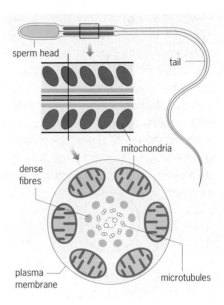

Fig 1.29 A cross-section of the tail of a human sperm showing the '9 + 2' arrangement of microtubules in its centre

6 EUKARYOTES: PLANT CELLS

Typical leaf cells from a plant, seen with the light microscope, are shown in Fig 1.30. These are **palisade mesophyll cells**.

Plant cells have several features that are not found in animal cells (see Table 1.5). A plant cell is surrounded by a cell wall made of cellulose.

Table 1.5 Comparing plant cells and animal cells		
Feature Cell:	**Animal**	**Plant**
Nucleus	✓	✓
Plasma membrane	✓	✓
Mitochondria	✓	✓
Rough ER	✓	✓
Smooth ER	✓	✓
Golgi bodies	✓	✓
Lysosomes	✓	✓
Cell wall	✗	✓
Plastids, eg chloroplasts	✗	✓
Vacuoles	✗	✓
Centrioles*	✓	✓*

*Centrioles are not present in the cells of higher plants

Fig 1.30 The light microscope shows some of the basic features of plant cells

A large proportion of the inside of the cell is taken up with a fluid-filled compartment known as the **vacuole**. Together, the wall and vacuole maintain the shape of the whole cell.

As you can see from Fig 1.31, plant cells have specialised organelles, the **chloroplasts**, which enable them to make their own food by photosynthesis. Chloroplasts bring together all the chemicals involved in photosynthesis, and provide sites for the chlorophyll molecules so that they can absorb the maximum amount of light.

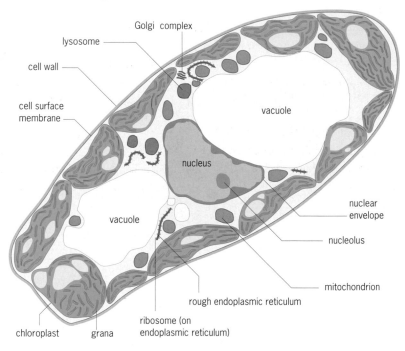

Fig 1.31 Transmission electron micrograph (above) and explanatory diagram (right) showing a palisade mesophyll cell from a soya bean leaf

Fig 1.32 The cellulose cell wall is a composite material. Fibrils containing thousands of cellulose molecules are cemented together by a 'glue', a complex mixture of compounds

THE CELLULOSE CELL WALL

Unlike animal cells, plant cells are enclosed by a **cell wall** that consists of many **cellulose fibres** cemented together by a mixture of other organic substances.

Cellulose is a **polysaccharide** (a sugar polymer) which consists of long straight chains of glucose molecules bonded to adjacent molecules by hydrogen bonds (see Chapter 3). In the cell wall, around 2000 parallel cellulose molecules are packed together to form **microfibrils**. These in turn are bundled together to form **fibrils**, as you can see in Fig 1.32. Like fibreglass, the cell wall has great strength because of the many strong fibres and the 'glue' that holds them together.

The main functions of the cell wall are:

- to provide rigidity and strength to the cell. Cell walls need to be strong enough to resist expansion to allow the cell to become turgid; see Chapter 4.
- to force the cell to grow in a certain way. A particular arrangement of fibrils causes the cell to assume a particular shape – for example a long, thin tube.

VACUOLES

Plant cells also have a **vacuole**, a large, fluid-filled cavity. In mature cells the vacuole can occupy over 80 per cent of the cell volume, and is filled with a fluid called **cell sap** (Fig 1.33). The sap consists of a complex mixture of sugars, salts, pigments and waste products in water.

Vacuoles have several functions:

- They absorb water by **osmosis** and therefore swell, pushing the cytoplasm against the cell wall. In this state, a plant cell is said to be turgid.
- They store food substances such as sugars and mineral salts.
- They store pigments that give colour to plant structures such as petals.
- They can accumulate waste products and by-products of metabolism. Sometimes these chemicals may be toxic or have an unpleasant taste and are used by the plant to make it less palatable to herbivores.

Fig 1.33 In a mature plant cell, the vacuole becomes so large that the cytoplasm is reduced to a thin layer around the cell

SEE QUESTIONS 6 AND 7

CHLOROPLASTS AND OTHER PLASTIDS

Chloroplasts are one of a group of plant cell organelles known as **plastids**. They are surrounded by a double membrane and contain an elaborate internal membrane system that houses the chemicals of photosynthesis (Fig 1.34).

lipid droplets

stroma

thylakoid

membrane-bound ribosomes

granum

outer membrane

starch grains

stroma thylakoid

free ribosomes

inner membrane

Fig 1.34 Chloroplasts are organelles in which all the chemicals associated with photosynthesis are brought together. Chlorophyll molecules are housed on membranes to maximise light absorption. In a flowering plant, most chloroplasts are found in the palisade cells of the leaves

The structure and function of chloroplasts is covered in detail in Chapter 34.

EXTENSION The origin of mitochondria and chloroplasts: the endosymbiont theory

Scientists continue to speculate whether mitochondria and chloroplasts originated from inside eukaryotic cells. Many researchers think that they were originally free-living prokaryotes, similar to bacteria and blue-green bacteria, which, at some point, became incorporated into larger cells.

Let's look at the evidence. Mitochondria and chloroplasts have several similarities to prokaryotes:

- They are similar in size to prokaryotes.
- They have their own DNA, as prokaryotes do, suggesting an ability to reproduce independently.
- The DNA of both organelles is circular, like that of bacteria.
- Both organelles have their own ribosomes, smaller than those in eukaryotic cells but identical to those in prokaryotes.
- All reproduce by binary fission.
- The biochemistry of mitochondria is closer to that of aerobic bacteria than that of the eukaryotic cell; chloroplasts and cyanobacteria are even more similar.

This theory was first suggested by a team led by Lyn Margulis at Boston University, USA. Assuming the theory is true, how could these organisms have become incorporated into a larger cell without being destroyed by lysosomes?

Imagine a heterotrophic, anaerobic organism – similar to an amoeba in structure and movement – that feeds by phagocytosis and respires without oxygen, though oxygen is present in its environment. If such a cell were to ingest aerobic bacteria that it did not digest, the bacteria could respire aerobically, and in doing so could provide the host organism with far more ATP than the host could produce on its own. The host cell in turn would give the bacteria material to respire – food.

Similarly chloroplasts, which may originally have been free-living blue-green bacteria, could also have become incorporated into larger cells. If these were ingested – but not digested – they would continue to photosynthesise within the host and so provide it with organic food and oxygen – an obvious advantage. The host, in turn, could provide the blue-green bacterium with carbon dioxide and nitrogen.

Associations in which an autotroph lives inside a heterotroph, or an aerobe lives inside an anaerobe, are both very convenient arrangements. One organism produces what the other needs. This is the key to the success of several other associations, for example, lichens (algae and fungi) and corals (algae inside animals). Both host and 'passenger' gain a survival advantage: the host gets food and the smaller organism is protected from predators.

The idea of incorporation by ingestion could also explain why both organelles have a double membrane. The inner membrane would represent the original prokaryotic membrane while the outer one could represent the food vacuole into which the organism was taken.

> ✔ **REMEMBER THIS**
>
> An **autotroph** is an organism that can make its own food: a green plant is an autotroph that make its own food by photosynthesis. A **heterotroph** is an organism that must obtain ready-made food: a cow is a heterotroph that obtains its food from grass.

SUMMARY

- All living organisms (except viruses) are composed of either **prokaryotic** or **eukaryotic** cells.
- **Bacteria** are prokaryotes; **plants**, **animals**, **protoctists** and **fungi** are eukaryotes.
- Prokaryotic cells are relatively small, with very little internal organization and few internal membranes.
- Eukaryotic cells are relatively large, with a high degree of internal organisation.
- Eukaryotic cells have a true, membrane-bound **nucleus** and many **organelles** – membrane-bound compartments in which particular chemical processes take place.
- The nucleus, **ribosomes**, **rough endoplasmic reticulum (ER)**, **vesicles** and **Golgi complex** in eukaryotic cells all take part in different stages in the making and packaging of chemical products. Some chemicals are used within the cell. Others – such as hormones or digestive enzymes – are released.

- **Mitochondria** are the site of production of most of the eukaryotic cell's **ATP**, which provides the energy for other cellular processes.
- Eukaryotic cells possess a **cytoskeleton** – a network of fine **tubules** that supports and shapes the cell.
- Plant cells contain structures not found in animal cells: **chloroplasts**, a **cell wall**. The **vacuole** in most plant cells is much larger than the vacuoles found in an animal cells.
- The cell can be studied using a variety of experimental techniques: the **electron microscope** reveals the structure of the cell, **cell fractionation** isolates individual organelles and, with the use of radioactive isotopes in **autoradiography**, the passage of materials through the cell can be followed.

? EXAM QUESTIONS

1

a) Explain concisely the main functions of the following structures in bacterial cells.
 (i) cell wall
 (ii) cell membrane
 (iii) DNA
 (iv) plasmids
 (v) ribosomes

b) Complete the table below to show the main differences between the structure of prokaryotic and eukaryotic cells.

prokaryotic	eukaryotic

OCR Biology Cells and Structures March 2001 Q2

2 The size of a cell can be measured using a microscope fitted with an eyepiece graticule. The eyepiece graticule has to be calibrated using a micrometer slide on the stage of the microscope. The figure below shows an eyepiece graticule against a micrometer slide, with a one millimetre line with subdivisions. A ×10 objective lens and a ×10 eyepiece were used on the microscope.

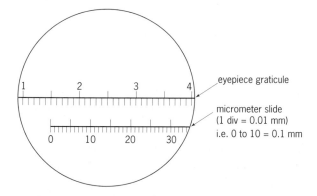

eyepiece graticule

micrometer slide
(1 div = 0.01 mm)
i.e. 0 to 10 = 0.1 mm

a) Calibrate the eyepiece graticule, stating the width, **in μm**, of the smallest division indicated on the graticule. Show your working.
 The figure below shows the same eyepiece graticule, using the same ×10 objective lens and the same ×10 eyepiece, with some cells in focus.

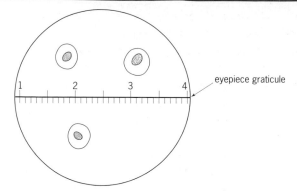

eyepiece graticule

b) **(i)** Measure the largest diameter of each of the three cells shown. Calculate the mean diameter of the three cells, in μm. Show your working.
 (ii) What would the **mean** diameter of these cells be if they were observed using the ×40 of this microscope?

OCR Biology Cells and Structures June 2000 Q5

3 The optical (light) microscope and the transmission electron microscope have both been of significant importance as tools in biological science. Complete the following paragraphs using appropriate words or phrases.

Tissues for use with both the light and the electron microscope are first … . This prevents the cells breaking down and prepares them for further treatment. The tissues are then dehydrated using … . Materials for use with the light microscope are embedded in wax but in the case of the electron microscope they are supported using … . Thin sections of material are produced using a microtome fitted with a sharp metal blade for the light microscope but a blade made of … , which gives much thinner sections, is used for the electron microscope.

The embedding material for the light microscope is removed and the sections can then be treated with stains. The presence of lignin can be demonstrated using the stain … and the stain … shows the presence of cellulose. Stains cannot be used on sections for use with the electron microscope but the sections are immersed in solutions of … . There is a differential absorption of these which gives contrast in the resulting image.

The energy source in the light microscope is … and in the transmission electron microscope is … .

Quite often students consider that the electron microscope is a more important research tool because of magnification but in reality it is the … of the electron microscope which is important.

OCR Biology Cells and Structures March 2001 Q1

? EXAM QUESTIONS

4

a) Complete the diagram of a mitochondrion and **label four** structures **clearly**.

b) **(i)** What process takes place in the mitochondrion?
(ii) Name a tissue where you would expect to find large numbers of mitochondria.
(iii) Explain why mitochondria are particularly important **in this tissue**.

WJEC Biology Module BI1 January 2001 Q2

5 The pie charts below show the relative amounts of types of membrane found in two types of cell.

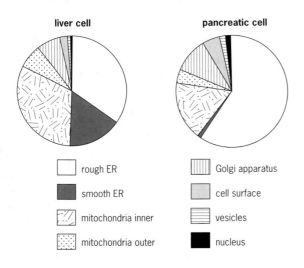

liver cell   pancreatic cell

- ☐ rough ER
- ■ smooth ER
- ▨ mitochondria inner
- ▧ mitochondria outer
- ▥ Golgi apparatus
- ▦ cell surface
- ▤ vesicles
- ■ nucleus

a) State which type of membrane is present in the greatest amount in the liver cell.

b) **(i)** Compare the relative amounts of mitochondrial membranes in the two types of cell.
(ii) Suggest **one** reason for the difference between the cell types in amount of mitochondrial membranes.

c) Using only the data in the pie charts, predict which cell type secretes more protein. Give **two** reasons for your answer.

d) Predict which type of membrane would be present in the largest relative amount in a prokaryote cell.

IBO Biology Paper 3 Standard level November 1999 QC1

6 State the name of the cell organelle, or structure, to which each of the following statements refers.

a) Fully permeable layer composed of a polysaccharide.
b) Series of membranes in the cytoplasm through which materials can be transported and in which proteins accumulate.
c) Site of ribosomal RNA synthesis.
d) Contains large quantities of chromatin.
e) Responsible for the translation of mRNA.
f) Partially permeable layer about 7.5 nm thick.
g) Structures responsible for producing the spindle apparatus.
h) Contains membrane stacks called grana.
i) Site of aerobic respiration.
j) Structure containing cell sap.
k) Proteins with a quaternary structure are assembled here.
l) Structures made of the protein tubulin.

OCR Biology Cells and Structures November 2000 Q2

7 Carefully study the drawings in the figure below made from observations of electron micrographs.

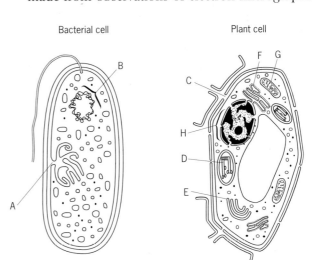

Bacterial cell   Plant cell

a) Cells are divided into two major groups according to their structures. To which group do the above cells belong?
b) Name the structures indicated by the letters **A**, **B** and **C** on the diagrams above.
c) State **one** difference between the structures labelled **B** and **H**.
d) State the **main** function of the following structures: **D**, **E**, **F** and **G**.

WJEC Biology Module BI1 June 2001 Q3

KEY SKILLS **ASSIGNMENT**

GETTING A SENSE OF PROPORTION

In this assignment we shall look at two skills that are commonly tested in examinations; scaling and matching structure to function.

Fig 1.A1 Electron micrograph showing epithelial cells from the lining of a mammalian small intestine (×1800). Cell A1: width, line a to b; diameter of the nucleus, line c to d

Fig 1.A2 Electron micrograph showing a palisade cell from a flowering plant (×1720)

MICROSCOPY AND SCALING

There are three aspects to scaling; the **actual size**, the **observed size** (that you can measure with a ruler) and the **magnification**. If you know two of these values, you can always work out the third one.

The formula is:

$$\text{Magnification} = \frac{\text{Observed size}}{\text{Actual size}}$$

This can be simplified even further to a triangle with O on top, M and A below. In exam questions, cover up the value you need to find and the calculation you need should be clear.

1 Use the scales given (and a ruler) to work out the actual size of the cells in μm.

2 Copy and complete the table below.

Organelle	Size on paper	Actual size in the cell
Cell A1:		
Nucleus		
Mitochondrion		
Cell A2:		
Nucleus		
Nucleolus		
Mitochondrion		
Chloroplast		

3 Why do you think that the endoplasmic reticulum and the cell's surface membrane have been left out of the above table?

Fig 1.A3 shows some red blood cells.

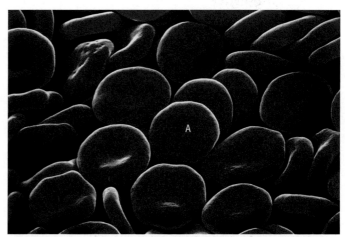

Fig 1.A3 Scanning electron micrograph of human red blood cells (×2150)

4 The magnification of Fig 1.A3 is ×2150. Calculate the diameter of the red cell labelled A. Give your answer in micrometres.

5 What evidence can you see from Fig 1.A3 that it was taken with a scanning electron microscope?

KEY SKILLS **ASSIGNMENT**

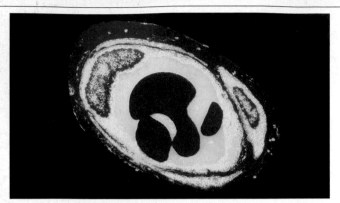

Fig 1.A4 Thin section of capillary showing different shaped red cells that have been cut across different planes (×1925)

6 Fig 1.A4 shows a blood capillary in the lungs. In order to prepare the slide for examination with the electron microscope, thin sections had to be cut. Explain how this results in the different shapes of red cells seen in the micrograph.

RELATING STRUCTURE TO FUNCTION

As we shall see in Chapter 2, cells inside an organism are specialised and perform particular functions. The structure of cells usually reflects their specialisation.

7 The **intestinal epithelial cell** (Fig 1.A1) allows rapid absorption of digested food into the bloodstream. The two mechanisms for the absorption of food molecules are diffusion and active transport. (You can look at digestion in more detail in Chapter 11.) Suggest reasons why this cell has large numbers of microvilli and mitochondria.

8 The **palisade cell** is on the upper surface of a leaf, the best location for photosynthesis. Remember that photosynthesis requires light, carbon dioxide and water and produces glucose and oxygen. Suggest how the following features help the palisade cell to perform its function:
a) the large numbers of chloroplasts,
b) the observation that chloroplasts circulate within the cell,
c) the deep, cylindrical shape of the cell.

You will gain a more complete picture of plant function when you study the structure of the whole leaf in Chapters 11 and 31.

9 **Red cells** are unusual because they lack a nucleus.
a) What is the main function of red cells?
b) Suggest an advantage of having no nucleus.
c) The cell would have a greater volume if it were spherical. Suggest an advantage of the biconvex shape.

02

Tissues and organs

Tissues and organs

Stephen Badylak, a senior research scientist in Purdue's Department of Biomedical Engineering, holds a sheet of scaffolding material harvested from the urinary bladder of a pig. The material holds promise as a scaffold for the repair or replacement of many different body parts, such as the trachea and oesophagus in human patients.

As we learn more about how tissues interact to form organs and organ systems, we come across new opportunities to create artificial tissues to use in medicine. Artificial heart valves and tubes to keep blood vessels open have been used for quite a while, but more recently a team of bioengineers in the USA have gone one better. They have developed a way of treating the lining from the small intestine, stomach, bladder and liver of pigs and other animals to make an organic mesh that can be used to repair internal organs such as vocal cords and damaged ligaments. This mesh is also proving useful for treating skin wounds and chronic sores.

The tissues are removed and sterilised, then processed into sheets, tubes and other shapes which are grafted into areas that need to be replaced or repaired. Because the material is obtained from living tissue, it not only contains collagen, the fibrous protein that has great strength and flexibility, but also growth factors, proteins that send messages to cells to tell them where and how to grow and differentiate.

Once implanted into the body, the growth factors send instructions to the surrounding cells. The resulting new tissue growth is essentially the same shape and performs the same functions as the damaged tissues.

1 THE COMPLEXITY OF LIVING ORGANISMS

Most human cells are microscopic. The largest cell in the human body is the egg cell, or **ovum**. This is just visible with the naked eye. It needs to be larger than other cells because it needs space to store food reserves (Fig 2.1).

Humans, like all large organisms, are **multicellular**. Our body systems consist of complex organs and tissues (Fig 2.2) that interact and co-operate to keep all the cells of the body in the best possible environment. Different groups of cells carry out different tasks and functions: they have **differentiated** and become **specialised**. Different types of cell develop to carry out specific functions that contribute to the success of the organism as a whole. This is known as **division of labour**.

Fig 2.1 This photograph shows the huge difference in size between an ovum (a human egg cell) and the sperm that will fertilise it. The cytoplasm is packed with food reserves that allow the cell to divide several times before it implants in the uterus. The DNA in a fertilised egg carries all the information necessary to make a complete human baby in just 9 months

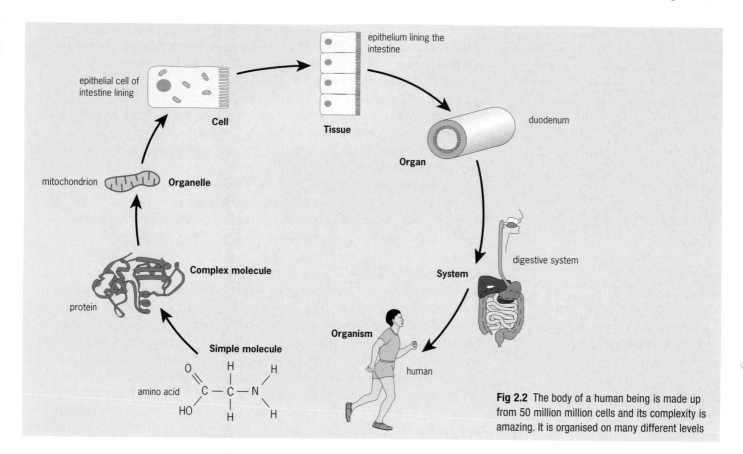

epithelial cell of intestine lining

Cell

epithelium lining the intestine

Tissue

duodenum

Organ

mitochondrion **Organelle**

digestive system

System

Complex molecule

protein

Organism

human

Simple molecule

amino acid

Fig 2.2 The body of a human being is made up from 50 million million cells and its complexity is amazing. It is organised on many different levels

2 TISSUES

A **tissue** is a collection of similar specialised cells that work together to achieve a particular function. Different tissues combine to form more complex structures called **organs**. An organ usually has a specific function in the organism, for example to detect light, to absorb food or to produce a hormone.

Studying the detailed structure of individual tissues is a lot more interesting if you look at how the structure of a tissue relates to its function in the body of the organism. Getting familiar with normal tissues is particularly important in medicine; recognising what is normal and abnormal in a particular type of tissue can mean the difference between life and death.

HUMAN TISSUES

In humans, as in all mammals, each organ of the body is made up from a combination of these four basic tissue types:

- **Epithelial tissue** forms thin sheets that line and cover body structures. Epithelial tissue lines the intestines, for example.
- **Connective tissue** is tough and fibrous and forms structures that hold the body together. Ligaments and tendons are mainly connective tissue.
- **Muscular tissue** can contract to produce movement.
- **Nervous tissue** has the ability to conduct impulses, allowing communication between different parts of the body.

EPITHELIAL TISSUE

Epithelial tissues form continuous sheets that line or cover most structures and cavities in the body. Different types are classified according to the size and shape of their cells, and the number of cell layers they contain. For example, simple **squamous epithelium** contains only one layer of cells

REMEMBER THIS

A **tissue** is a collection of similar specialised cells.

Histology is the study of cells, tissues and organs, using microscope techniques.

Mitosis is a type of cell division (see Chapter 23).

and is therefore thin and ideally suited to be a membrane where substances are exchanged, such as at the surface of alveoli. **Stratified squamous epithelium** is several cells thick and can be replaced continuously, making it perfect as a protective covering to the skin.

Epithelial tissues have very little mechanical strength; they are supported and attached to underlying connective tissue by a **basement membrane**, a continuous sheet consisting of **collagen** and other proteins.

Epithelial tissues form barriers that keep different body systems separate. They also have many other functions. For instance, the epithelial cells which line the mammalian respiratory tract are **ciliated**: they form a 'carpet' of tiny hair-like processes which trap inhaled dust and sweep it away from the lungs and out towards the throat where it is swallowed. Other epithelial tissues, such as the lining of the mammalian intestine, secrete **mucus**.

CONNECTIVE TISSUE

As their name suggests, **connective tissues** usually connect structures together. Connective tissue cells are not packed closely together but are separated by a non-cellular **matrix** (a mesh) which they secrete. (Notice the contrast to epithelial tissue where the cells are tightly packed.) The nature of the matrix accounts for the property of the tissue. In bone, for example, the matrix is a hard mineral strengthened by fibres of collagen, a tough protein (see Chapter 3).

The main types of connective tissue found in the human body are listed in Table 2.1. Cartilage, tendons, ligaments, blood and bone are all connective tissues. Each one is adapted to perform a particular function. All except blood have a structural role in the body. Blood is described as a fluid connective tissue because plasma, a fluid, separates and surrounds the cells. Because connective tissues are mainly structural, their metabolic demands (their need for food and oxygen) are relatively small.

 REMEMBER THIS

Connective tissue consists of cells in a matrix that usually contains strengthening fibres.

Table 2.1 The main types of mammalian connective tissue

Connective tissue	Function
Bone	Forms the skeleton. Protects, supports the main organs of the body. Anchors the muscles
Tendon	Attaches muscles to bones
Ligament	Attaches bones to bones and provides support at joints
Cartilage	Smoothes surfaces at joints. Prevents collapse of trachea and bronchi
Adipose tissue	Stores fat and provides insulation
Blood	Transports substances around the body (see Chapter 10)
Areolar tissue	Protects organs, blood vessels and nerves. Gives strength to epithelial tissue. General 'packing tissue' arrangement

 REMEMBER THIS

Fibrous cartilage is a good shock absorber because it is compressible (it can withstand pressure without being damaged) and it can absorb the impact of a rapid jolt which would damage a more brittle structure such as a bone.

 REMEMBER THIS

Tendons are made from fibrous connective tissue that contains many closely packed collagen fibres. In some situations, such as in the legs during vigorous exercise, the tendons withstand enormous forces.

CARTILAGE

Cartilage is a tough, smooth and flexible connective tissue. There are three types of cartilage:

- **Hyaline cartilage**. This is a compressible and elastic tissue found in the nose and trachea, and at joints where it covers the ends of bones to provide smooth surfaces.
- **White fibrous cartilage**. This is found between the vertebrae and acts as a shock absorber.
- **Yellow elastic cartilage**. This is a very elastic tissue found in the ears, pharynx (throat) and epiglottis (the flap that closes over the airway during swallowing).

! SCIENCE FEATURE Corneal transplants with a difference

Mesh made from the lining of animal intestines is being used to repair damaged ligaments and other tissues (see opener at the start of the chapter) and artificial skin is widely used to treat burns and slow-to-heal skin wounds. It may also soon be commonplace for people to have their damaged cornea replaced without using donated human corneas.

The technique, which has taken over 10 years to perfect, involves taking a few corneal stem cells (cells that grow to become corneal cells) from the patient's healthy eye, or from a related donor, in a painless procedure that lasts about 5 minutes. These cells are then grown in laboratory dishes under special conditions until they produce a fragile film, just one cell thick. This delicate sheet is then transferred onto a mesh made from an amniotic membrane that has been sterilised. The amniotic membranes are donated by mothers after the birth of their babies. On the mesh, the corneal cells grow into a fatter layer, between 5 and 10 cells thick. The end product is elastic like the amniotic membrane but has the biological properties of corneal tissue. The membrane is then placed over the patient's eye, and held in place with delicate stitches.

In a small trial carried out in 2000, 10 out of 14 patients who had damaged corneas, and whose traditional corneal transplants had failed, gained better eyesight with this new type of transplant. One man of 78, whose cornea became severely scarred after he suffered a rare reaction to medicated eye drops, could not drive or see a computer screen before his operation in 1998. Now, with the extra help of a contact lens, his vision is near normal in one eye and he can now drive, read and use his computer again.

In the UC Davis technique, the corneal stem cells are harvested from the patient's healthy eye or from the eye of a donor.

The tissue is separated to extract the living corneal cells.

This bioengineered composite tissue is then stitched onto the patient's eye, after the abnormal corneal tissue has been removed.

The cells are grown in laboratory dishes, where the stem cells produce a fragile film of corneal cells just one cell thick.

On the matrix, the corneal cells grow into a layer 5 to 10 cells thick, forming a sturdy composite tissue that combines the elasticity and resilience of the amniotic membrane with the biological properties of corneal tissue.

Scientists transfer the corneal cells, including the surviving stem cells, to the surface of a matrix of sterile amniotic membrane.

Some cells also can be frozen and banked for possible later use.

Fig 2.3 The technique of corneal transplantation using stem cells grown on a membrane from a placenta

MUSCULAR TISSUE

Muscular tissues are contractile: they contain protein fibres, actin and myosin, which produce movement when they slide over each other (see Chapter 17).

As Fig 2.4 shows, there are three types of muscle: skeletal muscle, smooth muscle and cardiac muscle.

Fig 2.4 Micrographs of muscle: left, skeletal; middle, smooth; right, cardiac. See Table 2.2 for descriptions

The properties of the three types are summarised in Table 2.2.

? TEST YOURSELF

B Which type of muscle would you expect to have the fewest mitochondria? Explain.

C What features of skeletal muscle make it unsuitable to form the muscles of the heart?

Table 2.2 Comparing the three types of mammalian muscle

Muscle:	Skeletal	Smooth	Cardiac
Other names	striped, striated, voluntary	unstriped, unstriated, involuntary, visceral	heart
Site in the body	attached to skeleton	tubular organs; gut, reproductive system, glands, bronchioles	heart
Function	movement and maintenance of posture	usually controls movement of substances along tubes	heartbeat
Control	voluntary	involuntary (not under conscious control)	myogenic (self-generating)
Speed of contraction	rapid	slow, usually known as peristalsis (rhythmic squeezing movements) (see Chapter 8)	rapid
Speed of fatigue	rapid	slow	slow

The muscles of most vertebrates make up a large proportion of their body mass. As you might expect, active muscle tissue has a very high metabolic rate, and many mitochondria are packed between the muscle fibres.

NERVOUS TISSUE

Nervous tissue contains neurones, specialised elongated cells that transmit electrical impulses from one end to the other. Nervous tissue controls and coordinates the activities of the body. Changes in its internal and external environment act as stimuli that are detected by receptor cells in sense organs. From here, impulses travel along sensory neurones to the central nervous system (CNS). The CNS consists of the brain and spinal cord, both also made up mainly of neurones. The brain processes the information it receives and decides which response to make, often taking into account previous experience (memory). Impulses are sent out along motor neurones to effector organs: the muscles and glands.

Control and co-ordination in the nervous system is covered in Chapters 15 and 16.

PLANT TISSUES

Like animals, plants are made up of different tissues that, together, form a whole functioning organism. Of course, there are basic differences between plants and animals, and plant tissues reflect these differences. Tissues of a multicellular plant are adapted for the following functions:

- photosynthesis
- transport
- support
- storage
- protection
- reproduction

The process of photosynthesis is covered in Chapter 34.

3 ORGANS AND BODY SYSTEMS IN ANIMALS

An **organ** is a collection of tissues that work together to perform a particular function. For example, think about that most delicate and sophisticated of organs, the eye.

- The iris contains muscular tissue, as do the suspensory ligaments that control the lens.
- The retina is made from nervous tissue that detects light and sends information to the brain.
- Epithelial tissues make up blood vessels and other structures such as the capsule around the lens, and the conjunctiva that covers the front of the eye.
- The whole organ is held together and protected by connective tissue. The sclera, in particular, forms a tough, protective outer coating.

Fig 2.5 The human eye. The eye is a sphere, about 2.5 cm across. Its outer wall has three layers: an inner layer, the retina, a middle layer, the choroid, and a tough outer layer, the sclera

blind spot – where the optic nerve leaves the brain

vitreous humour – a transparent jelly behind the lens

aqueous humour – a clear liquid in front of the lens

optic nerve – carries nerve impulses to and from the brain

conjunctiva – protects the front surface of the eye

fovea – has a high density of the cells involved in colour vision

iris – controls the amount of light entering the eye by varying the size of the pupil

lens – adjusts the focusing of the eye for near and distant vision

retina – contains light-sensitive cells

choroid – contains the blood supply to the eye

cornea – a transparent window that refracts light

sclera – the tough, white outer coat of the eye

eye muscles – move the eye

ciliary muscles and suspensory ligaments – these alter the shape of the lens to focus on near and distant objects

In a whole organism, organs form **systems** that carry out life processes such as digestion, excretion, reproduction. The combined systems of the body also need to create a stable internal environment to keep cells bathed in fluid that has the optimum temperature and composition.

? TEST YOURSELF

D Which body system in the list on this page does not contribute directly to the maintenance of internal conditions in the body?

Here is a short summary of the main systems in the human body together with their primary functions:

● The **digestive system** extracts from food the simple molecules that cells need (Fig 2.6).
● The **excretory system** removes the cell's waste products from the body.
● The **respiratory system** provides oxygen and removes carbon dioxide.
● The **circulatory system** transports all necessary substances to the cells, and removes waste products.
● The **muscles**, **skeleton** and **nervous system** combine to ensure that the organism obtains food and avoids danger.
● The **reproductive system** enables us to produce offspring and to pass on our genes.

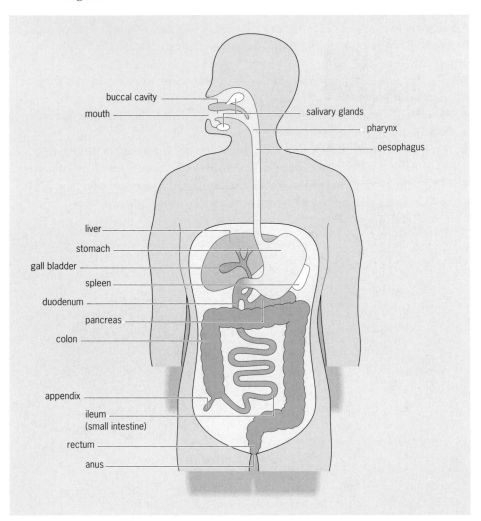

buccal cavity
mouth
salivary glands
pharynx
oesophagus
liver
stomach
gall bladder
spleen
duodenum
pancreas
colon
appendix
ileum (small intestine)
rectum
anus

Fig 2.6 The digestive system. Many organs form the system that enables us to process our food

SUMMARY

When you have studied this chapter, you should know and understand the following:

■ The cells that make up the human body are specialised for a variety of different functions. This is called **division of labour**.

■ Specialised cells form **tissues**, tissues form **organs** and organs form **organ systems**.

■ A tissue is a collection of similar specialised cells.

■ The tissues of the human body can be classified into four broad groups: **epithelial**, **connective**, **muscular** and **nervous**.

■ The whole organism consists of all the systems working together to produce an individual that is able to control its internal conditions.

KEY SKILLS **ASSIGNMENT**

BURNS AND TISSUE GRAFTS

The severity of a burn can be measured on a three-point scale:

Fig 2.A1

First degree burns are minor or 'partial thickness' burns (part-way through the skin), often caused by a mild scald or over-exposure to sunlight. The main symptom is redness of the skin. Though painful, they usually heal quickly and need minimal medical treatment.

Second degree burns are partial thickness burns caused by severe scalding, contact with flames or very severe sunburn. The dermis of the skin is damaged as well as the epidermis, and redness is accompanied by blistering. This type of burn is extremely painful and usually needs medical attention.

Third degree burns are severe, full thickness burns. Though they may not be painful, third degree burns completely destroy the skin and often some of the underlying tissue such as fat, muscle and bone.

1 Why might a patient with third degree burns often be in less pain than one with less severe burns?

Estimating the damage

The extent of a burn is expressed as a percentage of the total body area the burn covers. As a rough guide, medical staff use the rule of nines, shown in Fig 2.A2. One of the main problems that occurs once a large area of skin has been damaged, is that the body loses its ability to regulate its temperature and the person's energy requirements rise to that of a marathon runner. They need huge amounts of energy to try and deal with the fluid loss, the need to keep their body temperature normal, and to fight infection.

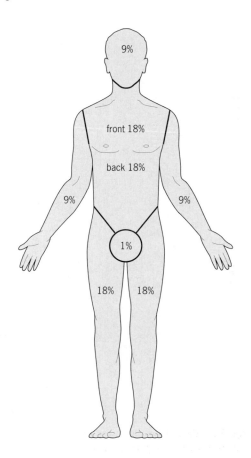

Fig 2.A2 The extent of a burn can be estimated by using the 'rule of nines'

2 **a)** A patient has burns to the whole of his back and all of both legs. What percentage of his body surface is this?
b) From your knowledge of the functions of skin, list some of the problems that will result from the loss of a large area. Refer to Chapter 12 if you need a reminder about skin function.
c) Before modern treatments were available, patients with large burns who survived the initial shock often died weeks later from starvation. Can you explain why?

KEY SKILLS **ASSIGNMENT**

Treatment

For many years, the main treatment available for burns, where the damage was not too extensive, was a **skin graft**. Skin from another region of the body was transferred onto the damaged area where it would eventually grow and provide an acceptable, if often unsightly, working replacement. Burns covering up to 40 per cent of the skin surface were usually treated by grafting techniques.

Fig 2.A3 New culture techniques allow large areas of skin to grow from a few healthy cells. Here, the cultured skin is sliced to stretch over three times its area

3 Why is it desirable to use a patient's own skin rather than a graft from another person?

During the 1990s, it became possible to culture skin outside the body (Fig 2.A3), and this is still being used and further developed. It is thought to be the most effective treatment for severe burns in the long term. Initially, however, the damage has to be covered, and this used to be done using either pig skin or skin from dead people (cadaver skin). This would be rejected within a few days, but kept the patient going until the cultured skin could be produced in large enough sheets. More recently, however, the development of artificial skin has revolutionised the treatment of burns and the death rates from the most serious burns have been reduced dramatically.

4 What advantages do you think artificial skin provides over animal or cadaver skin, or cultured skin?

Artificial skin is now produced commercially and is available in burn centres throughout the world. When a patient arrives with large area, third degree burns, surgeons have learned that the best treatment is to excise any damaged skin still left. This is a horrific operation in which the patient is 'peeled' with an instrument that does look quite like a potato peeler. The healthy underlying tissue is then covered as quickly as possible with artificial skin. This stops loss of fluid from the huge wound and also prevents microorganisms getting in to cause infection. The artificial skin also acts as a matrix for the patient's own skin to regenerate into and patients now suffer much less scarring than previously. The patient is kept sedated for long periods, so that they can recover with as little psychological trauma as possible.

5 Use the internet to research the survival rates from burns that cover between 70% and 90% of the body. How have the survival rates improved over the past 20 years or so?

The future

Researchers continue to make advances in wound healing and infection control. Newer skin substitutes, including those made from cultured cells of infant foreskins (a previously useless product of circumcision), are already or soon to be on the market. A new, aerosol spray-on suspension of the patient's own skin cells that takes only a few days to grow has also undergone trials in Australia. This method can also reduce scarring and is particularly useful for areas that are often hard to graft, such as the soles, underarms, and the palms of the hands. In Australia, more than 1200 patients have been treated using this method, but in Britain only a handful of patients have been treated so far.

03

The chemicals of life

03 The chemicals of life

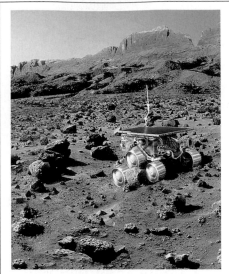

Was there ever life on Mars? In 2003, instruments on a Mars lander will look for minute traces of molecular fossils as some scientists believe that similar conditions to those on Earth may have once produced organic molecules. Finding amino acids on Mars could finally solve the mystery of whether life of any sort did once exist on Mars and it could also help us understand life's origins on our own planet

All new organisms come from other organisms – they don't just arise spontaneously. So how did life first come about? Could it have developed from non-living chemicals?

Some experiments, carried out in the US in the 1950s by Stanley Miller, showed that this is a definite possibility. In his laboratory, Miller recreated the conditions thought to have been present on Earth 3500 million years ago. Then, the atmosphere contained no oxygen. Instead, it was probably a mixture of methane, ammonia, carbon dioxide, hydrogen sulphide and hydrogen. There were frequent electrical storms, and no ozone layer to protect the surface of the Earth from the Sun's ultraviolet rays.

When these conditions were recreated in the laboratory, the chemical reactions in the mixture produced several biological molecules: sugars, amino acids and bases. The findings were important because these molecules are the building blocks of nucleic acids and proteins – the chemicals on which all life is based. Since these groundbreaking experiments, further work over the past half century has suggested that one of the earliest self-replicating 'life-forms' may have been RNA. An 'RNA world' may have persisted for a long time, until conditions became milder and DNA was able to take over. However, we still have many jigsaw pieces to find before we work out how even these complex molecules made the giant step to become life.

1 STUDYING THE CHEMICALS OF LIFE

The study of the chemicals and reactions that take place inside living organisms is a science in its own right – **biochemistry**. Biochemists aim to understand how living organisms work by studying how the molecules within their cells interact. In this chapter we look at **carbohydrates**, **lipids**, **proteins** and **nucleic acids**, and describe the characteristics that make them the chemicals of life.

SOME CENTRAL THEMES IN BIOCHEMISTRY

The chemical systems that make up living things seem incredibly complicated, but there are some simple underlying patterns:

- Living organisms contain a huge range of **macromolecules**, but these large molecules are built from a small number of simple molecules.
- These simple building blocks are very similar in all organisms, suggesting that all life had a common origin.
- The characteristics of an organism are determined by the information contained in its DNA.
- The DNA contains information that the cell can use to make proteins. Many proteins are **enzymes**, which control the physical and chemical activities of an organism.
- The chemical activities that go on inside an organism can be given the general term **metabolism**.
- Metabolic reactions can be divided into two general categories: **anabolic** and **catabolic**. Anabolic reactions build up large molecules from smaller ones, while catabolic reactions do the reverse, breaking down larger molecules.

Photosynthesis

Respiration

Fig 3.1 Organic molecules are normally made in plants via the process of photosynthesis. These molecules are then available to the rest of the food chain

Fig 3.2 All organisms, including plants, need energy. They obtain energy by breaking down organic molecules (food). This process is respiration

- Anabolic reactions usually involve **condensation reactions** in which building-block molecules are joined together and a water molecule is released.
- Catabolic reactions, such as those that occur during digestion, usually involve hydrolysis reactions in which larger molecules are split as they react with water.
- In photosynthesis (Fig 3.1), plants use the energy from sunlight to build up organic molecules such as sugars from simple ones such as carbon dioxide and water.
- All organisms need a supply of energy, which they obtain via respiration (Fig 3.2). In respiration, organic molecules are oxidised into simpler molecules, usually carbon dioxide and water. The resulting energy is used to fuel the many energy-requiring processes within the organism.

LIFE ON EARTH IS CARBON-BASED

Why is life on Earth based on carbon? It is because this element can bond with itself repeatedly, to produce an infinite variety of molecules. Living things are composed mainly of organic molecules: proteins, carbohydrates, lipids and other molecules that have 'skeletons' made from carbon.

THE CARBON ATOM

Carbon has an atomic number of 6: it has a nucleus containing six protons and six neutrons, orbited by six electrons, two in an inner shell, the other four in an outer shell (Fig 3.3). Carbon acquires a full, and therefore stable, outer shell of eight electrons by sharing four more. So, carbon forms four covalent bonds – bonds with electrons shared between the atoms. We say it has a **valency** of 4.

Carbon can form covalent bonds with other carbon atoms to form stable chains or rings. C–C bonds are the most common; compounds that contain single carbon bonds are said to be **saturated**. C=C bonds are also frequent and C≡C are possible. Compounds with double or triple bonds are said to be **unsaturated**. This means that they are not saturated with hydrogen (more hydrogen atoms could be added at their multiple bonds). This is particularly relevant with respect to the fatty acids, as we see later in this chapter.

 REMEMBER THIS

Water is essential to humans, as it is to all living organisms. Life evolved in water, most biochemical reactions take place in solution and all organisms that live on land – like ourselves – can do so because they have their own aquatic environment inside them.

All of the living cells in the human body contain water and are surrounded by it; in fact, most of the human body is made of water. Even the most complex organ – the human brain – is about 85 per cent water.

You can find more information about this important substance in the Extension box in Chapter 4.

REMEMBER THIS

Atoms have electrons in shells. Two electrons make the first shell full; eight electrons fill the second. Atoms with a full outer shell are stable. Atoms with a less than full outer shell are reactive and combine with others in order to fill the outer shell.

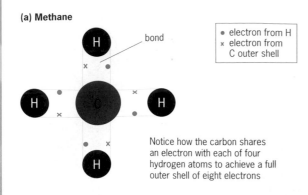

(a) Methane

H — bond

- electron from H
× electron from C outer shell

Notice how the carbon shares an electron with each of four hydrogen atoms to achieve a full outer shell of eight electrons

(b) Ethene

bond

bond

H

H

The two carbon atoms in this compound achieve a full outer shell of eight electrons by sharing electrons with hydrogen atoms and by forming a double bond

Fig 3.3 Dot and cross diagrams showing two simple carbon compounds

Carbon atoms commonly form covalent bonds with hydrogen, nitrogen, oxygen, phosphorus and sulphur. These elements make up the vast majority of organic molecules.

WHAT IS AN ORGANIC MOLECULE?

The word *organic* is much used today, but the 'supermarket' use of the word to mean 'no added chemicals' is not a scientific one. Originally, the word organic meant 'of living origin', back in a time when scientists thought that the chemicals that made living things had a divine origin and could not be made by people. When scientists started to make organic molecules in the laboratory (urea was first) this idea was abandoned.

For our purposes, we can say that organic molecules are ones with carbon backbones – at least one carbon bonded to another one. Some organic molecules are enormous, with relative molecular masses in the millions. Table 3.1 shows some common examples of simple organic and inorganic compounds.

In this chapter, we look at four types of large organic molecules commonly found in living things: carbohydrates, lipids, proteins and nucleic acids.

Table 3.1 Some simple organic and inorganic compounds	
Inorganic	**Organic**
Water (H_2O)	Ethane (C_2H_6)
Nitrogen gas (N_2)	Ethanol (C_2H_5OH)
Ammonia (NH_3)	Ethanoic acid (CH_3COOH)
Carbon dioxide (CO_2)	Glucose ($C_6H_{12}O_6$)
Hydrogen sulphide (H_2S)	Amino acid ($CHRNH_2COOH$)
Nitrate ion (NO_3^-)	

✔ REMEMBER THIS

A **monomer** is a single molecule. A **dimer** is formed when two of the same molecules link together. A **polymer** is formed when many of the same molecules link together.

2 CARBOHYDRATES

Carbohydrates are organic molecules that contain three elements: carbon, hydrogen and oxygen. Carbohydrates include **sugars** and **starches** (Fig 3.4 and Table 3.2). There are three basic types of carbohydrate molecule, named according to their structure and size:

- **Monosaccharides** are single sugars.
- **Disaccharides** are double sugars (made from two monosaccharides).
- **Polysaccharides** are multiple sugars (polymers of many monosaccharides).

Carbohydrates are the first molecules made in photosynthesis. Lipids, proteins and nucleic acids are formed from carbohydrates.

Fig 3.4 The term carbohydrate covers a range of chemicals which include sugars, starches and cellulose. This picture shows food containing complex carbohydrates – these foods are staples of the human diet

MONOSACCHARIDES AND DISACCHARIDES

Monosaccharides and disaccharides are classed as sugars and usually have names ending in **-ose**, such as **sucrose** and **lactose**. In monosaccharides, the three elements carbon, hydrogen and oxygen are always present in the same ratio and they have the basic formula $(CH_2O)_n$. Monosaccharides are classified according to the number of carbon atoms they have: 3, 5 and 6 are the most usual (Table 3.3). In glucose, for example, n is 6, so its formula is $(CH_2O)_6$ or $C_6H_{12}O_6$.

Table 3.2 Some common carbohydrates

CHO type	Compound	Sub-units	Occurrence in living things
Monosaccharide	glucose		widespread
	fructose		sweet fruits
	galactose		milk
Disaccharide	maltose	2 × glucose	germinating seeds
	sucrose	glucose + fructose	fruit
	lactose	glucose + galactose	milk
Polysaccharide	starch	glucose	plants (storage)
	glycogen	glucose	animals (storage)
	cellulose	glucose	plant cell walls

Table 3.3 Common monosaccharides

Number of carbon atoms	Name	Common examples
3	triose	glyceraldehyde (see Chapter 34)
5	pentose	ribose, deoxyribose (in RNA and DNA)
6	hexose	glucose, fructose, galactose

GLUCOSE

It is useful to begin a study of carbohydrates with **glucose**: it is the main source of energy for many organisms – including humans – and most of the common polysaccharides are glucose polymers. Glucose is a hexose (6-carbon) sugar that has the formula $C_6H_{12}O_6$. All other hexose sugars, such as **fructose** and **galactose**, have the same formula (Fig 3.5).

SEE QUESTION 1.

Fig 3.5 Variations on a common theme: common hexose sugars. A molecule of a glucose is shown (left) with all the atoms in place. This can be simplified to allow us to focus on the functional parts of the molecule – the OH groups on carbons 1 and 4. These groups react most commonly with other sugars. Where H on its own is joined to a C, the H can be omitted in the simpler form. The other three monosaccharides shown share the formula $C_6H_{12}O_6$ but, as you can see, they all have a slightly different structure. They are said to be **isomers** of each other. You might think these differences are too small to affect the molecule, but they do greatly affect properties such as taste (sweetness), digestibility and the nature of the polymers formed when the monomers join together

BIOCHEMICAL TESTS

Foods are usually mixtures of carbohydrates, lipids and proteins; one way to find out what type of molecules a food contains is to do biochemical tests. Some of the common ones you will use at A level are shown in Table 3.4 – these are basically the food tests you learned at GCSE level.

Table 3.4 Some simple biochemical tests

Nutrient	Reagent used	How test is carried out	Positive result
Reducing sugar	Benedict's solution	Add Benedict's solution to sample in a test-tube. Heat in a water-bath	orange-red precipitate
Non-reducing sugar	hydrochloric acid, and Benedict's solution	Once the reducing sugar test has proved negative, boil with dilute acid. Add sodium hydrogencarbonate to neutralise. Carry out reducing sugar test, as described above	orange-red precipitate
Starch	iodine solution	Add a few drops of iodine solution to the sample	blue-black staining
Lipid	ethanol	Shake the sample with ethanol in a test-tube. Allow to settle. Pour clear liquid into water in another test-tube	cloudy-white emulsion
Protein	Biuret solution	Add biuret solution to sample in a test-tube. Warm very gently	lilac/mauve colour

CHROMATOGRAPHY

This is useful when only tiny amounts of each substance is present and may not be detectable by one of the biochemical tests. It can also distinguish between substances that give the same result in biochemical tests. For example, the Benedict's test gives a positive result for both glucose and fructose because both are reducing sugars but chromatography can separate them very easily.

DOING CHROMATOGRAPHY

Fig 3.6 shows a basic **chromatography tank**. The paper has a pencil line at the bottom – the samples are dotted onto the paper along this line at regular intervals. Each sample is allowed to dry before adding the next, to avoid spreading and merging of the substances under test. When all the samples have been put on the paper, the bottom end is lowered into the solvent at the bottom of the tank. The lid is put on to prevent evaporation of the solvent, and the apparatus is then left for a few hours to allow the solvent to soak up the paper.

Different substances dissolve in a particular solvent to different degrees – a substance that dissolves very well will move further up the paper, perhaps to near the top. A substance that hardly dissolves at all remains near the original spot placed on the paper. If you are separating out pigments, say, for example, from a leaf extract, it is easy to see where the different substances are on the paper. However, if the test substance is colourless, you just end up with a piece of white paper – this must be treated with a stain to reveal the spots. For amino acid or protein mixtures, the paper is sprayed with **ninhydrin** which shows up the spots. This is highly toxic, so spraying is always carried out in a fume cupboard.

Fig 3.6 Chromatography in a beaker showing separation of pigments

IDENTIFYING YOUR SUBSTANCE

Having run the chromatogram, substances can be identified by working out their R_f value.

$$R_f = \frac{\text{Distance moved by spot}}{\text{Distance moved by solvent front}}$$

R_f values vary between 0 (didn't move at all) and 1 (went as far as the solvent front). A particular compound will always have the same R_f value in

(a)

A   B   C

Sample spot
of mixture

Sample spots
of known
substances

(b)

P

Q

y

x

R

$$R_f = \frac{\text{Distance moved by spot}}{\text{Distance moved by solvent front}} = \frac{x}{y}$$

Fig 3.7 Chromatograms

a particular solvent. So, for instance, every time you run lysine in the solvent propanone, it will give you exactly the same R_f value. Fig 3.7 shows how an R_f value is calculated.

CHROMATOGRAPHY: A WORKED EXAMPLE

Paper chromatography was used to find out which amino acids were present in polypeptide X (a polypeptide is a small section of a protein). First, enzymes were used to digest the polypeptide into its amino acids. Chromatography was then carried out using a specially coated plastic sheet (this has exactly the same role as blotting paper, but gives clearer results). After running with solvent Y, the dried chromatogram was sprayed with ninhydrin.

● EXAMPLE

Q Fig 3.8 shows the chromatogram produced by polypeptide X; Table 3.5 gives R_f values of some amino acids. Which amino acids are present in polypeptide X?

A The R_f of an amino acid is

$$\frac{\text{Distance moved by the spot}}{\text{Distance moved by the solvent front}}$$

Spot A has moved x mm, the solvent front has moved y mm. Therefore the R_f value is: $x/y = 48/50 = 0.96$. This amino acid is therefore probably proline.

Spot B has moved x mm, the solvent front has moved y mm. Therefore the R_f value is: $x/y = 45/50 = 0.90$. This amino acid is therefore probably leucine.

Spot C has moved x mm, the solvent front has moved y mm. Therefore the R_f value is: $x/y = 33/50 = 0.66$. This amino acid is therefore probably tyrosine.

Spot D has moved x mm, the solvent front has moved y mm. Therefore the R_f value is: $x/y = 25/50 = 0.50$. This amino acid is therefore probably glycine.

Spot E has moved x mm, the solvent front has moved y mm. Therefore the R_f value is: $x/y = 19/50 = 0.38$. This amino acid is therefore probably glutamic acid.

? TEST YOURSELF

A Fig 3.7a shows that mixture A contains substance B. How many substances does mixture A contain? Does the mixture contain substance C? Give the reason for your answer.

B In Fig 3.7b, the R_f value of substance P is 40/50, which is 0.80. Calculate the R_f values of substances Q and R. Which substance is least soluble in the solvent used?

✔ REMEMBER THIS

If you work out an R_f value to be more than 1, you've probably got the formula upside down.

Table 3.5 R_f values of some amino acids

Amino acid	R_f value
Alanine	0.70
Arginine	0.72
Glutamic acid	0.38
Glycine	0.50
Tyrosine	0.66
Leucine	0.91
Proline	0.95

Solvent front

A
B

C

D

E

Original spot

Fig 3.8 Amino acid chromatography

TWO-WAY CHROMATOGRAPHY

Sometimes, two substances will have the same R_f values in a particular solvent, so they won't be separated out. The problem is overcome by using two-way chromatograms. The first chromatogram is run in the usual way but then the paper or plastic sheet is turned through 90° and run again with a different solvent. As Fig 3.9 shows, this reveals extra spots not distinguishable in the first chromatogram.

Fig 3.9 Two-way chromatography

Fig 3.10 The Benedict's test can be used to estimate the amount of reducing sugars present in foods such as fruit juice. Samples with no reducing sugars remain blue, as left; those with a low concentration produce a green suspension, those with more produce yellow and orange suspensions, and juices very rich in reducing sugars produce an orange precipitate, as shown right

EXAMPLE

Q Fig 3.9 shows a one-way and two-way chromatogram run with a mixture of sugars. How many sugars were present in the mixture?

A The original chromatogram suggests only 4, but the two-way chromatogram reveals that there were actually 7. Three of the sugars must have very similar solubility in the first solvent used.

Q Which spots in chromatogram 1 contained more than one substance?

A Spots A and C. Spot A contains E and F. Spot C contains K, J and H.

Q Which substance was insoluble in the second solvent?

A Substance K because it has not moved towards the second solvent front.

? TEST YOURSELF

C In the Benedict's test, why does sucrose give a negative result before hydrolysis but a positive result after hydrolysis?

WHAT IS A 'REDUCING SUGAR'?

The term 'reducing sugar' reflects the fact that some sugars can reduce other chemicals. This basically means that the sugar can donate electrons to other substances. For a fuller explanation of reduction–oxidation reactions (redox reactions), see Chapter 34.

A standard test for a reducing sugar involves boiling the sample with Benedict's solution, a blue solution that contains copper sulphate. If a reducing sugar is present, the Cu(II) ions in copper sulphate are reduced to Cu(I) ions, resulting in an orange-red precipitate (Fig 3.10). Glucose, fructose, galactose, maltose and lactose are all reducing sugars, but sucrose is not. However, after sucrose is boiled with dilute acid to hydrolyse (split) it into its monosaccharides, it does produce a positive result.

MONOSACCHARIDES LINK BY MEANS OF GLYCOSIDIC BONDS

When two monosaccharides join together in a condensation reaction, the bond between them, a **glycosidic** link, centres around a shared oxygen atom (Fig 3.11). Two α-glucose molecules join together to make one molecule of **maltose**. Sucrose, the familiar sugar we buy in bags, consists of one molecule of glucose and one of fructose. Lactose, the main sugar in milk, is a disaccharide that contains glucose and galactose. The structures of sucrose and lactose are shown in Fig 3.12.

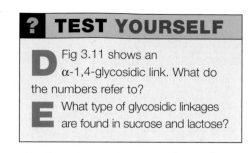

? TEST YOURSELF

D Fig 3.11 shows an α-1,4-glycosidic link. What do the numbers refer to?

E What type of glycosidic linkages are found in sucrose and lactose?

SEE QUESTION 2.

Fig 3.11 Two glucose molecules join to form maltose. Like many anabolic (building-up) reactions, this is a condensation reaction which involves the production of water

Fig 3.12 The structure of sucrose and lactose. Sucrose is made by the condensation of glucose and fructose. Lactose is made by the condensation of glucose and galactose

? TEST YOURSELF

F If a storage compound was soluble, what would it do to the cytoplasm of the cells in which it was stored?

? TEST YOURSELF

G Which component of starch – amylose or amylopectin – would the plant be able to build up and break down the quickest? Explain your answer.

POLYSACCHARIDES

STARCH

Starch is the most abundant storage chemical in plants (Fig 3.13) and it is the single largest provider of energy for most of the world's population.

Starch has the three properties that are necessary for a storage compound. It is:

- compact
- insoluble
- readily accessible when needed.

Starch is a mixture of two compounds, **amylose** and **amlyopectin**. Amylose is an unbranched polymer in which glucose monomers are joined by α-1,4-glycosidic linkages. These bonds bring the monomers together at a slight angle and, when they are repeated many times, a spiral molecule is produced. In amylose there are six glucose residues in a turn of the spiral.

The glucose chains of amylopectin have α-1,4-glycosidic linkages and α-1,6-glycosidic linkages. This allows branching (Fig 3.14).

Fig 3.13 In many plants, starch is the main storage carbohydrate. Foods such as rice, pasta and potatoes all have a high starch content. When treated with an iodine/potassium iodide solution, a blue-black starch–iodide complex is formed. This reaction has been used here to demonstrate the presence of starch in a potato

Fig 3.14 Amylose and amylopectin, the two different polymers that make up starch. When a plant needs to break down starch to provide glucose for respiration, it removes the terminal (end) units of amylose and amylopectin to release glucose. Since the branched amylopectin molecule has more terminal glucose units that can be removed simultaneously, it can be broken down more quickly than amylose

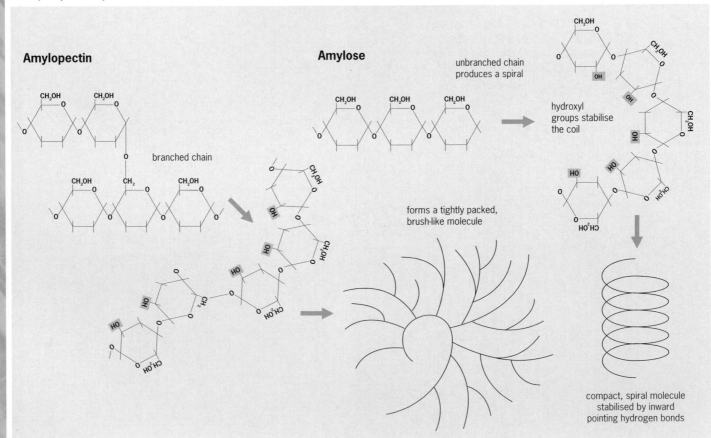

Amylopectin

branched chain

Amylose

unbranched chain produces a spiral

hydroxyl groups stabilise the coil

forms a tightly packed, brush-like molecule

compact, spiral molecule stabilised by inward pointing hydrogen bonds

GLYCOGEN

In humans, **glycogen** is the main storage carbohydrate. Its structure is similar to amylopectin, but it is even more frequently branched. In humans, glycogen is stored in large amounts in the liver and the muscles. During prolonged exercise, when the immediate supply of glucose is used up, the body restores its supplies by breaking down glycogen. If an average person goes without food, his or her glycogen stores last for about a day, but prolonged exercise such as marathon running can use all of the body's glycogen in less than two hours. When glycogen runs out, the body turns to using its lipid stores. This
is why eating less while taking more exercise is the quickest way to lose weight.

One of the major changes associated with improving fitness is an increase in the amount of the enzyme glycogen synthetase in the muscles. This allows glycogen to be built up faster after it has been used (Chapter 14).

CELLULOSE

Cellulose is a structural polysaccharide: it gives strength and rigidity to plant cell walls. Individual cellulose molecules are long unbranched chains containing many β-1,4-glycosidic linkages (Fig 3.15). The molecules are straight and lie side by side, forming hydrogen bonds along their entire length. This results in strong bundles of chains called **microfibrils**.

Cellulose is probably the most abundant structural chemical on Earth, but few animals can digest it because they do not make the necessary enzyme, **cellulase**. Herbivorous animals, whose diet contains large amounts of cellulose, can deal with it because they have cellulase-producing microorganisms in their digestive system. Humans cannot digest cellulose, but we make good use of it in other ways – its strength is used in products such as paper, cotton, Lycra and nail varnish.

> **? TEST YOURSELF**
>
> **H** Why is it an advantage to an animal to have a storage chemical – glycogen – that is even more branched than amylopectin?

> **? TEST YOURSELF**
>
> **I** Why do cellulose-based materials such as cotton take a long time to rot?

microfibril (enlarged)

cellulose chains

microfibril

cellulose β-1,4 linkages

Fig 3.15 Cellulose and the structure of the plant cell wall

3 LIPIDS

Lipids are a varied group of compounds that include the familiar **fats** and **oils**. As they are non-polar molecules, most lipids are insoluble in water but soluble in non-polar solvents such as alcohol and ether. Important exceptions are phospholipids, which have polar heads. The **emulsion test** for lipids, shown in Fig 3.16, is based on the solubility of lipids in ethanol.

Lipids contain the elements carbon, hydrogen, oxygen and sometimes phosphorus and nitrogen. They are intermediate-sized molecules that do not achieve the giant sizes of the polysaccharides, proteins and nucleic acids.

Fig 3.16 The emulsion test for lipids. The food to be tested is broken up into very small pieces, mixed with pure ethanol and shaken vigorously. Any lipid present dissolves in the alcohol. This top layer is poured off and mixed with water. If lipid is present, the mixture turns white as an emulsion, a suspension of fine lipid droplets forms. If no lipid is present, the mixture remains clear

LIPID STRUCTURE AND FUNCTION

The **triglycerides**, which act mainly as energy stores in animals and plants, are a large important group of lipids.

Triglycerides consist of one molecule of **glycerol** and three **fatty acids**. Fig 3.17 shows how the four components join by condensation reactions to form an E-shaped molecule.

SEE QUESTIONS 3 TO 5.

Fig 3.17 A triglyceride is formed from a condensation reaction between one molecule of glycerol and three fatty acids. The bonds formed are called ester bonds. The bottom right diagram shows the final shape of the triglyceride: glycerol is combined with three fatty acids to form an E-shaped molecule

The glycerol molecule is common to all triglycerides and so the properties of different triglycerides depend on the nature of the fatty acids. Fatty acids vary in the length of their chain and in the degree of **saturation** they show.

Table 3.6 shows the chain length of some common fatty acids. Chains of about 14 to 16 carbon atoms are the most usual, but they range from 4 to over 28. A **saturated fatty acid** has the maximum amount of hydrogen and therefore has no double bonds. **Monounsaturated fatty acids** possess one C=C bond and **polyunsaturates** contain more than one.

REMEMBER THIS

Fatty acids are organic acids. They contain a carboxyl (—COOH) group.

Table 3.6 Some common fatty acids

Fatty acid	No. of carbon atoms	No. of double bonds	Abundant in	Melting point/°C
Palmitic acid	16	0	palm oil	63.1
Stearic acid	18	0	cocoa	69.6
Lauric acid	12	0	coconut, palm oil	44.2
Oleic acid	18	1	olive, rapeseed	13.4
Linoleic acid	18	2	sunflower, maize	−5.0

? TEST YOURSELF

J Look at Table 3.6. Which fatty acids are unsaturated?

K Of the fatty acids listed in Table 3.6, which are liquid at room temperature (20 °C)? What do they have in common?

! SCIENCE FEATURE Fat cells have hidden depths

Fig 3.18 Fat cells

Fig 3.19 Leptin-deficient rat

People used to think that fat cells were just little blobs of fat surrounded by a membrane – which is certainly what they look like under the scanning electron microscope (Fig 3.18). However, we are now beginning to realise that fat cells are an important tissue in their own right, and that they help to regulate our metabolism.

Fat cells produce and secrete a wide range of proteins that act on other cells. Particularly interesting are the hormones such as leptin that act on centres in the brain that tell us to stop eating when we have had enough. (But how often do we listen?) The importance of leptin was demonstrated in the 1980s when rats that don't make any leptin were first bred. These animals look normal at birth, but they gain weight rapidly because they don't know when to stop eating. When they are a few months old, they become grossly fat, resembling spheres of fur with a head, four feet and a tail (Fig 3.19).

Other hormones produced by fat cells that also help control appetite and feeding behaviour have since been found. The interactions between them are so complex that we don't yet understand how they fit together. Even so, researchers are confidant that progress will be made and that we may have that elusive weight control pill in the next decade or so.

Fig 3.20 Lipids are vital chemicals in all living organisms. In humans, triglycerides occur mainly in adipose (fat storage) tissue, which forms under the skin and around internal organs. As well as storing energy, adipose tissue insulates against the cold, protects organs against physical damage, and contributes to a person's overall shape. The female body shape is, to a large extent, determined by adipose tissue distribution. This painting by Reubens shows the fuller female figure that was fashionable in previous centuries

Fig 3.21 There has always been considerable debate about the ideal body; it often appears to be dictated by fashion and culture, rather than common sense. The body shape of this model might by considered 'ideal' by many but it is a shape that few girls can achieve. It is also debatable whether such a low level of body fat is actually healthy

PHYSICAL AND CHEMICAL CHARACTERISTICS OF DIFFERENT TRIGLYCERIDES

Triglycerides, which contain longer chain fatty acids and saturated fatty acids, generally form 'hard' fats such as lard and suet that are solid at room temperature. Animal tissues contain a higher proportion of saturated fats than plants. Plant lipids contain shorter chain, unsaturated or polyunsaturated fatty acids and so are 'light' oils that are liquid at room temperature. Unsaturation leads to a lowering of the melting point because double bonds produce kinks in the carbon chain. This increases the distance between molecules by stopping chains from lying parallel and so reducing weak intermolecular forces – look ahead to Table 3.8. This makes the lipid more 'fluid'.

TRIGLYCERIDES AS ENERGY STORES

Many animals store energy in the form of triglycerides: gram for gram, they yield more than twice as much energy as proteins or carbohydrates. Triglycerides are **highly reduced** compounds; they contain many C–H bonds which can yield energy during respiration. If you look at the formula for a lipid, such as $C_{14}H_{26}O_2$ you can see that there is a far greater ratio of hydrogen to oxygen than in a carbohydrate such as sucrose, which has the formula $C_{12}H_{22}O_{11}$.

PHOSPHOLIPIDS

Phospholipids have a similar structure to triglycerides but one of the fatty acids is replaced by polar **phosphoric acid** (Fig 3.22). This gives the molecule a polar head and a non-polar tail. When placed in water, phospholipids arrange themselves with their hydrophobic ('water-hating') tails pointing inwards and their hydrophilic ('water-loving') heads pointing outwards. This is vitally important because it results in double layers called bilayers. Phospholipid bilayers form the basis of all biological membranes (see Chapter 4).

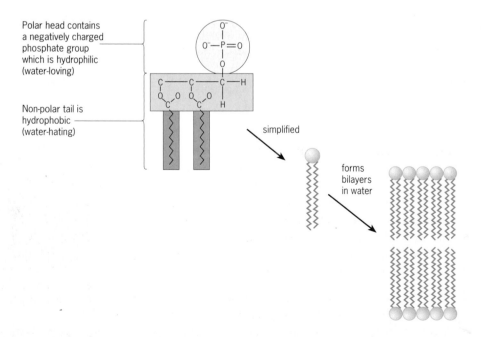

Fig 3.22 The structure of a phospholipid. The phosphoric acid gives the molecule a polar head and a non-polar tail. Cell surface membranes – and other membranes in the cell – form automatically due to the behaviour of phospholipids. In water, the hydrophilic polar heads face outwards while the hydrophobic tails point inwards

CHOLESTEROL

Many people associate **cholesterol** with heart disease, but this lipid is a perfectly normal constituent of every cell in our body (Fig 3.23 and see Chapter 4). As well as eating food that contains cholesterol, we can also synthesise cholesterol in the liver. The more there is in the diet the less the liver needs to make: vegans, who eat no animal products, are easily able to make all the cholesterol they need.

STEROID HORMONES

Steroid hormones have a similar structure to the cholesterol from which they are made. They include testosterone, progesterone and the oestrogens. The role of these hormones is discussed in Chapter 18.

WAXES

Waxes are lipids that are often used to waterproof surfaces, so preventing water loss. The cuticle of a leaf and the protective covering on an insect's body are both waxes. Waxes consist of a very long chain fatty acid joined to an alcohol molecule (not glycerol as in triglycerides). They have no nutritional value because they cannot be digested by lipases (lipid-digesting enzymes).

cholesterol

Fig 3.23 Cholesterol molecules are found between phospholipid molecules in membranes. Cell surface membranes are particularly rich in cholesterol, while organelle membranes contain less cholesterol

---

! SCIENCE FEATURE Steroids in athletics

Anabolic steroids are synthetic compounds that are similar in structure to **testosterone**, the male sex hormone. They increase anabolic (building-up) reactions, which enhance muscle size.

Athletes, notably weightlifters, began injecting testosterone in the 1950s to gain extra strength.

When steroids were banned, athletes looked for other natural substances that would be undetectable in blood tests. **Human growth hormone** (HGH, or **somatotropin**) and also **erythropoietin** are other illegal but potentially performance-enhancing substances. HGH increases muscle size in adults and erythropoietin increases the oxygen-carrying capacity of the blood. See the Assignment in Chapter 10.

HOW ANABOLIC STEROIDS WORK

Steroid hormones are lipid soluble and so can pass through the cell surface membrane. Studies have shown that anabolic steroids pass into the nucleus, where they force the cell to 'read' the genes that code for the muscle proteins actin and myosin. As well as increasing muscle size, athletes can train for longer, and the greater aggression that they tend to develop may also give them an added competitive edge.

The use of anabolic steroids is unfair and dangerous. To achieve a muscle-enhancing effect, athletes need to inject 10 to 40 times the normal dose, and this can have irreversible side effects. Symptoms of steroid abuse include acne, impotence, sterility, diabetes, heart disease and even liver cancer. Studies of male steroid abusers have shown that their testes may

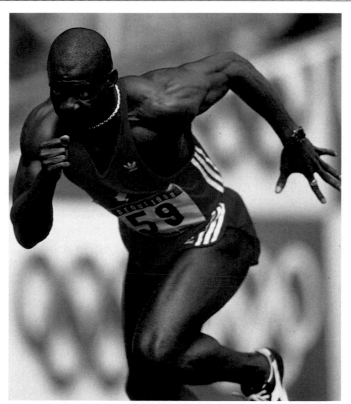

Fig 3.24 Athletes such as Ben Johnson have risked the shame of being banned from their sport by taking steroids to enhance their performance

decrease in size by as much as 40 per cent. Their sperm count is correspondingly lowered. Female athletes who use steroids tend to have fewer periods and can develop masculine features such as excessive body hair and a deeper voice.

Blood clotting factors.
Many components in the complex chain reaction of blood clotting are proteins

Keratin gives strength to skin, hair, nails, claws, hooves, etc

Tubulin forms microfilaments which make up the cytoskeleton of cells

Antibodies have a central role in defence against disease

Albumen is a blood plasma protein essential for normal circulation

Enzymes are globular proteins which control metabolism

Some **hormones,** eg insulin, are proteins

Membrane proteins are vital to the functioning of the plasma membrane, eg forming pores or acting as carrier molecules in active transport

Actin and **myosin** filaments produce movement, eg in muscles

Collagen gives strength to connective tissues – see the Feature box on page 57

Fig 3.25 After water, proteins are the major constituent of our bodies

Fig 3.26 The basic structure of an amino acid

Fig 3.27 Five of the 20 amino acids found in the human body. The R groups are highlighted in red, sulphur is shown in blue. A further 200 or more amino acid structures are possible, but these need to be synthesised (made artificially). They are never found in living systems

4 PROTEINS

Proteins play a central role in the structure and metabolism of all living organisms. Protein molecules have a huge variety of shapes and sizes (the structure of carbohydrates and lipids is relatively limited by comparison). This versatility of shape is the key to the role of proteins in the cell.

THE IMPORTANCE OF PROTEINS TO LIVING ORGANISMS

We can appreciate the extent to which organisms use proteins by looking at their distribution in the human body (Fig 3.25). Consider some of the jobs that proteins need to do:

● Each cellular metabolic reaction must be catalysed by a different **enzyme**.

● Each substance that passes across a cell membrane requires a different **carrier molecule**.

● A different **antibody** is needed to combat the chemicals produced by the many (and also constantly changing) disease-causing organisms such as bacteria and viruses.

Enzymes, carrier molecules and antibodies are all proteins that are 'tailor-made' to fulfil all these requirements.

THE STRUCTURE OF PROTEINS

Proteins are large and complex molecules (Table 3.7). If a water molecule (relative molecular mass = 18) was the size of a brick, proteins would be whole buildings.

In addition to the elements carbon, hydrogen and oxygen, proteins always contain nitrogen and sometimes sulphur. The building blocks of proteins are called **amino acids**. Fig 3.26 shows the basic structure of an amino acid. As the name suggests, all amino acids contain an amino group ($-NH_2$) and a carboxylic acid group ($-COOH$). Both of these groups are attached to a central carbon atom, known as the α-carbon. The 'backbone' of an amino acid is made up from the three atoms C–C–N.

The R group varies between different amino acids. Fig 3.27 shows five of the different R groups in the 20 amino acids that make up all of the proteins in the human body.

Table 3.7 The relative molecular masses of several proteins	
Protein	**Relative molecular mass**
Insulin	5 700
Haemoglobin	64 500
Myoglobin	16 900
Hexokinase	102 000
Glycogen phosphorylase	370 000
Glutamine synthetase	592 000

AMINO ACIDS IN SOLUTION

Amino acids are readily soluble in water. When in solution they can resist a change in pH by mopping up or releasing both hydrogen ions (H^+) and hydroxyl ions (OH^-), acting as buffers, that is, regulators of pH. In whole proteins, the buffering effect is largely due to the R groups. Maintaining a constant pH is an important aspect of **homeostasis** (see Chapters 11 to 14).

PEPTIDES, POLYPEPTIDES OR PROTEINS?

It is easy to be confused about the precise meaning of the terms **peptide**, **polypeptide** and **protein**; there isn't a universally accepted rule to follow. This is how we use them in this book.

When amino acids are assembled into a short chain a peptide is formed. Longer chains are known as polypeptides. The term protein is reserved for the finished, functional molecule. Some proteins consist of one polypeptide, others consist of two or more than two. Haemoglobin, for example, contains four polypeptides.

HOW AMINO ACIDS FORM PROTEINS

Amino acids join together in long chains to form proteins by means of **peptide bonds**. Fig 3.28 shows how two amino acids join to form a dipeptide. This is an example of a **condensation reaction**.

All proteins are complex molecules and biochemists look at their structure at four different levels: **primary**, **secondary**, **tertiary** and **quaternary**.

PRIMARY STRUCTURE

The primary structure of a protein refers to the sequence, or order, of amino acids in that protein (Fig 3.29).

A simple primary structure of a tiny protein could be shown as:

alanine–histidine–phenylalanine–glutamine–cysteine

Real proteins usually consist of a lot more than five amino acids. The hormone insulin, for example, a relatively small protein, has 51 amino acids.

The code for the primary structure of any protein is contained in the gene or genes that code for that protein. This code determines the precise order in which amino acids are assembled. This order then dictates the way they will twist and turn to produce the precise three-dimensional shape that allows the protein to carry out its specific function in the body. The first level of three-dimensional twisting is described as the secondary structure of the protein.

SECONDARY STRUCTURE

When combinations of amino acids join together in a chain they tend to fold into particular shapes and patterns (such as spirals) in some places. These shapes form because the amino acids twist around to achieve the most stable arrangement of hydrogen bonds. The secondary structure of the protein refers to the patterns contained within the amino acid chain. Such patterns exist in different places in different proteins, producing an almost infinite variety of molecular shapes.

REMEMBER THIS

No matter how many amino acids are in the chain, if there is a —COOH group on one end, there will be an —NH_2 group on the other.

Fig 3.28 Two amino acid molecules react together in a condensation reaction (a reaction that releases water) to produce a dipeptide

Fig 3.29 When amino acids join, there is a repeating sequence of C—C—N—C—C—N— etc. This backbone runs throughout the length of the protein

SEE QUESTION 6

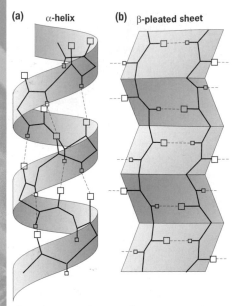

(a) α-helix **(b)** β-pleated sheet

Fig 3.30 In an α-helix, the polypeptide is coiled into a helix that is held in position by hydrogen bonds between different groups in the backbone. In a β-pleated sheet, the amino acid chains lie side by side, forming a sheet that is held together by hydrogen bonds between adjacent parts of the polypeptide

✔ **REMEMBER THIS**

A **residue** is the name given to the remains of a monomer that has become incorporated into a polymer.

The main types of secondary structure in proteins are:

● The α-**helix**, a spiral, is the most common type of secondary structure (Fig 3.30a). Amino acid **residues** in the spiral twist on their axis, each residue forming a hydrogen bond with another residue four units along. These hydrogen bonds stabilise the α-helix.

● The β-**pleated sheet**, a flat structure that consists of two or more amino acid chains running parallel to each other, linked by hydrogen bonds (Fig 3.30b).

The secondary structure of a protein depends on its amino acid sequence: some amino acids (or combinations of amino acids) tend to produce α-helices, others usually make β-sheets. A few amino acids tend to produce a sharp bend in the chain, a vital function that allows the chain to fold back on itself. Fig 3.31 shows the secondary structures within the enzyme lysozyme.

Fig 3.31 The secondary structures in the enzyme lysozyme: the α-helices and β-pleated sheets are clearly visible. The α-helices are shown as spiral ribbons, the β-pleated sheets as broad, flat arrows. This enzyme is present in tears and sweat, where it catalyses the breakdown of some bacterial cell walls

α-helix

β-sheet

β-bend

cystine S–S disulphide linkage

TERTIARY STRUCTURE

The tertiary structure of a protein is its overall three-dimensional shape and is produced as a result of the following:

● The sequence of amino acids that produces α-helices, β-sheets and bends at particular places along the chain.

● The hydrophobic nature of many amino acid side chains. Globular proteins are surrounded by water and so the hydrophobic side chains tend to point inwards.

Tertiary structure is maintained by attractive forces that arise when the amino acid chain folds (Table 3.8) and by **disulphide bridges**, covalent bonds that form when two **sulphur-containing cysteine residues** react. Disulphide bridges occur most often in structural proteins, where they contribute to strength.

We cannot over-emphasise the importance of tertiary structure for protein function. Functional proteins, such as enzymes and antibodies, must have an exact shape – and sometimes the ability to change shape – to fulfil their role in the organism. Many structural proteins depend on their tertiary structure for strength. The large number of disulphide bridges in keratin, for example, makes body structures such as hair and nails very tough.

If a protein consists of only one polypeptide, the tertiary structure

Table 3.8 The different types of bond that maintain the secondary, tertiary and quaternary structure of a protein

Type of bond	Formed between	Relative strength
Hydrogen bonds	H and an electronegative atom, usually O or N	weak, very common
Ionic bonds	oppositely charged ions	weak
van der Waals forces	non-specific, between nearby atoms	weak
Disulphide bridges	two SH-containing cysteine residues	strong, contribute to strength of fibrous proteins

is the final shape of the molecule. If, however, the protein has more than one, it has a further, quaternary level of structure.

QUATERNARY STRUCTURE

Many proteins consist of more than one polypeptide chain and sometimes also have non-protein **prosthetic** groups, often vital to the function of the protein. The quaternary structure refers to the three-dimensional structure, or **conformation**, produced when all the sub-units combine to give the final, active molecule (Fig 3.32).

Fig 3.32 A computer-generated model of the protein molecules of collagen, the most abundant protein in the human body

FIBROUS AND GLOBULAR PROTEINS

The final three-dimensional structure of proteins results in two main classes of protein – **fibrous** and **globular**. Fibrous proteins contain polypeptides that bind together to form long fibres or sheets. They are physically tough and are insoluble in water (Fig 3.33).

Fig 3.33 This diagram shows how the fibrous protein keratin is arranged inside a human hair

Globular proteins are usually individual molecules with complex tertiary and quaternary structures. They are spherical, or globular in shape, hence the name. Most are soluble in water and they tend to have a biochemical rather than a structural function (Fig 3.34).

HOW STABLE ARE PROTEINS?

As the final shape of globular proteins is maintained by relatively weak molecular interactions such as hydrogen bonds, proteins are very sensitive to temperature increases and other changes in their environment. As the temperature goes up beyond 40 °C, molecular vibration increases and bonds that are holding together the tertiary or quaternary structure break, changing the shape of the molecule. This is known as **denaturation**. Different proteins are denatured at different temperatures. Some begin to be denatured after about 40–45 °C or even below, but many are not totally denatured until 60 °C or even higher. It is an oversimplification to say 'organisms die at temperatures over 44 °C because their proteins become denatured'. In practice organisms die because of a metabolic imbalance, caused when enzymes work at different rates – see Chapter 5.

Proteins can also be denatured by adverse chemical conditions. Chemicals that affect weak bonds alter the overall structure; even a slight change in protein shape can mean loss of function. Some proteins are particularly sensitive to changes in pH.

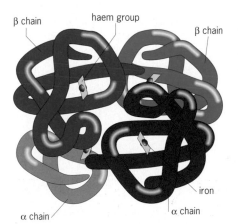

Fig 3.34 Haemoglobin is an example of a globular protein. This complex molecule picks up oxygen from the lungs and releases it to respiring tissues. Each molecule consists of four polypeptide chains held together by disulphide bonds. There is a prosthetic group, the haem group, at the centre of the each polypeptide chain. The function of haemoglobin is discussed more fully in Chapter 9

! SCIENCE FEATURE When diabetes is not helped by insulin

Like many other protein molecules with a biochemical function, such as enzymes and hormones, the hormone insulin is a globular protein (Fig 3.35). Produced by the pancreas, it is a major regulator of metabolism. When blood sugar levels are high, insulin is secreted into the blood and it quickly travels all around the body. All cells of the body have insulin receptors on their surface and when insulin binds, this causes glucose channels to change shape slightly, particularly in muscle cells and fat cells. This then allows these cells to take up glucose from the blood and store it as glycogen or fat.

Type I diabetes, which we cover in more detail in Chapter 11, arises when the cells in the pancreas that produce insulin become damaged. It usually occurs in young people (under 20) and so is sometimes called juvenile-onset diabetes. It is treatable by insulin injections. However, there is also another type of diabetes – called Type II. This affects much older people – in their 40s and 50s and beyond – and often does not respond to insulin injections. The problem here is that muscle and fat cells become resistant to the effect of insulin – one reason is that they don't have enough insulin receptors.

One of the hormones produced by fat cells has recently been linked to this type of diabetes. Mice that don't make any adiponectin show the same symptoms as people with Type II

diabetes. This hormone is now being considered as a future treatment as it is able to increase cells' sensitivity to insulin, possibly by increasing the number of insulin receptors made.

Fig 3.35 Although it looks like a blob, each insulin molecule contains two sub-units. Each sub-unit has two chains and is bound to a single zinc atom. Atoms and molecules that bind to proteins are called ligands. Ligands are bound to the proteins by weak bonds and are usually more easily removed from the protein than are prosthetic groups. Proteins can have many different types of ligands and these are often important for the function of the protein.

ANALYSING PROTEINS

We can find out if a sample contains protein using the **Biuret test** (Fig 3.36). More complex biochemical techniques can be used to tell us more about proteins in living systems. We can:

- find out which proteins are present in a mixture, such as plasma,
- work out the exact three-dimensional shape of a protein using techniques such as X-ray diffraction and nuclear magnetic resonance,
- find out which amino acids are present in a particular protein,
- analyse the exact amino acid sequence of a protein,
- use computers to predict the three-dimensional shape of a protein, using only its amino acid sequence.

food sample Biuret solution sample goes lilac
(CuSO₄ + NaOH) if protein is present

Fig 3.36 The Biuret test is a simple laboratory test that detects the presence of peptide bonds

IDENTIFYING INDIVIDUAL PROTEINS

The proteins present in a body fluid such as blood plasma can be separated from each other by electrophoresis. Each protein carries a particular overall electrical charge. Electrophoresis uses this fact to separate individual proteins, as Fig 3.37 explains.

Fig 3.37 Electrophoresis can be thought of as chromatography using electricity. Instead of separating chemicals according to their solubility, electrophoresis causes individual molecules to move to the positive or negative terminal at speeds that depend on their mass and overall charge. This is also the principle behind DNA fingerprinting – see the Assignment at the end of the chapter

Myosin (200 000)
Phosphorylase a (92 000)
Bovine serum albumin (68 000)
Ovalbumin (43 000)
Glyceraldehyde 3-phosphate dehydrogenase (36 000)
Carbonic anhydrase (29 000)
Myoglobin (17 200)
Cytochrome c (11 700)

> ⚠ **SCIENCE FEATURE Collagen**
>
> Collagen is probably the most widespread structural protein in the human body. It is a fibrous protein that gives strength to tissues such as tendons, ligaments, bone and skin. Fig 3.38 shows the structure of collagen. A single collagen molecule contains three polypeptides – each of about 1000 amino acids – intertwined to form a triple helix. This arrangement has great strength, mainly due to the large number of hydrogen bonds that occur along the length of the polypeptides.
>
> Collagen is secreted in an unassembled form because the instant formation of large fibres would damage the cell that made it. Complete collagen is produced as enzymes act on the individual polypeptides, causing them to twist together to form very long fibres (sometimes several millimetres long). These have the tensile strength of steel and are used to strengthen bone in much the same way as metal rods reinforce concrete. Brittle bone disease (Fig 3.39), is a genetic disorder which causes a fault in the bonding between collagen and the mineral component of bone.
>
> The primary structure of collagen is very regular. It consists of a repeating sequence of glycine and two other amino acids, often proline and hydroxyproline. These amino acids do not cause the chain to gain the normal α-helix or β-sheet structure. Instead, they form long separate chains that allow the collagen triple helix to form. This is a good example of how the primary structure is ultimately responsible for the shape and properties of the whole protein.
>
>
>
> **Fig 3.38** The structure of collagen
>
>
>
> **Fig 3.39** The legs of a small child suffering from 'brittle bone' disease – an unusually severe case. Frequent fractures of the leg result in the deformities

WORKING OUT THE THREE-DIMENSIONAL SHAPE OF A PROTEIN

X-ray diffraction is useful here. When a beam of X-rays is fired at a crystal of purified protein, the atoms in that protein diffract (bend) the X-rays, producing a specific pattern on a photographic plate. The shape of a protein is not immediately obvious from the information produced, but a trained scientist with a computer can produce an accurate three-dimensional model of the molecule.

IDENTIFYING WHICH AMINO ACIDS ARE PRESENT

Protease (protein-degrading) enzymes break up the protein and then the constituent amino acids are identified using chromatography or electrophoresis. These techniques do not reveal the sequence of amino acids.

DETERMINING THE AMINO ACID SEQUENCE

The protein is digested into manageable chain lengths of amino acids and each length is analysed in turn. The polypeptide is treated with a chemical that binds to the N-terminal (N-end) amino acid but not to any of the others. The 'tagged' terminal amino acid is removed from the rest of the chain and identified by chromatography. Frederick Sanger, who won a Nobel Prize in the 1950s, was the first scientist to sequence a complete protein (insulin). Today, the process is fully automated and has become a standard technique.

COMPUTER MODELLING

As biochemists accumulate knowledge about the structure of many individual proteins, they produce computer software to help work out the shape of other complete proteins using their amino acid sequences. This is incredibly useful

SEE QUESTIONS 7 AND 8.

in genetics. Sometimes a new gene is sequenced and no one knows its function. The DNA's sequence can be given to a computer which then works out the sequence of amino acids that would be produced if the gene were expressed in a cell. It can then 'build' the final protein and compare it to other proteins with known functions.

5 NUCLEIC ACIDS

Nucleic acids are so called because they are slightly acidic molecules, and it because was thought originally that they occurred only in the nucleus. The two types of nucleic acid, **DNA** and **RNA**, both contain carbon, hydrogen, oxygen, nitrogen and phosphorus.

THE STRUCTURE OF NUCLEIC ACIDS

Nucleotides are the building blocks of nucleic acids. A nucleotide consists of three units (Fig 3.40):
- a **sugar** (ribose or deoxyribose),
- a **phosphate** group,
- a nitrogen-containing **base**.

As the names imply, **deoxyribo**nucleic acid has nucleotides in which the sugar is **deoxyribose**, while **ribo**nucleic acid contains the sugar **ribose**.

DEOXYRIBONUCLEIC ACID (DNA)

Deoxyribonucleic acid (DNA) is the macromolecule that carries the **genetic code**, the information for making the cell's proteins. Most of the DNA in a eukaryotic cell is in the nucleus (Fig 3.41).

The nucleotides in DNA can contain any one of four nitrogenous (nitrogen-containing) bases: **adenine**, **guanine**, **cytosine** or **thymine**. When DNA **replicates** (copies itself), it makes new strands by adding nucleotides. These are available as free molecules in the cytoplasm. Generally, cells can synthesise their own nucleotides.

Fig 3.40 The basic structure of a nucleotide

SEE QUESTION 9.

? TEST YOURSELF

L If there are 46 chromosomes in a human cell, each with an average of 5 cm uncoiled length, estimate the length of DNA contained within each cell.

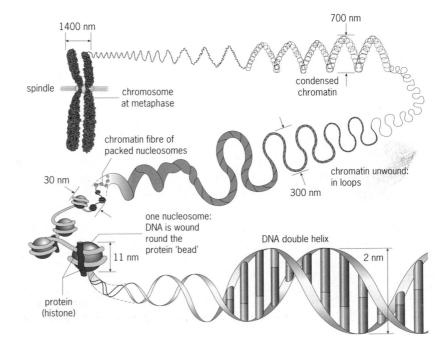

Fig 3.41 Studies of chromosomes have shown that each one is a giant, highly coiled DNA molecule. Its shape is achieved by supercoiling – coils within coils, condensing a huge amount of DNA into a tiny length. Each chromosome can contain 300 000 000 nucleotide units and, unravelled, the DNA within it would be between 5 and 10 centimetres long!

HOW DNA CARRIES THE GENETIC CODE

DNA has two remarkable characteristics:

● It is a store of genetic information.
● It can copy itself exactly, time after time.

Looking at the structure of DNA helps us to understand how it does this.

HOW THE BASES PAIR

Adenine and guanine belong to a group of chemicals called **purines** while thymine and cytosine are **pyrimidines**.

Because of the shape of the two types of molecule, each purine always bonds with only one pyrimidine. So, in DNA, adenine always bonds with thymine, and cytosine with guanine. In RNA, cytosine bonds with guanine and adenine bonds with uracil:

DNA: A=T   RNA: A=U
     G≡C   G≡C

The base pairs are held together by hydrogen bonds. There are two H-bonds between A and T (or U) and three between C and G.

Fig 3.42 shows how the base pairs within DNA fit together to form a double-stranded helix. The sides are formed by alternating sugar–phosphate units, while the base pairs form the cross-bridges, like the rungs of a ladder. Each base pairing causes a twist in the helix and there is a complete 360° turn every 10 base pairs.

There is more about how DNA stores information to act as the genetic code in Chapter 23.

SEE QUESTIONS 10 AND 11.

? TEST YOURSELF

M If one half of a DNA strand had the base sequence ATCGTTACC, what sequence would the other strand have?

✔ REMEMBER THIS

Hint: To remember which bases are pyrimidines and which are purines, concentrate on the letter Y: Thymine and cytosine are pyrimidines.
Adenine and guanine are purines.

? TEST YOURSELF

N List the four major differences between RNA and DNA.

Fig 3.42 Nucleotides condense to form DNA. There are two parallel strands of bases. For any gene, only one strand of the DNA – known as the sense strand – is used to make proteins. The other side serves to stabilise the molecule. The sense strand can alternate between one gene and the next; it is not always on the same side of the molecule

RIBONUCLEIC ACID (RNA)

Three of the bases in RNA – adenine, guanine and cytosine – are the same as those in DNA. The fourth is different: RNA contains uracil instead of thymine.

RNA molecules are much smaller than DNA molecules. DNA can consist of over 300 000 000 nucleotides; RNA usually consists of a few hundred. RNA is also less stable. DNA molecules are the permanent store for genetic information and last for many years. In contrast, RNA molecules have a short-term function and are easily replaced. There are three forms of ribonucleic acid (RNA) in the cell:

● **Messenger RNA** (mRNA) (Fig 3.43a) can be thought of as a mobile copy of a gene. Small lengths of mRNA are assembled in the nucleus using a single gene within the DNA as a template. When a complete copy of the gene has been produced, the mRNA moves out of the nucleus to the ribosome, where the protein is synthesised according to the code taken from the DNA.

● **Transfer RNA** (Fig 3.43b) is found in the cytoplasm and is a carrier molecule, bringing amino acids to the ribosomes for assembly into a new amino acid chain, according to the order specified on the mRNA code.

● **Ribosomal RNA** makes up part of the ribosome, a small organelle that brings together all the chemicals associated with protein synthesis.

More details about nucleic acids can be found in Chapter 23.

etc

Fig 3.43(a) Messenger RNA is a relatively delicate, short-lived molecule composed of a single strand of nucleotides. It carries the base sequence from the gene to the ribosome and provides a template for protein synthesis

hydrogen bonds

single nucleotide strand

Fig 3.43(b) Transfer RNA is a single folded strand of nucleotides that is stabilised by hydrogen bonds, producing what is often described as a 'clover leaf' structure. The tRNA brings amino acids to the ribosome so that they can be added on to the growing amino acid chain

SUMMARY

When you have finished this chapter, you should know and understand the following:

CARBOHYDRATES

- **Carbohydrates** contain the elements carbon, hydrogen and oxygen. They are the first products made by plants in photosynthesis.
- The term **sugars** describes **monosaccharides** and **disaccharides**. Their names end in the suffix **-ose**.
- Monosaccharides include **glucose**, **fructose** and **galactose**. These sugars are **isomers**: they have the same formula but their atoms are arranged in different ways. Monosaccharides are linked by **glycosidic bonds**. These are formed by **condensation reactions**.
- Disaccharides, such as **maltose**, **sucrose** (cane sugar) and **lactose** (milk sugar), consist of two monosaccharides linked together.
- Most **polysaccharides** are **polymers** of glucose. **Starch** is used for storage in plants, **glycogen** is used for storage in animals, and **cellulose** gives strength to plant cell walls.

LIPIDS

- **Lipids** contain carbon, hydrogen, oxygen, often phosphorus and occasionally nitrogen. Most are **non-polar** chemicals and therefore insoluble in water.
- Lipids are used for energy storage, protection and insulation.
- In living things there are two main types of lipid: **triglycerides** and **phospholipids**. Triglycerides are the familiar fats and oils. Phospholipids form cell membranes.
- Fatty acids vary in chain length, and may be **saturated** (with hydrogen) or **unsaturated**. These factors determine the properties of the triglyceride, such as its melting point and viscosity.
- Phospholipids are similar to triglycerides, but phosphoric acid replaces one of the fatty acids. They have a polar head, allowing them to form **bilayers** (membranes) in water.
- **Cholesterol** is a normal constituent of cell membranes.

PROTEINS

- **Proteins** consist of chains of **amino acids** linked by **peptide bonds**.
- There are 20 different amino acids in living things. All have a **carboxylic acid** group and an **amino group** but differ in their **R group**.
- **Fibrous proteins** often join to form large fibrils whose function is to provide strength or produce movement. **Globular proteins** – including enzymes, antibodies and some hormones – are usually individual molecules with a chemical function.
- The **primary structure** of a protein is the sequence of its amino acids.
- The **secondary structure** refers to the patterns and shapes formed within the **polypeptide** chain, for example, an **α-helix**.
- The **tertiary structure** refers to the three-dimensional shape of a polypeptide chain, which results from the interactions as the chain folds back on itself. If the protein consists of one polypeptide chain, the tertiary structure refers to the overall shape of the molecule.
- The **quaternary structure** refers to the overall three-dimensional shape of a protein that consists of more than one polypeptide chain.

NUCLEIC ACIDS

- There are two types of **nucleic acid**, **deoxyribonucleic acid (DNA)** and **ribonucleic acid (RNA)**. RNA itself has three forms, **messenger RNA**, **transfer RNA** and **ribosomal RNA**.
- DNA carries the genetic code. Its structure allows it to store information, pass information on to RNA so that proteins can be made, and to copy itself, allowing the genetic code to pass to new cells.
- Messenger RNA copies the genes, to allow them to be used as templates for protein synthesis. Transfer RNA brings amino acids to the ribosome during synthesis. Ribosomal RNA is a major structural component of the ribosome.
- DNA and RNA are composed of **nucleotides**, which themselves contain a **sugar**, a **phosphate group** and a **nitrogen-containing base**.
- DNA contains the bases A, C, T and G. RNA contains the bases A, C, U and G. A always pairs with T (or U) and C always pairs with G.

? EXAM QUESTIONS

1 The figure shows two possible structures for glucose.

α glucose   β glucose

Both molecules α and β have the same general formula $C_6H_{12}O_6$.

a) **(i)** What is the term used to describe compounds with the same formula but different structure?
(ii) Using the information in the above diagram distinguish between α and β glucose molecules.

b) Indicate, by placing a ring around the appropriate atoms on the diagram above, which atoms are lost when the two glucose molecules join together.

c) Name the bond that is formed when the molecules join together.

d) Name the sugar that is formed after the molecules have joined.

e) Starch and cellulose are both made up from a number of glucose molecules. Suggest **two** differences between the structures of starch and cellulose.

f) Give **one** function of **each** of the molecules in cells.
(i) Starch
(ii) Cellulose.

WJEC Biology Module BI1 June 2001 Q5

2
a) Write a chemical equation to summarise how two molecules of glucose are joined to form maltose. Structural formulae are **not** expected.

b) State **two** ways by which the reaction you have shown in **(a)** could be reversed.

c) **(i)** Maltose and sucrose both have the same empirical formula, but different structural formulae. What biochemical term is used to describe this?
(ii) Maltose is a *reducing sugar* but sucrose is a *non-reducing sugar*. If you were given a solution suspected to contain a mixture of these two sugars and asked to prove their presence, describe the procedure you would use.

OCR Biology Molecules and Life March 2000 Q1

3
a) **(i)** Draw and label a diagram to show the structure of a triglyceride.
(ii) Indicate, with an X on the diagram, a site where hydrolysis takes place.

b) Explain the differences in solubility between triglycerides and the products of their hydrolysis.

c) Suggest why triglycerides release twice as much energy on oxidation compared with an equivalent mass of carbohydrates.

OCR Cambridge Linear Biology Paper 3 June 2000 Q3

4 The figure shows diagrams that represent lipid and lipid-related molecules.

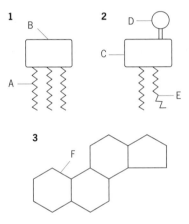

a) **(i)** Which are the **three** most common **elements** in these molecules?
(ii) Give the number of the molecule (1, 2 or 3) which contains a fourth element and name this additional element.

b) **(i)** Which molecule (1, 2 or 3) could most easily be hydrolysed to provide four molecules by boiling with alkali such as sodium hydroxide solution?
(ii) Name the products that would result from this procedure.

c) **(i)** What general name is given to molecules with a structure similar to diagram 3?
(ii) Suggest **one** important function in mammals for the type of molecule.

d) **(i)** Which **one** of the parts labelled **A** to **F** in the figure, would most likely be metabolised in order to provide energy if a person goes on a 'low calorie diet'?
(ii) State how this part is used in the metabolic pathway to provide energy.

? EXAM QUESTIONS

The nutritional information label on a tub of 'low fat' margarine states:

Fat … 38.0 g (per 100 g), of which, saturates = 10.6 g, unsaturates = 27.4 g.

e) **(i)** Describe briefly how molecules of **unsaturated** fats are different in structure from molecules of saturated fats.

(ii) How does this difference in chemical structure affect their physical properties?

Lipids as energy providers contain about $592\,000\,\text{kJ}\,\text{kg}^{-1}$. The equivalent figure for glycogen is $3\,800\,\text{kJ}\,\text{kg}^{-1}$.

In a human male weighing 70 kg, on average, 20% of body mass is fat stored around internal organs and under the skin.

f) **(i)** Calculate the amount of chemical energy contained in the fat.

(ii) Suggest **one** reason why humans do **not** use carbohydrates as long term energy stores.

OCR Biology Molecules and Life June 2000 Q4

5 The figure shows the structure of a lipid (triglyceride) molecule.

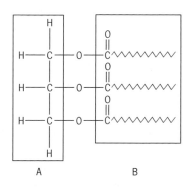

(a) Name the parts **A** and **B**.

(b) **(i)** What type of chemical reaction takes place to form the bonds between **A** and **B**?

(ii) What is the other product of this reaction besides the lipid molecule?

(c) State **one** function of this type of lipid in living organisms.

WJEC Biology Module BI1 January 2001 Q3

6 The figure represents the structure of lysozyme, a protein consisting of a single polypeptide, found in egg white.

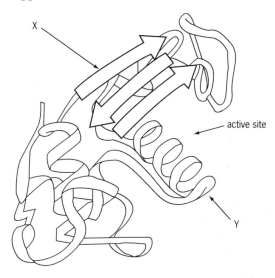

a) State the name given to the shape of this type of protein.

b) State what is meant by the primary structure of a protein.

c) In the regions labelled **X** and **Y two** different types of secondary structure are found.

(i) Identify each type of secondary structure.

(ii) State the type of bonding that is used to stabilise these structures.

d) Explain the importance of the tertiary structure of this protein to its function.

IBO Standard level November 1998 Q3

7 The following list shows some chemical compounds found in living organisms. Match the appropriate letter to one of the following descriptions. Use each letter once only.

A Glucose, **B** Cellulose, **C** Lipid,
D Protein, **E** Starch.

(i) is destroyed by heating to 65 °C …

(ii) has a storage function in plants …

(iii) is soluble in water …

(iv) has a support function in plants …

(v) does not mix with water …

WJEC Biology Module BI1 January 2001 Q1

? EXAM QUESTIONS

8 The following table lists some features of biological compounds. Complete the table by ticking (✓) in the appropriate column(s) if the feature is found in carbohydrates, lipids or proteins. You can tick one, two or three columns for each feature.

Feature	Carbohydrate	Lipid	Protein
can be saturated or unsaturated			
contains peptide bonds			
contains the elements carbon, hydrogen and oxygen			
can contain disulphide bonds			
cellulose and glycogen are examples			

WJEC Biology Module BI1 June 2001 Q2

9

a) **(i)** In the spaces below, write the **names** of the nitrogenous bases which pair together in DNA.
… pairs with …
… pairs with …
(ii) What type of chemical bond holds these base pairs together in a DNA molecule?

b) There are two types of nitrogenous bases. Name the two types.

c) State **three** ways in which the structure of RNA differs from DNA.

OCR Biology Molecules and Life June 2000 Q1

10 The diagram shows the basic structure of amino acids:

$$\begin{array}{c} COOH \\ | \\ R-C-H \\ | \\ NH_2 \end{array}$$

a) State what is represented in the diagram by the letter R.

b) Draw a simple diagram to show how two amino acids are linked together.

c) Amino acids are linked together to form polypeptides at special sites in the cytoplasm of both prokaryotic and eukaryotic cells. Compare the sites where polypeptides are formed in prokaryotic cells with the sites in eukaryotic cells.

IBO Standard level November 1999 Q3

11

a) The figure shows the molecular structure of a dipeptide which has been formed by joining two amino acids together.

(i) Give the formula of the chemical group present at position **X** on this molecule.
(ii) Draw a circle round the chemical bond which has been formed as the result of the condensation reaction between the two amino acids.
(iii) The table shows the chemical structure of the R-group in a number of different amino acids.

Amino acid	Structure of R-group
Alanine	CH_3
Asparagine	CH_2CONH_2
Aspartic acid	CH_2COOH
Glutamine	$(CH_2)_2CONH_2$
Serine	CH_2OH

Use the information in the table to name the two amino acids from which the dipeptide was formed.

b) The relative molecular mass (M_r) of a molecule is a measure of its size.
(i) The mean M_r of an amino acid is 110. The M_r of a particular protein is 40 000. Calculate the number of amino acids that make up this protein. Show your working.
(ii) Explain how it is possible for different proteins to have the same relative molecular mass.

AQA Processes of Life June 2000 Q3

KEY SKILLS **ASSIGNMENT**

DNA FINGERPRINTING

In 1987, Robert Melias made legal history in the UK when he was convicted of rape based on evidence obtained by DNA fingerprinting. This technique – more accurately called DNA profiling – was developed at Leicester University by Alec Jeffreys in 1984 and is proving to be an invaluable tool in forensic medicine.

The principles behind DNA profiling

Every human body cell contains 46 chromosomes. Each one consists of a single elaborately coiled piece of DNA which, if stretched out, can be as long as 5 cm. The structure of DNA can be used to identify individuals.

Between the many genes that occur along the DNA molecule are regions that code for nothing at all. Within this non-coding DNA are hypervariable regions, so called because they vary enormously in length from person to person. Hypervariable regions consist of particular base sequences called core sequences, which are repeated again and again. Different people have different numbers of repeats and so have differently sized hypervariable regions. When these are labelled and separated according to size, a pattern is produced. Each pattern is unique to each individual person and can be used as the basis of DNA profiling (Fig 3.A1).

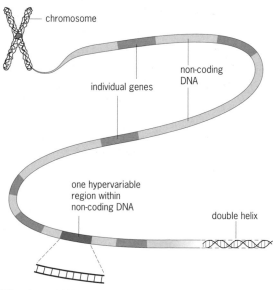

Within a hypervariable region a core sequence is repeated a certain number of times: different people have different numbers of repeats and therefore different sized hypervariable regions

Fig 3.A1 Although each chromosome contains several thousand genes, they only account for about 10 per cent of the length: the rest is non-coding DNA. Within these segments are hypervariable regions. These are unique to each individual and, when they are separated out, the pattern they form provides the basis of DNA profiling

1 **a)** What is a gene?
b) What is the name given to a fault which sometimes occurs when DNA is copied?
c) Faults accumulate in the hypervariable regions more frequently than they do in the genes. Why do you think this is?

Getting a sample of DNA

All body cells contain the same DNA, and so virtually any tissue sample can be used for DNA profiling. The amount of tissue required is very small: $0.05 \, cm^3$ of blood, $0.005 \, cm^3$ of semen or one hair root!

2 **a)** Which blood cells contain DNA?
b) Why do forensic scientists need a larger sample of blood than of semen?

Processing the DNA and separating the fragments

Once you have a DNA sample, the next stage is to cut the molecule using restriction enzymes, which act rather like molecular scissors (see Chapter 24). These enzymes cut DNA at specific base sequences, giving a complex mixture of DNA fragments, some of which contain the hypervariable regions.

Electrophoresis (see page 56) separates the fragments. The fragment mixture is placed in a trough in some gel. When a current is applied, the negatively charged DNA fragments move towards the positive terminal, or anode. The fragments move at different speeds according to their size: small ones move faster than larger ones. The fragments separate out into bands, as shown in Fig 3.A2 for blood.

Fig 3.A2 The major steps in the preparation of a DNA profile

KEY SKILLS **ASSIGNMENT**

The bands of DNA are transferred from the gel onto a nylon membrane by Southern blotting, a process which works by capillary action. At this stage the bands are still invisible, and must be stained so that the hypervariable regions can be seen.

Labelling the fragments

Within hypervariable regions are core sequences that are common to all humans. It is the number of times the sequences are repeated which varies from person to person.

Pieces of DNA complementary to these core sequences have been isolated and are produced in bulk for use as genetic probes. They are labelled with a marker chemical, commonly

the enzyme alkaline phosphatase which fluoresces (produces light) when a particular substrate is added.

When the probes, complete with enzyme, are added to the DNA sample, they attach to the core sequences, thus marking the hypervariable regions. Excess probe is washed off, substrate for the enzyme is added, and bands that contain hypervariable regions fluoresce. If the blot is exposed to an X-ray film, dark lines appear wherever bands in the blot have emitted light, forming the familiar DNA profile.

3 **a)** Why do the DNA bands fluoresce?
b) Do all the DNA bands fluoresce? Explain your answer.

Case studies

You have covered the basic theory of DNA profiling: now you can put it into practice. Here are two case studies for you to interpret.

CASE 1: WHODUNNIT?

M   C   Mr Z

Fig 3.A3 The DNA profile of six suspects compared with the sample taken from the tissues

In 1989, a man robbed a bank at gunpoint while suffering from a particularly heavy cold. The mucus-filled tissues found on the scene provided enough DNA for a DNA profile.

Look at Fig 3.A3.

4 On the basis of the evidence, which suspect robbed the bank?

CASE 2: WHOSE BABY?

Bands in an individual's DNA profile must have come from either the mother or the father. When a child's DNA profile is compared to the mother's, it is easy to see which bands have been inherited. Any bands in the child's profile which did not come from the mother must have come from the father. Use this information to solve the next case study.

In this case study, the mother M is claiming that Mr Z, a famous rock musician, is the father of her 14-year-old daughter C. She is seeking a large maintenance settlement. Predictably, Mr Z denies all knowledge of this and his lawyer demands that a DNA profile be carried out. Fig 3.A4 shows the results.

5 **a)** Which of the daughter's DNA bands were inherited from her mother?
b) Do the remaining bands match those from Mr Z, showing that he is the father?

Fig 3.A4 The DNA profiles from Case Study 2, for the mother (M), child (C) and alleged father (Mr Z)

Questions for discussion

6 What would be the advantages and disadvantages of compulsory DNA testing for the whole UK population, so that the police have comprehensive files?

7 Would you like details of your DNA to be available to businesses, in the way that financial records are today? What problems could this cause?

04

Movement in and out of cells

04 Movement in and out of cells

A kidney waiting to be transplanted. The pale pink colour changes to a much darker red when the new blood supply is connected

The list of people awaiting an organ transplant is growing. Improving surgery and reducing organ rejection rates has made transplants increasingly successful, leading to a greater demand for kidneys, hearts, livers and other body parts.

Surgeons face problems: an organ can become available in Dundee, but the most suitable recipient might be in Norwich. So, the organ usually has to be transported. But cells in tissues that are separated from their blood supply cannot carry out their normal metabolic functions if they are unable to exchange materials with blood. They run out of oxygen, accumulate waste and can die within minutes.

The solution is to cool the cells so that their metabolic demands are lowered. Organs packed in ice can survive a journey of several hours. During the transplant, the kidney is warmed to normal body temperature and connected without delay to the recipient's blood system.

1 THE CELL AND ITS ENVIRONMENT

Each living cell is a dynamic system that can exchange a volume of fluid several times bigger than its own volume every second! This happens only because cells are very small: if they were any larger the nutrients could not be taken up quickly enough to satisfy demand and waste could not be expelled efficiently enough to prevent poisoning of the cell. All organisms larger than about 1 mm have developed strategies to increase the exchange of materials, to meet the needs of all their cells.

Material passes into or out of cells by these basic processes:
- **diffusion** (and **facilitated diffusion**),
- **osmosis**,
- **active transport**,
- **endocytosis** and **ectocytosis**.

In this chapter, we look at each of these in detail but, since they all involve the cell surface membrane, we first take a detailed look at this important structure.

2 THE CELL SURFACE MEMBRANE

The **cell surface membrane**, often called the **cell membrane** and sometimes known as the **plasma membrane** or **plasmalemma**, is the boundary between a cell and its surroundings. It has little physical strength, but it plays a vital role in regulating the materials that pass in and out of the cell. Many of the organelles of the eukaryotic cell are also made up of membranes (Fig 4.1) with the same basic structure as the cell surface membrane.

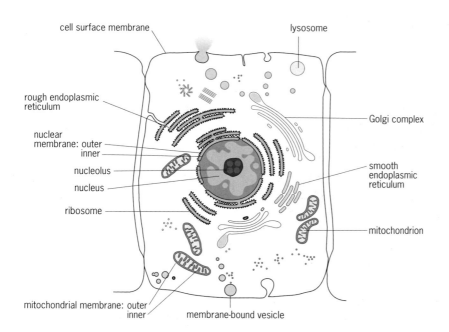

cell surface membrane

lysosome

rough endoplasmic reticulum

Golgi complex

nuclear membrane: outer
inner

nucleolus

smooth endoplasmic reticulum

nucleus

ribosome

mitochondrion

mitochondrial membrane: outer
inner

membrane-bound vesicle

Fig 4.1 Whenever you draw a eukaryotic cell, nearly every line you draw represents a membrane. In this diagram of an animal cell, all the membranes are drawn as red lines

THE STRUCTURE OF THE CELL SURFACE MEMBRANE

As we saw in Chapter 1, the cell surface membrane is basically a double layer of **phospholipid** molecules about 7 to 10 nm thick (Fig 4.2) It cannot be seen with a light microscope and so its structure could not be studied directly until the electron microscope was developed. Instead, early cell biologists deduced its structure by investigating its properties.

phospholipid molecule

7–10 nm

head

tail

Fig 4.2 The phospholipids in a membrane are arranged tail to tail, forming a bilayer

THE FLUID MOSAIC THEORY

In 1972, aided by electron microscope studies and evidence from other techniques, Singer and Nicholson put forward the **fluid mosaic theory** (Figs 4.3 and 4.4). Today, most scientists accept it as the model that best represents the structure of living cell membranes.

Fig 4.3 The fluid mosaic model of the structure of a cell surface membrane. The membrane has been described as 'protein icebergs in a lipid sea'.

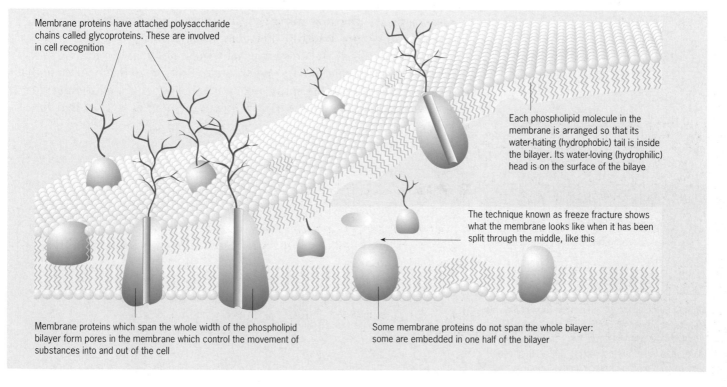

Membrane proteins have attached polysaccharide chains called glycoproteins. These are involved in cell recognition

Each phospholipid molecule in the membrane is arranged so that its water-hating (hydrophobic) tail is inside the bilayer. Its water-loving (hydrophilic) head is on the surface of the bilaye

The technique known as freeze fracture shows what the membrane looks like when it has been split through the middle, like this

Membrane proteins which span the whole width of the phospholipid bilayer form pores in the membrane which control the movement of substances into and out of the cell

Some membrane proteins do not span the whole bilayer: some are embedded in one half of the bilayer

Fig 4.4 A simplified diagram of a cell surface membrane, showing the features described in the fluid mosaic theory. This level of detail should be enough to enable you to explain the essential properties and functions of the membrane

SEE QUESTIONS 1 TO 3

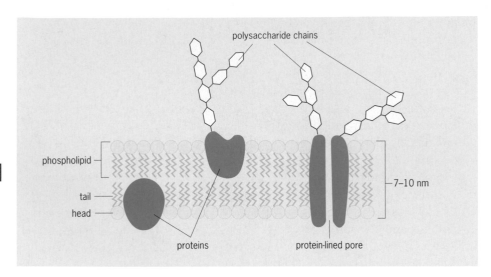

REMEMBER THIS

Water is a **polar molecule** (it has regions of positive and negative charge) and is a solvent for **polar substances** such as sugars, charged ions (Na^+, Cl^-, Ca^{2+}, K^+), B and C vitamins and amino acids. Polar substances do not dissolve in lipid and so can cross a cell membrane only by going through pores.

Most fats, oils and lipids are **non-polar molecules** (they do not have charged regions) and do not dissolve in water. Other non-polar substances (such as vitamins A, D, E and K) can dissolve in lipids and so can cross cell membranes without going through pores.

? TEST YOURSELF

B Explain the terms hydrophilic and hydrophobic. How do these terms relate to a phospholipid molecule?

According to the fluid mosaic model, the cell membrane consists of a double layer of phospholipid molecules, known as a **lipid bilayer**. This is studded with proteins and other molecules. The name **fluid mosaic** is used because the bilayer is a very fluid structure (the phospholipid molecules are in constant sideways motion – a bit like table tennis balls on the surface of a stretch of water) and it contains a 'mosaic' of protein molecules.

THE CHEMICAL MAKE-UP OF CELL MEMBRANES

Cell membranes contain **phospholipids**, **proteins**, **cholesterol** and **polysaccharides**.

Phospholipids are a major constituent of cell membranes. They naturally form membranes in water because they automatically arrange themselves into a bilayer that is virtually impermeable to water and to anything that is water-soluble (see Fig 4.2). So the membrane keeps the cell contents in and it keeps everything else out – except it's not as simple as that. Some substances, mainly water-soluble chemicals, need to get in and out of cells all the time – so how does the cell manage this?

The answer is – **membrane proteins**. These protein molecules act as **hydrophilic pores**, water-filled channels that allow water-soluble chemicals to pass through (Fig 4.5). Pores are usually small and highly selective; they allow only specific molecules or ions through. Proteins in the membrane that form pores usually span the entire membrane, but other proteins with other functions can occur only in the top or only in the bottom layer of **lipids**.

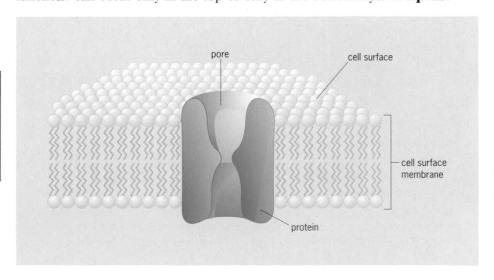

Fig 4.5 Some of the proteins in the membrane form pores. These allow particles that cannot dissolve in lipid to enter and leave the cell. Membrane pores are not simply 'holes'; they can control what passes through them

Membrane proteins have several functions. They:
- form pores through which water and water-soluble chemicals can pass,
- act as carriers in active transport,
- form receptor sites for hormones,
- are important in cell recognition.

Cell surface membranes can also contain **cholesterol** molecules, which fit in between the phospholipids (see Fig 4.16). Cholesterol is a vital constituent of animal cells, despite being notorious for its connection with heart disease (Fig 4.6).

Polysaccharides, branched polymers of simple sugars, stick out from the outer surface of some membranes like antennae (see Figs 4.3 and 4.4). They attach to lipids, forming **glycolipids**, or to proteins, forming **glycoproteins**. Glycolipids and glycoproteins help cells to recognise each other – allowing the immune system to tell the difference between body cells and invading bacteria, for example.

Fig 4.6 A narrowed artery. High cholesterol levels are associated with the development of coronary heart disease. The fatty deposit, atheroma, builds up under the endothelium of the blood vessel, narrowing the lumen

3 DIFFUSION

Diffusion is basically a 'mixing of molecules'. In a gas or liquid the molecules or ions move continuously, randomly bumping into each other and changing direction. In this way, particles tend to diffuse, or spread, so that they are evenly spaced rather than being concentrated in one place (Fig 4.7).

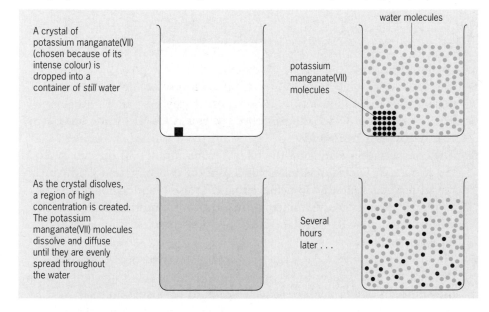

A crystal of potassium manganate(VII) (chosen because of its intense colour) is dropped into a container of *still* water

As the crystal disolves, a region of high concentration is created. The potassium manganate(VII) molecules dissolve and diffuse until they are evenly spread throughout the water

water molecules

potassium manganate(VII) molecules

Several hours later . . .

Fig 4.7 If you drop a crystal of potassium manganate(VII) (chosen for its visibility) into still water, it dissolves and the ions diffuse until evenly spread. This takes several hours (depending on temperature) but can be speeded up by stirring. Diffusion is a passive process which requires no input of energy: it does however depend on the kinetic energy (energy of movement) of the molecules in a gas or a liquid

A useful working definition of diffusion is:

Diffusion is the movement of particles within a gas or a liquid from a region of high concentration to a region of lower concentration.

Diffusion mainly moves substances over short distances. It is too slow to move substances efficiently over distances much greater than a fraction of a millimetre. In the mammalian lung, for example, oxygen diffuses through the thin epithelium of the alveolus and into the blood, a journey of usually less than a hundredth of a millimetre (10 µm). It is then carried away to other parts of the body by the circulatory system. The rapid movement of materials around an organism in a stream of fluid is called **mass flow**.

See the Assignment at the end of the chapter, which deals with diffusion in organs of the human body.

DIFFUSION AND ENERGY

The difference in concentration of a substance between two regions is called a **concentration gradient**. Particles that are free to move have **kinetic energy**. The region of a gas or liquid with the highest concentration of particles of a particular substance has the highest kinetic energy for that substance. Particles move down a concentration gradient by diffusion, until they are spread evenly.

The most important point to remember is that diffusion is a **passive process**: it requires no input of energy from the cell. It follows that movement of a substance against a concentration gradient – **active transport** (see page 79) – requires energy.

FACTORS THAT AFFECT THE RATE OF DIFFUSION

The rate at which molecules of a substance diffuse from one region to another has a great influence on the design of cells, organs and whole organisms (Fig 4.8).

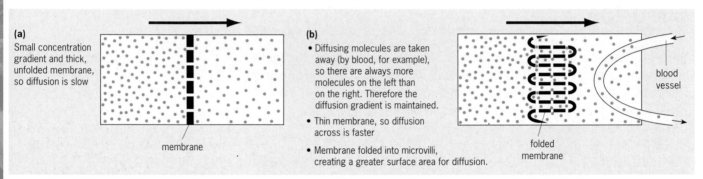

(a) Small concentration gradient and thick, unfolded membrane, so diffusion is slow

membrane

(b)
- Diffusing molecules are taken away (by blood, for example), so there are always more molecules on the left than on the right. Therefore the diffusion gradient is maintained.
- Thin membrane, so diffusion across is faster
- Membrane folded into microvilli, creating a greater surface area for diffusion.

blood vessel

folded membrane

Fig 4.8 A schematic diagram to show some of the features which increase the rate of diffusion. An efficient exchange of materials requires a large surface area, maintenance of a diffusion gradient and a thin cell surface membrane. Increasing the temperature would also speed up diffusion, but this is not an option in living systems

Several factors affect the rate of diffusion:

- The surface area between the two regions. The greater the surface area, the greater the rate of diffusion.
- The distance over which diffusion occurs. This is known as the **length of the diffusion pathway**.
- The concentration gradient – the relative concentration of the substance in the two areas. Diffusion is more efficient if the concentration gradient can be maintained. One way to achieve this is to use blood to transport the substance away from the immediate area once it has diffused. Another way is to combine the substance with another chemical to prevent it from diffusing back.
- The size and nature of the particles. Fat-soluble substances can diffuse through the lipid bilayer of the membrane. Water-soluble substances must pass through the protein pores, which tend to be small and selective. Generally, very large molecules cannot diffuse into or out of cells at all.
- The temperature at which the process takes place. At higher temperatures, molecules have more kinetic energy and so diffuse more quickly.

CALCULATING THE RATE OF DIFFUSION

The rate at which substances diffuse can be estimated from a simple formula that takes into account the factors that affect diffusion. Rate of diffusion is proportional to:

$$\frac{\text{surface area} \times \text{concentration difference}}{\text{length of diffusion pathway}}$$

This relationship is known as **Fick's law**. For efficient diffusion, the values on the top line of this equation should be as large as possible and the value on the bottom line should be as small as possible.

? TEST YOURSELF

D Phosphorylated chemicals (those with a phosphate group added) cannot pass easily through cell membranes. Suggest why glucose molecules are phosphorylated as soon as they enter a cell.

DIFFUSION IN THE HUMAN BODY

Diffusion is not fast enough to be effective over distances greater than a few millimetres. The human body has many organs that are adapted for the exchange of materials – the lungs, intestines and kidneys are obvious examples. However, these organs are of little use without a circulatory system that transports the exchanged materials to all other areas of the body. The blood system is an example of what is known as a **mass flow system**.

The main function of this system is to bring materials to within diffusing distance of living cells, and to take away cell products. In the human body, no cell is more than a few micrometres away from a capillary (see Chapter 10).

? TEST YOURSELF

E Name some organs where exchange of materials is the main function. (Note: Think about plants too.)

4 FACILITATED DIFFUSION

Some substances enter and leave cells much faster than you would expect if only diffusion occurred. We now know that some membrane proteins assist, or **facilitate**, the diffusion of some substances across the cell membrane. Two types of protein are responsible for **facilitated diffusion**:

- Specific **carrier proteins** take particular substances from one side of the membrane to the other.
- **Ion channels** are proteins that open and close to control the passage of selected charged particles.

CARRIER PROTEINS

Until the 1970s, cell biologists thought carrier proteins worked by rotating within the membrane, like turnstiles. Newer research points to a different explanation (Fig 4.9). As soon as the diffusing molecule binds to the carrier protein, the protein undergoes a change in shape so that the diffusing molecule ends up at the other side of the membrane, where it is released.

Like diffusion, facilitated diffusion involves movement *down* a concentration gradient and requires no *input* of metabolic energy.

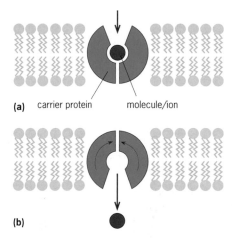

(a) carrier protein   molecule/ion

(b)

Fig 4.9 Facilitated diffusion using a carrier protein. The diffusing molecule interacts with the carrier protein, causing a change in shape which 'squeezes' the molecule through the channel

ION CHANNELS

Ion channels are proteins with a central 'hole' lined with polar groups (Fig 4.10). Ion channels facilitate the diffusion of charged particles such as Ca^{2+}, Na^+, K^+ and Cl^- ions. Many are **gated**, so can open or close. Cells use ion channels to control the movement of ionic substances between themselves and other cells, and to regulate the ionic composition of their cytoplasm.

(a) Open

Protein spans the membrane

Ions such as Ca^{2+} pass through the protein pore lined with polar groups

(b) Closed

Protein changes shape, so pore becomes too to allow ions through

Fig 4.10 An ion channel. Because they are charged particles, ions cannot easily pass through the non-polar lipid bilayer. Specific membrane proteins form polar pores through which ions can pass. These channels are usually specific for one type of ion and can open and close according to the needs of the cell. When fully open, over 1 million ions per second can flow through a single channel

---

! SCIENCE FEATURE Diabetes and carrier proteins

The concentration of our blood sugar is kept relatively constant by controlling how much glucose passes from the blood into cells. After a meal, when blood glucose levels are high, the hormone insulin is released. This hormone activates a mechanism in the cell membrane that facilitates the diffusion of glucose into the cell, so lowering the blood sugar concentration.

One form of diabetes in humans is caused by a gene mutation that alters the structure of glucose carrier proteins in cell surface membranes. An affected person cannot get enough glucose into their cells. Unlike diabetes that results from the inability to make insulin (see Chapter 11), this form of the disease is difficult to treat, because injecting insulin has little effect.

A difference in concentration of ions may lead to a net positive or negative charge in a particular region. This type of concentration gradient is called an **electrochemical gradient**. Charged particles move towards regions of opposite charge. This is important in the process of transmission of a nerve impulse (see Chapter 15) and in the process by which red blood cells exchange materials with the tissues (see Chapter 10).

5 OSMOSIS

In biology, we usually talk about the diffusion of substances that are dissolved in water. But what about the water molecules themselves? Does water diffuse down a diffusion gradient? The answer is yes, and the diffusion of water is known as **osmosis**.

A working definition of osmosis is:

> **Osmosis is the diffusion of water only. It is the net movement of water molecules from a region of their higher concentration to a region of their lower concentration, through a partially permeable membrane.**

Water, water everywhere, and not a drop to drink… The concept of osmosis may be easier to grasp if you think of a real life-and-death situation. Imagine a ship-wrecked survivor floating in a dinghy on the open sea, the sun burning down on him for 15 hours a day, no land or rescue in sight. After a few hours, he gets so thirsty, he gives in and drinks sea water. However, because sea water contains salt, its sodium chloride concentration is about three times that found in human blood. The survivor is already dehydrated, and the salt causes water to move out of his blood and into his stomach by osmosis, causing further dehydration of his body cells. Death from dehydration is more likely than if he had drunk nothing.

IMPORTANT FACTS ABOUT WATER

Understanding some of the properties of water is the key to understanding osmosis, so before going into detail about this process, we need to look at what happens when a substance dissolves in water.

As you can see from the information in the Extension box on water (see page 76), water molecules are polar – they carry a tiny positive and a tiny negative charge. In pure water, all the water molecules present are able to move around freely, not associating with each other or with anything else – think of them as 'free' water molecules. However, when a substance is placed in the water that is soluble, things change. As each of the solute molecules dissolves, it becomes charged and attracts the charges on the water molecules. It soon becomes surrounded by a shell of water molecules. When this happens, the water molecules that form the 'shell' are no longer 'free' as they were before: they have been 'tied up' by the solute molecules. This reduces the concentration of 'free' water molecules in the solution.

AN OVERVIEW OF OSMOSIS

Now think about what happens if you have two solutions, each with different concentrations of solute, on either side of a semi-permeable membrane.

Fig 4.11(a) shows what happens. There are more solute molecules in solution B (on the right of the membrane) than in the solution A (on the left). In solution B, more water molecules are 'tied up' in shells around the solute molecules, so there are fewer water molecules in solution B than in solution A. There is therefore a concentration difference between the two solutions, both of solute molecules and of water molecules.

The pores in the membrane are too small to allow solute molecules through, so their concentration has to stay the same. What can change is the concentration of 'free' water molecules on either side of the membrane. These water molecules can pass through the pores and water can diffuse down its concentration gradient. Water molecules move from the left, where there is a higher concentration of water molecules, to the right, where there is a lower concentration of water molecules (Fig 4.11b). This is osmosis.

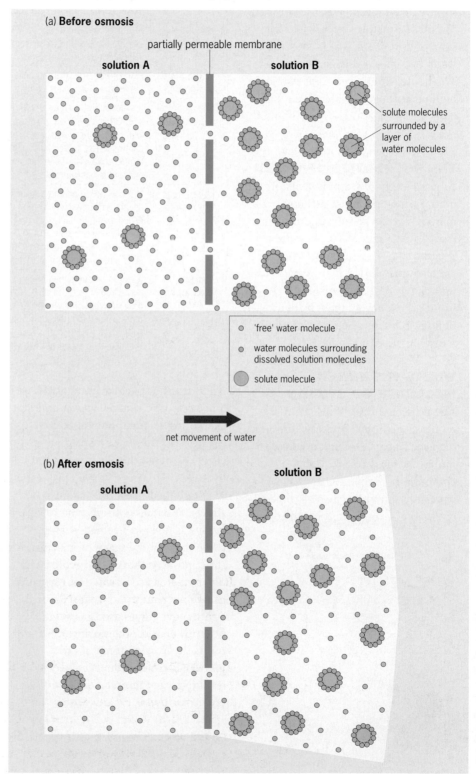

Fig 4.11(a) A 'weak' solution is one with a low concentration of solute molecules but a high concentration of water molecules. It has more 'free' water molecules than the 'concentrated' solution on the right. In osmosis, water molecules move from a region of their higher concentration to a region of their lower concentration – in this case from the solution on the left to the solution on the right

Fig 4.11(b) As osmosis occurs, the number of water molecules on either side of the membrane starts to equalise. Eventually, both solutions will have the same concentration of solute molecules relative to water. Because water has moved from left to right, the solution on the left will have a much lower volume than the one on the right

There is no life without water: life almost certainly originated in water. Most biochemical reactions take place in solution and all organisms that live on land can do so only because they have their own internal aquatic environment. Some organisms can survive dehydration as spores or seeds, but these structures simply allow a state of dormancy, which ends when water becomes available once again. The bodies of all organisms contain a high percentage of water. Even that most complex of organs, the human brain, is 85 per cent water.

THE PROPERTIES OF WATER

All cells are tiny compartments of fluid. Living cells must exchange materials with the surroundings, and this can only take place in solution. So, as well as being full of water, cells are also surrounded by it. Water has unique physical and chemical properties, many of them significant in biology. Some of the most important are summarised in Table 4.1.

WHY IS THE WATER MOLECULE SO STICKY?

The water molecule consists of two atoms of hydrogen and one atom of oxygen. This gives water a relative molecular mass of 18. Many compounds of this size are gases: water is a liquid because its molecules are 'sticky' – they cling to each other.

Table 4.1 Some properties of water

Property	Significance in living systems
Cohesion	Hydrogen bonding causes water molecules to stick to each other. This is important, for example, when water evaporates from the leaves of a tall tree: water is so cohesive that a continuous column can be sucked up from roots which might be over 100 metres below. This cohesion causes surface tension. Water has a very high surface tension, and several organisms, such as pond skaters and rat-tailed maggots, take advantage of this property
The polar nature of water	This makes water a very good solvent – it has been called the 'universal solvent'. Though obviously not true since it doesn't dissolve everything, it does act as a solvent for all ionic and polar compounds
Melting and boiling point	Water freezes at 0 °C and boils at 100 °C and is therefore a liquid over most of the temperature range found on Earth
Specific heat capacity	A gram of water absorbs a lot (4.2 joules) of heat before its temperature rises by 1 °C. Organisms have a high water content and so can absorb a lot of energy (or lose a lot) before their temperature changes significantly. This makes water a useful thermal buffer. Most organisms cannot tolerate great changes in temperature, and the water in their bodies protects them from rapid changes in external temperature
Latent heat of vaporisation	The hydrogen bonding between water molecules makes it difficult for them to separate from each other and vaporise (evaporate) to form a gas. When water does evaporate, the escaping molecules take a lot of energy with them, and so vaporisation is a very efficient cooling process. Anyone who has ever stepped out of the shower and stood in a draught will know just how efficient! The processes of sweating, panting and transpiration all involve cooling by evaporation
Thermal conductivity	A naked person sitting in a room at 25 °C would find it comfortable. However, if immersed in water at 25 °C, the same person would feel cold. This is because the thermal conductivity of water allows heat to be transferred away from the body more quickly through the water than through air. Materials with a low thermal conductivity are good insulators. Fur is a good insulator because it traps air, but the insulating properties are lost if the fur gets wet

Electrons are negatively charged. Since opposite charges attract and like charges repel, the lone pairs of electrons of the oxygen atom effectively repel the electrons of the hydrogen – oxygen bonds. This results in a 'v-shaped' molecule, as Fig 4.12 shows. The region around the oxygen atom has a slight negative charge, while the two hydrogen atoms have a slight positive charge. Molecules with different areas of positive and negative charge are said to be **polar**.

Attractive forces exist between water molecules: the positive charge on the hydrogen atoms of one molecule attracts the negative charge on the oxygen atoms of another molecule. These attractive forces are

Fig 4.13 Weak attractive forces exist between the positive areas of one water molecule and the negative areas of adjacent ones. Many of the unique properties of water are due to this hydrogen bonding

called hydrogen bonds (Fig 4.13). In effect, each water molecule acts as a 'mini magnet'. As well as 'sticking' to each other, water molecules also cling to other charged particles.

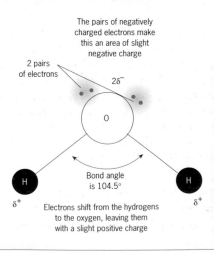

The pairs of negatively charged electrons make this an area of slight negative charge

2 pairs of electrons

$2\delta^-$

O

Bond angle is 104.5°

Electrons shift from the hydrogens to the oxygen, leaving them with a slight positive charge

H δ^+ H δ^+

Fig 4.12 Water is a polar molecule: it has distinct areas of positive and negative charge

OSMOSIS IN LIVING CELLS

All cells contain cytoplasm, a complex solution separated from its surroundings by a partially permeable cell surface membrane. So, all cells have the potential to gain or lose water by osmosis. The overall tendency for water to enter or leave a cell is determined by **water potential**.

WATER POTENTIAL

If no other factors are involved, a cell that is placed in a solution of the same concentration will neither gain nor lose water. Water molecules pass equally in either direction and there is no net change in the cell volume: Fig 4.14(a).

A cell placed in a solution that contains less solute than the cytoplasm will gain water by osmosis: Fig 4.14(b). If the surrounding solution contains more solutes than the cytoplasm, the cell will lose water by osmosis: Fig 4.14(c).

The water potential is the potential of any system to absorb water molecules by osmosis.

Fig 4.14 Factors that determine whether water enters or leaves a cell by osmosis
(a) When the solution surrounding the cell contains the same amount of solute as the cytoplasm (ie is **isotonic**), the cell neither gains nor loses water
(b) When the cell's cytoplasm contains more solutes than the surroundings (ie is **hypertonic**), the cell gains water by osmosis
(c) When the cell's cytoplasm contains fewer solutes than the surroundings (ie is **hypotonic**), the cell loses water by osmosis

water | solute molecules | cell surface membrane

! SCIENCE FEATURE How water molecules cross cell membranes

We have already said that the cell surface membrane is virtually impermeable to water. In fact, it is not 100 per cent impermeable. A tiny amount is able to pass through a lipid bilayer. But recent studies have shown that water passes through purified phospholipid bilayers, with all the proteins removed, at rates approximately 100 to 1000 times faster than you might expect.

The exact reason for this is not known, but scientists think the constant sideways motion of the phospholipids, coupled with the flexing of the fatty acid tails, creates temporary holes (Fig 4.15) through which the water molecules can slip. These holes appear in only one half of the membrane at a time, so the water molecules 'wait' at the halfway point until a hole appears in the other side, rather like crossing a busy road, one lane at a time.

Membranes with a high proportion of cholesterol are less permeable to water,

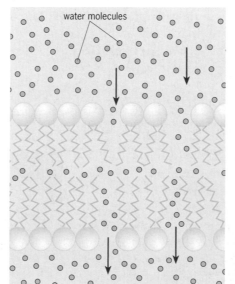

Fig 4.15 A possible explanation for the fact that lipid bilayers are permeable to water. Temporary holes, created by the random sideways movement of the lipid molecules, act as channels through which water can pass

and to simple ions such as sodium and chloride. It could be that cholesterol has a stabilising effect on the phospholipids.

Fig 4.16 Cholesterol is an important constituent of the membranes of many types of cell. The effect of cholesterol is to reduce leakage of water and ions through the lipid bilayer, probably by reducing the sideways motion of the lipid molecules

By slotting in between them, the cholesterol molecules minimise the creation of temporary holes (Fig 4.16).

? TEST YOURSELF

F Preserves such as jam and marmalade have a high sugar content. Suggest why bacteria do not thrive when they land on these foods.

Water potential is measured in units of pressure, kilopascals, kPa, and is a negative scale. For example, a relatively weak solution of sucrose can have a water potential of –200 kPa while a more concentrated solution can have water potential of –500 kPa. Pure water has the highest possible water potential: zero.

If the two solutions described above were separated by a partially permeable membrane, water would move from the more dilute solution into the more concentrated one. So, if there are no other factors involved, such as physical pressure, water will move from a region of high water potential to a region of lower water potential. This is from the solution with the water potential of –200 kPa to the solution with the more negative value of –500 kPa. If you have a problem with negative scales, as many people do, compare it to temperature. Of two solutions, one at –40 °C and the other at –60 °C, it is obvious which is the colder. If the two solutions were placed in contact with each other, heat would pass from the warmer to the colder until they were both about equal. It's the same with the movement of water in osmosis.

? TEST YOURSELF

G Why is it incorrect to say that 'sea water is a hypertonic solution'?

SEE QUESTION 4

OSMOSIS IN ANIMAL CELLS

If you place a human cell, or any animal cell in distilled water, it absorbs water by osmosis and swells up. As cell surface membranes have virtually no physical strength, the cell often bursts.

Fig 4.18 shows what happens to red blood cells placed in **hypertonic**, **isotonic** and **hypotonic** solutions. When blood plasma becomes hypertonic to red blood cell cytoplasm, as it does during severe dehydration, water is lost from the red blood cells, which shrink and become crinkled, or **crenated**. At the other extreme, hypotonic plasma causes red blood cells to swell and burst, leaving 'ghosts' of cell surface membrane. This process in known as **haemolysis**: literally 'blood splitting'.

In order to protect our cells from the effects of osmosis, we must maintain our body fluids at the same solute concentration as our cell cytoplasm. We do this by **osmoregulation** (see Chapter 13).

! SCIENCE FEATURE Osmosis explains why AIDS doesn't spread by kissing

AIDS is a sexually transmitted disease caused by the human immunodeficiency virus (HIV). Semen, breast milk and vaginal secretions are known to contain sufficient HIV to set up an infection in another person. Scientists studying how the disease spreads could detect the virus in these body fluids but saliva seemed to be the odd one out. And there was no evidence that the virus could be passed to another person by kissing. The answer to this puzzle is that saliva is very low in salt compared to other body fluids. It is actually less concentrated than the inside of blood cells that naturally find their way into the mouth, because of wear and tear of the cheeks and tongue. In a person infected with HIV, these cells could carry a high load of virus – and if passed on through kissing or sneezing, could theoretically infect the person on the receiving end. Fortunately, because cells are hypertonic compared to saliva, water passes from the saliva into the cells by osmosis, causing them to lyse and die, giving no chance of transmission of HIV.

Fig 4.17 No risk of HIV transmission...

Fig 4.18(a) All animal cells are prone to the same effects as these red blood cells when placed in solutions of different concentrations

(b) Red blood cells become crenated in a hypertonic solution

(c) Red blood cells take in water and burst in a hypotonic solution (some red cells have not yet burst)

6 ACTIVE TRANSPORT

We have seen that diffusion is movement of molecules down a diffusion gradient:

> ### Active transport is movement of molecules *against* a concentration gradient.

Cell membranes have evolved specialised carrier proteins that transport a particular ion or molecule across the membrane, often from a region of low concentration to a region of much higher concentration. Active transport requires an input of energy from cell respiration, usually provided by ATP. These processes all involve active transport:

- Absorption of amino acids from the gut into the blood.
- Reabsorption of glucose and other useful molecules in the kidney.
- Pumping sodium ions out of cells.
- The exchange of sodium and potassium ions which allows conduction of nerve impulses.

Fig 4.19 An example of an active transport mechanism. The H⁺–ATPase pump removes hydrogen ions from a cell against a diffusion gradient. It is found, for example, in the membrane of the kidney tubule cells, where it contributes to the production of acidic urine

✔ REMEMBER THIS

The membrane proteins responsible for facilitated diffusion and active transport are similar to enzymes. Carrier proteins have binding sites for specific chemicals, they can only work at a particular rate, becoming saturated at high concentrations of transported substance (Fig 4.20), and they can be inhibited. They do, however, differ from enzymes in one important respect: they do not alter the substances they transport.

Step 1 The carrier molecule binds to and hydrolyses ATP. The terminal phosphate group converts the molecule into a 'high energy' complex

in: low [H⁺]

out: high [H⁺]

Step 2 The high energy complex combines readily with an H⁺ ion

Step 3 The carrier protein undergoes a conformational (shape) change which exposes the H⁺ to the other side of the membrane

Step 4 The conformational change lowers the affinity of the carrier for H⁺. The phosphate group is also lost and this induces a final conformational change, back to Step 1

Fig 4.20 The effect of increasing concentration on rate of uptake. Note that the rate of active transport increases to the point at which all the carrier proteins are working at full capacity. If this graph was showing diffusion, it would not level off

The mechanism of active transport is similar to that of facilitated diffusion (Table 4.2). Cell membranes contain specific carrier proteins that combine with the substance to be transported, enabling it to pass rapidly across the membrane. The key difference is that, while facilitated diffusion is a passive process, the carrier protein complex in active transport is activated by an input of energy (Fig 4.19).

The rate of active transport is closely linked to the rate of respiration. We can assume that active transport is taking place if investigations show that:

- movement of particles takes place against a diffusion gradient,
- the rate of transport increases as the rate of respiration increases,
- the process is inhibited by respiratory poisons such as cyanide.

Table 4.2 Comparing diffusion, facilitated diffusion and active transport

	Simple diffusion	Facilitated diffusion	Active transport
Type of membrane molecule involved	lipids	proteins	proteins
Force driving the process	concentration gradient	concentration gradient	ATP hydrolysis
Direction of transport	with concentration gradient	with concentration gradient	against concentration gradient
Specificity	non-specific	specific	specific
Saturation at high concentration of transported molecules	no	yes	yes

SEE QUESTIONS 5 AND 6

7 ENDOCYTOSIS AND EXOCYTOSIS

The transport processes covered so far involve the movement of individual molecules or ions across the cell membrane. There are other processes, **endocytosis** and **exocytosis**, which transport larger volumes of material (solids and liquids) into or out of a cell. Fig 4.21 shows the main features of these two processes and highlights the difference between them.

ENDOCYTOSIS – BRINGING MATERIAL INTO THE CELL

In endocytosis, the cell surrounds material with a cell surface membrane, bringing it into the cell within a small vesicle. There are two main types of endocytosis:

- **Phagocytosis**. Cells **phagocytose** (take in) solid particles such as bacteria and old red blood cells in order to destroy them.
- **Pinocytosis**. A process identical to phagocytosis but which takes in fluid (Fig 4.21a).

PHAGOCYTOSIS AND PINOCYTOSIS

In the process of phagocytosis, a cell's cytoplasm flows round solid material. When this is completely surrounded, a segment of cell membrane is pinched off and a vesicle is formed within the cytoplasm. The membrane and the solid material inside the vesicle form a **food vacuole**. Lysosomes fuse with the vacuole membrane, allowing digestive enzymes to enter the vacuole and begin to digest the contents. Soluble products, such as amino acids and sugars, are absorbed into the cytoplasm, while undigested remains are discharged from the cell by exocytosis.

Examples of phagocytosis include the removal of foreign matter by white cells (see Fig 1.24 on page 16) and the removal of old red blood cells from circulation by the Kupffer cells of the liver.

EXOCYTOSIS – RELEASE OF MATERIAL FROM THE CELL

In exocytosis (Fig 4.21b), materials enclosed in vesicles are expelled from the cell when the vesicle membrane fuses with the cell surface membrane. When the process involves a liquid it is called reverse pinocytosis and this is how the cell secretes many of its synthesised products, such as digestive enzymes, mucus, hormones and the components of milk.

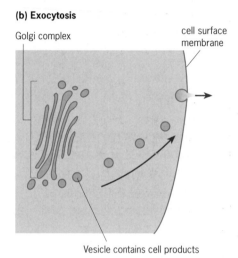

(a) Endocytosis (pinocytosis)

An invagination (pocket) in membrane fills with fluid

Membrane closes round droplet, forming vesicle

Labelled vesicles pass to various parts of cell according to the nature of their contents

(b) Exocytosis

Golgi complex

cell surface membrane

Vesicle contains cell products

Fig 4.21 Endocytosis and exocytosis. These processes appear to be the reverse of each other, but endocytosis must be selective and the contents must be 'labelled' by the cell so that they can be processed in the correct way

---

⚠ SCIENCE FEATURE Pus

When someone has a heavy bacterial infection, such as a cut or infected sweat gland, white blood cells gather in great numbers and go into what can be described as a 'phagocytosis frenzy'. They engulf so many bacteria and dead cells that they become literally 'full'. Toxins secreted by the bacteria then kill the white cells that form pus. The lysosomes of dead white cells tend to liquefy the material surrounding them, forming a smelly, runny fluid.

Fig 4.22 This person's acne is caused by an oversecretion of sebum, a greasy substance that blocks sebaceous glands and hair follicles, trapping bacteria underneath. Pus is formed when white cells do their best to deal with the problem

SUMMARY

When you have finished this chapter, you should know and understand the following:

■ **Diffusion** is the movement of particles down a concentration gradient until they are evenly distributed.

■ In order to maximise the process of diffusion, organisms have a large **surface area**, make the exchange surface as thin as possible and maintain a **diffusion gradient**.

■ **Facilitated diffusion** is diffusion enhanced by specific proteins in cell membranes.

■ **Osmosis** is the diffusion of water from a region of high water potential to a region of lower water potential.

■ **Water potential** is the tendency for water to move by osmosis. It has a negative scale. Pure water has a water potential of zero. More concentrated solutions have a more negative water potential.

■ Diffusion, facilitated diffusion and osmosis are passive processes: they do not require energy.

■ **Active transport** is the movement of particles across cell surface membranes against a diffusion gradient. This process requires energy, usually in the form of ATP. It is achieved by carrier proteins in cell membranes.

■ **Endocytosis** is the passage of droplets of fluid (**pinocytosis**) or solid particles (**phagocytosis**) into the cell by becoming enveloped in cell surface membrane.

■ **Exocytosis** is the passage of material (usually a cellular secretion) out of the cell.

? EXAM QUESTIONS

1 Give an **illustrated** account of the detailed structure and functions of the plasma membrane.
WJEC Biology Module BI1 January 2001 Q9(a)

2 The figure below shows a section through part of the fluid mosaic model of the cell surface membrane with a Na⁺/K⁺ pump protein.

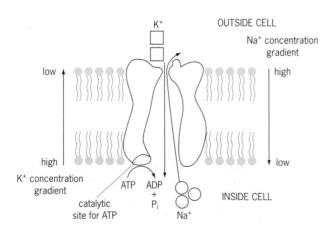

a) Explain why the cell surface membrane is described as a fluid mosaic.

b) Describe how the channel surface of the protein differs from its surface next to the phospholipid tails.

c) Explain why Na^+ and K^+ cannot pass freely across the phospholipid bilayer.
Cholesterol and glycolipids are associated with cell surface membranes.

d) Suggest **one** function of each compound in membranes.
OCR Biology paper 3 June 2000 Q2

3 The figure below represents the fluid mosaic model of a cell membrane.

a) **(i)** State the name of the molecule labelled I.
(ii) Label the diagram to show which part of molecule I is hydrophobic and which part is hydrophilic.

b) **(i)** Identify whether molecule II is an intrinsic or an extrinsic protein.
(ii) Describe the part played by molecule III in active transport.
IBO Standard level paper 2 May 1999 Q2

? EXAM QUESTIONS

4 Consider the information given below and then answer the questions which follow.

Some A level Biology students carried out an investigation involving the extraction of DNA from yeast cells. The method for extracting and purifying the DNA involved the principles of diffusion, osmosis and dialysis.

The procedure they followed is summarised below.

1. Mix yeast cells with sodium hydrogen carbonate solution in a flask and maintain at 40 °C for about 24 hours in order to autolyse (break open) the cells.
2. Filter the mixture to remove cell debris such as cell walls, retain the filtrate.
3. With the help of a syringe, transfer the filtrate into a length of cellulose (Visking) tubing which has previously been sealed at one end. When full, tie a knot in the end of the Visking tubing, which acts as a partially permeable membrane.
4. In order to concentrate and purify the filtrate, place the filled tubing in a beaker containing concentrated sucrose solution and leave for about 1 hour, until the filtrate has reduced (by exosmosis) to about half of its original volume.
5. Remove the Visking tubing from the beaker and put the concentrated filtrate into a boiling tube. Add ice-cold ethanol carefully down the side of the tube. A white, gelatinous precipitate of DNA forms at the surface.

The figure shows the apparatus used.

Concentration of the filtrate occurs due to osmosis, but dialysis is occurring at the same time.

a) Explain, in molecular terms, what is happening during dialysis.

b) **(i)** Name **two** substances (molecules or ions) which would pass out of the Visking tubing during the experiment.

(ii) Name **two** types of molecule which would be retained inside the Visking tubing during the experiment.

OCR Molecules and Life paper 6910 June 2000 Q2

5

a) Give **two** differences between active transport and facilitated diffusion.

b) Describe how ions are transported through a cell surface membrane by active transport.

c) The table shows the concentration of three ions inside and outside a cell of an alga that lives in pond water.

Ion	Concentration of ion in cytoplasm/mmol dm⁻³	Concentration of ion in pond water/mmol dm⁻³
Chloride	58.0	1.3
Potassium	93.0	0.1
Sodium	51.0	1.0

(i) By how many times have the potassium ions been concentrated in the cytoplasm compared with the pond water?

(ii) The ions enter the cell of the alga by active transport. What does the information in the table suggest about the active transport of these three ions through the cell surface membrane? Explain your answer.

AQA AS level January 2001

6 Some proteins in membranes are used for active transport.

a) List **three** other functions of proteins in membranes.

b) Explain the reasons for large numbers of vesicles often being found near to the Golgi apparatus of cells.

IBO Standard level paper 3 May 1999 Q3

KEY SKILLS **ASSIGNMENT**

EXCHANGE PROCESSES IN THE HUMAN BODY

All organisms exchange materials with their surrounding. The amount of material an organism needs is proportional to its volume – the mass of living tissue it possesses. However, the amount of material it is able to exchange is proportional to its surface area – the area of tissue in contact with the environment. Humans are, biologically speaking, huge demanding organisms. The aim of this Assignment is to develop an appreciation of the ways in which the human body is designed to maximise the exchange processes.

The surface area and volume problem

Imagine a cube-shaped multicellular organism that exchanges materials over its surface. (Obviously, no cubic organisms exist, but this keeps the calculations simple.)

1 Copy and complete the following table:

Size of each side of the organism (units)	1	2	5	10	?
Surface area (units^2) for a total of 6 sides	6	24	?	600	60 000
Volume (units^3)	1	8	125	?	?
Surface area: volume ratio	6:1	?	1.2:1	?	?

The important principle shown by the table is that as organisms get larger, their volume increases faster than their surface area.

The consequence of this is that organisms must develop ways of increasing their surface area so that they can exchange more materials. So larger organisms like ourselves can only survive if they evolve ways to optimise the exchange of materials in spite of this. Organisms can:

● increase their surface area.
● make membranes/barriers as thin as possible.
● maintain a diffusion gradient.

As soon as materials are absorbed they are moved away from the site of absorption, often by a circulatory system to ensure that an equilibrium is never reached.

Fig 4.A1 Trypanosomes in blood

2 Fig 4.A1 shows a simple unicellular organism called a trypanosome, the parasite responsible for sleeping sickness. Does this organism need special adaptations to increase its surface area to volume ratio? State a reason for your answer.

Fig 4.A2 shows some of the human organs designed to maximise the exchange of materials, including the lungs, digestive system and placenta.

3 **a)** Briefly state the function of the lungs, digestive system and placenta in Fig 4.A2.
b) List the features common to these three organs, and explain how each helps the exchange process.
c) The human circulatory system is described as a mass flow system. What does this mean and why are such systems necessary?

1 Life depends on the exchange of materials between different cells and between cells and their surroundings. Here, in the plant, the processes of photosynthesis, respiration and transport of materials all involve diffusion.

2 Oxygen which diffuses out of the plant during periods of active photosynthesis enters the lungs of this woman as she breathes in. The lungs are adapted to maximise the process of diffusion: they have moist thin surfaces with a very large surface area.

3 From her lungs, oxygen enters the blood system, a mass flow system which transports large volumes of material over distances too large for diffusion to cope with.

4 Her digestive system relies on diffusion as well as active transport to absorb food molecules.

5 The baby growing in her womb exchanges all its oxygen, food and waste across the placenta, another organ that is well adapted for efficient diffusion.

Fig 4.A2 Many of the vital life processes depend directly on diffusion

05

Enzymes and metabolism

It is easy to see how confectioners make chocolates with hard centres: they simply pour molten chocolate over the centre and wait for it to set. But what about the soft centres? Surely it isn't possible to pour liquid chocolate over a liquid centre and still keep the shape. So, how do they make the runny, yolklike inside to chocolate eggs?

Chocolate lovers everywhere may be surprised to learn that the answer is: use an enzyme. To start with, the centre is solid and contains an enzyme and a polysaccharide. After the chocolate coating has set, the enzyme breaks down the long polysaccharide chains, turning the hard centre into the familiar runny filling. Increasingly, enzymes are being used in medicine, industry and biotechnology.

1 WHAT ARE ENZYMES?

Enzymes are complex chemicals that control reactions in living cells. They are biochemical **catalysts**, speeding up reactions that would otherwise happen too slowly to be of any use to the organism. This definition is a bit of an oversimplification. An active enzyme may speed up a particular reaction, but living organisms do not need all reactions to be going at the maximum rate all of the time. The key word is control. It is more accurate to say that enzymes interact with other molecules to produce an ordered, stable reaction system in which the products of any reaction are made when they are needed, in the amount needed.

THE ROLE OF ENZYMES IN AN ORGANISM

The metabolic pathway chart, part of which is shown in Fig 5.1, shows some of the large number of different but interconnected chemical reactions in a living cell. Charts like this are too complex to learn but they illustrate several important points:

- Many of the complex chemicals that living organisms need cannot be made in a single reaction. Instead, a series of simpler reactions occurs, one reaction after another, forming a **metabolic pathway**. A single pathway may have many steps in which each chemical is converted to the next. A specific enzyme controls each reaction.

- Each individual step in the chart represents one of the simple chemical reactions. At first glance, there seem to be a great variety of reactions, but look more closely and you will see that the same types of reaction occur again and again.

- Enzymes control cell metabolism by regulating how and when reactions occur. Using this very simple pathway as an example,

$$A \rightarrow B \rightarrow C \rightarrow D$$

the final product is substance D, the chemical needed by the living organism. The pathway needs three different enzymes and, when D is no longer needed, or if too much has been produced, one of the three enzymes is 'switched off'.

Fig 5.1 Part of a much larger metabolic chart showing some of the biochemical reactions that occur in living cells. Some reactions carry a number: this identifies the enzyme that catalyses a particular reaction

ONE ENZYME, ONE REACTION, BUT WHAT A DIFFERENCE!

When carbon dioxide dissolves in water, a small proportion of the carbon dioxide molecules combine with water to form carbonic acid.

$$CO_2 + H_2O \rightarrow H_2CO_3 \rightarrow H^+ + HCO_3^-$$

It is a rather slow reaction, but, in the presence of the enzyme **carbonic anhydrase**, it is speeded up about 10^7 (ten million) times. One molecule of carbonic anhydrase can convert 600 000 molecules of carbon dioxide into carbonic acid every second. In living cells, this reaction allows other processes to occur much faster:

- In red blood cells, the enzyme speeds up the production of acid, which in turn causes oxyhaemoglobin to give up its oxygen. Without the enzyme, delivery of oxygen to the tissues would be much slower.
- In certain cells in the stomach lining, the activity of carbonic anhydrase in a pathway allows hydrochloric acid to be secreted rapidly. This creates the acidic conditions necessary for digestive enzymes in the stomach to work properly.
- In the cells of the kidney tubule, carbonic anhydrase speeds up excretion of excess acid, and so helps to maintain the pH of the body at the correct level.

2 THE CHEMICAL NATURE OF ENZYMES

Enzymes are globular proteins. They have a complex tertiary structure (see Chapter 3) in which polypeptides are folded around each other to form a roughly spherical, or **globular** shape (Fig 5.2). This overall three-dimensional shape of an enzyme molecule is very important: if it is altered, the enzyme cannot bind to its substrate and so cannot function. Enzyme shape is maintained by hydrogen bonds and ionic forces (Fig 5.3) and their function can be affected by changes in temperature and pH.

> **? TEST YOURSELF**
>
> **A** By what process are enzymes made in the cell?

> **✔ REMEMBER THIS**
>
> The active site of an enzyme is part of the enzyme molecule and not part of the substrate.

Fig 5.2 The three-dimensional shape of ribonuclease A, an enzyme that helps break up mRNA in the cytoplasm of bacteria. The **active site**, the 'pocket' into which the substrate fits, is clearly visible

✔ **REMEMBER THIS**

An enzyme acts on a chemical known as its substrate. The name of an enzyme often comes from substituting or adding -ase in the name of the substrate, so, for example, lactose is the substrate of the enzyme lactase.

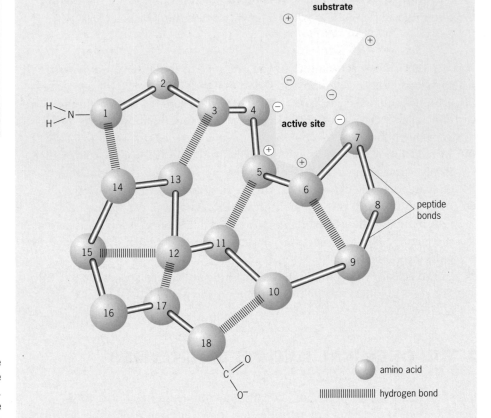

Fig 5.3 This schematic diagram shows the three-dimensional structure of an enzyme molecule. Hydrogen bonds are shown in red. The shape and the electrical charges on the substrate closely match those of the active site

SEE QUESTION 1

? **TEST YOURSELF**

B Predict substrates for the following digestive enzymes: protease, lipase, nucleotidase.

C If the enzyme amylase is specific, how can it catalyse the digestion of bread, potatoes and rice?

Enzymes have several important properties:
● Enzymes are **specific**: each one catalyses only one reaction.
● Enzymes combine with their substrates to form temporary enzyme–substrate **complexes**.
● Enzymes are not altered or used up by the reactions they catalyse, so can be used again and again.
● Enzymes work very rapidly and each has its own **turnover number** (see Table 5.1).
● Enzymes are sensitive to temperature and pH.
● Enzyme function can be slowed down or stopped by **inhibitors**.

THE SPECIFICITY OF ENZYMES

Taking digestive enzymes, you might find it difficult to believe that each enzyme catalyses only one reaction. Trypsin, for example, can begin the digestion of a wide variety of foods rich in protein: eggs, pork, chicken and soya, for example. But when you look at how trypsin works at the molecular level, you can see that this enzyme *is* **specific**. Trypsin cuts an amino acid chain at a point between two particular amino acids, arginine and lysine, and nowhere else. Most proteins have these two amino acids next to each other at some points in their polypeptide chain, and so can be partly digested by trypsin.

Several scientists in biochemistry have devised mechanisms – often called **models** – to explain how enzymes work. In coming up with ideas, they have had to take account of enzyme specificity. Two models that are used to explain how enzymes work are the **lock and key hypothesis** and the **induced fit hypothesis**.

THE LOCK AND KEY HYPOTHESIS

This idea assumes that enzyme function depends on an area on the molecule known as the **active site**, visible in Figs 5.2 and 5.3. The active site is a groove or pocket in the surface of the enzyme into which the substrate molecule fits. Typically, the active site is formed by 3 to 12 amino acids. The size, shape and chemical nature of the active site corresponds closely with that of the substrate molecule, so they fit together like a key fits into a lock (Fig 5.4) or, perhaps more realistically, like two pieces in a three-dimensional jigsaw.

Although this model helps us to understand some of the properties of enzymes, it is now generally accepted that a modified version, known as the induced fit hypothesis, better represents what happens when an enzyme catalyses a reaction.

Fig 5.4 The lock and key hypothesis. The active site is a particular shape (the lock) into which only one substrate (the key) will fit. The enzyme and substrate combine for an instant to form an enzyme–substrate complex. The formation of this complex brings about the desired chemical reaction, converting substrate into product(s)

THE INDUCED FIT HYPOTHESIS

Experimental evidence suggests that the active site in many enzymes is not exactly the same shape as the substrate, but moulds itself around the substrate as the enzyme–substrate complex is formed (Fig 5.5). Only when the substrate binds to the enzyme is the active site the correct shape to catalyse the reaction. As the products of the reaction form, they fit the active site less well and fall away from it. Without the substrate, the enzyme reverts to its 'relaxed' state, until the next substrate molecule comes along.

Both models show why enzymes are not altered by the reactions that they catalyse: they bind to a substrate momentarily, allowing a reaction to happen, but do not themselves undergo any chemical change.

SEE QUESTIONS 2 AND 3

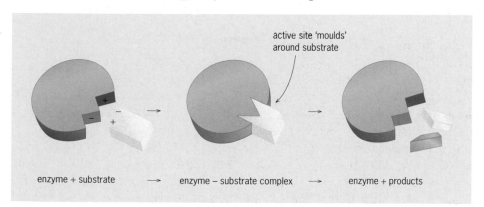

Fig 5.5 The induced fit hypothesis. Before substrate binding, the enzyme's active site is 'relaxed'. When the substrate binds, the active site is pulled into the correct shape by molecular interactions between the two molecules, and an enzyme–substrate complex forms. As the products fall away from the active site, the molecule becomes 'relaxed' again

THE CHEMISTRY OF ENZYME ACTION

In biology almost all reactions are **reversible**. If they were not, there would be no recycling of molecules because once large molecules had been built up they could not be broken down again (see Chapter 35).

A simple reversible reaction can be expressed as:

$$A + B \rightleftharpoons C + D$$
$$\text{reactants} \qquad \text{products}$$

In this reaction, the **reactants** (A and B) combine to give the **products** (C and D). If this reaction were to take place in a test tube, a proportion of the products would react to form A and B again. Eventually, an **equilibrium** would be reached in which the relative proportion of reactants and products would remain the same.

In theory, an enzyme allows a reversible reaction to reach equilibrium more quickly. An enzyme can speed the reaction in either direction; the way it does depends on whether there are more reactants or products present at the start. In living organisms, however, enzymes usually speed up reactions in one particular direction. This is because, in any organism, an enzyme acts only if the product of the reaction that it catalyses is needed. And, as the product is used as soon as it is made, equilibrium is rarely reached.

ENZYMES WORK BY LOWERING ACTIVATION ENERGY

For any chemical reaction to take place, bonds must be broken before new ones can form. The energy needed to break these bonds, and so set the reaction in motion, is the **activation energy**.

Many reactants need a large amount of energy to push them to a state where they can take part in a reaction, so many reactions take place only at high temperature. Many substances burn, for instance, but only after the

Fig 5.6 In order for a reaction to happen, activation energy must be supplied. Then the rest of the reaction proceeds, just as the boulder rolls down the hill once the energy has been supplied to push it to the top. Catalysts such as enzymes work by lowering the activation energy

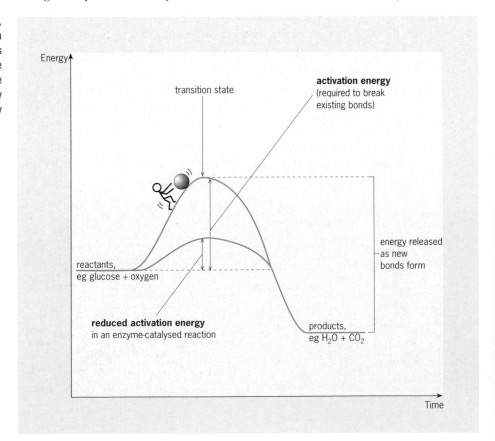

initial activation energy has been supplied, perhaps by lighting a match. In the presence of enzymes, the activation energy is greatly lowered and this allows reactions to take place at the relatively low temperatures normally found in living organisms. The 'halfway' point in a reaction is called the **transition state**, and this is represented by the top of the curve in Fig 5.6. The transition state represents the stage when the old bonds have been broken in order to allow new ones to form. Enzymes lower the activation energy by making it easier to achieve the transition state.

Enzymes speed up reactions in a number of ways:

- They can hold the substrates close together at the correct angle – this would otherwise have to occur by chance collision.
- They can position any charged groups on the active site to help the reaction to occur.
- By acid or base catalysis: the active site of the enzyme can behave as an acid or a base, donating or accepting protons (hydrogen ions, H^+) from the substrate.
- Through structural flexibility: the flexible shape of the enzyme can change during catalysis. This ensures that the substrates are brought together in the correct sequence for the reaction.

Many metabolic reactions require energy in addition to the presence of the relevant enzyme. This energy is supplied by ATP made by cell respiration (see Chapter 34). Reactions that need energy are made to happen by coupling them with ATP breakdown, and so are called **coupled reactions**.

HOW FAST DO ENZYMES WORK?

The speed at which an enzyme works is expressed as its **turnover number**. This is usually defined as the number of substrate molecules turned into product in one minute by one molecule of enzyme. Values range from less than a hundred to many millions. Some examples are given in Table 5.1.

NAMING AND CLASSIFYING ENZYMES

Older enzyme names such as **pepsin**, **catalase** and **trypsin** give no clues about the nature of the reaction they catalyse. To cope with the rapidly expanding number of new enzymes, the International Union of Biochemistry has developed a scheme for naming and classifying enzymes. Very generally, enzymes are named be adding the suffix ase to the name of their substrate. The rest of the name attempts to indicate the nature of the reaction taking place. Alcohol dehydrogenase, for example, catalyses the removal of hydrogen from alcohol (ethanol). Further examples are given in Table 5.2.

? TEST YOURSELF

D The enzyme lysozyme helps to split bacterial cell walls. Where would you expect this enzyme to be present in the human body?

Table 5.1 Some turnover numbers

Enzyme	Turnover number
Carbonic anhydrase	36 000 000
Catalase	5 600 000
β-galactosidase	12 000
Chymotrypsin	6 000
Lysozyme	60

Source: *Biochemical society guidance notes 3, Enzymes and their role in biotechnology.*

Table 5.2 Some common enzymes and their substrates

Enzyme	Substrate	Reaction catalysed
Maltase	maltose	hydrolysis of maltose to glucose
Amylase	starch	hydrolysis of starch to maltose
Alcohol dehydrogenase	alcohol (ethanol)	removal of hydrogen from alcohol
DNA ligase	DNA	joining together two DNA strands
RNA polymerase	nucleotides that make RNA	synthesis of mRNA on a DNA molecule
Glycogen synthetase	glucose	polymerisation of glucose into glycogen
ATPase	ATP	synthesis or splitting of ATP

? **TEST YOURSELF**

E People with a great deal of physical stamina have been found to have a lot of **glycogen synthetase** in their muscles. Suggest what this enzyme does.

F Suggest which type of bonds are broken when an enzyme is denatured. (Look at Fig 5.3 for help.)

✔ REMEMBER THIS

Heating denatures enzymes; it does not kill them.

✔ REMEMBER THIS

The factors that affect enzyme activity also affect the functions of the cell and, ultimately, the organism.

TYPES OF ENZYME

Although there are many different enzymes, they can be put into one of six main categories according to the type of reaction they catalyse:

- **Oxidoreductases** These catalyse oxidation and reduction (redox) reactions. In aerobic respiration, most of the cell's ATP is generated by redox reactions – see Chapter 34.
- **Transferases** These catalyse the transfer of a chemical group from one compound to another, such as the transfer of an **amino group** from an amino acid to another organic acid in the process of **transamination**.
- **Hydrolases** These catalyse **hydrolysis** (splitting by use of water) reactions. Most digestive enzymes are hydrolases.
- **Lyases** These catalyse the breakdown of molecules by reactions that do not involved hydrolysis.
- **Isomerases** These catalyse the transformation of one isomer into another, for instance the conversion of glucose 1,6 diphosphate into fructose 1,6 diphosphate. This is one of the first reactions in **glycolysis**, the first stage of respiration (Chapter 34).
- **Ligases** These catalyse the formation of bonds between compounds, often using the free energy made available from ATP hydrolysis. DNA ligase, for example, is involved in the synthesis of DNA.

FACTORS THAT AFFECT ENZYME ACTIVITY

Enzymes are proteins and their function is therefore affected by:

- Temperature
- pH
- Substrate concentration
- Enzyme concentration
- Inhibitors

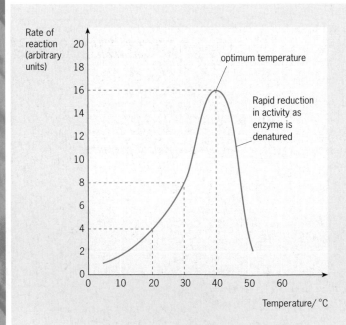

Fig 5.7(a) Up to about 40 °C, the rate of enzyme-controlled reactions increases with temperature. The optimum temperature for many enzymes is about 40 °C, although the activity of an individual enzyme may increase up to about 50 °C or beyond. However, as the temperature passes 43–44 °C, most enzymes lose their activity

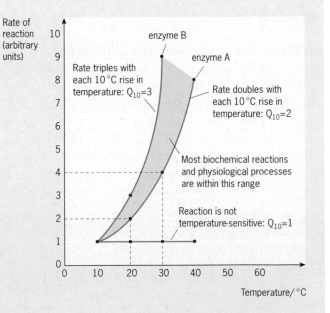

Fig 5.7(b) Not all enzymes have the same Q_{10} values. This variation is important because enzymes, even those in the same pathway, can vary significantly in their sensitivity to temperature. Outside an organism's normal temperature range, the enzymes may begin to work at different rates. This causes a metabolic imbalance that may be lethal, and may explain why some organisms die at temperatures that would seem to be relatively mild. Some Antarctic fish, for example, live in water that remains very constant at around −2 °C, and die if placed in water above 6 °C

TEMPERATURE

For a non-enzymic chemical reaction, the general rule is: the higher the temperature, the faster the reaction: Fig 5.7(a). This same rule holds true for a reaction catalysed by an enzyme, but often only up to about 40 to 45 °C. Above this temperature, enzyme molecules begin to vibrate so violently that the delicate bonds that maintain tertiary and quaternary structure are broken, irreversibly changing the shape of the molecule. When this happens the enzyme can no longer function and we say it is **denatured**.

The effect of temperature on a reaction can be expressed by the temperature coefficient, commonly known as the Q_{10}.

Where t is the chosen temperature, the formula for the Q_{10} is:

$$\frac{\text{rate of reaction at } t + 10\,°C}{\text{rate of reaction at } t\,°C}$$

Fig 5.7(b) shows how to calculate the Q_{10}. To avoid denaturing the enzyme, the values for living organisms need to fall in the range 4 to 40 °C, so we have chosen t as 20 °C.

$$Q_{10} = \frac{\text{rate at 30 °C}}{\text{rate at 20 °C}} = \frac{6}{3} = 2$$

In practice, most enzymes have a Q_{10} between 2 and 3. A value of 2 means that the rate of reaction doubles with a 10 °C temperature rise, 3 means that it triples.

Some organisms have enzymes that are less sensitive to heat than those found in mammals. For example, certain bacteria can survive in hot volcanic springs and deep sea hydrothermal vents (see Fig 5.8) at temperatures of over 90 °C, so their enzymes must be active at these extreme temperatures.

SEE QUESTIONS 4 TO 6

? TEST YOURSELF

G Egg albumen, when heated, goes solid and white. What do you think has happened to the proteins in the egg white, and why does the white not become runny again when the egg is cooled?

? TEST YOURSELF

H Why is it an advantage for humans to have a constant body temperature of around 37 °C?

! SCIENCE FEATURE Stable enzymes for washing powders

Enzymes are unstable, particularly at high temperature, so their commercial usefulness is limited. Many industrial processes need to take place in 'unnatural' environments, at high temperature and extremes of pH. But there are organisms that can thrive at high temperatures. Heat-loving bacteria have thermostable enzymes that are also more resistant to extremes of pH and other unfavourable conditions, such as organic solvents.

Genes that code for some thermostable enzymes have been transferred into bacteria that are easy to grow in large quantities, using recombinant DNA technology (see Chapter 27). As the bacteria that have received a particular gene multiply, the gene is translated into protein and large amounts of the enzyme are produced for commercial use.

An example of a thermostable enzyme is the alkaline protease **subtilisin**, the famous stain digester in biological washing powders. This enzyme is produced on a grand scale by the bacterium *Bacillus subtilis*, and is active in alkaline environments, and so is compatible with the other ingredients in washing powder. It is active at temperatures up to 60 °C, allowing it to be used in a wide variety of wash programmes.

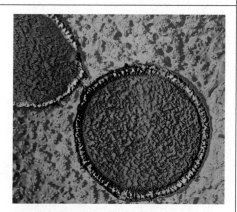

Fig 5.8 This bacterium *Staphylothermus marinus* lives near deep sea hot-water vents. It thrives at temperatures of up to 98 °C. Such heat-loving bacteria have thermostable enzymes (enzymes that are very resistant to heat damage). These have great potential in industry

? TEST YOURSELF

Chymotrypsin is found in the small intestine at pH 8. What would happen to this enzyme if it found its way into the stomach?

SEE QUESTION 7

pH

Like other proteins, enzymes are stable over a limited range of pH. Outside this range, at the extremes of pH, enzymes are denatured. Free hydrogen ions (H^+) or hydroxyl ions (OH^-) affect the charges on amino acid residues, distorting the three-dimensional shape and causing an irreversible change in the protein's tertiary structure.

Enzymes are particularly sensitive to changes in pH because of the great sensitivity of their active site. Even if a slight change in pH is not enough to denature the molecule, it may upset the delicate chemical arrangement at the active site and so stop the enzyme working (Fig 5.9).

Most enzymes are **intracellular**: they work inside cells, and their optimum pH is around 7.3 to 7.4. Most organisms have buffer systems that resist changes in pH, and many are able to excrete excess acid or alkali.

Fig 5.10 shows some enzymes that have optimums of pH that are distinctly acid or alkaline. These include digestive enzymes that normally work in the stomach or inside lysosomes. We describe enzymes that work outside cells as **extracellular**.

At correct pH

substrate

Charges match: enzyme–substrate complex forms

Too acidic

System flooded with H^+ ions: enzyme + substrate repel

Too alkaline

System flooded with OH^- ions: enzyme + substrate repel

Fig 5.9 The effect of pH on enzyme activity

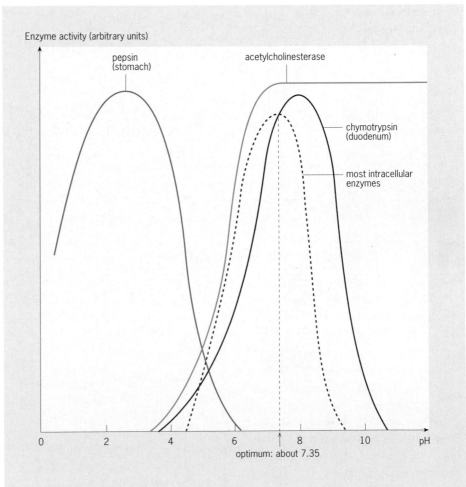

Fig 5.10 The optimum pH of some different enzymes

SUBSTRATE CONCENTRATION

The rate of an enzyme-controlled reaction increases as the substrate concentration increases, until the enzyme is working at full capacity. At this point the enzyme molecules reach their turnover number and, assuming that all other conditions such as temperature are ideal, the only way to increase the speed of the reaction even more is to add more enzyme (Fig 5.11).

ENZYME CONCENTRATION

In any reaction catalysed by an enzyme, the number of enzyme molecules present is much smaller than the number of substrate molecules. Look back at Table 5.1, which shows the turnover numbers of some enzymes, and you can see that one molecule of an enzyme can convert millions of substrate molecules into products every minute.

When there is an abundant supply of substrate, the rate of reaction is limited by the number of enzyme molecules. In this situation, increasing the enzyme concentration increases the rate of reaction.

INHIBITORS

Inhibitors slow down or stop enzyme action. Usually, enzyme inhibition is a natural process, a means of switching enzymes on or off when necessary. Inhibition tends to be **reversible**, and the enzyme returns to full activity once the inhibitor is removed. Many 'external' chemicals, such as drugs and poisons, can inhibit particular enzymes. This type of inhibition is often **non-reversible**. Reversible inhibitors are either **competitive** or **non-competitive**.

The different types of inhibitor are summarised in Fig 5.12.

COMPETITIVE INHIBITORS

Competitive inhibitors compete with normal substrate molecules to occupy the active site. The inhibitor molecules must be a similar size and shape to the substrate to fit the active site, but cannot be converted into the correct product. Effectively, competitive inhibitors 'get in the way' (Fig 5.13) and reduce the number of interactions that can happen between enzyme and substrate.

Fig **5.11** The effect of substrate concentration on the rate of enzyme action

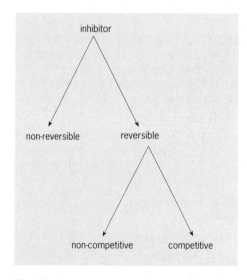

Fig **5.12** A summary of the different types of enzyme inhibitor

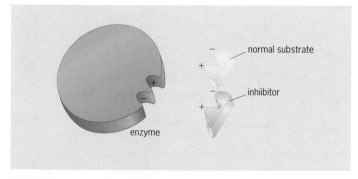

Fig 5.13(a) A competitive inhibitor fits into the active site of the enzyme, preventing the real substrate from gaining access. The inhibitor cannot be converted to the products of the reaction and so the overall rate of reaction is slowed down. If an inhibitor is present in equal concentrations to the substrate, and if both types of molecule bind to the active site equally well, the enzyme can only work at half its normal rate

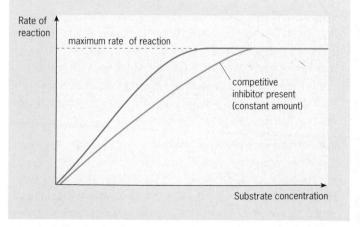

Fig 5.13(b) The effect of a competitive inhibitor on the rate of reaction. Note that the effect of the inhibitor can be overcome by adding more substrate

NON-COMPETITIVE INHIBITORS

Non-competitive inhibitors bind to the enzyme away from the active site but change the overall shape of the molecule, modifying the active site so that it can no longer turn substrate molecules into product (Fig 5.14a).

Non-competitive inhibition has this name because there is no competition for the active site. The presence of a non-competitive inhibitor has the same effect as lowering enzyme concentration: all inhibited molecules are taken out of action completely. Fig 5.14(b) shows the effect of a non-competitive inhibitor.

Fig 5.14(a) Non-competitive inhibitors attach to enzyme molecules and alter the overall shape, so that the active site cannot function. Although the substrate can still bind, forming an enzyme–inhibitor substrate complex, the substrate cannot be turned into product. When the inhibitor molecule is removed, normal function is restored

Fig 5.14(b) Non-competitively inhibited enzymes show no activity at all, but unaffected ones work normally. Thus, the maximum rate of reaction is lowered, as it would be if we used less enzyme

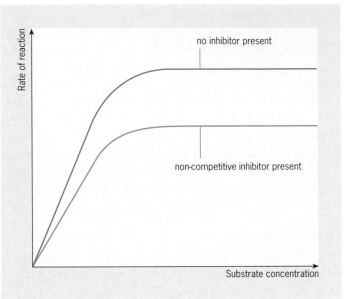

! SCIENCE FEATURE How enzymes help babies digest cow's milk

When babies are first born, their intestines don't behave at all like those of an adult. So that they can absorb and make use of important antibodies in their mother's breast milk, they have intestinal walls that have a structure that resembles a string vest. It has lots of large pores through which whole proteins can pass, without being digested into their component amino acids.

Everything is fine, as long as the baby is fed breast milk, but some mothers cannot breast feed for very good reasons. What happens then? A hundred years ago, babies that couldn't feed from their mother usually fell behind in their growth, sickened and often died. Cow's milk does not contain the same proteins as human milk, and it also has a completely different balance of minerals such as sodium. Feeding tiny babies on untreated cow's milk leads to a generalised immune response against the cow proteins and can also damage the delicate kidneys, which are not able to cope with large amounts of sodium.

Today, many babies are fed on infant formula milk (Fig 5.15) and they thrive just as well as babies that are breast fed (although the natural method is still considered the best). Enzyme technology has had a lot to do with this – enzymes have been used in the production of infant formula milk from cow's milk for over 50 years. Manufacturers used proteases to digest the milk proteins

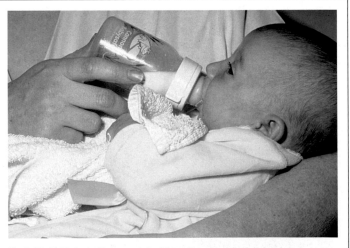

Fig 5.15 Babies can thrive on infant formula milk, which is produced using enzymes

into short peptides and individual amino acids, so preventing whole cow proteins from entering the baby's system. This 'pre-digested' cow's milk is also adjusted for fat and mineral content, and today's formula milks are as close to human milk as possible. Whether biotechnology will ever manage to add all the antibodies, cells, human enzymes and other components of naturally produced human milk remains a question for the future.

IRREVERSIBLE INHIBITORS

Irreversible inhibitors bind permanently to the enzyme, rendering it useless. For obvious reasons organisms rarely produce this type of inhibitor for their own enzymes, but they are splendid weapons to use against other organisms. A wide variety of natural toxins are irreversible inhibitors, as are many pesticides.

HOW INHIBITORS HELP TO CONTROL METABOLISM

Many metabolic pathways are self-controlling: when a substance is needed, a particular pathway is activated to produce it. When enough has been produced, the pathway is deactivated.

This happens because some enzymes in a metabolic pathway are inhibited by the end product. As Fig 5.16 shows, if too much product begins to accumulate, this inhibits one of the enzymes in the pathway. When the product is once more in short supply, the inhibition is lifted and the pathway becomes active again. This self-regulation is an example of **negative feedback**. This is a fundamental principle important in homeostasis (see Chapter 11).

REMEMBER THIS

Cyanide is an irreversible inhibitor of **cytochrome oxidase**, one of the enzymes involved in respiration. Organisms poisoned with cyanide die because they are deprived of ATP, their immediate energy source.

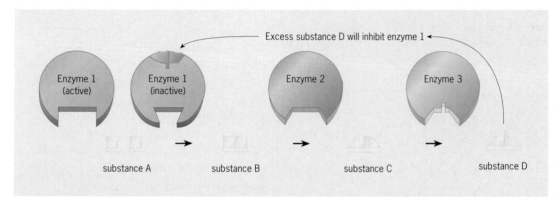

Fig 5.16 A metabolic pathway can be self-regulating by having the end product act as a non-competitive inhibitor on one of the enzymes in its production pathway

SUMMARY

After reading this chapter, you should know and understand the following:

■ Enzymes are **globular** proteins with a precise, but delicate, three-dimensional shape maintained by **ionic** and **hydrogen bonds**.

■ During a reaction, the substrate fits into a region on the enzyme surface called the **active site**.

■ Enzymes are **specific**: each enzyme catalyses one particular reaction.

■ Enzymes speed up reactions by lowering the **activation energy** needed to get the reaction started.

■ Between 4 °C and 40 °C, the rate of an enzyme-controlled reaction increases between two- and threefold for every 10 °C rise in temperature. Increase in temperature beyond 40 °C usually **denatures** the enzyme. Activity is lost when its three-dimensional structure is destroyed.

■ Most enzymes have an optimum pH. For **intracellular** enzymes this is usually about 7.35. **Extracellular** enzymes, such as those found in digestive juices, may have optimums of extreme pH values.

■ The rate of an enzyme-controlled reaction is limited by the supply of substrate, enzyme or the enzyme cofactor.

■ **Inhibitors** are substances that slow down or stop enzyme activity. They may be **reversible** or **non-reversible**. Reversible inhibitors may be **competitive** or **non-competitive**.

■ Competitive inhibitors tend to be similar in structure to the substrate and compete for the active site. Non-competitive inhibitors do not bind to the active site itself but alter the shape of the enzyme so that the active site is no longer functional.

❓ EXAM QUESTIONS

1 Define the term 'enzyme' and relate its function and properties to its structure.
WJEC Biology Module BI1 June 2001 Q8

2 Describe the *induced fit* model of enzyme action.
IBO Standard level paper 3 November 1999 QC2

3
a) Outline how enzymes in the cytoplasm of cells are produced.
b) Compare the induced fit model of enzyme activity with the lock and key model.
c) Explain, using **one** named example, the effect of a competitive inhibitor on enzyme activity.
IBO Higher level paper 2 May 1999 Q6

4 The graph shows the results from a number of experiments to determine the effect of temperature on the rate of a reaction catalysed by a mammalian enzyme.

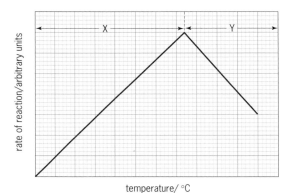

a) Add the scale and numbers on the temperature axis for a typical mammalian enzyme.
b) **(i)** Explain the shape of the curve in region **X**.
(ii) Explain the shape of the curve in region **Y**.
c) If you were carrying out this experiment state **two** factors which you would need to keep constant in order to obtain reliable results.
WJEC Biology Module BI1 January 2001 Q5

5 The total amount of product formed in an enzyme-controlled reaction was investigated at two different temperatures, 55 °C and 65 °C. The results are shown on the graph.
a) **(i)** Explain how you would calculate the rate of the reaction at 55 °C over the first 2 hours of the investigation.
(ii) Explain why the initial rate of this reaction was faster at 65 °C than it was at 55 °C.

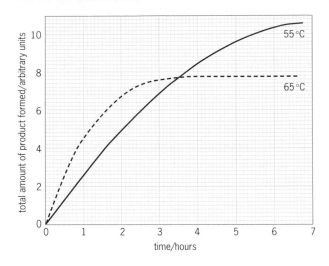

b) Use your knowledge of enzymes to explain the difference in the two curves between 4 and 6 hours.
c) In this investigation, the enzyme and its substrate were mixed in a buffer solution. What was the purpose of the buffer solution?
AQA Biology/Human Biology Molecules, Cells and Systems January 2001 Q1

6 *Bacillus licheniformis* is a bacterium that secretes an extracellular enzyme which digests protein. This enzyme is used in biological washing powders.
a) Describe how the protease enzyme might be produced commercially. Start with a culture of *Bacillus licheniformis* and a suitable medium.
b) **(i)** Biological washing powers containing protein-digesting enzymes are used to remove 'biological stains'. Suggest a 'biological stain' which could be removed by a protein-digesting enzyme.
(ii) The instructions on a packet of biological washing powder stated 'Pre-soak at 40 °C for 1 hour or leave overnight in cold water'. Explain why the recommended soaking time is longer if cold water is used.
AQA Making Use of Biology January 2001 Q3

7 One of the first enzymes to be purified and studied in the 1930s was urease, which catalyses the breakdown of urea to carbon dioxide and ammonia. The reaction may be summarised thus:

$$CO(NH_2)_2 + H_2O \xrightarrow{\text{urease}} CO_2 + 2NH_3$$

It is known that urea molecules will slowly break down to CO_2 and NH_3 in the absence of a catalyst. This spontaneous (uncatalysed) breakdown rate can be speeded up if acid (H^+ ions) is added. However,

❓ EXAM QUESTIONS

the rate of breakdown is very much faster in the presence of the enzyme urease. This is shown in the figure below, which is an energy profile diagram.

a) State, with a reason, whether this reaction is endergonic or exergonic.

b) **(i)** What does **x** represent?
(ii) What does **y** represent?
(iii) What does level **z** represent?

Experiments were performed to investigate the influence of pH on the activity or urease, using phosphate buffer solutions.

Enzyme activity is determined by measuring the amount of urea broken down to CO_2 and NH_3 per unit time. The table shows the results of the investigation.

pH	Enzyme activity (arbitrary units)
4.6	0.06
5.0	0.12
5.6	0.32
6.0	0.50
6.4	0.70
6.8	0.88
7.2	0.84
7.6	0.62
8.0	0.40
8.4	0.14

c) **(i)** Construct a graph of these data on graph paper.
(ii) With reference to your graph, what is the most likely optimum pH value for urease?
(iii) If the urease enzyme solution is made **more** acidic than the optimum by 1 pH unit, by what percentage is the reaction rate lowered?
(iv) Explain why a change in pH affects the activity of an enzyme.
(v) Suggest a possible mechanism, based on the results of this investigation, whereby cells could regulate enzyme activity.
(vi) State **two** factors which must be kept constant during the investigation described above.

OCR Molecules and Life March 2000 Q5

KEY SKILLS **ASSIGNMENT**

ENZYMES AT WORK

Enzymes have enormous commercial potential. Since they can catalyse particular reactions at relatively low temperatures, they are more versatile and much cheaper than inorganic catalysts. Once a suitable enzyme is found, it is made on a large scale and then purified.

There is a vast worldwide demand for sweeteners, mainly for confectionery and soft drinks. Traditionally, manufacturers have used sucrose extracted from sugar beet or cane but, in recent years, **high fructose syrup**, a cheaper sugar beet product has become more common. Fructose is sweeter than sucrose, and can be made from the starch in sugar beet. About 40 million tonnes are now produced each year.

1 What sort of carbohydrate is starch?

Fructose is produced by an enzyme-catalysed process: the starch is first broken down into glucose and then glucose is converted to fructose.

2 What type of enzyme catalyses the conversion of:
a) starch to glucose?
b) glucose to fructose?

Three enzymes are involved in the production of high fructose syrups: **bacterial β-amlyase**, **fungal amyloglucosidase** and **bacterial glucose isomerase**. In the first step, the starch paste is heated to 105 °C. β-amylase is then added to reduce the **viscosity** (thickness) of the paste. The process results in a mixture of **maltodextrins** – branched sugars.

3 **a)** How do we describe an enzyme that works at high temperatures?
b) What is the advantage of higher temperatures in industrial processes?
c) Why does the action of the enzyme reduce the viscosity of the paste?

The next step is the hydrolysis of the maltodextrins to glucose. The substrate is cooled to 55–60 °C and acidified to pH 4.5. The enzyme amyloglucosidase, obtained from the fungus *Aspergillus niger*, is added. This is an **exoenzyme**: it removes the terminal glucose units from the maltodextrins.

4 **a)** What is the difference between an endoenzyme and an exoenzyme?
b) In the hydrolysis of polymers like starch, why is it an advantage to add an endoenzyme before using an exoenzyme?
c) Sketch graphs to show the likely effect of variations in temperature and pH on the activity of amyloglucosidase.

Amyloglucosidase acts on maltodextrins to produce glucose syrup. Although not as sweet as sucrose, it still has its uses in the food industry. The production of the extrasweet fructose needs the enzyme glucose isomerase, which is obtained from bacteria (*Bacillus* or *Streptomyces* sp.). This enzyme is fixed on rigid granules and packed into a column. The glucose syrup then flows between the granules and the enzyme converts the glucose into fructose.

5 What is the advantage of using a fixed (immobilised) enzyme?

The end product is high fructose syrup – a mixture of about 42 per cent fructose and 55 per cent glucose. In these proportions it has the same sweetening power as sucrose.

An alternative sweetener is **aspartame**, a dipeptide that is 180 times sweeter than sucrose. The commercial production of aspartame again involves enzymes – particularly **aspartase**. The ease and cost of aspartame production affects the demand for high fructose syrups.

6 What is the main dietary advantage of using aspartame instead of sucrose?

7 All of the enzymes mentioned in this assignment come from organisms such as yeast and bacteria. Why are these organisms particularly suitable for large-scale enzyme production?

8 Imagine that a scientist isolates a human gene that codes for an enzyme that catalyses a reaction giving a very valuable product. List, and perhaps discuss, the steps that would have to be taken in order to get this product into commercial production.

Fig 5.A1 Many soft drinks contain a high concentration of fructose syrup

06

Biotechnology

Biotechnology

Scientists in the USA have identified new strains of bacteria that could revolutionise how we deal with pollution. One 'eats' benzene, a toxic chemical found in petroleum-based fuels. Fuel spills and leaking storage tanks mean that benzene is a common contaminant of groundwater supplies and, since long-term exposure to benzene can lead to cancer in humans, it must be dealt with before it can affect soil and drinking water.

Benzene breaks down when exposed to the atmosphere but, without oxygen, this degradation process slows to a crawl. Removing the compound from airless environments is therefore difficult. However, *Dechloromonas* strain RCB and *Dechloromonas* strain JJ apparently oxidise benzene to carbon dioxide without the help of oxygen. These organisms are creating a lot of interest because they can speed up the break down of benzene in anaerobic conditions by up to 10 times. What once took 70 days now happens in 7 days.

Another bacterial species digests toluene, another common but toxic groundwater contaminant that comes from fuels, adhesives and household solvents. These bacteria also work in the absence of oxygen and research has already identified four genes that code for enzymes in the metabolic pathway responsible.

Using these organisms, or other bacteria given pollution-busting genes, to clean up contamination would have several advantages over current methods. Not only would they be cheaper, but also the bacteria could be brought to the contaminated area, which would avoid the environmental scarring caused by the removal of tonnes of contaminated soil for dumping or cleaning. The toluene-eating species also might be used to prevent pollution by treating industrial waste before it ever reaches the environment.

Culture of oil-degrading bacteria. Genetically engineered microbes may become more widely used to extract oil from the ground and valuable metals from factory wastes

1 WHAT IS BIOTECHNOLOGY?

A definition of biotechnology is quite difficult to pin down. Broadly speaking, it is the manipulation of biological organisms, systems or processes for the benefit of people, in areas such as agriculture, food production and medicine.

The oldest biotechnologies include the **fermentation process**, which has been used to make bread, beer, wine and cheese for centuries, and also **plant and animal breeding**. More recent biotechnology techniques include the **manipulation of genes** to make **transgenic** plants and animals, using **monoclonal antibodies** to target cancer cells, and using advanced chemistry to **immobilise** enzymes to control and improve manufacturing processes.

Elsewhere, you will also see biotechnology defined more narrowly – as the manipulation of DNA and genes to make organisms that are commercially useful. In this book, we consider both aspects of biotechnology. This chapter covers the wider aspects, with sections on the use of biotechnology in farming and food production, in industry and in medicine. We also consider some of the social and ethical implications of biotechnology. The manipulation of

DNA and genes is also here, particularly when it is relevant to farming and food production. However, gene technology and topics such as human cloning and the human genome project are dealt with in more detail in the section on genetics (Chapters 22–27).

2 BIOTECHNOLOGY AND FARMING

MANIPULATING FARM ANIMALS

Some people might argue that farming animals is manipulating them too much already, but biotechnology is now used increasingly as a tool to develop more productive animals. Not surprisingly, many of the developments are controversial. All have their supporters and their opponents, and in the sections that follow we have tried to present a factual balanced view. Delve into it deeper if you want, and then make up your own mind.

USING HORMONES TO INCREASE MILK PRODUCTION

In order to produce milk at all, cows need to produce a calf each year – they are constantly either pregnant or lactating, or both. With careful timing of one pregnancy after another, a dairy cow can produce a steady supply of milk. Its calf is usually taken away and hand reared, or is used for food. The quality and quantity of this supply depend on the genetic make-up of the cow, how old she is, how much food she is given and the quality of that food.

For some years now, farmers in many developed countries such as the UK and USA have been enhancing the level **bovine somatotrophin** (**bST**) in dairy cows. Also known as **bovine growth hormone** (**BGH**), this is a natural protein hormone produced by the **pituitary gland** of all cattle.

Biotechnology has enabled scientists to produce a **recombinant** form of this protein, called **rbST**. This means that they have isolated the gene for rbST from cows, and then inserted it into the genome of bacteria. These are then cultured and produce the hormone is vast quantities in industrial vats. The industrial production of recombinant growth hormone is summarised in Fig 6.1.

When this readily available hormone is injected into cows on a regular basis, it reliably improves their efficiency as milk producers. The mammary glands of such dairy cows take in more nutrients from the bloodstream and produce more milk.

It is very unlikely that bST, whether recombinant or natural, has any biological effect on humans. It already exists naturally in milk, so there is just a bit more of it in the milk from rbST-treated cows. However, the human digestive juices break down the protein into the amino acids that make it up, so it disappears, except in babies who, in any case, should not be fed on cow's milk. It cannot gain entry to the human body as a complete protein. And even if it did, the hormone is **species-specific**. The form found in cows does not fit into any receptor in the human body; it is a completely different shape from human growth hormone, for example.

Cows that are given rbST increase their overall milk production between 5 and 10 per cent but they don't need to eat 5 to 10 per cent more food, so overall this is a cost-effective technique. Those in favour argue that it allows dairy farmers to be more efficient and that it enables more milk to be produced in areas where food is needed urgently. They propose that nations suffering from hunger and malnutrition could benefit from rbST.

However, those against hormone use say that milk produced by treated cows should be labelled so consumers can decide whether to drink it or not.

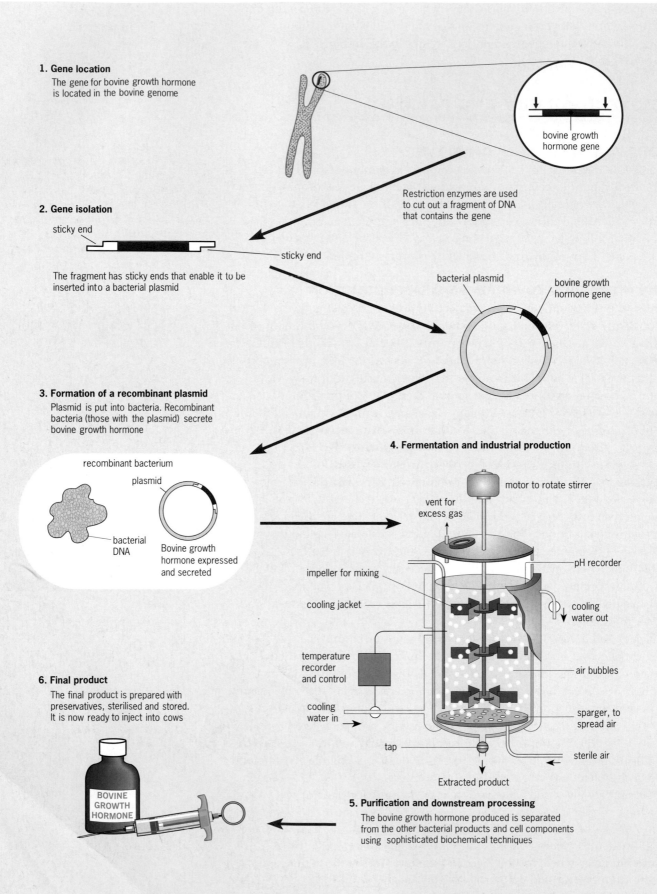

1. Gene location
The gene for bovine growth hormone is located in the bovine genome

bovine growth hormone gene

Restriction enzymes are used to cut out a fragment of DNA that contains the gene

2. Gene isolation

sticky end

sticky end

The fragment has sticky ends that enable it to be inserted into a bacterial plasmid

bacterial plasmid

bovine growth hormone gene

3. Formation of a recombinant plasmid
Plasmid is put into bacteria. Recombinant bacteria (those with the plasmid) secrete bovine growth hormone

recombinant bacterium

plasmid

bacterial DNA

Bovine growth hormone expressed and secreted

4. Fermentation and industrial production

motor to rotate stirrer

vent for excess gas

pH recorder

impeller for mixing

cooling jacket

cooling water out

temperature recorder and control

air bubbles

cooling water in

sparger, to spread air

tap

sterile air

Extracted product

6. Final product
The final product is prepared with preservatives, sterilised and stored. It is now ready to inject into cows

BOVINE GROWTH HORMONE

5. Purification and downstream processing
The bovine growth hormone produced is separated from the other bacterial products and cell components using sophisticated biochemical techniques

Fig 6.1 Steps in the production of recombinant bovine growth hormone. This diagram is not drawn to scale

Fig 6.2 Cow's udder being injected
with antibiotic

Fig 6.3 Embryo duplication allows up to 16
embryos to be obtained from one mating between
a valuable female and a valuable male. The calves
are carried by surrogate mothers. The cow that is
the biological mother does not carry a calf; she is
inseminated again soon afterwards to obtain another
embryo for splitting between another set of
surrogates. In this way, one pair of prime animals
can be used to generate dozens of calves every year

This may lead to an unsold surplus of milk, which may need
to be bought by governments to prevent farmers going out
of business. They also say that cows that are given growth
hormone suffer more udder infections than untreated cows,
and that the greater quantity of antibiotics used to treat them
then gets into the human food supply.

EMBRYO DUPLICATION

The technique of **artificial insemination** – the use
of sperm from a bull or ram with characteristics that are
particularly sought after in the farming world – has been used
to breed 'better' animals for several years. In the last 15 years
of the twentieth century reproductive technology became a lot
more sophisticated and it became possible to make several
female animals pregnant with identical embryos. When these
are produced by mating a prime female with a prime male, the
resulting embryos can be split, sometimes into two, sometimes
more, and implanted into any healthy female of the same
species. This means that a whole herd of prime stock animals
can be produced from one breeding pair, whose embryos are
carried by **surrogate mothers**.

SYNCHRONISING SHEEP BREEDING

In Britain, the changing day-length of the seasons dictates
the breeding behaviour of sheep. They synchronise their
reproductive cycle to produce lambs in early spring
(March/April) when new grass is plentiful. This also gives
the young as much time as possible to mature before the next
winter. The gestation period in sheep is 21 weeks, and ewes
and rams must mate in October/November to get the timing
just right.

Normally, left to their own devices, sheep rely on their
hormones to do this. In the autumn, the stimulus of shortening
days affects the **hypothalamus**, which responds by secreting
a hormone called **gonadotrophin-releasing factor**. As its
name suggests, this acts on the pituitary gland to cause the
release of a gonadotrophin hormone – **follicle stimulating**

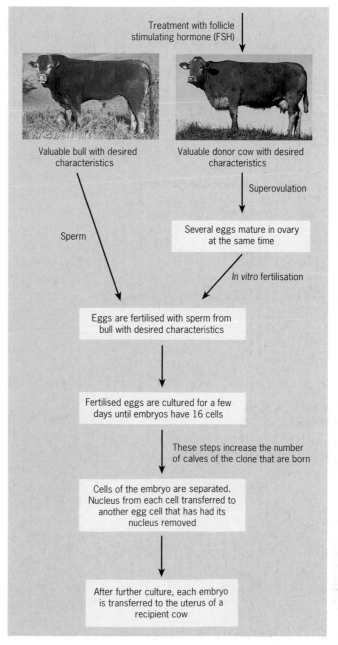

Treatment with follicle
stimulating hormone (FSH)

Valuable bull with desired
characteristics

Valuable donor cow with desired
characteristics

Superovulation

Sperm

Several eggs mature in ovary
at the same time

In vitro fertilisation

Eggs are fertilised with sperm from
bull with desired characteristics

Fertilised eggs are cultured for a few
days until embryos have 16 cells

These steps increase the number
of calves of the clone that are born

Cells of the embryo are separated.
Nucleus from each cell transferred to
another egg cell that has had its
nucleus removed

After further culture, each embryo
is transferred to the uterus of a
recipient cow

Fig 6.4 When ewes come into season, the farmers release rams into their field for a night. In the morning, all the females should have been impregnated. Just to make sure none has been missed, some farmers take the precaution of taping a block of dye to each ram's chest. This rubs off during mating. Any ewes that do not have dyed bottoms by morning have to stay with the rams for another night

❓ TEST YOURSELF

A Why could a castrated ram not be used as a teaser when a farmer is trying to synchronise mating in his flock of ewes?

hormone (FSH). FSH stimulates the development of follicles in the ovaries; these secrete oestrogen, and this causes the ewe to come into season. As she begins to give off vaginal secretions that contain **pheromones**, nearby rams are aroused and mating occurs.

The problem with this for farmers of today is that the mating process, and therefore the lambing, can take place over a period of about a month. This makes it very difficult to care for all the ewes and lambs, particularly if the flock is very large. To enable farmers to concentrate their effort and resources (such as vets on standby), they try to ensure that lambing actually takes place over a 5-day period. To do this, they must ensure that most ewes mate on the same day.

Achieving this takes some extraordinary measures:
● Ewes that are to be mated must be kept away from all sexually active males during the time they are weaning their previous lambs. Sexually active males include not only rams, but male lambs and billy goats, which can secrete some of the hormones that attract the female.
● This absence of males allows the next stage to have its maximum effect. Vasectomised but thoroughly randy rams are then put with the ewes four weeks before the intended mating day to bring them into season. These rams are termed 'teasers', for obvious reasons.
● Two weeks after the introduction of the teasers, the ewes are separated again and small sponges impregnated with the hormone **progesterone** are inserted just inside the vagina of each one. The progesterone prevents ovulation in the ewes until it is removed 14 days later, at which point all the ewes should be at the same stage of their oestrous cycle.
● Sometimes, another hormone, called **pregnant mare serum gonadotrophin** (**PMSG**), is injected into the ewes just after sponge removal. This slightly increases the litter size of the ewes, and may also make the timing more precise.
● 36 to 48 hours after sponge removal, ewes come into heat. Tonight's the night!
● The rams are then introduced to the ewes and, after few hours of frenzied activity, the farmer achieves his or her objective. Alternatively, the farmers use **artificial insemination** – important if they want to use a particular ram to father as many lambs as possible.
● If all has gone well, the ewes should lamb 145 days later, with some lambs born 2 days or so either side of the expected date.

USING BIOTECHNOLOGY IN CROP DEVELOPMENT

Farmers have been practising a crude form of biotechnology for centuries by **cross-breeding** plants to introduce and maintain desirable characteristics. More recently, plant technologists have bombarded plant cells or small plants with radiation that damages DNA, causing random changes that sometimes result in 'better' plants. However, the newest, truly biotechnological techniques involve **genetic engineering** and produce the **genetically modified** (**GM**) plants and crops that are so often in the news.

HOW ARE PLANTS GENETICALLY MODIFIED?

Genetic modification usually involves the transfer of a single gene, together with an obvious marker so that the plant breeder knows which plants have taken up the gene. The gene chosen usually codes for a protein that gives the plant a higher rate of growth, or makes it resistant to drought, pests or weedkillers, or that makes part of the plant more useful for the food industry (see the Flavr Savr tomato on page 110).

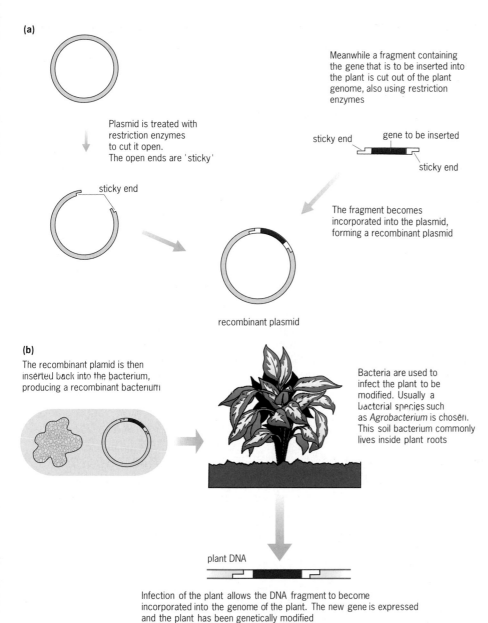

(a)

Plasmid is treated with
restriction enzymes
to cut it open.
The open ends are 'sticky'

sticky end

Meanwhile a fragment containing
the gene that is to be inserted into
the plant is cut out of the plant
genome, also using restriction
enzymes

sticky end   gene to be inserted

sticky end

The fragment becomes
incorporated into the plasmid,
forming a recombinant plasmid

recombinant plasmid

(b)
The recombinant plamid is then
inserted back into the bacterium,
producing a recombinant bacterium

Bacteria are used to
infect the plant to be
modified. Usually a
bacterial species such
as *Agrobacterium* is chosen.
This soil bacterium commonly
lives inside plant roots

plant DNA

Infection of the plant allows the DNA fragment to become
incorporated into the genome of the plant. The new gene is expressed
and the plant has been genetically modified

Fig 6.5 (a) Production of a recombinant bacterial
plasmid containing a gene for transfer into a plant
(b) Bacteria such as *Agrobacterium* sp. are forced
to take up the recombinant plasmid and are then
introduced into the plant. The plasmid enters plant
cells where it opens up to release the gene. This is
then incorporated into the DNA of the plant. As the
plant genome is expressed, so is the new gene.
This diagram is not drawn to scale

Fig 6.5(a) shows how a gene is first incorporated into a **plasmid**
(a small circular piece of DNA) and is then introduced into cells of the plant
that is being modified (Fig 6.5b). Sometimes bacteria can infect a whole
plant; other techniques introduce the bacteria carrying the plasmid into
individual plant cells. These are then cultured to produce plant tissue and
then whole plants that carry the 'foreign' gene.

THE ADVANTAGES OF GENETICALLY ENGINEERED PLANTS

The main advantage of such a specific transfer of one or perhaps a handful
of genes is that the plant breeder can avoid transferring lots of other genes,
which may not be identified. The only changes that will occur in the
modified plants are the changes that result from the known gene. Genetic
engineering is therefore more predictable and precise than 'traditional' plant
breeding techniques.

SEE QUESTIONS 1 AND 2

AND THE DISADVANTAGES

The major problem that is preventing more widespread use of genetically
modified crops and foods is public suspicion and hostility.

✔ **REMEMBER THIS**

Genetically modified plants could also be called **transgenic plants**, but this tends to be a term used more commonly for animals that have been bred to carry specific genes.

Opponents of GM foods are particularly vocal in the UK (Fig 6.6). In other countries in Europe, and in the USA, people have accepted them more readily.

Here are some of the arguments made by people opposed to GM foods:

● GM foods are bad for your health: swapping genes from one organism to another is unnatural, and therefore dangerous. What might be the effect of eating 'foreign' genes from bacteria?

● Genes from GM plants might escape and get into other plants, creating 'superweeds'. Bacteria might pick up the plasmids that have been used to carry genes into plant cells, and then infect other plants. The genes may spread like wildfire, making weeds resistant to all known weedkillers.

● GM plants might breed with traditional varieties, producing hybrids that are no use. Organic farmers might find that crops that have taken years to establish become contaminated with genes from GM crops.

● Growing GM foods resistant to weedkillers and pesticides will increase the use of such chemicals, and insects and other useful wildlife will be lost.

● The technology behind GM crops is only available to the multinational, large agrobusinesses, not the small, independent farmer. Organic farmers may find their produce outsold by GM foods, which can be produced at a much lower cost.

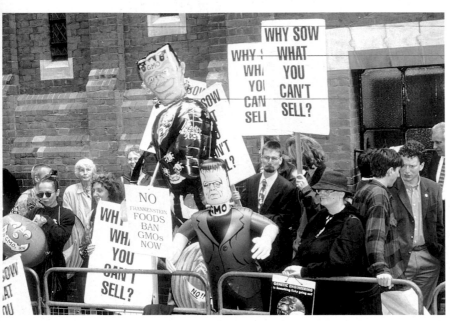

Fig 6.6 The use of genetic engineering in food production has provoked strong feelings in many people

EXAMPLES OF GM FOODS AND CROPS

THE POMATO

One of the earliest GM foods produced was the 'pomato', a laboratory-generated combination of two members of the *Solanaceae* family, the potato and the tomato. This could grow into a plant but could not breed. In 1994, however, researchers managed to cross the pomato **hybrid** with an ordinary potato but not without difficulty. Today, groups around the world are still trying to improve the pomato, and are also attempting to use the hybrid to transfer frost-resistance genes from the tomato plant to the potato.

Tomato is better able to withstand frost than potato. Dutch potato growers sow seed potatoes in April. If there is frost at night the leaves freeze and wither. The potato survives but yields are considerably reduced. Farmers who produce seed potatoes are most troubled by this problem. They want to sow as early as possible so that they can also harvest early in order to avoid viruses.

The plan is to cross the hybrid pomato with a potato. Each time sex cells are made through a process of **meiosis** (see Chapter 23), some of the chromosomes disappear in the hybrid. This forces single tomato chromosomes to pair up with a potato partner. The partner chromosomes then exchange parts and segments of the tomato chromosome become permanently built into a potato chromosome. However, this transfer of tomato chromosome parts to the potato does not happen as frequently as the researchers had expected, and success has so far eluded them.

? **TEST YOURSELF**

B Look at the list of objections made about GM crops and foods. From a scientific point of view, say which ones you think are valid, and which are not. Explain how you came to your decision.

TRANSGENIC TOMATOES COULD HAVE AN EXCITING FUTURE

Although the Flavr Savr tomato is no longer available (see the Science Feature on page 110), other transgenic tomatoes are being developed all the time. Here are a few of the developments we might see in the next few years:

- Transgenic tomatoes may be produced containing genes that help them tolerate salty growing conditions, drought, flood and extreme temperatures. One gene that regulates the plant's ability to cope with cold weather has already been identified.
- A vitamin-rich tomato has been produced that could help to prevent heart disease and cancer. This tomato has about four times the normal levels of beta-carotene, which the body uses to make vitamin A, and twice the levels of lycopene (a close cousin of beta-carotene), the compound that helps make tomatoes red. Eating lots of lycopene may reduce the risk of some cancers, particularly prostate cancer, so transgenic tomatoes may be eaten as part of a cancer prevention diet.
- Herbicide-resistant, virus-resistant and pest-resistant tomatoes are in development. *Bacillus thuringiensis* (Bt) is a naturally occurring soil bacterium that produces a protein that kills a range of common insects once it is ingested. The *Bt* gene has been isolated and inserted into crops including tomatoes.
- Tomatoes may be used as edible vaccines. Tomatoes are being engineered to carry genes that can act as vaccines for diseases such as rabies.

 REMEMBER THIS

Seed potatoes may only be sold if they are certified 'virus-free'.

! **SCIENCE FEATURE** Transgenic fish

As you will see in more detail in Chapter 27, the DNA that carries a specific gene can be introduced into an organism of the same or a different species by injecting it directly into embryo cells using a tiny glass pipette. Several species of transgenic animal have, for example, been bred to carry an extra growth hormone gene. As soon as the organism starts to develop, and this gene is expressed, it is exposed to a much higher concentration of growth hormone than usual, sometimes with fairly predictable results. Northern pike (a marine pike) with the growth hormone gene are two to six times bigger than fish without the extra gene. In one case, a fish 13 times larger than normal was created.

However, the effect that these fish have on their environment is less predictable. If this is a fish farm, they can grow as big as they can, and then be 'harvested'. However, if the fish are allowed into the open sea, there is a real danger that they could out-compete their smaller siblings and so disrupt the delicate balance of their ecosystem. Perhaps there is a case for introducing other genes as well as the growth hormone gene so that any transgenic fish are unable to breed, and therefore unable to take over.

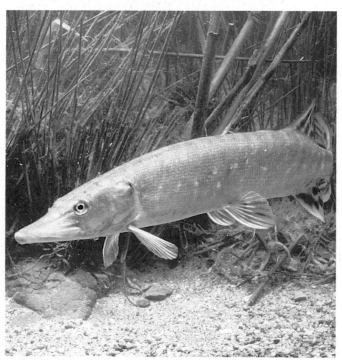

Fig 6.7 Transgenic northern pike, *Esox lucius*

SCIENCE FEATURE The Flavr Savr tomato

The Flavr Savr tomato, the world's first transgenic tomato, was produced by an American company called Calgene in the late 1980s. It makes use of something called anti-sense technology (Fig 6.8). Basically, a gene is introduced into the tomato plant that produces messenger RNA. This blocks the transcription of a gene in the tomato that codes for an enzyme called polygalacturonase. This enzyme breaks down the pectin that normally holds cell walls together. When it is present, it causes the tomato to go soft and squishy, and the fruit rots. The messenger RNA produced by the anti-sense gene stops the enzyme being made, and the fruit stays much firmer. Other parts of the ripening process are not affected so the flavour of a ripe tomato still develops.

The major advantage of the Flavr Savr tomato was that it could be left on the vine to ripen for longer, producing tomatoes with a better flavour, but that did not become mush when transported.

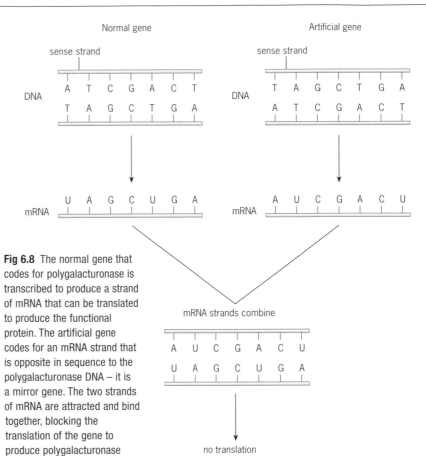

Fig 6.8 The normal gene that codes for polygalacturonase is transcribed to produce a strand of mRNA that can be translated to produce the functional protein. The artificial gene codes for an mRNA strand that is opposite in sequence to the polygalacturonase DNA – it is a mirror gene. The two strands of mRNA are attracted and bind together, blocking the translation of the gene to produce polygalacturonase

Fig 6.9 GM tomato on vine

Release of the Flavr Savr was reviewed by the Food and Drug Administration in the USA in the spring of 1994. It was found to be as safe as conventionally produced tomatoes. This is the first time a whole food produced by biotechnology had been evaluated for its public safety. However, concern still surrounded Flavr Savr's introduction to the market. People worried about whether the tomato was a danger to health, whether it would give them food allergies, whether it contained 'unnatural' toxins. It was termed a 'Frankenfood' and in 1997 it was withdrawn from the market by the company that developed it.

Fig 6.10 These tomatoes are the same age and variety. The deterioration of those on the right is normal. The tomatoes on the left have been genetically modified and stay fresher for longer

3 BIOTECHNOLOGY AND FOOD

LACTIC ACID FERMENTATION

MAKING YOGHURT

Yoghurt is basically milk (cow, goat or ewe) that is fermented by bacteria. Although it originated in Western Asia and Eastern Europe, it is now eaten all over the world. To make yoghurt, a starter culture of the lactic acid bacteria *Lactobacillus bulgaricus* and *Streptococcus thermophilus* is added to whole or skimmed milk. The bacteria multiply and ferment lactose, the disaccharide sugar in the milk, producing lactic acid. As you will know if you have ever added lemon juice to milk, acid causes the milk to curdle. Separation of the acidified milk into solid curds gives yoghurt its taste and texture.

Yoghurts can be pasteurised to destroy the fermentation bacteria but if this is not done, the result is 'live' yoghurts. Many supermarkets now sell yoghurts they describe as 'bio-yoghurts'. These have been produced in the usual way, left unpasteurised, and then extra bacteria are added. Bio-yoghurts usually contain live *Lactobacillus acidophilus* and *Bifidobacterium bifidum*. Various health claims are made for these yoghurts, but whether they actually do keep you healthy or not is still an open question (see the Assignment on page 127 of this chapter).

CHEESE-MAKING: ONE OF THE OLDEST BIOTECHNOLOGIES

Cheese-making dates back at least 5000 years and the basic method used has changed very little over that time. Three main ingredients are needed:

- Milk.
- A source of the protein-degrading enzyme **chymosin**, which causes the milk proteins to clump together. Traditionally, this was added in the form of calf **rennet** (chymosin can also be called **rennin**). Non-animal sources of this enzyme are now available and can be used in the manufacture of cheese suitable for vegetarians.
- A starter culture of lactic acid bacteria.

In many ways, cheese-making is similar to yoghurt-making. The lactic acid starter culture is added to the milk to convert the lactose to lactic acid, which helps to preserve the cheese. Chymosin causes the milk to set, producing hard cheese. Once bacterial fermentation stops, the maturation of the cheese then begins. As the bacteria die, they are digested by their own enzymes – a process called **autolysis**. This process produces more enzymes, including **peptidases**, that produce the strong flavours that we associate with a 'good' cheese. Some strains of bacteria produce carbon dioxide during maturation – this makes the holes in cheeses such as Gouda and Emmental.

A fungus of the Penicillin genus (*Penicillium roquefortii*) is used to inoculate fermented curd to produce blue cheeses such as Stilton and Danish Blue. This fungus grows through the cheese during the maturation process, producing the characteristic blue-green veins.

✓ **REMEMBER THIS**

Anaerobic bacteria that can ferment carbohydrates to lactic acid are widely used in the food industry. They lack the enzymes of the citric acid cycle and the respiratory chain, and so ferment carbohydrates into lactic acid – this process is called **fermentation**.

✓ **REMEMBER THIS**

Pasteurisation of milk involves heating it to 60 °C so that most of the bacteria in it are killled, but the taste is not unduly affected. Sterilised milk has been heated to 100 °C to kill all the bacteria, which is why it keeps much longer. However, at this temperature, some of the proteins in it are denatured and this changes the taste of the milk many people dislike the new taste and sterilised milk has never been popular for drinking.

Fig 6.11 The pictures show the holes in Gruyere (below) and a blue cheese, Roquefort, being probed during its maturation to control the development of the blue veins (above)

FERMENTATION USING YEAST

Yeast is an essential microorganism in the production of bread, wine and beer. Under the right conditions, yeast will ferment carbohydrates to release carbon dioxide and alcohol in a series of chemical reactions.

BREAD-MAKING

Bread is basically a mixture of flour, water and a little fat that is baked. Some breads in the world are still produced like this (pitta bread, for example) but yeast fermentation is usually used to cause the dough to rise, so that a lighter, less dense loaf can be made. As the flour, water and yeast are mixed, enzymes called amylases that are present in the flour break down the starch into the disaccharide, maltose. Other enzymes break this down into glucose, which is then fermented by the yeast to carbon dioxide and alcohol. In bread, the alcohol evaporates during baking, but the carbon dioxide bubbles become trapped in the sticky dough, causing the baked loaf to have lots of tiny pockets.

WINE AND BEER

In wine-making, the fermentation of grape juice is a complex process that involves yeasts, bacteria and filamentous fungi – but the yeasts have the starring role. During fermentation, yeasts use the sugars and other components of grape juice and convert them into ethanol, carbon dioxide and other end-products that contribute to the chemical composition and taste of wine. Fermentation stops when the yeast becomes poisoned by its own waste – the alcohol – which is why drinks made by fermentation only are no stronger than about 14 per cent alcohol by volume. Spirits such as whisky and gin, which have a much higher alcohol content, are **distilled** from a fermented mixture.

(a) Batch production

sterile air for oxygenation

nutrient culture + bacterial starter culture added at one go

nutrient culture and bacteria

(b) Continuous production

sterile air for oxygenation

fresh nutrient culture

valve to control how much fresh medium flows in

nutrient culture and bacteria

Fig 6.12 (a) Batch production: when fermentation is complete, the entire contents of the tank are released for processing. The tank is cleaned and then the whole process is repeated. Each production is a different batch
(b) Continuous production: fermentation is continuous and can carry on indefinitely as long as fresh medium is added and product is removed and the conditions for optimum growth are maintained. Continuous production runs a higher risk of contamination than batch production, so great care has to be taken to keep the incoming medium sterile, apart from the microorganism in the culture

Most wine-makers are more interested in the alcoholic by-product of yeast fermentation but a brewer producing beer needs both the alcohol and the carbon dioxide made by the yeast. Carbon dioxide gives the beer its characteristic fizziness.

The brewing industry uses barley as the source of the food that the yeast ferments to make the alcohol in beer. However, barley stores food in the form of starch – which yeast cannot use directly. In order to solve this, the brewer allows the barley grains (seeds) to germinate. During germination, enzymes in the barley convert the starch into maltose sugar, which the yeast can ferment. This process is called **malting**.

Beer production is made as efficient as possible by providing the best possible conditions for yeast to grow and ferment. This means that the temperature, oxygen supply and amount of glucose must be carefully controlled, and unwanted microorganisms must be kept out. The easiest way to make sure that this happens throughout the fermentation is to set up all the conditions at the start, together with the raw materials, and then to leave the whole system closed and untouched until the fermentation is complete. This has the disadvantage that production is only possible in batches rather than a continuous process. The main features of batch processing and continuous processing are shown in Fig 6.12.

USING OTHER MICROORGANISMS AS A FOOD SOURCE

Biotechnology has been used in the past few years to make foods that are as rich in protein as meat, but that are much cheaper. Pruteen™ is one example (Fig 6.13), mycoprotein (for example, Quorn™) is another. Both can be eaten by vegetarians. Mycoprotein has the following nutritional advantages over meat:

- It has no animal fat, little overall fat and no cholesterol.
- It has a high protein content (as high as that of skimmed milk protein).
- It is high in fibre.
- It contains useful amounts of trace elements and B vitamins.

Mycoprotein is produced by the fungus *Fusarium graminearum*, which is related to mushrooms and truffles. This fungus can use the carbohydrate present in cheap, readily available carbohydrates – wheat in the UK, potato in Ireland, and cassava, rice or sugar in tropical countries, for example. Mycoprotein contains 45 per cent protein and 13 per cent fat, a composition similar to that of grilled beef. It is also high in fibre and has a complete amino acid content.

TEST YOURSELF

C What is the difference between fermentation and glycolysis? (See Chapter 34).

REMEMBER THIS

In exam questions on fermentation you will be expected to know about the basics of respiration, particularly glycolysis, and also be able to apply your knowledge of the properties of enzymes.

SEE QUESTIONS 3 AND 4

REMEMBER THIS

Mycoprotein is made by a filamentous fungus – one that produces lots of thin strands called **hyphae**.

REMEMBER THIS

Pruteen™ is a microbial protein produced by bacteria that can break down methanol.

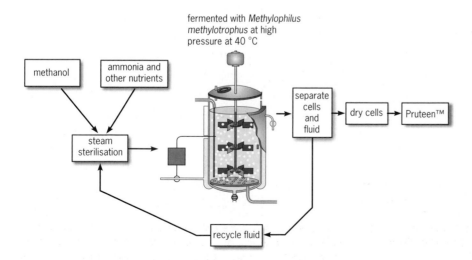

Fig 6.13 Pruteen™ production. Pruteen™ is the trade name for microbial protein produced by growing *Methylophilus methylotrophus* bacteria on methanol, which is derived from methane or natural gas. Since the product is a low-value commodity, large quantities have to be manufactured to make profits. Usually very large (3000 m³) airlift fermenters are used with a diameter of 7 m and a height of 60 m. These can be run for 100 days at a time

✔ REMEMBER THIS

The following fact shows why the potential for using microorganisms in food production is so interesting in the commercial world: with a generation time of 20 minutes, and with no limiting factors, a single bacterium, with a biomass of only 10^{-12} g, would produce 22×10^{25} tonnes of biomass in two days. That's 4000 times the mass of the planet Earth.

The industrial fermenters currently being used to manufacture mycoprotein are 40 metres high (similar in height to Nelson's Column). They run continuously for 6 weeks, after which there is a 2-week period for cleaning and preparing the fermenter for the next run. After production, the mycoprotein looks like pastry, but it is then mixed with binding agents and colours to form a product that looks like different kinds of meat.

OTHER FERMENTED FOOD PRODUCTS

SAUERKRAUT

Sauerkraut is a German delicacy that is now popular in other parts of the world, particularly the USA. It is made by fermenting cabbage that has been salted. The main bacteria responsible for the fermentation are lactic acid bacteria. *Leuconostoc mesenteroides* ferments the cabbage juices to produce lactic acid, acetic acid, carbon dioxide and ethanol. Two others, *Lactobacillus plantarum*, which produces lactic acid only, and *Lactobacillus brevis*, are used to complete the process.

TOFU, TVP AND SOY SAUCE

Biotechnology is also used to process soya beans to give the following food products:

- **Textured vegetable protein (TVP)** – defatted soya flour that has been processed and dried to give a substance with a sponge-like texture that resembles meat. Soya beans are **dehulled** (their skins are taken off) and their oil is extracted before they are ground into flour. This flour is then mixed with water to remove soluble carbohydrate and the residue is textured by either **spinning** or **extrusion**. Extrusion involves passing heated soya residue from a high-pressure area to a reduced pressure area through a nozzle – this causes the soya protein to expand. The soya protein is then dehydrated and may be either cut into small chunks or ground into granules.
- **Tofu** – soya bean curd made from coagulated soya milk. Soya beans are soaked, crushed and heated to produce soya milk to which a coagulating agent such as calcium sulphate or calcium chloride is added. The resulting soya curd is then pressed to give tofu. Tofu is sometimes known as soya cheese, and is sold as blocks packaged in water.
- **Tempeh** – a fermented soya bean paste made by inoculating cooked soya beans with the fungus *Rhizopus oligosporous*. This forms a mycelium holding the soya beans together and is responsible for the black specks in tempeh. Tempeh has a chewy texture and distinctive flavour and can be used as a meat substitute in recipes. It may be deep-fried, shallow-fried, baked or steamed.
- **Miso** – a fermented condiment made from soya beans, grain (rice or barley), salt and water. Miso production involves steaming polished rice, which is then inoculated with the fungus *Aspergillus oryzae* and left to ferment to give an end-product called **koji**. Koji is then mixed with soya beans that have been heated and extruded to form strands, together with salt and water. This is then left to ferment in large vats. Miso varies widely in flavour, colour, texture and aroma. It is used to give flavour to soups, stews, casseroles and sauces.
- **Soy sauce** – true soy sauce, called **shoyu**, is made by fermenting soya beans with cracked roasted wheat, salt and water. **Tamari** is similar but slightly stronger and made without wheat (and so is gluten-free).

Fig 6.14 Soya products: dried soya beans, soya milk (blue bowl) and two types of tofu or bean curd – dried tofu (upper right) and fresh tofu (brown bowl)

Fermentation for shoyu and tamari takes about one year. Much of the soy sauce available in supermarkets is not true soy sauce but is made by chemical hydrolysis from defatted soya flour, caramel colouring and corn syrup without any fermentation process.

4 BIOTECHNOLOGY IN INDUSTRY

Enzymes were first used on a large scale by the textile industry. Thread used for weaving is protected by a coating of starch paste – useful during the weaving process to make the thread more 'slippy' so that weaving was easier, but totally useless afterwards. The starch used to be removed by acids, alkalis or oxidising agents but these often discoloured the cloth. In the early part of the twentieth century, the industry started using enzyme extracts from malt, and later from an animal pancreas. A thermostable bacterial amylase was discovered in 1917, but its use in the mass production of cloth took a few years to perfect. At around the same time, the leather industry also started to experiment with enzymes to reduce costs. Protease enzymes began to replace calcium hydroxide and sodium sulphide for removing animal hair from skins.

These early attempts at using enzymes seem crude by today's standards and it took until after 1945 for large-scale production and use of enzymes to be commonplace in industry.

ENZYMES IN THE INDUSTRIES OF TODAY

The ability of enzymes to catalyse specific chemical reactions at body temperature makes them very useful tools in the commercial world. Table 6.1 outlines some of industrial applications of enzymes. This is a rapidly changing field, and new applications of enzyme technology appear all the time.

✔ **REMEMBER THIS**

Thermostable enzymes, those that work outside the normal physiological range of the human body, are particularly useful in industry. Many have been discovered in bacteria and other organisms that live in the extreme environments of deep ocean vents. The enzymes in these organisms are not denatured at high temperatures – and so can be used to catalyse industrial processes at temperatures approaching 100 °C. An application of thermostable enzymes in washing powders is in Chapter 5 (page 93).

Table 6.1 Some applications of enzymes			(Source: Biochemical Society Guidance Notes 3, Enzymes and their role in biotechnology)
Enzyme	**Reaction**	**Source of enzyme**	**Application**
Industrial applications			
α-amylase	breaks down starch	bacteria	converts starch to glucose in the food industry
glucose isomerase	converts glucose to fructose	fungi	production of high-fructose syrups
proteases	digest protein	bacteria	washing powder
rennin	clots milk protein	animal stomach linings; bacteria	cheese-making
catalase	splits hydrogen peroxide into $H_2O + O_2$	bacteria; animal livers	turns latex into foam rubber by producing gas
β-galactosidase	hydrolyses lactose	fungi	in dairy industry, hydrolyses lactose in milk or whey
Medical applications			
L-asparaginase	removes L-asparagine from tissues – this nutrient is needed for tumour growth	bacteria (*E. coli*)	cancer chemotherapy – particularly leukaemia
urokinase	breaks down blood clots	human urine	removes blood clots, eg in heart disease patients
Analytical applications			
glucose oxidase	oxidises glucose	fungi	used to test for blood glucose, eg in Clinistix™, used by people with diabetes
luciferase	produces light	marine bacteria; fireflies	binds to particular chemicals indicating their presence, eg used to detect bacterial contamination of food
Manipulative applications			
lysozyme	breaks 1–4 glycosidic bonds	hen egg white	disrupts bacterial cell walls
endonucleases	break DNA into fragments	bacteria	used in genetic manipulation techniques, eg gene transfer, DNA fingerprinting

COMMERCIAL PRODUCTION OF ENZYMES

Table 6.2 shows some important industrial enzymes and their sources. As you can see, the majority of enzymes used in industry are **extracellular**. This type of enzyme is easier to isolate, as it is secreted into the growth medium by the source organism, and can be extracted relatively easily. **Intracellular** enzymes are more tricky, as their removal requires the cultured cells to be broken open, and then the enzyme must be separated from all the other proteins and other molecules. Fig 6.15 provides a flow chart to show how a protease for use in a stain remover is produced on a large scale from a bacterial culture.

Table 6.2 Some important industrial enzymes and their sources

	Enzyme	Source	Intracellular/ extracellular	Industry/ industrial use
Animal enzymes	catalase	liver	I	food
	chymotrypsin	pancreas	E	leather
	lipase	pancreas	E	food
	rennet	abomasum	E	cheese
	trypsin	pancreas	E	leather
Plant enzymes	actinidin	kiwi fruit	E	food
	α-amylase	malted barley	E	brewing
	β-amylase	malted barley	E	brewing
	bromelain	pineapple latex	E	brewing
	β-glucanase	malted barley	E	brewing
	ficin	fig latex	E	food
	lipoxygenase	soya beans	I	food
	papain	pawpaw latex	E	meat
Bacterial enzymes	α-amylase	*Bacillus*	E	starch
	β-amylase	*Bacillus*	E	starch
	asparaginase	*Escherichia coli*	I	health
	glucose isomerase	*Bacillus*	I	fructose syrup
	penicillin amidase	*Bacillus*	I	pharmaceutical
	protease	*Bacillus*	E	detergent
	pullulanase	*Klebsiella*	E	starch
Fungal enzymes	α-amylase	*Aspergillus*	E	baking
	aminoacylase	*Aspergillus*	I	pharmaceutical
	glucoamylase	*Aspergillus*	E	starch
	catalase	*Aspergillus*	I	food
	cellulase	*Trichoderma*	E	waste
	dextranase	*Penicillium*	E	food
	glucose oxidase	*Aspergillus*	I	food
	lactase	*Aspergillus*	E	dairy
	lipase	*Rhizopus*	E	food
	rennet	*Mucor miehei*	E	cheese
	pectinase	*Aspergillus*	E	drinks
	pectin lyase	*Aspergillus*	E	drinks
	protease	*Aspergillus*	E	baking
	raffinase	*Mortierella*	I	food
Yeast enzymes	invertase	*Saccharomyces*	I/E	confectionery
	lactase	*Kluyveromyces*	I/E	dairy
	lipase	*Candida*	E	food
	raffinase	*Saccharomyces*	I	food

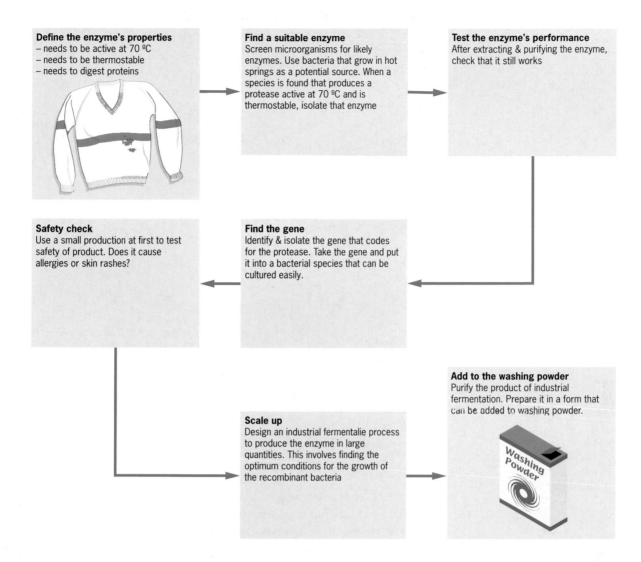

Define the enzyme's properties
– needs to be active at 70 ºC
– needs to be thermostable
– needs to digest proteins

Find a suitable enzyme
Screen microorganisms for likely enzymes. Use bacteria that grow in hot springs as a potential source. When a species is found that produces a protease active at 70 ºC and is thermostable, isolate that enzyme

Test the enzyme's performance
After extracting & purifying the enzyme, check that it still works

Safety check
Use a small production at first to test safety of product. Does it cause allergies or skin rashes?

Find the gene
Identify & isolate the gene that codes for the protease. Take the gene and put it into a bacterial species that can be cultured easily.

Scale up
Design an industrial fermentalie process to produce the enzyme in large quantities. This involves finding the optimum conditions for the growth of the recombinant bacteria

Add to the washing powder
Purify the product of industrial fermentation. Prepare it in a form that can be added to washing powder.

Washing Powder

Fig 6.15 Production of protease from *Bacillus* bacteria on a large scale for use as a stain remover

SEE QUESTION 5

IMMOBILISED ENZYMES

In an industrial process controlled by an enzyme, the enzyme is usually the component that costs the most. Like all catalysts, enzymes can be used several times, so it is a shame to waste enzyme. In a process in which the enzyme is simply added to the substances that you want to react, a lot of enzyme is wasted because not all of it can be removed from the products.

To solve this, biotechnologists have developed methods to **immobilise** enzymes (see also Fig 6.16):

- **Cross-linkage**: enzyme molecules can be linked together to make much larger molecular structures that are easy to extract from the products. The enzyme can also be chemically attached to a supporting material.

- **Entrapment**: the enzyme can be mixed with ingredients that form a gel. When the gelling reaction is complete, the enzyme becomes trapped. Gaps in the gel are large enough to let the substrate in, but not let the enzyme out. Enzymes can also be trapped in a compartment behind a semipermeable membrane that, again, lets reactants in, but doesn't allow the enzyme molecules out.

- **Adsorption**: the enzyme can be stuck temporarily to various surfaces – usually this method is only used for experimental studies, not full-scale industrial manufacture.

Fig 6.16 Immobilising enzymes

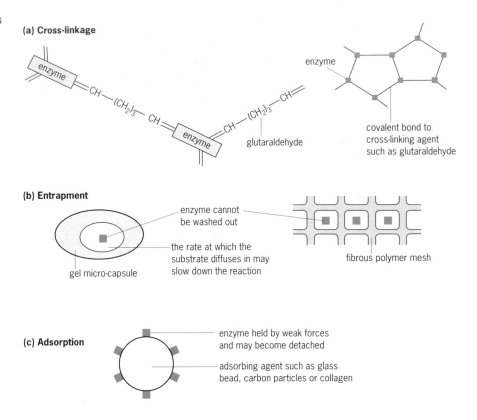

(a) Cross-linkage

enzyme

CH—$(CH_2)_3$—CH

enzyme

glutaraldehyde

enzyme

covalent bond to cross-linking agent such as glutaraldehyde

(b) Entrapment

enzyme cannot be washed out

the rate at which the substrate diffuses in may slow down the reaction

gel micro-capsule

fibrous polymer mesh

(c) Adsorption

enzyme held by weak forces and may become detached

adsorbing agent such as glass bead, carbon particles or collagen

The first major commercial use of immobilised enzymes took place in the 1970s, when immobilised **glucose isomerase** was first used to make high-fructose syrups from starches. Immobilised enzymes are now widely used in the food industry and the pharmaceutical industry. Immobilised **penicillin amidase** is, for example, used to prepare different forms of penicillin from the type produced by fungi.

Enzyme immobilisation is also important in **biosensors** (see page 120). These often require immobilisation of an enzyme into an electrode tip or into a paper strip (see the Science Feature on the Clinistix™ test).

BIOFUELS

Biotechnology is also used industrially to produce fuel. A biofuel is defined as a gaseous, liquid or solid fuel that contains energy that has come from a biological source. An example of a biofuel is rapeseed oil, which can be used in place of diesel fuel in modified engines. The methyl ester of this oil, **rapeseed methyl ester (RME)**, can be used in unmodified diesel engines and is sometimes known as **biodiesel**. Other biofuels include **biogas** and **gasohol**.

GASOHOL

Gasohol is produced by fermentation of sugar cane by the yeast *Saccharomyces cerevisiae* (commonly known as 'baker's yeast'). It has been produced on a large scale in Brazil, where it is manufactured in large distilleries and provides an alternative source of energy for motor vehicles. In Brazil all cars are adapted so as to be able to use gasohol or a mixture of gasohol and petrol.

Fig 6.17 Filling up with petrol made from sugar

Fig 6.18 A biogas digester in India

BIOGAS

Biogas is a mixture of methane and carbon dioxide that is produced by the anaerobic decomposition of such waste materials as domestic, industrial and agricultural sewage. The decomposition is carried out by **methanogenic bacteria**, anaerobes that respire to produce methane, the main component of biogas. Methane is collected and used as an energy source for domestic processes, such as heating, cooking and lighting. The production of biogas is carried out in special **digesters**, which are widely used in China and India.

⚠ SCIENCE FEATURE The Clinistix™ test for glucose in urine

Enzymes are both specific and sensitive: this makes them ideal for use in analysis, which often involves very small samples. One such application is in the analysis of glucose – an important technique in both medicine and industry. You may have heard of Clinistix™ (Fig 6.19) – the sticks used to test for the presence of glucose in urine. Clinistix™ contain two enzymes: glucose oxidase and peroxidase.

Glucose oxidase catalyses the following reaction:

glucose + oxygen $\rightarrow$ gluconic acid + H_2O_2 (hydrogen peroxide)

In a simple, visible test, the production of peroxide is coupled to the production of a coloured dye or chromagen.

The second enzyme, peroxidase, catalyses the following reaction:

$$DH_2 + H_2O_2 \rightarrow 2H_2O + D$$

D stands for the chromagen, a colourless hydrogen donor. When it loses its hydrogen it becomes coloured. The intensity of colour indicates the amount of glucose present.

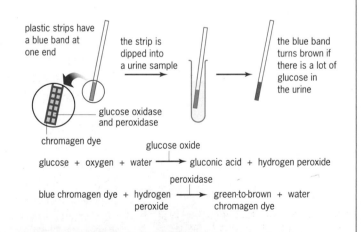

plastic strips have a blue band at one end

the strip is dipped into a urine sample

the blue band turns brown if there is a lot of glucose in the urine

glucose oxidase and peroxidase

chromagen dye

glucose oxide

glucose + oxygen + water $\longrightarrow$ gluconic acid + hydrogen peroxide

peroxidase

blue chromagen dye + hydrogen peroxide $\longrightarrow$ green-to-brown chromagen dye + water

Fig 6.19 The urine glucose test for diabetes. Glucose oxidase, peroxidase and DH_2 are fixed on a cellulose fibre pad. When this is added to a sample of urine, the colour reaction gives a quantitative measure of glucose

SEE ALSO THE ASSIGNMENT IN CHAPTER 11 – LIVING WITH DIABETES.

Fig 6.20 The Glucowatch® Biographer

SEE QUESTION 6

Fig 6.21 How the Glucowatch® Biographer detects the level of glucose in the blood

Fig 6.22 Daily variations in blood glucose level in a person with diabetes

5 BIOTECHNOLOGY IN MEDICINE

BIOSENSORS

Biosensors are being developed for many applications in medicine. One such device that is already available is the Glucowatch® Biographer (Fig 6.20), a watch that is able to measure blood glucose levels in diabetics without puncturing the skin to take a blood sample.

The watch looks like an ordinary watch; it is worn on the wrist and it has an LCD dial. however, people with diabetes use it to measure their blood glucose level, not the time.

HOW THE GLUCOSE BIOSENSOR WORKS

The back of the Glucowatch® Biographer that comes into contact with the skin has two circular flat plates, as shown in Fig 6.21. The two plates generate a low electric current that is directed into the skin. This causes positive and negative ions to move out of the skin, pulling glucose molecules from the blood with them. The glucose collects in gel discs that contain glucose oxidase. This enzyme catalyses a reaction that generates a small electrical signal. The size of the signal is proportional to the glucose level in the blood. The Glucowatch® Biographer uses this signal to calculate a reading, which is then displayed.

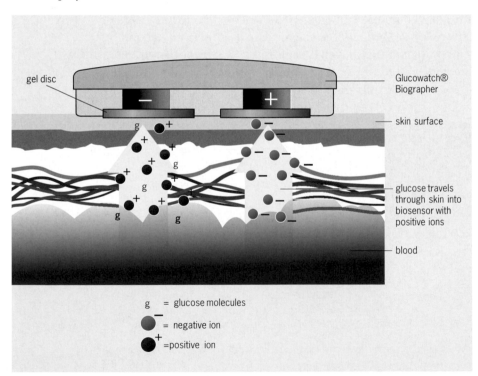

HOW IS THE BIOSENSOR USED?

Usually, people with diabetes who control their condition by insulin injections measure their blood glucose using the standard finger-prick method. They do this three or four times a day; before each meal and last thing at night. The disadvantage of this is shown in Fig 6.22. The curve shows how the blood glucose level actually changes during a 12-hour period. This can be detected by the Glucowatch® Biographer because it takes readings every 20 minutes, rather than by much less frequent finger-prick tests. These show the glucose levels at points A, B and C, lulling the person into thinking that their blood glucose is stable.

MONOCLONAL ANTIBODIES

Monoclonal antibodies are antibodies that can be produced in large amounts and that bind only to one, very specific target. This is usually part of a molecule, not even a whole molecule.

HOW ARE MONOCLONAL ANTIBODIES PRODUCED?

The method used to create monoclonal antibodies is called **hybridoma technology**. It was first devised in the 1980s by Cesar Milstein, working in laboratories in Cambridge, UK. It basically involves fusing a cell that produces antibodies with one that is **immortal**. Fig 6.23 summarises the process of making a monoclonal antibody.

The first step in the production of a mouse monoclonal antibody is to immunise a mouse with the molecule against which we want the monoclonal antibody to react. White blood cells in the mouse produce a range of different antibodies, directed against different parts of the molecule.

In the mouse, cells called B lymphocytes that are responsible for making these antibodies are isolated from the mouse's spleen. These cells are then fused with tumour cells that divide continuously and rapidly. This creates a single, large cell called a **hybridoma**. These hybridomas show properties of both 'parent' cells; they produce antibodies and they divide continuously.

One hybridoma can divide to produce a whole clone of identical hybridomas, all secreting the same antibody molecule. Once a clone has been established and is growing well in culture, the antibody produced by this collection of cells is called a **monoclonal antibody**.

MONOCLONAL ANTIBODIES USED IN DIAGNOSIS

Monoclonal antibodies can be linked to a radioactive isotope and, in this form, they have been used for diagnosing and monitoring disease since the early 1980s. Some forms of cancer can be diagnosed using monoclonal antibodies directed against molecules found only on the surface of cancer cells. When injected into patients, such a monoclonal antibody will carry its radioactive marker straight to the tumour, where it can be detected using **CT (computerised tomography)** scanning (Fig 6.24).

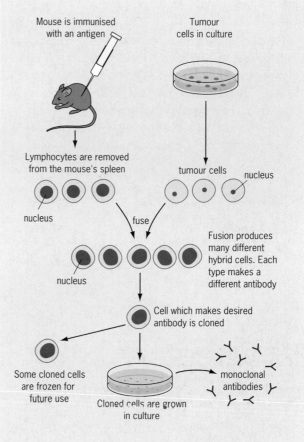

Fig 6.23 How a monoclonal antibody is made. Tumour cells, which divide without control, are fused with lymphocytes that produce antibody. The hybrid cell produced shares the properties of the two cells that formed it. It divides endlessly (it is immortal) and it produces an antibody. When one cell is then used to form a whole culture, all the cells are clones. They all produce the same antibody, a **monoclonal** antibody

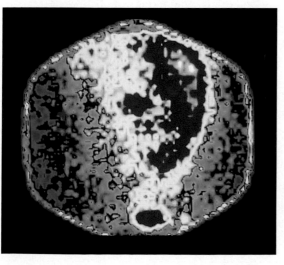

Fig 6.24 A cancer of the colon. Radioactivity from the increased concentration of monoclonal antibody in the colonic cancer cells is shown by the red, pink and white areas at the centre-right

REMEMBER THIS

Monoclonal antibodies also form the basis of modern pregnancy testing kits (see Science Feature box overleaf).

⚠ SCIENCE FEATURE The pregnancy test

By the time a pregnant woman would have been due for her next period, the embryo will have implanted and would be starting to secrete human chorionic gonadotrophin (hCG) Enough of this hormone passes into the urine for it to be detected by a modern pregnancy testing kit. This kit makes use of monoclonal antibodies that are specific for hCG and which are attached to a coloured chemical.

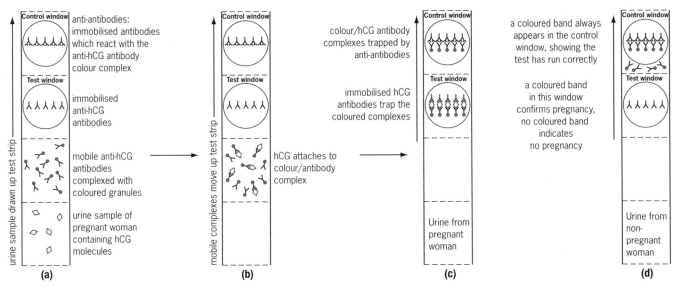

Fig 6.25 How a home pregnancy test kit works for a pregnant woman (a, b and c) and for a woman who is not pregnant (d)
(a) Urine is drawn up the test strip by capillary action
(b) hCG attached to the colour/antibody complex moves up the test strip

(c) In the urine of a pregnant woman, hCG attached to the colour/antibody complexes moves up to dock with the immobilised (fixed) hCG antibodies in the test window. All the complexes are stopped at one place, forming a visible coloured band and confirming pregnancy

(d) For the urine of a non-pregnant woman, the colour/antibody complexes have no attached hCG, so all the complexes move past the immobilised hCG antibodies in the test window and attach to the anti-antibodies in the control window, showing that the test has worked

TREATING DISEASE WITH MONOCLONAL ANTIBODIES

Many different diseases can now be treated with monoclonal antibodies. The antibody is made against a key molecule on the diseased cell, such as a protein only expressed by cancer cells, and is then linked with a drug molecule, or a radioactive isotope. The monoclonal antibody is injected into the patient and it 'homes in' on the diseased cells, binding to them and releasing its toxic passenger.

A few examples of how monoclonal antibodies are used to treat disease are shown in Table 6.3.

Table 6.3 Treatment with monoclonal antibodies

Disease	Monoclonal Ab	How it works
Non-Hodgkin's lymphoma	Rituximab	this antibody binds to molecules of B cells and kills them directly
Non-Hodgkin's lymphoma	^{131}I Tositumomab	also binds to molecules on B cells, but also carries a radioactive isotope that it carries to the tumour
Breast cancer	Trastuzumab	targets a protein found in one-third of women with breast cancer, binds to it, and slows down tumour growth
Cardiovascular disease	Abciximab	targets clotting proteins to prevent clot formation in patients undergoing angioplasty (see Chapter 31)
Crohn's disease	Infliximab	binds to a protein that causes inflammation in the bowel, preventing it from causing further damage
Rheumatoid arthritis	Etanercept	binds to a protein in the body that causes inflammation, preventing it from causing further damage

RECOMBINANT PROTEINS AND THEIR USE IN MEDICINE

RECOMBINANT HUMAN INSULIN

We saw earlier in this chapter that bovine growth hormone can now be produced in bacteria as a recombinant protein. Other proteins of medical importance can also be manufactured in large quantities in this way. An important example is human insulin.

ANIMAL VERSUS HUMAN INSULIN

The daily doses of insulin that people with diabetes use to manage their glucose levels can come from several sources, including the pancreas of a pig or cow. Because pig insulin is nearly identical to human insulin, pigs used to be the most common source. However, some people reacted badly to 'foreign' insulin from pigs; researchers argued that they might be better off with human insulin.

Pharmaceutical companies now use genetic engineering to produce large quantities of human insulin for people around the world.

The common bacterium *Escherichia coli* (*E. coli*) is used to produce insulin using recombinant DNA technology. The gene for producing insulin is inserted into DNA of *E. coli*. These genetically engineered bacteria are turned into tiny insulin-producing factories. When grown in bulk in large **industrial fermenters**, they can easily produce insulin in the vast amounts needed to supply people with diabetes worldwide.

Human insulin is now widespread, and the use of purified pig insulin and insulin from other animal sources is being phased out. However, some people are unhappy about that. A few have had bad reactions to human insulin, although they had no problems with animal insulin. They are arguing that both products should be made available, so that they can have a choice if needed.

> ### ✔ REMEMBER THIS
>
> The hormones insulin and glucagon work antagonistically to control blood glucose levels. Insulin is produced if blood glucose rises beyond the normal range; it lowers blood glucose by making cells in the liver and muscles more permeable to glucose. Glucagon is produced when blood glucose falls; it activates enzymes inside cells to convert glycogen into glucose. (See Chapter 11 Homeostasis.)

SEE QUESTION 7

⚠ SCIENCE FEATURE Farming or pharming?

A new and highly experimental use of genetic engineering is 'pharming'. This involves farming, but its objective is not to produce food. The idea is to genetically engineer crop plants and animals usually used for food so that they can produce medicines. The technique is difficult and expensive because it requires a transgenic organism that is bred to carry a human gene that makes a particular protein – it is possible, for example, to genetically engineer a cow to produce human haemoglobin in its milk. Once transgenic embryos have been produced in the laboratory, these are then implanted into surrogate mothers and carried to term, resulting in transgenic cows that secrete human haemoglobin in their milk.

Milk from a cow is relatively easy to obtain in large amounts; the secreted haemoglobin is then extracted and purified. This process has been used to produce other blood components, human growth hormone and large quantities of human proteins needed for research.

Fig 6.26
Transgenic sheep. These sheep have a human gene which makes them produce α-1-antitrypsin (AAT) in their milk. People who lack AAT suffer from emphysema

SUMMARY

- Biotechnology is defined generally as the **manipulation** of biological organisms, systems or processes for the benefit of people, in areas such as agriculture, food production and medicine.
- Biotechnology is widely used in agriculture. For example, bovine growth hormone injections are used to increase milk production in cows; artificial insemination of a prime cow using semen from a prime bull is used to generate multiple embryos carried by surrogate mothers (ordinary cows); hormones are used to control oestrus and mating in sheep to synchronise lambing in a large herd.
- The technology that creates **transgenic animals** is advancing rapidly. Some animals are bred with genes that make them more productive for the food industry (fish with an extra growth hormone gene); other animals are given genes that allow them to produce human proteins used in medicine.
- Biotechnology is also used to **modify plants** used in food production. Genetically modified foods offer both advantages and disadvantages. Some pressure groups and some ordinary people in the UK are very suspicious of the development and use of GM foods. Others are in favour but don't know anything about them. Knowledge, evaluation and caution are important.

- Biotechnology is not new. The process of **fermentation** has been used to make yoghurt, cheese, beer, wine and bread for thousands of years.
- New biotechnology techniques are used to make edible protein from fungus (Quorn™) and bacteria (Pruteen™).
- Other **fermented food products** include sauerkraut, and products made from soya beans (eg tempeh and soy sauce).
- **Enzymes** are used in industry, particularly in the food and manufacturing sectors. Enzymes allow chemical processes to occur quickly and at relatively low temperatures, so are useful for keeping costs down.
- Most of the enzymes used in industry are extracellular. These are secreted by bacteria into the growth medium and are easy to separate and purify. Some intracellular enzymes are used but extracting them involves breaking up the bacterial cells first.
- Enzymes are often immobilised so they can be reclaimed and reused. The main methods of immobilising enzymes are **cross-linkages**, **entrapment** and **adsorption**.
- Immobilised enzymes are important in **biosensors** and in **diagnostic tests**.

? EXAM QUESTIONS

1 The following diagram is a simplified representation of a technique used in genetic engineering. Look carefully at the diagram and answer the following questions.

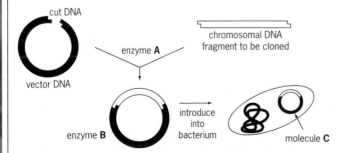

cut DNA

vector DNA

enzyme **A**

chromosomal DNA fragment to be cloned

enzyme **B**

introduce into bacterium

molecule **C**

a) Give the name of enzyme **A** and enzyme **B**.

b) Suggest why the same enzyme (enzyme **A**) is used to cut both the vector DNA and the chromosomal DNA fragment.

c) What is the function of enzyme **B**?

d) What name is normally given to the molecule **C**?

e) Describe **two** ways in which the chromosomal DNA fragment can be obtained.

f) What can be added to molecule **C** to aid in selection and identification of the cells that have successfully taken up the fragment?

g) Suggest a possible medical use for this technology.

WJEC Biology Module BI1 June 2001 Q7

2 The bacterium *Bacillus thuringiensis* has a gene on its plasmid coding for the production of a protein toxic to many insect pests. The bacterial cell can be made to synthesise copies of these plasmids which can be introduced into tomato cells as shown below. Tomato plants grown from these cells are resistant to a wide range of insect pests.

a) (i) Name the chemical that makes up the bacterial plasmid.

(ii) What structure has been removed from the treated plant cell in stage 2 which would otherwise prevent uptake of the plasmid?

? EXAM QUESTIONS

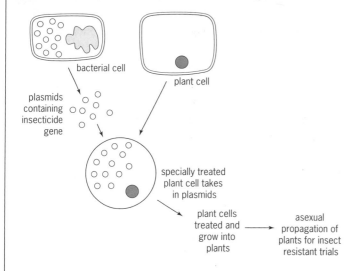

bacterial cell

plant cell

plasmids containing insecticide gene

specially treated plant cell takes in plasmids

plant cells treated and grow into plants

asexual propagation of plants for insect resistant trials

(iii) Why is it important that the entire process only involves asexual reproduction?

(iv) Suggest the **two** main advantages to farming of introducing a gene for this toxic protein into the plant.

b) What important factor would need to be determined so that genetically engineered tomatoes could be grown commercially?

WJEC Biology Module BI1 January 2001 Q7

3 Yeast cells produce ethanol under certain conditions.

a) State the conditions that cause yeast to produce ethanol.

b) Outline how glucose is converted into ethanol in yeast.

IBO Standard level Paper 3 November 2000 QC2

4 At the start of the brewing process, a small 'starter culture' of yeast is added to the fermenter under aerobic conditions. The yeast cells feed, grow and divide, leading to a growth in the yeast population.

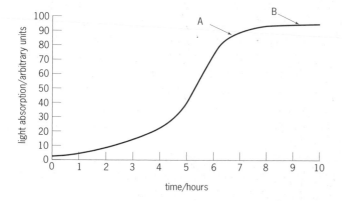

When the population of yeast is large, air is no longer allowed into the fermenter, the conditions become anaerobic and the yeast cells stop growing and dividing but are still alive. Ethanol is produced under these conditions. The growth of the yeast population in a fermenter under aerobic conditions was measured using turbidimetry. The results are shown on the graph.

a) Explain why the rate of population growth decreases after 6 hours.

b) As a result of this experiment, it was decided to turn off the air supply in future fermentations at Time **A**, rather than waiting until Time **B**. Suggest **one** commercial reason for this decision.

c) Explain why turbidimetry is **not** the best method for finding the concentration of live yeast cells in the fermenter at Time **B**.

AQA Microorganisms and Biotechnology June 2000 Q6

5

a) In industrial processes enzymes are often immobilised. Give **two** advantages of immobilising enzymes and for each explain how the industrial process is improved.

b) Chymotrypsin is an enzyme that breaks down proteins. It can be isolated from microorganisms and used industrially. Experiments were carried out on the effects of temperature on the activity of chymotrypsin. In one experiment, the enzyme was in solution, and in another it was immobilised on silica beads.

(i) Suggest reasons for the differences in the results of the experiments.

(ii) In an industrial process chymotrypsin is used at 45 °C. Suggest whether chymotrypsin is better used in solution or immobilised at this temperature. Explain the reason for your answer.

AQA Microorganisms and Biotechnology March 2000 Q7

❓ EXAM QUESTIONS

6 The figure below shows a possible structure for a biosensor that uses immobilised enzymes to detect glucose.

a) What are immobilised enzymes?

b) Suggest what happens when the glucose in the solution reaches the immobilised enzymes.

c) Suggest the form in which the information about the concentration of glucose is transmitted to the display.

d) **(i)** Why should a buffer be added to the substrate solution?

(ii) Suggest **one** other variable that should be kept constant.

e) Competitive (X) and non-competitive (Y) inhibitors can be used to slow down or to stop enzyme controlled reactions. Label the following figure to suggest where these inhibitors are most likely to act. Use X for the competitive inhibitor and Y for the non-competitive inhibitor.

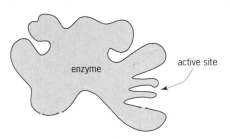

f) Describe how each of these inhibitors works.
X............... Y...............

g) Give an example of a competitive inhibitor.

WJEC Biology Module BI1 June 2001 Q5

7 In 1987 an attempt was made to produce human insulin from genetically engineered yeast. Genes for human insulin were inserted into small loops of DNA called plasmids. These plasmids were then used to try to carry the insulin genes into yeast cells.

a) Describe how an insulin gene could be removed from human DNA and inserted into the plasmid DNA.

b) Explain why a microorganism such as yeast might be particularly useful for producing insulin on an industrial scale.

c) The yeast used to produce insulin was a mutant strain that did not have the gene for a enzyme needed by the yeast in respiration. The missing enzyme is called triose phosphate isomerase. The DNA of the plasmids, however, did contain the gene for triose phosphate isomerase. Explain the importance for insulin production of using plasmids that have the gene for triose phosphate isomerase and a mutant yeast that does not have this gene.

AQA Microorganisms and Biotechnology March 2000 Q9

KEY SKILLS **ASSIGNMENT**

YOGHURT:
IS IT ALL IT'S MADE OUT TO BE?

People have been making yoghurt for thousands of years. It is easy to make at home, with very little equipment – just put some warm milk in a vacuum flask with a couple of tablespoons of live yoghurt, and leave it overnight. In the past, people probably did the same thing, but used a jug by the fire. Today, there is a multi-million pound international industry that has built up around yoghurt-making, which is done on a huge scale.

Fig 6.A1 outlines the major steps in the production of a commercial yoghurt.

Fig 6.A1

Production

1 **Stages in yoghurt making**
Mixing the ingredients
Milk, sugar, glucose and filtered
sterilised water are mixed together

2 **Sterilisation**
The mixture is heated to destroy any bacteria
that are not needed

3 **Fermentation**
The sterilised mixture is transferred to a fermentation
tank that may hold 6000 litres. The transfer is carried
out under sterilised conditions via a system of pipes
and valves that are closed to the external environment.
Bacterial starter cultures are added and the contents
of the tank are kept at about 37 °C for several hours
to several days

Post-production processing

4 **Flavouring**
If flavoured yoghurt is being made,
flavourings are added here

5 **Storage and dilution**
The concentrated yoghurt mixture is stored in
another large tank that is cooled to 2 °C. Sterile
water is used to dilute the yoghurt to the correct
consistency before packaging

6 **Packaging**
The finished yoghurt is put into the cartons
or bottles, sealed, labelled and stored below 6 °C.
It is then distributed for sale

KEY SKILLS **ASSIGNMENT**

1 **a)** What is the essential raw material used in the initial mix during yoghurt manufacture?
b) Give two reasons why the mixture is heat-treated.

2 **a)** Two spieces of bacteria that can be used in yoghurt production are *Loctobacillius bulgaricus* and *Streptococcus thermophilius*. Describe two ways by which you could distinguish between different spieces of bacteria such as these.
b) Why are the organisms incubated around 37 °C.

3 Describe what happens during the formation of yoghurt. Include details of the microorganisms present.

4 How are the nutritional requirements of the bacteria in the yoghurt culture satisfied by the milk used in the process?

5 Give an example of the downstream processing of yoghurt.

Is yoghurt always good for you?

A lot of advertising claims that yoghurt is the ultimate in healthy eating. Many people think that all yoghurts are non-fattening and can be used when trying to lose weight, or at least when trying not to gain any extra kilos. Those with live bacteria are represented as a vital food for a healthy digestive system. But how much truth is there behind these two assumptions?

The following table shows different nutritional contents of two types of yoghurt; A is a fruit-flavoured yoghurt with added cream; B is made with sweetener and fruit flavouring.

Quantity per 100 g	A	B
Energy (kJ/kcal)	462/110	245/58
Protein (g)	3.7	4.1
Fat (g)	3.9	0.8
Carbohydrate (g)	15	1.9
Salt (g)	0.1	0.1
Taste	sweet and fruity	sweet and fruity
Consistency	thick and creamy	thick and creamy

A gram of fat yields 9 calories; a gram of protein yields 4 calories and a gram of carbohydrate yields 4 calories.

6 **a)** How many calories in the two different yoghurts come from the fat it contains?
b) What percentage of the total energy content is this?
c) What percentage of the total energy content of both yoghurts comes from protein? And from carbohydrate?

7 **a)** Which of these yoghurts do you consider more healthy and why?
b) How do you think these two types of yoghurt are marketed?
c) Do you think the marketing of yoghurt is sometimes misleading?
d) Is there such a thing as 'an unhealthy yoghurt'?

Live yoghurts are not heated to a high temperature to kill the bacteria present before they are packaged for sale. This could have a disadvantage in that contaminant bacteria could live and grow in the yoghurt and cause it to 'go off' in storage. However, the health advantages of live yoghurt are thought to be worth it. Manufacturers claim that the bacteria in the yoghurt populate the large intestine, and set up a correct balance of 'friendly' bacteria in the bowel.

8 **a)** Using your knowledge of digestion, and the conditions present in the large bowel, do you think this claim might be true?
b) What study would you need to carry out to determine if eating live yoghurt actually did change the bacteria in the bowel?

07

Nutrition

In the increasingly popular cosmetic treatment of liposuction, the fatty adipose tissue is literally hoovered out of 'problem' areas

In the affluent countries of the western world a lot of people eat too much. The number of overweight people is on the increase, and at the same time, others have made dieting an obsession.

We know that to lose weight we have to eat less food and exercise more, but the trick is to find the self-discipline to do it. One glance at the newsagent's shelves shows that there are many magazines devoted to slimming. Many of the articles are basically the same – diets, exercise regimes and inspirational stories such as: I lost ten stones and so can you.

If will-power fails, there are more drastic methods. Liposuction removes excess fatty tissue, usually from the buttocks and thighs, while 'beer bellies' can be removed in the same way. To cut down on intake, desperate people have even had their jaws wired up or part of their stomach surgically stapled so that it cannot hold much food.

REMEMBER THIS

Inorganic substances are generally made from simple molecules. Water, mineral ions (sodium chloride, etc) and gases such as carbon dioxide are all inorganic.

Organic substances – those based on carbon – include carbohydrates, lipids, proteins and nucleic acids; see Chapter 3.

1 WHY DO WE NEED TO EAT?

All living things need food to obtain energy, to grow, and to repair damaged cells and tissues. This chapter is about nutrition – the process of obtaining food. There are two types of nutrition, **autotrophic nutrition** and **heterotrophic nutrition**. For most human biology courses, you need only be aware of a broad definition of each:

Autotrophic nutrition involves taking in simple **inorganic** molecules and using them to form complex substances. Nutrition in plants is **autotrophic**. Their method of nutrition is more commonly known as **photosynthesis**, a process in which they build up large complex molecules from carbon dioxide and water (see Chapter 34).

Heterotrophic nutrition involves taking in complex organic molecules, and then breaking them down into simpler molecules that are absorbed into the body. This process of breaking down food is called **digestion**, and is covered in Chapter 8.

Humans are heterotrophic organisms, or **heterotrophs**. All the food we eat comes ready-made. Heterotrophs can be divided into three groups: **herbivores**, **omnivores** and **carnivores**. Herbivores, such as cows and sheep, obtain all their energy and nutrients directly from plants; carnivores, such as tigers and polar bears, eat only other animals. Humans are omnivores; they eat both plants and other animals.

The old cliché 'we are what we eat' is perfectly true. As developing babies, we were built when molecules from the food our mothers ate were assembled according to the genes we inherited, and ever since we have been taking in the food we need to build and power our bodies.

In Chapter 3 we looked at the chemicals of life: carbohydrates, proteins, lipids and nucleic acids. This chapter looks at how we provide our bodies with these and other essential substances.

2 WHAT DO WE NEED TO EAT?

You will have heard much talk of a **balanced diet**. Many assume that to achieve a balanced diet, all you need to do is eat a little bit of everything. However, as we shall see, like many things in life, it's just not that simple.

THE CONCEPT OF A BALANCED DIET

A diet must be balanced in terms of energy and in terms of its chemical content.

Whether a person is fat, thin, or 'just right' depends on the balance of their energy intake versus their energy expenditure. Put simply, this means that if you take in more energy in the food you eat than you use up in your everyday activities, you will start to put on weight. Someone who is **obese** eats more than they need over long periods of time (Fig 7.1a). If you take in less energy that you use up, you lose weight. Someone who is **starving** cannot meet their energy demands, and they start to digest their own body fabric (Fig 7.1b). A diet that is balanced in terms of energy, enables us to maintain a constant, healthy weight.

It is also possible to eat a balanced diet in terms of energy, but then suffer from **malnutrition**. This is an imbalance in the chemical composition of the food we eat. If, for example, you ate only chocolate, you could easily maintain your body weight but your body would suffer from a range of deficiencies after a time – chocolate contains lots of sugar and fat, but few vitamins, minerals, complex carbohydrates and no fibre.

THE ENERGY IN FOOD

As you sit here reading this book, you might think that you are not using much energy, but even at rest, your body has a steady energy demand. Think about some of the processes going on:

Your heart is beating.

You are breathing.

Food is being pushed along your intestines.

Food is being absorbed from your intestines into your blood.

● Urine is moving from your kidneys to your bladder.

Nerves are taking information from your sense organs to your brain, keeping you informed of changes in your environment.

You are thinking: your brain is processing information coming in from the sense organs, sorting out what is important and what can be ignored.

● Unless you are somewhere very hot, you are probably losing heat to your surroundings. You need to replace the heat you lose in order to maintain a constant internal body temperature.

● Some of the molecules from digested food, such as amino acids, are being built up into molecules that will help to make new cells and tissues.

All these processes (and many others) require energy that is released from food molecules by respiration, a process that goes on constantly in all of our cells.

The amount of energy a person needs depends on:

● their age,

their level of activity,

● their size,

their genetic background (some people inherit a very high or a very low metabolic rate).

Fig 7.1(a) When the body takes in more energy from food than it needs, the excess is stored as fat under the skin and around the internal organs. The distribution of body fat is determined by the sex hormones; males tend to accumulate fat around the waist, while females accumulate fat more on the buttocks and thighs

Fig 7.1(b) When the body does not get enough energy from food, it has to respire stored fat and then protein. The actual fabric of the body – such as the muscles – start disappearing

SEE QUESTION 1

? TEST YOURSELF

A The size of an organism affects the amount of food it needs per unit of body weight. Explain.

The energy and dietary requirements of people at different times of life are covered in detail on page 140.

Ideally, we need to eat enough food to maintain an energy balance so that:

INTAKE	EXPENDITURE		
energy transfer from carbohydrates and fats	energy necessary for essential activities	energy for physical activity	energy for metabolising food

with: INTAKE = energy necessary for essential activities + energy for physical activity + energy for metabolising food

The energy required for the body's essential activities, that is, for the processes that take place when we are at rest, is called the **basal metabolic rate** (abbreviated to BMR).

SCIENCE FEATURE Measuring the energy in food

The SI unit of energy is the joule (J) and the amount of energy in food is measured in joules. When we talk of the energy requirement of people, or of the energy content of food, we talk about kilojoules.

$$1000 \text{ J} = 1 \text{ kJ}$$

The energy value of food can be estimated by burning a known amount of the food in oxygen and measuring the energy released as heat. This process is called **calorimetry**. The information on energy content given on food packaging has been calculated in this way.

Fig 7.2(a) shows one very simple method of measuring the energy contained in a food sample. The sample is weighed and set alight. The energy released as the sample burns is used to heat a known volume of water. We can then calculate the total amount of energy contained in the food sample since we know that 4.18 joules of energy will raise the temperature of 1 cm³ of water by 1 °C.

Unfortunately, although this method is simple, it is not very reliable:

● A lot of the energy escapes and does not go into heating the water. In particular, heat is lost to the surroundings.

● Food samples are very difficult to burn. There is usually a lot of unburned material left at the end of the experiment.

The apparatus shown in Fig 7.2(b) is much more accurate.

The calorie is an older unit for energy in food, but is still widely used. In popular nutrition guides, the calories referred to are actually kilocalories, which should be abbreviated Cal, with a capital C.

$$1000 \text{ cal} = 1 \text{ kcal} = 1 \text{ Cal}$$

A calorie is defined as the energy needed to raise the temperature of 1 cm³ of water by 1 °C.

$$1 \text{ calorie} = 4.18 \text{ joules}$$
$$1 \text{ Calorie} = 4.18 \text{ kilojoules}$$

Fig 7.2 (c) Packaging for muesli shows the energy given in kilojoules and kilocalories

a)
- Test tube contains water
- Tin lid contains dried food sample

b)
- Stirrer keeps water at an even temperature
- thermometer
- Large volume of water produces a small rise in temperature. Little heat is lost to the atmosphere
- Heating coil sets fire to food sample in crucible
- Oxygen supply ensures food sample burns completely

Fig 7.2 (a) A simple calorimeter. **(b)** A more accurate calorimeter

NUTRITION INFORMATION

		Typical value per 100g	Per 40g Serving with 125ml of Semi-Skimmed Milk
ENERGY	kJ	1350	800*
	kcal	310	180
PROTEIN	g	8	8
CARBOHYDRATE	g	66	33
(of which sugars)	g	(31)	(19)
(starch)	g	(35)	(14)
FAT	g	1.5	2.5 *
(of which saturates)	g	(0.3)	(1.5)
FIBRE	g	13	5
SODIUM	g	0.7	0.4

> ## ! SCIENCE FEATURE Starving in the midst of plenty
>
> In Chapter 4, we saw how someone adrift in the sea can die from dehydration even though surrounded by water. Some people in our society can also starve, even though they have access to plenty of food. Unlike the shipwreck victim, their problem is psychological rather than physiological – they have the eating disorder **anorexia nervosa**.
>
> Anorexia usually affects teenage girls, but the number of boys affected is rising. A diet usually triggers the condition. Restricting food to lose a bit of weight gets out of control and the affected teenager becomes obsessed with food. They lose confidence as well as weight, and they often try to control the eating patterns of those around them. Some anorexics will, for example, make large meals for others, and encourage them to overeat, but eat virtually nothing themselves. Treatment does help, but an unlucky few do die from anorexia. Once the body weight of a teenage girl drops below about 7 stones (44.5 kg), her periods stop. Below 6 stones, she is unable to carry out ordinary activities like going to school, and once her weight has dropped to below 5 stones, she is in real danger of sudden death because of heart or other organ failure.
>
> A related eating disorder, **bulimia nervosa**, also tends to affect young adults, mainly girls, and can also have devastating effects. Bulimia can also be triggered by dieting, but instead of under-eating, a bulimic will binge on food, eating vast quantities until overfull (Fig 7.3a). Often, she will then make herself vomit to get rid of the uncomfortable feeling, and because she feels shame and self-disgust (Fig 7.3b). The vomiting leads to further negative feelings, and is then followed by another bingeing
>
> session. This cycle can continue for hours, or may only happen now and again. The body weight of a bulimic does not change drastically, and if the bingeing and vomiting is carried out in secret, friends and family may not know about the problem.
>
>
>
> **Fig 7.3 (a)**
> Bulimic girl bingeing
>
> **Fig 7.3 (b)**
> Vomiting to get rid of the feeling of fullness

BALANCING THE DIET

What constitutes a balanced diet varies from animal to animal; in a big cat like a tiger, a balanced diet is a large meal of antelope every two days. Humans, because they are omnivores, need to eat a wider range of foods and the healthiest diets seem to be those based mainly on carbohydrates, with relatively low amounts of fat, some protein, and plenty of fresh fruits, vegetables and other plant products such as cereals and grains. The typical 'western' diet tends to have too much fat and protein, and too few fresh fruits and vegetables. Taking in a high proportion of fat in the diet leads to obesity because fat contains almost twice as much energy per unit mass than either carbohydrate or protein:

- 1 g carbohydrate produces in the body approximately 17 kJ;
- 1 g protein produces approximately 18 kJ;
- 1 g fat produces about 39 kJ.

A balanced diet for most people provides more than 55 per cent of its energy in the form of carbohydrates. Any diet should also include non-energy rich components – vitamins, minerals and fibre (see above). These compounds are needed only in tiny amounts, but they are vital components of the enzymes and other molecules that are involved in life processes.

SEE QUESTION 2

? **TEST YOURSELF**

B How do you test for the presence of a simple sugar in a food? (Hint – you may need to revise Chapter 3, page 45)

CARBOHYDRATES

Carbohydrate is the most important energy provider among the food types, accounting for between 40 and 80 per cent of total energy intake of most human diets. It is also necessary for the metabolism of proteins and fats. Carbohydrates can be divided into three groups:

- **Sugars** such as sucrose and glucose – sweet foods and drinks contain large amounts. Sugars can be thought of as 'instant' fuels. The fastest way to get energy into our bodies is to have a glucose drink: the glucose does not have to be digested and so passes quickly into the bloodstream where it can be carried to the working muscles.
- **Starches**, abundant in many staple foods such as rice, potatoes, pasta and bread.
- **Non-starch polysaccharides** (NSPs) such as cellulose which collectively make up much of the fibre in our diet.

Although almost all types of carbohydrates provide similar amounts of energy, the nature of the carbohydrate can have a significant affect on our health. For instance, sugars can be classified in the following way:

- **Intrinsic sugars** are those contained within the cell walls of food, eg sugars within the cells of a banana.
- **Extrinsic sugars** that are not contained within cells, such as those found in processed foods such as sweets and soft drinks.
- **Milk sugars** occur in milk and milk products. Lactose is the primary milk sugar.

A major health problem caused by extrinsic sugars is tooth decay, or dental caries, probably the most widespread disease in the western world. In the UK, about 13 per cent of the total dietary energy comes from extrinsic sugars – efforts are being made to reduce this to below 10 per cent to prevent tooth decay.

Generally, nutritionists disapprove of eating too many simple carbohydrates, advising instead to base your diet on starchy foods. In planning a meal such as breakfast, for example, you should eat a wholegrain, unsweetened cereal, or wholemeal bread with fruit and maybe low-fat yoghurt instead of a doughnut or Danish pastry. The energy from complex carbohydrates is released slowly over a few hours and this stops you being hungry until the next meal, so you can avoid 'snacking' on more high-sugar foods.

This makes sense because it keeps your blood glucose level from rising and falling rapidly – a major factor in hunger is blood glucose level. When it is low we feel hungry. Foods containing a lot of extrinsic sugars, or processed foods that allow the rapid digestion of starch, both enable glucose to be released rapidly, and it enters the blood. This triggers the body's insulin-based control mechanism, which reduces the glucose level again, causing hunger to return with a vengeance.

THE IMPORTANCE OF FIBRE

Fibre, the indigestible component of carbohydrate-rich foods, provides bulk to the digestive tract, stretching the gut wall and stimulating a faster through-put of food. Generally there is very little fibre in animal products; the vast bulk of indigestible material comes from plant tissues. If we eat a highly processed diet (one based on white bread, sweet foods, high-fat foods with little fresh fruit, vegetables or whole grains) we do not take in enough fibre. In the short term, this means chronic constipation and it increases our risk of developing gallstones and cancer of the colon later in life.

SEE QUESTION 3

SCIENCE FEATURE What makes for a healthy colon?

Many large studies involving thousands of people have shown that a diet low in fibre and high in red meat and fat makes it more likely that someone will develop cancer of the colon or rectum later in life. It is one of the clearest cases of a lack of a food type being associated with a particular cancer.

One large study, the European Prospective Investigation into Cancer and Nutrition (EPIC), published some of its findings in 2001. This study started in 1993, and monitored over 400 000 people across nine different countries to investigate the links between diet and cancer. It concluded that eating a diet high in fibre – at least 18 g per day – cut the risk of developing colon cancer by 40 per cent.

It is, however, difficult to explain why fibre is so important for keeping the colon healthy. It may be that a higher fibre diet increases the speed at which food passes through the alimentary canal. If food has less time to be in the colon, there is less chance that cancer-causing components in food can affect the cells in the lining.

PROTEINS

The human body is made largely from protein; it accounts for more than half of the dry mass of our bodies – see Chapter 3. The proteins in our diet are digested to amino acids that then travel to our cells to be reassembled in a different order to make the proteins of the body. Proteins are needed for growth and repair of tissues, and not surprisingly are an important part of the diet of bodybuilders (Fig 7.4). Table 7.1 shows the protein content of some familiar foods.

Although there are many different types of protein, there are only 20 different amino acids. These can be divided into two dietary groups: the **non-essential amino acids** are those that the body can make, the **essential amino acids** are those that the body cannot make (Table 7.2). These must be obtained from food. We make non-essential amino acids in the liver – see Chapter 11.

As a general rule, proteins of animal origin contain all the essential amino acids, while different plants contain only some of them. If you are a vegetarian, you need to combine proteins from different plants in the same

? TEST YOURSELF

C How do you test for the presence of protein in a food? (Hint – you might need to revise Chapter 3, section 4.)

Fig 7.4 In order to make the extra muscle tissue, bodybuilders must have a high protein diet

Fig 7.5 Each one of these eggs provides enough daily protein for an adult human. When proteins are heated they are denatured and often coagulate. This accounts for the change in colour and texture when an egg is cooked. However, the amino acid content is usually unchanged

Table 7.1 The protein content of some familiar foods	
Food	**Protein content/ g per 100 g**
Milk (whole, cow's)	3.2
Cheddar cheese	25.5
Beef	20.2
Chicken	17.6
Turkey	20.6
Cod	17.4
Baked beans	5.2
Red kidney beans (dry)	22.1
Peanuts	25.5
Eggs	12.3
Apples	0.4
Cornflakes	7.9
Potatoes	2.1

Source: MAFF reference book 342, Manual of nutrition

Fig 7.6 This child is suffering from kwashiorkor, a Swahili word which means 'displaced child', referring to the fact that the condition often develops after breast-feeding has stopped. A lack of protein in the blood means that water cannot be drawn back from the tissues, resulting in the swollen abdomen

Table 7.2 Essential and non-essential amino acids	
Essential	**Non-essential**
Isoleucine	Alanine
Leucine	Arginine
Lysine	Aspartic acid
Methionine	Asparagine
Phenylalanine	Cysteine
Threonine	Glutamic acid
Tryptophan	Glutamine
Valine	Glycine
	Proline
	Serine
	Tyrosine
	Histidine (essential in infants)

> ✔ **REMEMBER THIS**
>
> Plasma proteins are found in the blood, and one of their vital functions is to draw water back into the blood from tissue fluid. A lack of blood protein leads to the swollen abdomen seen in cases of starvation (Fig 7.6).

SEE QUESTION 4

meal so that you take in the full range of amino acids. This is not as difficult as it sounds: a simple meal of beans on wholemeal toast (pulses plus grains) contains all the essential amino acids.

It may come as a surprise to know that a healthy adult human needs only about 60 g of protein per day, about the size of a large egg (Fig 7.5). The human body cannot store protein. If we eat too much, the excess amino acids that are produced are **deaminated** in the liver. The resulting ammonia is combined with carbon dioxide to form urea, which is excreted in the urine. Although we need only a small amount of protein, this tends to be the most expensive part of the diet and is therefore often lacking (Fig 7.6).

FATS

What we usually think of as 'fat' is mainly triglyceride, but it also includes edible oils, fats, waxes and related compounds. Approximately 95 per cent of the lipids in the human diet are triglycerides (three fatty acids attached to a glycerol molecule, see Fig 3.17). The remaining 5 per cent comprise mainly cholesterol and the phospholipids. The body can synthesise many of its fatty acids from carbohydrates and proteins: only a few, the essential fatty acids, must come from food.

The fat content of most vegetables is low (seeds are an exception). In general, animal products have a higher fat content as animals store fat in their bodies – even lean meat is quite high in fat. Animal sources include mammals (beef, pork, lamb, etc) and oily fish such as herring and pilchard. Milk and milk products, such as cheese, are also rich in fats: milk contains 4.5 per cent fat by mass, and cheese is 30 per cent fat.

VITAMINS

Vitamins are essential organic compounds needed by the body in small amounts (Table 7.3). For, example, the recommended daily amount of vitamin A for a young adult male is 1000 µg (only one thousandth of a gram). This means that the average human needs to take in only about 25 g of vitamin A (a mouthful) during an entire lifetime.

Obviously, we get most of our vitamins from our food, but there are other sources. Gut bacteria synthesise vitamin K (essential for blood clotting). In addition, humans can make vitamin A from carotene, a substance found in most plant foods, particularly carrots, and ultraviolet radiation from sunlight acts on a compound in our skin called **ergosterol**, converting it to vitamin D.

> ✓ **REMEMBER THIS**
>
> Vitamins help to regulate cell chemical processes, often through interaction with important enzyme systems. Lack of a particular vitamin in the diet leads to a **deficiency disease**.

SEE QUESTION 5

Table 7.3 Vitamins important in the human diet

Vitamin	Name	Minimum daily requirement	Rich food source	Function	Deficiency disease
Fat-soluble vitamins					
A	Retinol	1000 µg	Fish liver oils, dairy products, liver. Most leafy vegetables and carrots contain carotene that can be converted into retinol.	Needed for healthy epithelial cells. Needed for regeneration of rhodopsin in rod cells of the eye.	Dry (keratinised) skin. Night blindness.
D	Calciferol	10 µg	Fish oils, egg yolk, butter. It can be made by the action of sunlight on skin.	Promotes absorption of calcium from intestines. Necessary for formation of normal bone and reabsorption of phosphate from urine.	Rickets in children ('soft' bones, bend easily). Painful bones in adults.
E	Tocopherol	10 µg	Vegetable oils, cereal products. Found in many foods.	In rats, formation of red blood cells, affects muscles and reproductive system. Unclear in humans. Because the vitamin is widely found in foods, deficiency is rare in humans.	Mild anaemia and sterility in rats.
K	Phylloquinone	70 µg	Fresh, dark green vegetables. Gut bacteria.	Formation of prothrombin (involved in blood clotting).	Delayed clotting time. May occur in newborn babies before the establishment of gut bacteria.
Water-soluble vitamins					
The B vitamins form a vitamin complex. They are chemically unrelated, but tend to occur together and have similar functions.					
B$_1$	Thiamine	1.5 mg	All plant and animal tissues contain thiamine but rich food sources are yeast, cereals, nuts, seeds, pork.	Coenzyme in cell respiration, necessary for complete release of energy from carbohydrates.	Accumulation of lactic acid and pyruvate. Beriberi (severe muscle wastage stunted growth, nerve degeneration).
B$_2$	Riboflavin	1.8 mg	Liver, milk, eggs green vegetables.	Coenzyme in cell respiration. Precursor of FAD.	Cracked skin, blurred vision.
B$_3$	Niacin	20 mg	Liver, yeast, whole cereals, beans.	Coenzyme in cell respiration. Precursor of NAD/NADP.	Pellagra (dermatitis, diarrhoea, dementia).
B$_6$	Pyridoxine	2 mg	Meat, fish, eggs cereal bran, some vegetables.	Interconversion of amino acids – transamination.	Dermatitis, nerve disorders.
B$_{12}$	Cyano-cobalamin	2 µg	Liver, milk, fish, yeast. None in plant foods.	Maturation of red blood cells in bone marrow. Maintenance of myelin sheath.	Pernicious anaemia. Nerve disorders.
	Folic acid	200 µg	Liver, raw green vegetables, yeast, gut bacteria.	Formation of nucleic acids. Formation of red blood cells.	Anaemia, especially during pregnancy.
C	Ascorbic acid	60 mg	Blackcurrants, potato, rose hips, citrus fruits.	Formation of collagen and intercellular cement.	Scurvy. Poor wound healing.

! SCIENCE FEATURE Vitamin K injection and new-born babies

Vitamin K is made in the body by helpful bacteria in the large intestine. Everybody has them, so no-one should be short of vitamin K. However, one very special group of people are at risk of vitamin K deficiency. All babies are born with sterile guts – not a bacteria in sight – and it takes about a week for the normal collection of different bugs to build up. During that first 7 days of life, this makes new babies prone to a disease called **haemmorrhagic disease of the new-born**. Vitamin K is vital for blood clotting, and a lack of vitamin K can lead to bleeding all over the body. This can occur if the skin is broken, at the navel, but also in the gut, the bowel and the brain. Brain bleeding is rare but is highly dangerous. Minor bleeding seems to happen in about one baby in 400, while more serious bleeding is very rare, about one baby in 10 000 suffers.

Fortunately, haemorrhagic disease of the new-born can be prevented very simply by giving vitamin K either by injection or by mouth on the first day of life (Fig 7.7). Recently, there has been some controversy about doing this because some studies suggested that babies given vitamin K by injection were more prone to leukaemia. Other studies have shown no such link, and most babies in the developed world now receive the vitamin K supplement.

Fig 7.7 A new-born baby being given vitamin K

? TEST YOURSELF

D Early Arctic explorers learned from the resident Eskimos not to eat the liver of polar bears. It causes headache, drowsiness, vomiting and extensive peeling of the skin. Since polar bear liver may contain up to 600 000 µg vitamin A per 100 g, what is the likely explanation for this observation?

Fig 7.8 Vitamin C is needed for healthy gums; the inflammation seen in this photograph is caused by vitamin C deficiency – otherwise known as scurvy

SEE QUESTION 6

Vitamins can be divided into two groups: fat-soluble vitamins (A, D, E and K), which contain only carbon, hydrogen and oxygen, and water-soluble vitamins (vitamin C and the B complex of vitamins), most of which also contain nitrogen. Fat-soluble vitamins can be stored in the body (especially in the liver) but water-soluble ones cannot, and are excreted in the urine. Water-soluble vitamins need to be taken into the body every day, and it is virtually impossible to 'overdose' on them. Taking too much of fat-soluble vitamins can cause problems: they can build up in fatty tissues until they reach toxic levels.

MINERALS

The minerals essential to human health are listed in Table 7.4. Minerals have four functions in the body:

- They are raw materials for the construction of body tissues.
- They have a homeostatic role, providing the correct chemical environment for cells.
- They are metabolites for various cell processes.
- They are essential partners for enzymes and vitamins controlling cell activity.

Generally, the minerals in the first two categories are needed in relatively large amounts and are known as **macronutrients**. Minerals in the last two categories are needed in specific chemical reactions and in tiny amounts. For this reason they are known as **micronutrients**.

The presence of other vitamins and minerals can affect the actual amount of a mineral that the body takes up. For example, calcium, the most abundant mineral in the body, is taken up more quickly in the presence of vitamins C and D, and iron is better absorbed in the presence of vitamin C.

Table 7.4 Minerals essential in the human diet

Mineral	Major food source	Function	Total body content/g
Major minerals (macronutrients)			
Calcium	Milk, cheese, bread, watercress	Needed for muscle contraction; nerve action; blood clotting and the formation of bone	1000
Phosphorus	Cheese, eggs, roast peanuts, most foods	Bone and tooth formation; energy transfer from foods; DNA, RNA and ATP formation	780
Sulphur	Dairy produce, meat, eggs, broccoli	Formation of thiamin, keratin and coenzymes	140
Potassium	Potatoes, meat, chocolate	Muscle contraction; nerve action; active transport	140
Sodium	Any salted food, meat, eggs, milk	Muscle contraction; nerve action; active transport	100
Chlorine	Salted foods, seafood	Anion/cation balance; gastric acid formation	95
Magnesium	Meat, green vegetables	Formation of bone; form coenzymes in cell respiration	19
Trace elements (micronutrients)			
Iron	Liver, kidney, cocoa powder, watercress	Formation of haemoglobin, myoglobin and cytochromes	4.2
Fluorine	Water supplies, tea, sea food	Resistance to tooth decay	2.6
Zinc	Meat, liver, legumes	Enzyme activator; carbon dioxide transport	2.3
Copper	Liver, meat, fish	Enzyme, melanin and haemoglobin formation	0.072
Iodine	Seafood, iodised salt, fish	Constituent of thyroxine	0.013
Manganese	Tea, nuts and spices, cereals	Bone development; enzyme activator	0.012
Chromium	Meats, cereals	Uptake of glucose	>0.002
Cobalt	Meat, yeast, comfrey	Component of vitamin B_{12}; formation of red blood cells	0.0015

WATER

Just under 70 per cent of the human body is water: two-thirds of this is found within the cells and one-third outside the cells in tissue fluid and blood plasma. It is essential in the diet and can come from three main sources:

- as a drink,
- as food (lettuce, for example, contains 94 per cent water),
- as metabolic water: water is released from cellular chemical reactions, especially cell respiration.

An individual must take in enough water to balance what is lost in urine, sweating, breathing out and in faeces. Otherwise, they lose water and can become dehydrated. Severe dehydration leads to an increase in heart rate and blood urea concentration and, eventually, death.

The chemical properties of the water molecule are discussed in Chapter 4, the role of the kidneys in water balance is covered in Chapter 13.

? TEST YOURSELF

E To what extent does a meal of brown bread, butter and cheese satisfy dietary requirements for an adult man? What does the meal provide, and what does it lack?

? TEST YOURSELF

F Why do experts recommend that a combination of dieting and increasing the amount of exercise is the most effective way to lose weight if you are overweight? Why not just diet or do more exercise?

3 A DIET FOR LIFE

The nutritional requirements of a human change according to age, sex, level of physical activity and state of health. Within groups of people there are still variations, but it is possible to say broadly what the general food and energy requirements of the groups are.

Fig 7.9 shows the energy requirements for different people at different stages of life. The Tour de France cyclist is included as a contrast to the 'ordinary' lifestyles of people who are very active. This is one of the most demanding sporting events in the world, requiring peak output for several hours at a time, often over several days.

Fig 7.9 The approximate energy needs of different people

Table 7.5 Composition of human milk

Component	Amount present per 100 ml
Carbohydrates	
Lactose	7.3 g
Oligosaccharides	1.2 g
Proteins	
Caseins	0.2 g
Lactalbumin	0.2 g
Lactoferrin	0.2 g
Secretory IgA	0.2 g
β-Lactoglobulin	None
Milk lipids	
Triglycerides	4.0%
Phospholipids	0.04%
Minerals	
Sodium	5.0 mM
Potassium	15.0 mM
Chloride	15.0 mM
Calcium	8.0 mM
Magnesium	1.4 mM

THE FIRST YEAR OF LIFE

When a baby is first born, its mother's breasts produce a fluid called **colostrum**. This is rich in antibodies, which pass whole into the blood of the baby through large pores in its intestine. Generally, the colostrum also has about three times more protein than the breast milk produced later. There is a gradual transition between colostrum and early breast milk over the first week. As the baby suckles, it first takes in **foremilk** – a more dilute breast milk that quenches its thirst before richer, fattier **hindmilk** is produced towards the end of the feed.

Breast milk produced at about the third month of lactation has 75 calories per 100 ml; its main components are shown in Table 7.5.

During the first 4–6 months of life growth and development are rapid. Breast milk (or infant formula) contains all the nutrients required during this period; babies don't really need any solid food before the age of 4 months.

As the baby grows, the fat content of breast milk decreases. After 4–6 months, milk is no longer enough and the baby starts taking in other foods – this is called **weaning** (Fig 7.10). Iron requirements increase rapidly and it is important that the diet given during weaning contains enough iron. Requirements for protein, thiamin, niacin, vitamin B_6, vitamin B_{12}, magnesium, zinc, sodium and chloride also increase between 6 and 12 months.

TODDLERS

Once babies start to toddle, walk and then run, they need a lot more energy from food. Growth and development are very rapid at this stage and, although protein requirements do not increase much, children between the ages of 1 and 3 years need more of all the vitamins, except vitamin D (this is synthesised in the skin in sunlight) and all the minerals, except zinc. Slightly lower amounts of calcium, phosphorus and iron are needed.

Children at this stage of life need energy-dense diets. The usual rules of what makes a healthy diet for an adult just do not apply. They should drink whole milk, not skimmed or semi-skimmed, and they shouldn't eat too much fibre. If the diet is too bulky, there is a possibility that a child cannot absorb enough energy from the food he or she eats. After the age of 2 years, children can switch to semi-skimmed milk, as long as the rest of the diet contains enough energy. Children shouldn't switch to completely skimmed milk until they are at least 5 years old.

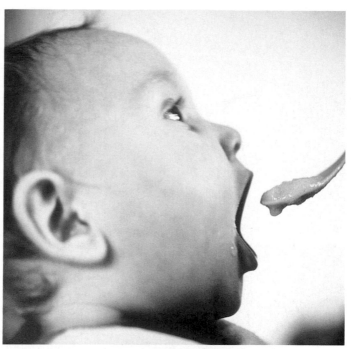

Fig 7.10 He won't stay this clean for long. Modern baby foods are well balanced and nutritious, giving a great start in life. It's not until he reaches the age of the children shown below that he'll start nagging for high-fat, high-salt or high-sugar foods. Broccoli? No chance.

Fig 7.11 Young children are growing rapidly and are very active, so they require lots of energy from their food

CHILDHOOD

From the age of 5 until about 11, growth is still rapid and children of this age are very active (Fig 7.11):

- 5–7 years: children need more energy, protein, more of all the vitamins (except C and D) and more of all the minerals.
- 7–10 years: there is a marked increase in requirements for energy and protein. Children of this age need no more thiamin, vitamin C or vitamin A than they did at the age of 5 but their need for the other vitamins and minerals increases.
- 11–14 years: the energy needed continues to increase and young teenagers need 50 per cent more protein than they did at the age of 10. By the age of 11, the vitamin and mineral requirements for boys and girls start to differ. Boys of 11–14 need more of all the vitamins and minerals. Girls need no more thiamin, niacin, vitamin B_6, but they need more of all the minerals. When periods start, girls need more iron than boys to make up the iron lost in menstrual blood.
- 15–18 years: boys need more energy and protein and more thiamin, riboflavin, niacin, vitamins B_6, B_{12}, C and A, and more magnesium, potassium, zinc, copper, selenium and iodine. In girls, the need for energy, protein, thiamin, niacin, vitamins B_6, B_{12} and C, phosphorus, magnesium, potassium, copper, selenium and iodine all increase, but girls continue to need slightly less energy than boys of the same age. Boys and girls have the same requirement for vitamin B_{12}, folate, vitamin C, magnesium, sodium, potassium, chloride and copper. Girls still need more iron.

ADULTHOOD

As young adults stop growing, their need for energy and some vitamins and minerals stabilises, and usually decreases. The energy requirements of adults depend mainly on their lifestyle – a very active person will need more energy (see Figs 7.9 and 7.12).

- 19–50 years: adult men and women both need less energy in their diet than when they were teenagers and their requirements for calcium and phosphorus fall as they are no longer making large amounts of new bone. There is also a reduced requirement in women for magnesium, and in men for iron. An adult's need for protein and most of the vitamins and minerals remain virtually unchanged from the teenage years.

Fig 7.12 Tour de France cyclist being pushed to the limit

- 50+ years: energy requirements decrease gradually after the age of 50 in women and age 60 in men (Fig 7.13). Protein requirements decrease for men but increase slightly in women. Vitamin and mineral intake should continue at the same level as younger adults, for both men and women. The only exception is that once women have gone through the menopause and their periods have stopped, they need the same amount of iron as men of the same age.

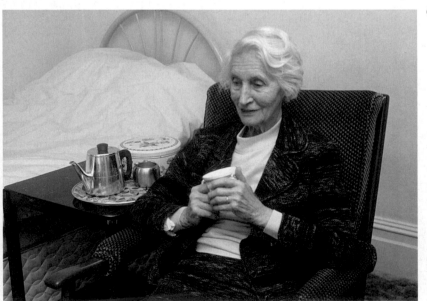

Fig 7.13 As we age, our metabolic rate slows and we tend to become less active. This can lead to an increase in weight but elderly people often lose interest in food and eating, and can become very thin and frail

THE SPECIAL CASES OF PREGNANCY AND BREAST-FEEDING

Women are now advised to take folic acid supplements to reduce the risk of spinal defects in the months when they are planning to get pregnant, and for the first 3 months of pregnancy. Despite the old saying that pregnant women are 'eating for two', a pregnant woman does not need additional energy or other food components until the last 3 months of pregnancy (Fig 7.14). Her mineral requirements do not increase at all.

Once the baby is born and the woman starts to breast-feed, she needs to take in more energy, protein, all the vitamins (except B_6), calcium, phosphorus, magnesium, zinc, copper and selenium than a new mother who does not breast-feed. The actual amount of extra energy needed depends on how much milk is produced, the size of the fat stores that have accumulated during pregnancy and how long the woman chooses to breast-feed.

Fig 7.14 Eating for two? A developing baby is fed via the placenta, an organ which takes the molecules from food directly out of the mother's blood. When the infant is born, the mother will need even more extra food in order to produce milk

---

► EXTENSION Dietary reference values (DRVs)

In the UK, estimated requirements for particular groups of the population are based on advice given by the Committee on Medical Aspects of Food and Nutrition Policy (COMA). The COMA panel reviews scientific evidence and makes proposals that are used by the government. In 1991, the COMA report 'Dietary Reference Values for Food Energy and Nutrients for the United Kingdom' was published. This estimated the nutritional requirements of the age groups shown in Table 7.6.

Dietary reference values for energy intake exist for all different age groups, but dietary reference values for specific food groups (carbohydrates and vitamins, for example) only apply to people over 5 years old.

Source: Department of Health. Dietary Reference Values for Food Energy and Nutrients for the United Kingdom. London: HMSO, 1991.

Table 7.6 How energy requirements change during life

	Estimated average requirements for energy			
	Megajoules per day		Kilocalories per day	
Age	Males	Females	Males	Females
0–3 months	2.28	2.16	545	515
4–6 months	2.89	2.69	690	645
7–9 months	3.44	3.20	825	765
10–12 months	3.85	3.61	920	865
1–3 years	5.15	4.86	1230	1165
4–6 years	7.16	6.46	1715	1545
7–10 years	8.24	7.28	1970	1740
11–14 years	9.27	7.72	2220	1845
15–18 years	11.51	8.83	2755	2110
19–50 years	10.60	8.10	2550	1940
51–59 years	10.60	8.00	2380	1900
60–64 years	9.93	7.99	2380	1900
65–74 years	9.71	7.96	2330	1900
74+ years	8.77	7.61	2100	1810
Women only pregnancy		8.9		1140

SUMMARY

When you have read and studied this chapter, you should know and understand the following:

■ Food supplies us with the energy and raw materials we need for repair, growth and development of body tissues.

■ Humans need a **balanced diet** which contains carbohydrate, protein, fat, **water**, **fibre**, vitamins and minerals.

■ **Carbohydrate** foods (sugars and starches) should be the main source of energy in the diet, although **lipids** provide more than twice as much energy per gram.

■ When intake of 'energy' foods exceeds requirements, the body will store the excess as glycogen and fat. When intake is too low, the body will turn to the lipid stores and increasingly use **protein** when the other stores run low.

■ **Vitamins** are either water soluble (A, D, E and K) or fat soluble (B complex and C). Lack of particular vitamins or **minerals** in the diet can lead to **deficiency disease**s.

? EXAM QUESTIONS

1 Humans vary considerably in the amount of energy that they need in their diet. Two factors which affect the energy requirement are body mass and physical activity. The amount of physical activity carried out can be measured and expressed as a numerical value, PAL (physical activity level). In teenage boys PAL values range from 1.4 (inactive) to 2.0 (extremely active).

The nomogram below can be used to estimate the energy requirements of boys from 10 to 18 years old if their body mass and PAL are known.

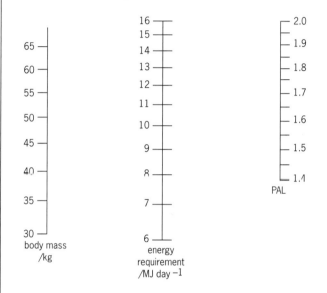

a) Estimate, using the nomogram, the energy requirement for a boy with:
(i) a mass of 60 kg and PAL of 1.4,
(ii) a mass of 40 kg and PAL of 1.8.
b) State the relationship between:
(i) physical activity level and energy requirement,
(ii) body mass and energy requirement.
c) Suggest **one** reason for the relationship between:
(i) physical activity level and energy requirement,
(ii) body mass and energy requirement.
d) Predict how the energy requirement of a 10 to 18 year old boy would differ from that of:
(i) a 10 to 18 year old **girl** with the same mass and PAL as the boy,
(ii) an adult man with the same mass and PAL as the boy.
IBO Biology Standard Level Paper 3 May 1999 QA1

2 Outline **two** uses of carbohydrates in the human body.
IBO Biology Standard Level Paper 3 May 1999 QA2

3 An experiment was carried out to investigate the effects of fibre in the diet on formation of tumours (cancers) in the large intestine of rats. At the start of the experiment all of the rats received a treatment that causes tumour formation.

Immediately following the treatment the rats were divided into two groups. One group was fed a normal diet (control group). The other group was fed on a normal diet with added fibre in the form of wheat bran (experimental group).

The rats were examined every week to check for tumours in the large intestine. The results are shown below.

a) State the week when tumours were observed in more than 10% of the rats in:
(i) the control group;
(ii) the experimental group.
b) State when the difference in percentage of rats with tumours is greatest between the two groups.
c) Discuss the effectiveness of fibre in the experiment in preventing tumour formation.
IBO Biology Subsidiary Standard Level Paper 3 Nov 1998 Option A QA1

4
a) Define *malnutrition*.
b) Low levels in the amount of protein in the diet can result in a deficiency disorder. Suggest a reason why eating protein from only one source can cause similar disorders.
c) State **two** different sources of protein acceptable to vegans.
IBO Biology Subsidiary/Standard Level Paper 3 Nov 1998 QA2

5

a) Explain what is meant by the term *vitamin*.

b) Copy and complete the following table to show the biochemical importance of some vitamins and elements in animal nutrition.

Name of vitamin or element	Origin (important source)	One major biochemical reason for importance in organism
calcium		
	sea-foods	needed to form part of molecule of hormone thyroxine
iron	liver	
		precursor of rhodopson molecule in retina cells
D		
	leafy vegetables; eggs	needed to form coenzyme FAD in respiration
K	green vegetables; also synthesised by gut microbes	

OCR 6910 Molecules and Life March 2000 Q2

6 Surveys of the British population show that requirements for nutrients, such as iron, show a normal distribution. Three Dietary References Values (DRVs) are based on this normal distribution as shown in the diagram.

LRNI = lower reference nutrient intake
EAR = estimated average requirement
RNI = reference nutrient intake

With reference to the diagram, state the percentage of the population that has a requirement for iron between the LRNI and the RNI.

The table shows the DRVs for iron for males and females aged 17.

	Males			Females		
	LRNI	EAR	RNI	LRNI	EAR	RNI
iron intake /mg day^{-1}	6.1	8.7	11.3	8.0	11.4	14.8

b) With reference to the diagram and the table, explain why:

(i) nutritionists recommend that people consume the RNI for nutrients, such as iron;

(ii) the DRVs for iron for females are higher than those for males.

People with anorexia nervosa are likely to consume much less than the LRNI for iron.

c) Suggest the likely consequences for people with anorexia nervosa of consuming less than the LRNI for iron over a long time.

OCR Human Health and Disease June 2001 Q7

KEY SKILLS **ASSIGNMENT**

OBESITY AND BODY MASS INDEX (BMI)

Obesity is a term that means 'grossly overweight' and, despite more people understanding the problems of the condition, the proportion of overweight and obese people is on the increase.

Table 7.A1 The increasing incidence of obesity						
Country	Weight level	Age range	Year	Incidence/%		
				men	women	
England	obese	16–64	1980	6	8	
	(BMI >30)		1994	13	16	
England	overweight	16–64	1980	35	24	
	(BMI 25–30)		1991	40	26	
Germany	obese	25–69	1985	15	17	
	(BMI >30)		1990	17	19	
USA	obese*	20–74	1978	24	24	
	(white people)		1988–91	34	34	
USA	obese*	20–74	1978	26	45	
	(black people)		1988–91	32	49	

*In the USA survey the threshold for obesity was BMI >27.8 men, >27.5 for women
Sources: Obesity Research Information Centre, London, and Dept of Health Report 46 (1994) Nutritional Aspects of Cardiovascular disease, HMSO.

1 Given the present rate of increase, what percentage of men and women in England will be obese by the year 2008?

Body mass index (BMI)

2 Obesity could be defined in terms of weight alone. What would be wrong with this?

In 1869 a Belgian astronomer Quetelet worked out that people's body mass varied not in proportion to their height, but in proportion to the square of their height.

So the formula for body mass index is:

$$\frac{\text{mass}}{\text{height}^2}$$

For example, a man who is 1.83 m tall (6 feet) and weighing 82 kg would have a BMI of

$$\frac{82}{(1.83)^2} = \frac{82}{3.35} = 24.47$$

3 Work out your own BMI.

Fig 7.A1 This is Daniel Lambert (1770–1809), once thought to be the heaviest man in England. He weighed 336 kg (52 stone 11 lb) and was 180 cm (6 ft) tall

4 Work out the BMI for Daniel Lambert.

5 List some of the problems that a person as heavy as Daniel Lambert would face in their everyday life. Assume that they are alive today, not in the seventeenth century
a) What sorts of things would it be impossible for them to do?
b) Can you think of any advantages of being this weight?

Table 7.A2	
BMI/kg m^{-2}	Description
less than 20	underweight
20–24.9	normal
25–29.9	overweight
30–40	obese
over 40	severely obese

The BMI is now universally used to define underweight, overweight and obesity, as shown in Table 7.A2.

The problems of obesity

There was a time when fat people were looked upon just as happy, jolly folk. Nowadays, we know the problems associated with obesity, and obese people have to deal with the intolerance of others towards their size and body image.

One look at the population of an old people's home will support the blunt statement that 'old people aren't fat and fat people aren't old'. Obesity puts a great strain on the body that more often than not leads to severe medical problems and a premature death. Consider some of the problems:

Coronary heart disease. Obesity increases the levels of triglyceride and cholesterol in the blood (see Chapter 31).

KEY SKILLS **ASSIGNMENT**

6 Suggest why obesity puts a strain on the heart. Why does it have to work harder?

Diabetes. This is the inability to control blood sugar levels. In an obese person, insulin becomes less effective than normal, a problem known as insulin resistance.

High blood pressure.

Joint problems. Excess weight puts a lot of strain on some joints.

7 Suggest which joints will be affected most.

Accidents. Obese people tend to be clumsier than less heavy people, and unable to move quickly.

Depression and suicide. These could be responses to a poor body image.

Fig 7.A2 The United States has the highest proportion of obese people in the world

The causes of obesity

Many obese people claim that eating too much is not the problem; it's in their genes. Recent research suggests that there may be some truth in this, but we can only inherit a tendency towards obesity or 'leanness'. We must still eat the food which allows us to fulfil our potential, and studies suggest that social and environmental factors play a huge part. Children born to overweight parents tend to be given larger portions and generally follow the influence of their parents.

Lifestyle

People are generally eating more and exercising less, using the car for even the shortest journeys and taking part in leisure activities with minimal energy needs.

8 **a)** Suggest how the leisure activities of children have changed over the past few decades.
b) Imagine you are a family doctor. How would you persuade an obese couple to change their lifestyle?

Drugs

Research has shown that adipose (fat storage) tissue gives off a hormone, leptin (Greek, leptos = slender/thin) which suppresses the appetite centre of the brain. There are artificial appetite suppressants available, but their effect is only temporary. Clearly, there would be a huge market for an effective drug.

9 What properties would the perfect weight control drug have?

Surgery

Adipose tissue can be surgically removed; the process of liposuction is one of the cosmetic surgeons' biggest moneyspinners. However, all surgery carries some risk and, without a long-term change in eating and exercise habits, the problem will return.

08

Digestion

Fructose is present in many everyday foods

Humans are omnivores; they have a digestive system that is adapted to eat a range of foods, from both animals and plants. But that doesn't mean that all foods suit everybody. Some people suffer food allergies or are intolerant to specific foods. Lactose intolerance is well known (see page 157) but researchers are uncovering a problem with sugar digestion – an intolerance to fructose.

Everyone experiences indigestion or stomach upsets now and again, but some people have chronic digestive problems – this is commonly known as irritable bowel syndrome (or IBS), and is extremely common. Its main symptoms are abdominal discomfort, a bloated feeling and experiencing diarrhoea and then constipation for no apparent reason. In the autumn of 2001, researchers in the USA revealed that about 10 per cent of IBS cases could be explained by an intolerance to fructose. Fructose is a simple sugar found in honey and many fruits.

When people with IBS symptoms were given fructose, tests showed that over three-quarters of them produced breath that contained hydrogen and methane. This indicated that they were not metabolising fructose normally, and these abnormal by-products were being produced in the gut.

1 THE HUMAN DIGESTIVE SYSTEM

The overall function of the **digestive system** is to break down the larger molecules in food, such as protein and starch, into their monomers – amino acids and sugars – so that they can be absorbed easily into the body.

The **alimentary canal**, or **gut**, can be thought of as a long coiled tube that runs through the body. The food we eat is subjected to 'conveyor-belt' food processing as it passes through the different regions of the gut. A common misconception is that the food we eat is 'inside' us, when really it travels through a cavity in the middle of the body. If we cannot digest the food, it cannot be absorbed into the body and passes straight through.

The alimentary canal, or 'gut', is a muscular tube that leads from the mouth to the anus. The overall process of nutrition can be divided into several stages:

- **Ingestion:** this is taking in food. In humans food is put into the mouth and chewed. **Swallowing** takes it down through the **oesophagus** into the stomach. Food is propelled through the alimentary canal by **peristalsis**, rhythmic contractions of the gut wall.
- **Mechanical breakdown:** the breaking up of food material into smaller pieces, mainly by the action of chewing or the churning action of the stomach.
- **Digestion:** the chemical breakdown of complex food molecules such as carbohydrates, proteins and fats into simpler ones.
- **Absorption:** the passage into the bloodstream of simple food molecules such as amino acids, sugars and fatty acids, vitamins, minerals and water.
- **Egestion:** the elimination of undigested food material from the body.

✔ REMEMBER THIS

Humans are **heterotrophs**; we cannot make our own food and so must obtain it in the form of large organic molecules from other organisms. All these molecules were originally made by green plants.

2 STRUCTURE OF THE HUMAN DIGESTIVE SYSTEM

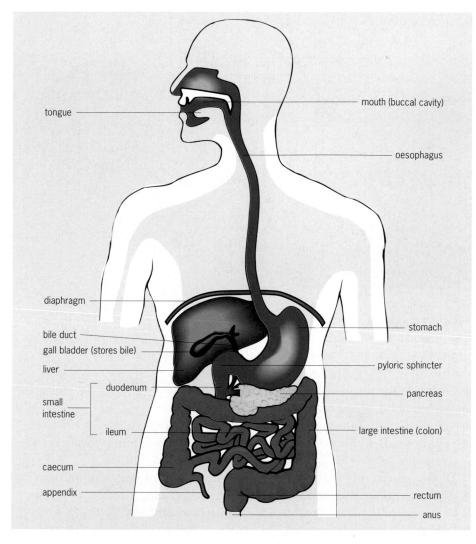

Fig 8.1 The human digestive system

The human digestive system consists of the alimentary canal and its associated glands, the salivary glands, the liver and the pancreas (Fig 8.1).

The alimentary canal begins at the mouth and ends at the anus. Between the two openings is a long convoluted tube organised into several distinct regions. The **oesophagus** carries food from the mouth to the **stomach**, a muscular bag or sac that stores food and is the first significant site of digestion.

Beyond the stomach is the **small intestine**. This has two main parts, the **duodenum** and the **ileum**. These are the main sites of digestion and absorption. The large intestine includes the **appendix**, the **colon**, whose main function is to absorb water, and the **rectum**. It ends at the **anus**.

BASIC FEATURES OF THE GUT WALL

The wall of the gut is not the same all the way along the alimentary canal. Each region has specific features, which can be related to the function of different parts of the gut.

The gut wall is divided into three main layers:

- an outer **muscle layer**, protected by a thin coating of fibres;
- a middle layer, the **submucosa**;
- an inner layer, the **mucosa**.

SEE QUESTION 1

THE OESOPHAGUS

The main function of the oesophagus is to push food from the mouth to the stomach. The structure of the wall of the oesophagus is shown in Fig 8.2. The endoscope image shows that the inside surface has large ridges, but it is fairly smooth.

Fig 8.2 The oesophagus

THE STOMACH

The stomach has deep ridges called **rugae** that help with the mechanical breakdown of food.

Overall, the stomach:

- mixes food with gastric juice by muscular action;
- retains food, giving enzymes time to act;
- digests proteins through the action of the enzyme pepsin;
- curdles milk with the enzyme rennin, absorbs some simple chemicals such as water, salts as ions, and alcohol.

The muscle layer is thick; three different layers of muscle run in different directions. This gives the stomach the power to contract and relax to grind food up. Glands in the **gastric pits** produce a strong acid that reduces the pH of the stomach contents. The stomach also produces large quantities of **mucus** to protect its own cells from damage from the acid.

REMEMBER THIS

Note that all regions of the gut have two layers of muscle, apart from the stomach, which has three. The extra *oblique* layer runs at 45° to the other two, and helps churn the food.

Fig 8.3 The stomach

THE DUODENUM AND ILEUM

The small intestine is about 5 metres long and is made up of two main parts, the duodenum and the ileum. The duodenum is the first 25 cm of the small intestine, the ileum forms the rest of it. While the duodenum is the main site of digestion, most absorption takes place in the ileum.

The small intestine:

- moves food from the stomach to the large intestine by peristalsis. Food moves through the intestine at about 1 cm per minute;
- secretes the large amounts of water necessary for most of the chemical reactions involved in digestion;
- completes the digestion of carbohydrates, proteins and fats;
- absorbs most of the small soluble food molecules produced by digestion.

The pancreatic duct and bile duct join with the duodenum, releasing

Fig 8.4 The duodenum

epithelium

mucosa

blood vessel

serosa; fibrous coat consisting of connective tissue

longitudinal muscle

submucosa

circular muscle

Fig 8.5 The ileum

mucous gland

epithelium

mucosa

blood vessel

serosa; fibrous coat consisting of connective tissue

longitudinal muscle

submucosa

circular muscle

digestive enzymes and bile into the lumen. This neutralises stomach acid, and creates a pH that is optimal for digestion. The wall of the small intestine looks less folded than that of the stomach, but looking at it under a light microscope reveals the tiny microscopic infoldings that increase the surface area. The tiny folds are called villi; there are 4–5 million of them in the ileum. They allow intimate contact between the wall of the intestine and the products of digestion inside this part of the gut, ensuring that the absorption of digestion products is as efficient as possible.

THE COLON

The ileum joins the large intestine at the caecum, near the appendix (refer back to Fig 8.1). The **colon, rectum** and **anus** make up the final section of the digestive tract. Sodium ions leave the lumen of the colon and re-enter the blood by active transport; water follows by osmosis. The colon therefore absorbs water from the undigested food that reaches it. What is left in the colon is called **faeces**. As well as undigested food, faeces contain dead cells, bacteria and bile pigments. These wastes are then **egested** from the anus during **defecation**.

Fig 11.6 The colon

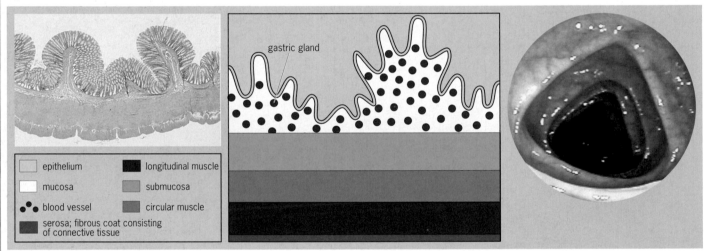

☐ epithelium	■ longitudinal muscle
☐ mucosa	▨ submucosa
•°• blood vessel	▨ circular muscle
▨ serosa; fibrous coat consisting of connective tissue	

gastric gland

3 INGESTION AND MECHANICAL BREAKDOWN OF FOOD

Whatever your table manners are like, you ingest food by getting it into your mouth. Your lips and tongue then mix the food with saliva and the chewing action of grinding the food between your teeth breaks up the food into small pieces. The technical term for this is **mastication**.

Adding saliva to food softens it, making it easier to swallow. Salivary glands (Fig 8.7) produce saliva at the rate of about 1 to 1.5 litres per day. It is produced constantly, but more is released when we see, smell, taste or even just think about food.

Saliva is mainly water (99.5 per cent) with some dissolved substances (0.5 per cent) including:

- **mineral salts** such as phosphates and hydrogencarbonates;
- **salivary amylase**, a starch-digesting enzyme that breaks molecules of starch into maltose;
- **mucin**, a slimy glycoprotein lubricant;
- **lysozyme**, an enzyme that kills bacteria.

Salivary amylase, the starch-splitting enzyme in saliva, begins the process of chemical breakdown. However, the speed at which most people chew and swallow their food means that salivary amylase has little chance to act, which is probably why people who lack the gene to make this enzyme are rarely aware of their 'genetic deficiency'.

SWALLOWING

Swallowing is not a voluntary action, but it is under some voluntary control because you can decide to swallow if you want to. When you are eating, the act of moving food to the back of your throat is the voluntary action but swallowing is a reflex response to food or liquid touching the back of your throat.

Theoretically, when the tongue forces food or liquid to the back of the throat, the food is able to travel in one of four directions:

- back out of the mouth;
- into the nasal cavity;
- into the trachea, or windpipe;
- into the oesophagus (where it should go).

Normally, when we swallow, the first three options are closed, and food or liquid is forced into the oesophagus. If you have ever been laughing, eating crisps and drinking a fizzy drink, you may have experienced fizzy liquid going 'up your nose' or crisps going 'down the wrong way'.

The oesophagus is about 25 cm long and 2 cm in diameter. When you swallow food, smooth muscles in the oesophageal wall contract rhythmically to propel it, by **peristalsis**, from the **pharynx** to the stomach (Fig 8.8). Elastic tissue in the walls enables the oesophagus to expand and, as in the mouth, stratified epithelium protects against friction damage.

Mechanical breakdown continues in the stomach, as the churning action breaks the food up into yet smaller pieces.

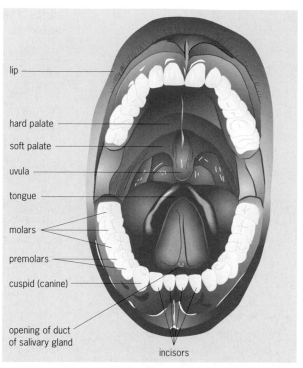

Fig 8.7 There are three pairs of salivary glands in the mouth and one pair can easily be seen. If you look under the tongue and dry the area, you will be able to see saliva oozing out

---

✔ REMEMBER THIS

The intestine, along with many other tubular organs in the body (ureters, vas deferens, uterus) are made from smooth muscle whose main function is the slow rhythmic contraction known as peristalsis.

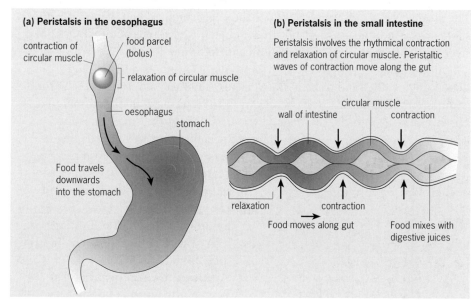

(a) Peristalsis in the oesophagus

contraction of circular muscle

food parcel (bolus)

relaxation of circular muscle

oesophagus

stomach

Food travels downwards into the stomach

(b) Peristalsis in the small intestine

Peristalsis involves the rhythmical contraction and relaxation of circular muscle. Peristaltic waves of contraction move along the gut

circular muscle

wall of intestine

contraction

relaxation

contraction

Food moves along gut

Food mixes with digestive juices

Fig 8.8 Food is propelled through the gut by peristalsis. The squeezing action of peristalsis moves semi-solid food from the mouth to the stomach in about 4 to 8 seconds. Food moves through the intestines at a rate of about 1 cm/min

4 DIGESTION

Digestion is usually thought of as the chemical breakdown of complex food molecules into simpler ones that can be absorbed and used by the body. Enzymes are key players in this process; they control all of these breakdown reactions. Table 8.1 gives an overview of the main enzymes involved in human digestion.

In the first part of this section, we look at where the different enzymes act, and what they do; in the following section we see how the release of these secretions is controlled.

Table 8.1 A summary of the main human digestive enzymes

Secretion	Enzymes produced	Site of production	Site of activity	pH	Substrate	Products
Saliva	salivary amylase	salivary glands	mouth	6.5–7.5	starch	maltose
Gastric juice	pepsin	stomach	stomach	2.0	proteins, polypeptides	
	rennin				caseinogen (milk protein)	
Pancreatic juice	trypsin	secretory cells of the pancreas (acini)	duodenum	7.0	proteins, polypeptides	short polypeptides
	chymotrypsin				proteins	polypeptides
	carboxypeptidase				polypeptides	dipeptides, amino acids
	pancreatic amylase				starch	maltose
	maltase				maltose	glucose
	sucrase				sucrose	glucose, fructose
	lactase				lactose	glucose, galactose
	lipase				fats and oils	fatty acids, glycerol

DIGESTION IN THE STOMACH

The stomach is a muscular sac under the diaphragm. When empty, it is the same size as a large sausage, but it can stretch to the size of a melon. The stomach is the first area of any significant digestion.

Food remains in the stomach for between 30 minutes and 4 hours depending on the type of meal (fatty meals stay there the longest). The food is digested mechanically and chemically. The resulting semi-liquid material, **chyme**, passes into the duodenum, the first part of the small intestine.

Specialised groups of cells inside **gastric pits** in the mucosa secrete **gastric juice** (Fig 8.9). Each type of cell produces a specific secretion:

- **Oxyntic cells** secrete a solution of hydrochloric acid which brings the pH of gastric fluid down to between pH 2.0 and 3.0.
- **Zymogen cells** (peptic cells) secrete the enzyme pepsinogen which is later converted into the protein-splitting enzyme, pepsin.
- **Mucous cells** secrete the mucus that protects the stomach lining from the digestive action of its own secretions.

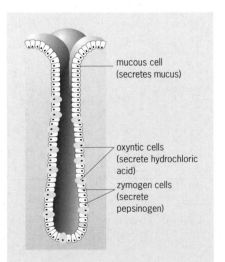

mucous cell
(secretes mucus)

oxyntic cells
(secrete hydrochloric acid)

zymogen cells
(secrete pepsinogen)

Fig 8.9 A gastric pit from the mucosa of the stomach showing the specialised cells that secrete gastric juice

THE IMPORTANCE OF STOMACH ACID

The hydrochloric acid in gastric fluid:
- provides the optimum pH for **pepsin** and **rennin**;
- denatures proteins and helps to soften tough connective tissue in meat;
- is a strong **bactericide** (it kills bacteria) and so protects the body from some of the harmful microbes that might enter the body in food.

PROTEIN DIGESTION IN THE STOMACH

One of the main functions of the stomach is to begin to digest proteins. **Pepsin**, is a powerful **endopeptidase enzyme**, digests proteins. This enzyme breaks peptide bonds in the middle of the protein chain, turning protein molecules into polypeptides (see Chapter 3, page 52). Protein digestion is completed when **exopeptidase** enzymes remove amino acids from the ends of the polypeptides.

The stomach avoids digesting its own tissue by secreting pepsin in an inactive form, **pepsinogen**. This is converted to pepsin in the lumen of the stomach only after contact with hydrochloric acid. The hydrogen ions in the acid cause the pepsinogen to unfold and become pepsin, the active form of the enzyme.

MILK DIGESTION

Caseinogens, the proteins in milk, are water soluble. They are valuable nutrients (milk is the sole source of food for young mammals; see the Science Feature on this page) but, if they remained in their soluble form, they would leave the stomach before protein digestion had finished. To avoid this, the stomach produces **rennin**. This curdles milk, converting soluble caseinogen into insoluble **casein**. Like pepsin, rennin is also secreted in an inactive form, **prorennin**. Like pepsinogen, prorennin is converted to its active form by contact with stomach acid.

! SCIENCE FEATURE Lactose

Like all mammals, humans feed their young on milk (Fig 8.10). The main carbohydrate in breast milk is the disaccharide lactose. But why should a mother go to the trouble of combining two monosaccharides into lactose when the baby will simply have to break it down again to use it as an energy source?

The answer lies in the size of the lactose molecule. In the mother, milk is made in the mammary gland and is stored there between feeds. Since it is a disaccharide, lactose stays in the milk, rather than diffusing into the surrounding tissue as smaller monosaccharides would tend to do. In the baby, lactose remains in the gut for the same reason and is broken down only gradually to form glucose. So there is a steady absorption of glucose into the infant's blood, rather than a sudden surge at feeding time.

Several medical conditions can arise because of a failure to digest lactose. In adult life, some people stop making the enzyme lactase and so cannot break down lactose. Undigested lactose cannot be absorbed and it accumulates in the gut, encouraging the growth of bacteria that produce large amounts of carbon dioxide and lactic acid. The result is diarrhoea and wind. Affected people are said to be **lactose intolerant**.

A type of lactose intolerance that results in far more serious problems is found in people with an inherited condition called **galactosaemia**. Affected people can break down lactose into glucose and galactose but their liver cannot convert galactose to glucose. Galactose builds up to dangerous levels and the condition can be fatal.

Fig 8.10 Milk contains all of the food chemicals that a new baby needs. The main carbohydrate in milk is lactose

DIGESTION IN THE SMALL INTESTINE

The small intestine is the major site of digestion, which is brought about mainly by enzymes and secretions that flow into the duodenum from the pancreas and liver. These neutralise stomach acid and also contain a variety of enzymes that break down proteins, carbohydrates and fats into their monomers, ready for absorption.

PANCREATIC JUICE

Pancreatic juice enters the duodenum via the pancreatic duct. Over a litre of alkaline pancreatic juice is secreted every day. The fluid contains several enzymes that are produced by secretory cells in the acini of the pancreas (see page 206, for the structure of the pancreas):

? **TEST YOURSELF**

E Why do the protein-splitting enzymes trypsin and chymotrypsin need to be produced in an inactive form?

SEE QUESTIONS 2 AND 3

- **Trypsin**: this is a powerful **endopeptidase enzyme** that breaks down proteins into polypeptides. It is secreted in an inactive form called **trypsinogen**. Inactive trypsinogen is turned into active trypsin by the action of **enterokinase**, an enzyme released by the intestinal wall:

$$\text{trypsinogen} \xrightarrow{\text{enterokinase}} \text{trypsin}$$

- **Chymotrypsin** continues the digestion of proteins into polypeptides. This enzyme is secreted in an inactive form called chymotrypsinogen, and is converted to active chymotrypsin by the enzyme trypsin:

$$\text{chymotrypsinogen} \xrightarrow{\text{trypsin}} \text{chymotrypsin}$$

- **Carboxypeptidase** is a polypeptide-digesting enzyme, which converts polypeptides into smaller peptides and amino acids. Carboxypeptidase is also activated by trypsin in the lumen of the small intestine:

$$\text{procarboxypeptidase} \xrightarrow{\text{trypsin}} \text{carboxypeptidase}$$

- **Pancreatic amylase:** this completes the breakdown of starch to maltose started by salivary amylase in the mouth. A second amylase enzyme is needed as salivary amylase is inactivated by stomach acid.

- **Pancreatic lipase:** this continues the breakdown of fats into fatty acids and glycerol.

Hydrogencarbonate ions make pancreatic juice slightly alkaline (pH = 7.1 to 8.2), allowing it to neutralise the stomach acid and helping to create optimum conditions for the intestinal digestive enzymes.

The pancreas and liver both have other functions that are not connected directly with digestion. The pancreas produces the hormones insulin and glucagon (Chapter 11, Section 2) and the liver has many regulatory – homeostatic – functions (also Chapter 11, Section 3).

BILE

Liver cells called **hepatocytes** produce thick yellow-brown or olive green bile. This contains:

- water;
- **bile salts** (sodium glycocholate and sodium taurocholate);
- **bile pigments** (breakdown products of red blood cells, for example bilirubin) and some mucus;
- cholesterol.

Liver cells produce around 0.8 to 1.0 litre of bile daily. Secretions from individual cells pass into tiny canals called **bile cannaliculi**. These lead to the gall bladder, a small sac-like organ, which stores the bile until it is needed. Bile is released into the duodenum when the muscle wall of the gall bladder contracts. Bile reaches the duodenum through the bile duct. Bile:

- **emulsifies** fats (breaks large fat or oil droplets into an emulsion of microscopic droplets). This process massively increases the surface area available for fat-digesting enzymes;
- neutralises the (acidic) chyme from the stomach and creates the ideal pH for intestinal enzymes;
- stimulates peristalsis in the duodenum and ileum;
- allows the excretion of cholesterol, fats and bile pigments.

CONTROL OF DIGESTION

Since we do not eat food continuously, it is important that we produce digestive enzymes only when food is in the gut. If large quantities of the acids and protein-attacking enzymes were released into an empty stomach, there would be a real danger that you would digest yourself.

The release of digestive secretions is controlled by nerves and hormones working together. Fig 8.11 shows an overview of what happens.

Fig 8.11 The digestive tract is supplied by nerves from the autonomic nervous system (see Chapter 16). Particularly important is the vagus nerve, which stimulates the secretion of digestive juices. Hormones travel to the digestive tract in the blood.

Information coming into the body from food (about its appearance and smell) travels to the hypothalamus and cerebral cortex in the brain. These send signals back to the body to stimulate the salivary glands, stomach, liver and pancreas to start releasing their secretions. As food enters the stomach, gastrin is released and this encourages more gastric juice secretion. Similarly, the presence of food in the duodenum causes the release of gastrin, secretin and CCK-PZ (cholecystokinin-pancreozymin).

We do not carry on eating until we burst because food intake is controlled by hunger and satiety centres in the hypothalamus. A full stomach and high blood glucose levels stimulate the satiety centre, so that we do not feel hungry any more

Fig 8.12 Secretion of gastric juice is partly controlled by a nerve (the vagus nerve) and partly by a hormone (gastrin). Gastrin stimulates gastric glands, causing them to release their secretions. It increases movement in the tract and relaxes the sphincters that control the movement of food into and out of the stomach

SEE QUESTION 4

CONTROL OF GASTRIC SECRETIONS

Nerves and hormones control the secretion of gastric juice in a process that has three distinct phases (Fig 8.12):

- **Nervous phase.** The sight, smell or taste of food initiates a nerve reflex in which impulses from the brain trigger gastric glands to release their secretions.
- **Gastric phase** (hormonal). Food in the stomach stimulates the lining to secrete the hormone **gastrin**. Gastrin increases gastric juice secretion through direct action on the gastric glands.
- **Intestinal phase** (hormonal). The duodenal lining, stimulated by partially digested food, produces a second hormone, **enteric gastrin**. This hormone also acts on the stomach's gastric glands, producing further small amounts of gastric juice.

Pressure sensors and chemical sensors in the stomach detect stomach stretching and the presence of chyme. The resulting nerve impulses, together with the action of gastrin, direct the stomach to start emptying its contents into the duodenum.

CONTROL OF PANCREATIC SECRETIONS

Pancreatic secretion, like gastric secretion, is controlled by both nervous and hormonal mechanisms. Impulses travel from the brain down the **vagus nerve** to trigger the secretion of pancreatic juice.

The lining of the duodenum secretes two hormones in response to acidic chyme arriving in the lumen of the duodenum. **Secretin** stimulates the pancreas and liver to secrete pancreatic juice rich in hydrogencarbonate ions. **CCK-PZ (cholecystokinin-pancreozymin)** stimulates the gall bladder to release bile and the pancreas to release digestive enzymes.

CONTROL OF BILE SECRETION

The hormone secretin acts together with nervous stimulation by the vagus nerve to increase the rate of bile secretion. The acidity of chyme in the duodenum and the hormone CCK-PZ stimulate the gall bladder to contract.

! **SCIENCE FEATURE** Feeling sick? Blame the brain

Lots of different things make you feel sick – eating food that is 'off', travelling on a boat or in a car or aeroplane, having a bad cold or having a serious illness such as cancer, even when the digestive system isn't affected directly. Getting rid of the stomach contents if you have eaten something bad is an understandable response by the body – but why do these other experiences cause nausea?

Travel sickness results when the inner ear is telling the brain something different from the messages being sent from the eyes. The fine hairs in the inner ear detect motion and if this is in conflict with the visual sensation of movement, you start

to feel yukky. Keeping a long-distance view on things helps – looking at near objects, such as reading a book on a long car journey, is asking for trouble.

But what about other illnesses like having a cold, or more serious illnesses such as cancer? Recently, it has become clear that, when you have an illness, your body responds by producing different chemicals that stimulate the immune system. One of these is called tumour necrosis factor. One of its functions, as its name suggests, is to clobber cancer cells, However, it also acts on neurones in the brain, in a part of the brain stem that is concerned with digestion and vomiting. When tumour necrosis factor binds to

them, these neurones cause the stomach to stop all movement. This dramatic stomach relaxation is perceived as nausea and occurs just before a bout of vomiting.

Fig 8.13

5 ABSORPTION OF THE PRODUCTS OF DIGESTION

Food absorption takes place mainly in the ileum of the small intestine. The small soluble molecules produced by digestion in the stomach and duodenum can be easily absorbed through the gut lining. The epithelial lining has a much greater surface area than that of an equivalent smooth-sided tube; an average 70 kg man has approximately 100 square metres of absorbing surface in his small intestine. This is possible because of three important structural features:

- folds in the inner surface of the intestinal wall;
- movable projections (0.5–1.0 mm in length) called **villi** that are present on the folded surface of the wall;
- microscopic projections called **microvilli** on the cell surface membranes of epithelial cells that line the villi.

The epithelial cells absorb amino acids, monosaccharides, fatty acids and glycerol, water and other substances including vitamins, nucleic acids, ions and trace elements (Fig 8.14). The type of transport process used varies depending on the substance, but **diffusion**, **facilitated diffusion** and **active transport** are all involved (see Chapter 4).

ABSORPTION OF CARBOHYDRATES

Glucose is actively transported into the cells of the small intestine and across the walls of blood capillaries. Active transport is necessary because diffusion alone would be too slow to supply the body's needs. Also, there would be diffusion out of the epithelial cells if the concentration of food in the intestines were very low. Energy for the transport is obtained from the hydrolysis of ATP.

ABSORPTION OF PROTEINS AND FAT

Amino acids and small peptide molecules (di- and tripeptides) are actively transported into epithelial cells of the small intestine.

Bile salts and lipase enzymes break up complex lipid molecules into monoglycerides and free fatty acids. The monoglycerides and some of the fatty acids combine with bile salts to form microscopic droplets called **micelles**. Micelles are soluble in water and diffuse easily into epithelial cells, along with the remaining fatty acids. Inside the cells, the fatty acids and glycerol recombine, forming triglycerides, which acquire a protein coat (that stops them sticking together) and form particles called **chylomicrons**. It is in this state that lipids leave epithelial cells and enter, not a blood capillary this time, but a branch of the lymphatic system called the **lacteal** (see Fig 8.14).

Fig 8.14 The absorption of the products of digestion through the lining of the intestine

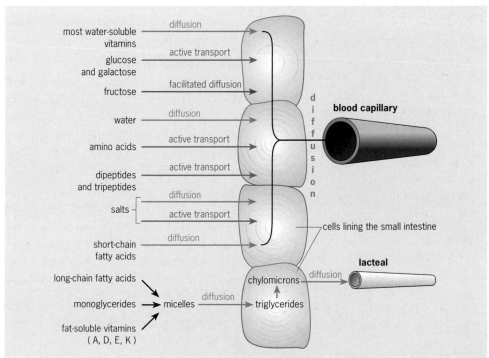

TRANSPORT OF ABSORBED FOOD PRODUCTS

Amino acids, dipeptides and simple sugars diffuse out of epithelial cells in the small intestine, directly into blood capillaries within the villi. From here they pass to the liver via the **hepatic portal** vein. The liver processes absorbed food, converting some into storage products. Glycogen, copper, iron and vitamins A, D, E and K can all be stored. The liver breaks down other food products, including excess amino acids which are **deaminated** – their amine groups are removed. These breakdown products are then passed to the kidneys for excretion. See Chapter 11 for more about liver function and Chapter 13 for details on excretion.

The lymphatic system plays a major role in the transport of absorbed lipids. Chylomicrons that have entered the lacteals remain in suspension, giving the lymph fluid a milky-white appearance. From here, the chylomicrons move into larger lymphatic vessels that eventually drain into a large duct that empties into the blood. In the blood, these complexes are broken down into fatty acids and glycerol and enter cells to be used in the synthesis of complex lipids.

● EXAMPLE

Q Explain how the structure of the ileum relates to its function of absorbing the products of digestion.

A Relating structure to function is one of the basic skills a biologist must master. The structure of the small intestine is a classic example.

When faced with this question, many students will simply state a well-learned GCSE list of 'large surface area, good blood supply, thin walls, etc', and expect full marks. At A-level a little more depth is required. Why is a large surface area vital? What does a good blood supply do?

Large surface area

The basic idea behind a large surface area is that a lot of intestine is in contact with a lot of digested food molecules. These are absorbed by diffusion or active transport and so the more membrane there is, the more pores and carrier proteins there are available. The fact that the intestine is long is the most obvious way to increase the surface area, but there are also the villi and the microvilli.

Good blood supply

Like most organs which are adapted for the exchange of materials, there is also a way of maintaining the diffusion gradient so that there is always more digested food on one side of the absorbing membrane than the other, thus ensuring that absorption is continuous and as rapid as possible. As soon as the digested food molecules are absorbed into the blood, they are taken away and replaced with blood containing fewer of these molecules.

Thin walls

Fick's law (Chapter 4) states that the speed of diffusion is inversely proportional to the diffusing distance. Put simply, the thinner the membrane, the faster the diffusion, and so the barrier between food and blood is as thin as possible, being as little as a few micrometres – the thickness of the epithelial cell and the cell lining the capillary.

The small intestine also has another feature of note: the epithelial cells contain many mitochondria which make the ATP needed to power the active transport mechanisms.

6 EGESTION

Egestion of undigested food from the body occurs after it has been through the large intestine. This is about 1.5 m long and extends from the ileum to the anus. It has four sections, the **caecum**, the **colon**, the **rectum** and the **anus**. The large intestine:

- absorbs water;
- is the site of manufacture of certain vitamins (microorganisms within the colon of some animals produce vitamin K and folic acid);
- forms and expels undigested food residue as faeces.

The caecum receives material from the ileum. In humans, the bottom of the caecum is attached to a small blind tube, the **appendix**. This tube is twisted and coiled and is about 8 cm long. The human appendix plays no part in digestion, but can become inflamed, causing appendicitis.

Much of the water that is poured on to food during its passage through the digestive tract (as saliva, gastric juice, pancreatic juice, for example) is reabsorbed. This reabsorption is vital: the digestive system pours up to 8 litres of fluid into the gut each day. If we lost all this fluid we could dehydrate. Most water absorption occurs in the small intestine, but of the litre or so that enters the large intestine, all but 100 cm³ is reabsorbed by the colon. Minerals (as ions) can also diffuse or be actively transported into the bloodstream from this segment of the large intestine.

Diseases such as cholera and dysentery that cause severe diarrhoea can result in potentially lethal dehydration within a very short time because the toxins produced by the bacteria that cause these diseases reverse the reabsorption of water. To combat this, oral rehydration therapy can be given, a mixture of glucose and salts in water – see the Science Feature box on page 232.

The final sections of the large intestine, the rectum and anus, are concerned with the compaction of faeces and the act of **defecation**.

? TEST YOURSELF

F Why are the nutrients produced by colon bacteria (in the large intestine) of less use to us than nutrients provided by bacteria in the small intestine?

✔ REMEMBER THIS

Faeces contain some water, mineral salts, undigested food, bile pigments, products of bacterial decomposition and epithelial cells that have become detached from the digestive tract.

SUMMARY

- The human **alimentary canal** is a tube that leads from the mouth to the anus. It is associated with glands such as the **salivary glands**, **pancreas** and **liver**.
- The alimentary canal is also called the **gut** or **digestive system**. It is responsible for the **ingestion**, **digestion**, **absorption** and **egestion** of food.
- Food is broken down **mechanically** by the chewing action of teeth and by the muscular churning of the alimentary canal.
- Digestive juices are poured on to food as it travels down the alimentary canal. These soften and lubricate the food and also contain **digestive enzymes** that break the food down **chemically**.
- The muscles of the alimentary canal 'squeeze' the food – a process called **peristalsis** – and push it along from one end to the other.

- Large food molecules are broken down to their building blocks: carbohydrates are digested to form simple sugars, proteins are broken into amino acids, and fats are converted to fatty acids and glycerol.
- Digestion of food in the small intestine is due to the actions of bile and pancreatic juice, and the process is completed on the surface of the epithelial cells themselves.
- Digested food products are absorbed into the body across the wall of the alimentary canal. Most enter the bloodstream, but the products of fat breakdown enter the **lymphatic system**.
- Most digestion occurs in the **duodenum**; most absorption occurs in the **ileum**.
- A sophisticated control system ensures that digestive enzymes are released in the right place at the right time.
- Water is absorbed from undigested food that enters the colon, turning it into **faeces** that are egested from the body through the **anus**.

? EXAM QUESTIONS

1 The diagram shows a section through the stomach wall.

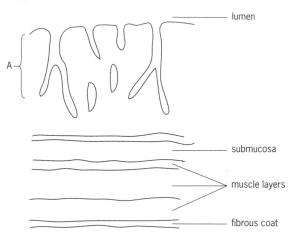

Name the part labelled **A**.
Give **two** ways in which the structure of the wall of the stomach differs from the wall of the oesophagus.
Give **one** way in which the stomach wall is adapted:
(i) to churn food;
(ii) to make food poisoning by bacteria less likely;
(iii) to prevent enzymes produced in the stomach wall digesting the surface of the stomach.

AQA Biology (B) BYB1 Core Principles June 2001 Q1

2 Read through the following account of the digestion of carbohydrates by humans, then copy it out and insert the most appropriate word or words to complete the account.
One of the enzymes present in pancreatic juice is ___ which hydrolyses starch to form ___ . Disaccharides are hydrolysed to their constituent monosaccharides by various enzymes secreted by cells in the ___ . These enzymes include sucrase, which hydrolyses sucrose into a mixture of ___ and ___ .

Edexcel B2 June 1999 Q3

3 Read through the following passage, which refers to hormonal coordination of the functioning of the alimentary canal in humans, and then copy it out and insert the most appropriate words or words to complete the account.
The presence of partially digested food in the stomach causes the secretion of the hormone ___ which stimulates the secretion of gastric juice. On entering the duodenum, the acidic chime stimulates the secretion of ___ . This travels in the ___ and causes the release into the duodenum of an alkaline pancreatic secretion. The release of an enzyme-rich pancreatic secretion is caused by the hormone ___ which also causes contraction of the ___ .

Edexcel HB3 June 1999 Q4

4
a) Gastric juice is released by glands in the stomach lining and has an important role in digestion.
(i) Describe the processes involved in stimulating the release of gastric juice.
(ii) Describe the role of gastric juice during digestion.
b) Explain how the small intestine is adapted to its function in the absorption of the products of digestion.

AQA (NEAB) BY03 Physiology June 2000 Q9

09

Gas exchange

9 Gas exchange

Diseases of the lung such as asthma, bronchitis and emphysema can be investigated at first hand thanks to the endoscope, a camera on the end of a flexible tube. This technology allows doctors to investigate parts of the body – such as the lungs, the stomach and the intestines – without the need for surgery. Not only does it relay enlarged pictures to a TV screen, but tools on the end of the device can take samples of tissue (such as tumours), remove blockages and even perform small operations.

Having inserted an endoscope into a person's airway, the doctor's first decision is whether to turn into the right or left lung. Here, he or she could look for obstructions, tissue damage or tumours. In practice, an endoscope can be used only for the larger branches, the trachea and the bronchi. If the camera were small enough to go right into a terminal alveolus, the endoscope operator would face another 23 decisions of left or right, such is the extent of the branching of the airways.

In total, there are about 2^{23} terminal bronchioles that lead from the trachea to the alveoli of the human lung. The number of alveoli is even larger, about 300 million. The combined surface area for gas exchange is huge, about the size of a badminton court.

An endoscope picture that reveals the top of a Martini stirrer stuck at the top of a child's airway

1 GAS EXCHANGE

Humans, like all living organisms, need energy to carry out the processes of life. In particular, we need energy for movement, growth and repair and to keep our body temperature stable. We obtain this energy from the oxidation of organic molecules such as glucose in the process of **respiration** (see Chapter 3). Respiration uses glucose and oxygen and produces carbon dioxide, water and energy in the form of adenosine triphosphate (ATP). This goes on constantly in every living cell of our bodies and, to keep the process going, we must obtain a constant supply of oxygen and expel the carbon dioxide as waste. We humans are also large warm-blooded organisms, and so we need a massive amount of oxygen to supply our metabolic demands.

Fortunately, air contains a lot of oxygen, and it is relatively easy for us to move a large amount of air into and out of our bodies in a short time. Our highly developed lungs are adapted to bring enough air sufficiently close to the blood to allow rapid diffusion to take place.

In this chapter we look at the structure of the lungs, the mechanism of ventilation, the process of gas exchange and, finally, the way in which our breathing rate changes according to the varying demands of the body.

✔ REMEMBER THIS

A summary of the overall process of respiration:

glucose + oxygen

↓

carbon dioxide + water + energy
(in ATP)

2 STRUCTURE OF THE LUNGS

The intestines, kidneys and lungs, organs that are adapted for exchange of materials, all show several features that maximise the exchange process. The **lungs** are responsible for exchanging gases between the blood and the air outside the body. Oxygen enters the body by diffusion, and carbon dioxide leaves in the same way. The lungs increase the efficiency of diffusion by having:

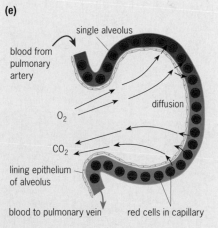

Fig 9.1 The structure of the human gas exchange system

(a) The overall structure

(b) A resin cast of human lungs

(c), **(d)** and **(e)** The fine structure of the alveolus and the mechanism by which gas is exchanged

- **A large surface area**. The huge surface area for gas exchange is the result of many little air pockets called alveoli, which are surrounded by many blood capillaries.
- **Thin, permeable walls**. The gases have only to pass through two thin cell layers, thus ensuring that the diffusing distance is as short as possible.
- **A good blood supply**. The lungs receive as much blood per minute as the rest of the body put together. Blood whisks incoming oxygen away quickly to a different part of the body. This ensures that a diffusion gradient is always present: there is always less oxygen in the blood that flows through the lungs than there is in the air.
- **Efficient ventilation**. This is a physical pumping mechanism, usually called breathing, that continually brings fresh air to the gas exchange surfaces in the alveoli and keeps the diffusion gradient as large as possible.

✓ **REMEMBER THIS**

Alveoli don't collapse when we breathe out because their surface is covered by an anti-sticking chemical called surfactant (see the Science Feature box on page 171).

Fig 9.2 The mechanism of breathing
(a) Inspiration (breathing in) happens when the external intercostal muscles contract and pull the ribcage upwards and outwards, away from the spinal column. At the same time, the diaphragm contracts and flattens, pushing down on the abdominal organs such as the liver. These movements increase the volume and therefore lower the pressure in the thorax. As the pressure in the thorax falls below that of the atmosphere, air is forced into the lungs to equalise the pressure
(b) Expiration (breathing out) is normally a passive process: it uses no energy. When inspiration is over, the muscles in the thorax relax and breathing out follows due to a combination of gravity, the elastic recoil of the connective tissues of the lungs and the pressure exerted by the abdominal organs. We can consciously speed up expiration by forcing air out of our lungs using our internal intercostal muscles. This happens, for instance, when we blow up a balloon or play a wind instrument
(c) The pressure changes associated with breathing

3 MECHANISM OF VENTILATION

The lungs themselves are spongy – they cannot move on their own – so how do we breathe? Lungs **inflate** and **deflate** because they are *expanded* and *compressed* by movements of the ribcage and diaphragm. This is possible because two **pleural membranes** hold the lungs to the ribcage and diaphragm. The outer membrane lines the **thoracic cavity**: the inner membrane encloses the lungs. Between the two is a very narrow space, the **pleural cavity**, filled with **pleural fluid**. This allows the pleural membranes to slide smoothly over each other during breathing, and prevents them from separating. Fig 9.2 shows how we breathe in and out.

To understand why two pleural membranes continue to stick together during breathing, imagine two wet pieces of glass pressed together. They can easily slide over each other, but it is virtually impossible to pull them apart without introducing air into the middle. If air is introduced into the pleural cavity, after a stab wound for example, the lung collapses, a situation known as a **pneumothorax**. The pleural cavity is filled with a foamy mixture of air and blood, while the elastic lungs shrink back from the thorax wall. The ribs can still move (though it is painful) but they don't inflate the lungs.

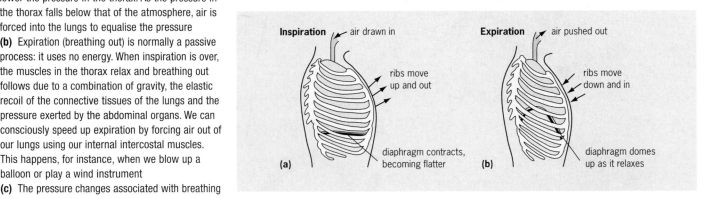

Inspiration → air drawn in
ribs move up and out
diaphragm contracts, becoming flatter
(a)

Expiration → air pushed out
ribs move down and in
diaphragm domes up as it relaxes
(b)

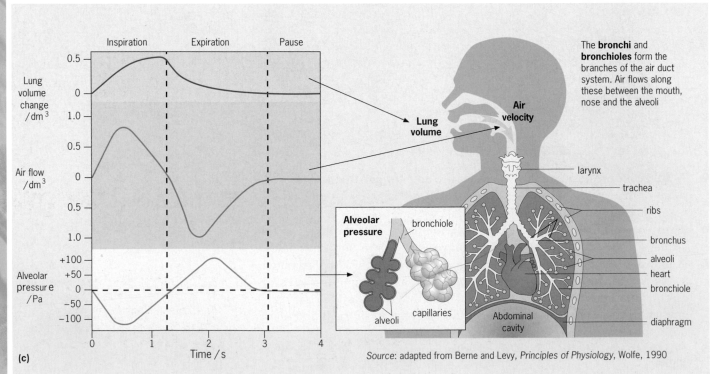

The **bronchi** and **bronchioles** form the branches of the air duct system. Air flows along these between the mouth, nose and the alveoli

Lung volume

Air velocity

Alveolar pressure
bronchiole
alveoli
capillaries

larynx
trachea
ribs
bronchus
alveoli
heart
bronchiole
Abdominal cavity
diaphragm

(c)

Source: adapted from Berne and Levy, *Principles of Physiology*, Wolfe, 1990

USING A SPIROMETER TO DETERMINE LUNG VOLUME

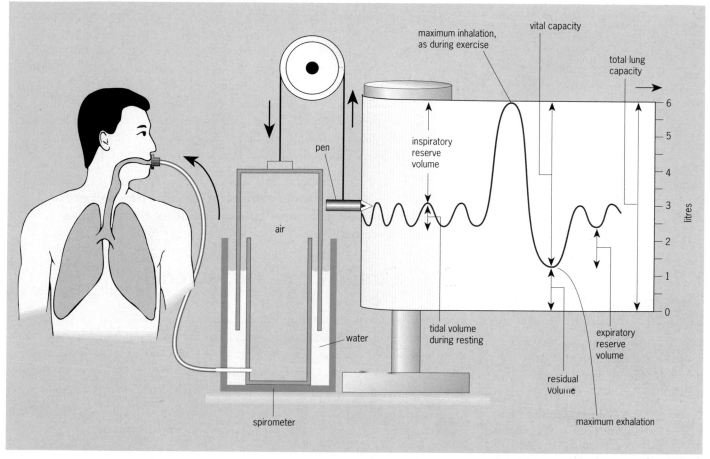

Fig 9.3 A spirometer and the trace it produces

Fig 9.3 shows a **spirometer**, a device for measuring the volumes of air that a person breathes in and out. The trace that this apparatus produces tells us a lot about lung volume.

First of all, your lungs have a **total capacity**, the maximum amount of air that the lungs can hold during the deepest possible breath. This is not the total volume of the lungs, however, because the lungs are never totally emptied. Even when you have exhaled as far as possible, there is still some air in the alveoli, and in the bronchi and trachea, which are held open permanently by rings of cartilage. The volume of air that remains in the lungs after breathing out is called the **residual volume**. The maximum usable lung volume (the total lung capacity minus the residual volume) is called the **vital capacity**. So if you breathe in as far as you can, then exhale as much as you can, that's your vital capacity. The average vital capacity is 4.5 to 5 litres for men and 3.5 to 4 litres for women.

Secondly, during normal breathing, the volume of air that moves in and out of the lungs in each breath is called the **tidal volume**. In a normal adult at rest, this is about 0.5 litres.

Thirdly, the trace shows that after breathing in at rest, the person could inhale an extra 3 litres, a value known as the **inspiratory reserve volume (IRV)**. He or she could also breathe out another litre or so: the **expiratory reserve volume (ERV)**. These are the extra volumes of air that we breathe in and out during exercise.

Finally, we can use the trace to work out the **ventilation rate**: the volume of air taken into the lungs in one minute. To do this, multiply the number of breaths taken in a minute by the tidal volume.

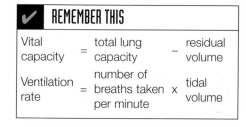

✓ REMEMBER THIS

Vital capacity = total lung capacity − residual volume

Ventilation rate = number of breaths taken per minute × tidal volume

SEE QUESTION 1

? TEST YOURSELF

A If the average breathing rate is 15 breaths per minute, and the tidal volume is 0.5 litre, calculate the ventilation rate.

Fig 9.4 In addition to lung volume, the spirometer can also be used to measure the volume of oxygen used. If the carbon dioxide is absorbed as soon as it is exhaled, the total volume of air falls, showing how much oxygen is being used. We can use the measurements to determine how much oxygen is used at rest, and to see how this changes with exercise

❓ TEST YOURSELF

B The alveolar membrane is 0.2 μm thick. What fraction of a millimetre is that?

(a)

Fig 9.5 (a) In this micrograph of mammalian lung tissue, the alveoli and capillary networks that surround them can clearly be seen
Fig 9.5 (b) Gas exchange at the alveolar surface. In the human lung, the alveolar endothelium, the membrane that lines the air-sacs, is only 0.2 μm thick. This minimises the distance oxygen has to diffuse to enter the blood. Breathing maintains a fresh supply of air and the circulatory system takes away oxygenated blood so that constant gas exchange is possible

SEE QUESTION 2.

USING A SPIROMETER TO DETERMINE OXYGEN CONSUMPTION

We can use a spirometer to estimate a person's rate of oxygen consumption. If the person re-breathes the air in the closed system of the spirometer, the composition of that air changes: the oxygen level decreases and the carbon dioxide level increases. Soda lime placed in the apparatus absorbs the carbon dioxide and so the volume of air in the spirometer decreases in volume as the oxygen is used up. Fig 9.4 shows a trace produced in this way.

To calculate the amount of oxygen used, we measure the volume decrease in a given time. In our example, the trace shows that the air volume fell from the 1600 cm^3 mark to the 1300 cm^3 mark in one minute. So, this person used up 300 cm^3 of oxygen in one minute.

WARNING!! Re-breathing your own air can be dangerous. You should never do investigations such as this without close supervision.

4 GAS EXCHANGE AT THE ALVEOLI

Gas exchange between air and blood occurs at the alveoli (Fig 9.5). These tiny air-sacs create a huge surface area: 1 cm^3 of frog lung tissue has a surface area of about 20 cm^2, but the corresponding figure for a mouse lung is over 800 cm^2. Other mammals have a similar value. The total surface area of one human lung is about 70 to 100 m^2.

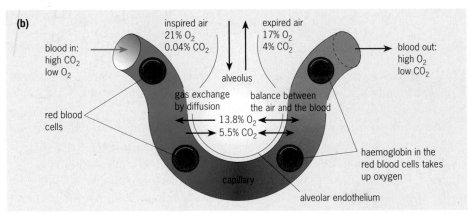

(b)

As we breathe in, fresh air enters the lungs and passes into individual alveoli. Oxygen diffuses rapidly through the walls of the alveoli and into the blood. Here, most of it combines with haemoglobin in red blood cells (see page 193). At the same time, carbon dioxide diffuses out of the blood and into the alveoli. It is breathed out during the next few expirations.

Table 9.1 shows the composition of atmospheric, alveolar and exhaled air. Each value for exhaled air is an average of the values for inhaled and alveolar air because exhaled air is a mixture of the two.

Table 9.1 The composition of inhaled, alveolar and exhaled air

| | Percentage of total volume | | | | |
	O_2	CO_2	N_2 + inert gases	H_2O vapour	Temperature (°C)
Atmospheric	21	0.04	79	variable	variable
Alveolar	13.8	5.5	80.7	saturated	37
Exhaled	17	4	79.6	saturated	37

> ⚠ **SCIENCE FEATURE** The lungs of premature babies

Alveoli are minute bubble-like air-sacs lined with moisture, and are liable to collapse because of surface tension. To prevent this, the alveolar epithelium secretes a surfactant, a mixture of phospholipids, which greatly reduces the surface tension and keeps the alveoli open.

Without surfactant the lungs cannot function effectively and severe breathing problems can develop. An unborn baby does not start to secrete surfactant until about the 22nd week of pregnancy and the lungs have not accumulated enough surfactant to cope with breathing until about the 34th week.

Any babies born before this have immature lungs and suffer from a condition called **respiratory distress syndrome**. The effort needed to inhale and inflate the collapsed alveoli becomes too great and, without medical help, the baby can die from exhaustion and suffocation. Surfactant can now be made artificially. It is introduced into the lungs of premature babies to help them to begin breathing. This major

breakthrough means that babies as young as 23 weeks (17 weeks premature) now have more of a chance of survival.

Fig 9.6 This baby was born prematurely, but survived owing to the treatment of her lungs with surfactant. Only a tiny amount (0.5 cm³) of surfactant is needed, enough to line the alveoli with a layer one or two molecules thick

The composition of exhaled air varies during the course of a single expiration. The first air to emerge has a very similar composition to atmospheric air because it has been nowhere near the alveoli – it has simply filled the **dead space** in the trachea and bronchi. As the exhalation continues, air that has been deep inside the alveoli is breathed out. The composition of this air has been altered by gas exchange: it contains more carbon dioxide and less oxygen than atmospheric air.

Fig 9.7 The air we exhale is saturated with water vapour. When you wake up on a cold day to see condensation on the windows, it is likely that a significant amount of that water came from your lungs

> ❓ **TEST YOURSELF**
>
> **C** Why would it be impractical to breathe under water using a 6 metre snorkel?

THE EFFECT OF SMOKING ON GAS EXCHANGE

Smoking is the largest cause of preventable death in the western world, and is responsible for about 100 000 deaths per year in the UK. In addition to lung cancer and coronary heart disease (see Chapter 31), smoking causes bronchitis and generally reduces the gas exchange capacity of the lungs.

Research has shown that there are two distinct types of smoke. **Mainstream** smoke passes into the smoker through the filter, **sidestream** smoke comes out of the burning end of the cigarette. In a room of smokers, about 85 per cent of the smoke is sidestream, which means that passive smokers don't get the benefit of the filter. Cigarette smoke contains three substances that have a direct effect on the lungs: nicotine, carbon monoxide and tar.

Nicotine is not the most harmful chemical in smoke, but it is highly addictive, and has a powerful effect on the nervous system (see page 273). It is thought to increase the stickiness of blood platelets, increasing the risk of blood clots forming inside arteries and veins. If these clots break away they can cause a heart attack or a stroke (see Chapter 31).

Fig 9.8 The lungs have an efficient cleaning system: goblet cells (orange) secrete mucus that traps dust, while a carpet of cilia (tiny hairs, green) waft the mucus to the throat where it is swallowed (or to the nose, from where it is blown). Smoking disrupts this cleaning mechanism, and often leads to chronic bronchitis

! SCIENCE FEATURE Climbing Mount Everest — an explanation of partial pressures

As you study the workings of the lungs and blood in more detail, you will come across the term partial pressure. To explain this, it helps to think about someone climbing Mount Everest (height 8848 metres).

At sea level, there is a lot of air pushing down on us, and this atmospheric pressure has a value of about 100 000 pascals (Pa) or 100 kPa. We can therefore say that the barometric pressure is 100 kPa. Dry air is 20.9 per cent oxygen at sea level so the partial pressure of oxygen (P_{O_2}) is 20.9 per cent of 100, which is about 20.9 kPa.

As our mountaineer progresses up the mountain, the atmospheric pressure becomes less because there is less air pushing down on him. The partial pressure of oxygen decreases accordingly (Fig 9.9).

Oxygen passes into the lung tissues because of the different concentration in the alveolar air and the blood, and at higher altitudes the difference is smaller, making it difficult to take in enough oxygen to meet demands. At 5300 metres the P_{O_2} is only half of that at sea level, and by the time the summit

is reached it is only about 8 kPa – just over one-third that at sea level. This is why most mountaineers who attempt Everest do so with pressurised oxygen containers, and to conquer the mountain without the help of additional oxygen – as has been done – is a remarkable feat.

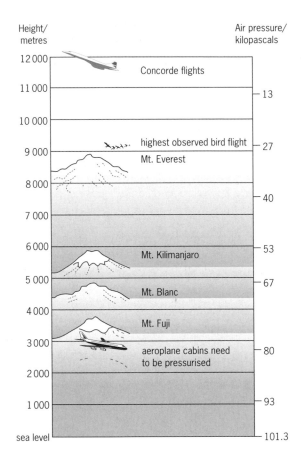

Fig 9.9 The effect of altitude on the partial pressure of oxygen. Aircraft cabins must be pressurised and mountaineers usually need the help of pressurised oxygen canisters. The human body is able to acclimatise to lower partial pressures of oxygen by increasing the amount of haemoglobin in the blood – see the Chapter 10 Assignment for more details

Carbon monoxide binds to haemoglobin molecules, preventing them from carrying oxygen. Smokers therefore have a lower oxygen-carrying capacity than non-smokers.

Tar is a complex mixture of chemicals that paralyses the cilia but stimulates the goblet cells (Fig 9.8), resulting in a build-up of mucus that traps dust and pathogens. This results in 'smoker's cough' and increases the smoker's chance of developing chronic bronchitis.

In the long term, the epithelia become damaged and replaced by scar tissue, cilia are destroyed, and some of the smaller airways become blocked by mucus. Chronic bronchitis develops with its persistent cough that produces large quantities of phlegm (mucus, bacteria and white cells). Tar also contains carcinogens and heavy smokers have a greatly increased chance of developing lung cancer.

5 THE CONTROL OF BREATHING

Control of breathing is **involuntary**: we don't have to think continually about breathing in and out, and our breathing rate is automatically matched to our needs. Breathing rate changes as our bodies detect the physical and chemical variations that occur when we carry out different activities.

SETTING A REGULAR PATTERN

Breathing is controlled by the **respiratory centre**. This is located in the brain in an area called the **medulla oblongata** (Fig 9.10).

Regular nerve impulses travel down **efferent** (outgoing) nerves that pass from the respiratory centre out to both the external intercostal muscles and the diaphragm. These muscles then contract, initiating inhalation. As air enters the lungs, stretch receptors in the airways start firing and feed information to the brain about how the inflation of the lungs is progressing. The more the lungs inflate, the faster the stretch receptors feed back impulses. When the lungs are sufficiently inflated, signals from the respiratory centre stop for a short time and exhalation follows automatically.

CHANGING BREATHING RATE TO MEET DEMAND

Ensuring that the body has a constant supply of oxygen is an important aspect of homeostasis (see Chapter 11). But, surprisingly, the body is relatively insensitive to falling oxygen levels. It is much more sensitive to an increase in carbon dioxide and so this is a much more reliable indicator of the need for oxygen. The levels of oxygen in the arterial blood vary little, even during exercise, but the carbon dioxide levels vary in direct proportion to the level of exertion. The heavier the exercise, the greater the carbon dioxide concentration of the blood. Lactate levels also increase during exercise. Any increase in carbon dioxide or lactate concentration in the blood lowers its pH.

SEE QUESTIONS 3 AND 4

> ✔ **REMEMBER THIS**
>
> The terms 'lactic acid' and 'lactate' are often used interchangeably. In water, lactic acid forms lactate ions and hydrogen ions (H^+), so in this chapter we refer to lactate. The same goes for other organic acids, such as pyruvic acid/pyruvate.

> ✔ **REMEMBER THIS**
>
> Together, the carotid and aortic bodies are known as the peripheral chemoreceptors. These cells sense changes in carbon dioxide and pH levels and, to a lesser extent, they are also sensitive to changes in oxygen levels.

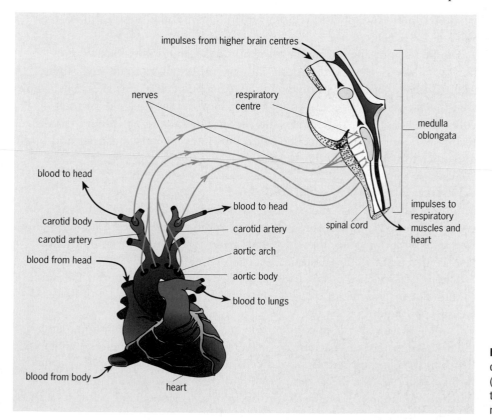

impulses from higher brain centres

nerves

respiratory centre

medulla oblongata

blood to head

blood to head

carotid body

carotid artery

carotid artery

aortic arch

blood from head

aortic body

spinal cord

impulses to respiratory muscles and heart

blood to lungs

blood from body

heart

Fig 9.10 Regulation of the levels of oxygen and carbon dioxide in the blood. Chemoreceptors (carbon dioxide/pH sensitive cells) occur in both the carotid and aortic bodies, as well as in the respiratory centre itself

Fig 9.11 In 1875, three French physiologists decided to investigate the effects of low oxygen levels on the human body. The easiest way to get oxygen-poor air was to go up in a hot-air balloon to observe the effects of the rarefied atmosphere on each other. At first there were no obvious effects and they happily continued to throw out ballast, going up to 8000 metres. At this point they all fainted. The balloon eventually came down on its own, and one of the scientists woke up to find the other two dead

SEE QUESTION 5

Chemoreceptors (see Fig 9.10) are extremely sensitive to the composition of the blood that flows past them, and can detect very small changes in pH. They occur at three sites:

- **Central receptors** in the medulla oblongata. These are sensitive to the carbon dioxide concentration in the blood that flows through this region of the brain.
- The **carotid bodies** in the wall of the **carotid artery**.
- The **aortic bodies** situated on the **aortic arch**, just above the heart.

When chemoreceptors register a change in carbon dioxide level or pH, they send nerve impulses to the respiratory centre in the brain. This responds by sending more frequent impulses to the external intercostal muscles and diaphragm. When this happens, our ventilation rate increases: you breathe harder and faster. Heart rate also increases (see Chapter 10) and so the body automatically increases oxygen delivery at the same time as removing the extra carbon dioxide.

The control of breathing rate is very similar to the control of heart rate (see page 186) with one important difference: we can control our breathing rate by thinking about it. This suggests that the higher, 'conscious' centres of the brain are more closely linked to the respiratory centre than they are to the cardiovascular centre, the part of the brain that controls heart rate. Also, research shows that pulse and ventilation rate change dramatically during exercise, even before the concentration of blood gases has a chance to change. It is as if the body predicts what is about to happen. How this works is not understood and is an active area of research.

THE EFFECTS OF OXYGEN DEPRIVATION

In some situations, at high altitudes for example, oxygen levels can fall without carbon dioxide levels increasing. In a rarefied or artificial atmosphere, normal breathing flushes carbon dioxide out of the lungs but there may not be enough oxygen to replace it. When this happens, the chemoreceptors often fail to register a problem, and the brain can become starved of oxygen.

The first symptoms of oxygen starvation are feeling ridiculously happy, having impaired senses and lacking judgement. When mountaineers, fighter pilots or deep-sea divers start giggling and making stupid mistakes, it is a sure sign that they are not getting enough oxygen. It is also a signal for their colleagues to act fast, if they can, and provide them with emergency oxygen. If they don't get help quickly, they soon lapse into unconsciousness, and brain damage and death follow (Fig 9.11).

ATHLETES AND V_{O_2}(MAX)

Many physical activities, such as jogging, swimming and team sports, rely on energy released by the aerobic pathway of cell respiration (see Chapter 34). The level of performance an athlete can achieve is largely governed by how fast oxygen gets to the muscles.

The rate at which a person uses oxygen is called the V_{O_2} and is measured in terms of the volume of oxygen consumed (cm^3), per kilogram of body weight, per minute.

The V_{O_2}(**max**) is the maximum rate at which oxygen is consumed and is the amount of oxygen that can be delivered to the tissues when the lungs and heart are working as hard as possible. Athletes use a knowledge of V_{O_2}(max) in their training, as a measure of how hard they are working. A training schedule, for example, might require an athlete to work at 65 per cent of their V_{O_2}(max) for a set length of time.

The Example shows how to calculate V_{O_2} and $V_{O_2}(\text{max})$ for an average person. (You can work out your own, if you are able to measure your tidal volume, the number of breaths you take per minute and your weight in kg.) You can see that the amount of oxygen consumed during exercise increases tenfold or 1000 per cent from 4.28 to 51.42 cm^3 O_2 kg^{-1} min^{-1}.

● EXAMPLE

Q What is the rate of oxygen consumption of a normal 70 kg adult at rest? Assume that the tidal volume of a normal adult at rest is 0.5 litres and that he or she takes 15 breaths per minute.

A We can work out the ventilation rate from the formula:

$$\text{Ventilation rate} = \frac{\text{number of breaths}}{\text{taken per minute}} \times \text{tidal volume}$$

$$\text{Ventilation rate} = 15 \times 0.5 \text{ litres min}^{-1} = 7.5 \text{ litres min}^{-1}$$

Atmospheric air is one-fifth oxygen and from Table 9.1 we know that only one-fifth of the available oxygen is absorbed. So:

amount of oxygen used per minute

$$= 7.5 \times 0.2 \times 0.2 \text{ litres min}^{-1}$$

$$= 0.3 \text{ litres min}^{-1}$$

And the V_{O_2} for this person is:

$$\frac{0.3}{70} = 0.004\,28 \text{ litres } O_2 \text{ kg}^{-1} \text{ min}^{-1},$$

or: 4.28 cm^3 O_2 kg^{-1} min^{-1}

Q What is the $V_{O_2}(\text{max})$ for the same adult? Assume that the volume of air taken in during each breath during strenuous exercise is 3 litres and that the number of breaths per minute increases to 30.

A Knowing that the tidal volume of the same adult during exercise is 3 litres and that he or she takes 30 breaths per minute, we can work out the ventilation rate:

$$\text{Ventilation rate} = \frac{\text{number of breaths}}{\text{taken per minute}} \times \text{tidal volume}$$

$$\text{Ventilation rate} = 30 \times 3 \text{ litres min}^{-1} = 90 \text{ litres min}^{-1}$$

Atmospheric air is one-fifth oxygen and from Table 9.1 we know that only one-fifth of the available oxygen is absorbed. So:

amount of oxygen used per minute

$$= 90 \times 0.2 \times 0.2 \text{ litres min}^{-1}$$

$$= 3.6 \text{ litres min}^{-1}$$

And the $V_{O_2}(\text{max})$ for this person is:

$$\frac{3.6}{70} = 0.051\,42 \text{ litres } O_2 \text{ kg}^{-1} \text{ min}^{-1},$$

or: 51.42 cm^3 O_2 kg^{-1} min^{-1}

SUMMARY

After reading this chapter, you should know and understand the following:

- We must **exchange gases** to provide oxygen for cell respiration, and to take away the waste carbon dioxide that this process produces.
- Lungs are **respiratory organs** that increase the surface area for gas exchange. These organs have a large surface area and thin membranes. Ventilation and a good blood supply help to maintain a high diffusion gradient.
- Our breathing movements serve to ventilate air-sacs called **alveoli**, bringing in a continuous supply of fresh oxygen and flushing out waste carbon dioxide.
- The lungs cannot move on their own: they inflate when pulled outwards by the **external intercostal muscles** and the **diaphragm**. The lungs are attached to the inside of the ribcage by attraction between the two pleural membranes and the pleural fluid between them.

- At rest, our lungs inflate mainly by movement of the diaphragm. The amount of air moved in and out at rest is called the **tidal volume** and the **ventilation rate** can be worked out by multiplying the tidal volume by the number of breaths taken per minute. The total amount of air that we can inhale, after exhaling as much as possible, is called the **vital capacity**.
- The rate of breathing is controlled by the **respiratory centre** in the **medulla oblongata** of the brain. This sets a regular breathing pattern that is modified according to information received from **chemoreceptors**. These detect the rise in carbon dioxide in the blood and the fall in pH that indicate the body's need for oxygen.

? EXAM QUESTIONS

1 The graph shows some of the changes which take place in a man during breathing.

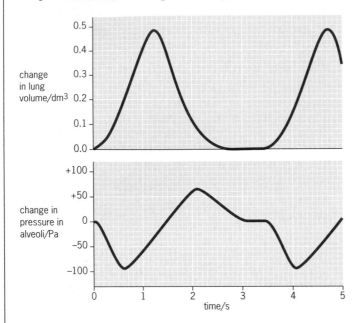

a) (i) Use the information in the graph to calculate this man's rate of breathing in breaths per minute. Show your working.

(ii) The volume of air in the man's lungs after he had exhaled was 2400 cm³. What is the volume of air in his lungs immediately after he had inhaled?

b) Explain how muscles cause the change in alveolar pressure between 0 and 0.5 s.

The table shows the composition of atmospheric air and gas samples obtained while this man was breathing normally.

Sample	Percentage O_2 of total volume	Percentage CO_2 of total volume
Atmospheric air	21.0	0.0
Air from alveoli	13.8	5.5
Exhaled air	16.9	4.1

c) Explain why the percentage of carbon dioxide in atmospheric air is given in the table as zero.

d) (i) Not all the air that a person breathes out has been in the alveoli. Explain why.

(ii) Use your answer to part **(i)** to explain the difference in carbon dioxide concentration between exhaled air and air from alveoli.

e) Describe how the structure of the lungs and the red blood cells enable efficient diffusion and transport of oxygen.

AQA Biology (A) BYA1 Molecules, Cells and Systems June 2001 Q8

2 a) In the lungs oxygen passes from the alveoli to the blood.
Describe and explain the features that make this process rapid and efficient.

b) The graph shows the pressure changes in the lungs during the period of one breath (inspiration + expiration) in a person at rest.

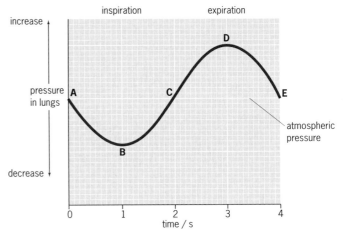

Use your knowledge of breathing to explain the changes in pressure during inspiration and expiration. The letters on the graph are to help you to refer to different parts of the curve.

AQA Biology (B) BYB1 Core Principles January 2001 Q7

3 The diagram shows some of the nerve pathways involved in the control of ventilation in humans.

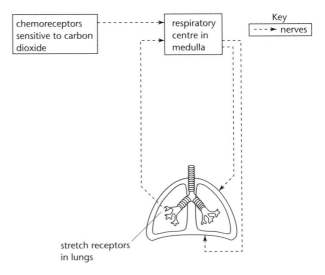

a) Give **one** place where chemoreceptors sensitive to carbon dioxide are found.

b) Describe the role of the stretch receptors in the lungs in the maintenance of breathing.

? EXAM QUESTIONS

c) A person was asked to hyperventilate while his pattern of breathing was being recorded. The trace shows the changes in the pattern of breathing before, during and after hyperventilation.

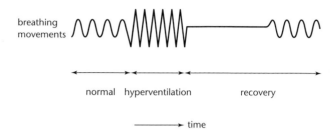

breathing movements

normal   hyperventilation   recovery

time

(i) Describe how the pattern of breathing changed when the person was hyperventilating.
(ii) Suggest an explanation for the long delay after hyperventilation before breathing started again.
AQA Biology (B) BYB3 Physiology and Transport June 2001 Q5

4 a) Both muscle and bones are made up of cells. Explain why muscle cells may use oxygen at a faster rate than bone cells do.
b) Describe what causes air to enter the lungs.
c) Describe the role of the nervous system in:
(i) maintenance of regular breathing whilst the body is at rest;
(ii) increasing the rate of ventilation of lungs during exercise.
NEAB Biology: Human Systems Module Test June 1998 Q9

5 At high altitudes, atmospheric pressure is lower than at sea level. This means that people living at high altitude have less oxygen available to them compared with those living at low altitude. The table below shows the results of three measurements made on blood of normal individuals resident at each of three different altitudes.

Altitude/m	Percentage saturation of arterial blood with oxygen	Oxygen content of arterial blood/ cm^3 per 100 cm^3	Haemoglobin in blood/g per 100 cm^3
150	94.2	18.3	15.5
3700	88.2	19.9	17.4
4375	83.5	20.9	18.6

a) (i) How does the percentage saturation of arterial blood with oxygen change with increasing altitude?
(ii) Using the information in the table, explain how people living at high altitude have adapted to the low oxygen availability.
(iii) State one feature, other than those shown in the table, which enables high altitude residents to overcome the problem of low availability of oxygen.
b) State three environmental factors, other than availability of oxygen, which vary between sea level and high altitude sites at the same latitude.
c) A visitor to high altitude from sea level may suffer from mountain sickness. State three symptoms of mountain sickness.
Edexcel Human Biology Module Test HB2 June 1998 Q6

KEY SKILLS **ASSIGNMENT**

ASTHMA

Asthma is a common childhood ailment, affecting at least one in ten children, and a large number of adults. The number of asthma cases has risen dramatically in recent years, fuelling speculation that an increase in air pollution is to blame.

To fully understand why asthma happens, you need to be familiar with the fine structure of the lungs – see page 167.

1 **a)** What is the difference in structure between the bronchi and terminal bronchioles?
b) What type of muscle lines the bronchi and terminal bronchioles?

What is asthma and what causes it?

Put simply, asthma is a difficulty in breathing caused when the smooth muscles of the bronchioles contract, narrowing the airways that lead to the alveoli. Asthma sufferers then find it hard to breathe and have to make far more effort to deliver a normal amount of air to the lungs.

Many factors can cause asthma, including an allergic reaction (an 'over-reaction' of the immune system to a substance that, in non-sufferers, has no effect). Common allergens are house dust, fur, feathers and pollen.

2 Suggest what measures could be taken to reduce the problems caused by the house dust mite shown in Fig 9.A1.

Treating asthma

Several approaches are used together to help control asthma. A doctor's first priority with all but very young children is to explain what is happening inside the lungs. Sufferers will be

less frightened and can to come to terms with their problem. The worst physical effects of the condition are then controlled by a combination of drugs and prevention.

3 What advice would you give to a patient suffering from pollen-induced asthma?

Generally, two types of drugs are prescribed to asthma sufferers: **bronchodilators** and **steroids**. Bronchodilators give instant relief from chest tightness and wheezing because they contain chemicals which relax the bronchiole walls. Many asthma sufferers carry an inhaler around with them. It contains the hormone adrenaline which relieves asthma. This is very useful in the treatment of severe attacks, but is unsuitable for regular use.

4 **a)** What are the effects of the hormone adrenaline on the body? (You might need to look at page 320.)
b) Suggest what happens when an asthma sufferer takes an adrenaline-based drug.
c) Suggest why adrenaline is not suitable for use in long-term asthma treatment.

Steroids (see page 313) such as **Becotide**™ act by reducing the degree of inflammation of the bronchioles. Steroids are preventative: asthma sufferers take regular doses morning and night to reduce the problem of over-reaction. The amount of steroids taken is minimal but they are effective if taken according to instructions.

5 **a)** Suggest why it is an advantage to inhale steroids rather than inject them.
b) The girl in Fig 9.A2 is using an inhaler. To get the full benefit from each dose – and not just coat the inside of her mouth – she must learn the correct technique. Find out what this is.

6 Describe how you would you investigate the hypothesis that air pollution is linked with the measured increase in asthma cases.

Fig 9.A2 Using an inhaler is easy, once you have mastered the right technique. Many people have had the wrong technique for years, with the result that their inhalers are not as effective as they might be

Fig 9.A1 House dust mites, *Dermatophagoides farinae*, are normal inhabitants of our mattresses. Feeding on the large deposits of skin which accumulate there, the faeces they produce can cause an allergic reaction that leads to asthma in some people

10

The circulatory system

During open-heart surgery, the job of the heart and lungs is performed by a heart–lung machine. This sophisticated device also has the ability to lower the core temperature of the body by about 10 °C. This reduces the body's demand for oxygen, giving the surgeons approximately twice as long to perform the operation

When the blood vessels supplying the heart muscle become narrowed by fatty deposits, the heart can become starved of oxygen. The patient will suffer heart pains on exercise – called angina – and is very susceptible to heart attacks. Often, the only effective remedy is surgery.

One of the available operations – sometimes known as a 'cabbage' (CABG = coronary artery bypass graft) – takes a team of two or three surgeons about five hours to complete. The overall aim is to take pieces of vein from the patient's leg and graft them from the aorta, over the blocked artery and into the heart muscle itself. But how do you sew small blood vessels onto an organ that won't keep still?

The answer is that you don't. The heart has to be stopped and this is only possible due to a heart–lung machine – a device that can take blood from the patient, oxygenate it and return it to the body at the right temperature and pressure. Once this machine has been connected the heart can be stopped, and this is done simply by pouring iced water over it. Once the heart is still the surgeon can sew on the new pieces of blood vessel. When complete, the heart is re-started by means of two electrodes placed on the heart itself.

This standard operation can greatly improve the quality of life for patients, as well as prolonging their life expectancy.

1 WHY DO WE NEED A CIRCULATORY SYSTEM?

 REMEMBER THIS

The human circulatory system is an example of a **mass flow system**, in which large volumes of fluid are carried to all parts of the organism.

The human body contains specialised organs such as intestines, lungs and kidneys whose function is the exchange of materials such as digested food, oxygen and waste. However, these organs could not function without a **circulatory system**: a network of tubes filled with fluid that can deliver vital materials to all the cells of the body, and then take away their waste.

The circulatory system is so extensive that every living cell is only a few micrometres from a capillary. The blood that passes nearby provides the cell with the materials it needs, takes away its waste and generally bathes it in favourable conditions.

The human circulatory system has three components:
● A fluid, **blood**, which flows in the system, carrying materials around the body.
● A system of tubes or vessels – **arteries**, **veins**, **capillaries** – which carry the fluid.
● A pump, **the heart**, which keeps the fluid moving through the vessels.
In this chapter, we look at these three components in detail.

A DOUBLE CIRCULATION
Fig 10.1(a) shows that humans, like other mammals, have a double circulation: a **pulmonary circulation** that carries blood between the heart and the lungs and a **systemic circulation** that carries blood between the heart and the rest of the body.

To understand the advantage of a double circulation, you might find it helpful to consider the situation in fish (Fig 10.1b). Fish have a single circulation and a two-chambered heart. Blood first travels to the gills where it passes through thin systemic capillaries, picking up oxygen but losing pressure. Blood continues to travel around the body of the fish, back to the heart, but more slowly because it is at low pressure. This situation works for fish but is no use in warm-blooded animals such as ourselves; the slow single circulation would not be able to supply all our cells with sufficient amounts of the freshly oxygenated blood they need.

With a double circulation, the right side of the heart receives deoxygenated blood from the body and pumps it to the lungs. Here, the blood gains oxygen and again loses pressure. However, the blood returns to the heart, which gives it a boost so that it can reach all the body parts quickly. The volume of blood going around each circulation in any given time is the same, but the systemic circulation is pumped under greater pressure by the heart because the blood has to reach all the extremities of the body.

Fig 10.1 Double and single circulations compared **(a)** Humans have a double circulation and a four-chambered heart. Deoxygenated blood, coloured blue in the diagram, passes into the right side of the heart and is pumped to the lungs where it picks up oxygen and releases carbon dioxide. Oxygenated blood, coloured red, returns to the heart to be pumped to all parts of the body except the lungs **(b)** The single circulation of the fish is a simple circuit in which blood loses pressure as it passes through the gills and so moves slowly around the rest of the body

2 THE HUMAN CIRCULATORY SYSTEM

The human heart beats over 100 000 times a day, creating the pressure to force blood through more than 80 000 kilometres of arteries, veins and capillaries. Fig 10.2 illustrates the main vessels that make up the human circulatory system.

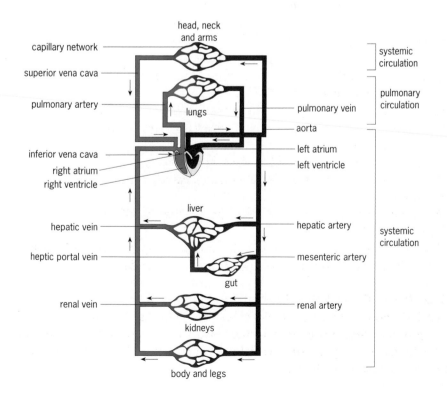

Fig 10.2 An overview of human circulation. Deoxygenated blood (coloured blue) passes into the right side of the heart through the superior vena cava and the inferior vena cava. The right ventricle pumps it into the pulmonary circulation and it picks up oxygen and releases carbon dioxide as it passes through the lungs.
Now oxygenated (coloured red), the blood returns to the heart. The left ventricle then pumps it around the systemic circulation and it travels to all parts of the body except the lungs

STRUCTURE OF THE HEART

The heart is composed mainly of **cardiac muscle**, a specialised tissue that contracts automatically, powerfully and without fatigue, throughout our lives (Fig 10.3).

The thickness of the walls in the different heart chambers reflects their function. The atria are thinly muscled: they pump blood the short distance to the ventricles directly below them. The right ventricle is more heavily muscled than either of the atria: it has to force blood a much further distance to the lungs. The left ventricle has the thickest wall: it has to push blood all around the body.

Fig 10.3(a) A longitudinal section of the heart
Fig 10.3(b) The heart is a slightly twisted, asymmetrical organ: the best way to appreciate its three-dimensional structure is to dissect a pig's or sheep's heart or to study a model

Fig 10.3(c) The coronary circulation contains all the blood vessels that supply the heart muscle. In the photograph, these blood vessels are easy to see because they have been filled with a dye that is opaque to X-rays
Fig 10.3(d) The fine structure of cardiac muscle. The rapid spread of impulses through cardiac muscle is possible because individual cells are connected by specialised junctions called intercalated discs. This system of communication ensures that a wave of contraction passes through the cardiac muscle, so that the chambers contract at the right time

It is important that blood flows through the heart in one direction only. Two sets of valves close to prevent backflow:

- The **atrio-ventricular valves** (**AV valves**) lie between the atria and the ventricles to prevent blood from returning to the atria when the ventricles contract. The **tricuspid valve** on the right side has three **cusps**, or flaps. The **bicuspid valve**, or **mitral valve**, on the left has two cusps. Both atrio-ventricular valves are subjected to great pressure and ultra-tough tendinous cords, arrowed in Fig 10.3(b), prevent them turning inside-out.
- The **semilunar** valves guard the openings to the pulmonary artery and the aorta and prevent backflow of blood into the ventricles. Both sets of valves have three semilunar ('half moon'-shaped) cusps.

HOW THE HEART BEATS

The sequence of events in a single heartbeat is known as the **cardiac cycle** (Fig 10.4). The cycle involves **systole**, or contraction, and also **diastole**, or relaxation, of the atria and ventricles. The cycle has four overlapping stages:

- **Atrial systole**. Both atria contract, forcing blood into both of the ventricles. This stage lasts 0.1 seconds.
- **Ventricular systole**. Both ventricles contract, forcing blood through the pulmonary artery to the lungs and through the aorta to the rest of the body. This takes 0.3 seconds.
- **Atrial diastole**. The atria relax, although the ventricles are still contracted. Blood enters the atria from the large veins coming from the body. This takes about 0.7 seconds.
- **Ventricular diastole**. The ventricles relax, and become ready to fill with blood from the atria as the next cycle begins. This takes about 0.5 seconds.

Given an average heart rate of 75 beats per minute, each cycle takes 0.8 seconds.

> ✔ **REMEMBER THIS**
>
> Valves are simply strong flaps of tissue: they cannot move on their own. A common mistake is to credit valves with the active control of blood flow. As blood begins to flow back, the valve is forced shut and this prevents any further backflow.

> ✔ **REMEMBER THIS**
>
> The heart continues to beat when removed from the body. Individual heart muscle cells grown in culture beat on their own! Because of this, we say that the heart is **myogenic**: the stimulus which drives it to beat originates in the muscle itself. (Some animals have hearts that are **neurogenic**: they beat only when stimulated by external nerves.)

Fig 10.4 The timescale of the cardiac cycle at rest. During exercise the heart rate can increase to over 150 beats per minute, so the cycle only takes half the time

CONTROL OF THE CARDIAC CYCLE

Heartbeat must be carefully controlled so that each chamber contracts only when full of blood. To achieve this, the events of the cardiac cycle are carefully co-ordinated (Fig 10.5).

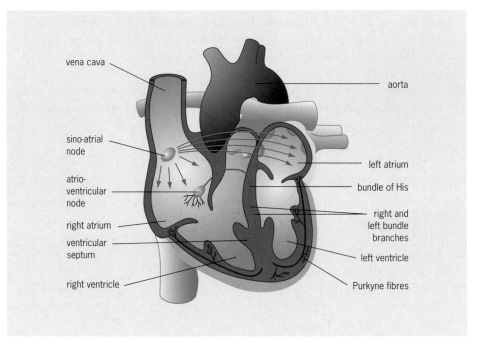

Fig 10.5 The conduction system of the human heart initiates and controls a heartbeat. Cells in the sino-atrial node act as a pacemaker, initiating impulses which spread through the walls of the atria and, after a delay, through the walls of the ventricles via the atrio-ventricular node

A single heartbeat starts with an electrical signal from a region of specialised tissue called the **sino-atrial node** (**SAN**), on the wall of the right atrium. This is the 'pacemaker', or heartbeat regulator. The electrical signal spreads out over the walls of the atria, causing them to contract.

From there, the signal does not pass directly to the ventricles. If it did, the ventricles would begin to contract before they had filled with blood. Instead, the impulse is delayed slightly. A second node, the **atrio-ventricular node** (**AVN**) picks up the signal and channels it down the middle of the **ventricular septum** through a collection of specialised cardiac muscle fibres called the **bundle of His**. From here the signal spreads throughout the wall of the ventricles, through the **Purkyne** (or **Purkinje**) fibres, and the ventricles contract after they have filled with blood.

ELECTRICAL EVENTS AND VOLUME/PRESSURE CHANGES IN THE CARDIAC CYCLE

We have looked at what happens to the blood that passes through the heart during the cardiac cycle, and at the electrical events that co-ordinate the contraction and relaxation of the atria and ventricles. Fig 10.6 shows how these events correspond to pressure and volume changes in the heart and blood vessels, and also to the sounds that you hear when you listen to someone's heart beating. You will need to refer to this figure as you read on.

During **atrial systole**, the atria fill with blood from the vena cava and the pulmonary vein. Some of the blood that enters the atria flows straight into the ventricles, without any need for contraction. Atrial systole is initiated when the SAN sends out an electrical signal that spreads out over both the atria, causing contraction and forcing the remainder of the blood into the ventricles. There are no heart sounds. Atrial systole is very short, little more than a 'twitch'. Atrial relaxation, or **atrial diastole**, lasts for the remainder of the cycle. All the rest of the 'action' is in the ventricles.

✔ REMEMBER THIS

The QRS complex is associated with the electrical impulse passing through the AVN, the bundle of His and the Purkyne fibres.

SEE QUESTIONS 1 AND 2

Fig 10.6 The cardiac cycle.

Key:

■ atrial systole
■ atrial diastole
■ ventricular systole
□ ventricular diastole
— arterial pressure
— atrial pressure
--- ventricular pressure

Ventricular systole begins when the ventricles have filled with blood. The AVN picks up the signal from the SAN and then conducts impulses down through the bundle of His and on through the Purkyne fibres in the walls of the ventricles. This stimulates the ventricles to contract. Blood is forced upwards, forcing the semilunar valves open and the AV valves shut. The closing of the AV valves causes the 'lub' of the 'lub-dup' heart sound. Pressure in the arteries rises sharply as blood is forced into them.

During **ventricular diastole**, the ventricle walls relax, arterial pressure falls and blood begins to flow back into the ventricles. This reversal of the flow causes the semilunar valves to shut, causing the second, 'dup', heart sound. Meanwhile, the atria have been filling with blood and, as the ventricles relax, blood flows from the atria into the ventricles, forcing the AV valves open again.

STROKE VOLUME AND CARDIAC OUTPUT

The volume of blood pumped by the heart during one cardiac cycle is the **stroke volume**. A typical stroke volume in an adult is about 80 cm³: every time the heart beats, 80 cm³ of blood is forced through the pulmonary artery to the lungs and 80 cm³ is forced into the aorta to the body. So, the heart pumps over 8500 litres of blood per day. Stroke volume increases during exercise, and regular exercise results in a permanent resting increase to 110 cm³, or more.

The volume of blood pumped in one minute is called the **cardiac output**. It is calculated by multiplying the stroke volume by the heart rate and is expressed in litres of blood per minute.

Fig 10.6 The cardiac cycle. See Fig 10.11 for an explanation of the ECG trace. Many exam questions use this diagram, without labels, to test understanding of the events of one heartbeat. It is important to remember what is cause and what is effect, ie:
• Pacemaker cells initiate systole (contraction)
• The squeezing of the muscle walls reduces the volume and so increases the pressure in the chambers, forcing blood in a particular direction
• The direction of blood flow causes the valves to open or close. The valves ensure that blood flow through the heart is in one direction only

✔ **REMEMBER THIS**

Cardiac output =
stroke volume × heart rate
(measured in litres:
1000 cm³ = 1 litre)

❓ **TEST YOURSELF**

A Calculate the cardiac output:
a) at rest when the stroke volume is 80 cm³ and the heart rate is 75 beats per minute;
b) during vigorous exercise when the stroke volume is 100 cm³ and the heart rate is 150 beats per minute.

cardiovascular centre in medulla

decelerator nerve (parasympathetic nerve)

sino-atrial node

heart

accelerator nerve (sympathetic nerve)

Fig 10.7 The nerves that connect the cardiovascular centre in the brain to the heart. The accelerator and decelerator nerves are part of the autonomic nervous system

CONTROL OF HEART RATE

Heart rate is modified according to the needs of the body. It increases during physical exercise to deliver extra oxygen to the tissues and to take away excess carbon dioxide. At rest, a normal adult human heart beats about 75 times per minute: during very strenuous exercise it might beat 200 times per minute. Heart rate is controlled by the sino-atrial node. The rate goes up or down when the SAN receives information via two autonomic nerves which link the sino-atrial node with the **cardiovascular centre** in the brain (Fig 10.7):

- A **sympathetic** or **accelerator nerve** speeds up the heart. The synapses at the end of this nerve secrete noradrenaline.
- A **parasympathetic** or **decelerator nerve**, a branch of the vagus nerve, slows down the heart. The synapses at the end of this nerve secrete **acetylcholine**.

A **negative feedback system** controls the level of carbon dioxide and, indirectly, the level of oxygen in the blood. During exercise, the blood level of carbon dioxide starts to rise. This is detected by **chemoreceptors** (cells sensitive to chemical change) situated in three places: the carotid artery, the aorta and the medulla. Nerve impulses travel from these receptors to the cardiovascular centre. In response, the cardiovascular centre sends impulses down the sympathetic nerve to increase the heart rate.

When the carbon dioxide level drops, the cardiovascular centre responds by sending impulses down the parasympathetic nerve, and the heart rate returns to normal. The control of heart rate is closely related to the control of breathing (see Chapter 9).

Several factors affect heart rate:

- Secretion of **adrenaline** in response to stress, excitement or other emotions. Adrenaline is the hormone that prepares the body for action, and one of its effects is to increase heart rate.
- Movement of the limbs, as in exercise. It is thought that stretch receptors in the muscles and tendons relay information to the brain, telling the cardiovascular centre that oxygen levels will soon fall and that carbon dioxide will soon build up. This initiates signals that speed up heart rate and breathing rate.
- The level of respiratory gases in the blood, as described above.
- Blood pressure. When this gets too high, a fail-safe mechanism prevents any further increase in heartbeat.

3 BLOOD VESSELS

SEE QUESTION 3

There are three types of blood vessels: **arteries**, **veins** and **capillaries**. The structure of each is closely related to its function (see Fig 10.8 and Table 10.1).

Arteries carry blood away from the heart towards other organs of the body. Arteries branch into smaller **arterioles**, which branch into tiny capillaries. Capillaries are **permeable** ('leaky') vessels whose walls are one cell thick, allowing exchange of materials between blood and nearby cells. Blood flows from capillaries into venules, which drain into larger veins.

As Fig 10.8 and Table 10.1 show, the walls of arteries and veins have the same three layers: the **tunica interna**, the **tunica media** and the **tunica externa**. The relative thickness and composition of these layers vary according to the function of the vessel. Arteries, which have to withstand pulses of high pressure, have a thick tunica media containing smooth muscle and elastic fibres. Veins have a thinner tunica media and a larger lumen and carry slower flowing blood at low pressure. Valves prevent blood flowing backwards.

✔ REMEMBER THIS

Arterial blood is under high pressure and surges occur every time the ventricles contract. The walls of the arteries are able to absorb and smooth out these pulse waves so that by the time blood reaches the arterioles, it is flowing steadily.

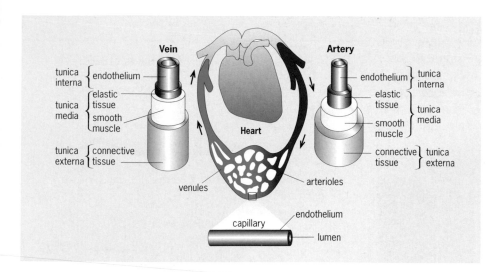

Fig 10.8 The structure of blood vessels and their distribution in the circulatory system

CAPILLARY CIRCULATION

The circulatory system keeps all cells bathed in tissue fluid. The composition of tissue fluid stays reasonably constant because permeable capillaries allow exchange of materials between tissue fluid and the blood.

HOW TISSUE FLUID IS FORMED

Fig 10.9 shows what happens in living tissue. Blood from the arterioles is under high **hydrostatic pressure**. When blood enters the capillary, substances begin to leak out through the permeable capillary wall. The capillary walls act as filters and a proportion of all chemicals below a particular size is squeezed out, forming tissue fluid. The composition of tissue fluid is very similar to that of plasma, but tissue fluid lacks most of the large proteins that cannot pass through the capillary wall. Plasma proteins remain in the blood where they play an important part in the drainage of tissue fluid.

> ✔ **REMEMBER THIS**
>
> All living cells are surrounded by tissue fluid. This is also known as **interstitial** or **intercellular fluid**.

> ✔ **REMEMBER THIS**
>
> Hydrostatic pressure is physical pressure. The contraction of the ventricles of the heart creates hydrostatic pressure in the blood vessels.

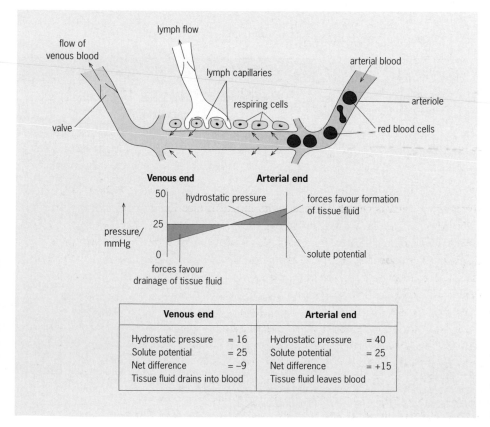

Venous end	Arterial end
Hydrostatic pressure = 16	Hydrostatic pressure = 40
Solute potential = 25	Solute potential = 25
Net difference = −9	Net difference = +15
Tissue fluid drains into blood	Tissue fluid leaves blood

Fig 10.9 The drainage and formation of tissue fluid. At the arterial end of the capillary, the high hydrostatic pressure forces water and small molecules out of the blood, providing the tissues with a fluid that contains nutrients and oxygen. The hydrostatic pressure decreases as blood flows along the capillary. When it falls, water drains back into the blood by osmosis, taking with it wastes such as urea and carbon dioxide

⚠ SCIENCE FEATURE The electrocardiogram (ECG)

The electrical events that control the cardiac cycle are recorded by placing electrodes on certain parts of the body – Fig 10.10. A normal, healthy heartbeat produces a distinctive trace – upper Fig 10.11. Certain heart defects produce a modified trace and this makes the ECG a useful diagnostic tool – lower Fig 10.11.

Fig 10.10 The electrocardiogram: electrodes taped to the chest pick up the electrical events of the cardiac cycle as they pass through the body. People with heart problems can now transmit their ECG trace down the telephone to a specialist who can detect problems at a distance

Features of a normal electrocardiogram (ECG) trace

R shows contraction of ventricle muscles

T wave shows relaxation of ventricle muscles

P wave shows contraction of atrial muscles

P–R interval shows time taken for an electrical impulse to travel from sino-atrial node to Purkyne fibres

A normal (ECG) trace

Some abnormal ECG traces

Tachycardia (heart rate of over 100 beats/minute)

Ventricular fibrillation (uncontrolled contraction of the ventricles which pumps very little or no blood)

Heart block (ventricles are not always stimulated following contraction)

Fig 10.11 A healthy ECG trace and some abnormal traces, showing common heart defects that can be diagnosed using the ECG

Table 10.1 A comparison of arteries, veins and capillaries

Vessel	Cross-section	Direction of flow	Pressure	Oxygen content	Size of lumen	Presence of valves	Properties of wall
Arteries		away from heart	high	high*	relatively small	no	tough, powerful elastic recoil
Veins		back to heart†	low	low*	relatively large	yes	thin, distend easily
Capillaries	endothelial cell	through organs	medium	oxygen lost through wall	small, about the diameter of one red blood cell	no	one cell thick, permeable

*Except in pulmonary circulation, where oxygenated blood is carried in the veins and deoxygenated blood is carried in the arteries.

†Except portal veins, eg the hepatic portal vein, which carries blood between organs and not to or from the heart.

HOW TISSUE FLUID RETURNS TO THE BLOOD

Two main forces act on the blood as is passes along the capillary:

- The hydrostatic pressure of the blood: this tends to force water and solutes *out* of the capillary.
- The water potential of the plasma; this tends to draw water *into* the capillary by osmosis.

As blood flows along a capillary, fluid passes to the tissues, and the volume and hydrostatic pressure of the blood in the capillary decrease (Fig 10.9). As the water potential created by the large plasma proteins is relatively constant, water begins to drain back into the blood when hydrostatic pressure falls below the water potential. Cell waste, such as urea, and substances that have been secreted into the tissue fluid, diffuse into the capillary, contributing to the composition of venous blood.

BLOOD FLOW IN VEINS

Blood that drains into the venules is deoxygenated, under low pressure and contains many waste products and cell secretions. Several features allow blood to return to the heart:

- Valves in veins close and prevent backflow.
- Working muscles surrounding the veins squeeze blood along as they contract. The action of muscles, especially those in the legs, is so important that it has been called the 'secondary heart' or the 'venous pump' (Fig 10.13).
- Gravity helps blood to flow from organs 'above' the heart.
- The negative pressure created in the thorax during **inspiration** (breathing in) draws blood from surrounding veins.
- The action of the heart. After systole (contraction), the elastic walls of the chambers recoil, drawing blood in from the veins.

Fig 10.12 Capillary walls are about 1 μm thick. Most capillaries have a lumen that is a little larger than the diameter of a red blood cell. Note that cells in photos often seem to be strange shapes. There are two reasons for this. First, they are very flexible and will distort to squeeze along capillaries without causing blockages. Second, microscope slides take thin two-dimensional sections through three-dimensional objects, so they often give a false impression of shape

SEE QUESTION 4

Walking
When walking, our leg muscles contract and squeeze the leg veins:

The valve will open and then close to prevent backflow

Contraction of skeletal muscle forces blood upwards

Passive upright position
When a person is standing upright, the blood pressure at the base of the large veins in the leg rises

The blood is almost static due to the pressure caused by the height of the column of blood above

Fig 10.13 The action of the leg muscles squeezes blood along veins, and the valves ensure that flow can only be in one direction. This venous pump is very important to the circulation, and regularly standing still for any length of time can lead to problems. Millions of people suffer from varicose veins or from haemorrhoids (piles), damaged veins whose walls have been stretched by pools of accumulated blood

✓ REMEMBER THIS

Peripheral vasoconstriction refers to constriction of the blood vessels in the skin (periphery = outer region).

? TEST YOURSELF

B What effect do the following have on blood pressure?
a) vasoconstriction
b) vasodilation

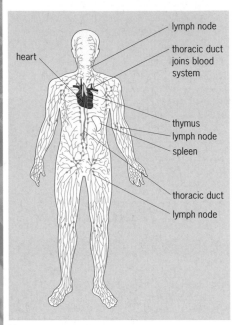

Fig 10.14 The major lymphatic vessels in the human body. The lymph system drains into the blood system in the upper thorax. A large lymph vessel, the thoracic duct, connects with the subclavian vein (sub-clavicle = under collarbone). Here lymph mixes with blood before entering the vena cava on its way to the heart

CONTROL OF BLOOD FLOW

The body often needs to alter blood flow to different areas, according to circumstances. For example, when we are hot, we lose excess heat because more blood flows to the surface of the skin. Blood flow can be modified by **sphincters** and **shunt vessels**, or by altering the diameter of the vessel itself.

A sphincter is a ring of muscle around a blood vessel which can contract, reducing the lumen size of the vessel and so reducing or preventing blood flow to a particular area. Sphincters can redirect blood into shunt vessels and these bypass a particular area. For instance, when we are cold, sphincters reduce blood flow to the skin surface, redirecting it along shunt vessels that keep the blood deeper in the body (see Chapter 12).

Many blood vessels, particularly arterioles, contain smooth muscle fibres that can contract to reduce blood flow through the vessel; this is **vasoconstriction**. Blood flow increases when the muscle fibres relax; this is **vasodilation**. Blood pressure can be regulated by altering the degree of vasoconstriction or vasodilation.

4 THE LYMPHATIC SYSTEM

The **lymphatic system** is part of the immune system (Chapter 30). It is also part of the circulatory system: it returns to the heart the small amount of tissue fluid that cannot be returned by the veins.

We can think of the lymph system as an extra set of veins (Fig 10.14). Flow through lymphatic vessels is slow, but important. Just how important is obvious when you see someone with the disease elephantiasis (Fig 10.15).

Lymphatic flow begins in the **capillary beds**, where small amounts of tissue fluid drain into tiny **lymphatic capillaries** (Fig 10.16). The walls of these vessels are more permeable than blood capillaries to lipids and large molecules such as proteins, and so lymph contains a high proportion of these substances. Many cells secrete substances that are too large to enter the blood directly, and so can only pass into the general circulation via the lymphatics. The lymph capillaries drain into larger lymph vessels that look like thin, transparent veins. These vessels have valves to prevent backflow. Lymph contains no red blood cells, and so is pale and clear. All lymph vessels flow towards the upper chest, passing through numerous lymph nodes, which filter out bacteria and cell debris.

Fig 10.15 Lymphatic drainage takes about 120 cm³ of fluid out of the tissues every hour. When a lymphatic vessel is blocked, the limb quickly swells until the pressure interferes with normal blood flow. This photograph shows someone suffering from elephantiasis, a condition caused by a parasitic worm, *Wuchereria bancrofti*, which blocks lymphatic vessels

Fig 10.16 A longitudinal section through a lymphatic capillary. The cells that form these capillary walls overlap, forming tiny valves that allow flow of lymph in one direction only. Lymph capillaries in the villi of the intestines are called lacteals, and are important in the drainage of lipids and large molecules

5 BLOOD PRESSURE

The term **blood pressure** refers to the physical pressure of blood flowing through the arteries. This depends on the force created by the pumping of the heart, the volume of circulating blood and the size of the blood vessels. The body must keep blood pressure inside fairly strict limits. It must be high enough to force blood through all the capillaries so that all cells of the body are well nourished, but not too high because this would make the heart work unnecessarily hard and risk damage to blood vessels. High blood pressure is a risk factor for **atherosclerosis** (hardening of the arteries; see Chapter 31), and it causes a variety of other health problems. Fig 10.17 shows how we measure blood pressure.

The **diastolic pressure** is used as an indicator for medical problems: a normal reading is between 60 and 80 mmHg, with anything over 90 regarded as high. Elsewhere in the book we talk about pressure in terms of kPa, but blood pressure is still measured in the old mmHg units.

CONTROL OF BLOOD PRESSURE

A negative feedback system

...... detects blood pressure

> **REMEMBER THIS**
>
> Have you ever stood up very quickly, and then felt faint or dizzy? This happens because the rapid change in posture has altered the body's blood distribution, causing a temporary lack of blood to the brain. Fortunately, body mechanisms quickly bring the blood flow back to normal.

Fig 10.17 Blood pressure is usually measured using a sphygmomanometer. An inflatable cuff is placed around the upper arm and inflated until all blood flow, in or out of the arm, stops. The blood flow in the brachial artery (at the elbow) is monitored using a stethoscope.

After inflation, there is no sound, but, as air escapes from the cuff, the pressure decreases until it falls just below that created by the heart as it contracts. At this point, blood is heard spurting through the constriction in the artery. This pressure is the **systolic value**.

Pressure in the cuff continues to drop until blood can be heard flowing constantly. This is the **diastolic value** and represents the pressure to which arterial blood falls between beats

> **REMEMBER THIS**
>
> Many mechanisms work together to keep blood pressure constant. The size of blood vessels, the heart rate and the volume of circulating blood can all change to increase or decrease blood pressure, according to the needs of the body. The kidneys are particularly important because they control how much fluid we lose. A drop in blood volume and pressure causes more fluid to be retained in the blood and less to be lost in the urine (see Chapter 13).

> **? TEST YOURSELF**
>
> **C** 'Blood pressure one twenty over seventy', shouts the doctor. What does this mean?

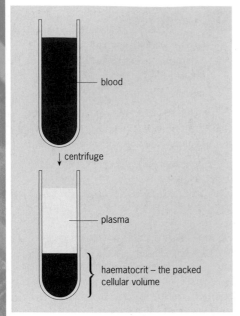

Fig 10.19 The percentage of blood volume taken up by cells is known as the haematocrit. A haematocrit of 39 would indicate that 39 per cent of the blood is composed of cells, mostly red cells. The average value for men is about 42 while women average 38. These values are affected by factors such as anaemia or high altitude

? TEST YOURSELF

D Suggest how **(a)** high altitude and **(b)** anaemia will affect the haematocrit.

6 THE BLOOD

Blood is the fluid that flows through blood vessels, and the sight of blood is a sure sign that a vessel has ruptured. Losing large amounts of blood has serious consequences: death can result if the ruptured vessels are not sealed and blood volume is not rapidly returned to normal. In an emergency, when there is no time to find out the patient's blood group, accident victims are given fluids called **plasma expanders**. These are isotonic fluids that contain no cells but still increase blood volume.

Blood is a complex mixture containing cells, cell fragments and a range of dissolved molecules. It does the following:

- Transports materials. Blood transports digested food and oxygen to respiring tissues. It also takes carbon dioxide and waste products away from respiring cells to the various organs that remove them. It carries hormones from endocrine glands to target organs.
- Distributes heat. The blood helps to keep body temperature stable by distributing heat from metabolically active organs, such as the working muscles, to the rest of the body (see Chapter 12).
- Provides pressure. Many organs of the body depend on the physical pressure of the blood to carry out their function. For example, filtration in the kidney, formation of tissue fluid and erection of the penis all depend on blood pressure.
- Acts as a buffer. The blood contains many proteins and ions which act as buffers, keeping the pH constant by 'mopping up' any excess acid or alkali. Haemoglobin in red blood cells is also an important buffer.
- Defends the body against infection (see Chapter 30).

WHAT'S IN BLOOD?

Fig 10.19 shows that centrifuged blood separates out into two distinct layers: a **cellular portion** called the **haematocrit**, and the **plasma**. The cellular portion contains red cells, white cells and platelets.

PLASMA

Plasma is the fluid part of blood. Its main constituents are shown in Table 10.2. The exact composition of the plasma varies greatly. For instance, in the hepatic portal vein, which leads from the intestines to the liver, the plasma is much richer in dissolved foods such as sugars, vitamins and amino acids than the plasma in any other vein.

Table 10.2 The main constituents of plasma		
	Components	**Function**
Proteins:	albumin	osmotic balance
	antibodies	immunity
	fibrinogen	blood clotting
Salts:	sodium, potassium, chloride, hydrogencarbonate, calcium	osmotic balance, conduction of nerve impulses, carriage of CO_2, buffering, blood clotting
Products of digestion:	glucose, amino acids, fatty acids, glycerol, vitamins	nourishment of cells
Hormones:	protein, eg insulin; lipid (steroids), eg testosterone	communication
Heat	distributed around body to maintain constant body temperature	
Oxygen	vital in aerobic cell respiration	
Waste products:	urea, CO_2	none: they must be removed by excretion (see Chapters 11 and 13)

Fig 10.20(a) If you look at a drop of blood under the microscope, the most obvious feature is the mass of red blood cells. There are also two white blood cells in this picture **Fig 10.20(b)** Red blood cells are biconcave discs filled with haemoglobin and other chemicals which help the loading and unloading of oxygen

RED BLOOD CELLS

In this chapter we study red blood cells only (Fig 10.20): white cells and platelets are covered in Chapter 30.

Also known as **erythrocytes**, red blood cells are by far the most numerous in the blood. A single cubic millimetre of blood contains around 5 million, sometimes more. Red blood cells have one function: they carry the respiratory gases. The **haemoglobin** (Hb) they contain picks up oxygen in the lungs and swaps it for carbon dioxide in the respiring tissues.

Red cells have no nucleus: they are packed with Hb and the enzymes and chemicals that allow the Hb to carry oxygen effectively. Having Hb inside red blood cells rather than in solution in the plasma gives several advantages:

- A much greater volume of Hb can be carried in cells than could be dissolved in plasma.
- Hb can be kept in a favourable chemical environment to allow faster loading and unloading of respiratory gases.
- Hb molecules of a particular age are kept together and can be easily replaced when old.
- Hb in cells does not affect the osmotic properties of the blood (free Hb would).
- Hb in cells cannot be lost by excretion.

Red blood cells have a regular shape, described as a biconcave disc. This shape is maintained by the **cytoskeleton**, a complex but flexible internal scaffolding made from protein fibres. When you think about what red blood cells do it is easy to see why they have this shape. They must have a large enough volume to carry useful amounts of oxygen but they also need a large enough surface area to load and unload it quickly. The biconcave shape is a good compromise between the maximum volume of a sphere and the maximum surface area of a flat disc.

Red blood cells are distorted when they have to squeeze through capillaries, which often have lumens slightly smaller than the diameter of the red cell. Red cells rarely cause blockages because they have a flexible shape and a smooth membrane.

BLOOD AS A TRANSPORT MEDIUM

Oxygen is carried in the blood in two ways: 98 per cent travels as **oxyhaemoglobin**, the oxygen–haemoglobin complex. The remaining 2 per cent is dissolved in the plasma. Carbon dioxide is carried in three ways: as HCO_3^- ions in the plasma (70 per cent), combined with haemoglobin as a **carbamino** compound (23 per cent) and in simple solution in the plasma (7 per cent).

> ✔ **REMEMBER THIS**
>
> Red blood cells are made in the bone marrow of the vertebrae, ribs and pelvis by specialised **stem cells**. The turnover of red cells is very rapid: we make millions of red cells every second. Red blood cells circulate for about 100 to 120 days before they are destroyed in the liver by the phagocytes called **Kupffer cells** and also in the spleen.

SEE QUESTION 5

? TEST YOURSELF

E Why do you think that haemoglobin has been described as the 'Robin Hood' molecule?

REMEMBER THIS

Conjugated means 'joined'.
A conjugated protein, such as
haemoglobin, consists of a protein
attached to a **prosthetic group**
(a non-protein). The prosthetic
group in haemoglobin contains
iron, so we need a regular supply
of iron in our diet to make this
vital chemical.

 REMEMBER THIS

The amount of oxygen in
a mixture of gases such as air
is often described in terms of
its **partial pressure**, or **tension**,
rather than its concentration –
see Chapter 9. The greater the
concentration of dissolved oxygen,
the higher the partial pressure.

SEE QUESTION 6

THE STRUCTURE OF HAEMOGLOBIN

The Hb molecule is a large conjugated protein with a molecular mass of about
64 500 kilodaltons. Each Hb molecule consists of four **globin** sub-units
consisting of a polypeptide chain and a prosthetic group called **haem**.
At the centre of each haem is an iron ion (Fe^{2+}) which combines with oxygen.
There are four haem groups, so the overall equation for the reaction is:

$$Hb + 4O_2 \rightarrow HbO_8$$
$$\text{haemoglobin} + \text{oxygen} \rightarrow \text{oxyhaemoglobin}$$

The polypeptide chains hold the haem groups in place (see Fig 3.34) and
help to load oxygen. When the first haem group combines with an oxygen
molecule, the Hb molecule alters its shape so that the next haem group is
exposed, making the loading of the second oxygen easier.

HAEMOGLOBIN IN ACTION

There are many substances that react readily with oxygen, but Hb is one of
the few that can combine with oxygen where it is abundant and then release
it when the concentration falls. This property is illustrated in a graph called
an **oxygen dissociation curve** (Fig 10.21). This is plotted by analysing
the percentage of Hb saturated with oxygen at different concentrations of
oxygen. The graph shows that Hb becomes fully saturated with oxygen at the
concentrations found in the lungs, and gives up a relatively large proportion
of its oxygen in the lower oxygen concentrations that occur in the tissues.

The dissociation curve in Fig 10.21 was plotted using mixtures of gases
in which the concentration of carbon dioxide was constant. Fig 10.22 shows
that, at higher levels of carbon dioxide, the curve moves to the right.

This is a vital point: it shows that carbon dioxide lowers the **affinity** of
Hb for oxygen. This means that when carbon dioxide concentration is higher,
haemoglobin does not hold on to its oxygen quite as well: Hb therefore tends
to give up oxygen in areas of high carbon dioxide – such as in the respiring
tissues which need it most. This lowering of affinity by carbon dioxide is
called the **Bohr effect**, or the **Bohr shift**.

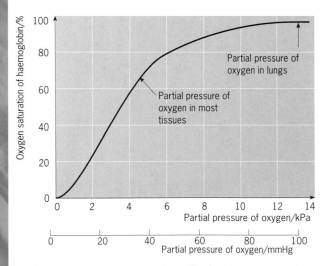

Fig 10.21 The oxygen dissociation curve. At the partial pressure found in the lungs,
haemoglobin becomes 97–99 per cent saturated with oxygen. Surprisingly, haemoglobin
releases only about 23 per cent of its oxygen in the respiring tissues, so the blood returning in
veins is still about 75 per cent saturated. This suggests that three out of the four haem groups
are still bound to oxygen. This allows great flexibility: if a tissue such as a working muscle
becomes particularly oxygen starved, the blood can release large amounts of extra oxygen

Fig 10.22 If the oxygen dissociation curve is plotted at
higher carbon dioxide concentrations, it moves to the right,
showing that haemoglobin has a reduced affinity for
oxygen: this is called the Bohr effect

FETAL HAEMOGLOBIN

Before birth, a fetus must obtain oxygen from its mother via the placenta. If fetal Hb had the same affinity for oxygen as the mother's Hb, no transfer of oxygen would be possible. But, as Fig 10.23 shows, fetal Hb has a higher affinity for oxygen than adult Hb. It can therefore pick up oxygen in the same conditions that cause the maternal blood to release it. After birth, the baby's body makes adult Hb which gradually replaces the fetal version.

UNLOADING OXYGEN AT THE TISSUES

The series of events that result in haemoglobin unloading its oxygen to supply respiring tissues is shown in Fig 10.24.

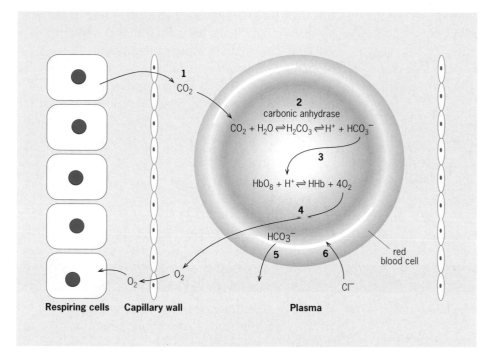

Fig 10.24 Unloading oxygen at the respiring tissues

Step 1 Carbon dioxide diffuses from the respiring tissue, through the capillary wall and the plasma, into the red blood cell

Step 2 Inside the red cell the enzyme carbonic anhydrase catalyses the conversion of carbon dioxide into carbonic acid (H_2CO_3)

Step 3 The hydrogen ions released by the carbonic acid makes the haemoglobin molecules less stable, causing them to release the four oxygen molecules

Step 4 The oxygen is free to diffuse into the respiring tissues

Step 5 The accumulating HCO_3^- ions diffuse out into the plasma, leaving the inside of the red cell with a net positive charge

Step 6 To maintain a neutral charge, Cl^- ions (the commonest negative ions in plasma) diffuse into the cell; this is called the chloride shift

MYOGLOBIN AND MUSCLES

Mammalian muscle contains the respiratory pigment, myoglobin (myo = muscle). The myoglobin molecule can be thought of as one-quarter of a haemoglobin molecule: it consists of one polypeptide chain and one haem, and combines with one oxygen molecule. Myoglobin has a higher affinity for oxygen than haemoglobin and so it can pick up oxygen from haemoglobin and store it until the muscle becomes short of oxygen. When this happens (during exercise, for example), the oxygen tension in the muscle drops. Haemoglobin has no more oxygen to deliver and myoglobin then releases its oxygen.

Fig 10.23 The dissociation curve (above) for fetal haemoglobin is to the left of the adult version. So, at the oxygen concentrations found at the placenta there is an efficient transfer of oxygen from mother to fetus (below)

| ✔ | REMEMBER THIS |

A human baby is known as an embryo up to about eight weeks of development. After this, it is known as a fetus for the rest of the pregnancy.

SUMMARY

When you have read this chapter, you should know and understand the following:

- Humans have a double circulation: the **pulmonary circulation** to the lungs and the **systemic circulation** to the rest of the body.
- The human **heart** is a four-chambered muscular pump made of **cardiac muscle**. This can contract powerfully and without fatigue.
- A single heartbeat, the **cardiac cycle**, consists of **atrial systole** (contraction) followed by **ventricular systole**, then **atrial** and **ventricular diastole** (relaxation).
- The heart is **myogenic**: the electrical impulses that control the cardiac cycle arise from the muscle itself.
- From the heart, blood passes into vessels in the following order: **arteries**, then **arterioles**, then **capillaries**, then **venules**, then **veins**.
- The **lymphatic system** acts as an extra set of veins, helping to drain fluid from the tissues back into the bloodstream.
- Blood consists of **red cells**, **white cells**, **platelets** and **plasma**.
- Red blood cells, or **erythrocytes**, contain **haemoglobin**, a conjugated protein that can pick up oxygen where it is abundant (lungs) and release it where it is needed (respiring tissues).
- Blood carries carbon dioxide in three ways: as hydrogencarbonate (HCO_3^-), combined with haemoglobin and as simple solution in the plasma.

? EXAM QUESTIONS

1 The diagrams show the left side of the heart at two stages in a cardiac cycle.

Diagram **A**

Diagram **B**

a) Name the structure labelled **X**.
b) Describe **two** pieces of evidence in Diagram **B** which indicate that the ventricle is emptying.
c) The cardiac output is the volume of blood that one ventricle pumps out per minute. The resting heart rate of an athlete often decreases as a result of training, even though the cardiac output remains the same. Suggest an explanation for this.

d) During exercise the rate of blood flow to heart muscle increases from $270\,\text{cm}^3$ per minute to $750\,\text{cm}^3$ per minute.
(i) Calculate the percentage increase in rate of blood flow to heart muscle during exercise. Show your working.
(ii) Explain the advantage of the increase in the rate of blood flow to heart muscle during exercise.

AQA (B) BYB3 Physiology and Transport June 2001 Q4

2 The diagram shows variations in certain features of blood vessels in different parts of the circulatory system.

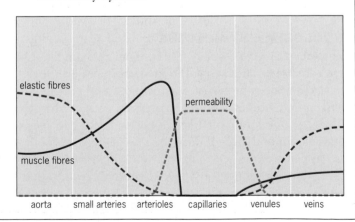

? EXAM QUESTIONS

a) Describe the difference in the ratio of elastic fibres to muscle fibres between the aorta and the arterioles.

b) Explain the importance of this ratio in
   (i) the aorta;
   (ii) the arterioles supplying the skin.

c) Explain how the structure of the capillaries is related to their function.

AQA (NEAB) BY03 Physiology June 2000 Q6

3 The graph shows changes in the volume of blood in the left ventricle as the heart beats.

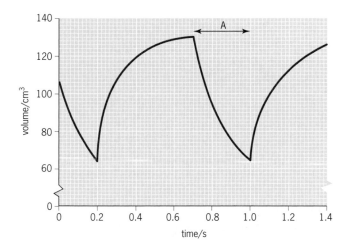

a) **(i)** The horizontal line labelled **A** on the graph shows when blood is leaving the ventricle. Explain, in terms of blood pressure, why blood does not flow back into the atrium during this period.
   (ii) Draw a horizontal line on the graph, to show the period in one cardiac cycle when the muscle in the wall of the ventricle is relaxed. Label this line with the letter **B**.

b) **(i)** Draw a horizontal line on the graph to show one complete cardiac cycle. Label this line with the letter **C**.
   (ii) Use line **C** to calculate the number of times the heart beats in one minute. Show your working.

c) The table shows the blood flow to different parts of the body at rest and during a period of vigorous exercise.

Part of the body	Rate of blood flow/cm3 minute−1	
	at rest	during exercise
Brain	750	750
Heart muscle	300	1*200
Gut and liver	3*000	1*400
Muscle	1*000	16*000
All other organs (except lungs)	1*550	1*550

(i) Use the figures in the table to calculate the cardiac output at rest.
(ii) Give **two** ways in which cardiac output is increased during a period of vigorous exercise.

d) Describe the parts played by the sinoatrial node (SAN) and the atrioventricular node (AVN) in controlling the heart beat.

AQA (A) BYA1 Molecules, Cells and Systems January 2001 Q8

4 The diagram shows the relationship between different body fluids.

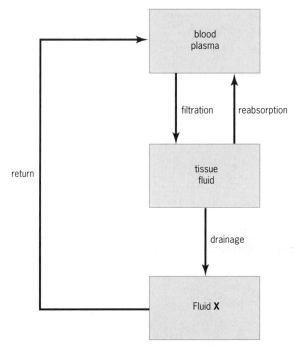

a) Name fluid **X**.

b) Give **one** way in which the composition of tissue fluid differs from that of plasma.

c) Filtration of blood plasma is partly due to high blood hydrostatic pressure. How is the high blood hydrostatic pressure produced?

d) Children who have insufficient protein in their diet develop a swollen abdomen. This is due to the accumulation of tissue fluid. Explain the link between insufficient protein in the diet and the accumulation of tissue fluid.

AQA (B) BYB3 Physiology and Transport June 2001 Q2

❓ EXAM QUESTIONS

5

a) Explain why blood may be described as a tissue.

b) The diagram shows how the different types of cells found in human blood may be distinguished from each other.

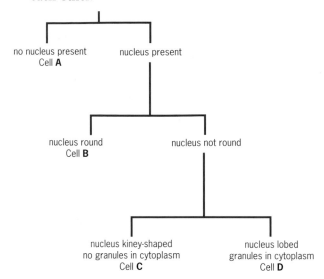

Name:

(i) Cell **C**;

(ii) Cell **D**.

c) The drawing has been made from an electronmicrograph. It shows a red blood cell in a capillary from a human lung.

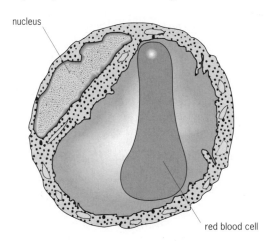

The diameter of the red blood cell is approximately the same as the diameter of the inside of the capillary. Suggest **two** ways in which this increases the efficiency with which oxygen diffuses from the alveoli into the red blood cell.

The red blood cell is a uniform colour. The nucleus has patches of light and dark colour. Explain this difference in appearance.

AQA (A) BYA1 Molecules, Cells and Systems January 2001 Q6

6 The graph below shows oxygen dissociation curves for human haemoglobin at low and high partial pressures of carbon dioxide. The effect of high partial pressure of carbon dioxide on haemoglobin dissociation is known as the Bohr effect.

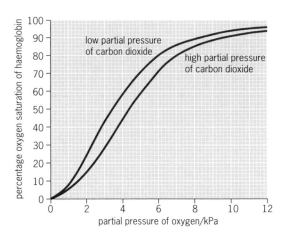

a) Explain how the shape of the dissociation curve at a low partial pressure of carbon dioxide is related to the ability of haemoglobin to transport oxygen from the lungs to respiring tissues.

b) **(i)** From the graph, calculate the difference in percentage oxygen saturation of haemoglobin at low and high partial pressure of carbon dioxide, at 4 kPa partial pressure of oxygen. Show your working.

(ii) Suggest why the Bohr effect is of importance to tissues when they increase their activity.

c) The respiratory pigment myoglobin is found in muscle tissue. The dissociation curve for myoglobin lies to the left of the curves for haemoglobin and it is 97% saturated with oxygen at an oxygen partial pressure of 1 kPa. Explain the function of myoglobin in muscle.

Edexcel B3 June 1999 Q7

KEY SKILLS **ASSIGNMENT**

ALTITUDE TRAINING

This Assignment links aspects of breathing (Chapter 9) and aspects of the circulatory system that we have covered in this chapter. So to answer the questions, you will need to refer to both chapters to find the relevant text.

Staying power

Which sport is the most demanding? Is it football? Or marathon running? It might surprise you, but these two sports are just a stroll compared with the most gruelling events that you can do. Top of the list are cross-country skiing and long-distance bike racing (such as the Tour de France).

Fig 10.A1 Cross-country skiers and long-distance cyclists top the table for maximum oxygen consumption

But how do we compare the demands of different sports and events? One way is to measure oxygen consumption. In Chapter 9 we saw that a person's oxygen consumption can be expressed in terms of V_{O_2} and V_{O_2}(max).

1 a) Suggest what does the term V_{O_2} means.
  b) What is the difference between V_{O_2} and V_{O_2}(max)?
c) What are the average values for V_{O_2} and V_{O_2}(max), as shown in Chapter 9?

Studies show that competitors in cross-country skiing and long-distance cycling can have oxygen consumptions as high as 90 cm^3 per minute per kilogram.

In events covering many kilometres over many hours, just a few seconds' advantage can mean the difference between success and failure. Not surprisingly, there is a search for ever more effective training methods that can increase athletes' oxidative capacity – their ability to get oxygen to their muscles. For several decades, one of the favoured methods of increasing the oxygen-carrying capacity of the blood has been to train at altitude.

Altitude training

In Chapter 9, we saw that mountaineers who ascend without oxygen apparatus experience problems as the air gets thinner and the partial pressure of oxygen decreases.

2 a) Explain what is meant by partial pressure of oxygen.
  b) Explain what happens to the partial pressure of oxygen as you go up a mountain.

The human body is able to *acclimatise*; it adapts to low oxygen levels if given time – usually several weeks. The count of red cells in the blood increases from about 5 000 000 per cubic millimetre to around 7 000 000.

3 a) What percentage increase is this?
  b) What is the advantage of this change?

Acclimatisation

The mechanism of this acclimatisation not completely understood, but this is what we think happens. The kidney detects the lowered oxygen levels in the blood and responds by releasing a hormone, **erythropoietin** (EPO), which targets cells in the bone marrow. In response, the bone marrow increases red cell production.

There are several ways of measuring the oxygen-carrying capacity of the blood. In addition to the number of red cells per cubic millimetre, there is the haematocrit and the amount of haemoglobin per litre. The haematocrit is the percentage of the blood taken up by cells (average values 41–42 per cent for males, about 38 per cent for females). The average amount of haemoglobin per litre is about 16 g for males and 14 g for females.

4 a) Write a flow diagram to summarise the mechanism of acclimatisation.
b) A person has trained at altitude and has returned to sea level. What effect will this have on his or her resting pulse rate? Explain your answer.
c) What do you think happens to the red blood cell count and resting pulse rate in the long term, if the athlete remains at sea level?

For a long time, athletes have realised the great benefit of training at altitude: if you go up into the mountains to train, the increased oxygen-carrying capacity of your blood gives you a distinct advantage over those competitors who have remained at sea level.

However, a problem comes from the fact that exercise done at altitude is not as beneficial as that done at sea level – you can't work the muscles as hard. So there's a problem; how do you 'live high and train low'?

KEY SKILLS **ASSIGNMENT**

There are two ways to achieve this. One is to have 'the Alps in a caravan' and live at low altitude in a mobile home with a controlled atmosphere, so that the partial pressure of oxygen is lowered to simulate high-altitude conditions.

The other solution is the highly illegal, unfair and dangerous practice of taking genetically engineered erythropoietin (RhEPO – recombinant human erythropoietin), which will mimic the effects of altitude training without the time and expense.

5 Outline the steps involved in making genetically engineered human erythropoietin. (You will need to look at Chapter 27 to answer this.)

In the past decade, the deaths of several top cyclists have been linked to RhEPO abuse and the resulting ultra-high haematocrit. The main danger comes from the increased risk of an embolism, a blood clot that circulates before becoming lodged in a vital blood vessel, such as in the heart, lungs or brain.

6 Suggest how an increased haematocrit can lead to an embolism.

7 Read the following extract and then answer the questions that follow.

Human proteins are boosting performance: the evidence

Genetic engineers have found a way to mass-produce EPO, which is a hormone and a front-line drug for treating anaemia. Use of EPO boosts red blood cell production and consequently blood haemoglobin levels.

The increase in haemoglobin is a powerful lure for athletes in endurance events. Evidence for the use of EPO has come from comparing two cross-country skiing events at Lahti and Thunder Bay. EPO is detectable in the body for only six to eight hours but its effect lasts as long as a red blood cell – around 120 days.

(Reproduced with permission from an article in the *Daily Telegraph*)

a) Explain how the graph of Fig 10.A2(a) provides evidence that EPO has been used by some cross-country skiers.

b) Explain how increased haemoglobin concentration might lead to increased performance in endurance events.

The graph of Fig 10.A2(b) shows the oxygen dissociation curves for haemoglobin as blood passes through capillaries in the lungs and in the skeletal muscles of an athlete.

c) Explain how features of the oxygen dissociation curves for haemoglobin in the lungs and in the skeletal muscles benefit an athlete.

Fig 10.A2(a) Comparing the haemoglobin concentration in the blood of cross-country skiers from two events, one in 1989 and one six years later, in 1995

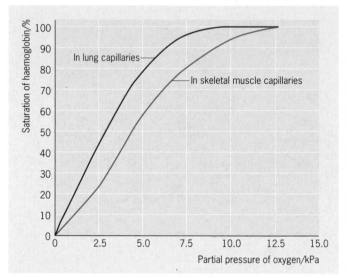

Fig 10.A2(b) The oxygen dissociation curves for haemoglobin in blood passing through capillaries in the lungs and in the skeletal muscles of an athlete
[Question 7 is an original exam question: NEAB Biology Physiology Module Test March 1998 Section B Q8]

Catching the cheats

So how can you tell when an athlete has cheated and used RhEPO? It is a natural hormone and so you would expect to find it in the blood, especially in those who have trained at altitude. The authorities have decided that for the present the only practical way to detect abuse is to measure the effect of the RhEPO, that is, the amount of haemoglobin in the blood. The assumption is that unnaturally high levels are indicators of abuse.

Recently the cycling authorities set the level at a haematocrit of 50 per cent, and the International Skiing Federation set haemoglobin levels at 18.5 grams per litre for males and 16.5 for women. Predictably, athletes have complained that it is perfectly possible to have levels this high naturally, and the debate looks set to continue.

11

Homeostasis

For Catherine Hayes, the discovery of insulin came just in time. Daily injections of insulin allowed the cells of her body to absorb glucose effectively, so nourishing her tissues and lowering her blood sugar

Millions of people worldwide are unable to control their blood glucose level. They have a condition known as diabetes mellitus. Early symptoms include a raging thirst, extreme fatigue and excessive urination. If sufferers are not treated, they lose weight and eventually die.

Until 1922, doctors regarded the symptoms of diabetes mellitus as a death sentence, especially when they occurred in children. Physicians confirmed a diagnosis by tasting the patient's urine to find out if it contained sugar (today, there are better methods). One noted, 'In children the disease is rapidly progressive, and may prove fatal within a few days... As a general rule, the older the patient at the onset, the slower the course.'

In 1923 a 16-year-old Canadian girl, Catherine Hayes, wrote, 'Only a year ago I was a human skeleton. Can you imagine being 5 feet 4 inches and weighing only 55 pounds [less than 4 stone; 24 kg]? My doctor had placed me on a starvation diet, the only available treatment at the time. I became too weak to engage in any physical activity, even walking, and I eventually lost most of my muscle tissue. My skin became so dry that it flaked and peeled. It was not only painful, but embarrassing, and I wondered how my school friends could bear to look at me. Many of them could not, and in the last year, I spent much of my time alone.

'However, what was even harder to bear was the knowledge that my life could end at any time. It was very difficult to live with the thought that if I took a turn for the worse I could lapse into a coma and die.

'Then came the news about the discovery of insulin...'

1 THE CONCEPT OF HOMEOSTASIS

In this part of the book we focus on the basic concepts and mechanisms of homeostasis, and then look at some examples including blood sugar, temperature (Chapter 12) and the workings of the liver and kidney (Chapter 13). Finally, we investigate exercise physiology (Chapter 14) – the short-term and long-term changes that happen to our bodies when we exercise. The fact that our muscles can work 20 times as hard and yet our body can still maintain acceptable internal conditions is one of the most impressive feats of homeostasis.

The word **homeostasis** means 'steady state'. Homeostasis describes how the body regulates its processes to keep its internal conditions as stable as possible. Homeostasis is necessary because cells, especially those of humans and other higher animals, are efficient but very demanding. To function properly they need to be bathed in tissue fluid that can provide the optimum conditions. Nutrients and oxygen must be delivered and waste needs to be removed. The concentration, temperature and pH of the fluid between cells must also be kept at levels that guarantee efficient cell functioning.

However, the phrase 'steady state' is a bit misleading. The conditions inside our bodies are not constant, but are kept within a narrow range. Some factors, such as core temperature and blood pH, fluctuate only slightly, while others, such as blood glucose, vary considerably throughout a normal day without producing any harmful effects.

In this chapter we look at the control of blood glucose and at the role of the liver. Other aspects of homeostasis are covered elsewhere in the book. Table 11.1 shows you where to find them.

Table 11.1 Homeostatic mechanisms covered in this book	
Homeostatic mechanism	**Covered in**
Temperature	Chapter 12
Blood glucose	Chapter 11
Blood pressure	Chapter 10
Solute concentration of blood	Chapter 13
Blood pH	Chapter 13
Blood volume	Chapter 13
Blood hormone levels	Chapter 18
	Chapter 20

THE MECHANISM OF HOMEOSTASIS

When you start to study how the body controls a physiological factor such as temperature, blood glucose or blood pressure, it is important to organise your thoughts by asking the following questions:
- What conditions bring about a change in the factor being considered?
- What detects the change?
- How is the change reversed?

You will soon notice a pattern. Whenever a physiological factor changes, the body detects the change and then, by using nervous or hormonal signals, or both, it reverses the change. The extent of the correction is monitored by a system called **negative feedback** (Fig 11.1). This makes sure that, as levels return to normal, corrective mechanisms are scaled down.

? TEST YOURSELF

A Why are homeostatic mechanisms often described as detection–correction systems?

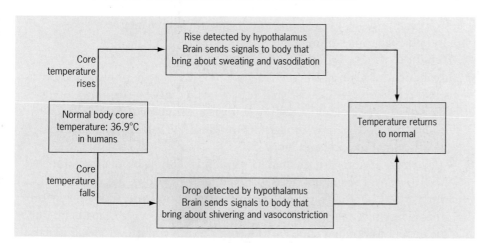

Fig 11.1 The mechanism of homeostasis as illustrated by temperature control. All other examples follow the same general pattern

Control of body temperature illustrates this mechanism well (see Chapter 12). When we are in a very hot environment or when we have been doing strenuous exercise, our body temperature rises. The brain detects this and sends signals to the body to bring the temperature down, using various corrective mechanisms such as sweating and increased blood flow to the skin. As the body cools, the drop in temperature is monitored by the brain, which begins to send out fewer signals. Sweating then decreases.

The opposite of negative feedback is **positive feedback**. In this situation, a change is amplified rather than returned to normal. Positive feedback in living systems is rare, but there are a few examples. Damage to a blood vessel causes a cascade reaction that brings about blood clotting: a few molecules of a substance become activated and each one then activates many more (see Chapter 12). The result is a tangled mesh of protein fibres that plugs the hole in the blood vessel.

Positive feedbacks also occur in abnormal situations and when normal homeostatic mechanisms get out of control. Elderly people whose sensory systems have deteriorated can suffer from hypothermia (see the Assignment at the end of Chapter 12). When they start to become cold, their body systems fail to respond, and the drop in body temperature goes uncorrected. As they grow colder, their metabolic rate decreases still further, they produce less heat and so in cold conditions they continue to cool down at an ever-increasing rate. Death occurs when their core temperature falls to about 25 °C.

HOMEOSTATIC MECHANISMS AT THE MOLECULAR LEVEL

We can also look at homeostasis at the level of molecules. In Chapter 5 we saw that enzymes control metabolic pathways, and that particular products are made, step by step, in a series of carefully controlled reactions. In a metabolic pathway, any product is made only as fast as it is needed. Excess product often prevents further quantities being made by inhibiting one of the enzymes in the pathway. When the excess has been used up, the inhibition is lifted and production continues. The process is self-regulating and so we describe it as a **homeostatic mechanism**.

2 THE CONTROL OF BLOOD GLUCOSE

In normal circumstances, we obtain most of our energy by respiring glucose. Cells therefore need a regular supply of this simple sugar. Some vital organs, notably the brain, cannot do without it even for a short time: lack of glucose can cause brain damage. Other cells and tissues, such as muscles, can respire lipids or even proteins for a short time if glucose is unavailable.

Blood glucose comes from:
- Digestion of carbohydrates in the diet.
- Breakdown of **glycogen** (see Chapter 3). This storage polysaccharide is made from excess glucose in a process called **glycogenesis**. Glycogen is particularly abundant in liver and muscle cells. When needed, glycogen can be broken down quickly to release glucose in the process of **glycogenolysis**.
- Conversion of non-carbohydrate compounds. Following deamination, the acid part of the amino acid is converted to glucose. Pyruvate and lactate (see Chapter 3) can also be converted to glucose. The conversion process in either case is called **gluconeogenesis**, which means literally 'the generation of new glucose'. During prolonged fasting, blood glucose is maintained by conversion of the body's protein and lipid stores.

The blood of a healthy person contains between 80 and 90 mg of glucose per 100 cm^3. This normal value is maintained even during prolonged fasting. But the value rises to around 120 to 140 mg per 100 cm^3 shortly after a meal, when carbohydrate digestion is in full swing. Feedback mechanisms bring the levels back to normal in about two hours. Under normal conditions, the kidney is able to reabsorb all of the blood glucose passing through it, preventing any from being lost in the urine.

 REMEMBER THIS

Glucose is one of the most abundant substances in our diet. Plant material contains starch and cellulose, and meat contains glycogen. All three are glucose polymers. During digestion, both starch and glycogen are broken down into glucose, which then passes into the blood in large amounts. Humans cannot digest cellulose, and this constituent of plant material forms much of our dietary fibre.

REMEMBER THIS

The terms used to describe glucose metabolism – glycogenesis, glycogenolysis and gluconeogenesis – can be confusing. Try remembering the origins of the words that make them up:
Glyco, gluco = sugar
lysis = splitting
neo = new
genesis = generation, or formation

THE MECHANISM OF BLOOD GLUCOSE CONTROL

SEE QUESTIONS 1 TO 3

The control of blood glucose level is a good example of homeostasis. A negative feedback mechanism operates to detect and correct the level of blood glucose, maintaining it within 'safe' limits. The pancreas plays a central role in the control of blood glucose. The digestive functions of the pancreas are covered in Chapter 8 but in this section we are concerned with its **endocrine role**: how it produces the hormones **insulin** and **glucagon** to control blood glucose (look ahead to Fig 11.3).

The pancreas itself detects any change in the level of blood glucose. If blood glucose becomes too high, β **cells** in the **islets of Langerhans** respond by releasing insulin. This hormone travels to all parts of the body in the blood, but exerts an effect mainly on cells in muscles, liver and adipose (fat storage) tissue. Insulin lowers blood glucose by making cell surface membranes more permeable to glucose. It activates transport proteins in the membranes, allowing glucose to pass into cells. Insulin also activates enzymes inside the cells. Some of these enzymes convert the glucose to glycogen, others increase protein and fat synthesis.

If the levels of blood glucose get too low, α **cells** in the islets of Langerhans secrete glucagon. This hormone fits into receptor sites on cell surface membranes, and activates the enzymes inside the cells that convert glycogen to glucose. The glucose then passes out of the cells and into the blood, raising blood glucose levels.

WHEN CONTROL OF BLOOD GLUCOSE FAILS

People with the disease **diabetes mellitus** are unable to control the level of glucose in their blood. This produces a range of symptoms. Blood glucose levels can get too high because the affected person produces little or no insulin. Without insulin, glucose cannot pass into cells and remains in the blood. The solute concentration of the blood increases, interfering with effective circulation and making the individual very thirsty. The cells are starved of their main fuel and are forced to respire lipids and proteins, leading to weight loss and eventual starvation. Glucose appears in the urine because blood sugar levels are so high that the kidney cannot reabsorb it all.

Diabetes has a variety of causes and varying degrees of severity. In the UK, about 25 people in every thousand suffer from diabetes in one form or another. That means that there are over a million sufferers. About a third of these have Type I diabetes, roughly two-thirds have Type II (see below).

TYPES OF DIABETES MELLITUS

There are two main types of diabetes: **Type I** and **Type II**.

Type I diabetes, also known as 'early onset diabetes', occurs when the body cannot make insulin. This is often caused by an auto-immune reaction which attacks and destroys cells in the islets of Langerhans. This form of the disease usually appears before the age of 20, and the onset is sudden. Sufferers have this condition for the rest of their lives, but they can be treated by regular injections of insulin (Fig 11.2) matched to their glucose intake (in diet) and expenditure (in exercise). Before insulin was available, untreated Type I diabetes was usually fatal within a year of diagnosis. Most people with diabetes today need insulin injections, but they lead normal lives.

Type II diabetes, also called 'late onset diabetes', is more common than Type I, accounting for about 70 per cent of the cases in the UK. It tends to begin during middle age and is more common in those who are overweight. This form of diabetes is due to a decline in the efficiency of islet cells, or to a failure of the cell surface membranes to respond to insulin. Fortunately, in many cases it can be controlled by regulating the diet.

✔ **REMEMBER THIS**

Diabetes mellitus is describes an inability to control blood glucose levels. It is not the same as diabetes insipidus. This form of diabetes is caused by a lack of anti-diuretic hormone in the body (see page 238).

Fig 11.2 This girl has diabetes and has to inject insulin every day. For many years, insulin for human treatment was derived from cows or pigs, but both are slightly different from human insulin and can cause an immune reaction, reducing their effectiveness. Today, diabetics use human insulin produced by genetically engineered bacteria (Chapter 6)

! SCIENCE FEATURE The discovery of insulin

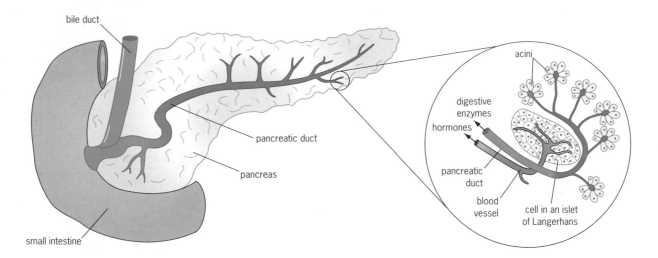

In the early days of science, one of the most direct ways to find out an organ's function was to remove it surgically and then observe the effects of its loss on the organism. If the symptoms could be relieved by injecting a ground-up extract of the same gland, then the organ in question was an endocrine gland – a gland that releases hormones into the bloodstream.

By the early years of this century, this effective but less-than-subtle approach had been used to clarify the role of several organs, including the thyroid gland. But, although researchers suspected that the pancreas was an endocrine gland, they ran into problems when they tried to use this method to demonstrate that the pancreas made a hormone responsible for controlling glucose metabolism. When the pancreas of a dog was removed, the animal developed the symptoms of diabetes. But injecting an extract of ground-up pancreas failed to relieve the symptoms, a result that recurred in several experiments.

Frederick Banting, a Canadian doctor, first had the idea that the digestive enzymes also made by the pancreas could be destroying any other active substances that were being made there. He persuaded the head of the physiology department at the University of Toronto, John J R Macleod, to let him have a laboratory, ten dogs and an assistant – Charles Best.

The dogs were subjected to one of two treatments. The pancreatic ducts of the dogs in group 1 were tied so that the animals could not produce pancreatic juice. Over the course of several weeks the acini, the cells that make pancreatic juice, degenerated, leaving just the islets of Langerhans functional. These dogs did not develop diabetes. The pancreases of the group 2 dogs were removed, and these animals did become diabetic.

Banting then made a pancreatic extract from the group 1 dogs and injected it into the diabetic dogs of group 2. The result was

Fig 11.3 The major part of the pancreas makes a juice containing digestive enzymes, but small patches of cells, called islets of Langerhans, produce the hormones insulin and glucagon

Fig 11.4 Banting and Best. In 1923 the Nobel Prize was awarded to Banting and Macleod, but Banting shared his prize with Best, and Macleod did likewise with Collip

dramatic – an instant reduction of blood sugar that could be achieved repeatedly in several different animals. There was great excitement, and then disappointment as the extracts were tried in humans and found to be too impure (they produced fever). Banting and Best enlisted the help of the biochemist James Collip. He went on to make a purer extract than had previously been possible, which could be used for people.

Some people feel that animal experiments are a bad idea, but the few dogs used in these early experiments enabled Banting's team to change a progressive and fatal illness into a chronic, manageable condition. As a direct result of these experiments, insulin became available to the world's diabetics. Today there are over 300 000 insulin-dependent diabetics in the UK, and as many as 30 million worldwide. Without insulin, they would not be alive.

3 THE LIVER PLAYS A CENTRAL ROLE IN HOMEOSTASIS

The liver receives blood from two sources: from the intestines via the hepatic portal vein, and from the hepatic artery (see Fig 11.5b). The composition of blood that flows into the liver can fluctuate greatly, depending on factors such as the timing and nature of the last meal, but the content of blood leaving the liver is remarkably constant.

THE STRUCTURE OF THE LIVER

Before we go on to look at the functions of the liver in more detail, it is important to understand its structure (see Fig 11.5 on the next page).

Blood is brought into the liver by two blood vessels. The 30 per cent that arrives in the **hepatic artery** is oxygenated blood, while the 70 per cent delivered by the **hepatic portal vein** contains relatively little oxygen but is rich in nutrients that have been absorbed from the intestines. The **hepatic vein** removes blood from the liver.

The liver consists of hundreds of thousands of **lobules**, each about 1 mm in diameter, which surround branches of the hepatic vein. Channels called **sinusoids** radiate out from each central vein. These are surrounded by rows of liver cells called **hepatocytes**. These apparently unspecialised cells perform the majority of the liver's functions, and the composition of the blood changes as it flows along each sinusoid towards the central vein.

Dotted along the sinusoids are numerous white cells called **Kupffer cells** (see Fig 11.5e). These are phagocytes (see Chapter 30) that engulf bacteria and debris. Parallel to the sinusoids are fine channels called **bile canaliculi** (singular: canaliculus). Hepatocytes secrete the constituents of bile into the canaliculi, and this drains into the gall bladder. Bile is covered in detail in Chapter 8.

THE MAIN FUNCTIONS OF THE LIVER

The thousands of different chemical functions performed by the liver contribute greatly to the overall composition of the blood. They can be grouped under a few basic headings:

- Control of blood glucose levels
- Control of amino acid levels
- Synthesis of plasma proteins
- Synthesis of fetal red blood cells
- Destruction of red blood cells
- Detoxification
- Production of bile
- Control of lipid levels
- Storage of vitamins
- Cholesterol formation

THE CONTROL OF AMINO ACID LEVELS

The human body cannot store proteins. Every day an adult needs a minimum amount (40 to 60 grams, about the weight of one egg) to provide the amino acids the body needs to repair and grow new cells. Most people take in more than this in their diet, and the liver breaks down any amino acids that are not used. Obviously, growing children and pregnant or breast-feeding mothers need more protein.

? TEST YOURSELF

B During their experiments, Banting and Best noticed that flies congregated around the urine of the diabetic dogs. Suggest an explanation for this.

? TEST YOURSELF

C Hepatocytes have many mitochondria and microvilli. What does this suggest about their function?

Fig 11.5 Structure of the liver

(a) A liver prepared for organ transplant sugery. This liver has been cut to fit onto the recipient

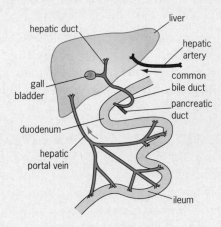

(b) The blood supply to the liver and associated organs. The liver weighs about 1.5 kg in a normal adult

(c) The liver consists of about a half to one million cylindrical blocks of cells called lobules

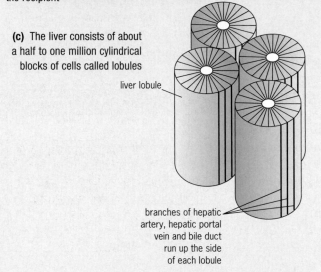

liver lobule

branches of hepatic artery, hepatic portal vein and bile duct run up the side of each lobule

Sinusoids: spaces through which blood flows as it passes from the hepatic portal vein and hepatic artery to the central vein

Kupffer cell

Bile canaliculus: tiny tube into which bile is secreted

bile duct

branch of hepatic portal vein

branch of hepatic artery

central vein

Liver cells are arranged in vertical plates. These act rather like the oil filter of a car

(d) A micrograph of liver lobules. Radiating sinusoids drain into the central veins

(e) The fine structure of the liver. Blood is delivered to the liver cells in branches of the hepatic artery and the hepatic portal vein. As blood flows along the sinusoids, some chemicals are removed while others are added by secretion. The liver secretes bile into the canaliculi. These drain into small branches of the bile duct

Fig 11.6 The ornithine cycle is a series of reactions in which toxic ammonia is combined with carbon dioxide to form urea. Effectively, a molecule of urea is assembled on the ornithine. When complete, the urea will become detached, leaving ornithine to continue the cycle. The basic steps are as follows:

1 One molecule of CO_2 and one of NH_3 are added to the molecule of ornithine, forming citrulline

2 A second molecule of NH_3 is added to citrulline, forming arginine

3 Arginine is split to produce one molecule of urea and one molecule of ornithine to continue the cycle

(a) Deamination
Amino group and hydrogen removed from amino acid as ammonia

$$\frac{1}{2}O_2 + NH_2 - C - COOH$$

(amino acid)

Two molecules of ammonia enter the ornithine cycle

$$NH_3 + \quad C - COOH$$

Organic acid can enter Krebs cycle and be respired

(b) Ornithine cycle

lungs
liver
gall bladder
kidney
colon
ureters
bladder
urethra

ornithine

1 → citrulline

carbon dioxide from respiration

2

arginine

ornithine regenerated

3 arginine is hydrolysed to release urea

one molecule of urea produced

urea (contains two amino groups)

Amino acids, like many other digested foods, reach the liver via the hepatic portal vein. They can be:

- **Deaminated.** This process removes the amino group (NH_2) of an amino acid and forms ammonia (NH_3): Fig 11.6(a). The organic acid residue is usually respired, while the toxic ammonia is quickly converted into a more harmless substance, urea, via the ornithine cycle: Fig 11.6(b). The kidneys remove urea from the body (see Chapter 13).

- **Transaminated.** There are 8 essential amino acids (10 in children) that must be present in the diet. The remaining 12 are termed non-essential amino acids because they can be made in the liver by transamination. This process involves the transfer of an amino group from an amino acid to an acid (derived from carbohydrate metabolism), thereby making a new amino acid.

- Used to synthesise plasma proteins such as fibrinogen.

- Released unchanged into the general circulation (most cells need a supply of amino acids for protein synthesis).

DETOXIFICATION AND HORMONE BREAKDOWN

The liver concentrates and **detoxifies** (breaks down) many harmful chemicals. Some, such as hydrogen peroxide, are produced by the body itself. Others, alcohol and food additives for example, come from outside. Drinking large amounts of alcohol over a long period of time can cause the death of liver cells, followed by replacement with connective tissue ('scar' tissue). This is known as **cirrhosis** of the liver. (Fig 11.7)

The liver also breaks down many circulating hormones. Removal of hormones from the blood is an important aspect of the control process as it ensures that hormones act only for as long as they are needed. Insulin, for instance, is rapidly metabolised by the liver and has a half-life of about 10 to 15 minutes.

? TEST YOURSELF

D Why can't we take in a week's supply of protein in a single meal by eating one large steak or omelette?

Fig 11.7 The liver of an alcoholic who developed cirrhosis of the liver. In the liver, alcohol (ethanol) is converted into ethanal by the enzyme alcohol dehydrogenase. Ethanal is toxic and seems to be responsible for much of the long-term alcoholic liver damage, although the exact mechanisms involved are unclear

STORAGE

The liver stores relatively large amounts of vitamins A, D and B_{12}, enough to supply the body for several months. Other vitamins, such as most of the B complex, and the minerals copper and iron are also stored but in smaller amounts. This high vitamin and mineral storage capacity explains why eating liver is good for you, even though some people don't find it a pleasant experience.

MANUFACTURE OF BLOOD PROTEINS AND BLOOD CELLS

The liver makes many important blood proteins, including the most abundant plasma protein, **albumin**, fibrinogen and other substances involved in blood clotting (see Chapter 12). Given a supply of vitamin K, the liver can make prothrombin and other blood clotting factors.

The liver of a fetus manufactures red blood cells, but this complex process is taken over by the bone marrow after birth.

SUMMARY

When you have studied this chapter, you should know the following:

- The word **homeostasis** means 'steady state'. Homeostatic processes keep conditions in the body within narrow limits.
- Homeostasis is usually maintained by **negative feedback mechanisms**. The body detects a change in a particular internal factor, such as core body temperature, and then activates a corrective mechanism to reinstate the normal level.
- Blood glucose levels are monitored by the **islets of Langerhans** in the pancreas. If levels get too high, β cells in the islets secrete **insulin**. If blood glucose gets too low, α cells secrete **glucagon**.

- **Insulin**, a peptide hormone, increases the permeability of cell surface membranes to glucose. It also appears to activate intracellular enzyme systems that convert glucose to glycogen, fat and protein.
- **Glucagon** promotes the breakdown of **glycogen**, releasing more free glucose into the blood.
- The liver is a large organ that plays a major role in controlling the composition of the blood.

? EXAM QUESTIONS

1 An experiment was carried out to determine how the level of glucose in the blood varies in a normal person. Samples of blood were taken at hourly intervals for a period of 18 hours, from 06.00 to 24.00 hours. During this period, three meals were taken: breakfast at 7.00, lunch at 13.00 and tea at 16.00.

The results are shown in the graph below.

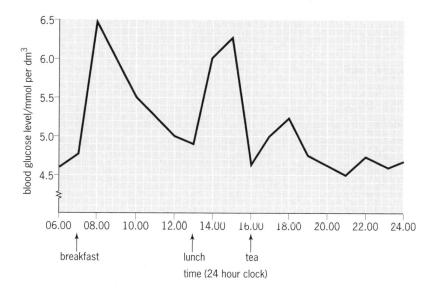

a) **(i)** Explain why the blood glucose level rises sharply between 7.00 and 8.00 hours.
(ii) Explain how the changes in blood glucose level between 08.00 and 12.00 hours are brought about.

b) Describe and explain two ways in which the changes in the blood glucose level after lunch (13.00 hours) and tea (16.00 hours) differ from those after breakfast.

c) Explain why it is important that the level of glucose in the blood is controlled.

Edexcel HB6 (Synoptic) June 1999 Q5

2 Use appropriate terms, or phrases, to complete the following passage. Each term or phrase may only be used once.

The importance of maintenance of constancy both within cells and within organisms was first appreciated by the French physiologist Claude Bernard. This maintenance of constancy is referred to as _____ . Many examples of this are carried out by the liver, for example the maintenance of constant blood glucose levels. Two other examples in the liver are the maintenance of _____ and _____ .

If the blood sugar level rises, the ß cells in the _____ , which are found in an organ called the _____, release a hormone called _____ . This decreases the blood sugar level by causing glucose to be converted into a polysaccharide called _____ which is stored in the liver. If the blood glucose levels fall, the hormone _____ is produced which causes this polysaccharide to be converted into glucose. Control mechanisms that involve a corrective procedure which returns levels to the norm, or set point, are referred to as examples of _____ . Control of a target organ by hormones, transported in the blood from the endocrine gland in which they are produced, is an example of chemical communication at the level of the organism. The binding of hormones with receptors on the cell surface membrane causes the release of secondary messengers within the cytosol which influence enzyme activity. This is an example of chemical communication at the _____ .

OCR 6911 Cells and Structures June 2000 Q2

❓ EXAM QUESTIONS

3 Read the following passage.

Diabetes

Diabetes mellitus is a group of disorders that all lead to an increase in blood glucose concentration (hyperglycaemia). The two major types of diabetes mellitus are type I and type II. In type I diabetes there is a deficiency of insulin. Type I diabetes is also called insulin-dependent diabetes mellitus because regular injections of insulin are essential. It most commonly develops in people younger than age twenty.

Type II diabetes most often occurs in people who are over forty and overweight. Clinical symptoms may be mild, and the high glucose concentrations in the blood can often be controlled by diet and exercise. Some type II diabetics secrete low amounts of insulin but others have a sufficient amount or even a surplus of insulin in the blood. For these people, diabetes arises not from a shortage of insulin but because target cells become less responsive to it. Type II diabetes is therefore called non-insulin-dependent diabetes mellitus.

a) Describe how blood glucose concentration is controlled by hormones in an individual who is not affected by diabetes.

b) Suggest how diet and exercise can maintain low glucose concentrations in the blood of type II diabetics.

c) Glucose starts to appear in the urine when the blood glucose concentration exceeds about 180 mg dm^{-3}. Explain how the kidney normally prevents glucose appearing in urine.

AQA (NEAB) BY03 Physiology June 2000 Q8

KEY SKILLS **ASSIGNMENT**

LIVING WITH DIABETES

Having diabetes doesn't mean you can't live life to the full, but unlike other chronic conditions such as asthma, someone with diabetes cannot just take their medication and forget there was ever a problem. Instead, they must become experts in the management of their own condition, matching up dietary intake with insulin doses and exercise regimes in order to keep their blood glucose within definite limits.

This is an account of the way in which one person, Anna, deals with her diabetes. It should be stressed from the start that the condition affects different people in different ways. A blood sugar level that would have one person fainting on the floor may produce no symptoms in another. Bear this in mind as you read on.

1 Explain how a non-diabetic person responds to a rise in blood sugar level. Your answer should include the effect of insulin on cells.

2 What is the difference between a chronic and an acute condition?

Types of diabetes

There are two basic types of diabetes: Type I and Type II. Type I is also known as 'early onset' or 'insulin-dependent' as it appears in early life and controlling it usually requires insulin. At the root of the problem is thought to be an auto-immune reaction in which the body's antibodies destroy its own insulin-producing cells. As these cells cease to produce insulin, less insulin is released into the blood after a meal. Without sufficient insulin, most of the glucose that enters the blood from the intestines after digestion cannot get into cells. Glucose accumulates in the blood, and the cells, which are starved of their main fuel, turn to alternatives – lipid and protein.

Five of the major symptoms of Type I diabetes are:
- excessive thirst;
- excessive urinating;
- weight loss;
- glucose in the urine;
- the fruity smell of ketones on the breath (a by-product of lipid metabolism).

3 Using the information given above, suggest explanations for each of the five symptoms.

The story of Anna

Anna, who is now in her forties, was diagnosed as a diabetic at the age of 18 months. This is early, but her brother had also developed diabetes (when he was 4) and so her mother was acutely aware of the symptoms.

4 It is thought that people can inherit a susceptibility to diabetes, but it needs an environmental trigger of some sort, such as exposure to a virus. Does Anna's story support this idea?

Anna has come to terms with her diabetes and has found that management is easiest within a routine. Overall, the aim is to keep her blood glucose within the range of 4 to 11 millimoles per litre. (Non-diabetics maintain their blood glucose to within about 4 to 9 millimoles per litre.)

When she wakes up in the morning, Anna tests her blood sugar. This is done by pricking her finger and putting a drop of blood on a test strip, similar to that shown on page 119. The level of blood glucose is then obtained by either matching up the colour to the chart or by placing the strip into a machine which takes the reading automatically. Anna would expect her glucose levels to be about 3 at this time in the day. Anything less than 4 is regarded as hypoglycaemic, although Anna gets no symptoms until the value approaches 1.

5 **a)** What fraction of a mole is a millimole?
b) Suppose Anna has a blood glucose reading of 10 millimoles. If the relative molecular mass of glucose is 180, how many grams of glucose are in one litre of her blood?

After taking the reading, Anna injects herself with insulin. Two types of insulin are now available; fast-acting and slow-acting. Anna injects 2 units of fast-acting and 14 of slow-acting. Fast-acting is normal, soluble insulin and works straight away. Slow-acting insulin is attached to a retarding agent that releases the insulin slowly over the next few hours.

Fig 11.A1 For diabetics, there is now an alternative to syringes. Special pen-like injection devices such as this Novopen are widely available. They are very accurate and discreet to use

KEY SKILLS **ASSIGNMENT**

6 **a)** Suggest the advantage to diabetics of having slow-acting insulin.

b) Suggest why you must have some fast-acting insulin.

Until relatively recently the insulin was obtained from animal pancreas. This was less than perfect because the non-human insulin slowly brought about an immune reaction. Eventually, the 'foreign' insulin had little effect. With modern genetic engineering, however, it is now a relatively simple task to produce human insulin on a large scale – see Fig 11.A2.

Fig 11.A2 Human insulin is made on a large scale by genetic engineering and is supplied in different doses

7 Why can't insulin be taken orally?

After breakfast, Anna goes to work. In her handbag are glucose tablets and some biscuits as well as her insulin and testing equipment. Anna has a sweet drink and a couple of biscuits at break.

8 It is frequently said that 'diabetics can't eat sugar' but this is obviously not true. Suggest why Anna needs the mid-morning biscuits.

Going 'hypo' and 'hyper'

A Type I diabetic who takes no insulin experiences the five symptoms listed on the previous page – they have gone 'hyper'. In Anna's case, the early warning signs are headaches and feeling tired or lethargic.

On the other hand, once a diabetic has taken insulin they must take care to keep their blood glucose levels up or they might go 'hypo'. For Anna, symptoms of too little blood sugar are similar to going hyper: tiredness, irritability and an inability to concentrate. Often, other people notice the symptoms first.

9 The terms hypo and hyper are short for hypoglycaemia and hyperglycaemia. What do these words mean?

10 If Anna finds that she is going hypo, suggest what she can do.

Anna has a normal lunch and another snack in mid-afternoon. At tea time she takes her blood sugar again – it is normally about 8 millimoles. After this she takes more insulin, 2 units of short-acting and 10 of slow-acting. If she is going out for a meal, and will eat more than normal, she will take an extra two units of short-acting insulin. She must also be careful what she drinks, because alcohol has a powerful effect on glucose metabolism.

Before bed time, Anna has a sweet drink, usually chocolate, and checks her blood sugar again. It needs to be around 8 millimoles because she will not eat for another 8 hours or so.

11 **a)** Why is it important that Anna does not go 'hypo' in the night?

b) Suggest what Anna does if her reading is, say, 6 units?

12 From what you have read of Anna's daily routine, sketch a graph of her blood glucose levels over a 24 hour period.

Type II diabetes

Many diabetics have Type II diabetes, This is also known as 'late onset' and the treatment depends on the underlying cause. In most Type II cases, the condition is due to a combination of one or more of the following:

● Muscle and fat cells do not respond to insulin.

● The body can't produce enough insulin to meet demand.

● Liver cells release too much glucose from their stores.

For many Type II diabetics, a diet and exercise plan may be enough to keep blood sugar levels within normal limits.

12

Temperature regulation

Babies are at home underwater – they spend nine months there. At this age the diving reflex is well developed.

There are many recorded cases of 'near drowning', incidents in which people have been plucked from water and resuscitated, and subsequently restored to full health. Rescuing the victim quickly – within minutes – lessens the chance of brain damage as a result of oxygen starvation.

Things are different, however, when the water is very cold. People have been pulled from freezing water – having fallen through thin ice, for example – and have later recovered completely, despite being under water for as long as one hour. How is this possible?

It seems that two factors contribute. Firstly, a remarkable response known as the diving reflex occurs in people (especially infants) who are suddenly immersed in cold water. Heartbeat and breathing decrease and blood is redirected so that the organs at the core of the body – the heart, lungs and brain – receive much more blood than the limbs. In this way the vital organs get what little oxygen is available.

Secondly, the rapid cooling effect of the water, known as immersion hypothermia, reduces the oxygen demand of the brain and other organs. This means that people can survive for several minutes, perhaps even as long as an hour after the heart and breathing have stopped.

The same principle is used in surgery. Heart operations often involve stopping the heart and then connecting the patient to a heart–lung machine to supply the body with oxygenated blood. The heart itself, however, receives no blood and so to prolong the time available to operate, the core temperature of the patient's body is lowered to 27 °C. This reduces the oxygen demand sufficiently to double the time the surgeon has before the heart needs to be reconnected.

1 THERMOREGULATION: SOME BACKGROUND

Different animals control their body temperature in different ways. Many animals (fish, insects, amphibians, etc) have a body temperature that is more or less the same as their surroundings. Humans, other mammals and birds usually maintain a constant body temperature, despite the external temperature.

These two categories of animal are commonly described as **cold-blooded** and **warm-blooded**. But these terms are not really satisfactory: a cold-blooded animal, for example, is often not cold. In bright sunshine on a hot day, the core temperature of a crocodile can be several degrees higher than that of a mammal (Fig 12.1). Nor are the terms **poikilothermic** (cold-blooded) and **homoiothermic** (warm-blooded) strictly accurate. The word poikilos means changeable, while homeo means constant. However, many cold-blooded animals, especially aquatic ones, have a remarkably constant body temperature, due to the stability of their surroundings. Hibernating mammals, on the other hand, show a dramatic drop in their core temperature during their dormancy.

Fig 12.1 This crocodile is gaping to keep cool. The external temperature is over 45 °C and so its blood is far from cold. Evaporation from its large mouth helps it to cool down

To overcome these problems, we now use two other terms: **ectotherm** ('heat from outside') and **endotherm** ('heat from within'):

- **Ectothermic organisms** can control their body temperature only by changing their behaviour. All animals except mammals and birds are ectotherms.
- **Endothermic organisms** maintain a stable core body temperature using physiological and behavioural means. Mammals and birds are the only endotherms. Humans are therefore endotherms.

HUMANS AND HEAT

Some mammals, such as polar bears, are obviously adapted to withstand the cold. In contrast, humans almost certainly evolved in Africa, and, as a species, we are adapted to a warm climate. We have a thin covering of insulating fat and our body hair is sparse. Most hairs are tiny and no use for insulation. We are one of the few animals to be covered in sweat glands, and our skin can make melanin, a dark pigment that blocks out harmful ultraviolet light.

It seems that we were able to spread to the colder areas of the planet only because of our ability to control our environment. We could build shelters, make clothes and control fire. These skills more than made up for the fact that we had few physical features to allow us to cope with the cold.

2 THERMOREGULATION: THE BASICS

Humans, like most mammals, can maintain a constant core body temperature, despite changes in the surroundings.

The core temperature of the human body remains reasonably constant at around 37 °C. It can fluctuate by a degree or so (more during fever) but it is generally very stable. Fig 12.2 is a thermal image of an adult male, showing the definition of the body core; this includes the trunk, the head and the upper part of the arms and legs. Body temperature is taken by placing a clinical thermometer in sites that detect the temperature of the core (Fig 12.3).

Heat is produced inside the body as a by-product of metabolic reactions. Heat production occurs throughout the body, and is especially high in working muscles. It is often stated that the liver is a particular source of heat, but tests have shown that the temperature of blood in the hepatic vein is no higher than in other vessels, suggesting that heat production occurs more generally in the organs of the body. Since body heat is a by-product of metabolism, the amount of heat produced depends on the metabolic rate.

Fig 12.2 This is a thermal image of a man taken with a camera sensitive to infrared radiation. Areas of higher skin temperature look red or orange, cooler areas look blue or purple. The temperature of the skin can vary greatly, but the core temperature remains more or less stable

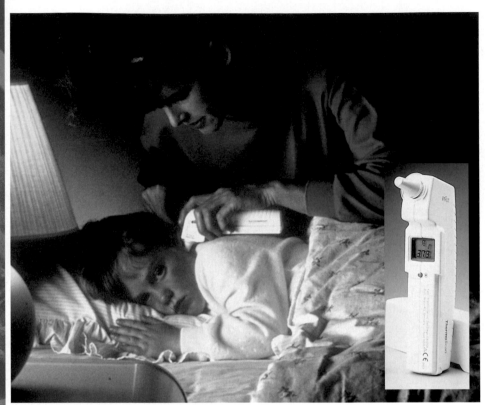

Fig 12.3 Measuring core body temperature. A clinical thermometer provides a good estimate of core temperature when it is placed in the mouth, armpit or rectum. This more recent thermal probe allows an instant reading of the temperature of the blood near to the brain, without damaging the ear

This can be increased by doing more exercise and by the secretion of hormones such as thyroxine or adrenaline (Chapter 18).

However, the principal way we control our body temperature is by increasing or decreasing heat loss to the environment, according to our needs. In humans, the basic mechanism that underlies temperature control, or **thermoregulation**, involves part of the brain called the **hypothalamus**. This acts as a thermostat. It can detect the temperature of the blood that passes through it and, if the temperature of the blood increases or decreases even slightly, the hypothalamus initiates corrective responses such as sweating or shivering.

When we encounter a particularly warm or cold environment, temperature receptors in the skin inform the hypothalamus. They also stimulate the higher, voluntary, centres of the brain. This means that we 'feel' hot or cold and decide to do something about it such as changing position (Fig 12.4), changing our clothing or turning the heating up or down. Often, this behavioural response corrects the situation without the need for any physiological response.

Before we can look at temperature control in detail, we need to understand a little about the processes of heat transfer, and to study the structure of the human skin.

Fig 12.4 When it's cold, we may automatically assume a fetal position in bed to keep warm: when the arms and legs are drawn into the body this reduces the surface area that can lose heat. When it's hot we do the opposite, spreading out as much as possible

3 HEAT: WHERE DOES IT COME FROM? WHERE DOES IT GO?

For any organism to maintain a stable body temperature, its heat loss must equal its heat gain. **Homoiotherms** such as ourselves, who live in a temperate climate, produce our own internal body heat to keep warm but we also control the amount of heat we lose to the environment.

Heat can be gained or lost in four different ways:
- conduction
- evaporation
- convection
- radiation

CONDUCTION

Conduction involves the transfer of heat between two objects that are in contact with each other. Heat is always conducted from a region of higher temperature to a region of lower temperature. When you sit on a cold seat, for example, heat is conducted from your body into the seat, until both are approximately the same temperature.

Fig 12.5 When the Titanic sank in the North Atlantic, the water temperature was similar to that of the air – a little above freezing. However, as we now know, those who remained dry in the lifeboats had a far greater chance of survival. The high thermal conductivity of the water drew the heat out of the unfortunate victims in the sea, most of whom died from hypothermia

The efficiency with which a material conducts heat is called its thermal conductivity. Different materials have different conductivities. Air has a low thermal conductivity, water has a much higher one (Figs 12.5 and 12.6).

A clothed human walking in air at 15 °C could maintain body temperature comfortably. If immersed in water at that temperature, it would not be long before their core temperature dropped and hypothermia began to set in. The heat loss into the water would be much greater – water can 'draw out' heat approximately 25 to 30 times faster than air at the same temperature.

Materials with a low thermal conductivity are very good insulators. Animals with fur keep warm largely by trapping air between the hairs. Humans rely more on fatty (adipose) tissue for insulation. This has a lower thermal conductivity than other body tissues and so conducts heat more slowly to the surroundings.

Fig 12.6 Most divers wear wet suits, even in tropical waters, because the body loses heat rapidly to any water that is below 29 °C. The wet suit traps a layer of water between the skin and the rubber. This warms up quickly as a result of conduction from the skin. The warm water layer cannot escape and so further heat loss is slow.
For very cold water, dry suits are available. These trap a layer of air next to the skin. This provides even better thermal insulation

CONVECTION

Convection is heat transfer due to currents of air or water. A person immersed in cold, absolutely still water would be able to heat up the water immediately next to the skin, and could reduce their heat loss to some extent by not moving.

However, this situation does not happen in real life. For a person who has fallen into the sea or a river, there are usually strong currents that continually move the water over the skin, causing water to be lost quickly by convection. Similarly, fast-moving air causes greater heat loss by convection. You have probably heard of the **wind chill factor**. A cold windy day feels a lot worse than a cold still day, even when the actual air temperature is the same.

In air, our clothes reduce heat lost by convection by trapping a layer of air next to the skin. In water, divers reduce the risk of heat loss by convection by wearing wet suits and dry suits (see Fig 12.6).

EVAPORATION

Evaporation is the change in state from a liquid to a gas. The evaporation of water uses up a large amount of energy, and this is known as the **latent heat of evaporation**. What this means for everyday life is that evaporation of water from a surface has a great cooling effect. Anyone who has ever stepped out of the bath and stood in a draught will have felt the power of this cooling. It is also why sweating is usually an effective way of losing heat when we get too hot. Even hot air – such as from a hair dryer – can have a cooling effect provided that the skin is wet so that evaporation can take place.

RADIATION

Radiation is the loss of infrared heat into the surroundings. A person sitting in a room at 20 °C radiates heat into the surrounding air, particularly from the exposed skin on the head, neck and hands. Under normal situations, at rest at comfortable room temperature, most of our heat loss is due to radiation. The heat that we radiate is in the infrared range. This is why infrared cameras can be used to search out humans and other warm-blooded animals from their colder surroundings.

In conduction, convection and evaporation, heat transfer occurs as a result of the movement of molecules. Radiation is fundamentally different: it does not depend on molecular movement. This explains why radiated heat can pass through a vacuum.

4 THE PROBLEMS OF TEMPERATURE EXTREMES

Humans can maintain a stable body temperature over a wide range of external temperatures. But what happens when we are no longer able to thermoregulate effectively, and our internal temperature changes? What happens to people stranded in the desert under a baking sun, or to those trapped in freezing water?

Experiments have been carried out in which a naked man is asked to sit in a room while the temperature around him is gradually raised or lowered. Fig 12.7 shows a generalised graph that plots what happens.

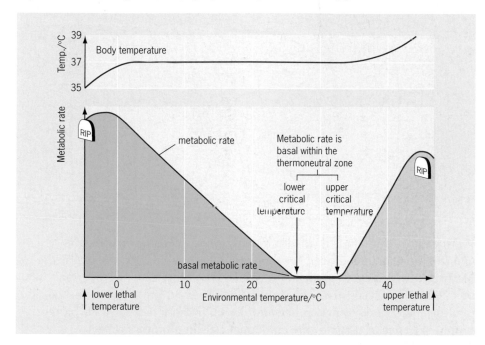

Fig 12.7 The effect of temperature on metabolic rate

Throughout the range 15 to 30 °C the naked man is able to maintain a constant core temperature. A point is reached, however, when his sweating and vasodilation are no longer able to keep him cool and he is said to have reached the **upper critical temperature**. After this, his metabolic rate starts to rise, but the experiment is always stopped at this point. If it were not, a positive feedback would begin. The man's enzymes would work faster and consequently make even more heat, raising his temperature further. At about 42 to 44 °C, the **upper lethal temperature** would be reached and he would die.

Why do humans die when the upper lethal temperature is reached? A common theory says that vital enzymes become denatured, but this cannot be the whole story. It is more likely that death from excess heat is due to a metabolic imbalance, caused by enzymes working at different rates. As the temperature increases, some enzymes work faster than others: some intermediate chemicals build up while some vital end-products become scarce. The end result is that cells, notably those in the brain, become irreversibly damaged.

If the same type of experiment is done, but temperature is lowered, the room eventually becomes so cold that the body is no longer able to maintain a constant core temperature: the **lower critical temperature** has been reached. At this point the metabolic rate begins to rise, and this helps the situation by generating more heat. Again, the experiment is always halted at this stage but, if heat loss were allowed to continue, a **lower lethal temperature** would be reached and the subject would die. Find out more about this in the Assignment at the end of the chapter.

? **TEST YOURSELF**

A Why does the upper critical temperature vary with humidity?

SEE QUESTION 1

5 THE STRUCTURE OF HUMAN SKIN

The skin is in direct contact with the external environment and so plays a central role in maximising or minimising heat loss. Fig 12.8 shows the structure of mammalian skin.

Fig 12.8 The structure of human skin. Human skin differs from that of most other mammals by being quite bare. We do not necessarily have fewer hairs, but most are tiny and no use for protection or insulation. We are also covered in sweat glands, and can lose heat by evaporation over most of our body surface. In contrast, most mammals have far fewer sweat glands, because evaporation from wet fur is not effective as a cooling mechanism

? TEST YOURSELF

B Are tattoos placed above or below the Malpighian layer? Explain.

Structurally, the skin is divided into two layers, the outer **epidermis** and the **dermis** underneath. Forming the boundary between the two is the **Malpighian layer**, the function of which is to produce new epidermal cells by mitosis. These cells make the protein keratin, and once they are pushed up into the epidermis they flatten, die and dry up because they have no blood supply. We are constantly losing epidermal cells in a process called **desquamation**. House dust is mainly desquamated skin cells, and a significant proportion of the mass of our mattresses and pillows is due to accumulated skin cells (one estimate says that skin adds 1 per cent to the weight of your mattress every year).

The dermis is much thicker than the epidermis, and contains many different structures such as nerve endings, hair follicles and blood vessels, held together with elastic connective tissue. Beneath the dermis is the **hypodermis** that usually contains at least some **subcutaneous fat**. This fat storage tissue, called **adipose tissue**, determines human body shape to a large extent.

Human skin has several functions:

- It detects stimuli using cells that are sensitive to heat, cold, touch, pressure and pain.
- It prevents excessive water loss or gain.
- It plays a role in thermoregulation by adjusting heat loss according to the circumstances.
- It prevents entry of microorganisms.
- It secretes hair, fingernails and toenails.
- It allows subtle forms of communication. We secrete natural scents, **pheromones**, from modified sweat glands (see Chapter 18).

6 THERMOREGULATION

PHYSIOLOGICAL RESPONSES TO COLD

There are four main physiological responses when the hypothalamus detects
a drop in blood temperature:

- We **shiver** when muscles contract and relax rapidly. Shivering muscles
  give out four or five times as much heat as resting muscles.
- **Vasoconstriction** occurs when the arterioles that lead to the capillaries
  in the surface layers of the skin **constrict** (narrow), so reducing blood
  flow to the skin (Fig 12.9a). This cuts down the amount of heat lost
  through the skin. Vasoconstriction is controlled by sympathetic nerves that
  pass from the **vasomotor centre** in the brain (Chapter 16). This centre,
  in turn, receives information from the hypothalamus.
- **Piloerection** means literally 'erection of hairs' and involves a reflex.
  In most mammals, piloerection makes the fur 'thicker', so that it traps
  more air to provide extra insulation. In humans, the **erector pili**
  muscles in the skin (see Fig 12.8) pull our tiny hairs upright, but only
  succeed in creating goose pimples.
- **Increased metabolic rate**. The body secretes the hormone adrenaline
  in response to cold. This raises metabolic rate and therefore increases heat
  production. People who live in cold conditions for a period of several
  weeks or months show a more permanent increase in metabolic rate due
  to the secretion of thyroxine (see Chapter 18).

PHYSIOLOGICAL RESPONSES TO HEAT

There are two main physiological responses to heat:

- **Vasodilation**. This occurs when arterioles that lead to skin capillaries
  dilate and shunt vessels are closed off, resulting in a greatly increased
  blood flow to the skin (see Fig 12.9b). As a result, more heat can be lost
  to the environment. Coupled with sweating, vasoconstriction
  is a very effective cooling mechanism.
- **Sweating**. Sweat is a salty solution made by sweat glands. Evaporation
  of sweat from the skin's surface leads to cooling. The efficiency of
  sweating depends on the humidity. In dry air, humans can tolerate
  temperatures of 65 °C for several hours. In humid air, however, when
  the sweat cannot evaporate, temperatures of only 35 °C (lower than
  the core body temperature) cause overheating.

SEE QUESTION 2

? TEST YOURSELF

C Which of the responses to cold
will be the least effective in
humans? Explain your answer.

✓ REMEMBER THIS

A common misconception is that
blood vessels move to the surface
of the skin. Blood vessels are fixed.
It is the blood flow through them
that can be changed.

? TEST YOURSELF

D Why do you think that people
in the tropics are recommended
to take salt tablets regularly?

Fig 12.9 Blood flow to the skin can be altered by
controlling the flow of blood in the arterioles leading
to the surface capillaries
(a) When tiny sphincter muscles in the arteriole
walls contract to prevent blood flow to the surface,
blood is forced through shunt vessels and away
from the skin, so less heat is lost
(b) When the sphincter muscles relax, the arteriole
walls relax, allowing more blood to flow into the
skin to increase heat loss

! SCIENCE FEATURE Freezing humans

Can we keep people alive by freezing them? Can people suffering from incurable diseases be put into deep-freeze until such time as a cure is found, and then revived? The simple answer is no.

Interestingly, human life can survive sub-zero temperatures in a metabolically inactive dormant state. Sperm, eggs and even embryos kept in liquid nitrogen at −196 °C can be thawed out and used successfully in infertility treatment (see Chapter 20). However, it is not possible to freeze people and bring them back to life. One of the main problems is that the formation of ice crystals destroys the cells, but a critical factor in maintaining life seems to be the timescale involved: ice crystals take time to develop. When very small samples are immersed in liquid nitrogen, the freezing process happens rapidly: all the molecules simply 'lock' in position and ice crystals do not form. The cells and organelles are not damaged and, when thawed, function normally. This technique, however, only works with tiny samples of tissue, which freeze instantly and entirely.

Fig 12.10 If we developed a way of instantaneously freezing all the cells of a fully developed human, could they be thawed out and live to tell the tale? Only time will tell

7 SKIN COLOUR AND SUNLIGHT

The colour of human skin is determined genetically. This is an example of continuous variation (Chapter 22), which suggests that there are several genes operating that give rise to a whole range of skin colour from the palest to the darkest. The darkness of skin is due to the amount of **melanin** present (Fig 12.11). This pigment is able to absorb ultraviolet light, thus protecting the DNA in skin cells from damage (Fig 12.12).

Fig 12.11 Skin contains cells called melanocytes that produce structures called melanosomes, which contain the melanin. Dark skin contains no more melanocytes than pale skin, but the melanosomes are larger and more evenly spread. Each melanocyte can protect the nuclei of several neighbouring cells. All but the darkest skins will respond to sunlight by producing more melanin

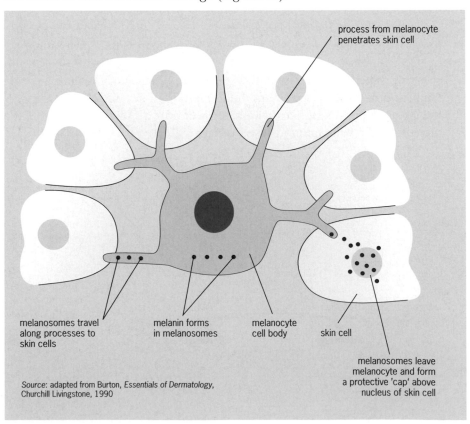

process from melanocyte penetrates skin cell

melanosomes travel along processes to skin cells

melanin forms in melanosomes

melanocyte cell body

skin cell

melanosomes leave melanocyte and form a protective 'cap' above nucleus of skin cell

Source: adapted from Burton, *Essentials of Dermatology,* Churchill Livingstone, 1990

Fig 12.12 (a) Black skin colour is due to the pigment melanin, a substance that protects skin from the damage by ultraviolet light. Skin cancer is rare in dark-skinned races. Fair-skinned people are not adapted to cope with strong sunlight **(b)** This is not a tan, but a first degree burn! When the redness dies down the skin begins to make melanin, giving the desired tan. However, the tan wears away in a couple of weeks. People with pale skin lack the quantities of melanin that provide protection against the genetic damage that ultraviolet light can cause and so run a high risk of developing skin cancer if they sunbathe excessively

Humans originated in Africa, and needed deeply pigmented skin to protect them from the fierce tropical sun. Under such circumstances, those with darker skin would have had an advantage over paler skinned individuals. It is more difficult, however, to see why white-skinned races evolved. Scientists think that the reason for this may be to do with the ability to synthesise vitamin D.

At the Equator, the sunlight was strong enough to penetrate dark skin and enable the synthesis of enough vitamin D for health. At lower light intensities, however, dark-skinned individuals could not make enough vitamin D. In these circumstances, the paler skinned individuals had a selective advantage and so natural selection brought about the development of pale skin in northern or southern latitudes (Fig 12.13).

Fig 12.13 As a general rule, the further you go from the Equator, the paler the skin colour. The map shows the relationship between skin colour and ultraviolet light intensity. (Adapted from Jolly and Plug, *Physical Anthropology and Archaeology*, 2nd edn, Knopf 1979, and Goldsby, *Race and Races*, Macmillan 1971)

lightest ☐☐☐☐☐☐ darkest   contours show UV light intensity in watt-seconds cm⁻¹

⚠ SCIENCE FEATURE The effect of sunlight on skin

The sun produces electromagnetic radiation of many different wavelengths. Some ultraviolet radiation is filtered out by the ozone layer, but some reaches the Earth's surface. Ultraviolet light is a potential problem because it is mutagenic – it damages DNA.

Ultraviolet radiation (Fig 12.14) is split into three categories:

● UVA (long wavelength, 315–400 nm) is the least likely to cause sunburn, and it cannot stimulate melanin production. However, it can damage DNA and lead to gene mutations that can give rise to skin cancer (Fig 12.15).

● UVB (medium wavelength, 295–314 nm) causes sunburn and stimulates melanin production.

● UVC (short wavelength, 200–294 nm) is very damaging to skin but fortunately much of it is filtered out by the ozone layer.

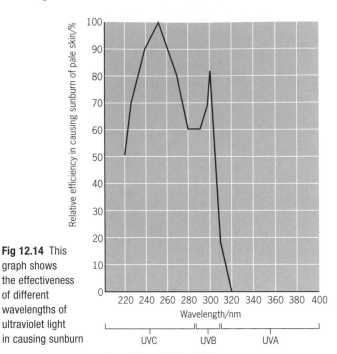

Fig 12.14 This graph shows the effectiveness of different wavelengths of ultraviolet light in causing sunburn

Fig 12.15 Skin cancer: this is a malignant melanoma, a common cancer especially where light-skinned people are exposed to strong sunlight. If the cancer can be removed before it spreads to other parts of the body, the chances of survival are over 90 per cent

SUMMARY

After studying this chapter, you should know and understand the following:

■ Animals are divided into two groups according to their ability to thermoregulate: **endotherms** ('heat from within') maintain a stable core temperature by behavioural and physiological means; **ectotherms** maintain a relatively steady body temperature using behaviour only.

■ Humans, like all mammals, are ectotherms. We also say that mammals are warm-blooded, or **homoiothermic**. This means that they can maintain a stable core body temperature despite changes in the external temperature.

■ To maintain a stable body temperature, heat loss must equal heat gain.

■ Unless the climate is very hot, we regulate body temperature by producing heat as a result of metabolic reactions, and by controlling how much of it we release to the environment. When we are too hot, we maximise heat loss; when we are cold, we try to lose as little as possible.

■ Heat is transferred in four ways: **conduction**, **convection**, **evaporation** and **radiation**.

■ Temperature receptors in skin provide an early warning system for temperature change, allowing us to thermoregulate by our behaviour (altering position, clothing, etc). This often prevents the need for a physiological response.

■ The **hypothalamus** can detect the temperature of the blood in the core of the body. If the blood temperature drops, the hypothalamus initiates corrective responses such as **vasoconstriction** and **shivering**. It the blood temperature increases, we respond by **vasodilation** and **sweating**. These are physiological responses.

■ Extremes of both heat and cold can cause death. Death as a result of cold occurs as a result of **hypothermia**. This is common in people who have fallen into freezing water.

■ The pigment **melanin** protects the skin from ultraviolet radiation. This type of radiation is **mutagenic** and exposure to some types of UV light can cause skin cancer.

? EXAM QUESTIONS

1 The diagram below shows how the body reacts to environmental conditions which may lead to lowering of the body temperature.

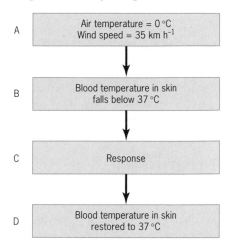

A — Air temperature = 0 °C
Wind speed = 35 km h⁻¹

B — Blood temperature in skin falls below 37 °C

C — Response

D — Blood temperature in skin restored to 37 °C

a) (i) Name the cooling effect caused by the conditions given in box A.
(ii) Suggest one response in box C which would lead to a rise in the blood temperature.
b) Explain what might happen to the exposed parts of the body if the blood temperature in the skin falls dramatically.
c) Name the condition in which the core temperature of the body may eventually fall below 35 °C.

Edexcel HB2 June 1999 Q3

2
a) Describe how blood vessels in the skin help regulate body temperature in response to hot external conditions.
b) Explain why regulation of body temperature is described as a negative feedback system.

IBO 410 Higher Paper 3 November 1998 QH2

KEY SKILLS **ASSIGNMENT**

HYPOTHERMIA: NO ONE IS DEAD UNTIL WARM AND DEAD

The definition of hypothermia is a drop in core temperature to 35 °C or below. This condition is very common, and is regularly seen in elderly people or in people who have been immersed in cold water – such as those caught in shipwrecks or people who fall through ice.

People can be revived from the most extreme states of hypothermia, even when all the vital signs point to death. People who are cold and blue, showing no heartbeat or breathing and with fixed and dilated pupils, have been revived and have later recovered, hence the above title.

Fig 12.A1 A mountain rescue team uses an insulated bag to restore the core temperature of an accident victim

1 **a)** What do we mean by core temperature?
b) What temperature is normal core temperature?
c) Why are people more likely to suffer hypothermia when wet or immersed in water than when they are just out in the cold air?

THE SIGNS OF HYPOTHERMIA

There are several stages of hypothermia:

Impending hypothermia

The core temperature drops to 36 °C. The person begins to feel uncomfortable and the skin may become pale, numb and waxy. Muscles become tense and can shiver if the patient is not moving. Fatigue and signs of weakness begin to show.

Mild hypothermia

Core temperature drops below 35 °C. Uncontrolled, intense shivering begins. Victims are alert and aware of the situation and so may still be able to help themselves. Movements become clumsy and the cold causes pain and discomfort.

Moderate hypothermia

Core temperature drops below 33 °C. Shivering slows or stops, muscles begin to stiffen and mental confusion and apathy begin. Speech becomes slow and difficult to understand. Drowsiness and eccentric behaviour can follow.

Severe hypothermia

Core temperature falls below 31 °C. The skin is cold and often bluish-grey. The eyes are dilated and victim may appear drunk and deny that there is a problem. There is a gradual loss of consciousness. In extreme cases the victim can appear dead, being rigid and not breathing.

2 **a)** Describe the relationship between temperature and enzyme activity (refer to page 93 in Chapter 5).
b) What happens to the oxygen demand of the body as the temperature gets lower?

TREATING HYPOTHERMIA

The basic rationale behind hypothermia treatment is stop the victim's heat loss, allowing the core temperature to return to normal gradually. The basic goals of early care are to stabilise the core temperature and to prevent cardiac arrest. One of the body's first responses to cold is peripheral vasoconstriction and one of the first measures with a hypothermia victim is to wrap the victim in a blanket and allow them to warm up gently while preventing excessive heat loss. Surprisingly, heating the skin – by placing them near a fire, for example – is not a good idea. This will cause vasodilation that will cause cold blood to return to the body core, resulting in an afterdrop. This can be fatal as it can cause the heart to stop.

3 **a)** What is peripheral vasoconstriction, and how does it help to conserve heat?
b) Chemicals such a caffeine and alcohol often give a person a warm 'glow' and this is because they cause vasodilation. Explain why coffee or alcohol should not be given to someone suffering from hypothermia.

Fig 12.A2 Elderly people whose sensory systems have deteriorated can suffer from hypothermia. When they start to become cold, their body systems fail to respond and the drop in temperature goes uncorrected. As the temperature drops, their metabolic rate decreases and they produce less heat, and so they cool down even more quickly. Death occurs when their core temperature reaches about 25 °C

13

Excretion and water balance

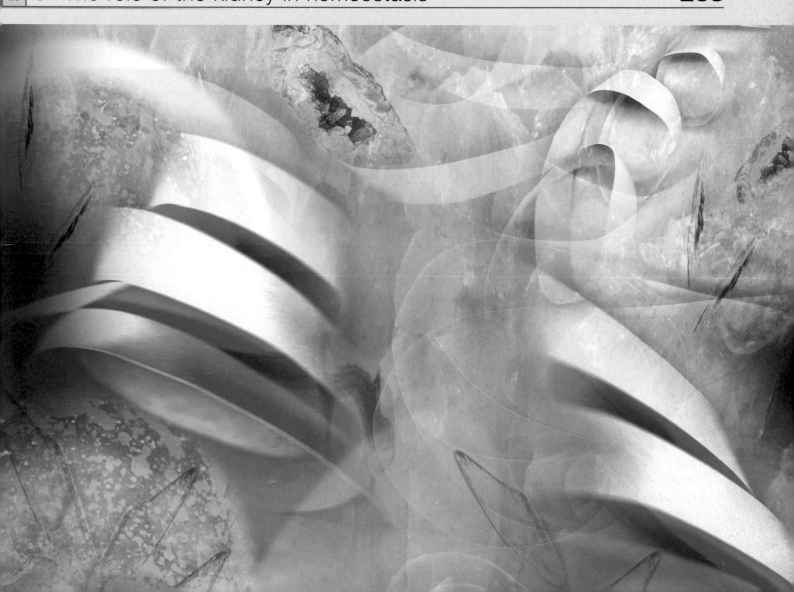

13 Excretion and water balance

This patient is undergoing dialysis. For several hours, blood flows from the patient's forearm into the machine. Here, much of the waste, together with excess salt and water, is removed by filtration. Find out more about this process in the Assignment at the end of this chapter

If one of your kidneys were to stop working because of injury or disease, you could probably still lead a normal life. However, people whose kidneys both fail are faced with a crisis: water, urea and potassium build up rapidly in their bodies. They may continue to pass some urine, but they cannot get rid of all the waste produced by normal cell processes.

Most people suffering from kidney failure hope for a transplant, but there is a shortage of donors. Until a suitable organ becomes available, patients have to rely on dialysis to filter their blood and balance their fluid intake. As the photograph shows, dialysis is uncomfortable and inconvenient. To reduce the time they need to spend in dialysis, kidney patients must stick to the strict Giovanetti diet. This comes of something of a shock. They must limit their fluid intake to only half a litre a day, about a quarter of the amount a human adult would normally drink. They must also control their protein intake to about 30 to 40 grams per day, the amount of protein in a small egg.

But perhaps the biggest problem is the need to regulate potassium. This ion is a normal constituent of the body, but large amounts cause serious problems, including heart failure. Potassium-rich foods include citrus fruits, bananas, instant coffee, peanuts, treacle and – a big blow for many people – chocolate. In this strange diet, carbohydrates are not a priority. Few patients feel like eating anyway, and this new restricted diet makes the task of finding appetising food even more difficult.

1 WASTE AND WATER CONTROL

The human body consists mainly of water. A person weighing 65 kg contains about 40 litres of water, of which 28 litres is **intracellular** (inside cells). The rest, the **extracellular** fluid, is made up from about 9 to 10 litres of tissue fluid and 2 to 3 litres of blood plasma.

Fig 13.1 When we drink a large amount of fluid, our kidneys have the job of getting rid of the excess water, to prevent body fluids from becoming too dilute. This man is drinking beer and the alcohol it contains also affects his kidney function – more of that later

The kidneys play a major role in regulating the volume and composition of these body fluids. They excrete or conserve water and salt so that the volume and composition of the blood and body fluids remain more or less constant. This is a vital aspect of homeostasis.

The kidneys also ensure that waste products do not build up by filtering the blood. Waste is allowed to pass through and then out of the body, while important substances such as glucose are reabsorbed and conserved. The removal of metabolic waste from the blood is called **excretion**.

Urine is the end-product of all processes that occur in the kidney.

? TEST YOURSELF

A If a person has 3 litres of blood plasma, and their haematocrit (the volume occupied by cells in the blood) is 40 per cent, what is their total blood volume?

Fig 13.2 Adult humans need about 40 to 50 grams of protein per day, and so there is enough here for over a week! However, the body cannot store protein and so the daily excess is broken down in the liver. The resulting urea is excreted in the urine

Most excreted waste leaves the body in the urine, but some waste is also lost in **sweat** and in air that we breathe out. Sweat contains mainly salt and water, while **exhaled air** contains carbon dioxide and water vapour.

In this chapter we concentrate on **nitrogenous excretion**, the removal of waste compounds that contain nitrogen. The main nitrogenous compound excreted by humans is **urea**. This is made in the liver following the breakdown of excess amino acids. Enzymes in the liver remove the amine (NH_2) group from amino acids in the process of **deamination**. The ammonia formed is a highly toxic compound that must not be allowed to build up. It immediately enters a series of reactions called the **ornithine cycle**, which produce the relatively harmless compound urea:

$$2NH_3 \quad + \quad CO_2 \quad \rightarrow \quad CO(NH_2)_2 \ + \ H_2O$$
$$\text{ammonia} \ + \ \text{carbon dioxide} \ \rightarrow \quad \text{urea} \quad + \ \text{water}$$

See Chapter 11 for more details on the ornithine cycle. Urea is carried in the blood from the liver to the kidneys to be excreted.

In this chapter we look at the role of the kidneys in homeostasis. The kidneys selectively eliminate water and solutes, such as sodium, potassium and chloride ions, so that the water and solute balance of the body is kept at the correct level. The need to balance the solute concentration of body fluids is called **osmotic regulation**, or **osmoregulation**.

Osmoregulation is the maintenance of a constant solute concentration within the body. In Chapter 4 we saw why this is important: if we place animal cells in a *hypertonic* solution, they lose water by osmosis and shrivel. If we put them in a *hypotonic* solution, they gain water and may burst. Both situations are harmful to cells, and so it is important that we maintain the solute concentration of our body fluids within narrow limits.

✔ **REMEMBER THIS**

Excretion is the removal of chemical waste from the body, waste that is produced by the metabolic processes within cells and that would be toxic if allowed to accumulate. You should not confuse excretion with **egestion** or **defecation**, the removal of undigested food and other debris from the intestine.

SEE QUESTION 1

✔ **REMEMBER THIS**

Different fluids are often compared with respect to their solute concentrations. A solution that has a higher solute concentration than another is said to be **hypertonic**. A solution that has a lower solute concentration than another is **hypotonic**. A solution that has exactly the same solute concentration is **isotonic**.

So, for example, we can say that seawater is hypertonic to human blood plasma, because seawater has a much higher concentration of dissolved salts.

Most people will have experienced the unpleasant symptoms that accompany a bout of mild food poisoning due to a dodgy late-night kebab, an undercooked chicken leg at a barbecue, or those prawns that lurked in the fridge just a little too long.

Vomiting is caused by bacterial toxins that irritate the gut lining. In the intestine, the frequency of peristalsis increases and the contents move along the gut rather more rapidly than usual. This doesn't give the large intestine enough time to absorb water from the waste and the result is **diarrhoea**. Fortunately, for most of us, the symptoms are short lived.

However, both vomiting and diarrhoea can be deadly if severe and prolonged. Dysentery and cholera can rapidly lead to death by dehydration as they cause the body to lose fluid faster than it can be replaced.

Every day the average person consumes about 2 to 3 litres of fluid in one form or another, and we also pour a huge volume of fluid (over 8 litres) into our intestines in the form of digestive juices. Diarrhoea does not allow for the efficient reabsorption of these fluids. Also, vital ions such a sodium, potassium and chloride, collectively known as **electrolytes**, are lost. If untreated, this loss can lead to muscle spasms, cramps, coma and heart failure.

Oral rehydration therapy can be used to treat dehydration. This does not involve expensive drugs, simply a mixture of glucose and salt in water. In cases where the patient cannot keep anything down, the rehydration solution can be given directly into the bloodstream via a drip. This simple treatment has saved millions of lives in places where dysentery and cholera are very common.

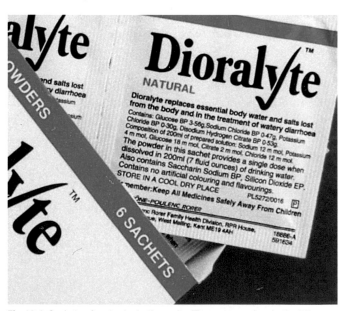

Fig 13.3 The baby in this picture has bacterial dysentery that he developed after drinking water contaminated with faeces. He risks death from dehydration if he does not receive rehydration therapy.
An oral rehydration mixture is easy to make up: one level tablespoonful each of glucose and salt dissolved in a pint of boiled and cooled water

Fig 13.4 Sachets of oral rehydration salts. Those we can buy in the UK contain more than the essential ingredients

---

2 AN OVERVIEW OF THE HUMAN URINARY SYSTEM

✔ REMEMBER THIS

The urinary system is also known as the **renal system**.
If you place your hands on your hips, your thumbs show the position of your kidneys.

The human urinary system is shown in Fig 13.5. The kidneys lie at the back of the abdominal cavity, just below waist level, where they are protected to some extent by the spine and the lower part of the ribcage. Usually, the left kidney is slightly above the right.

The kidneys receive blood from the two **renal arteries** that branch off the aorta (see Chapter 10). The kidneys receive the largest blood supply of any organ, per gram of tissue. About 1200 cm^3 of blood flows to each of them every minute. Incoming blood must be at high pressure to ensure proper

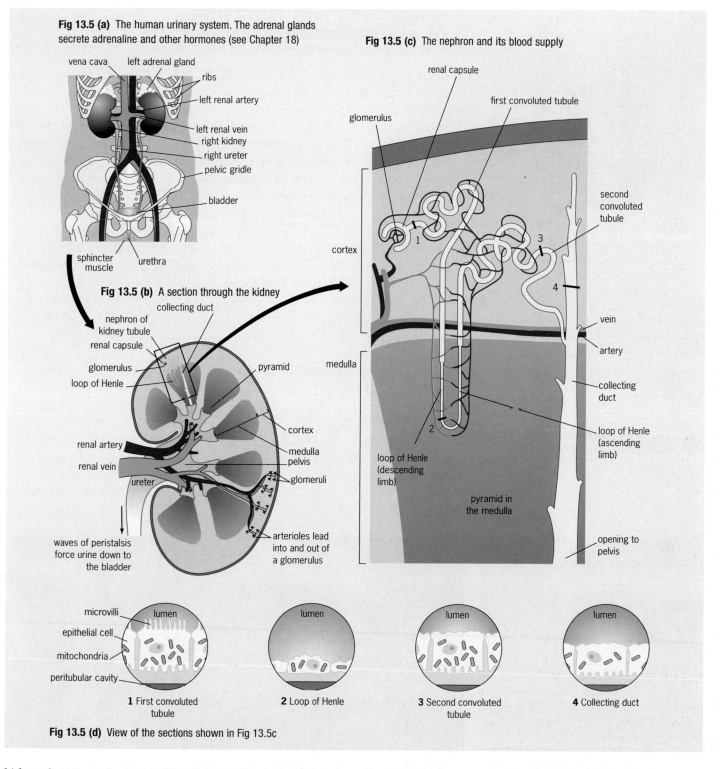

Fig 13.5 (a) The human urinary system. The adrenal glands secrete adrenaline and other hormones (see Chapter 18)

vena cava
left adrenal gland
ribs
left renal artery
left renal vein
right kidney
right ureter
pelvic gridle
bladder
sphincter muscle
urethra

Fig 13.5 (b) A section through the kidney

collecting duct
nephron of kidney tubule
renal capsule
glomerulus
loop of Henle
pyramid
renal artery
renal vein
ureter
cortex
medulla
pelvis
glomeruli
waves of peristalsis force urine down to the bladder
arterioles lead into and out of a glomerulus

Fig 13.5 (c) The nephron and its blood supply

renal capsule
glomerulus
first convoluted tubule
second convoluted tubule
cortex
medulla
loop of Henle (descending limb)
pyramid in the medulla
vein
artery
collecting duct
loop of Henle (ascending limb)
opening to pelvis

microvilli
epithelial cell
mitochondria
peritubular cavity
lumen

1 First convoluted tubule
2 Loop of Henle
3 Second convoluted tubule
4 Collecting duct

Fig 13.5 (d) View of the sections shown in Fig 13.5c

kidney function: they can't filter blood effectively if the pressure drops. Blood leaves the kidneys in the **renal veins**.

Urine made by the kidneys is pushed down muscular tubes, the **ureters**, by peristalsis (rhythmic muscular contractions). The ureters empty into the **bladder**, a muscular bag that stores urine until it is convenient to release it. The capacity of the human bladder varies from 400 to 700 cm³ or more. When it begins to get full, stretch receptors in the walls inform the brain of the urgency of the situation. **Urination**, or **micturition**, happens when the **sphincter muscles** relax, allowing urine to pass out of the body through the **urethra**.

? TEST YOURSELF

B A person has 5 litres of blood, and the kidneys filter 1.2 litres per minute. How many times on average does the total volume of blood in the body pass though the kidneys every hour?

3 STRUCTURE AND FUNCTION OF THE MAMMALIAN KIDNEY

The structure of the kidney is shown in Fig 13.5(b). The functional unit of the kidney is the **nephron** (or **kidney tubule**). It is important to know the position of the nephrons in relation to the overall plan of the kidney. The outer **cortex** of the kidney contains the **renal capsules** (also called **Bowman's capsules**) and the **first convoluted** and **second convoluted tubules** (also called the **proximal** and **distal tubules**), while the **medulla** houses the **loops of Henle** and the **collecting ducts**. Bundles of collecting ducts form **pyramids** that deliver urine into an open space called the **pelvis**. From here, urine flows down the ureters to the bladder.

Kidney function involves two processes: **ultrafiltration** and **active transport** (Fig 13.6). Ultrafiltration is filtration under pressure: blood is 'squeezed' to form a fluid called **glomerular filtrate** (usually just '**filtrate**'). Active transport then modifies the filtrate, secreting some substances into it and reabsorbing others from it, according to the needs of the body. The end result is that blood flows back into the body without much of its harmful waste. This waste, a solution containing urea, salts and various other chemicals, is the urine.

Let's now look at how individual parts of the kidney contribute to this overall process.

THE NEPHRON

Each human kidney contains about a million nephrons, together with a maze of blood vessels and some connective tissue (Fig 13.7). In this section we deal with the function of each region of the nephron in sequence, but you must remember that the nephron functions as a whole: the activities of one region are essential to the effectiveness of others.

Fig 13.6 A simple summary of kidney function. At the near end of the nephron, blood is filtered to produce a fluid that is virtually identical to tissue fluid. The filtrate is then modified by active transport. This involves active secretion of substances into the filtrate and active reabsorption of substances into the blood. These processes are possible only because of the close association between the nephron and blood system (see Fig 13.5c)

SEE QUESTION 2

Fig 13.7 The fine structure of the kidney
(a) Under the microscope, five renal capsules (containing glomeruli) can be seen, surrounded by sections of tubules. An underwater dissection allows some of the fine nephrons to be teased out and viewed individually without the aid of a microscope. Each nephron, although only 60 μm in diameter, can be over 14 cm long when uncoiled
(b) A scanning electron micrograph of a glomerulus with part of the torn renal capsule (whitish) round it

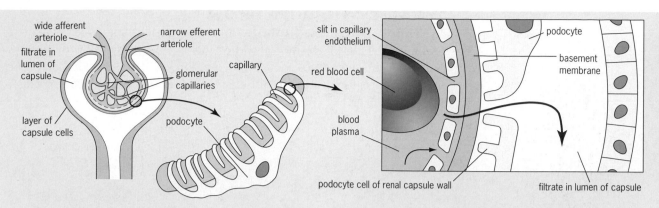

The walls of capillaries in the renal capsule are much more permeable than those of normal capillaries: the cells do not fit tightly together, but have thin slits in between, through which all the constituents of the plasma can pass	The renal capsule is lined with unique cells called **podocytes** ('foot-cells'). These cells, like those of the capillary, do not fit tightly together, but form a network of slits that fit over the capillary	Between these two relatively coarse filters is the continuous basement membrane. This finer filter prevents the passage of all molecules with a relative molecular mass greater than about 68 000 kilodaltons, so the larger molecules (mainly proteins) remain in the blood

ULTRAFILTRATION IN THE RENAL CAPSULE

Fig 13.8 shows the **renal capsule** in detail. It is shaped rather like a wineglass, with a central knot of blood vessels called the **glomerulus**. This area of the nephron filters blood by ultrafiltration (filtration under pressure). Obviously, this requires two things: a means of creating pressure and a filter.

The kidneys receive blood from the first branch off the aorta, so the blood is already under pressure when it reaches the nephron. This pressure is maintained and enhanced because the **afferent arteriole**, the blood vessel that takes blood into the glomerulus, is short and has a larger diameter than the longer **efferent arteriole** that takes blood away. This physical or **hydrostatic pressure** forces blood against a filter that consists of three layers:

● the lining or **endothelium** of the glomerular blood vessels,
● the **basement membrane**,
● the cells of the renal capsule itself.

Fig 13.8 shows how these membranes are arranged. The middle basement membrane acts as a fine filter and is therefore mostly responsible for the chemical composition of the filtrate. At this stage the filtrate is identical to tissue fluid (see Chapter 10).

The rate of filtrate production is high: about 125 cm³ per minute. Obviously, we don't produce anything like this volume of urine or we would be constantly in the loo and would dehydrate rapidly. On average we produce about 1 cm³ of urine per minute. The rest, over 99 per cent of the filtrate, is reabsorbed. In fact, after the renal capsule, the rest of the nephron is concerned with adjusting the volume and composition of the filtrate. Necessary substances are reabsorbed; toxic compounds and excess solutes and water are removed.

Several forces act on the fluids in the renal capsule, opposing or encouraging the filtration process. The high hydrostatic pressure of blood is the dominant force but it is opposed by the hydrostatic pressure and solute concentration of the filtrate (Table 13.1).

Fig 13.8 The fine structure of the renal capsule, a region of the kidney adapted for ultrafiltration of the blood. Note the difference in size between the afferent and efferent arteriole. The blood is filtered through three layers of cells: the endothelium of the capillaries, the basement membrane and the capsule wall

✓ REMEMBER THIS

Examiners like you to talk in terms of **water potential** when explaining osmosis and water movement. This is a negative scale – pure water has a water potential of zero, and the higher the solute concentration the lower the water potential. See Chapter 4 for details.

? TEST YOURSELF

C If an individual lost a large amount of blood in an accident, what effect would this have on kidney function?

Table 13.1 Summary of the forces acting in the renal capsule		
Force acting	**Opposes or encourages filtrate formation**	**Approximate value/kPa**
Hydrostatic pressure of blood	encourages	8.0
Hydrostatic pressure of filtrate in capsule	opposes	−2.4
Solute concentration of blood	opposes	−4.3
Overall filtration pressure	encourages	1.3

THE FIRST CONVOLUTED TUBULE

In the **first convoluted tubule** (Fig 13.9), many solutes such as glucose and amino acids are totally reabsorbed into the blood by active transport. Normal urine should not contain glucose. Only when blood glucose becomes very high, because of diabetes for example, does the reabsorption mechanism fail (Fig 13.10).

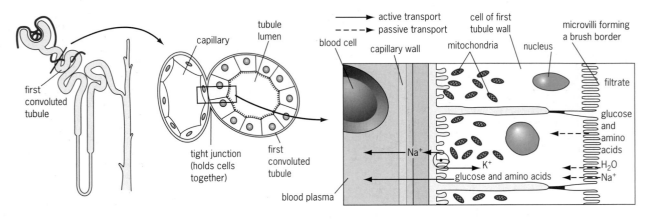

Fig 13.9 Most of the filtrate formed in the renal capsule is reabsorbed from the first convoluted tubule. Amino acids, glucose and sodium are removed from the tubule by active transport, and water follows passively by osmosis

Fig 13.10 Normally the kidneys are able to reabsorb all of the glucose into the blood. Only when blood sugar exceeds a threshold of about 350 mg per 100 cm³ does glucose begin to appear in the urine

In addition, this part of the tubule is very close to blood vessels that carry blood away from the glomerulus. Blood in these vessels has a low hydrostatic pressure but a relatively high solute potential due to the plasma proteins that remain there because they could not pass through the filter. This allows the blood to reabsorb a large percentage of the water from the first convoluted tubule by osmosis.

As you can see from Fig 13.9, the cells that line the first convoluted tubule show all the classic adaptations to active transport: a large surface area, provided by microvilli, and many mitochondria to provide ATP to power the process.

THE LOOP OF HENLE

The **loop of Henle** is a long U-shaped region of the nephron that descends deep into the medulla and then returns to the cortex. The loop creates a region of high solute concentration (low water potential) in the medulla. The collecting ducts pass through this region, and the osmotic gradient between the inside of the collecting duct and the outside draws water out of the duct by osmosis. Consequently, the urine becomes more and more concentrated (compared with body fluids) as it passes down the duct. So the loop is a vital adaptation that benefits humans and other land-living organisms: it allows us to get rid of waste without losing too much water.

Fig 13.11 How the loop of Henle works. The numbers refer to the solute concentration in mg per 100 cm³

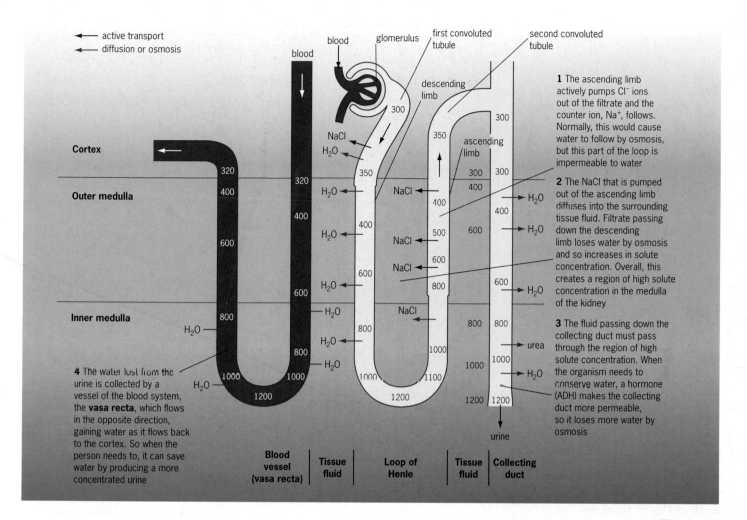

Fig 13.11 outlines how the loop of Henle works. Fluid in the two limbs of the loop flows in opposite directions. We describe this sort of arrangement as a **countercurrent system**. As fluid travels up the ascending limb, sodium chloride (NaCl) is transported actively out of the limb into the surrounding area. This causes water to pass out of the descending limb by osmosis. The net result is that the solute concentration at any one level of the loop is slightly lower in the ascending limb than in the descending limb. The longer the loop, the more chance there is for this mechanism to build up a high sodium chloride concentration. If the loop in Fig 13.11 were only half the length shown, sodium chloride would accumulate to only about 600 units.

THE SECOND CONVOLUTED TUBULE

While the first convoluted tubule is reabsorbing most of the filtrate, the **second convoluted tubule** 'fine tunes' the remaining fluid, according to the immediate needs of the body. This tubule plays an important role in the regulation of pH, salt and water balance.

? TEST YOURSELF

E People on survival courses are taught to assess their level of dehydration by looking at the colour of their urine. Explain how they would do this.

SEE QUESTION 3

4 THE ROLE OF THE KIDNEY IN HOMEOSTASIS

The kidney contributes to several vital homeostatic mechanisms. One of the most important is regulation of water content and blood volume.

WATER BALANCE IN HUMANS

Table 13.2 shows a typical water balance sheet for an average person, assuming normal activity and a comfortable external temperature. On average, we get almost two-thirds of our water from drinks and a third from food. We obtain a small but important proportion of water from metabolic reactions, notably cell respiration.

Some of the water loss shown in Table 13.2 is unavoidable. Metabolic waste must be removed in solution, and so some water loss in urine is inevitable. Similarly, water is always lost from the lungs as we breathe out. A significant amount of water is also lost by diffusion through our skin (this is not the same as sweating).

Table 13.2 The water balance sheet for a 24-hour period				
Water gain	**Volume/cm^3**		**Water loss**	**Volume/cm^3**
Food and drink	2100		through skin	350
Metabolic water	200		sweat	100
			in breath	350
			urine	1400
			faeces	100
Total	2300		Total	2300

THE MECHANISM OF WATER BALANCE

Like most homeostatic mechanisms, maintenance of water balance involves a negative feedback loop that consists of a **detector** and a **correction mechanism** (see Chapter 11).

A part of the brain, the **hypothalamus**, contains **osmoreceptor cells** that are sensitive to the solute concentration of the blood. When the solute concentration rises, indicating that water loss has exceeded intake, the hypothalamus responds in two ways:
● It stimulates the thirst centre in the brain.
● It stimulates the pituitary gland to release **anti-diuretic hormone** (**ADH**). Fig 13.12 summarises the mechanism of ADH action. ADH acts on the kidney to reduce the volume of urine produced. It achieves this by increasing the permeability of the second convoluted tubule and the collecting duct to water. The action of ADH causes more water to leave the tubule and re-enter the blood. Much more concentrated urine is produced and vital water is conserved.

Conversely, when fluid intake exceeds loss, the blood becomes more dilute. When the hypothalamus detects this, it reduces ADH production. The action of ADH on the kidneys lessens, resulting in less water reabsorption and the production of larger volumes of dilute, or **insipid**, urine.

People with the disease **diabetes insipidus** cannot produce ADH because they have a faulty pituitary gland. Once known as the 'pissing evil', this condition results in the constant production of dilute urine, leaving the sufferer permanently thirsty and unable to venture very far from the toilet. Today, it can be treated by giving extracted or synthesised ADH.

SEE QUESTION 4

CONTROL OF BLOOD VOLUME AND PRESSURE

Since it regulates water reabsorption, ADH also regulates blood volume. A drop in blood volume leads to a drop in blood pressure that is detected by stretch receptors in the walls of the aorta and carotid arteries. Impulses from these detectors pass to the hypothalamus, which then triggers the secretion of more ADH. This acts on the kidneys and causes them to retain more water, so increasing blood pressure.

Fig 13.12 When the solute concentration of the blood rises, the osmoreceptor cells in the hypothalamus stimulate secretion of ADH from specialised nerve cells in the posterior lobe of the pituitary gland. ADH makes the second convoluted tubule and the collecting duct more permeable to water, and allows more water to pass from the filtrate and into the blood. In this way, the blood becomes more dilute and blood volume increases

> **! SCIENCE FEATURE Hangovers!**
>
> Sooner or later, most people experience the unpleasant 'morning after' feeling which tends to follow a bout of drinking too much alcohol. Many hangover symptoms are due to dehydration, rather than to the toxic effects of the alcohol or other ingredients. Research has shown that alcohol inhibits the production of ADH, causing water that the body needs to be lost in the urine. Many of the symptoms disappear when the body is rehydrated.

Fig 13.13 Had he known about the dehydrating effects of alcohol, this man could have minimised his headache by having a long drink of water before he went to bed (assuming, of course, that he could find the tap)

SUMMARY

After reading this chapter, you should know and understand the following:

■ The kidneys remove metabolic waste and control water and solute levels in the body. As a result of these functions, the kidney also plays a vital role in the control of blood volume and pressure.

■ Each kidney is made from around one million **nephrons**, narrow tubules closely entwined with blood vessels.

■ At one end of the nephron, the **renal capsule**, the kidney filtrate is formed by ultrafiltration (pressure filtration) of the blood.

■ The first filtrate has the same composition as tissue fluid. As it passes along the nephron, the composition of filtrate is altered by various active transport mechanisms that reabsorb some substances while allowing others to pass into the urine.

■ A large amount of filtrate is formed, but over 99 per cent of it is reabsorbed in the **first convoluted tubule**, mainly by active transport mechanisms and osmosis. Usually, all glucose and amino acids are reabsorbed into the blood.

■ The movement of solutes from the **loop of Henle** creates a region of high solute concentration in the medulla, through which the collecting ducts must pass. As filtrate (now urine) flows along the collecting ducts, water leaves it by osmosis. The resulting urine is **hypertonic** (more concentrated than body fluids).

■ The **second convoluted tubule** is involved in several homeostatic mechanisms including the regulation of salt, water and pH levels.

■ When the water potential of the blood drops (eg when we dehydrate) it is detected by the **hypothalamus** which stimulates **ADH** release from the **pituitary**. ADH makes the second convoluted tubule and the collecting duct more permeable to water, so water leaves the filtrate and enters the blood.

? EXAM QUESTIONS

1

a) Describe how urea is formed in the liver.

b) The table gives information on the nitrogenous excretory products of turtles and tortoises from different habitats.

Habitat	Percentage of total nitrogenous excretory product		
	Ammonia	Urea	Uric acid
Water	40–50	40–50	10
Land and water	12	80–90	10
Land	12	30–50	70
Desert	8	10–15	80–90

(i) Describe how the relative percentages of nitrogenous products change with water availability in the habitat.

(ii) Explain the advantage of uric acid as the main nitrogenous excretory product in the desert.

AQA (NEAB) BY03 Physiology June 2000 Q7

2 The micrograph shows a section through part of the kidney cortex as seen under the high power of a

light microscope.

a) Name the structures labelled **A** to **C**.

b) Draw a simple sketch of the micrograph and show, by labelling with **X**, where ultrafiltration occurs.

Part of the kidney tubule is adapted for the reabsorption of glucose.

c) State **two** structural adaptations of this region of the kidney tubule and explain how each one assists reabsorption from the glomerular filtrate.

Kidney failure may be treated by use of a kidney dialysis machine.

d) **(i)** Explain how dialysis differs from ultrafiltration in the kidney.

(ii) Suggest two advantages of kidney transplantation over dialysis as a treatment for kidney failure.

OCR 9264 Biology Linear Paper 3 June 2000 Q1

3 The diagram below shows the loop of Henle from a nephron in the mammalian kidney.

The numbers on the diagram show changes in the relative concentration of solutes in the filtrate as it passes through this part of the nephron.

a) **(i)** Explain how, in the ascending limb, the relative concentration of solutes in the filtrate changes from 700 to 100.

(ii) Explain how, in the descending limb and the base of the loop, the relative concentration changes from 300 to 1200.

b) The ability of the nephron to reabsorb water from the filtrate is important in maintaining the water balance of the body.

Describe how the concentration changes in the loop of Henle assist in water reabsorption from the filtrate back into the body.

Edexcel B3 June 1999 Q5

4 Copy and complete the table summarising the role of each of the following hormones.

Hormone	Site of secretion	Part of kidney tubule affected	Effect of hormone
ADH			
Aldosterone			

AQA (NEAB) BY03 June 2000 Q3

KEY SKILLS **ASSIGNMENT**

TREATING KIDNEY FAILURE

In many cases of **chronic kidney failure** there is a gradual decline in kidney function, giving the patient plenty of warning. Acute renal failure is a crisis in which all kidney function effectively stops. There are many causes of **acute renal failure**, but generally they can be placed into the following categories:

● Sudden loss of large amounts of fluid (blood or tissue fluid).

● Inadequate blood flow to the kidneys.

● Bacterial infection in the kidneys.

● Effect of toxins.

● A blockage in the urinary tract caused, for example by damage to the ureter.

1 Why would a reduced blood flow to the kidneys interfere with kidney function, even though enough blood could get through to provide the kidney cells with the nutrients and oxygen they need to stay alive?

As we saw at the start of this chapter, the immediate problems of kidney failure are a build-up of fluid, urea and potassium. To minimise these problems, patients must follow the Giovanetti diet.

2 **a)** What are the essential features of this diet? See Opener.

b) Suggest why protein intake needs to be limited.

Dialysis and kidney machines

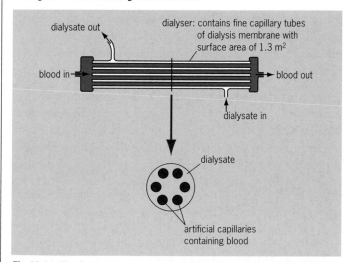

Fig 13.A1 The fine tubes seen in this dialysis filter are artificial capillaries. Blood flows through the middle of these tubes, while the dialysing fluid flows along the outside in the opposite direction. Each filter is an expensive piece of precision engineering, but it can be used for only a few dialysis sessions and must then be discarded

Dialysis is a method of separating small molecules from larger ones using a partially permeable membrane. Blood dialysis, or **haemodialysis**, separates the smaller constituents of plasma, such as urea and solutes, from the larger ones, such as proteins.

Blood is taken from the patient, usually from a vein in the forearm, and passed into the machine, where it runs through minute artificial capillaries. These are made from a partially permeable plastic that filters the blood. While blood flows inside the artificial capillaries, a special fluid, the **dialysate**, flows round the outside in the opposite direction.

In dialysis, molecules are exchanged between the blood and the dialysate. The composition of the dialysate is carefully controlled so that there is a net movement of urea, water and salts out of the blood.

3 **a)** By what physical process do solutes enter or leave the blood during dialysis?

b) Why do the blood and dialysate flow in opposite directions?

c) Suggest two problems that might occur if the dialysate was pure water.

d) Why must the dialysate contain glucose, amino acids and salt?

e) Why is there no urea in the dialysate?

Fig 13.A2 shows the circuit taken by the blood as it passes through the dialysis machine.

Fig 13.A2 The essential features of a haemodialysis machine

KEY SKILLS **ASSIGNMENT**

4 a) Calculate the volume of blood that is processed by the dialyser in four hours.

b) Why is heparin, an anti-coagulant, added to the blood?

c) Why is heparin not given in the last hour of dialysis?

d) Why is a filter included in the blood circuit?

e) Suggest why the omission of the bubble trap could prove dangerous to the patient.

f) What would a positive reading on the haemoglobin sensor indicate?

g) The dialysis fluid is maintained at approximately 40 °C. Suggest two reasons for this.

h) Excess water can be removed from the blood in a number of ways. One way is to increase the amount of glucose in the dialysing fluid. Explain how this method would work.

i) Another method involves partially clamping the blood tube at the region shown in Fig 13.A2. Explain the principle involved in this method of removing water.

CAPD

CAPD stands for **continuous ambulatory peritoneal dialysis**. In this fairly new treatment, individuals with kidney failure can use one of their own membranes, the **peritoneum**, as a dialysing membrane (Fig 13.A3). Over 3000 people in the UK are presently using this technique because of the advantages it offers over conventional dialysis.

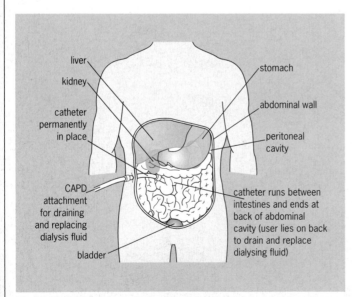

Fig 13.A3 The peritoneum is a semipermeable membrane that lines the abdominal cavity and organs such as intestines. Dialysate introduced into this cavity draws waste and excess water out of the blood

The basic principle is simple: patients have a hole, or **stoma**, made in their abdomen wall near the navel, through which a large volume of dialysing fluid is introduced through a tube or **catheter**. The patient is then free to walk around (hence the ambulatory part of the name) while dialysis occurs across the peritoneum between the blood and the dialysate. Every four to six hours the dialysate is replaced. This is a relatively simple exchange procedure that patients can carry out themselves after some basic training.

5 a) Suggest the biggest problem with the exchange procedure.

b) Explain why it would be no use leaving the dialysate inside the body for longer than the recommended time.

Kidney transplants

When a donor kidney becomes available, it is a relatively simple operation to transplant it into another body. Surprisingly, the old kidneys are left in place: they are rather inaccessible and so are difficult to remove, but they do no harm. The new kidney is placed in the lower abdomen. Surgeons choose this site because the new kidney can be attached easily to a large artery (the **femoral artery**, supplying the leg) and is usefully right next to the bladder.

6 What is the main problem with a transplant once it has been carried out?

Finding a suitable donor for an organ transplant is difficult. Although several hundred thousand people die each year in the UK, only a tiny fraction can provide organs for transplant. For example, accident victims can become organ donors only if their injuries have not affected the organ itself. Owing to road safety improvements, the number of serious accidents is decreasing. This is good news, but it means that the number of organs available for transplant is getting ever smaller.

A further complication is that only a minority of people carry donor cards, and permission to used body parts from recently deceased people has to be given by distressed relatives, who often say no. The problems of transplants are covered in more detail in Chapter 30, but you might want to find out more.

7 a) Discuss or jot down some ideas about what could be done to encourage more people to become organ donors.

b) It has been suggested that everyone should automatically be an organ donor, unless they take the decision to opt out of the scheme. Discuss whether this idea would work or not.

14

Exercise physiology

A variety of new techniques can speed up the healing process. In this controversial example a patient is being treated in a hyperbaric chamber that supplies oxygen at a pressure 2.2 to 2.4 times greater than atmospheric. This causes rapid uptake of oxygen into the blood and body tissues, speeding up the healing process in a variety of injuries. Studies suggest that this can speed up the healing process by as much as one third.

At an athletics meeting in Oxford in 1954, Roger Bannister became the first man to run a mile in under 4 minutes. In 1999, The Moroccan Hithcem El Guerrouj ran the same distance in 3 minutes 43 seconds. Genetically, humans today are no different from their predecessors. Yet year upon year, games upon games, athletes get better and world records keep tumbling. How is this being achieved?

Firstly, there are advances in science and technology. A substantial support team now surrounds top athletes: sports physicians, sports physiotherapists, sports dieticians, etc. Training can now be can be tailored to an individual's needs. Their muscle types can be analysed so that they do just the right balance of the different training methods and intensities. Training machines can isolate a particular muscle group and/or reproduce movement patterns such as a javelin throw or tennis serve. Movement can be filmed and analysed with a computer so that a biomechanics expert can target areas for improvement.

Diet can also play a big part both before and after the event. Endurance athletes prepare with a mainly carbohydrate diet so that their glycogen stores are topped up. Following an event, athletes need to rehydrate and replace lost fuels so that then can recover and begin training again as soon as possible.

Another important aspect in preparing a world-class athlete is the prevention of injury. Modern sports physiotherapists have an array of tools at their disposal to diagnose potential weaknesses. If an injury does occur, there is a variety of techniques available to speed the healing process (see the photo).

Lastly, and crucially, there has been a change from an amateur to a professional status. Roger Bannister was a doctor who fitted his athletics around his medical school training. Many of today's young athletes see their talent as a route to fame and fortune. They can devote all of their energies into excelling at their particular event or sport, knowing that success can set them up for life.

WHAT IS FITNESS?

What do we mean by **fitness**? Overall, it refers to a person's ability to perform well in their chosen activity. But there are several different aspects to the concept of 'fitness'.

There is strength and speed. Both aspects depend largely on the power in the muscles. Some men can lift 600 kilograms – about the weight of a small car – above their heads, and there are top sprinters who can cover 100 metres in less than 10 seconds – by running at a speed of about 36 kilometres per hour. A person's strength depends to some extent on the cross-sectional size of their muscles, but other factors such as the shape of their skeleton, good technique and motivation, are all important. So is skill. Some people are naturally gifted; their brain is able to control their muscles in just the right way to perform a particular activity. Of course, skill levels are further improved through coaching, training and practice.

Yet another factor is suppleness, the flexibility of the body. Often an underestimated aspect of fitness, most sports require a degree of suppleness, and performance can be improved by incorporating suppleness work into a training programme.

In this chapter we mainly focus on a fourth element of fitness: **stamina** – the body's 'staying power'. We look at the way in which the body gets its energy, and the different types of 'energy currency' that are important in different sports.

1 GEARING UP FOR EXERCISE

As you sit and read this book, your body is ticking over nicely. Oxygen demand is low, and can be met comfortably by relatively shallow breathing and low pulse rate. Blood is delivering oxygen and glucose to your cells and waste products are being taken away. The levels of these chemicals remain relatively constant, and homeostasis, that all-important *steady state* (see Chapter 11), is being maintained with relatively little fuss. There is, literally, 'no sweat'.

All of this is rudely interrupted when you begin to exercise (Fig 14.1). The metabolic rate of the muscles increases by up to 20 times, or by 2000 per cent. To fuel this frantic activity, and to maintain some sort of stability, your body must adapt. The overall response of the body to exercise is an excellent example of how the different systems work together to carry on maintaining homeostasis.

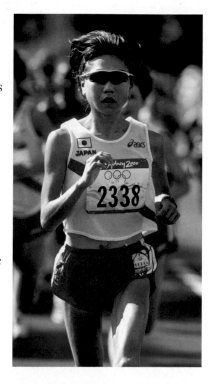

Fig 14.1 This is Naoka Takahashi who can run a marathon in a time of 2h 23min 14s. Throughout the 26 miles of this race, her body is able to keep her muscles supplied with oxygen and fuel (glucose and lipid), while ensuring that the waste products are eliminated before they build up. This is a magnificent feat of homeostasis

HOMEOSTASIS AND EXERCISE PHYSIOLOGY

Exercise physiology is the study of the responses of the body to exercise. The effects are familiar: in the short term we sweat, pant and go red, while in the long term we 'get fit', improving our muscle tone, strength, stamina and general well-being. This chapter builds on your knowledge of the human body, pulling together many different areas to show you that homeostasis is a whole-body process involving many different systems. Table 14.1 summarises the relevant topics that are covered elsewhere in the book.

Before we look at the ways in which the body responds to exercise, we need to set the scene by looking at some basic principles:

- The process of **cell respiration** (see Chapter 34) releases the energy in organic molecules such as glucose and lipids, and transfers the energy to a chemical called **adenosine triphosphate**, ATP (Fig 14.2).
- ATP provides the energy for muscular contraction; it allows the fibres to slide over each other (see Chapter 17).
- When ATP splits by **hydrolysis** into ADP and P_i (phosphate), energy is released (Fig 14.3).
- The purpose of cell respiration is therefore to resynthesise ATP from ADP and phosphate.

REMEMBER THIS

The word **homeostasis**, meaning 'steady state', refers to the fact that conditions within the body need to be kept within certain limits. Homeostatic mechanisms include the control of blood gases (oxygen and carbon dioxide), blood pH, body temperature and blood glucose levels.

adenosine

three phosphate groups

pentose sugar
(ribose)

simplified

adenosine

ribose P P P

Fig 14.2 The structure of ATP. In the computer graphics image, adenosine is blue, pentose is white and the three phosphate groups are red. This molecule is present in all cells and is used to provide energy. The purpose of respiration is to resynthesise ATP as fast as it is made, although during strenuous exercise this is not possible

- ATP splitting is a **coupled** reaction. When we exercise, ATP hydrolysis is coupled with muscular contraction.
- When ATP is split, the some of the energy is used to power the muscle but, as no energy transfer is 100 per cent efficient, some is always lost as heat. This is why vigorous exercise produces large amounts of heat that must escape from the body.

Fig 14.3 When the ATP molecule is split (hydrolysed), the terminal phosphate is removed and combined with water. This reaction releases energy, which is used to drive energy-requiring reactions such as muscular contraction

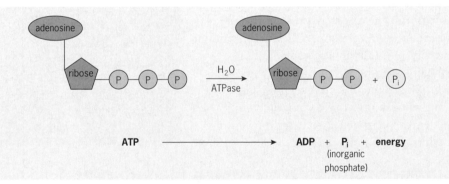

$$ATP \longrightarrow ADP + P_i + energy$$
(inorganic phosphate)

Table 14.1 Aspects of exercise physiology in this book

Topic	Found in
Homeostasis and negative feedback systems	Chapter 11
The cardiac cycle and its control	Chapter 10
Breathing an its control	Chapter 9
Respiration – the production of ATP	Chapter 34
Temperature control	Chapter 12
Movement of muscle/muscle types	Chapter 17
Electrocardiograms	Chapter 10
Artificial pacemakers	Chapter 31
Heart defects and their repair	Chapter 31
Blood transfusions	Chapter 28
Dehydration and water balance	Chapters 8 and 13
Diet and energy content of food	Chapter 7

ENERGY AND EXERCISE

If the movement of muscles requires ATP, it follows that the ability of an athlete to move his or her muscles for any length of time requires a continued supply of this essential chemical. Muscles have three sources of ATP:

- **The ATP already present**. This provides instant energy and allows us move on demand. When we contract our muscles as hard and as fast as possible, we use ATP far faster than it can possibly be made. So we rely on the ATP that has accumulated during periods of relative rest. During maximum effort there is only enough ATP for about three seconds, but there is a back-up chemical, **creatine phosphate**, CP. The energy in CP can be used to instantly resynthesise more ATP, allowing maximal exercise to continue for up to 10 seconds. This is called the **ATP/CP** system, or the **alactic anaerobic system**. Alactic means there is no build-up of **lactic acid**, or more accurately, **lactate ions**, and anaerobic means that this system does not require oxygen. All events that require explosive bursts of energy, such as weightlifting or short sprints, use the alactic anaerobic system.

- **The ATP provided by glycolysis, the first phase of respiration**. In Chapter 34 you can see that glycolysis provides two ATP molecules per glucose molecule. This might not seem a lot compared with the 36 or so available from complete respiration, but it has two big advantages: it is relatively quick and it does not need oxygen. Thanks to this system, exercise can continue at near-maximum levels for up to one minute. However, there is a price to pay – the accumulation of lactate ions. Lactate lowers the pH in the muscles, causing fatigue and interfering with enzymes in muscle cells. The body can tolerate only limited levels of lactate. This system is known as the **glycolytic** or **lactic anaerobic system**, and is the main energy source for events that last between 10 and 60 seconds, such as the 400 metres hurdles (Fig 14.4).

- **The ATP provided by aerobic respiration**. This is the complete breakdown of glucose, when each molecule yields about 36 molecules of ATP. The problem is that this system takes time to provide energy, and then it still has its limits. However, provided that the level of activity stays within those limits, the aerobic system can fuel exercise for a couple of hours of more. This is the main energy system for many sports: all those that last for longer than one minute. Prolonged exercise that increases heart rate and breathing rate, but that is sustainable for hours rather than minutes, is commonly known as aerobics (Fig 14.5).

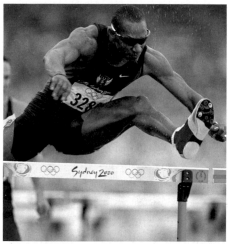

Fig 14.4 Sprinters and hurdlers rely on strength and are usually heavily muscled. To power their muscles during short races they rely on the ATP/CP system for up to about 10 seconds, and for the rest of the race the glycolytic system provides the energy

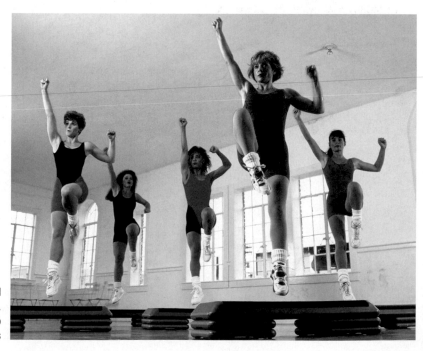

Fig 14.5 A familiar sight all over the western world, sustained aerobic exercise can strengthen the heart, lungs and circulation. As an added bonus, it also uses lipid (fat) as its main fuel, so aerobics are a vital part of many weight-loss programmes

THE ENERGY CONTINUUM

From what we have learned so far, it might appear that there are three separate energy systems. But, as we all know, we don't have to stop exercising after 10 seconds and then again after 1 minute to wait for the next energy system to cut in and give us some ATP. All three energy systems blend smoothly together, one taking over from the other. This phenomenon is known as the **energy continuum** (Fig 14.6).

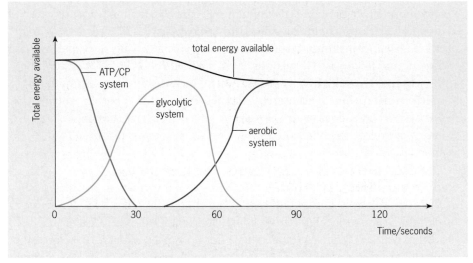

Fig 14.6 The energy continuum. The graph shows that one system gradually takes over from another, allowing us to move our muscles continuously

FUELS FOR EXERCISE

Respiration is the release of the energy contained in organic molecules such as glucose. All foods – carbohydrates, lipids or proteins – can be respired when the need arises. As a rough guide, the body uses the following fuels:

- **Glucose**. This supplies the normal, everyday energy needs when we are eating regularly (ie not dieting or starving). Anaerobic exercise always depends on glucose, because lipids cannot be used to fuel glycolysis.
- **Lipids**. Most lipid is stored in fat-storage, or **adipose** tissue. It takes a while to mobilise the lipid stores, but once the process has begun, this reservoir can fuel the body for as long as it lasts (for some of us almost indefinitely). Lipids cannot be used for short-term exercise, but they can be used to fuel aerobic exercise. This is why doing an aerobics class regularly can help to get rid of stored fat, if the exercise is done at the right intensity.
- **Protein**. The body usually gets around 5 per cent of its energy from protein. The proportion increases significantly only during starvation, when there is no other fuel available. Recent research suggests that protein can be used to fuel endurance events, but to what extent remains uncertain.

Overall, the main fuel used during exercise depends on the intensity of the activity. If you take running as an example, the main fuel for a long, gentle jog is lipid. But, as the intensity increases, so does the proportion of energy that comes from glucose. A summary of the energy systems used in exercise is shown in Table 14.2.

Table 14.2 A summary of the energy systems used in exercise

Energy system:	Anaerobic alactic	Anaerobic lactic	Aerobic alactic
Other names	ATP/CP; phosphate battery	glycolytic; lactic system	aerobic system
Energy comes from	ATP/CP already in muscles	glycolysis	complete respiration; glycolysis, Krebs cycle, ET chain
Nature of energy supply	allows maximum strength but is short lived	allows exercise to continue at near-maximum for up to 1 minute	allows long-term exercise at lower intensity
Timescale	up to 10 seconds	10 seconds to 1 or 2 minutes	anything longer than 1 to 2 minutes
By-product	no lactate	lactate	no lactate; carbon dioxide
Activity	explosive 'full strength'	longer sprint	endurance events such as cycling and marathons
Examples	sprints up to 100 metres; weightlifting and throwing events	200 to 400 metre sprints	800 metres, marathons
Training needed	to improve speed or strength	to improve lactate tolerance	to increase endurance capacity and to strengthen cardiovascular system

! SCIENCE FEATURE Diet for athletes

It has long been known that the correct diet can improve an athlete's performance, and in recent years the diets of sports stars such as footballers have been increasingly in the spotlight (Fig 14.7).

The best approach is a **holistic**, or *whole-body* one. The most dedicated athletes pay close attention to their diet, taking in the right amount and types of food at the right time. In addition, they ensure that they have the right amount of sleep and minimise their alcohol intake, especially before important events.

How do you plan a diet to maximise performance? A widely used technique is known as **carbo-loading** or **glycogen loading**. About six days before an event, the athlete goes on to a low-carbohydrate, high-protein diet. This depletes the body's glycogen reserves. Then, for the three days preceding the event the athlete goes on a high-carbohydrate diet, eating 8 to 18 grams of carbohydrate per kilogram of body weight per day. This significantly increases the amount of glycogen stored in the muscles, and this obviously helps during endurance events. It is of little help, however, if the event lasts for less than 90 minutes, and the effect wears off with repetition. Generally, athletes should aim for a diet in which 60 to 70 per cent of their energy comes from carbohydrate, increasing this figure in the day or two before an important event.

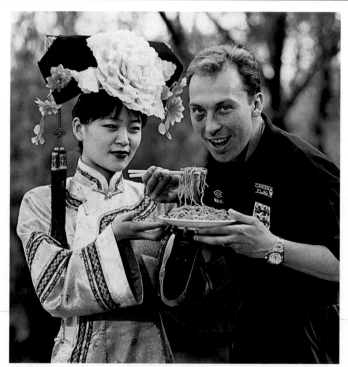

Fig 14.7 The prematch meal of footballers has traditionally been steak, but in recent years much more attention has been paid to their diet. Nutritionists recommend an easily digested high-carbohydrate diet, such as pasta, along with a little low-fat protein such as chicken or fish. This will leave the athletes feeling neither hungry nor bloated, and will top up their glycogen reserves

---

⚠ SCIENCE FEATURE Dehydration and isotonic drinks

Long-term exercise leads to prolonged sweating, and this can easily dehydrate an athlete. By 'dehydrate' we mean that the blood and body fluids become too concentrated. In the correct biological jargon we say that *the water potential is lowered*. The practical consequence of this is that the blood becomes more viscous and cannot flow as easily. In addition, sweating loses vital ions such as sodium, potassium and chloride, collectively known as electrolytes. A loss of electrolytes can lead to muscular cramps and a greatly reduced performance.

But, it is not too serious; all the athlete needs to do is to replace the lost water, electrolytes and glucose. This is easy enough if the athlete happens to be at home, but dehydration can strike when you are three-quarters of the way through a sporting event. Getting a drink that can do the trick takes a bit of planning.

The fastest way to combat dehydration is to drink pure water. This way, water is absorbed as fast as possible by osmosis. However, if you add glucose and salts to the drink you lower the water potential and therefore slow down the water uptake. In fact, if the drink becomes too concentrated it actually makes matters worse by drawing water out of the blood, in the same way as drinking seawater would.

The best compromise is to take an isotonic drink, a mixture that has the same water potential as body fluids. This way, all three components are absorbed as rapidly as possible. So do you have to spend money on expensive isotonic drinks? The simple answer is no. It is easy to make up an isotonic mixture on the same principle as the **oral rehydration therapy** given

to people suffering from diarrhoeal diseases (see page 232). For events lasting less than 90 minutes it is debatable whether isotonic drinks are any benefit at all. The fluid can be replaced by that cheapest of drinks, water, and the electrolytes and glucose are replaced later by any sensible meal.

Fig 14.8 Isotonic drinks contain a solution of electrolytes and glucose or short-chain polysaccharides, which can be easily digested and absorbed into the blood

---

2 SHORT-TERM RESPONSES TO EXERCISE

Exercise places great demands on the body. Think about the changes that occur when we exercise:

- oxygen levels fall,
- carbon dioxide and lactate levels increase,
- body temperature increases,
- blood glucose and glycogen levels fall,
- fluid and electrolytes (salts) are lost as we sweat.

All this presents a great challenge to the homeostatic mechanisms of the body. In an attempt to maintain some sort of stability, the body responds by:

- **Increasing heartbeat and breathing rate**. The increased respiration of the muscles raises the carbon dioxide level in the blood, and consequently lowers the pH (carbon dioxide is an acidic gas). These changes are detected by sensitive cells – **chemoreceptors** – situated in the **carotid** and **aortic bodies** in major blood vessels (see Chapter 9). In turn, these receptors inform the cardiovascular and respiratory centre in the brain, which responds by increasing the heart rate and the ventilation rate. The result is that gas exchange increases, as does the delivery of oxygen to the tissues and the removal of carbon dioxide.
- **Increased blood flow to the skin**. Muscle movement creates heat that must be lost. By increasing the diameter of the blood vessels that carry blood to the skin surface – a process called **peripheral vasodilation** – the blood can push heat to the body's surface where it can be more easily lost to the environment. This is why we 'go red' during and after exercise.
- **Increased sweating**. Sweat glands secrete a salty solution that evaporates from the skin, taking heat with it. Coupled with vasodilation,

 REMEMBER THIS

Ventilation rate is the amount of air taken in by our lungs in one minute. It can be calculated as depth of each breath (the tidal volume) × number of breaths per minute. At rest, this could be about 0.5 litres × 12 breaths = 6 litres per minute.

sweating is a very efficient means of removing excess heat (see Chapter 12 for more about heat transfer). The problem with excess sweating, of course, is that we also lose water and salts. If these are not replaced, the body becomes dehydrated and an athlete's performance can be impaired.

● **Increased mobilisation of glycogen**. In order to keep the muscles supplied with glucose, the glycogen stores in the muscles and the liver start to be mobilised. Glycogen is a highly branched polymer of glucose, which breaks down rapidly to release glucose (see Chapter 3). After prolonged exercise, glycogen stores are replenished from carbohydrate in the diet.

SEE QUESTION 1

THE RECOVERY PROCESS

After exercise, following the short-term changes listed above, the body does not immediately return to normal. Each system gradually returns to resting levels. Generally, the fitter the individual, more quickly the resting state is achieved.

Fig 14.9 shows the pulse rate of an athlete before, during and after a session on an exercise bike. Following exercise, pulse rate follows a classic pattern; a rapid initial fall followed by a slower return to normal. The shape of the curve results from two processes: the recharging of the ATP/CP system – this is known as the **alactacid component** of recovery – and the removal of lactate, called the **lactacid component** of recovery. Both occur because the body was not able to deliver enough oxygen to keep up with demand. This shortfall is known as the **oxygen debt** and is repaid after exercise stops. On a longer term basis, the body must also restore its glycogen levels. After a particularly gruelling exercise session, restoring the glycogen can take up to 48 hours and is not associated with raised heart or breathing rate.

We shall look at each of these components in turn.

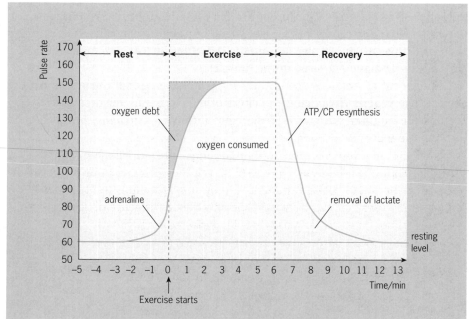

Fig 14.9 The effect of pulse rate on exercise. Note that the resting pulse is just under 60, normal for a trained athlete. The increase in heart rate just before exercise starts is due to anticipation, caused by the release of adrenaline: the more important the event, the more noticeable is the effect of adrenaline. When exercise starts, oxygen demand is greater than supply, and an oxygen debt builds up. This is paid off after the activity has finished

THE ALACTACID OXYGEN DEBT COMPONENT

This is the recharging of the ATP/CP system and takes one to two minutes to complete. Even fit people can't sprint continually, but after a short 'burst' it takes only a matter of seconds before there is enough ATP/CP to allow them to sprint again. In sports such as football, rugby, tennis or hockey, there are bursts of intense activity followed by periods of relative rest, during which the ATP/CP stores are restored by normal aerobic respiration.

Fig 14.10 This person has decided to improve his fitness by a little gentle jogging. He is comfortable at walking pace, but as soon as he starts to jog, he passes his anaerobic threshold, meaning that his muscles do not get oxygen quickly enough and lactate begins to accumulate, making his muscles ache. He will have to slow down until the lactate levels fall, then he will be able to run again. However, if he maintains regular exercise, he can raise his anaerobic threshold, allowing him to run for longer and/or faster without discomfort

THE LACTACID OXYGEN DEBT COMPONENT

This is the removal of the lactate ions that accumulated during the exercise. This acidic chemical lowers the pH of the blood, interfering with the enzymes in aerobic respiration and so reducing ATP supply to muscles. A significant build-up is painful, and there is only so much muscle fatigue we can take before we simply have to stop (Fig 14.10). Interestingly, one of the main effects of training is to increase our tolerance to lactate, so that we can continue to exercise despite higher lactate levels.

Lactate builds up in the muscles and diffuses into the blood. It then has five possible fates:

- 65 per cent is oxidised to carbon dioxide and water,
- 20 per cent is converted to glycogen,
- 10 per cent is converted to protein,
- 5 per cent is converted to glucose,
- a trace is excreted in urine and sweat.

The list clearly shows that the anaerobic system is rather wasteful in terms of energy. Only two ATP molecules are made per glucose, instead of a potential 36 (as in aerobic respiration). The production of two lactate molecules therefore represents a great waste of potentially useful energy: when each pair of lactate molecules are broken down in the liver, 34 molecules of ATP fail to be made available to the muscles.

The removal of lactate is speeded up by gentle exercise following the main activity. The **warm down** has the effect of reducing muscle soreness by keeping the capillaries dilated and therefore flushing oxygenated blood through the muscles.

Table 14.3 shows the recovery times that are recommended after exercise. This information is used by trainers to work out the optimum intervals between training sessions, and between training and events. The aim is to avoid chronic (long-term) fatigue that can have a drastic effect on an athlete's performance.

Table 14.3 Recommended recovery times after exhaustive exercise		
Aspect of recovery	**Recommended recovery time/min**	
	minimum	maximum
Restoration of muscle ATP and CP	2 minutes	3 minutes
Repayment of alactacid oxygen debt	3 minutes	5 minutes
Restoration of oxygen myoglobin	1 minutes	2 minutes
Restoration of muscle glycogen (after prolonged exercise)	10 hours	48 hours
Removal of lactic acid from muscle and blood	1 hour *	2 hours*
Repayment of lactacid oxygen debt	30 minutes	1 hour

* = speeded up by a warm down, ie if the muscles are kept working gently, the lactate is removed more quickly

3 LONG-TERM RESPONSES TO EXERCISE

If we exercise regularly, our bodies adapt and we 'get fit'. We feel better, look better and are able to cope easily with exercise that a few months earlier would have had us gasping in a heap on the floor. A remarkable feature of the human body is its ability to respond to exercise, making us more able to cope with our chosen activity.

SO WHAT IS ACTUALLY HAPPENING?

Observable long-term responses to exercise include changes to the heart, lungs and muscles, although the extent of the effects depends on the type of exercise done. For training to have an observable benefit, it must be above a certain intensity. Muscles must be **overloaded** before they begin to **adapt**.

For instance, if you were going to train for a rugby team, a brisk walk would be totally useless, because it would not overload your muscles. Rugby, like many sports, uses a combination of all the energy systems, and the training should reflect this balance. You would need to do endurance (cardiovascular) work as well as short and long sprints, along with training (eg with weights) that would exercise all of the relevant muscle groups. As the muscles adapt, the intensity of training must be increased continually, to ensure muscles are still overloaded. This is the concept of progressive **resistance** and ensures that the body continues to adapt and improve.

SEE QUESTION 5

CHANGES TO THE HEART

The heart responds to exercise like any other muscle; it enlarges, although the nature of the enlargement depends on the type of exercise done. Generally, there is an increase in the size of the **myocardium** (all the heart muscle) and therefore an increase in the size of the chambers. Consequently the **stroke volume** – the volume of blood pumped with each beat – increases. As a general guide, an untrained person has a stroke volume of about 90 cm^3, increasing to over 120 cm^3 after training.

When the heart can pump more blood per beat, it does not have to beat as often when the body is at rest. This is why getting fitter causes a decrease in resting pulse. This can go down from about 70 in the average untrained person to less than 50. Some of the world's top endurance athletes have a resting pulse of about 35 beats per minute.

Research also shows that exercise actually increases the strength of the blood vessels, allowing them to withstand higher pressures and reducing the risk of atherosclerosis (hardening of the arteries) later in life.

CHANGES TO THE LUNGS

The strength of the respiratory muscles (internal and external intercostals and the diaphragm) is increased, allowing a greater volume of air to be forced in and out. This means that **ventilation rate** improves. The total lung volume (the vital capacity) and surface area of the alveoli also increases, resulting in a greatly improved rate of gas exchange. Overall, the improvements in the circulation and gas exchange systems results in an increased V_{O_2}**(max)** (see Chapter 9 for a revision of these important terms).

REMEMBER THIS

Muscular contraction and the different muscle types are covered in Chapter 17.

REMEMBER THIS

A lot of damage can be done by over-training, and the effects of this can be worse than undertraining. Over-training, especially in young athletes who are still growing, can damage muscles, joints, ligaments and tendons. In the long term, the immune system can be damaged, leaving athletes susceptible to infection. For this reason training is usually cyclical. Intense periods of training are used to reach peak fitness for an important event, while at other times, the athlete simply maintains fitness.

CHANGES TO THE MUSCLES

The way in which the muscles adapt to exercise depends on the type of muscle fibres involved and the nature of the exercise. Here are some of the improvements that can occur:

- As a general rule, repetitive exercise with moderate resistance (such as aerobics, or light weights) improves muscle tone and stamina. Training with greater resistance (such as heavy weights) brings about an increase in muscle size, a process known as **hypertrophy**. Muscular hypertrophy involves an increase in the cross-sectional size of existing muscles. This is due to an increase in the number of myofibrils, the sarcoplasmic volume, and in the amount of connective tissue (tendons and ligaments).

- A general improvement in performance occurs because of an improved coordination of the motor units. A significant aspect of improving skill involves the antagonistic muscles. Studies have shown that as we practise particular movements, such as a tennis serve, we get smoother and more efficient. Not only do we train the active muscle to work better, we also learn to completely suppress the antagonistic muscle, preventing it from interfering with the action.

- With training, muscles increase their **oxidative capacity** – their capacity for respiration. This is achieved by an increase in the number of mitochondria in the muscle cells, an increased supply of ATP and CP and a rise in the quantity of the enzymes involved in respiration. The ability of the muscles to store glycogen, the amount of myoglobin and the ability to use lipid as an energy store are also increased.

- The density of the capillaries running through muscles increases in response to exercise, a process called **capillarisation**. This allows a more efficient exchange between working muscles and the blood.

SUMMARY

By the end of this chapter you should know and understand the following:

- Exercise places great metabolic demands on the body. Homeostatic mechanisms must cope with a much greater oxygen demand as well as a build-up of carbon dioxide and heat. In the long term, there may be a significant loss of water and salts (electrolytes) through sweating.

- Movement of muscles requires **adenosine triphosphate**, **ATP**, the chemical that supplies the muscles with the energy released in respiration.

- There are three sources of ATP: these comprise three energy systems. The most immediate is the **ATP/CP**, or **alactic anaerobic** system: ATP already present in the muscles is backed up by creatine phosphate, which is used to make more ATP instantly. The second system that kicks into action is the **glycolytic** or **lactic anaerobic** system: ATP is supplied by glycolysis and this system results in a build-up of lactate ions. The third is the **aerobic** system in which energy comes from the complete oxidation of glucose or lipid.

- The duration of an event determines the type of energy system used: up to 10 seconds, the ATP/CP system; from 10 to 60 seconds, the glycolytic system; and more than 60 seconds, the aerobic system.

- In the short term, the body responds to exercise by maintaining homeostasis. There is an increase in heartbeat and ventilation rate, coupled with vasodilation and sweating.

- During exercise we can build up an **oxygen debt**. This is paid off after exercise has finished, or during rest periods within the activity. This is why pulse and ventilation rate remain higher than normal even when activity has stopped. Extra oxygen is needed to replace the ATP/CP stores and to oxidise accumulated lactate.

- The long-term responses of the body to exercise are generally those that improve the body's fitness. These include strengthening of the cardiovascular system (heart and blood vessels) and making the lungs and muscles more efficient. The overall result is a body that is better able to cope with the chosen activity.

? EXAM QUESTIONS

Parts of questions marked * require a knowledge of training not covered in this chapter.

1 An investigation was carried out into effects of exercise on the heart rate and breathing rate of a student. His heart rate and breathing rate were first measured at rest. He then cycled at 5 km h⁻¹, using an exercise bicycle, for exactly two minutes. His breathing rate and heart rate were measured immediately after this exercise. After resting for five minutes, he cycled at 10 km hr⁻¹ for two minutes, then measurements of heart rate and breathing rate were taken as before. This procedure was repeated for cycling speeds of 15, 20 and 25 km h⁻¹. The results are shown in the table below.

Cycling speed / km h⁻¹	Heart rate / beats min⁻¹	Breathing rate / breaths min⁻¹
0 (rest)	64	12
5	70	13
10	80	14
15	100	17
20	140	20
25	180	27

a) Describe the relationship between cycling speed and heart rate.

b) (i) Describe and explain the changes that occurred in breathing rate during this investigation.
(ii) The volume of air breathed per minute is referred to as the *minute volume*. Suggest changes that are likely to occur in the student's minute volume during this investigation.

c) Suggest why the student rested for five minutes between each period of exercise.

d) During this investigation, the pH of the student's blood was found to have decreased from 7.4 (at rest) to 7.2. Explain how this change occurred.

Edexcel HB6 synoptic June 1999 Q4

2 Some athletes find that they recover faster after strenuous exercise by continuing to exercise slowly. This is sometimes called 'warming down'. An investigation was carried out to see if this was due to the removal of lactate from the blood.
A cyclist exercised vigorously on an exercise bicycle for six minutes and then rested for 34 minutes. When fully recovered, the same cyclist repeated the procedure, but this time cycled at a slow speed following the six minutes of fast cycling. Blood samples were taken at intervals and analysed for the concentration of lactate. The results are shown in the graph.

a) With reference to the graph,
(i) describe the changes in lactate concentration in the blood when the cyclist exercised and then rested during recovery.
(ii) explain the changes you have described in **(i)**;
(iii) suggest how cycling at slow speed during the 'warming down' period helps to lower the concentration of lactate in the blood more quickly than resting completely.
Trained athletes can exercise at much higher levels than untrained people before their blood lactate concentration starts to increase. This is due to changes in the cardiovascular system and in muscles.

b) Describe the changes that occur in the cardiovascular system and in muscles **during training**.

OCR 2802 Human Health and Disease June 2001 Q2

3
a) Define the term anaerobic metabolism.
b) The figure shows the relationship between oxygen consumption and time, before, during and after maximal exercise.

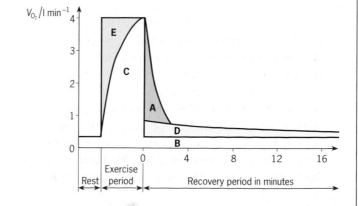

❓ EXAM QUESTIONS

Using the letters A, B, C, D and E which label each section, identify:

(i) the lactacid oxygen debt recovery;

(ii) the oxygen debt;

(iii) the alactacid oxygen debt recovery.

c) Describe the process of **ATP production** which restores the oxygen debt.

d) A student performs an interval training session during which the rate of **muscle phosphagen levels** during the recovery period was recorded. The results from this training session are given in the table below.

Recovery time/seconds	Muscle phosphagen restored
10	10%
30	50%
60	75%
90	87%
120	93%
150	97%
180	99%
210	101%
240	102%

(i) Using the results shown in the table, plot a graph of recovery time against the percentage of muscle phosphagen restored.

(ii) What resting interval would you recommend for a full recovery?

(iii) What would be the effect of restarting the exercise after 30 seconds?

e) (i) Part of the recovery mechanism after anaerobic exercise involves **myoglobin**. Explain the function of myoglobin during the recovery process.

(ii) Explain the importance of the cool down in the assistance of lactacid oxygen debt recovery and in the avoidance of muscle soreness.

(iii)* How could information on oxygen debt recovery be of use to an athlete and coach in the design of training sessions?

AEB Physical Education Paper 2 Section A Physiology of Exercise June 1995 Q1

4 An investigation was carried out to find the effect of exercise on the concentration of lactate in the blood. Four treadmills were placed at different gradients, ranging from 2 per cent to 14 per cent. A different athlete ran at a constant speed on each treadmill for five minutes. Three minutes after each run a blood sample was taken from each athlete and the concentration of lactate was measured. The graph shows the results.

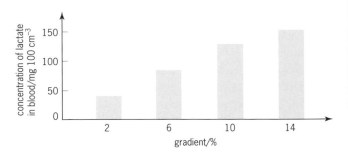

a) Name the process that produces lactate during exercise.

b) Describe the relationship between the intensity of exercise and the concentration of lactate in the blood.

c) Suggest **two** ways in which the design of the investigation could be improved to give more reliable results.

d) Suggest why the blood samples used to measure lactate concentrations were not taken from the athletes until three minutes after completion of a run.

e) The athlete running on the treadmill at a gradient of 14 per cent suffered from muscle fatigue. Explain what causes muscle fatigue.

AQA (B) BYB3 Physiology and Transport June 2001 Q3

5 During exercise, physiological changes take place within the body.

a) (i) What is the difference between a short-term physiological response and a long-term physiological adaptation? Give examples to support your answer.

(ii) If an athlete completed a 12-week programme of intense aerobic training, what physiological adaptations would you expect to take place within the cardiovascular system?

b)* The Multi-stage fitness test is commonly used by athletes to assess their level of endurance. State what this test measures and discuss the advantages and disadvantages of this method of assessment.

c) Myoglobin is found in the sarcoplasm of the muscle cell. Explain why an increase in myoglobin is beneficial to an endurance athlete.

UCLES Physical Education Paper 2 Optional Topics Section A: Scientific topics, Topic 4: Exercise Physiology June 1998 Q1

KEY SKILLS **ASSIGNMENT**

ENERGY SYSTEMS AND SPORT

In this chapter we have seen that the stamina of an athlete depends, to a great extent, on their ability to supply their muscles with sufficient oxygen to make ATP by aerobic respiration. (You might need to revise respiration by looking at Chapter 34 before attempting this Assignment.)

1 **a)** What is the essential difference between aerobic and anaerobic respiration?
b) How many molecules of ATP are produced, per molecule of glucose:
(i) from anaerobic respiration (glycolysis);
(ii) from complete aerobic respiration (glycolysis, link reaction, Krebs cycle and electron transport chain)?

For athletes, it would seem that the aerobic pathway is the most important, but we must take into account the time taken to complete the process. Complete aerobic respiration takes between 50 and 60 seconds, but glycolysis takes only about 10 seconds to produce ATP. Glycolysis is therefore very useful to an athlete because it is quick and requires no oxygen, so can still provide ATP even when the circulation can't supply oxygen quickly enough for aerobic respiration.

If it takes 10 seconds for glycolysis to produce ATP, how can we move instantly? The answer is that there is already ATP in the muscles: it was stored up during periods of inactivity. During strenuous exertion, such as when we sprint flat out or lift a very heavy weight, we use our ATP reserves in about 3 seconds. This leaves 7 seconds, which we can get through, thanks to a back-up chemical called creatine phosphate (CP). CP is similar to ATP but yields even more energy when broken down, and can be used to resynthesise ATP instantly, allowing maximum exertion to continue for another 6 or 7 seconds.

2 **a)** Suggest why we have enough ATP and CP to last for about 10 seconds.
b) Name the chemical produced as a consequence of anaerobic respiration. Explain its effect on the muscles.

The duration of a sport determines the main source of ATP used. The ATP/CP system is used for events lasting up to about 10 seconds, while the anaerobic lactic system (glycolysis) is used for events lasting between 10 and 60 seconds. Beyond about 60 seconds, the aerobic system provides most of the ATP.

3 Copy and complete the table below, identifying the energy system which provides most of the ATP for that particular activity.

Sport/Event	ATP/CP	Anaerobic	Aerobic
Javelin			
200 m sprint			
800 m run			
Shot put			
Marathon			
100 m sprint			
400 m sprint			
5000 m run			
Soccer			

Sports such as soccer, hockey and rugby use a mixture of all the systems because the activity is prolonged but not constant. There are periods of intense activity followed by periods of relative rest.

Fig 14.A1 Sprinting is an anaerobic activity which relies on strength. Sprinters therefore tend to be heavily muscled

Fig 14.A2 Prolonged exercise relies on the aerobic system to provide a continual supply of ATP. Generally, fatigue sets in only when fuels such as glycogen begin to run out. Endurance specialists such as this marathon runner tend to be of slight build, with small muscles

KEY SKILLS **ASSIGNMENT**

4 a) Suggest why the 400 metres (world record under 40 seconds) is thought by athletes to be a particularly painful event?

b) What use would the information in the completed table (above) be to someone who wanted to improve his or her fitness for a particular event?

c) Suggest why sprinters usually have large muscles while endurance athletes are generally quite thin.

d) An amateur footballer wishes to get fit for the new season, so he jogs 5 miles each day. Explain why this would be an insufficient preparation and suggest the additional training he would need to do.

e) You can occasionally see joggers out for a run with their dogs, although it is rare for a cat to do the same. Cats in the wild hunt their prey by pouncing or using a short sprint, but hunting dogs rely on stamina in a long chase. Which is the predominant energy system in dogs and cats? Explain why dogs need more regular exercise than cats.

Fig 14.A3 Wild dogs using stamina to run down wildebeest

Aerobics and weight loss

Many people want to improve their overall fitness. But can exercising help them to lose weight as well? To understand more about aerobics and weight loss, we need to think about the type of fuel being used in respiration. The way we use respiratory fuels has been likened to the way in which we use money.

We all need spending power, and a common form of payment is cash. When we have no cash, we have to go to the cashpoint or bank and get some more out instantly. When our bank account is low, we can turn to our savings account. This may take a while to get at, but it provides a useful cushion. When we are completely broke, we will have to sell our belongings and property just to survive. Eventually, with no injection of cash, we will be declared bankrupt.

5 Rewrite the last paragraph, applying the analogy to human energy sources. Include the terms lipid, protein, glucose, dead, energy, food and glycogen. When you have done this, write down or discuss the limitations of this analogy.

It takes time to run down our glycogen store and start to use fats. This is a shame, because otherwise you could simply go for a run and return slimmer, if a little baggy. In practice, it takes several sessions – and often a reduced calorific intake – to bring about an improvement. Generally, exercise is beneficial because it strengthens the cardiovascular system and increases metabolic rate, so you use up more calories all the time, not just while exercising.

6 a) What is meant by **(i)** metabolic rate and **(ii)** cardiovascular system?

b) Imagine you are a GP faced with a rather fat couple in their forties. Write down or discuss how you would try to persuade them to take exercise. What exercise would you recommend and what precautions would you include?

15

Nerves and impulses

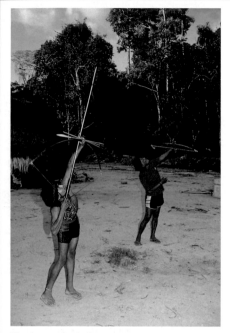

Some Amazonian Indians have a very effective way of hunting monkeys and other animals. They tip their blowpipe arrows in a preparation made from the bark of one of the local trees. The substance, known as curare, is a powerful neurotoxin.

For many centuries, the exact nature of curare was a mystery. In 1814, in an attempt to confirm his suspicions, the explorer Charles Waterton injected a donkey with curare. Within a few minutes, the donkey appeared dead. Waterton cut a small hole in her throat and inflated her lungs with a pair of bellows. According to his account, the donkey regained consciousness and looked around. This crude method of artificial respiration was continued for a couple of hours until the effects of the curare had worn off.

Analysis of curare has since shown that it prevents nerve impulses getting through to the muscles, causing paralysis. In addition to affecting the movement of skeletal muscles, curare interferes with heartbeat and breathing, although the exact effects depend on the dosage and whether the curare is taken orally or by injection. If heartbeat and breathing are severely affected, the result is usually fatal.

The Amazonian Indians, not being analytical chemists, assess the strength of their preparation by how long it takes a monkey to fall out of the tree. 'One-tree curare' is the most powerful; the monkey only escapes to the next tree before the paralysing toxin brings it tumbling to the forest floor. 'Two-tree curare' is obviously weaker and 'three-tree curare' is the weakest preparation that is useful. After that, the monkey travels too far and, although it still dies from the action of the poison, it's usually impossible to find

In 1939, the active ingredient in curare was isolated, and in 1943 doctors started to use it as an anaesthetic. This gift from the rainforest has proved to be a vital tool in surgery, relaxing muscles and generally making the surgeon's job a lot easier. It is also useful for treating conditions in which the muscles go into spasm – polio, tetanus, epilepsy and cholera. Today, synthetic analogues of the active ingredient in curare, such as d-tubocurarine, are used widely in medicine.

1 IT'S ALL ABOUT COMMUNICATION

The body of a human is incredibly complex. To ensure that the different parts work together effectively, there needs to be communication to register changes in internal and external conditions. In this section of chapters we look at the role of hormonal and nervous communication.

The nervous system is made up of specialised cells that allow the different parts of the body to communicate. These cells are called neurones (Fig 15.1). They carry information from one part of the body to another. We concentrate on the structure and function of these remarkable cells in this chapter.

cell body

nucleus

dendrites

axon

connections to other neurones or effector cells

Fig 15.1 A simple diagram of a neurone

OVERVIEW OF THE NERVOUS SYSTEM

The nervous system as a whole ensures that the body responds appropriately to the external conditions at any given time. It means we can do the following:

- Gather information. Sense organs called **receptors** detect **stimuli** from the internal and the external environment (see Chapter 19).
- Transmit sensory information to the **central nervous system** by means of the **sensory nerves** (see Chapter 16).
- Co-ordinate information. Incoming information travels to the brain via the spinal cord. The brain then decides what to do. Decisions are often based on memory, the result of our past experience.
- Transmit the information to the **effectors**: the muscles and glands. Impulses pass from the central nervous system to the effectors via **motor nerves** (Fig 15.2).

The structure of the mammalian/human nervous system is covered in detail in Chapter 16, the structure of muscles and their role in movement is covered in Chapter 17 and hormonal communication is the topic of Chapter 18. This section of chapters finishes with a look at senses and behaviour (Chapter 19).

HOW INFORMATION TRAVELS AROUND THE BODY

Information passes along **neurones** in the form of electrical signals called **nerve impulses**. A nerve impulse, known as an **action potential**, is not a message, nor is it an electrical current (a flow of electrons). It is more a change in ion balance in the nerve cell, which spreads rapidly from one end to the other, like the fire travelling along a burning fuse. It is little more than an electrical 'blip' – but the brain can make sense of the blips because they vary in frequency and arrive down specific nerves (see Chapter 16).

So what happens when the nerve impulse reaches the end of the neurone? It connects with other neurones at junctions called **synapses**. A nerve impulse crosses a synapse usually by means of a **chemical transmitter**. The whole of the nervous system therefore communicates by a mixture of electrical and chemical signals. This allows information to travel around an organism with far greater speed and precision than if only chemical signals were used.

Fig 15.2 Flow diagram summarising nervous system function

✔ **REMEMBER THIS**

The entire human body is made up from just four basic types of tissue (see Chapter 2). Epithelial tissue forms coverings and linings, connective tissue holds other tissues together, muscle tissue has the ability to contract and nervous tissue is excitable – it can transmit impulses. All of our organs are formed from these four basic tissue types.

? **TEST YOURSELF**

A What is the difference between a neurone and a nerve?

Fig 15.3(a) A nerve is a bundle of axons, together with connective tissue and blood vessels. The micrograph shows a cross section of a nerve

Fig 15.3(b) The brown threads (axons) and their cell bodies (see Fig 15.8) form bundles that make up nerves

NERVES

Neurones rarely act alone. They are bundled together into larger, visible structures called **nerves** (Fig 15.3). Nerves form a complex network throughout the body. Sensory nerves take information received from receptors, or sense organs, on the outer parts of the body – the eyes, ears, tongue and nose and touch receptors in the skin – into the central nervous system. Processing of that information happens in the brain, and then effector nerves take the information from the brain to effectors such as muscles and glands to cause the body to take some action. Think what happens if someone puts, say, a glass of water in front of you. Sensory nerves take the information from the eyes to the brain. If you are thirsty, the brain may then send information down effector nerves to your arm and hand and you would reach for the glass, pick it up and drink. If you aren't thirsty, information might travel down effector nerves to the muscles in your face and throat for you to say 'No thanks'.

In the reflex arc (see Chapter 16), the sensory and effector nerves are linked by a neurone in the spinal cord, rather than by neurones in the brain.

2 HOW DO NEURONES CARRY INFORMATION?

Neurones have two properties that enable them to carry information. They are **excitable** – that is, they can detect and respond to stimuli – and they are **conductive** – that is, they can transmit a signal from one end to the other. Before we look at how a neurone transmits information, let's find out what is going on in the neurone before information arrives.

THE NEURONE AT REST

At any given moment a neurone needs to be ready to conduct impulses. This state of readiness is called the **resting potential**. At this point, the axon membrane is **polarised**: the fluid on the inside is negatively charged with respect to the outside. This difference in charge, about –70 mV, results from an unequal distribution of ions known as an **electrochemical gradient**.

THE RESTING POTENTIAL

What causes the resting potential, and how is it maintained? Like most cells, neurones have selectively permeable membranes that contain specialist proteins. Some of these proteins act as passive ion channels; others are active transport mechanisms that pump ions. The steps that establish a resting potential are shown in Fig 15.4.

Fig 15.4 How the resting potential is established. In each square micrometre of axon membrane surface, there are between 100 and 200 sodium–potassium pumps. Each pump actively transports about 200 sodium ions out and about 130 potassium ions in, every second, causing positive ions to accumulate on the outside of the axon membrane. Potassium ions diffuse out through passive ion channels more readily than sodium ions can diffuse in. So overall, an electrochemical gradient develops in which the outside of the axon is positive relative to the inside

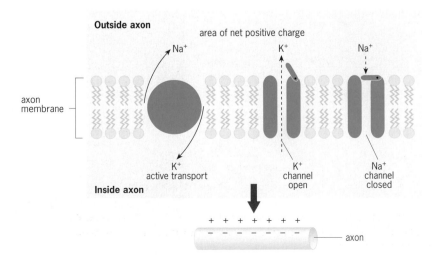

The first important thing that happens is that an active transport mechanism in the neurone membrane swaps sodium ions for potassium; Na$^+$ ions are moved out, while K$^+$ ions go inwards. On its own, this mechanism simply swaps one positively charged ion for another and so does not create the resting potential.

The second important point is that a neurone also has passive Na$^+$ channels and K$^+$ channels. In a resting neurone the K$^+$ channels are generally open while the Na$^+$ channels stay shut. Thus potassium is able to diffuse out faster than sodium can diffuse in, and diffusion via the passive ion channels is unable to balance the active transport mechanism. Positive ions accumulate on the outside of the membrane and this then becomes positively charged compared to the inside.

Overall, the resting potential results from an active transport mechanism, which uses the energy in ATP to create an ionic imbalance. If the active transport mechanism failed, the potential would quickly be lost, and the neurone would no longer be excitable.

THE NEURONE IN ACTION

So what exactly is a nerve impulse? In short, it is a temporary reversal of the resting potential; a change that spreads rapidly along the axon. A schematic diagram explaining how this happens is shown in Fig 15.5. Fig 15.6 shows an oscilloscope trace of an **action potential**, sometimes referred to as a 'spike' because of its shape.

Area of depolarisation

Wave of depolarisation moves along the axon

Fig 15.5 The basic concept of the nerve impulse – a wave of depolarisation which spreads along the axon. The active transport mechanism immediately re-establishes the resting potential as soon as the action potential has passed

SEE QUESTION 1

✔ REMEMBER THIS

Once the action potential starts, the wave of electrical activity travels, or is **propagated**, along the axon at great speed.

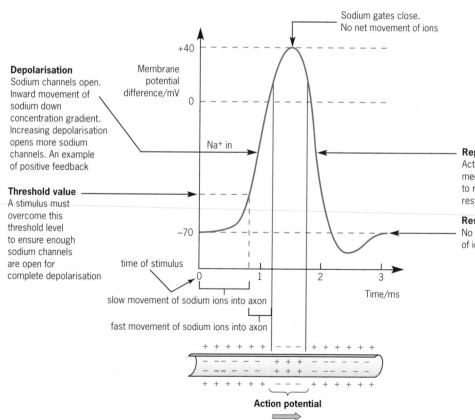

Depolarisation
Sodium channels open. Inward movement of sodium down concentration gradient. Increasing depolarisation opens more sodium channels. An example of positive feedback

Threshold value
A stimulus must overcome this threshold level to ensure enough sodium channels are open for complete depolarisation

Membrane potential difference/mV

+40

0

Na$^+$ in

−70

time of stimulus

Sodium gates close. No net movement of ions

Repolarisation
Active transport mechanism begins to restore resting potential

Resting potential
No net movement of ions

0   1   2   3

Time/ms

slow movement of sodium ions into axon

fast movement of sodium ions into axon

+ + + + + + – – – + + + + + +
– – – – – – + + + – – – – – –
– – – – – – + + + – – – – – –
+ + + + + + – – – + + + + + +

Action potential

Direction of nerve impulse

Fig 15.6 An action potential as recorded by an oscilloscope, and how it relates to the events in the axon. This is a key exam diagram that you should use when you are aiming to explain the basic sections of the trace

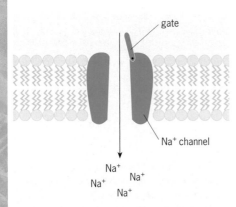

Rapid influx of Na$^+$ ions reverses resting potential

Fig 15.7 A diagrammatic model to show an ion channel in the axon membrane. When the Na$^+$ channels open, the rapid influx of ions causes the action potential by reversing the resting potential.

When the Na$^+$ channels close and K$^+$ channels open, this allows a rapid outflow of K$^+$ and repolarises the membrane.

In the action potential region, only about one in a million Na$^+$ or K$^+$ moves through channels

DEPOLARISATION

When a stimulus reaches the resting neurone, the Na$^+$ ion channels open at the site of stimulation. The ions move in relatively slowly at first, making the membrane potential less negative. Then, at a **threshold** of about –50 mV, a much more rapid inrush of Na$^+$ ions is triggered, and the inside suddenly becomes positive (+40 mV) with respect to the outside. As the electrical charge across the membrane is reversed, we say that the membrane becomes **depolarised**. Depolarisation initiates a nerve impulse.

THE ACTION POTENTIAL

Why does the area of depolarisation flow along the axon? When a patch of the axon membrane depolarises, a flow of current is produced locally as electrically charged ions move. This current stimulates the next patch of membrane, where Na$^+$ channels also open, as shown in Fig 15.7. In this way the action potential moves along the axon in the direction away from the cell body.

The change in voltage (the amplitude) of the action potential is the same at all points along the axon. It does not increase or fall off as the signal travels. After a very brief moment – about 1 millisecond – the Na$^+$ channels close and the resting potential is established once again. The Na$^+$ channels cannot re-open until after the resting potential has been re-established.

THE REFRACTORY PERIOD

Nerves conduct messages by 'firing' repeated action potentials along the nerve fibres. The time delay between one action potential and the next is called the **refractory period**. This has two phases:

- The **absolute refractory period**. During this time, immediately after the sodium channels close, no further impulse can be conducted.
- The **relative refractory period**. During this time, the membrane begins to recover and becomes increasingly responsive. It is possible to initiate another action potential provided that the stimulus is greater than normal.

The refractory period imposes a limit on the frequency of nerve firing. Large nerve fibres recover in 1 millisecond and could theoretically propagate 1000 impulses per second. Small fibres take longer to recover – about 4 milliseconds – and so could propagate about 250 impulses per second.

The refractory period ensures that each action potential is separated from the next, with no overlapping of signals. We can think of the information the signal conveys as coded information. The refractory period also ensures that a nerve impulse flows in one direction only: the wave of depolarisation can only move away from the refractory region, towards the axon terminal, and therefore onwards to the next neurone in the pathway.

THE SPEED OF AN ACTION POTENTIAL

The speed at which a nerve impulse or action potential travels is known as its **conduction velocity**. In human nerve fibres, values range from 1 to 3 metres per second in unmyelinated fibres, and between 3 and 120 metres per second in myelinated fibres (see below for what myelin is and what it does). In general, conduction velocity depends on the following factors:

- **Axon diameter**. The larger the axon, the faster it conducts.
- **Myelination of the neurone**. A nerve impulse travels faster in a myelinated nerve than in an unmyelinated nerve.
- **Number of synapses involved**. Communication between neurones across the tiny gaps at the synapses involves chemical release and a brief time delay. The greater the number of synapses in a series of neurones, the slower the conduction velocity.

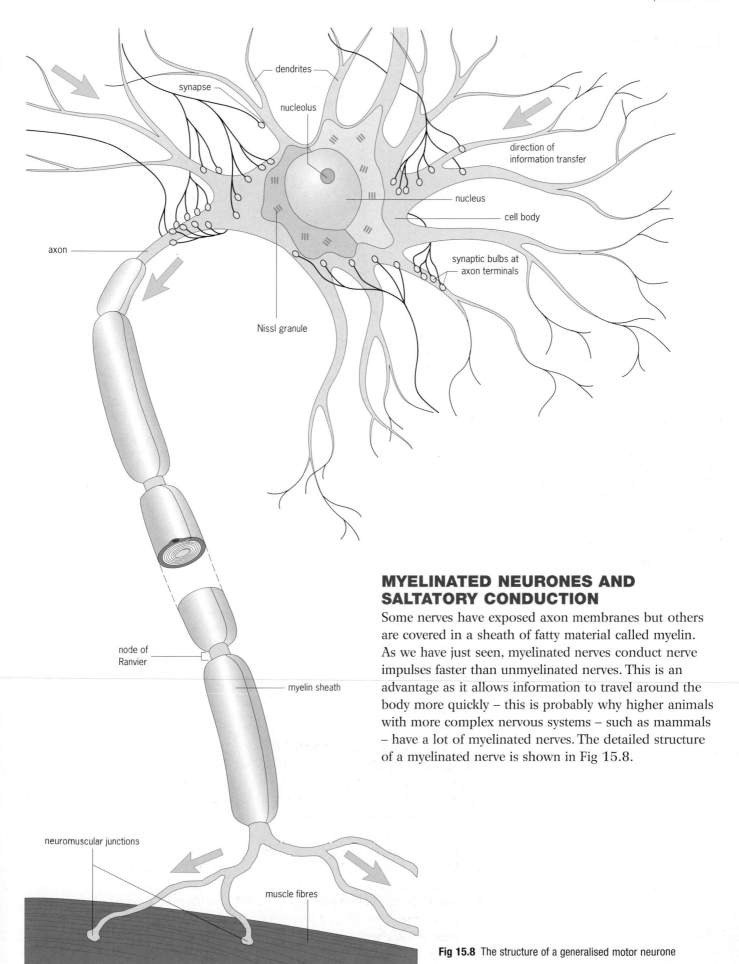

dendrites

synapse

nucleolus

direction of information transfer

nucleus

cell body

axon

synaptic bulbs at axon terminals

Nissl granule

node of Ranvier

myelin sheath

neuromuscular junctions

muscle fibres

MYELINATED NEURONES AND SALTATORY CONDUCTION

Some nerves have exposed axon membranes but others are covered in a sheath of fatty material called myelin. As we have just seen, myelinated nerves conduct nerve impulses faster than unmyelinated nerves. This is an advantage as it allows information to travel around the body more quickly – this is probably why higher animals with more complex nervous systems – such as mammals – have a lot of myelinated nerves. The detailed structure of a myelinated nerve is shown in Fig 15.8.

Fig 15.8 The structure of a generalised motor neurone

! SCIENCE FEATURE Myelin and multiple sclerosis

Multiple sclerosis is a chronic, often disabling disease of the central nervous system that is caused by a loss of myelin around neurones. Symptoms may be mild, such as numbness in the limbs, or severe – paralysis or loss of vision. Most people with MS are diagnosed between the ages of 20 and 40 but the unpredictable physical and emotional effects can be lifelong.

What starts MS off remains something of a mystery. In some cases, the immune system starts to recognise myelin as a 'foreign' protein and T cells (see Chapter 30) become active and start to destroy the myelin sheaths around many neurones. The resulting demyelination causes disruption to the transmission of nerve impulses within the brain and around the body. Some research suggests that this autoimmune response is provoked when a virus that has proteins similar in shape to myelin infects the body. As the body reacts against the virus, it also reacts against its own myelin.

One possible treatment for MS is a substance called copolymer-1, which is a random polymer of basic amino acids. This is structurally similar to myelin basic protein and researchers hope that it can 'mop up' some of the immune

cells and antibodies that would otherwise damage myelin. People with MS would need to have an injection of copolymer-1 every day but it has been shown to help and it decreases the frequency of relapses.

Fig 15.9 A coloured magnetic resonance image (MRI) scan of an axial section through the brain of a person with MS. The lesions due to MS are shown in red and yellow at the left and lower left

✔ REMEMBER THIS

The pale, creamy colour of myelinated nerves is due to the fatty (lipid) nature of the myelin that surrounds them.

THE IMPORTANCE OF MYELIN

But what is myelin and how does it form around neurones? Specialised cells called **Schwann cells** wrap themselves round the axons of some neurones as they develop in a growing embryo (Fig 15.10a) The Schwann cells form a thick, lipid-rich insulating layer called the myelin sheath (15.10b). This insulates the axon electrically, rather like the plastic layer round a copper wire in an electrical flex. Neurones with myelin sheaths are said to be myelinated.

Fig 15.10(a) Schwann cell growing round an axon

Fig 15.10(b) False-colour transmission electron micrograph of part of the myelin sheath (orange and green layers) of the human auditory nerve.

? TEST YOURSELF

D What effect does a myelin sheath have on conduction velocity?

SALTATORY CONDUCTION

Schwann cells cover most of the axon, but leave bare sections between the cells called **nodes of Ranvier** (see Fig 15.8). Nerve impulses that travel along a myelinated axon 'jump' between these gaps by **saltatory conduction** (Fig 15.11).

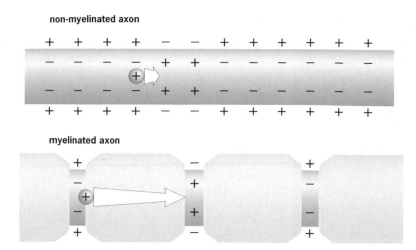

non-myelinated axon

myelinated axon

Fig 15.11 Myelin and conduction speed. Top: the impulse passes along the nerve fibre by each section of membrane depolarising the next, so transmission is relatively slow. Bottom: the impulse jumps from one node to the next, so only the membrane at the nodes is depolarised. This leads to rapid transmission

THE ADVANTAGES OF SALTATORY CONDUCTION

Saltatory conduction has two advantages:

- The conduction of a nerve impulse is fast. In human unmyelinated fibres, nerve impulses travel at 1 to 3 metres per second. Myelinated fibres conduct at speeds of up to 120 metres per second.
- Metabolically, saltatory conduction is quite economical, because fewer ions move across the membrane, so the ion pumps need less energy to restore the ionic balance.

> ✔ **REMEMBER THIS**
>
> In myelinated nerves, impulses can travel at up to 120 metres per second – the equivalent of a person running the length of a football pitch in one second, ten times quicker than the best sprinters can run.

3 COMMUNICATION BETWEEN NEURONES: THE SYNAPSE

When an action potential reaches the end of an axon it is passed on to the next neurone, or on to an effector cell such as a muscle or gland. The axon of one neurone does not usually make direct contact with the cell body of the next; the two cells are separated by a gap called a synapse (Figs 15.12 and 15.13).

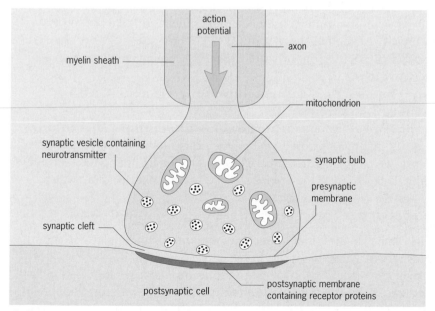

Fig 15.12 The basic structure of a synapse

Fig 15.13(a) False-colour scanning electron micrograph of the junction sites (synapses) between nerve fibres (purple) and a neurone cell body (yellow)

Fig 15.13(b) A synapse showing the mitochondria (top) and vesicles (centre) in the presynaptic neurone, and neurotransmitter passing across the synaptic cleft

REMEMBER THIS

The cell that carries a signal towards a synapse is a **presynaptic cell**; the cell carrying the signal away from the synapse is a **postsynaptic cell**. Presynaptic cells are always neurones but postsynaptic cells can be either neurones or effector cells.

REMEMBER THIS

Many drugs and poisons exert their effect because they interfere with the functioning of synapses – see the Assignment at the end of this chapter.

REMEMBER THIS

In some people, release of too much noradrenaline causes the heart to race. One way of treating this is to use drugs known as beta-blockers. These drugs have molecular shapes similar to noradrenaline.

THE MAIN FEATURES OF A CHEMICAL SYNAPSE

The axon terminal of a presynaptic neurone is swollen, and is often called the **synaptic bulb** or **synaptic knob** (see Fig 15.12). It meets the cell body of the next axon, leaving a gap or **synaptic cleft** of about 20 nm. This is small – you would have to split a hair 250 times to get it to fit sideways into this gap. Even so, synapses have a high electrical resistance, and this gap is too big to allow the action potential to simply jump from one neurone to the next.

HOW DO SYNAPSES WORK?

So how does the action potential get across? It is relayed by chemicals that diffuse across the gap and initiate an action potential in the neurone at the other side. The synaptic bulb contains many mitochondria, which provide energy for the manufacture of chemicals called **neurotransmitters**. Neurotransmitters are small molecules and they can diffuse easily across the synaptic cleft. Synaptic vesicles are temporary vacuoles (membrane-bound spheres) that store neurotransmitter chemicals, the most common being **acetylcholine**. Synapses that have acetylcholine as their transmitter are called **cholinergic synapses**.

TYPES OF NEUROTRANSMITTER

Hundreds of neurotransmitters have been identified and there are certainly more to find. There are four main groups:

- **Acetylcholine** (neurones that release acetylcholine are described as **cholinergic**).
- **Amino acids** such as **gamma-aminobutyric acid** (**GABA**), **glycine** and **glutamate**.
- **Monoamines** such as **noradrenaline**, **dopamine** and **serotonin** (neurones that release noradrenaline are described as **adrenergic**). Synapses that use noradrenaline affect heart rate, breathing rate and brain activity. This is similar to the effect of the hormone adrenaline, which prepares the body for emergencies.
- **Neuropeptides** (chains of amino acids) such as **endorphins**.

Acetylcholine also acts throughout the brain, modifying the activity of other neurotransmitters. Nerve pathways in which acetylcholine is a neurotransmitter seem to be involved in motivation and memory. A very fast-acting enzyme called **acetylcholinesterase** breaks down acetylcholine into acetic acid and choline. These substances are reabsorbed through the presynaptic membrane. ATP energy from mitochondria is used to provide the energy to resynthesise acetylcholine, which is then returned to the vesicles. The chemicals in some 'nerve gases' work by inhibiting acetylcholinesterase.

TRANSMISSION AT A SYNAPSE – A DETAILED EXPLANATION

As you read the next section, follow the stages of chemical transmission at a synapse numbered in Fig 15.14.

Fig 15.14 The sequence of events in chemical transmission at a synapse

1 An action potential arrives at the synaptic bulb.
2 Calcium channels open in the **presynaptic membrane**. As the Ca^{2+} ion concentration inside the bulb is lower than outside, Ca^{2+} ions rush in.
3 As the Ca^{2+} concentration increases, **synaptic vesicles** move towards the membrane.
4 The vesicles fuse with the membrane, releasing neurotransmitter into the synaptic cleft.
5 The short journey across the synapse takes about a millisecond, longer than an electrical signal takes to travel the same distance. This time is therefore called the **synaptic delay**.
6 At the **postsynaptic cell**, the neurotransmitter binds to receptors on the postsynaptic cell surface membrane.
7 Some neurotransmitters open sodium channels in the postsynaptic membrane, causing an inflow of Na^+ ions. This creates an **excitatory postsynaptic potential** (EPSP) in the membrane. This potential lasts for only a few milliseconds and can travel only a short distance, but it makes the membrane more receptive to other incoming signals.
8 Other neurotransmitters open chloride and potassium channels, causing Cl^- ions to flow into the cell and K^+ ions to flow out. This creates the **inhibitory postsynaptic potential** (IPSP) that makes the postsynaptic membrane less receptive to incoming signals. If more EPSPs than IPSPs are produced in the postsynaptic membrane, the change in potential can exceed the threshold potential needed to create a new action potential.
9 Once the neurotransmitter has acted on the postsynaptic membrane, it is immediately broken down by an enzyme on the postsynaptic membrane. If the transmitter remained, it could continue to stimulate the neurone, even without new impulses coming from the presynaptic cell.
10 The enzyme acetylcholinesterase splits acetylcholine into choline and ethanoic acid. These components then diffuse back into the presynaptic membrane, when they are resynthesised to acetylcholine using the ATP from the mitochondria.
11 An action potential is set up in the postsynaptic cell.

? TEST YOURSELF

E The synaptic cleft has a high electrical resistance but is very narrow. Suggest reasons for these two observations.

F Why can transmission occur only one way across synapses?

SEE QUESTION 2

✔ REMEMBER THIS

Receptor binding can also lead to the formation a second messenger (a transmitter substance) such as cyclic AMP (cAMP). This also changes the ionic permeability of the membrane, but it has a longer-lasting metabolic effect on the ion channels. Such long-term changes to brain neurones are thought to underlie memory.

SYNAPSES IN ACTION: FACILITATION AND SUMMATION

Does the arrival of an impulse at a synapse mean that an action potential is always generated on the postsynaptic neurone? The answer is no, because it would lead to chaos; all neurones would automatically connect to others. Synapses therefore have a vital role in **information processing**. Transmission of information across synapses is **graded**. They can amplify or damp down the information they receive. In many cases, they will not transmit it at all.

A neurone can be fed information by both excitatory synapses that produce EPSPs and inhibitory synapses that produce IPSPs (Fig 15.15). Whether the cell develops an action potential is determined by the sum of all the excitatory and inhibitory synapses at any particular moment. Put simply, impulses arriving at some synapses will 'excite' the cell, while others will 'calm it down'. Whether or not a neurone generates an action potential depends on the balance of the two types.

Facilitation occurs when many EPSPs reach the same area of the postsynaptic membrane at the same time, making it more likely that the threshold potential will be reached. When this happens, an action potential is propagated.

(a) Facilitation
If many EPSPs are set up at the same time in the same area of the postsynaptic membrane, the threshold potential may be reached and an action potential may result

(b) Summation
If the same number of EPSPs as IPSPs are set up at the same time in the same area of the postsynaptic membrane, the two potentials cancel each other out

Fig 15.15(a) Facilitation and **(b)** summation

Imagine a synapse discharging its transmitter on to a postsynaptic neurone. This will set up an EPSP, but if it is not big enough to reach the threshold, no action potential is generated. However, if other synapses discharge their transmitter at the same time, or shortly after, the EPSPs will add up, or **summate**, until an action potential is generated. Generally, there are two types of summation:

- **Temporal summation** – summation of two or more impulses that arrive rapidly one after the other down the same neurone.
- **Spatial summation** – summation of two or more impulses arriving down different neurones at the same time (spatial = related in space). One neurone can receive information from many others – this is synaptic convergence. It follows that the arrival of one excitatory impulse will leave the neurone more responsive to another one. This is known as facilitation, and results from the summation of two or more synapses discharging their transmitter substance at the same time.

As a simple example of this idea, imagine the touch receptors from one area of skin feeding into one sensory neurone. An action impulse down just one receptor is almost certainly an insignificant stimulus, and can be ignored. It will not create an EPSP large enough to generate an action potential in the sensory nerve. However, if several touch receptors are stimulated at the same time, they will summate and produce a sensory impulse.

WHY HAVE SYNAPSES?

Synapses are important because they allow the transfer of information in nerve networks to be controlled. Synapses:

- allow information to pass from one neurone to another;
- help ensure that a nerve impulse travels in one direction only;
- allow the next neurone to be excited or inhibited;
- can amplify a signal (make it stronger);
- protect nerve networks by not firing when over-stimulated. When this happens the synapse is said to be fatigued. Over-stimulation might damage muscle or gland tissue;
- can filter out low-level stimuli. For example, you fail to notice the sound of a clock ticking because synapses are 'filtering out' the signal of sound;
- aid information-processing by the action of summation (adding together the effect of all impulses received, see page 270);
- are modifiable and can form a physical basis for memory.

Overall, the significance of synapses cannot be over-emphasised. They allow us to select particular neural pathways. The process of learning is largely one of educating the synapses. People can play the violin or piano, or play tennis, because their synapses allow their brains to co-ordinate their senses and muscles in the right way. Your memories, too, have a basis in synapses choosing specific pathways. If you are asked, 'What's the capital of France?' your synapses will (we hope) select a pathway of neurones in your brain which will lead you to the answer 'Paris'.

A SPECIAL SORT OF SYNAPSE: THE NEUROMUSCULAR JUNCTION

When a motor neurone terminates on a muscle, it branches into many specialised synapses called **neuromuscular junctions**. Fig 15.16 shows a typical neuromuscular junction. These structures are wider than ordinary synapses and come into close contact with the surface membrane of the muscle, the sarcolemma. The area of sarcolemma in contact with the synapse is called the motor end-plate, and contains acetylcholine receptor sites. When an action potential arrives at a neuromuscular junction, vesicles of acetylcholine are released in the usual way. The transmitter changes the permeability of the motor end-plate to Na$^+$ and K$^+$, creating an end-plate potential (EPP) which results in an action potential passing along the sarcolemma. This impulse brings about the contraction of muscle fibres in that area.

REMEMBER THIS

Temporal = related in time.
Spatial = related in space.

SEE QUESTION 3

REMEMBER THIS

Facilitation is not a result of temporal summation, which is simply the accumulation of EPSPs because impulses arrive before the preceding EPSPs have died down.

SEE QUESTIONS 4 AND 5

REMEMBER THIS

Chapter 17 gives more detail about muscular contraction, and the Opener and Assignment of this chapter illustrate how chemicals can affect the neuromuscular junction.

Fig 15.16 Studies have shown that each vesicle contains about 10 000 acetylcholine molecules, and that about 100 vesicles need to be released before an action potential (leading to a muscle twitch) can be generated

4 DRUGS AND SYNAPSES

Transmitters are released in tiny amounts: only 500–1000 molecules from each synaptic knob are required to transmit an impulse. So, drugs that affect transmitters or their binding sites can have powerful effects when given in fairly small doses. Some chemicals, many of them from plants, have a dramatic effect on the nervous system.

NICOTINE AND ATROPINE

Nicotine is a substance found in tobacco. It has a similar shaped molecule to acetylcholine and so competes with acetylcholine to bind with its receptors. Once the nicotine has bound to the receptor, it opens the sodium channel that forms part of the receptor, causing nerve impulses to be generated. The effect of nicotine is covered in the Science feature on the next page.

Atropine also binds to acetylcholine receptors, but does not open the sodium channels. It blocks the receptors, and prevents acetylcholine binding to them. When this happens in motor neurones, it causes muscle paralysis. The binding of nicotine and atropine is summarised in Fig 15.17.

Fig 15.17 Drugs and transmitter binding sites

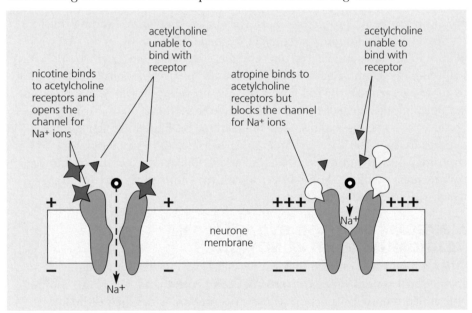

DRUGS AND NORADRENALINE

Beta-blockers have a similar shape to noradrenaline and so compete with it for receptors on the postsynaptic membrane. Amphetamines and cocaine also affect noradrenaline synapses, but they work in a different way. They prevent the reabsorption of noradrenaline from the synaptic gap. So, noradrenaline remains in the gap and the neurone keeps on firing. One of the effects of amphetamines is to make a person feel energetic and carefree, which is why they are often known as speed. Before the harmful effects were known, amphetamines were given to pupils with poor attention spans, to help them concentrate.

Amphetamines are psychologically addictive. Users become dependent on the drug to avoid the 'down' feeling they often experience when the drug's effect wears off. This dependence can lead a user to turn to stronger stimulants such as cocaine, or to larger doses of amphetamines to maintain a 'high'.

People who abruptly stop using amphetamines often experience the physical signs of addiction, such as fatigue, long periods of sleep, irritability and depression. How severe and prolonged these withdrawal symptoms are depends on the degree of abuse.

! SCIENCE FEATURE Nicotine — the drug that mimics acetylcholine

Do you wonder why people continue to smoke, even though they know the increased risk of lung cancer and heart disease associated with their habit? The answer lies partly with one of the components of tobacco: nicotine.

Nicotine is very addictive. It affects the brains of smokers, making them feel less stressed, better able to concentrate and less likely to eat sweet foods. Smokers become tolerant to nicotine over time, needing to smoke more to achieve the same effects. But how does nicotine cause addiction?

Studies carried out in the early 1980s using nicotine labelled with a radioactive tracer showed that it is taken into the brain very rapidly. Once there, it binds tightly to acetylcholine receptors, fooling postsynaptic cells into 'thinking' they are being stimulated. It also binds to other receptors which normally accept another neurotransmitter, dopamine.

The action of nicotine on both types of receptor in specific areas of the brain causes long-lasting changes to cell connections and may explain why it is addictive. We know that dopamine receptors, in particular, are involved in addictions to other substances such as amphetamines and cocaine.

Because nicotine is addictive but not carcinogenic (it is the chemicals in tobacco tar that have been shown to cause cancer),

smokers keen to kick the habit can get help. They can buy skin patches and gum that deliver nicotine to the brain, but not tar to the lungs.

Although patches and gum do help smokers to cut down or stop smoking altogether, they are only a partial solution. Nicotine affects the acetylcholine receptors in the parasympathetic nervous system that are involved with the constriction of blood vessels. Over time, circulatory problems and heart disease can result, and it is best to avoid these effects altogether.

Fig 15.18
Transdermal patches can be used almost anywhere on the skin. They slowly release substances such as nicotine into the bloodstream

CURARE

Curare is used as a muscle relaxant in abdominal surgery. It competes with acetylcholine at the neuromuscular junctions, preventing the nerve impulses that cause the abdominal muscles to contract. South American Indians first used curare as a poison. They used it mainly to catch animals for food by smearing arrow tips with it. In a poisoned animal it first affects the muscles of the toes, ears and eyes, then those of the neck and limbs, and, finally, those involved in respiration. The leads to death caused by respiratory paralysis.

PROZAC AND TRANQUILLISERS

Serotonin is a neurotransmitter normally active in the brain. Some forms of depression are caused by a reduced concentration of serotonin in the brain. The antidepressant drug Prozac is a known as a serotonin re-uptake inhibitor. Prozac alleviates depression because it competes with serotonin for the 'active' sites on the proteins that reabsorb serotonin, leading to a higher concentration of serotonin at synapses in the brain. It also competes for the active sites on the enzymes that break down serotonin at synapses.

Tranquillisers are drugs that reduce tension. Benzodiazepine tranquillisers, such as Valium, work by increasing the binding of inhibitory transmitters in the brain. Inhibitory transmitters hyperpolarise rather than depolarise the membrane of the next neurone. This makes the next neurone less excitable. Valium reduces stress and anxiety, but it can be addictive.

? TEST YOURSELF

H Why it is important that the muscles of the abdominal walls are relaxed during abdominal surgery?

SUMMARY

When you have read this chapter you should know and understand the following:

- The basic unit of the nervous system is the nerve cell, or neurone. This is a specialised cell with **dendrites** that take impulses into the cell body, and greatly elongated axons that take impulses away.
- The axon membrane is able to use an active transport mechanism to establish a **resting potential**. This is an electrical charge across the membrane caused by an unequal distribution of ions.
- The nerve impulse itself, called the **action potential**, is a momentary reversal in the resting potential, caused by a sudden rush of sodium ions into the axon. The action potential spreads rapidly along the axon.
- The action potential lasts for only a millisecond or so, after which the resting potential is re-established. When an action potential has passed, there is a brief period of time – the **refractory period** – during which it is impossible to generate another action potential.

- The nerve impulse passes from one nerve to another (or from a neurone to a muscle) by means of **synapses**.
- Synapses are vital in selecting some neural pathways and not others. As such, synapses play a vital role in memory, skill and co-ordination of the body's activities.
- Transmission across synapses occurs when a chemical, the neurotransmitter, is released by the **presynaptic membrane**. This diffuses across the gap and changes the permeability of the **postsynaptic membrane**, generating an **excitatory postsynaptic potential** (**EPSP**). If the EPSPs are sufficiently large, an action potential is generated in the next neurone.
- **Summation** means that the effect of several action potentials can add up to produce transmission at a particular synapse.
- Many drugs and poisons work by affecting synaptic transmission.

? EXAM QUESTIONS

1

a) Describe how the resting potential is maintained across the cell surface membrane of a neurone.

b) The graph below shows the change in potential difference across the cell surface membrane of a neurone during the propagation of a nerve impulse.

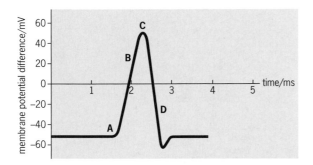

Explain, in terms of ion movements, the change in potential difference which takes place between:
(i) points **A** and **B**; **(ii)** points **C** and **D**.

c) In an investigation into action potentials, neurones were placed in different solutions. The concentration of sodium ions in solution **X** was the same as in tissue fluid. The concentration in solution **Y** was only 30% of that found in tissue fluid. The graph below shows the change in potential difference across the axon membranes after stimulation.

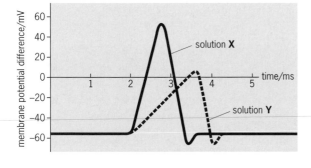

(i) Explain the difference in the size and shape of the curve obtained for the neurone placed in solution Y.

(ii) Explain why it was not possible to produce a series of action potentials when a respiratory inhibitor was added to solution X.

AQA (NEAB) BY04 Biological Basis of Behaviour March 2001 Q4

2 The graph shows changes in concentration of sodium ions inside the axon of a large neurone. The axon was stimulated at the points indicated on the graph. Between **A** and **B** on the graph the neurone was treated with dinitrophenol. This prevents the production of ATP.

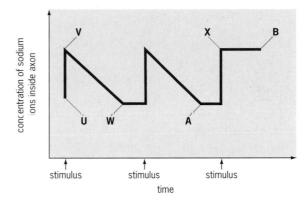

a) Explain the changes in sodium ion concentration
(i) between **U** and **V**;
(ii) between **V** and **W**.

b) Explain why the concentration of sodium ions did not change between **X** and **B**.

AQA (NEAB) BY04 Biological Basis of Behaviour June 2000 Q5

? EXAM QUESTIONS

3 The membrane potential of a neurone leading from a stretch receptor in a muscle was measured before, during and after a period in which the muscle was stretched. The results are shown in the graph.

a) Explain what causes the initial increase in membrane potential.

b) How do these results illustrate the 'all-or-nothing' signal?

c) **(i)** For how long was the muscle stretched?
(ii) Describe and explain the effect on the neurone of keeping the muscle stretched for this period.
(iii) Suggest **one** biological advantage of this effect on the neurone.

AQA (NEAB) BY04 Biological Basis of Behaviour March 1999 Q6

4
a) Describe the structures of a sensory neurone and of a motor neurone and explain their roles in a reflex arc.

b) Explain how a nerve impulse is transmitted across a synapse.

OCR 9264/3 Biology Linear Paper 3 June 2000 Q9

5 The diagram shows a neuromuscular junction.

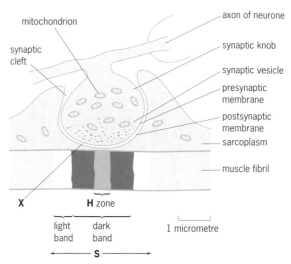

a) **(i)** Name the neurotransmitter that is released from the synaptic knob.
(ii) By what process does this neurotransmitter cross the synaptic cleft?
(iii) Calculate the width of the synaptic cleft at **X**. Show your working.

b) Suggest **two** functions of the energy released by the mitochondria in the synaptic knob.

c) Describe how the appearance of the section of the muscle fibril labelled **S** would change when the fibril is stimulated by the neurotransmitter.

AQA (NEAB) BY04 Biological Basis of Behaviour March 2000 Q3

KEY SKILLS **ASSIGNMENT**

TOXINS AND DRUGS

A huge variety of chemicals have an effect on the human body, and many act by interfering with the ways nerves or synapses work. In particular, they affect the transmission of impulses at synapses. In this Assignment we take a brief look at toxins and drugs.

1 Write a simple flow diagram outlining the sequence of events at a normal synapse.

Toxins

A toxin, by definition, is a poisonous substance of biological origin. The curare featured in the Opener is a toxin. In nature, animals and plants produce a huge variety of chemicals that organisms have developed to use against each other. Various species of spiders, frogs, jellyfish, snakes, molluscs, fish and plants – to name but a few – all make toxins that they use either to protect themselves or to kill prey.

Snake venom is of particular interest to neuroscientists. Venomous snakes have evolved poison glands as modifications of their salivary glands. Not only does the toxin subdue or kill the victim, preventing the snake from the serious injuries that might result from a violent struggle, some of them also start breaking down the prey's tissues. So when the snake gets round to eating it, the dead animal doesn't take so much digesting.

Fig 15.A1 The krait (*Bungarus multicinctus*). The krait's venom contains a powerful neurotoxin, bungarotoxin. Cobras produce similar substances in their venom

One of the first neurotoxic snake venoms was isolated from the krait (Fig 15.A1). In 1963, a small polypeptide consisting of about 70 amino acids was purified from the whole venom and was given the name bungarotoxin. An antibody to bungarotoxin was made and then attached to a fluorescent 'tag' so that its destination could be followed in the body of an animal injected with the krait's venom. The bungarotoxin was found almost exclusively at the neuromuscular junctions where it blocked the acetylcholine receptor sites on the postsynaptic membrane.

2 Predict the symptoms of krait bite. Explain your answer.

Bacterial toxins

Why is it that some bacteria cause disease while others are harmless? One of the key features is that pathogenic (disease-causing) bacteria produce an exotoxin that interferes with the host's metabolism in some way. A few of these toxins affect the nervous system. Botulism, a particularly nasty if rare form of food poisoning, is caused by the bacterium *Clostridium botulinum*. Botulinum toxin prevents the release of acetylcholine from the presynaptic membrane.

3 The symptoms of botulism include a weakness of the facial muscles, dilation of the pupils and difficulty in swallowing and breathing. Suggest why the botulinum toxin produces these effects.

Most people in the UK are vaccinated against tetanus (Fig 15.A2). Also known as lockjaw because it causes muscle paralysis in the face and neck, tetanus results from infection from the bacterium *Clostridium tetani*. The tetanus toxin that this bacterium produces prevents the release of the substance at the neuromuscular junction that breaks down neurotransmitter. As a result, the effects of acetylcholine persist much longer than required.

Fig 15.A2 A baby with muscle stiffness caused by tetanus. The toxins produced by the bacterium are causing the baby's muscles to go into spasm

4 What is the essential difference between the actions of tetanus toxin and botulinum toxin?

Drugs

Something that we touched on in this chapter, and will go on and study in more detail in the next, is the brain. This organ is incredibly complex. It contains billions of nerve cells, each with thousands of connections to other cells, allowing for an infinite variety of pathways. There are many different transmitter substances in the brain that mediate synaptic transmission through these pathways. It should be no surprise that chemicals that interfere with neurotransmitters in the brain often produce sensory effects such as hallucinations.

KEY SKILLS **ASSIGNMENT**

These effects can occur as a side-effect of a legal, prescription drug that could be used to reduce pain, to induce sleep, or to reduce tension or depression. There are also substances that, since they have been recognised to alter the way the brain processes information, have been abused. Alcohol and nicotine both act on the brain and can be abused, but they are legal. So-called 'recreational' drugs are illegal and often highly addictive. They interact with the brain and central nervous system and can have extremely dangerous side-effects.

5 Table 15.A1 includes the main classes of abused drugs, their mode of action and the common names of some individual drugs. However, they have been mixed up. Copy the table, matching up the correct examples and descriptions.

Table 15.A1 The main classes of abused drugs, the effect they have and some examples

Group	Effect	Examples
Depressants	interfere with brain function, Distort perception and induce a dream-like state	a variety of preparations of the plants *Cannabis sativa* and *Cannabis indica*. AKA hashish, marijuana, tac
Stimulants	similar to alcohol, they sedate the central nervous system	LSD, mescaline, magic mushrooms
Hallucinogens	pain killers	cocaine, ecstasy, amphetamine
Cannabis	keep the user awake, alert, excited	a variety of anti-anxiety drugs, sleeping tablets including benzodiazepines (eg Valium)
Analgesics	produce a relaxed, mild euphoria	morphine, heroin, methadone

Becoming dependent on drugs

If illegal drugs had a well-understood mode of action, were very short-lived in the body, had no side-effects and did not induce dependency or addiction, few people would worry. There are people who argue that drugs, if used 'sensibly', are not dangerous. However, the evidence is overwhelmingly against this.

With repeated use, many drugs can lead to addiction. This may be psychological, or physical. Psychological dependence

is the most widespread. The person develops a strong desire for the state induced by the drug – elation, euphoria, stimulation, sedation, hallucination – and the state becomes much more desirable than normality. In many cases, the person experiences tolerance to the drug. Any attempt to kick the habit can result in severe symptoms of depression.

6 How is a drug user most likely to respond when they start to become tolerant to the drug they are abusing?

Some drugs can, if used repeatedly, produce physiological changes and, after a time, the body cannot function 'normally' if the drug is not taken. Often a drug mimics one of the body's own products and this is no longer made as long as the drug is being taken. Therefore, when the drug is stopped, the body cannot respond immediately and start making the required product again. This situation occurs with heroin, which stops the body making its own endorphins. This is physical dependence and an attempt to stop will lead to dangerous and painful withdrawal symptoms.

7 Find out what some of the common withdrawal symptoms are.

8 Controversially, some drugs – notably alcohol and tobacco – are legal, while those listed in Table 15.A1 are not, or are available only with a prescription. What are the arguments for and against the legalisation of these drugs?

Fig 15.A3 Abuse of drugs brings many dangers. Drugs bought on the street might be mixed with a variety of other substances that can be lethal, and the sharing of needles can spread infections such as AIDS and hepatitis

16

The human nervous system

The human nervous system

16

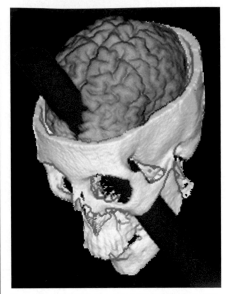

Ouch! Early knowledge about brain function was pieced together with the help of accidents such as that suffered by Phineas Gage. The metal rod missed the vital blood vessels and the crucial areas of the brain that keep us alive. It did, however, damage the frontal lobes, and the subsequent changes in Gage gave us important clues about the function of this part of the brain

The human brain is an organ of great mystery. In addition to being unbelievably complex, it is very difficult to investigate experimentally. We now know that different areas of the brain have particular functions, but how did we find out? Some of the earliest clues came from accidents in which people suffered damage to a particular part of the brain.

Consider the tale of Phineas Gage, a US railroad worker. He was a popular and reliable man, polite and responsible, and he had been made a foreman. In September 1848, he was jamming a stick of dynamite into a hole using a tamping iron. A spark from the iron ignited the dynamite, and the metal rod came out of the hole a lot faster than it went in.

The rod entered Gage's face below his left cheekbone, passing through the eyeball and through the top of his skull before landing several metres away. Gage fell back in a heap, as you would, but remarkably he did not die. He did not even pass out. He was driven by oxcart to a local physician, John Harlow. As the doctor stuck his fingers into the holes in Gage's head until the tips met, Gage asked when he would be ready to return to work. Within a couple of months he had recovered physically, but was no longer himself. Instead of being a gentle, honest, conscientious worker, Gage became 'a foul-mouthed and ill-mannered liar'.

He lived on as something of a celebrity for another 13 years before dying from an epileptic fit. On hearing of the death, Harlow managed to get Gage's body donated to medical research. He believed that the changes in Gage's behaviour had been caused by the damage to the frontal lobes of the brain. 'The equilibrium … between his intellectual faculties and animal propensities seems to have been destroyed', Harlow wrote.

Some 130 years on, scientists were able to use computer modelling to trace the damage done to Gage's brain. The metal rod missed the areas associated with language and motor function, but destroyed the ventromedial region on the left side of his frontal lobe. This is what made Gage so antisocial. People who have had tumours in this area of the brain have undergone the same sort of transformation.

1 GENERAL ORGANISATION

The human nervous system can be divided into two parts: the **central nervous system** (CNS) and the **peripheral nervous system** (Fig 16.1). The CNS consists of the **spinal cord**, which is protected by the vertebral column, and the **brain**, which is enclosed within the bony shell of the **cranium**. The peripheral nervous system brings information from the sense organs into the CNS, and then relays information out to the structures

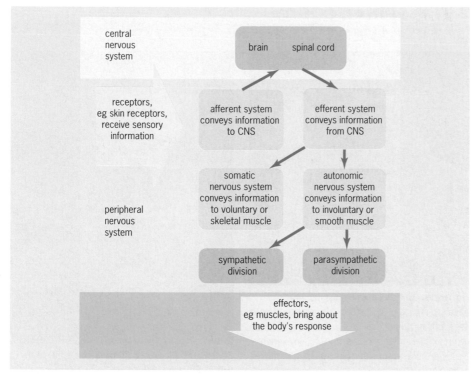

Fig 16.1 Overview of nervous system organisation

that bring about responses: the muscles and glands. The peripheral nervous system is divided into the **somatic** and **autonomic** nervous systems. The somatic system is responsible mainly for controlling voluntary actions, such as walking and talking, while the autonomic system keeps the involuntary body functions such as heartbeat, blood pressure and breathing ticking over. The autonomic system is further divided into the **sympathetic** system and the **parasympathetic** system.

 REMEMBER THIS

Neurone – a nerve cell (see the previous chapter).

Stimulus – a change in an organism's environment (internal or external) that can be detected by receptor cells.

Receptor – a specialised cell that detects a stimulus and initiates a nerve impulse.

Sensory neurone – a nerve cell that carries impulses from the receptor into the central nervous system.

Central nervous system (CNS) – the brain and spinal cord. The CNS processes the incoming information and decides on a response – often based on previous experience.

Motor neurones – nerve cells that carry impulses from the CNS to the effectors.

Effectors – the organs that bring about the response, usually muscles or glands.

 REMEMBER THIS

What is grey matter? Different areas of nerve tissue in the central nervous system are different colours. Grey matter contains nerve cell bodies; their nuclei are responsible for the grey colour. White matter consists largely of myelinated fibres (their fatty sheaths are creamy-white)

2 THE SPINAL CORD

Fig 16.2 Two images of the spinal cord showing a slipped, or prolapsed, disc

Fig 16.2 Two images of the spinal cord showing a slipped, or prolapsed, disc

Fig 16.2(a) This model illustrates the basic structure of the spinal cord. The delicate nerve cord itself runs down the middle of a channel in the centre of the vertebra. Between each bony vertebral disc is a shock-absorbing intervertebral disc made of cartilage. When the outer layer of the disc breaks or ruptures, part of the cartilage can push against the nerve cord, causing pain, numbness and – in severe cases – partial paralysis

Fig 16.2(b) A coloured 3D computed tomography (CT) scan image of the same problem. The slipped disc (yellow, lower centre) can be seen pushing against the nerve cord (purple). Compare this with the two normal discs above

REMEMBER THIS

Meningitis is a potentially fatal disease characterised by inflammation of the meninges – the protective membranes that surround the brain and spine.

SEE QUESTION 1

The spinal cord starts at the base of the brain and ends at the first **lumbar vertebra** (this is roughly at waist level). The spinal cord is enclosed within the **vertebral column** and has a diameter of about 5 millimetres (Fig 16.2). Further protection is provided by three layers of tough membranes called **spinal meninges**, and by the **cerebrospinal fluid** that cushions the cord, acting as a shock-absorber. Each vertebra has an opening on its right and left sides to let spinal nerves pass through. These nerves extend into the body, forming the **peripheral nervous system**.

3 THE REFLEX ARC

A reflex arc is the simplest example of co-ordination, so is a good place to start a study of how the nervous system works. The key feature of a reflex is that a particular stimulus leads to a fixed response – this is very rapid and can't be controlled because it does not pass through the conscious parts of the brain. An important feature of reflex arcs is that they contain as few synapses as possible. This speeds up the response and in many cases – such as the blinking reflex – avoids danger or minimises damage.

THE KNEE-JERK REFLEX

Fig 16.3(b) shows the knee jerk – a classic example of a reflex arc. This is a postural reflex, one of the many mechanisms we have to maintain our position and body control without having to constantly think about fine adjustments. Most reflexes involve either the constant fine adjustments to posture or have evolved to avoid danger.

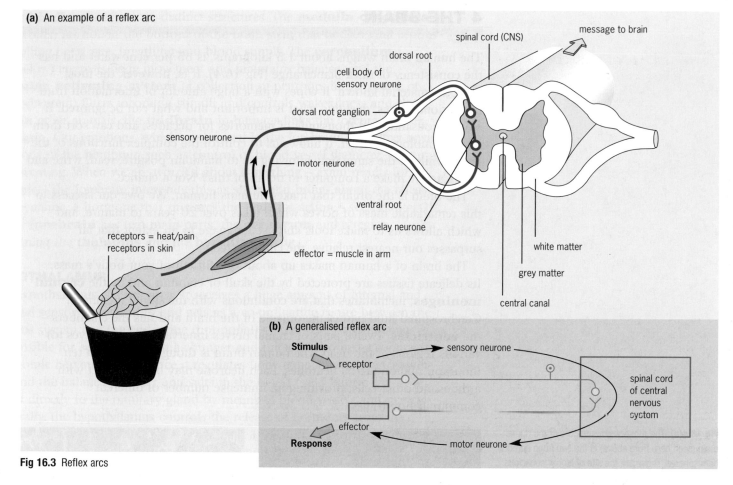

(a) An example of a reflex arc

spinal cord (CNS)

message to brain

dorsal root

cell body of
sensory neurone

dorsal root ganglion

sensory neurone

motor neurone

ventral root

relay neurone

receptors = heat/pain
receptors in skin

effector = muscle in arm

white matter

grey matter

central canal

(b) A generalised reflex arc

Stimulus

receptor

sensory neurone

spinal cord
of central
nervous
system

effector

Response

motor neurone

Fig 16.3 Reflex arcs

In the knee-jerk example the reflex is initiated by the stretch receptor in the patellar tendon – a tap on this tendon just below the knee has the same effect as the knee bending. Nerve impulses pass up the sensory neurone and into the spine. Here, the sensory neurones synapse with motor neurones and the nerve impulses pass straight out of the spine in the motor nerve, where they pass to the thigh muscle (the quadriceps). Contraction of the quadriceps straightens the leg. Impulses will also pass from the sensory nerves up the spine to the brain, but we are conscious of the stimulus only after the response has been initiated.

> **✔ REMEMBER THIS**
>
> In Fig 16.3 a synapse
> is represented as:
>
> direction of impulse
>
> synaptic knob

OTHER REFLEXES

Most reflexes contain more synapses than the knee-jerk example. Blinking when a foreign object enters the eye and the withdrawal reflex, or pulling your hand away from a hot pan (Fig 16.3a), involves a circuit containing sensory receptors, sensory neurones, spinal relay neurones (interneurones), motor neurones and effector muscles. The principle is the same: the heat of the pan stimulates receptors, and impulses pass along sensory neurones towards the spinal cord. Here, instead of making synapses with motor neurones, they pass their signals on to relay neurones. These connect with motor neurones that cause muscles to contract, moving your hand away from the pan.

You know that you have touched the hot pan because some impulses do travel to the brain, but the movement that is part of the reflex is involuntary, not under your conscious control. It is more difficult to persuade people that the swearing that accompanies such an event is also involuntary.

cerebrum of mammals is very large compared with the forebrains of other vertebrates and is certainly the dominant feature of the human brain. Folds in the surface of the cerebrum increase the surface area for centres of control, where incoming nerve impulses are interpreted, or integrated, in the light of information already stored in the brain. Overall, the function of the cerebral cortex can be summarised in three stages:

Vertical section

Cerebral cortex is 3 mm deep (grey matter is on the outside of the brain)

white matter

The brain is divided into left and right cerebral hemispheres above the cerebellum which is also divided into left and right sections

cell bodies and synapses

cerebral cortex (grey matter)

white matter

Axons only

Fig 16.6 The cerebral cortex

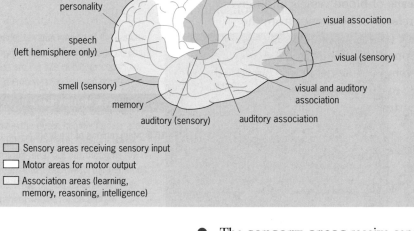

Localisation of function in the left cerebral hemisphere

premotor area (complex movement)

motor area

skin, touch and pressure (sensory)

memory, decisions, personality

muscle (sensory/ stretch receptors)

visual association

speech (left hemisphere only)

visual (sensory)

smell (sensory)

visual and auditory association

memory

auditory (sensory)   auditory association

☐ Sensory areas receiving sensory input
☐ Motor areas for motor output
☐ Association areas (learning, memory, reasoning, intelligence)

Fig 16.7 The key areas of the human cortex

● The **sensory areas** receive sensory information.
● The information is processed and interpreted in **association areas,** which decide on an appropriate response.
● Responses are initiated in the **motor areas** and modified by the **cerebellum**.

In this section we look at each of these areas in turn.

THE SENSORY AREAS

The sensory areas of the brain receive incoming impulses from the vast number of receptors that provide us with information about the world outside. The external receptors might tell us that it is cold, dark, raining and that the skin is getting wet. Your brain has to decide what to do.

The model shown in Fig 16.8 shows how our brains 'see' our body. There are many receptors in the skin but as you will know, some parts are more sensitive than others. The lips, tongue and fingertips have more receptors per unit area of skin than anywhere else. Fig 16.8 reflects this; the largest parts are the most sensitive. We think they ran out of plastic before they made the genitals.

Fig 16.8 No, you won't meet him on a dark night. Some parts of the body contain more receptors than others, and consequently a larger area of brain is devoted to them. In this distorted model the man's anatomy is proportional to the area in the brain that deals with the sensory information. Thus the hands, mouth and tongue are huge – they have many receptors per square centimetre – while other areas such as the torso and legs are relatively small. The eyes, however, are not in proportion in this model. These take up more of the human brain than the rest of the body put together
Fig 16.9 The different areas of the body are also mapped out on the motor area of the cortex

THE ASSOCIATION AREAS

The association areas interpret the sensory information and make decisions in the light of previous experience, ie our **memory**. If you are cold and wet, you may see a friend's house and your visual association areas will compare it with the images of houses stored away. Once you have recognised it, impulses pass to your frontal lobes, which then make a conscious decision to knock at the door.

The way information from the eyes is processed gives us a particularly impressive example of brain function. The millions of rods and cones in the retinas of both eyes generate a stream of impulses that pass down the optic nerves to the visual cortex at the back of the brain. The different information from the two eyes is processed in the **visual association area** into a single image in which we have a perception of depth as well as colour and detail.

THE MOTOR AREAS

The motor areas generate the impulses that cause the muscles to contract. This simple statement hides the fact that the process is incredibly complex and requires such a degree of computation that we are unlikely to be able to make lifelike 'android' robots for a long time to come.

Muscular contraction is not a matter of all or nothing. If it were, muscles would be either relaxed or fully contracted. Movement, if possible at all, would be very jerky. However, we can move our muscles with a great deal of precision, creating just the right tension for the action we are performing.

This is made possible by the process of **proprioception**, in which the brain receives constant feedback about the degree of contraction of the muscles and the position of the joints in relation to each other. Vital

SEE QUESTION 3

information also comes from the inner ear about our position with respect to gravity, direction of movement, acceleration and deceleration (see page 334). The process of movement control and postural maintenance is a matter of countless fine adjustments to alter the tension in the muscles. Interestingly, the left side of the body is controlled by the right motor area, and vice versa (Fig 16.10).

Fig 16.10 Two computer images of the left side of the human brain. These pictures were created by a process called magnetoencephalography, a technique that measures magnetic fields generated by neurone activity in the brain

Fig 16.10(a) Just before the subject moves her right index finger, neurones in the motor area can be seen to 'fire'

Fig 16.10(b) 40 milliseconds later, the hand sensory area processes feedback from the muscles in the finger, so that ongoing modifications can be made

There are three key components to the process of moving:
- selection of a **motor programme**;
- initiation of a motor programme – the muscles contract;
- modification of movement in the light of feedback.

Given the complexity of the task it is not surprising that several different areas of the brain are involved in the control of movement.

The process starts with the **basal ganglia**; a collection of structures lying deep in the forebrain. These are of fundamental importance because they select and initiate the motor programme. Fig 16.11 shows the pathway taken by motor impulses before they reach the muscles. From the basal ganglia impulses pass to the motor cortex.

The cerebellum is mainly concerned with learned skills, eg balance, posture, smooth movements

Impulses from association areas to motor area of opposite cortex

Impulses from motor area to cerebellum on the same side

Medulla *modulates* impulses before passing them on to muscles

Fig 16.11 Motor pathways and the cerebral cortex

! SCIENCE FEATURE Dyslexia and the cerebellum

Dyslexia is a very common condition that affects the learning ability of both children and adults. Commonly over-simplified as 'word blindness', dyslexia is a complex condition that involves an inability to concentrate and difficulty in co-ordinating, along with reading and writing problems. Dyslexics can become very frustrated and distressed with school and learning generally, to the extent that some become suicidal. One survey suggested that 70 per cent of convicted criminals have some sort of learning difficulty.

Now a new area of research is focusing on the cerebellum, an area previously thought to be solely concerned with movement, not with the higher intellectual processes. However, when dyslexics are given exercises to improve their balance and co-ordination, dramatic improvements are seen in their dyslexia within six weeks. Treatment is resulting in improvement time after time, to the extent that even the most sceptical of researchers are having to take notice. Centres offering this treatment are springing up all over the UK.

Fig 16.12 Exercises that improve balance and co-ordination may offer new hope for dyslexics

The motor cortex contains two major regions, the **primary motor cortex** and the **supplementary motor cortex.** The primary motor complex contains neurones that send impulses to skeletal muscles along nerve fibres passing down the brain stem and spinal cord.

The cerebellum is a complex structure lying on the posterior surface of the brain stem. Its function is to co-ordinate ongoing movements, thus producing smooth flowing movements. The cerebellum uses information from proprioceptors to decide whether the body's actions are going to plan or need modifying.

PARKINSON'S DISEASE

Parkinson's disease is a disease in which the death of a small number of cells in the basal ganglia leads to an inability to select and initiate patterns of movement. The exact symptoms vary, but generally Parkinson's sufferers have an increased rigidity in their muscles, usually accompanied by a tremor. This leads to slowness of movement, poor balance and speech problems. This shows the importance of the basal ganglia – without them the body is incapable of normal movement even though most of the motor system is intact.

Quite what triggers the onset of Parkinson's is still not clear. It appears to be due to a genetic predisposition and an as yet unidentified environmental trigger. What is known is that the underlying cause is an inability to produce **dopamine**, a neurotransmitter that has a number of functions, including enabling us to move smoothly and normally.

Some relief from Parkinson's has been achieved by treatment with **levadopa,** a precursor that is transformed into active dopamine in the brain. Dopamine reduces the symptoms – often spectacularly so – but is not a cure. Trials involving transplants of dopamine-producing cells from fetal or animal sources also show promise, but it is still early days.

⚠ SCIENCE FEATURE Epilepsy and the two sides of the brain

Epilepsy is a common disorder of the brain. Symptoms range in severity from a mild loss of concentration, known as an absence or petit mal, to full-blown convulsive fits (grand mal) in which the person blacks out and falls to the floor. These fits can be dangerous if the sufferer lashes out – injuring himself and others – or bites his own tongue.

The underlying cause of epilepsy is random, uncontrolled activity of some cells in the brain. This chaotic activity in both sensory and motor nerves causes patients to see and hear a variety of strange things, such as flashing lights and bells, while muscles jerk uncontrollably.

A diagnosis of epilepsy can be confirmed and studied using an EEG machine (see Fig 16.15a). Fig 16.13 shows a trace from a person during an epileptic seizure: you can compare it with the normal reading in Fig 16.15(b).

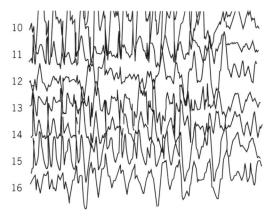

Fig 16.13 EEG traces taken during an epileptic seizure

Epilepsy can often be controlled successfully by drugs. However, in extreme cases, the condition is treated by brain surgery, and one such operation has given us a fascinating insight into the workings of the brain. The cerebral hemispheres have been described as two separate brains, and in order to work effectively as a whole, the two halves must communicate. The bridge between the two halves is known as the **corpus callosum** (Fig 16.14). Neuroscientists discovered that the corpus callosum was involved in the spread of epileptic seizures.

Fig 16.14 The corpus callosum is a broad, thick mass of nerve fibres connecting the right and left cerebral hemispheres. The left eye supplies information to the right hemisphere, and vice versa. The corpus callosum allows the two sides of the brain to communicate

In a seemingly drastic operation, surgeons sever most of the corpus callosum. This often causes the seizures to be less intense and dangerous. However, there are other amazing consequences. Initially, subjects appear to be perfectly normal: they can talk and read, and have no problems in recognising the world around them. However, if they close their right eye, and are given a familiar object such as a comb, they cannot put a name to it. Open the other eye, however, and, 'Ah, it's a comb!'

The same happens with words. If a word such as 'TIGER' is looked at it with the left eye only, the patient can't read it. If they open the right eye, they can read the word immediately. This is because the left eye supplies information to the right side of the brain, and that is not where the language centre is situated. The right eye supplies information to the left side of the brain, to the language-processing neurones.

From studies of split-brain patients, and other studies, it appears that different sides of the brain have different functions. The left hemisphere contains the language centre, and the three R's – reading, writing and 'rithmetic. The right side, in contrast, is responsible for our imagination and sense of humour. It can also appreciate form, geometry and music. It cannot, however, put words to things. If the right hemisphere needs a word, it has to put in a call to the left side, via the corpus callosum.

Split-brain patients do not experience the symptoms forever. Within a few months the right hemisphere develops more language skills and can function on its own. It has even been suggested that split-brain patients could read two books simultaneously, one with each eye!

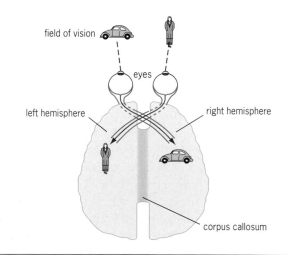

EEGS AND BRAIN STEM DEATH

In 1929 the German scientist Berger discovered that the brain produces electrical activity that can be measured by electrodes placed on the scalp. The machine that measures brainwaves is called an **electroencephalogram**, or EEG. It was discovered that the patterns of EEG activity change in different situations, particularly with levels of consciousness.

The four basic types of brainwaves, alpha, beta, theta and delta, are shown in Fig 16.15(b). The absence of β waves or any electrical activity from the brain of a patient indicates brain stem death. The absence of heartbeat, breathing movements and electrical brain activity is a clinical definition of death.

EEGs tell us little about how the brain works, but they are useful for locating and diagnosing the various types of epilepsy, sleep disorders and brain tumours.

Fig 16.15(a) Person wired up to EEG machine

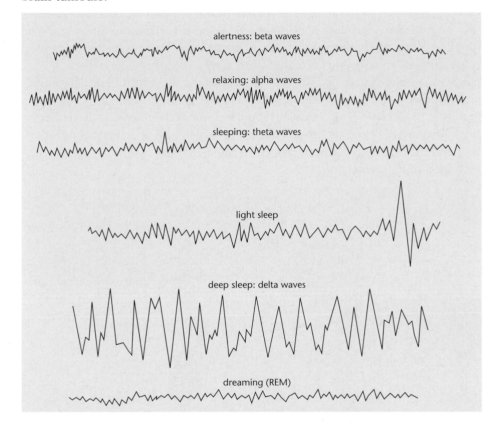

alertness: beta waves

relaxing: alpha waves

sleeping: theta waves

light sleep

deep sleep: delta waves

dreaming (REM)

Fig 16.15(b) EEG traces of the different brain waves

5 THE PERIPHERAL NERVOUS SYSTEM

The nerves of the peripheral nervous system behave like major road systems, carrying traffic in and out of the central nervous system. **Afferent nerves**, also called **sensory nerves**, carry information from sensory receptors into the CNS. **Efferent nerves**, also called **motor nerves**, carry information from the CNS out to effector organs. The efferent system can be further subdivided into the **somatic** and **autonomic** systems. These differ in their function, rather than their structure or position in the body.

THE SOMATIC NERVOUS SYSTEM

The somatic nervous system contains both afferent and efferent nerves. It receives and processes information from receptors in the skin, voluntary muscles, tendons, joints, eyes, tongue, nose and ears, giving an organism the sensations of touch, pain, heat, cold, balance, sight, taste, smell and sound. It also controls voluntary actions such as the movement of arms and legs.

THE AUTONOMIC NERVOUS SYSTEM (ANS)

The ANS consists of two sets of involuntary nerves that generally act antagonistically – they have opposing effects. The system is entirely motor, made up of efferent nerves only. It does not carry sensory information: feedback from muscles and glands travels via the somatic system.

 REMEMBER THIS

Afferent means 'incoming' while efferent means 'outgoing'. You can refer to 'afferent nerves' or 'efferent blood vessels', for example.

 REMEMBER THIS

There is a parallel between the nervous and muscular systems of the human body. In the same way as we have voluntary and involuntary components of our nervous system, we also have skeletal (striped) muscle to deal with voluntary muscle contraction, and visceral (smooth) muscle to deal with involuntary muscle contraction. The different muscle types are described in Chapter 17.

Fig 16.16 The autonomic nervous system, showing the antagonistic (opposing) effects of the sympathetic and parasympathetic division. Autonomic neurones, like others in the nervous system, release chemical transmitter substances where they communicate with other cells. Autonomic neurones can synapse with other neurones within the ANS or with effectors such as muscles and glands. Neurones are classified as cholinergic or adrenergic on the basis of the transmitter they release

The ANS controls basic 'housekeeping' functions such as heart rate, breathing, digestion and blood flow. Heart rate, for example, can increase or decrease. This and many functions like it are controlled by the two branches of the ANS, the sympathetic and the parasympathetic systems (Fig 16.16). These two sets of nerves have antagonistic (opposite) actions. So, for example, while the sympathetic system increases heart rate, the parasympathetic system lowers it.

Normally, the activity of both systems is balanced. But if the body is stressed, then the 'fight or flight' reactions of the sympathetic nervous system take over, causing an increase in heart rate, faster breathing, an increase in blood pressure and an increase in blood sugar level. This makes the body ready for sudden strenuous activity. When the emergency is over, the parasympathetic system takes over. It decreases the heart and breathing rates and diverts blood supply back to 'housekeeping' activities such as digestion and food absorption. The actions of the parasympathetic nervous system have been described as 'feed or breed' because parasympathetic stimulation leads to increased blood flow and peristalsis in the intestines, and sexual responses such as gaining an erection.

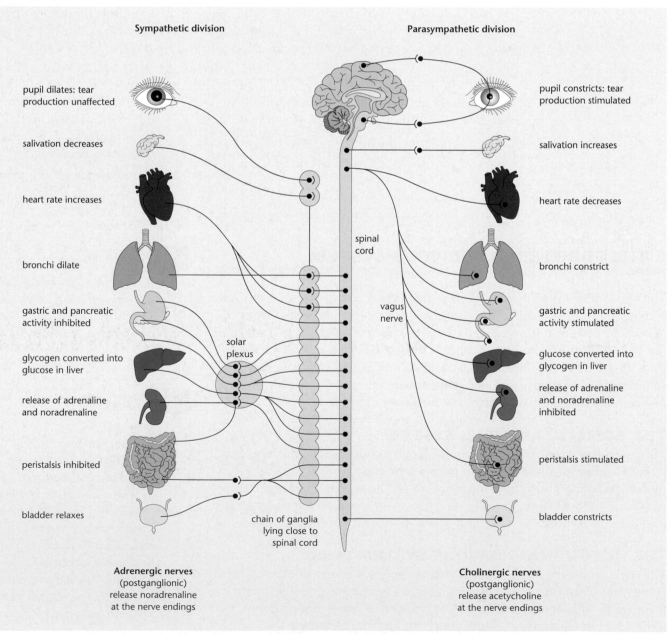

Sympathetic division

pupil dilates: tear production unaffected

salivation decreases

heart rate increases

bronchi dilate

gastric and pancreatic activity inhibited

glycogen converted into glucose in liver

release of adrenaline and noradrenaline

peristalsis inhibited

bladder relaxes

solar plexus

spinal cord

chain of ganglia lying close to spinal cord

Parasympathetic division

pupil constricts: tear production stimulated

salivation increases

heart rate decreases

bronchi constrict

gastric and pancreatic activity stimulated

glucose converted into glycogen in liver

release of adrenaline and noradrenaline inhibited

peristalsis stimulated

bladder constricts

vagus nerve

Adrenergic nerves
(postganglionic)
release noradrenaline
at the nerve endings

Cholinergic nerves
(postganglionic)
release acetycholine
at the nerve endings

The autonomic system was originally thought to be independent of the rest of the nervous system, hence the term autonomic, meaning 'on its own' or 'self-governing'. Now we appreciate that it is not autonomous, but is regulated by areas within the central nervous system, including the hypothalamus, cerebral cortex and the medulla oblongata.

SEE QUESTIONS 4 AND 5

DO WE HAVE ANY CONSCIOUS CONTROL OVER THE AUTONOMIC NERVOUS SYSTEM?

The answer is yes, in some cases. As children we became potty trained when we learned conscious control over the muscular valves (sphincters) in our bladder and rectum. There are people who have learnt conscious control over some other autonomic functions; those who are adept at advanced meditation and yoga techniques can voluntarily lower their heartbeat.

SUMMARY

When you have finished this chapter you should know and understand the following:

■ The **central nervous system** (CNS) consists of the brain and spinal cord.

■ The simplest co-ordinated response is the **reflex arc**. In the knee-jerk reaction, the pathway consists of the receptor in the tendon, the sensory nerve that connects directly to the motor nerve in the spine, and the effector, the thigh muscle.

■ More complex reflex arcs such as blinking are **polysynaptic**. These have more intermediate connections inside the CNS.

■ The human brain is divided into **hindbrain**, **midbrain** and **forebrain**. Generally, the midbrain and hindbrain are concerned with

'housekeeping' functions, while the **cerebral hemispheres**, which make up much of the forebrain, are responsible for the conscious functions such as language and memory.

■ The **peripheral** nervous system consists of the **somatic** and **autonomic** nervous systems.

■ The autonomic nervous system consists of the **sympathetic** and **parasympathetic** nervous systems. The two sets of nerves act antagonistically to control a variety of functions that are not under conscious control. Generally, sympathetic stimulation prepares the body for action while the parasympathetic system returns it to normal.

? EXAM QUESTIONS

1 The diagram below shows a transverse section of the spinal cord of a small mammal.

a) Name the substance present in the central canal.

b) The magnification of the diagram is ×25. Calculate the actual diameter of the spinal cord at the position between **X** and **Y**.

c) On the diagram, draw and label the neurones involved in a simple spinal reflex.

Edexcel B3 June 1999 Q2

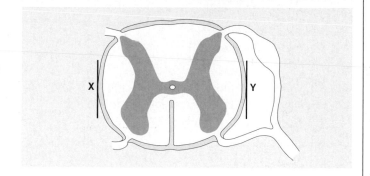

? EXAM QUESTIONS

2 The diagram shows a vertical section through a human brain.

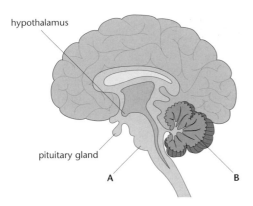

hypothalamus

pituitary gland

A   B

a) Give the name and one function of the part of the brain labelled **(i)** A; **(ii)** B.

b) The hypothalamus monitors the blood. It is also involved in homeostasis.
(i) Name one of the features of the blood monitored by the hypothalamus.
(ii) Explain what is meant by homeostasis.

AQA (NEAB) BY04 Biological Basis of Behaviour March 2001 Q1

3 The diagram shows a vertical section through a human brain.

a) **(i)** A person drinks a glass of lemonade. Name the parts of the brain to which the following descriptions would apply.

P receives impulses from touch receptors in the lips.
Q enables the person to coordinate the movements necessary to drink from the glass.
R contains receptors that respond to a change in concentration of the blood plasma.
(ii) Use the appropriate letter and a guideline to show the position of each of parts, P, Q and R on the diagram of the brain.

b) When the lemonade touches receptors in the throat, a swallowing reflex occurs. What are the effectors in the reflex action?

AQA (NEAB) BY04 Biological Basis of Behaviour March 1999 Q1

4
a) State the name of the neurotransmitter substance normally produced at the effector in:
(i) the sympathetic nervous system;
(ii) the parasympathetic nervous system.

b) Atropine is a drug that depresses the activity of the parasympathetic nervous system. Predict the effect of atropine on the heart rate.

c) Identify two other possible effects of atropine.

IBO Biology Higher Level Paper 3 November 1998 QE2

5 The sinoatrial node in the heart controls the heartbeat. The rate of heartbeat is coordinated by the brain.

a) Name:
(i) the two types of nerves that connect the brain to the sinoatrial node;
(ii) the part of the brain where these nerves originate.

b) Suggest:
(i) why it is necessary to have two separate nerve connections from the brain to the sinoatrial node;
(ii) how the two nerve connections are able to have different effects on the sinoatrial node.

AQA (NEAB) BY04 Biological Basis of Behaviour June 1999 Q1

17

Support and movement

17 Support and movement

It might look fun, but in zero gravity the body is under a lot of stress

Astronauts must often live and work in zero gravity. Weightlessness is not a restful state and it can put great stress on body systems, particularly the skeletal system.

Many changes occur in zero gravity. With less work to do, body muscles begin to break down and the bones start to lose calcium at an increased rate. The number of red blood cells in the body falls and there are dramatic shifts in fluid distribution: the face becomes puffy and the legs become thinner (astronauts call this condition 'bird's legs'). The lack of gravity also affects the spine. Without the constant downward force, the spine lengthens by as much as eight millimetres. This can lead to blocked nerves and back pain. Often, astronauts also lose their touch sensitivity.

NASA scientists are busy looking at forms of treatment and exercise that might prevent bone breakdown. Such information might also have practical benefits closer to home. By studying and finding ways to slow the accelerated changes produced in space, it may be possible to develop new treatments for bone diseases such as osteoporosis which occurs when bones become brittle due to a loss of calcium.

1 THE HUMAN SKELETON

The human skeleton has several functions:
- It acts as a framework that supports soft tissues.
- It allows free movement through the action of muscles across joints.
- It protects delicate organs and structures such as brain and lungs.
- It forms red blood cells in the bone marrow.
- It stores and releases minerals from bone tissue.

Fig 17.1 shows the structure and features of the human skeleton.

THE STRUCTURE AND PROPERTIES OF HUMAN BONE

Bone is one of the hardest tissues in the human body and is second only to cartilage in its ability to absorb stress. It is made up of bone cells called **osteocytes**, which sit in a bone **matrix**. The matrix consists of inorganic matter (mainly compounds of calcium and phosphorus) interwoven with **collagen fibres** (see page 57).

Bone needs to be tough and resilient. These properties are provided by two types of bone tissue: **compact bone** and **spongy bone** (Fig 17.2). **Compact bone** is deposited in sheets called **lamellae** arranged as cylinders inside cylinders. Nerves and blood vessels run in a central canal. The compacted structure of the lamellae gives immense strength. The lamellae of **spongy bone** are arranged in a criss-cross pattern, forming a spongy honeycomb. This structure has excellent shock-absorbing properties.

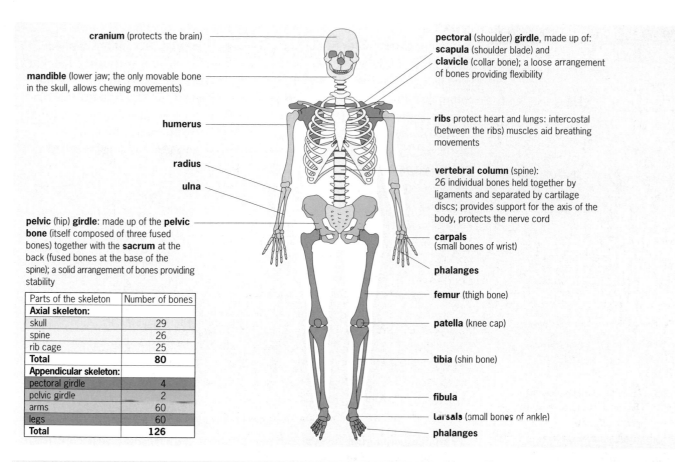

cranium (protects the brain)

mandible (lower jaw; the only movable bone in the skull, allows chewing movements)

humerus

radius

ulna

pelvic (hip) girdle: made up of the pelvic bone (itself composed of three fused bones) together with the sacrum at the back (fused bones at the base of the spine); a solid arrangement of bones providing stability

pectoral (shoulder) girdle, made up of: scapula (shoulder blade) and clavicle (collar bone); a loose arrangement of bones providing flexibility

ribs protect heart and lungs: intercostal (between the ribs) muscles aid breathing movements

vertebral column (spine): 26 individual bones held together by ligaments and separated by cartilage discs; provides support for the axis of the body, protects the nerve cord

carpals (small bones of wrist)

phalanges

femur (thigh bone)

patella (knee cap)

tibia (shin bone)

fibula

tarsals (small bones of ankle)

phalanges

Parts of the skeleton	Number of bones
Axial skeleton:	
skull	29
spine	26
rib cage	25
Total	**80**
Appendicular skeleton:	
pectoral girdle	4
pelvic girdle	2
arms	60
legs	60
Total	**126**

Fig 17.1 The human skeleton contains a total of 206 bones. It is divided into two parts: an axial skeleton that comprises the skull and vertebral column and an appendicular skeleton that is made up of the limbs and limb girdles.

All vertebrates show skeletal modifications related to their lifestyle. Since humans are bipedal (they walk on two legs), the hips and lower spine take most of the weight of the body and so the pelvic girdle (the hip) is larger and less flexible than the pectoral girdle (the shoulder)

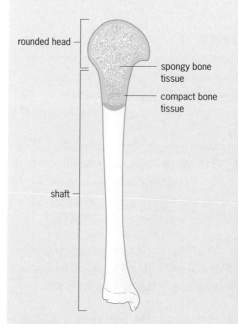

rounded head

spongy bone tissue

compact bone tissue

shaft

Fig 17.2(a) A human long bone. The part in section shows the distribution of compact and spongy bone

Fig 17.2(b) Compact bone makes up the shaft of the long bone where strength and rigidity are important

Fig 17.2(c) Spongy bone forms the rounded head of long bones which absorbs the shocks and jolts of movement

> ✔ **REMEMBER THIS**
>
> The human body also makes use of the supporting properties of water. The **amniotic fluid**, the fluid inside the uterus, surrounds and supports the developing fetus; and the **vitreous humour**, the jelly-like material inside the eyeball, supports the structures inside the eye.

SEE QUESTION 1

femur (thigh bone)

rectus femoris
(one of the
quadriceps muscles)

quadriceps
tendon

fatty pad

patellar
ligament

tendon of
gastrocnemius

posterior ligament
of knee joint

tibia
(shin bone)

fibula

gastrocnemius
(one of the
calf muscles)

Achilles tendon

calcaneum
(heel bone)

Fig 17.3(a) A diagram of the human leg showing the position of the main ligaments (green) and tendons (blue)

Fig 17.3(b) This micrograph shows a slice of tendon (yellow) connected to a piece of muscle (red). Tendons and ligaments look similar but their composition is different: tendons contain mainly **collagen fibres** (see page 57); ligaments are made up of fibres formed from the protein **elastin**

JOINTS

As in other vertebrates, the human skeleton is jointed. A joint is simply a place in the body where two bones meet. Most joints are **movable** but some, such as between the bones making up the **skull**, the **sacrum** (base of the spine) and the **pelvis**, are **immovable**, or **fused**. Elastic **ligaments** bind bones together, while tough inelastic **tendons** attach muscles to bone (Fig 17.3). The human knee joint is seen in Fig 17.4. Internally, the joint is lubricated by a viscous fluid, **synovial fluid**, secreted by the **synovial membrane**, the membrane that lines the joint.

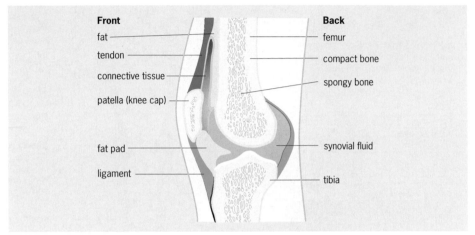

Front

fat

tendon

connective tissue

patella (knee cap)

fat pad

ligament

Back

femur

compact bone

spongy bone

synovial fluid

tibia

Fig 17.4 The features of a typical joint can be seen clearly in the knee joint

The ligaments and the synovial membrane together form a **joint capsule** which surrounds the end of the bones. Different types of joint allow for different kinds of movement (Fig 17.5).

HOW THE HUMAN BODY MOVES

A simple action, such as tapping your finger on the table, involves the skeletal system, the muscular system and the nervous system. This voluntary action begins as a stimulus in the cerebrum of the brain. Motor nerves carry impulses to muscles and cause them to contract, pulling on bones through the tough inelastic tendons.

Muscles can only pull, or contract; they cannot push. This means that muscles rarely act alone: most of the time they work in groups. Contraction of a muscle moves a bone at a joint, but a second (**antagonistic**) muscle returns the bone to its original position. Take movement in the arm, for example. The biceps muscle bends or **flexes** the arm; the triceps muscle straightens or **extends** the arm.

We all know that machines can help us to lift heavy loads using levers. The bones in our bodies also act as **levers**. The principle of leverage allows a muscle to overcome the **resistance** supplied by a weight. This happens in three ways:

- An arm bends because a force is applied between the **pivot** (the elbow) and the resistance (the weight to be lifted) (Fig 17.6a).
- We stand on tiptoes by pivoting the toes on the ground and using our calf muscles to raise the weight at the ankles. This is the wheelbarrow principle (Fig 17.6b).
- When standing or sitting, we can raise our head because muscles at the back of the neck tip the skull at the pivot at the top of the spine. This is the seesaw principle (Fig 17.6c).

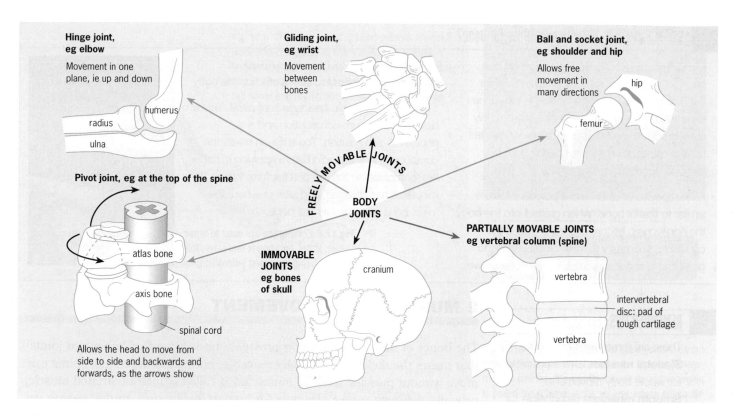

Hinge joint, eg elbow

Movement in one plane, ie up and down

humerus
radius
ulna

Gliding joint, eg wrist

Movement between bones

Ball and socket joint, eg shoulder and hip

Allows free movement in many directions

hip
femur

FREELY MOVABLE JOINTS

BODY JOINTS

Pivot joint, eg at the top of the spine

atlas bone
axis bone
spinal cord

Allows the head to move from side to side and backwards and forwards, as the arrows show

IMMOVABLE JOINTS eg bones of skull

cranium

PARTIALLY MOVABLE JOINTS eg vertebral column (spine)

vertebra
intervertebral disc: pad of tough cartilage
vertebra

The 'stuck finger' test demonstrates the effect of muscles working in groups (Fig 17.7). Each of your fingers has a separate tendon connecting it to muscles in the forearm. Try it. Curl up your middle finger and place the other four fingers on a hard surface as shown. Now lift up the other fingers one by one. You will find that you can lift all of your fingers except the ring finger. This is because these two fingers share the same tendon connection.

Fig 17.5 There are many different types of joint in the vertebrate skeleton. They are classified both by their shape and by their mobility

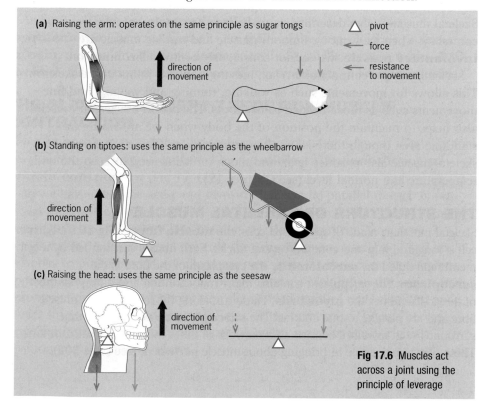

(a) Raising the arm: operates on the same principle as sugar tongs

△ pivot
← force
← resistance to movement

direction of movement

(b) Standing on tiptoes: uses the same principle as the wheelbarrow

direction of movement

(c) Raising the head: uses the same principle as the seesaw

direction of movement

Fig 17.6 Muscles act across a joint using the principle of leverage

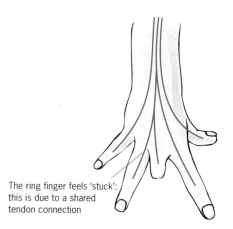

The ring finger feels 'stuck': this is due to a shared tendon connection

Fig 17.7 The stuck finger investigation

1 Muscle at rest

2 A nerve impulse reaches the muscle fibre, releasing calcium ions. This reveals myosin binding sites in the actin myofilament. The myosin heads use the energy gained from ATP hydrolysis to move towards the binding sites

3 As the myosin heads bind to the actin myofilament they tilt, pulling the actin myofilament past them

4 As the actin filament moves, the myosin heads detach and reattach to the next binding site along

Fig 17.11 The sliding filament theory to explain muscle contraction

THE ROLE OF CALCIUM IONS AND ATP IN MUSCLE CONTRACTION

Contraction of skeletal muscle does not simply occur 'out of the blue'. It must be initiated by a stimulus in the form of a nerve impulse (see Chapter 15) which arrives at the muscle fibre from the nervous system. So when a muscle contracts and then relaxes, the following events take place (refer again to Fig 17.11):

● The brain makes a decision that the body is to move. Nerve impulses travel from the brain, along the spinal cord and through the motor nerves, towards the muscle.

● The electrical signal reaches the **neuromuscular junction**, a specialised synapse. The arriving signal causes the release of acetylcholine (see Chapter 15). This chemical is a neurotransmitter: it travels across the gap between the end of the nerve and the muscle, and binds to receptors on the **motor end-plate**.

● This binding causes an electrical change in the muscle fibre which triggers the release of calcium ions from the sarcoplasmic reticulum into the myofilament.

● Calcium ions bind to troponin and tropomyosin, the proteins that are closely associated with actin filaments. This changes the three-dimensional shape of the troponin–tropomyosin–actin complex, revealing parts of the actin filament that were previously hidden. These are the active sites to which the heads of the myosin filaments will attach.

● The calcium ions also act directly on myosin, activating it so that it splits ATP into ADP and inorganic phosphate, releasing energy. This energy is used to move the heads of the myosin filaments towards the newly exposed binding sites on the actin filaments.

● As cross-bridges form between the actin and myosin filaments, the myosin heads tilt, pulling the actin myofilaments past them. As the myofilaments slide, the heads detach from one site and attach to the next. As many as 100 such attachments can occur every second.

● As the actin myofilaments from opposite ends of the sarcomere move towards each other, the muscle fibre contracts.

● When the stimulation of the muscle fibre stops (when the brain decides the body should stop moving), the calcium ions are pumped back into the sarcoplasmic reticulum.

● As the level of calcium ions falls, troponin and tropomyosin move back to their original positions, blocking once again the active sites available for myosin attachment. The myosin heads can no longer attach to the actin filament, and both sets of filaments return to their original position. When the antagonistic muscle contracts, the sarcomeres in the first muscle return to their normal length and the muscle relaxes.

FAST AND SLOW MUSCLE FIBRES

A nerve impulse is the trigger for muscle contraction, but the length of time a contraction lasts depends on how long calcium ions remain in the sarcoplasm. This time is different in different types of skeletal muscle fibres. **Fast twitch fibres** and **slow twitch fibres** are classified on the basis of their contraction times.

There are three important differences between fast and slow fibres:

- Slow twitch fibres have less sarcoplasmic reticulum than fast twitch fibres. This means that calcium ions remain in their sarcoplasm longer.
- Slow twitch fibres have more mitochondria which provide ATP for sustained contraction.
- Slow twitch fibres have significantly more myoglobin than fast twitch fibres. Myoglobin has a higher affinity for oxygen than the haemoglobin in blood and so is particularly efficient at extracting oxygen from the blood.

The overall result of these differences is that slow twitch fibres are responsible for sustained muscle contraction, such as that which maintains body posture, while fast twitch fibres are responsible for the shorter-acting but more powerful contractions important in locomotion (see the Assignment at the end of this chapter).

! SCIENCE FEATURE Time of death

When a body is discovered, a pathologist investigates and tries to estimate how long the person has been dead. This information can be important in a murder enquiry. The pathologist assesses the time of death, partly by taking the internal temperature, to see how much the body has cooled from the normal body temperature of 37 °C, and partly by looking at the state of the muscles.

When death occurs, ATP is no longer made. It is a short-lived chemical and so it runs out fairly quickly. This causes the muscles to lock into position as cross-bridges that formed between actin and myosin filaments before death can no longer be broken. This condition, **rigor mortis**, happens in all body muscles. It appears about four hours after death and lasts about 24 hours. After this time, muscle proteins are destroyed by enzymes within the cells and so rigor mortis disappears.

Fig 17.12 All suspicious deaths are investigated by a police pathologist: one of the first things he or she does is estimate the time of death

---

SUMMARY

After completing this chapter you should know and understand the following:

- The skeleton has three main functions; **support**, **protection** and **movement**. The human skeleton is movable because different bones meet at **joints**, across which **muscles** are attached. Because muscles can pull but not push, muscles work in **antagonistic pairs**.
- Muscles are **excitable**, **extensible** and **elastic**. **Skeletal muscle**, also called **striped**, **striated** or **voluntary muscle**, is responsible for human locomotion, maintaining body posture and generating body heat.

- Skeletal muscle is made of **muscle fibres** which contain bunches of **myofibrils** which in turn contain **myofilaments**. Thick **myosin** filaments lie alongside thin **actin** filaments.
- The **sliding filament hypothesis** explains muscle contraction at the molecular level.
- The events of muscular contraction can be summarised as follows. A nerve impulse arrives at the **neuromuscular junction**. This releases acetylcholine, which changes the ionic permeability of the **sarcolemma**. Calcium ions are released into the **myofilaments** where they bind to the tropomyosin–troponin–actin complex, and activate ATPase. Cross-bridges form between **actin** and **myosin**, the fibres slide over each other and the muscle contracts.

? EXAM QUESTIONS

1 The diagram below shows a section through a human synovial joint.

a) (i) Name the type of synovial joint shown in the diagram.
(ii) State **one** location of this type of joint in the human skeleton.
(iii) Describe the range of movement permitted at this type of joint.
b) Describe **one** function for each of the structures labelled **A** and **B**.
c) One form of arthritis, which is common in elderly people, causes the joints to become painful so that movement is difficult.
Suggest how the joints of a person affected by arthritis would differ in appearance from that shown in the diagram.
Edexcel HB3 June 1999 Q3

2 Copy the table which refers to striated and cardiac muscle. If the statement is correct, place a tick (✓) in the appropriate box and if the statement is incorrect, place a cross (✗) in the appropriate box.

Statement	Striated muscle	Cardiac muscle
Muscle cells (fibres) branched		
Interlaced discs present		
Cytoplasm contains myofibrils of actin and myosin		
Myogenic		

Edexcel B3 June 1999 Q1

3 The diagram shows part of a myofibril as seen through an electron microscope.

a) Name the main protein present in
(i) the thick filaments;
(ii) the thin filaments.
b) Describe the mechanism that brings about the change in position of the filaments when the myofibril contracts.
AQA (NEAB) BY04 Biological Basis of Behaviour June 2000 Q4

4
a) The diagram shows some of the muscles in a human leg.
(i) Describe what happens to the leg when: muscle **A** contracts; muscle **B** contracts.
(ii) Explain why muscles such as **C** and **D** are arranged in pairs.

? EXAM QUESTIONS

b) The photograph shows part of a muscle fibre.

Reproduced from An Atlas of Histology, W.H. Freeman and B. Bracegirdle by permission of Heinemann Ltd.

Give **three** differences in structure between a muscle fibre and an epithelial cell from the lining of the small intestine.

AQA (NEAB) BY04 Biological Basis of Behaviour March 2000 Q1

5 The diagram shows part of a myofibril from a skeleton muscle. It is in a relaxed state.

a) Name the protein present in the filament labelled **P**, **Q**.

b) (i) Name the organelle that provides the ATP required for muscle contraction.
(ii) Describe the changes that occur to the lengths of regions **R** and **T** when the muscle contracts.
(iii) Explain the mechanism of muscle contraction.

AQA (NEAB) BY04 Biological Basis of Behaviour March 2001 Q1

KEY SKILLS **ASSIGNMENT**

MUSCLE FIBRES AND ATHLETICS

This Assignment follows on from the Chapter 14 Assignment Energy systems and sport. You may need to revise this to answer all the questions here.

The different types of muscle fibre

You may have noticed that athletes are highly specialised for their particular event. Marathon runners can run great distances without fatigue, but they lack the power to run as fast as the sprinters. In contrast, sprinters fatigue far too quickly to run long distances. These features are in part due to the type of muscle fibres they have inherited, and partly to the training they have done. Research has shown that there are two basic types of muscle fibre: **fast twitch** and **slow twitch**. Fast twitch fibres can be further subdivided, giving us the three different muscle types:

Slow twitch oxidative (type I) fibres. These contain large amount of myoglobin, many mitochondria and a dense capillary network. They split ATP slowly and therefore can only contract slowly. These fibres are red and have a high resistance to fatigue.

Fast twitch oxidative (type IIA) fibres. These show similar features to slow twitch fibres: a lot of myoglobin, many mitochondria and blood capillaries; but they are able to hydrolyse ATP far more quickly and therefore to contract rapidly. They are relatively resistant to fatigue, but not as resistant as type I fibres.

Fast twitch glycolytic (type IIB fibres). These fibres have a relatively low myoglobin content, few mitochondria and few capillaries. They contain large amounts of glycogen which provides fuel for anaerobic ATP production via the process of glycolysis. They contract rapidly but fatigue quickly, owing to the build-up of lactic acid. Type IIB fibres appear whiter than the other two types.

1 Summarise the above information in the form of a table.

2 Movement of muscles requires energy which is provided by one or more of three energy systems: (i) the ATP/CP system, (ii) glycolysis and (iii) the aerobic system. Outline the essential features of these systems.

3 **a)** What is the function of myoglobin? (See page 303 for a reminder.)
b) Suggest why type IIB fibres have very little myoglobin.

Muscles are made up of motor units: a motor unit is a bundle of fibres controlled by a single motor nerve. Analysis of muscle tissue by biopsy shows that all the muscle fibres in a single motor unit are of the same muscle type, but there can be different muscle types in different units within the same muscle. The various muscles of the body have different proportions of the three types, according to their function.

4 Why is it an advantage to have all three types of muscle fibre in the same muscle?

The distribution of the different muscle types in the body

For the rest of this Assignment we will consider the two main types of muscle fibre: slow and fast twitch.

A central principle in physiology is that structure is closely related to function and this is illustrated clearly by the distribution of the muscle types in the body. For instance, the postural muscles – such as those in the neck and back – have a high proportion of slow twitch fibres. In contrast, the muscles that bring about fast, explosive movements such as running, jumping and throwing are packed with fast twitch fibres.

5 **a)** Where in the body would you expect the muscles to have a high proportion of fast twitch fibres?
b) Explain why slow twitch fibres are suitable for postural muscles.

The distribution of different fibre types in different athletes

The muscle fibre types of a selection of age-matched athletes were analysed, and the results are summarised in Table 17.A1. These are average values: there is considerable variation between different athletes.

Table 17.A1 Analysis of the average proportions muscle fibre types in selected muscles in different male athletes		
Type of athlete	**Approx % fast twitch**	**Approx % slow twitch**
Marathon runner	18	82
Swimmer	25	75
Cyclist	40	60
800 metre runner	52	48
Untrained person	55	45
Sprinter and jumper	62	38

6 What does 'age-matched' mean?

7 Explain in general terms how the proportion of fast and slow twitch fibres is related to the nature of an athlete's chosen activity.

8 What do you think the information about untrained people tells us?

18

Hormones

The tallest man who ever lived was Robert Wadlow, who stood a touch under 9 feet tall (2.7 metres) and had size 36 feet (US sizes). Born in Alton, Illinois, USA, Wadlow reached the height of over 7 feet at the age of 13 and continued to grow throughout his short life of 22 years. He might have continued to grow, but he died prematurely following an infected foot wound.

Wadlow's condition was caused by an over-productive pituitary gland that made too much human growth hormone, or somatotropin. Today people rarely suffer such extremes because the condition is diagnosed early. Blood tests done on a particularly large (or small) child can reveal hormone imbalances. Under-production can be remedied by injections of the hormone – either the real thing or a synthetic analogue (a chemical of similar structure that has the same effect). Conversely, over-production can be treated either by reducing the size of the gland, or by administering drugs that block the action of the hormone.

Many women find the symptoms of the menopause eased by hormone replacement therapy, and some women have even been able to conceive in their 60s if they have hormone treatment. Add this to the artificial hormones taken every day by women on the pill, and you can understand why making artificial hormones is big business.

Robert Wadlow made it into the record books as the tallest man who has ever lived. There might have been people who have grown taller – stories exist of men over 9 feet tall – but their existence, unlike Wadlow's, cannot be proved. Today, such extremes in size are rare because conditions can be diagnosed early and treated effectively

1 HORMONES ARE CHEMICAL SIGNALS

The human body communicates in two fundamentally different ways: by nerves and by chemicals. As Fig 18.1 shows, there are three main types of chemical signals:

Local signals, eg prostaglandins

Distance signals, eg hormones

Pituitary gland produces the hormone prolactin which stimulates the breasts to produce milk in a woman who has recently given birth

Prolactin travels from the pituitary to the breasts in the blood

pituitary gland

target cells

External signals, eg pheromones

Fig 18.1 A summary of the different types of chemical signalling that exist within and between organisms. Local signals include neurotransmitters such as acetylcholine (see Chapter 15) and local messengers such as prostaglandins, endorphins and histamines. Hormones allow communication between different organs, via the blood. Externally produced messengers allow communication between organisms

- Locally acting chemicals such as **endorphins**, **prostaglandins** and **histamines** (Fig 18.2). These act on cells that are close to their site of production.
- **Hormones**. These are produced by special tissues called **endocrine glands**. They pass into the blood and act away from their site of production. Hormones affect **target cells** and **target organs** in the body.
- **Pheromones** are chemical messengers that allow external communication: simple exchanges of information between people. Generally, pheromones are not an important means of communication in humans, despite the claims of some newspaper reports and advertisements.

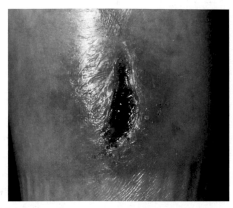

Fig 18.2 Histamine is a local messenger which is released by injured cells. It acts on the cells around an injury like the deep cut shown above, causing blood vessels near to the injury to dilate, bringing more blood into the area. This is an important part of the inflammation response (see Chapter 30)

In this chapter we focus mainly on hormones. The definition of a hormone in mammals is:

A chemical messenger that is produced by cells or tissues of the endocrine system and that travels through the blood to act on target cells or target organs, before being broken down.

The main part of the chapter describes the human endocrine system, and we finish with a brief overview of prostaglandins and endorphins.

2 THE HUMAN ENDOCRINE SYSTEM

In humans, more than a dozen tissues and organs produce hormones. Some, including the **pituitary gland**, the **thyroid gland**, the **parathyroid glands** and the **adrenal glands** are endocrine specialists: their major function is to secrete one or more hormones. Others, such as the pancreas, ovaries and testes, secrete hormones in addition to their other functions. Together, these glands make up the **endocrine system**, shown in Fig 18.3.

The endocrine system has four main functions:

- It maintains **homeostasis**, the balance of the body, by making sure the concentration of many different substances in body fluids are kept at the correct level. The control of blood sugar level, blood pH and water balance provides examples of homeostasis (see Chapters 11 to 14).
- It works with the nervous system to help the body respond to stress. The release of adrenaline in the **fight or flight** response is an example of this (see the Assignment at the end of this chapter).
- It controls the body's rate of growth.
- It controls sexual development and reproduction (see Chapter 20).

We know of more than 50 human hormones, and we divide them into two groups according to their origin: those made from fatty acids and those made from amino acids. The first group, which includes **steroids** such as oestrogen and progesterone, are lipid-soluble. Hormones based on **amino acids** are **modified amino acids**, **peptides** or **glycoproteins**: they are water-soluble and include insulin and adrenaline. Table 18.1 lists most of the hormones produced by the endocrine system and shows where they act and what they do. You might find it useful as a revision aid.

SEE QUESTIONS 1 AND 2

> ✔ **REMEMBER THIS**
>
> Fig 18.3 provides a useful reference point for the rest of this chapter, and also for the hormones described in Chapters 11, 13 and 20.

Table 18.1 Human endocrine glands and their principal secretions. You do not need to learn all of this, but you should find it useful for reference as you read the rest of the chapter

Gland	Hormone	Chemical structure	Effect
Pituitary 1 Posterior lobe	Anti-diuretic hormone (ADH)	Peptide	Reduces amount of water lost in urine. Raises blood pressure by constricting arterioles
	Oxytocin	Peptide	Contraction of smooth muscle during childbirth. Stimulates secretion of milk from mammary glands
2 Anterior lobe	Adrenocorticotrophic hormone (ACTH)	Peptide	Stimulates production and release of hormones from adrenal cortex
	Follicle stimulating hormone (FSH) also called ICSH in males	Glycoprotein	Controls the development of follicles in the ovary, and sperm cells in the testis
	Growth hormone	Protein	Promotes growth (especially of skeleton and muscles). Affects body metabolism
	Luteinising hormone (LH)	Glycoprotein	Stimulates ovulation and formation of the corpus luteum (stimulates testosterone production in males)
	Prolactin	Protein	Stimulates milk production and release during pregnancy
	Thyroid stimulating hormone (TSH)	Glycoprotein	Stimulates growth of thyroid gland; synthesis and production of thyroid hormones
Thyroid	Thyroxine + Triiodothyronine	Iodine-containing amino acids	Thyroxine Increases rate of cell metabolism, controls aspects of growth and development and controls basal metabolic rate (BMR)
Parathyroid	Parathormone	Peptide	Raises blood calcium levels by stimulating release of calcium from bone
Pancreas (Islets of Langerhans)	Insulin (produced by the β cells)	Protein	Lowers blood glucose levels by making cell membranes more permeable to glucose, increases glycogen storage in liver
	Glucagon (produced by the α cells)	Peptide	Raises blood sugar levels by stimulating glycogen breakdown in the liver
Adrenal 1 Cortex	Glucocorticoids, eg cortisol	Steroids	In response to stress, raises blood glucose
	Mineralocorticoids, eg aldosterone	Steroids	Concerned with water retention. Increases reabsorption of sodium chloride in kidneys, so important in control of blood volume and pressure
2 Medulla	Adrenaline	Modified amino acid	'Fear, flight and fight' reactions. Prepares the body for heightened activity. Mimics effects of the autonomic nervous system
	Noradrenaline	Modified amino acid	As adrenaline
Gonads 1 Ovary (follicle)	Oestrogens	Steroids	Female sex characteristics, rebuilding of uterus lining after menstruation, inhibits FSH
2 Ovary (corpus)	Progesterone	Steroids	Stimulates maturation of uterus lining, maintains pregnancy, inhibits FSH
3 Testis	Androgens, eg testosterone	Steroids	Support sperm production. Important in the development of male secondary sexual characteristics
Placenta	Human chorionic gonadotrophin (hCG)	Steroid	Causes corpus luteum to secrete progesterone, thus maintaining pregnancy

✓ **REMEMBER THIS**

There are two basic types of gland in the body: **exocrine glands** (eg salivary gland), whose secretions are released through a tube or **duct**; and **endocrine glands** (eg thyroid gland), which have no ducts, so they shed their secretions directly into the bloodstream for transport to their eventual destination.

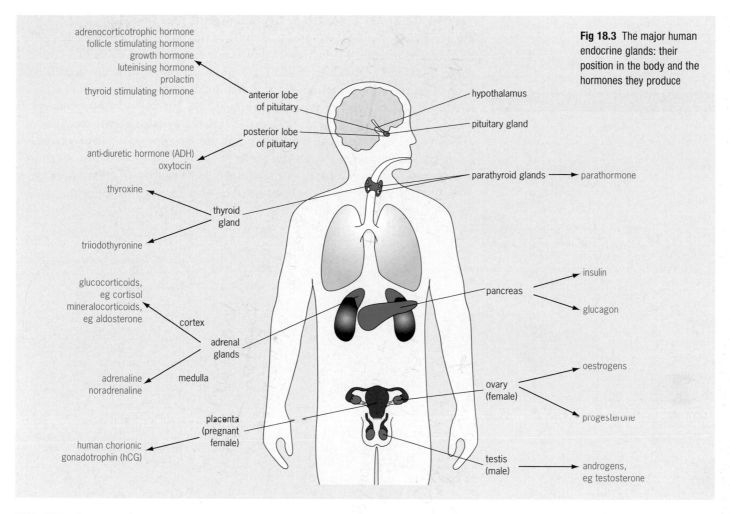

Fig 18.3 The major human endocrine glands: their position in the body and the hormones they produce

MAKING AND RELEASING HORMONES

Endocrine glands are tissues and organs that produce and release hormones. These glands have no ducts. The hormone is simply secreted into the blood and then travels to target organs or cells elsewhere in the body. A good example of an endocrine gland is the **adrenal gland**. Its structure is illustrated in Fig 18.4.

Hormones are formed inside endocrine cells. Some amino acid-based hormones, such as **insulin** and **glucagon**, are produced directly from a copy of a gene by the processes of transcription and translation (see Chapter 24). Most hormones, including the **steroids**, are made as an end-product of a series of chemical reactions. Each reaction makes minor changes to **precursor molecules**, until, eventually, the active hormone is produced. For the steroid hormones progesterone, oestrogen and testosterone, the precursor is cholesterol (see Chapter 3).

All endocrine glands contain some of the hormone they produce, at any given time. Although the hormone thyroxine is released continuously, most are not. **Negative feedback loops** control the release of hormones from many endocrine glands so that homeostasis is maintained. This concept is covered in Chapter 11.

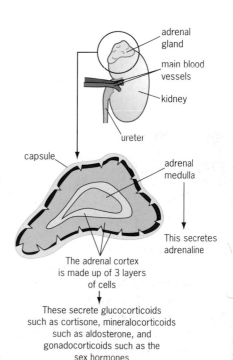

Fig 18.4 The adrenal glands sit on top of the kidneys. They consist of the adrenal medulla, the tissue in the centre of the gland which secretes adrenaline, and an outer layer called the adrenal cortex

THYROXINE: AN EXAMPLE OF NEGATIVE FEEDBACK

Metabolic rate is the rate at which all cells in the body carry out their biochemical reactions. It is a vital whole-body function that must be controlled within very strict limits. The hypothalamus in the brain detects even a small decrease in metabolic rate and responds by releasing more **thyrotropin releasing hormone** (Fig 18.5). This acts on the pituitary gland, causing it to release more **thyroid stimulating hormone**. This passes to the thyroid gland, which responds by secreting more **thyroxine**, the hormone that acts on individual cells to increase metabolic rate. As soon as metabolic rate gets back to normal levels, the hypothalamus responds by releasing less thyrotropin releasing hormone and, in a healthy person, homeostasis is maintained.

SEE QUESTION 3

Fig 18.5 Control of metabolic rate involves a negative feedback loop

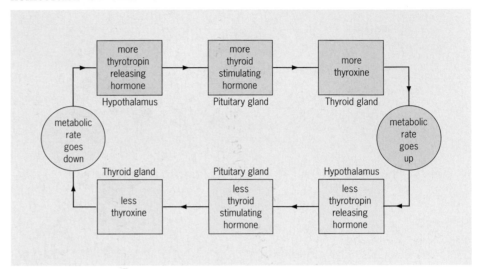

HOW HORMONES AFFECT TARGET CELLS

Hormones exert their effects in two different ways. Peptide hormones travel in the blood to all parts of the body. They do not, however, affect all cells in the body. Target cells or organs have specific proteins on their surface that act as receptor sites. The hormone fits into these sites as a key fits a lock, or like an enzyme fits its substrate (see Chapter 5). Once in place, the hormone–receptor complex brings about changes inside the cell, although the hormone itself never enters the cytoplasm. In contrast, steroid hormones pass easily into cells because they can get through the lipid bilayer. They bind to a specific receptor molecule inside the cytoplasm: this complex then causes biochemical changes inside the cell.

PEPTIDE HORMONES AND SECOND MESSENGERS

Hormones based on amino acids are water-soluble and vary in size from **thyroxine** (just 2 modified amino acids) to **growth hormone**, a protein made up of 190 amino acids. Because they are not lipid-soluble, they cannot get into cells through the lipid cell surface membrane. Instead, they act as first messengers, binding to target receptors on the outside of the cell surface membrane.

As Fig 18.6(a) shows, this binding event activates an enzyme, **adenylate cyclase**, which is located on the inside surface of the membrane. The activated enzyme converts ATP into a substance called **cyclic AMP**. This is the **second messenger** that moves about inside the cell, causing biochemical changes. The final effect on cell function depends on the type of hormone and the type of target cell involved.

Fig 18.6(a) Water-soluble hormones work via a second messenger which is formed inside the cell when they bind to receptors on the cell surface

Fig 18.6(b) When steroid hormones bind to a receptor inside the cytoplasm, they form a complex similar to an enzyme–substrate complex. This enters the nucleus of the cell and binds directly to the DNA, interfering with the cell's ability to read some of its genes. A steroid hormone can either switch on or switch off protein synthesis of particular genes. The overall effect of hormone action is different for each steroid hormone (see Table 18.1)

STEROID HORMONES

Steroids are small lipid-soluble molecules that can get through cell membranes easily. They enter cells and bind with **target receptors** inside the cytoplasm (Fig 18.6b). Cortisol, progesterone, oestrogen and testosterone all act in this way.

3 COMPARING THE NERVOUS AND ENDOCRINE SYSTEMS

Hormones do not act independently. They work together with other hormones and also with the nervous system. The nervous system and the endocrine system both control and co-ordinate the function of different parts of the body and they both rely on chemical messengers, but they have obvious differences (Table 18.2).

Table 18.2 Differences between the nervous system and the endocrine system in humans

Property	System	
	Nervous system	**Endocrine system**
Nature of signal	Nerve impulses are **electrical** signals, transfer of information across synapses, is chemical (see Chapter 15)	All hormones are **chemical** signals
Size of signal	**Frequency modulated** – determined by the frequency of nerve impulses sent along a nerve fibre, and the number of nerve fibres being stimulated	**Amplitude modulated** – determined by the concentration of hormone
Speed of signal	**Rapid**. Human nerves conduct nerve impulses at speeds ranging between 0.7 metres per second and 120 metres per second	**Usually slower**, by comparison. The release of insulin from the pancreas in response to a rise in blood sugar level takes several minutes
Effect in the body	**Localised** effect – each individual neurone links with only one or a few cells	More **general** effect – hormones can influence cells in many different parts of the body
Capacity for modification	**Can be modified** by learning from previous experience	**Cannot be modified**: no learning from previous experience

THE ENDOCRINE AND NERVOUS SYSTEMS ARE LINKED

Despite these differences, we now realise that the link between the endocrine system and the nervous system in humans is more definite than once thought. The main physical link between the two systems is between the **hypothalamus**, part of the base of the brain, and the **pituitary gland**, an endocrine gland just beneath it (Fig 18.7).

Fig 18.7(a) The position of the hypothalamus and pituitary gland in the brain can be identified from Fig 18.7(b)

Key to pituitary hormones
ADH - anti-diuretic hormone
ACTH - adrenocorticotrophic hormone
FSH - follicle stimulating hormone
ICSH - interstitial cell stimulating hormone
GH - growth hormone
LH - luteinising hormone
TSH - thyroid stimulating hormone

Fig 18.7(b) A vertical section through the human brain, showing the main areas. The human pituitary gland is a small outgrowth at the base of the brain. It is connected to the hypothalamus by the pituitary stalk. Both the posterior (back) and anterior (front) lobe secrete hormones

THE HYPOTHALAMUS AND PITUITARY: PARTNERS IN COMMUNICATION AND CONTROL

The pituitary has been called the 'master gland' because it controls the activity of most of the other endocrine glands, but it is itself actually under the control of the hypothalamus. The range of hormones produced by the pituitary and the way the hypothalamus controls pituitary function are shown in Fig 18.7(b).

The human pituitary gland has a **posterior lobe** and an **anterior lobe**. The **pituitary stalk** connects the posterior lobe to the brain. This direct physical link allows nervous communication between the pituitary and the hypothalamus. A rich network of blood vessels also links the hypothalamus with the anterior lobe, allowing hormonal signals to pass from the brain to the pituitary.

NEUROSECRETORY CELLS AND THE POSTERIOR PITUITARY

A **neurosecretory cell** is a modified neurone that synthesises a hormone. The hormone is produced in the cell body and is then transported down the axon and released into the blood through the synapse, which terminates on a capillary. The neurosecretory cells that arise in the hypothalamus and end in the posterior pituitary make two hormones: **anti-diuretic hormone** (**ADH**) and **oxytocin**. When the hypothalamus is stimulated by appropriate nerve impulses, it releases these hormones into the posterior pituitary. From here they pass into the blood and then travel around the body. ADH controls the reabsorption of water by the kidneys (see Chapter 13); oxytocin causes contraction of the uterus during childbirth (see Chapter 20).

HORMONAL CONTROL OF THE ANTERIOR PITUITARY

Blood vessels that connect the hypothalamus with the anterior pituitary carry hormones and chemical messengers. These either stimulate or inhibit the release of pituitary hormones, including growth hormone, prolactin, the gonadotrophins, adrenocorticotrophic hormone and thyroid stimulating hormone (see Fig 18.3).

NERVOUS AND HORMONAL CONTROL OF DIGESTIVE SECRETIONS

The body uses a combination of nerve communication and hormones to control the secretion of digestive juice to coincide with the presence of food in particular areas of the gut. Below is a brief summary of how this happens: for more detail see Chapter 8.

The taste or smell of food encourages the brain to signal the salivary glands and stomach to release saliva and stomach secretions. Food leaving the stomach stimulates the **vagus nerve**, which then triggers the release of bile and pancreatic juice. Digestive secretions can also be controlled by hormones. The hormone **gastrin**, produced by the stomach wall, travels in the bloodstream but exerts its effect locally, stimulating the production of both pepsinogen and hydrochloric acid.

Secretin and **cholecystokinin** control pancreatic and liver secretions. Both are formed by cells in the duodenal wall. Secretin causes the release of sodium hydrogencarbonate in the pancreas. In the liver, it increases the rate of bile formation. Cholecystokinin triggers the release of pancreatic enzymes such as lipase and trypsinogen.

THE ACTION OF ADRENALINE

As the nervous system and the endocrine system are closely linked, we should not be surprised that hormones can affect the way that we feel emotionally. The classic example of this is the **fear, fight or flight response** (see the Assignment at the end of this chapter).

REMEMBER THIS

Anterior means 'towards the front'.
Posterior means 'towards the back'.

SEE QUESTION 4

REMEMBER THIS

Fig 18.3 shows which endocrine glands produce these hormones. Table 18.1 shows what each hormone does.

? TEST YOURSELF

A What would happen if the production of digestive juices was not timed to coincide with the presence of food in the gut?

! SCIENCE FEATURE Acromegaly

Robert Wadlow, the tallest man who ever lived, had a condition called gigantism; he probably had a tumour that produced large amounts of growth hormone. The growth plates in his long bones didn't fuse as they normally do in the late teens, and he was still growing when he died at the age of 22. But what happens if someone develops a tumour that produces growth hormone later in life, after their growth plates have closed over? Tumours of the pituitary gland that cause excess production of growth hormone are rare, but they do happen. They cause a condition called acromegaly. The name acromegaly comes from the Greek words for 'extremities' and 'enlargement' and reflects one of its most common symptoms, the abnormal growth of the hands and feet. Soft tissue swelling of the hands and feet is often an early feature, with patients noticing a change in ring or shoe size. Gradually, bony changes alter the patient's facial features: the brow and lower jaw protrude, the nasal bone enlarges and spacing of the teeth increases. So, growth still occurs, but this does not lead to a change in height.

The disease does cause problems and is fatal if not treated. Overgrowth of bone and cartilage often leads to arthritis. When tissue thickens, it may trap nerves, causing numbness and weakness of the hands. Other symptoms of acromegaly include thick, coarse, oily, smelly skin; skin tags; enlarged lips, nose and tongue; deepening of the voice due to enlarged sinuses and vocal cords; snoring due to upper airway obstruction; excessive sweating, fatigue and weakness; headaches; impaired vision; abnormalities of the menstrual cycle and sometimes breast discharge in women; and impotence in men. Body organs, including the liver, spleen, kidneys and heart, can also enlarge. This is bad enough, but in the longer term, the most serious health consequences are diabetes mellitus, high blood pressure and increased risk of cardiovascular disease. Patients with acromegaly are also at increased risk for polyps of the colon that can develop into cancer.

Acromegaly is difficult to diagnose as the changes it brings about happen very slowly and can be mistaken for changes that result from the normal ageing process. However, once it is diagnosed, it can be treated and people with acromegaly can avoid the changes in their facial features that were common until about 25 years ago, when treatment became available. The first treatment is surgery to remove the pituitary tumour – this reduces the production of growth hormone significantly, but perhaps not down to normal levels. The slight excess must then be 'mopped up' using drug treatment. The main drugs in use are called octreotide and lanreotide. These are analogues of the hormone somatostatin – this dampens down the production of growth hormone by the pituitary. Early forms of the drug had to be injected under the skin every 8 hours but longer-acting preparations that last a month are now available. After surgery, a patient is given the short-acting form to judge which dose is most suitable for them, and then they move on to the once-a-month injections, which need to continue for the rest of their lives.

✓ REMEMBER THIS

The process of amplification seen in the enzyme cascade, which results in the fear, fight or flight response, allows one molecule of adrenaline to stimulate the production of millions of glucose molecules.

At the molecular level, adrenaline binds to the outside of a target cell and activates the enzyme **adenyl cyclase**. This converts ATP to cyclic AMP. Cyclic AMP, is a second messenger which sets off a **cascade**, a chain of reactions that convert glycogen to glucose.

As adrenaline binds to receptor molecules on the surface of a cell, changes in the membrane cause the production of **G proteins**. For every one molecule of adrenaline that binds, 10 molecules of G protein are produced. Each of those 10 molecules of G protein catalyses the production of 10 molecules of adenyl cyclase. Each molecule of adenyl cyclase stimulates the production of 10 molecules of cyclic AMP. This **amplification** effect continues along a chain of reactions that eventually breaks down glycogen into glucose.

4 PROSTAGLANDINS

Prostaglandins are a group of hormone-like compounds. They were originally thought to be produced by the prostate gland – hence the term prostaglandins. They are not true hormones as they are not produced by endocrine glands. Most mammalian cells synthesise prostaglandins, which act locally on surrounding cells.

Prostaglandins control cell metabolism, probably by modifying levels of cyclic AMP inside the cell, and are involved in a wide range of activities including blood clotting, inflammation and smooth muscle activity.

Prostaglandins are extremely powerful: one billionth of a gram produces measurable effects. The prostaglandins in human semen cause muscles of the uterus and oviduct to contract during female orgasm, helping the sperm on their journey towards the ovum. Synthetic prostaglandins can be used to stimulate labour in a pregnant woman whose baby is overdue.

5 ENDORPHINS

Endorphins are one of several morphine-like chemical messengers produced in the brain (endorphin = endogenous morphine). These polypeptide molecules mimic the effects of drugs such as morphine and heroin: like prostaglandins, they are involved in a wide range of activities, but their most important role seems to be in the management of pain. It is thought that endorphins bind to pain receptors and so block the sensation of pain. Long-distance runners who run until they collapse do so because abnormally high levels of endorphin block out the acute pain and discomfort.

Endorphins also seem to affect the 'pleasure centres' of the brain. Stimulation of these areas provides the intense feelings associated with orgasm. Scientists studying the chemistry of pleasure have shown that pleasurable sensations begin when the hypothalamus releases serotonin. This stimulates the release of endorphins, which turns on the supply of dopamine and simultaneously turns off the supply of GABA (an amino acid that suppresses dopamine). Dopamine stimulates the pleasure centre directly.

Drugs such as heroin work by mimicking endorphins. The brain of an addict stops making endorphins and so withdrawal from heroin results in a sudden lack of endorphin and a build-up of GABA, producing unpleasant withdrawal symptoms.

6 PHEROMONES AND BEHAVIOUR

Pheromones, sometimes called **ecto-hormones**, are substances that organisms release into the environment to communicate with members of the same species. Pheromones are small, volatile molecules that spread easily in the environment. They are active in very small amounts: the pheromone of the female gypsy moth causes a response in the antennae of the male moth at concentrations as low as one in a thousand million million molecules (Fig 18.8).

Pheromones are usually classified according to the type of response they produce:

- **Alarm pheromones** are produced by bees and ants when attacked. They excite other insects of the same species to swarm around the attacker.
- **Sex attractants** are released by moths, rats and possibly humans to attract members of the opposite sex. Humans do not have a particularly good sense of smell but there is some evidence that very young babies can recognise their mother by her characteristic smell. Some people also think that sexual partners can also recognise each other using their sense of smell, but there is less evidence to back this up.
- **Trail substances** are produced by ants to show other ants where to find sources of food.
- The **queen substance** is produced by the queen bee within a hive to suppress the production of other queens.

Fig 18.8 Sex attractants such as the pheromone bombykol are released into the air in minute quantities by female moths. The male is extremely sensitive to these pheromones, because his antennae are large and crammed with many specialist receptors. This male has about 17 000 receptors in each antenna that respond only to bombykol

SUMMARY

After reading this chapter, you should know and understand the following:

■ There are three basic types of chemical signal: **local signals** such as histamine, prostaglandins and endorphins; **hormones** such as insulin and adrenaline; external signals or **pheromones**.

■ The human **endocrine system** has four main functions. It controls growth, sexual development and **fear, flight and fight reactions**. It maintains body homeostasis.

■ Endocrine glands lack a duct: their secretions are delivered directly into the bloodstream.

■ Hormone levels are usually controlled by **negative feedback loops**.

■ There are two basic types of hormone: **peptide hormones** which are derived from amino acids (eg insulin) and **steroid hormones** which are made from fatty acids (eg oestrogen).

■ Peptide hormones bind to receptors on the outer membrane of target cells and achieve their effect via a **second messenger** such as cyclic AMP. Steroid hormones pass straight through the cell membranes and bind to target receptors inside the cytoplasm.

■ There is a close link between the nervous and endocrine systems, shown by the way in which, in the brain, the **hypothalamus** interacts with the **pituitary gland**.

■ Adrenaline, the hormone that causes the classic fear, flight and fight response, acts at the molecular level to bring about the production of millions of glucose molecules.

■ **Prostaglandins** and **endorphins** are local messengers which affect many different types of cell. Prostaglandins are best known for their effects on the female reproductive system: endorphins are chemicals that influence our perception of pleasure and pain.

EXAM QUESTIONS

1

a) Given below is a table of mammalian hormones and a list of hormonal functions **A–H.** Copy the table. From the list, choose function of each hormone and place the appropriate letter **A–H** in the table.

Hormone	Function
insulin	
anti-diuretic hormone	
glucagon	
oxytocin	
thyroxine	
progesterone	

List of hormonal functions

A Lowers the resting potential across the membranes of the muscle fibres in the wall of the uterus

B Causes hair follicles to produce hair in the lower abdominal regions of females

C Stimulates the hydrolysis of glycogen to glucose in the liver

D Increases the metabolic rate by affecting the rate of protein synthesis

E Inhibits the secretion of follicle-stimulating hormone during pregnancy

F Increases the rate of contraction of the cells in the sinoatrial node

G Renders the walls of certain areas of nephrons permeable to water

H Increases the ability of skeletal muscle to use glucose as a respiratory substrate

b) State three differences between hormones and enzymes.

c) State two differences between control using the endocrine system and control using the nervous system.

OCR 6918 Animal Studies March 2000 Q2

Refer to Chapter 11 page 205 to help you answer this question

2 Read the following passage about hormones involved in the regulation of blood glucose levels, then write add the most appropriate word or words to complete the passage.

When blood glucose levels rise above normal, ―― is released from the ―― cells of islets of Langerhans in the ――. This lowers the blood glucose levels by increasing the uptake of glucose into cells and promotes the conversion of glucose into ――. Other cells in the islets of Langerhans secrete ―― which raises the level of blood glucose.

Edexcel B2 January 2000 Q1

3 Thyroxine is a hormone secreted by the thyroid gland and it controls the basal metabolic rate. The diagram shows how the secretion of thyroxine is controlled.

a) Explain how the control of thyroxine secretion illustrates the principle of negative feedback.

b) In the formation of thyroxine, a protein-digesting enzyme detaches amino acid molecules from a globular protein called thyroglobulin. Thyroxine is formed from an amino acid molecule joined to iodine atoms. Name each of the following:
(i) The organelle where thyroglobulin is manufactured.
(ii) The type of bond that is broken when the amino acid is detached from thyroglobulin.
(iii) The process by which thyroxine is secreted from a cell in the thyroid gland.

c) Give three ways in which hormonal control differs from nervous control.

Edexcel B6 January 2000 Q5

4 Complete the table summarising the role of each of the following hormones.

Hormone	Site of secretion	Part of kidney tubule affected	Effect of hormone
ADH			
Aldosterone			

AQA (NEAB) BY03 Physiology June 2000 Q2

KEY SKILLS **ASSIGNMENT**

ADRENALINE AND STRESS

When we are nervous, familiar symptoms appear. We get butterflies in the stomach, our pulse increases, our hands become cold and clammy. This happens because our body is preparing for action by secreting **adrenaline**. This hormone:

- increases heartbeat,
- increases ventilation rate,
- causes vasoconstriction, resulting in redirection of blood from intestines and skin to brain and muscles,
- increases metabolic rate,
- dilates pupils.

1 **a)** Which organs secrete adrenaline?
b) What is the overall effect of adrenaline on the body?
c) Use your answer to 1b to explain why adrenaline leads to an increased blood pressure.

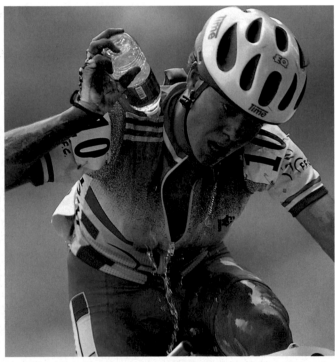

Fig 18.A1 Adrenaline helps this mountain biker to concentrate and compete to the best of her abilities. Her heart is pounding, blood pressure is raised and all of the capillaries in the muscles are open to maximise performance

The effects of adrenaline are very similar to the effects of the sympathetic nervous system (see Chapter 16). For example, secretion of adrenaline speeds up the heart, as does direct stimulation by a sympathetic nerve.

The difference is that adrenaline produces a general state of readiness for action, while the sympathetic nervous system is used for fine control. If your body needs to increase the heart rate a little, but does not need the other effects of adrenaline, it does so via the sympathetic nerves.

The effects of stress

Fig 18.A2 There are many stressful lifestyles: being a busy executive is just one. Stress happens when adrenaline is continually released, ready for action that is not taken. This leads to raised blood pressure which can increase the risk of heart disease or stroke

It is difficult to define stress. What is a stressful situation to some people may be sheer enjoyment to others. Certainly, some stimulation is desirable – it makes us feel more alive.

In the short term, the responses of the body to stress have a beneficial effect and leave the individual better able to cope with the crisis if and when it comes. In the long term, however, the effects can be far from beneficial. Continued secretion of adrenaline and general sympathetic stimulation bombard the body with stress chemicals. Table 18.A1 outlines some of the common symptoms of stress.

2 **a)** List some activities/sports that people enjoy because they stimulate release of adrenaline.
b) List some occupations or lifestyles that are commonly associated with stress.

3 **a)** Discuss or write down some ideas that might explain why there is more stress-related illness today than in previous generations.
b) Discuss or write down some strategies that people might use to minimise stress.

Table 18.A1 The effects of stress on bodily functions			
Normal state	**Adrenaline response**	**Short-term effect**	**Long-term effect (stress)**
Brain: normal blood flow	blood flow increases	think more clearly	headaches and migraines
Muscles: normal blood flow	blood flow increases	improved performance	muscular tension and pain
Heart: normal pulse and blood pressure	output and pressure increase	improved performance	hypertension and chest pain
Intestines: normal blood flow and peristalsis	blood flow decreases, peristalsis increases	slower digestion	abdominal pain and diarrhoea
General biochemistry: normal rate of oxygen use. Glucose and lipids liberated	oxygen, glucose and lipid use increases	more energy available quickly	rapid tiredness

19

Senses and behaviour

19 Senses and behaviour

Prosthetic legs enable people who have lost both lower limbs to lead fairly normal lives. New technology is now making use of the 'phantom limb' sensations that amputees experience, by developing limbs with electrical connections. These give the person more control over the movement of their artificial limbs

Losing a limb is a surprisingly common event. Amputation can follow the horrific injuries suffered in terrorist bomb attacks or in serious industrial or road accidents. However, the amputee wards in Britain's hospitals are more likely to be full of smokers whose circulation is no longer effective enough to keep their extremities alive.

When a person is unfortunate enough to lose an arm or a leg, you might expect that they would lose all the sensations that come from that part of the body. However, amputees can still feel pain or itching in the removed limb. This is an added misery for a person already seriously traumatised by their injury.

The phenomenon of the phantom limb arises because of the nature of the nervous system. Our brain interprets the environment by knowing where all the incoming nerves have originated. If someone stamps on your left foot, you know which foot hurts because impulses arrive via the left foot nerves. When a limb is removed, many of the nerves that connect it with the brain remain intact, and any stimulation of these nerves is interpreted by the brain as sensation in the missing limb.

1 RESPONDING TO THE ENVIRONMENT

We make sense of our surroundings by using our sense organs to gather information and relay it to the central nervous system. Table 19.1 shows the huge amount of sensory information that reaches the brain. Fortunately, we deal with most of it at a subconscious level, which is just as well, since otherwise we would have no time for thoughts. Overall, the brain copes with a huge range of sensory information and uses it to decide on appropriate courses of action.

In this chapter, we present an overview of how the sense organs work and then we examine the Pacinian corpuscle, the eye and the ear in more detail. You can find out more about taste and smell in Chapter 8, and touch and the skin are covered in Chapter 12. We conclude the chapter with a brief look at behaviour.

Table 19.1 Summary of the sensory information that goes to the brain	
External stimuli	**Internal stimuli**
Sound: volume, pitch, harmonics	Blood pressure
Vision: colour, shape, intensity, movement	Solute concentration of the blood: do we need a drink?
Touch: light or heavy pressure?	Internal pain: ulcer, wind, headache?
Texture: sandpaper or velvet?	Tension in muscles: posture?
Temperature: hot or cold?	Fatigue in muscles
Pain: mild or severe?	Tension in bladder or rectum
Movement: direction, acceleration or deceleration	
Direction of gravity	
Smell: which one of thousands of different chemicals?	
Taste: salt, sweet, sour or bitter?	

SOME IMPORTANT TERMS

Before going on to look at sensory systems in more detail, you will find it useful to become familiar with some important terms:

- A **stimulus** is a physical event, usually some form of energy change, that we can detect with our receptors. Examples include **mechanical stimuli** such as pressure, touch and movement of air, **thermal stimuli** such as heat and cold, **light** and **chemical stimuli** such as taste, smell, and the concentration of carbon dioxide or oxygen in the blood. There are many factors in our environment that are not stimuli – we cannot detect them because we do not have the appropriate receptors. We do not sense radio waves or high-pitched dog whistles, for example.

- A **sensation** refers to a general state of awareness of a stimulus. Aristotle first identified the five senses of hearing, sight, taste, smell and touch, but we now know that there are actually many more senses than this. The skin can experience the sensations of light touch or deep pressure, heat, cold and pain. There are also the general senses of balance and body movement. Internally, the body can sense changes in blood pressure and blood levels of chemicals; and we experience hunger and thirst when we need food or fluids.

- **Perception** is the interpretation of stimuli by the brain. Perception of a stimulus is different from just sensing that it is there – it allows us to assess the significance of the stimulus. Many stimuli can be ignored, such as the sensations from your clothes, to allow us to concentrate on more urgent stimuli (like the information in this book)

2 FEATURES OF SENSORY SYSTEMS

A sensory system must allow the following processes to take place:

- **Transduction**. This is the process by which a stimulus initiates a nerve impulse. Stimuli vary widely in their nature; they can be physical, light, heat or chemical, and so on. All, however, result in the generation of a nerve impulse (see Chapter 15). For example, when physical pressure is applied to the skin, sensory cells inside specialised receptors in the skin depolarise, and the physical stimulus is transduced, or changed, into the electrochemical event known as the action potential.

- **Transmission**. A nerve impulse is transmitted along a sensory neurone into the central nervous system (CNS) (see Chapter 16). The brain recognises the strength of a stimulus not by the size of the impulse but by its frequency. The stronger the stimulus, the higher the frequency of impulses.

- **Information processing**. The CNS processes information that it receives from several parts of the body and makes a decision about the best course of action to take. We have superb memories compared with other animals and so we are able to combine new sensory information with past experience before deciding on an appropriate course of action.

RECEPTORS

A receptor is part of the nervous system that has become adapted to receive stimuli. Some receptors are modified neurones while others are separate cells connected to neurones. Some receptors, such as those in the skin, occur alone or in small groups. Others are concentrated within a specialised sense organ such as the eye or ear.

Fig 19.1 Different species have different sense organs. Anyone who has taken a dog for a walk will know that their pet lives in a different world. They can drag you from lamp-post to lamp-post, obsessed with smells that we can't even begin to detect, even if we wanted to. Their hearing is much more acute, too. Dogs can hear much higher frequencies than we can. On the other hand, their colour vision is not as good

? TEST YOURSELF

A Look at the cells in Fig 19.2. What type of receptor cell are they? Classify them according to stimulus, location and complexity, as in Table 19.2.

Sensory receptors are vital. They sense the external environment and they also provide information to the CNS about the body's internal environment. Receptors that detect changes in the outside world are positioned at or near the body surface. These include the specialised sense organs, the ear and eye, the smell receptors of the nose and the taste receptors of the tongue. The human head, like that of most animals, is basically a set of sense organs in front of a mass of nerves (the brain) that is capable of processing all the incoming information. In contrast, receptors that respond to internal stimulation, such as stretch receptors in blood vessels or muscles, are found deep inside the body.

TYPES OF RECEPTOR

Receptors can be classified in three ways: by the type of stimulus they respond to, by their location within the body, or by their level of complexity (Table 19.2).

? TEST YOURSELF

B Look at Table 19.2. What type of receptors are the photoreceptors of the eye: are they exteroceptors, proprioceptors or interoceptors?

Fig 19.2 Taste buds in the human tongue

Table 19.2 Classification of receptors

Receptor name	Action	Example
Classified by stimulus		
Photoreceptor	responds to light	eye
Chemoreceptor	responds to chemicals	taste bud
Thermoreceptor	responds to changes in temperature	temperature receptors in skin
Mechanoreceptor	responds to mechanical (touch) stimuli	nerve fibres around hairs
Baroreceptor	responds to changes in pressure	baroreceptors in carotid artery
Classified by location		
Interoceptor	responds to stimuli within the body	baroreceptors in carotid artery
Exteroceptor	responds to stimuli outside the body	eye and ear
Proprioceptor	responds to mechanical stimuli conveying information about body position	muscle spindle
Classified by complexity		
General senses	single cells or small groups of cells	temperature receptors in skin
Special senses	complex sense organs	eye and ear

There are also two basic types of receptor, and both can be any of the classes of receptor described in Table 19.2. Simple or primary receptors consist of a single neurone whose axon carries a nerve impulse into the CNS. The temperature and pressure receptors in the skin are both examples of **primary receptors**.

Secondary receptors receive information from a cell that is not part of the nervous system and then pass the information on to a nerve cell for transmission to the CNS. Taste buds in the tongue are secondary receptors.

THE PACINIAN CORPUSCLE

When people shake hands, there is a subtle exchange of information. A firm, confident handshake communicates friendliness but 'I'll take no nonsense', while a feeble, cold, clammy, loose grip says 'I don't really want to be here'. This is only possible because of the simplest receptor in the human body – the **Pacinian corpuscle**. This is a **mechanoreceptor** that responds to pressure or any kind of mechanical stimulus.

The corpuscle itself is a nerve ending surrounded by several concentric capsules of connective tissue, separated by a viscous (stiff) gel. In its resting

REMEMBER THIS

The Pacinian corpuscle is named after Filippo Pacini, a nineteenth century Italian anatomist.

state, the Pacinian corpuscle in cross section resembles a dartboard with the nerve ending at the bull's eye (Fig 19.3). Touching the hands or feet makes this epidermal receptor deform into an oval cross section including the nerve ending. The viscous gel then moves and allows the nerve ending to resume its normal shape. If the pressure is released, the corpuscle as a whole returns to its original shape and the nerve ending is again deformed, signalling the end of the hand or foot pressure. The viscous gel flows back and everything returns back to the normal resting stage. The nerve sends the signal to the brain, alerting it to the exact location of the pressure. Pacinian corpuscles can detect vibrations and touch in terms of frequency, duration and intensity.

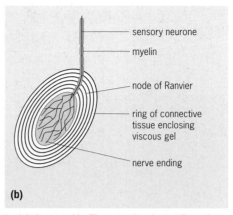

Fig 19.3(a) A micrograph of two Pacinian corpuscles (touch receptors) in human skin. They are abundant at fingertips. In this touch receptor, pressure is detected by the end of the neurone at the centre of the connective tissue capsule which is layered like an onion. When the neurone is stimulated, action potentials are propagated along the neurone. The brain interprets these impulses as pressure on that part of the body. **(b)** A simple diagram of a cross section through a Pacinian corpuscle

HOW DO RECEPTORS WORK?

How do cells detect stimuli? How, for instance, can we distinguish between a light and a heavy touch? The answer lies in the fact that primary receptors have proteins in their cell surface membranes that are sensitive to the type of stimulus. Whatever the stimulus, transduction follows the same basic pattern, which is summarised in Fig 19.4. A receptor protein is activated by a specific stimulus. In turn, the protein opens or closes specific ion channels in the cell surface membrane. This change in ion movement causes depolarisation of the membrane; this is the **generator potential**.

If the depolarisation goes beyond the threshold level for that cell, it triggers an action potential (see Chapter 15). All receptors therefore act as **biological transducers**: they create action potentials in neurones in response to physical and chemical stimuli from the environment.

The greater the stimulation of the sense cell, the greater the depolarisation. Once the generator potential exceeds the threshold level, an action potential is produced and is transmitted along the axon. It is important to remember that all action potentials are of the same magnitude – you can't have powerful and weak nerve impulses. The body is able to tell the difference between strong and weak stimuli by the frequency of the action potential created.

Receptors are very sensitive – they have to be in order to detect small changes in the environment. The human nose can detect ethanoic acid (acetic acid, the acid in vinegar) at a concentration of 5×10^{11} molecules per litre of air; a dog's nose, though, can detect the same smell at a concentration of about 2 million times less, 2×10^5 molecules per litre.

Body senses also **amplify** the incoming signal. A stimulus is often quite weak – just a few photons of light when viewing a distant star for example, or a few molecules of a chemical. But the action potential travelling from the eye to the brain, for example, has about 100 000 times as much energy as the few photons of light that triggered it.

The endocrine system also shows amplification. One molecule of the hormone adrenaline, for example, can release millions of glucose molecules in the cascade effect (see Chapter 18).

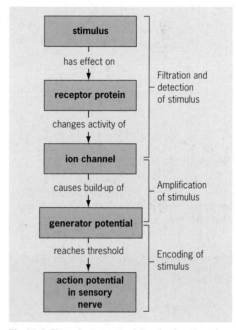

Fig 19.4 Flow chart summarising the function of receptor cells

✔ REMEMBER THIS

A stimulus alters the electrical properties of membrane of the receptor cell. The charge on the membrane is reversed: it goes from –70 mV to +40 mV. We say that it has depolarised. When a membrane depolarises there is the formation of a nerve impulse, or moving action potential, that is transmitted along a nerve fibre. Find out more about action potentials in Chapter 15.

3 VISION

The eyes are literally our window on the world (Fig 19.5). This most sensitive and intricate of sense organs has fascinated both scientists and poets for centuries. In this section we look at the basic structure of the eye before going on to study focusing and the correction of defects. Finally, we look at vision at the molecular level.

Fig 19.5 Mammalian eyes are contained in the orbit, a bony socket of the skull. They are protected by the bone and also by a pad of fat at the rear of each eyeball. Other features of the eye are also protective:
* The eyebrows prevent sweat running from the forehead into the eye
* The eyelashes prevent the entry of airborne particles such as flying insects, often adding to the efficiency of the eye blink reflex
* Tears constantly bathe the surface of the eye, removing the surface film of dust and dirt and killing bacteria
* The conjunctiva, a thin, transparent membrane which covers the cornea, also protects the eye from airborne debris

STRUCTURE OF THE HUMAN EYE

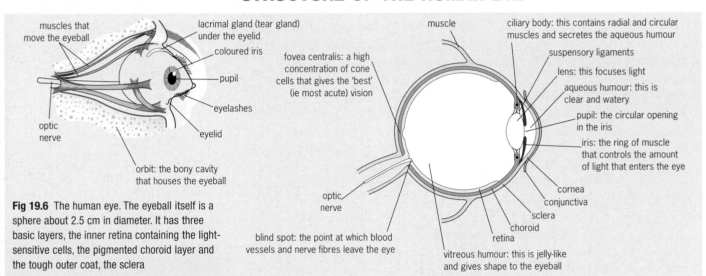

Fig 19.6 The human eye. The eyeball itself is a sphere about 2.5 cm in diameter. It has three basic layers, the inner retina containing the light-sensitive cells, the pigmented choroid layer and the tough outer coat, the sclera

Fig 19.7 As people get older, the lens in their eyes can become cloudy, interfering with normal vision. In this routine operation, an ultrasound probe is being used to break up a cloudy lens before a plastic lens is fitted

The eye is an important and delicate structure. The **orbits**, bony sockets in the skull, protect the eyes from physical damage, and a pad of fat behind the **eyeball** helps to cushion it from shocks. The other protective features of the eye are shown in Fig 19.6.

The main body of the eye is divided into two parts by a **biconvex lens**. This lens contains transparent **lens fibres** (long thin cells that have lost their nuclei) together with an elastic **lens capsule** made of glycoprotein. The lens is flatter at the front than at the back and is soft and slightly yellow. With age it becomes flatter, yellower, harder and less elastic; it can also become cloudy (Fig 19.7). There is no blood in the lens – substances diffuse in and out of it from the surrounding fluid.

The eyeball has three cavities: the region between the cornea and the iris, the region between the iris and the lens, and the cavity that fills all the space behind the lens. The first two cavities are filled with a fluid called the **aqueous humour**, which is thin and watery. The third cavity, the largest (it takes up about 80 per cent of the volume of the eyeball) is filled with a thicker fluid called the **vitreous humour**. This transparent gel has the consistency of egg white and contains 99 per cent water and 1 per cent collagen fibres and hyaluronic acid. The vitreous humour preserves the spherical shape of the eyeball and helps to support the **retina**.

THE RETINA

The retina is the light-sensitive layer at the back of the eye. It is a complex structure with a deep layer of light-sensitive cells called **rods** and **cones**, together with a middle layer of **bipolar neurones** which connect the rods and cones to a surface layer of **ganglion cells** (Fig 19.8). Fibres from the ganglion cells join up to form the **optic nerve**. Only the back of the retina is **photosensitive**. The front surface extends over the **choroid** and forms the inner lining of the **iris** and **ciliary body**.

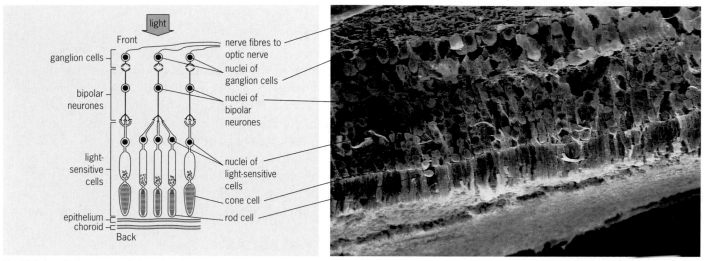

Fig 19.8 Detailed structure of the retina showing photoreceptors and their connections

The sensitive part of the retina has the area of a ten pence coin and acts as a projection screen on to which images are directed. Strangely, the nerves that come from the retina pass in front of the light-sensitive cells, but these do not interfere with vision.

THE CHOROID

The **choroid** contains many blood vessels. Blood supplies the cells of the eye with nutrients and oxygen and removes waste. The choroid is dark-coloured because of a high concentration of the pigment melanin in its cells. These pigmented layers prevent internal reflection of light rays and so prevent us seeing a confused and blurred image.

At the front of the eye, the choroid expands around the edge of the lens to form the ciliary body. The smooth muscles in the ciliary body alter the shape of the lens (Fig 19.9). The lining of the ciliary body secretes aqueous humour, the fluid that fills the front of the eye.

The iris is also an extension of the choroid, partially covering the lens and leaving a round opening in the centre, the pupil. Its function is to control the amount of light entering the eye. The iris contains **radial** and **circular muscles** that can alter pupil size. Iris muscles are controlled by the autonomic nervous system (see Chapter 16).

THE SCLERA

The **sclera** forms the 'white' of the eye. It is a tough external coat mainly of collagen fibres. Six exterior muscles attached to the sclera enable the eye to look up, down and side-to-side. The central one-sixth of the sclera is colourless and transparent, forming a clear window, the cornea, through which light gets into the eye.

? TEST YOURSELF

C The eye is often compared with a camera. Which part of the camera corresponds to the retina of the eye?

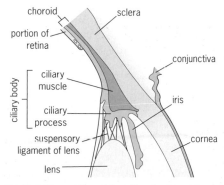

Fig 19.9 The position of the ciliary body and the suspensory ligaments in the eye. When we focus on objects that are distant from us, the ciliary muscles relax, increasing the tension on the suspensory ligaments. This pulls the lens into a thinner and flatter shape. When we look at objects closer to us, the ciliary muscles contract, the tension on the ciliary ligaments decreases and the lens becomes fatter

Bright light
Relaxation of radial muscles and contraction of circular muscles cause pupil to constrict

Dim light
Contraction of radial muscles and relaxation of circular muscles cause pupil to dilate

Fig 19.10 The eye is prevented from being dazzled by a speedy reflex that makes the pupil smaller. To achieve this **constriction** of the pupil, radial muscles in the iris relax and circular muscles contract. In dim light, the opposite happens: radial muscles contract and circular muscles relax, causing the pupil to enlarge

HOW THE EYE WORKS

The process of seeing is complex and involves five different stages:

- Light enters the eye.
- An image is focused on the retina.
- Energy in the light that makes up the image is transduced into an electrical signal.
- Nerve impulses carry information about the image into the brain.
- The brain decodes the information and perceives the image.

Let's look at each of the stages in more detail.

LIGHT ENTERS THE EYE

Light enters through the cornea and then passes through the pupil, the aqueous humour, the lens and the vitreous humour, before it reaches the retina. To operate well in conditions of different light intensity, the eye must control how much light reaches the retina. It does this by changing the diameter of the pupil. In bright light, the pupil **constricts** (gets smaller) to prevent overstimulation of the retina and the perception of 'dazzle'. In dim light, the pupil **dilates** (gets wider), letting in more light.

The size of the pupil is controlled by the muscles in the iris (Fig 19.10). These are themselves controlled by the autonomic nervous system and so pupil adjustment is a reflex response, not under the conscious control of the brain.

FOCUSING LIGHT ON THE RETINA

If you do not focus a camera correctly, the picture you take is blurred. Similarly, the eye must focus light to produce a high-resolution image on the retina. The eye focuses an image by refracting or bending light using the cornea and the lens, forming an upside down or inverted image on the retina (look ahead to Fig 19.12).

Because of the composition and curvature of the cornea, most of the refraction of light occurs in this structure. The lens is important in the fine focusing of light onto the retina. The cornea is fixed but the lens is adjustable, allowing **accommodation**, a reflex that makes the eye focus on objects that are at different distances from the eye.

Fig 19.11 The eye can focus on near and distant objects by altering the shape of the lens. This process is called accommodation

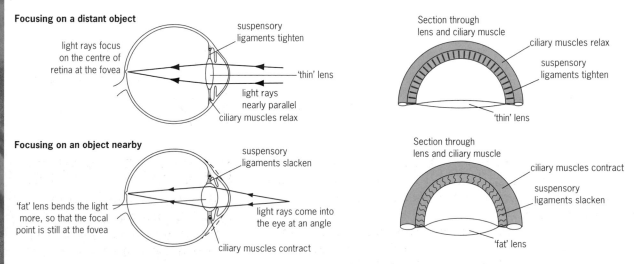

Focusing on a distant object

light rays focus on the centre of retina at the fovea

suspensory ligaments tighten

'thin' lens

light rays nearly parallel

ciliary muscles relax

Section through lens and ciliary muscle

ciliary muscles relax

suspensory ligaments tighten

'thin' lens

Focusing on an object nearby

'fat' lens bends the light more, so that the focal point is still at the fovea

suspensory ligaments slacken

light rays come into the eye at an angle

ciliary muscles contract

Section through lens and ciliary muscle

ciliary muscles contract

suspensory ligaments slacken

'fat' lens

ACCOMMODATION (FOCUSING)

Generally, when we open our eyes in the morning, they are not focused on nearby objects. At rest, the ciliary muscles have relaxed, pulling the lens flat. In this state we focus on distant objects.

When we focus on an object that is only a metre or so away, the ciliary muscles of the eye contract involuntarily (Fig 19.11). This reduces the tension on the ligaments that hold the lens in place, and the lens becomes 'fatter'. The focal length changes, and the image is focused. When we then look at an object much further away, the ciliary muscles relax, again automatically. The tension on the ligaments supporting the lens increases, and the lens becomes 'thinner' allowing our vision to **accommodate**.

SEE QUESTION 1

PROBLEMS WITH THE LENS SYSTEM

In some individuals the lens and cornea do not focus correctly (Fig 19.12). If you are short sighted, a condition known as **myopia**, you focus light from an object in front of the retina. This produces a blurred image. Short sight can result from an elongated eyeball or a thickened lens. It is corrected by using a **diverging** or **concave** lens.

Fig 19.12 In some people, an image forms in front of the retina (this is myopia) or behind the retina (this is hypermetropia). In these cases the eyeball is either too long or too short. Astigmatism can result from an irregular cornea or an irregular lens

If you are long sighted you are said to have **hypermetropia**, and you focus light behind the retina. Again, objects appear blurred. Long sightedness results from a shortened eyeball or a lens that is too thin, and is corrected using a **converging** or **convex** lens.

A more complicated visual defect is **astigmatism**. In astigmatism, the surface of the cornea is irregular, the object appears blurred because some of the light rays are focused and others are not (this is similar to the distortion produced by a wavy pane of glass). Astigmatism is corrected using a **cylindrical** lens that bends light rays in one plane only.

? TEST YOURSELF

D Imagine you are standing at a bus stop reading a book. What happens **(a)** to the ciliary muscle, **(b)** to the lens and **(c)** to the pupil of your eye as you look up to see the bus approaching in the distance?

Fig 19.13 Primates have both binocular and colour vision. Binocular vision allows animals to judge distance. This is vital when jumping from branch to branch. It is thought that colour vision allows food to be chosen with greater accuracy; many fruits use colour to signal when they are ripe. Humans, being primates, owe our vision to our ancestry

BINOCULAR VISION

Humans and some other mammals, notably primates and predators such as cats, have **binocular vision** (Fig 19.13). We have eyes at the front of our heads, so the field of vision seen by each eye is similar but not identical. Normally, the brain is able to compute the incoming signals so that we see only one image.

Binocular vision gives us a big advantage. When two sets of signals are decoded by the brain, we get the impression of distance and depth and of objects being three-dimensional. This **stereoscopic vision** allows us to judge distance and depth accurately. It is easy to see why primates need binocular vision; when living in the trees, any animal that cannot judge distance is likely to fall sooner rather than later.

VISION AT THE MOLECULAR LEVEL

When light reaches the back of the retina it is focused on to the **photoreceptors**, cells that are specialised to detect light. These cells contain pigment molecules that are bleached by light. Bleaching of the pigment results in a generator potential and, when the threshold is reached, action potentials are created and nerve impulses begin to travel towards the brain.

The two photoreceptors in the human eye are the rods and cones (Fig 19.14 and Table 19.3).

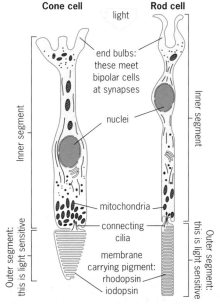

Fig 19.14 Rods and cones: the two types of light-sensitive cell in the human eye

Table 19.3 Comparing rods and cones		
Feature	**Rod cells**	**Cone cells**
Shape	rod-shaped outer segment	cone-shaped outer segment
Connections	many rods connect with one bipolar neurone	only a single cone cell per bipolar cell
Visual acuity	low	high
Visual pigment(s)	contain rhodopsin (no colour vision)	contain three types of iodopsin, responding to red, blue and green light
Frequency	120 million per retina – 20 times more common than cones	7 million per retina – 20 times less common than rods
Distribution	found evenly all over retina	found all over retina but much more concentrated in the centre, particularly the yellow spot or fovea centralis
Sensitivity	sensitive to low light intensities (used in dim light)	sensitive to high light intensities (used in bright light)
Overall function	vision in poor light	colour vision and detailed vision in bright light

PHOTORECEPTION IN ROD CELLS

Rods contain a reddish-purple compound called **rhodopsin**, or visual purple. This consists of two joined compounds – **opsin**, a protein, and **retinal**, a light-absorbing compound derived from vitamin A. When retinal is exposed to light and absorbs light energy, it changes shape (it is actually converted into an isomer) and this causes the two compounds of rhodopsin to break apart. The free opsin acts as an enzyme which sets in motion a series of biochemical events which leads to the hyperpolarisation of the rod cell surface membrane, that is, it becomes more negative. This generates action potentials in the connecting nerve cells. Nerve impulses then pass to the brain for decoding. In the absence of further light stimulation, the retinal molecule goes back to its original shape and then recombines with opsin to form rhodopsin. It is worth remembering that this resynthesis process takes time.

Rhodopsin is very sensitive to light and so rods can function effectively in dim light. In bright light our rhodopsin is broken down quicker than it can be re-formed, so that most of it is bleached for most of the time. In this state we are said to be light adapted. If we enter a dark room from a brightly lit one we cannot see anything because much of our rhodopsin is bleached. However, as we resynthesise the pigment and become dark adapted, things become clearer and we are able to see in dim light, even if it is only in shades of grey.

PHOTORECEPTION IN CONE CELLS

Fig 19.15(a) There are three types of cone responding (optimally) to three wavelengths of light. The colour we perceive depends largely on the proportion of the stimulation of the three types. If a single type of cone is missing, the individual will not be able to distinguish between certain colours – red–green is the commonest – and is said to be colour blind

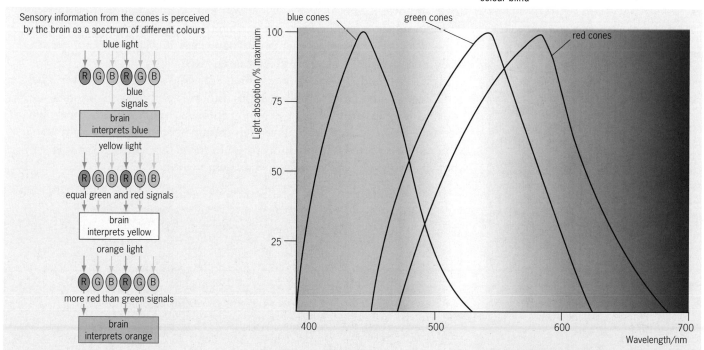

The function of the cones is to see colour and detail, although they can only do this in bright light. In dim light they do not function at all. Cone cells contain the pigment **iodopsin**, a photosensitive pigment made up of **photopsin**, a different protein from that found in rods, and retinal. The events that occur in cone cells stimulated by light are basically the same as those that occur in rod cells. There are, however, three different types of cone. Each one contains a slightly different pigment and responds to a different wavelength of light. One responds to red light, one to green light and one to blue light, so allowing us to see in colour (Fig 19.15).

Fig 19.15(b) Colour blindness can be diagnosed by means of Ishihara tests. A person with normal vision will see the pink and red loop, while a red–green colour-blind person will not.

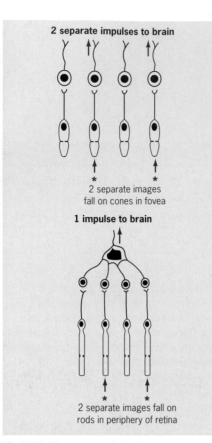

Fig 19.16 The concept of retinal convergence. Generally, the further out you go from the centre of the fovea to the edge of the retina, the greater the degree of convergence. Two separate stimuli (such as two dots or lines) may stimulate two separate cones, but not the one in between. Therefore, the brain will see these as two separate images. The same two stimuli falling on the rods will not stimulate separate neurones, because a group of rods all feed into the same neurone. In this case, the brain will perceive only one stimulus, and see only one line or dot

The colour that we 'see', or more accurately **perceive** (interpret in our brain), depends on which cones are stimulated. When all the cones are fully stimulated we see 'white'. When very few are stimulated we see black. Stimulation of separate types produces red, green or blue, and all colours in between are produced by combinations of different levels of stimulation of the three types together. This model of how three different types of cone can produce the range of colour vision that we experience is known as the **trichromatic theory of colour vision**.

VISUAL ACUITY: THE ABILITY TO SEE DETAIL

From what distance can you read a newspaper? Can you read the bottom line of the standard optician's eye-test board? Tests such as these are a measure of **visual acuity**, which is defined as the ability to resolve detail. For instance, two lines such as these ══════════ will obviously appear as two lines, but if you prop the book up and walk backwards, a point will be reached when they look like one grey line.

An interesting feature of visual acuity is that we only have a very narrow field of accurate vision. Focus on any particular word on this page and you see it in detail, but those on either side are unclear. We see only a very small proportion of our field of vision in detail – and that is the part that is focused on the fovea, and therefore that stimulates the cone cells.

Overall, rods are more sensitive than cones. Rods can function in poor light, but they cannot perceive detail; you can't read small print in dim light. The cones have a higher visual acuity than the rods, but they need bright light. Why is there such a difference? The answer lies largely in a phenomenon known as **retinal convergence**.

Figs 19.16 and 19.17 illustrate retinal convergence. There are over 120 million rods and about 7 million cones, but only about 1 million neurones in the optic nerve. Clearly, each rod and each cone cannot all have their own neurone. In fact, many rods converge into one neurone, while only a few cones are associated with each neurone. In the centre of the fovea it is thought that each cone actually has its own neurone.

Fig 19.17 Why can we see in detail only when we look straight at something? Taking the letter E as an example, it can be seen that if the image falls on the cones it will stimulate some neurones but not others. However, if it falls on the rods it will simply stimulate lots of adjacent neurones, and all we will see is a splodge, not a clear letter E

TRANSMISSION OF NERVE IMPULSES TO THE BRAIN

If you look back at Fig 19.8, you can see that the photoreceptor cells are linked to a set of nerve cells in the retina called bipolar cells. These link with a second type of nerve cell called ganglion cells. It is the axons of ganglion cells that take information from the eyes to the brain. Ganglion cell fibres are bundled together to form the optic nerve.

We cannot see an object whose image falls on the retina at the point where axons of ganglion cells leave to form the optic nerve. Since it contains no receptor cells, any light striking this small area is not sensed; hence its name, the blind spot (Fig 19.18).

Fig 19.18 To demonstrate the blind spot, hold your book at arm's length with the two symbols straight in front of your eyes. Close your left eye and concentrate on the cross with your right eye. Bring the book slowly towards you. Keep looking at the cross on the left: eventually, as the dot falls on the blind spot of your right retina, the image of the dot on the right will disappear

HOW THE BRAIN PERCEIVES AN IMAGE

SEE QUESTION 2

Action potentials do not differ very much from one another and so there are only two possible ways in which stimuli going into the brain can be coded. One is the rate at which action potentials arrive and the other is their destination in the brain.

The rate at which sensory neurones fire tells the brain the strength of the stimulus (see page 264): specific neurones tell the brain which part of the retina has been stimulated. The brain analyses this information to assess the nature of the original stimulus. This process, called **visual processing**, is immensely complex.

Specific areas of the cerebral cortex are associated with different sensory functions (see Chapter 16). The **visual cortex** in the **occipital lobes** deals with visual information (Fig 19.19). However, body senses must not be considered in isolation. Sensory and motor systems interact within the control regions of the CNS, and sensory information is processed at various points along the nerve pathway before it reaches the brain. In the brain it is processed further with information from other senses and stored memories.

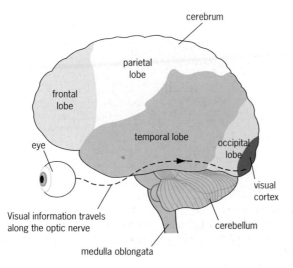

Fig 19.19 The location of the visual cortex in the human brain

SEE QUESTION 3

4 HEARING AND BALANCE

The ear is a miniature **receiver**, **amplifier** and **signal-processing system**. It is divided into three parts – an **outer ear**, **middle ear** and **inner ear** – and is connected to the brain by the auditory nerve (Fig 19.20).

Sound waves pass through the outer and middle ear and into the inner ear through the **oval window**. Pressure on the oval window squashes the fluid in the inner ear, and compression waves travel through the canals of the **cochlea**. The round window eventually receives these movements of fluid and dampens them down (this prevents them from being reflected back into the cochlea) by bulging into the middle ear. Sensitive hair cells in the cochlea fire off signals, sending action potentials along the auditory nerve to the brain.

Fig 19.20 The structure and function of the human ear

Ear ossicles: these are the smallest bones in the body. They form a system of levers which transmit vibrations from the air to the denser fluid in the ear

The eardrum has a surface area twenty times that of the oval window. The vibrations which compress the fluid in the cochlea are therefore amplified

Semicircular canals: movement of fluid inside these canals triggers hair cells in them. This triggers nerve cells to the brain. The three canals are at right angles to each other so that the brain can detect tilting in any direction

The fleshy part of the outer ear, the pinna, is cup-shaped and collects and amplifies sound waves which then pass along the ear canal to the eardrum

Oval window: this transmits vibrations to the cochlea

Cochlea: a fluid-filled spiral tube which transmits vibrations from the middle ear to the sensitive hair cells where movement is converted to a nerve signal

outer ear

middle ear

inner ear

brain

auditory nerves

hammer

anvil

ear canal

stirrup

Sounds waves travel towards the ear: a sound wave takes 20 milliseconds to pass through the ear

Sound waves strike the eardrum and cause the thin, stretched membrane to vibrate

Eustachian tube to throat: this equalises air pressure on either side of the eardrum

Round window: this acts as a 'dampener', reducing movement of fluid in the cochlea

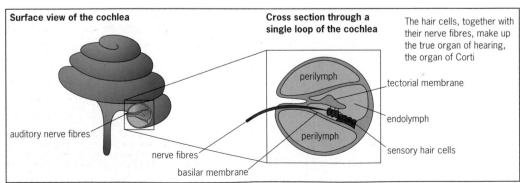

Surface view of the cochlea

Cross section through a single loop of the cochlea

The hair cells, together with their nerve fibres, make up the true organ of hearing, the organ of Corti

perilymph

tectorial membrane

endolymph

perilymph

sensory hair cells

auditory nerve fibres

nerve fibres

basilar membrane

5 BEHAVIOUR

SIMPLE BEHAVIOURS

One of the simplest types of behaviour is **orientation**. This enables organisms to move in order to find favourable environments and to avoid unfavourable ones. Animals such as insects show two types of simple orientation:

- **Taxes**: movements in a specific direction that are directed by a stimulus such as light or food.
- **Kineses**: more random movements that are not directed by a stimulus.

Taxes and kineses are usually regarded as **reflexes**. Reflexes also occur in higher animals – including mammals and humans; these are covered in more detail in Chapter 16.

SEE QUESTIONS 4 TO 7

TAXES

In many orientation behaviours, the animal moves directly towards or directly away from a stimulus. Maggots (fly larvae) have very simple eyes and can respond to the direction of light (Fig 19.21). Honey bees move towards a scent which is associated with food. These types of directional response are called taxes (singular: taxis).

A positive taxis involves movement towards a stimulus, a negative taxis involves movement away from the stimulus. As Table 19.4 shows, taxes are also classified according to the type of stimulus involved (light, water, gravity) or to the type of behaviour shown (straight-line responses, turning movements).

Maggot shows avoidance reaction when light is turned on

light

lamp is switched on

The maggot's head has simple light receptors

direction of locomotion

lamp is off

The maggot has to move its head from side to side to compare light levels on either side of its body

Fig 19.21 The maggot has simple eyes that can distinguish changes in light intensity but it cannot tell from which direction the light is coming. It moves its head from side to side to judge on which side the light is brightest

Table 19.4 Summary of some common taxes. These are normally classified either by the type of stimulus that triggers them or by the type of response they elicit

Name of taxis	Type of stimulus	Type of response
Telotaxis		fixing on a stimulus without scanning from side to side
Klinotaxis		moving head repeatedly from left to right to judge stimulus intensity
Geotaxis	gravity	
Thermotaxis	heat	
Phototaxis	light	
Anemotaxis	air currents	
Rheotaxis	water currents	
Chemotaxis	chemicals, eg salts	

KINESES

Imagine a ladybird, inactive during a cold night. It begins to stir at dawn. As the light level and temperature increase, the animal becomes more active and begins to move around and locate food. Conversely, woodlice are active in the dark and damp but become less active in brighter, drier conditions. These responses, which are not directional, are called kineses. Some common types of kineses are shown in Table 19.5.

Table 19.5 Some common kineses. Like taxes, these are normally classified either by the type of stimulus that triggers them or by the type of response they elicit

Name of kinesis	Type of stimulus	Type of response
Photokinesis	light intensity	
Chemokinesis	chemical gradients	
Hygrokinesis	humidity levels	
Klinokinesis		turns faster when stimulus is more intense
Orthokinesis		moves with greater speed when stimulus is more intense

MORE COMPLEX BEHAVIOURS

We can divide all animal behaviour into two categories: **instinctive behaviour,** or **innate behaviour**, which is inherited, and **learned behaviour**, which is modified by experience.

HUMAN BEHAVIOUR

Consider two examples of human behaviour:

- If children see pictures of small children's faces and adult faces they prefer the adult face. A change in preference occurs at 12 to 14 years of age in girls when they express a preference for baby pictures. Boys display the same trend two years later.
- If we sit too close to a stranger in a library then we tend to use physical 'barriers' such as books or bags to keep the stranger at a distance. If this person gets too close we often move seats.

These are typical of the types of behaviour studied in **human ethology** – the biology of human behaviour. It asks the question 'why do we behave the way we do?'

Ethologists view human behaviour in the same way as that of other animals: they think it is adaptive and has evolved to suit its purpose. In the two examples above, there is strong evidence to link children's behaviour with child rearing and adult behaviour with protection.

One goal of human ethologists is to look for patterns of behaviour shared by all peoples of the world. These are seen in examples of behaviour that involve body movements, rather than speech. Gestures such as smiling and raising the eyebrows, and behaviours such as grooming and hugging seem to be common to all human societies.

We all think that we have a free choice in our behaviour, but do we really? We may be under the control of our genes or be shaped by our surroundings. There is evidence for both points of view.

SUMMARY

After completing this chapter you should know that:

- A **sensation** is a general state of awareness of a **stimulus**; **perception** involves the **interpretation** of **sensory information**.
- A **receptor** is a **biological transducer**: it converts energy from one type of system (eg chemical) to another system (eg electrical). A receptor is classified according to the type of stimulus it receives (eg photoreceptor), its location (eg interoceptor) or its level of complexity.
- The **Pacinian corpuscle** is a simple receptor in the skin that detects pressure and vibration.
- The eyeball contains three outer layers, the **sclera**, **choroid** and **retina**.
- Incoming light rays are **refracted** (bent) by the cornea and by the **lens** of the eye.
- **Accommodation** is the ability to focus the eye automatically, to see near and distant objects.
- **Rod** and **cone** cells are stimulated by the effects of light on two pigments, **rhodopsin** and **iodopsin**. Breakdown of these compounds alters the electrical properties of the cell membranes.

- There are three types of cone cell pigment. Each responds to a different range of wavelengths of light. Humans have **trichromatic** vision: our cone cells detect three colours, blue, green, red.
- **Stereoscopic vision** enables us to perceive depth and distance.
- The human ear has two main functions, to detect sound and to maintain balance.
- Ethology is the study of animal behaviour under natural conditions.
- Patterns of behaviour are controlled by the nervous system; simple behaviours include **taxes** and **kineses**.
- A **taxis** is an orientation response related to the direction of the stimulus. A positive taxis describes movement *towards* a stimulus; a negative taxis, movement *away* from a stimulus.
- A **kinesis** is an orientation behaviour where the response relates not to the direction of the stimulus, but to its intensity.
- **Instinct** or **innate behaviour** is unlearned. It has survival value. There are also various types of learned behaviour.

? EXAM QUESTIONS

1

a) A person, who has been watching a distant view from a window, looks down and starts to read. Describe how a clear image is formed on the retina of this person's eye, and explain the changes that take place in the eye in order to produce a clear image when the person starts to read.

b) Explain the roles of the eye and brain in enabling humans to distinguish colours in a picture.

AQA (NEAB) BY04 Biological Basis of Behaviour March 1999 Q8

2 The diagrams show sections through the front of an eye during distance vision and near vision.

distance vision   near vision

a) Describe the mechanisms which bring about the changes to the lens and pupil as the eye changes from distance vision to near vision.

b) The diameter of the pupil may also change when light intensity changes. This is caused by a reflex action which involves the brain rather than the spinal chord. Describe the route taken by the nerve impulses between the receptor and the effector in this reflex action.

AQA (NEAB) BY04 Biological Basis of Behaviour June 2000 Q7

3 The following structures are all found in the mammalian ear. Explain concisely the function of each of the structures.

a) Tympanic membrane

b) Maleus, incus and stapes

c) Basement membrane (basilar membrane) in the organ of Corti

d) Tectorial membrane in the organ of Corti

e) Auditory nerve

f) Maculae in the walls of the utricle and saccule

g) Semicircular canals

h) Sensory hair cells in the cochlea

OCR 6918 Animal Studies March 2000 Q3

4 Chitons are small animals that live on the seashore. When the tide is out they are found on the lower surfaces of stones. When stones are turned over the chitons move to the new lower surface as shown in the diagram.

chiton

a) Suggest two advantages to the chitons of this response.

b) Give two factors, other than light, to which the chitons might be responding.

c) A student investigated the response of chitons to light. Three covered dishes, X, Y and Z, were arranged as shown in the diagram. One half of the top of each dish was painted black, the other half was transparent. The dishes were placed on a table outside at noon on a bright cloudy day. Ten chitons were placed in the light half of each dish. The number of chitons in each half of the dishes was recorded every five minutes for the next hour.

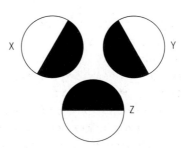

X   Y

Z

(i) Suggest why the dishes were arranged as shown in the diagram.

? EXAM QUESTIONS

(ii) The results of the investigation are shown in the table. What conclusions may be drawn from these results?

(iii) Explain how kinesis could account for the results shown in the table.

Time/ minutes	Number of chitons					
	Dish X		Dish Y		Dish Z	
	Light	Dark	Light	Dark	Light	Dark
0	10	0	10	0	10	0
5	9	1	7	3	8	2
10	6	4	7	3	7	3
15	5	5	7	3	6	4
20	3	7	5	5	5	5
25	3	7	4	6	5	5
30	3	7	3	7	5	5
35	3	7	3	7	5	5
40	2	8	3	7	4	6
45	1	9	3	7	4	6
50	1	9	2	8	4	6
55	1	9	3	7	2	8
60	1	9	1	9	1	9

AQA (NEAB) BY04 Biological Basis of Behaviour June 2000 Q1

5 The diagram shows apparatus used to investigate behaviour in rats. When a rat presses the lever a food pellet drops into the food tray.

The graph shows typical results from these investigations.

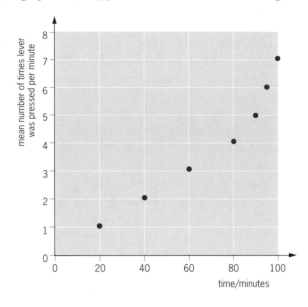

a) What type of learning is shown by the rat?

b) Explain why the rate of lever-pressings changes during the investigation.

c) Explain how similar principles may be used when training a dog.

AQA (NEAB) BY04 Biological Basis of Behaviour June 2000 Q3

6

a) Outline the process of learning by operant conditioning

b) Explain briefly the roles of rods and cones in perceiving visual stimuli.

IBO Biology Higher Level Paper 3 November 1998 QE1

7

a) **(i)** Describe one example of imprinting.
(ii) Give one biological advantage of the imprinting that you have described in part **(i)**.
(iii) Explain why imprinting is considered to be learned rather than innate behaviour.

b) Birds can be a danger to aircraft. One method used to keep them away from airfields is to play recordings of alarm calls which the birds make in response to predators. This works well for a time, but then becomes less effective.
(i) Explain why the alarm calls become less effective after a time.
(ii) Name the behavioural process involved.

AQA (NEAB) BY04 Biological Basis of Behaviour June 1999 Q2

20

Human reproduction

20 Human reproduction

Sometimes sex in humans does result in a new baby, but more often than not, it doesn't. Sex in humans is more than just making babies

Ask anybody what the purpose of sex is and, after a few funny looks, you'll probably get the answer 'to make babies'. Yet, although millions of acts of sexual intercourse take place across the globe every day, the vast majority of couples have no intention of making a baby. In practice, many of them use contraception to prevent pregnancy.

In the same way, if you ask what breasts are for, you might get the answer 'for breast-feeding'. However, when in close proximity to a topless woman, the average man probably does not spend much time thinking about how many children she could successfully breast-feed. All mammals produce milk, but humans are the only ones with permanently swollen hemispheres of fatty tissue.

Clearly, sex in humans is about more than just making babies. In fact, we seem rather obsessed with sex. Humans are one of the few species that have sex when there is no egg to be fertilised, and statistics tell us that the average human has sex several thousand times in their lifetime.

As you read through the account of human reproduction that follows, have a think about the function of sex in humans. If it is for more than basic reproduction, what is its purpose? Is it simply for pleasure? Or it is a form of 'social cement', strengthening the bond between people, encouraging them to fall in love and stay together? You decide.

1 HUMAN LIFE SPAN

As far as sex is concerned, humans are unique. We develop bodies capable of reproducing at around 12 or 13 years old – a good few years before most of us become independent from our parents. For most female mammals mating occurs only when there is an egg to be fertilised and most ovulations result in pregnancy. The female spends her life being pregnant or lactating, or both.

Fig 20.1 Different human cultures have different concepts of what makes a family. Here, four generations of one family have been photographed together; you can see the family resemblances and the effects of ageing quite easily

The ceaseless cycle of reproduction is brought to an end by death. Humans are quite different. The average woman in the UK ovulates about 400 times, but has only two children during her life. We take a detailed look at human reproduction in the rest of this chapter.

Old age is another part of life that is alien to most animals. Predators who get a bit slower catch fewer prey, get weaker and eventually starve. Ageing prey become easy targets for a predator having a lazy day. *Homo sapiens* is in a unique situation. When our physical capabilities start to fail, we find less demanding things to do. Top sports people, for example, might stop performing at the highest level in their mid-thirties but then they can progress into training, management or broadcasting. Eventually though, age catches up with us. We look at the human life span from birth onwards in Chapter 21, ending with a section on the effects of ageing.

As you read the next two chapters, you might like to reflect on the human animal; our attitudes to sex, a 20-year childhood, a retirement often longer than the working life and an ever-increasing wish to prolong its own life span. What a strange creature indeed.

2 THE NEED TO REPRODUCE

None of us is immortal and so, in common with all other types of organism, we must reproduce in order for our species to survive.

An interesting way of looking at the whole business of reproduction is to think of organisms as disposable containers for the genes they contain. None of us will live forever, but our genes, mixed with those from other people, might survive long after we are forgotten.

SEXUAL REPRODUCTION IN ANIMALS

Sexual reproduction involves the fusion of **gametes**, or sex cells. Female sex cells, called **egg cells** or **ova**, fuse with the male gametes, called **spermatozoa** or **sperm**, in a process known as **fertilisation**. The resulting cell, the **zygote**, develops into a new individual.

In sexual reproduction, the genetic material of two different individuals is mixed and combined to produce an individual that is genetically different from either parent. This produces variation within a population (see Chapter 25). All individuals, unless they have an identical twin, are **genetically unique**: different from all other individuals in the group.

REPRODUCTION IN MAMMALS

Mammals show a great range of reproductive strategies that reflect their circumstances, but the basic life cycle of most mammals follows a general pattern. After mating, the fertilised egg cell(s) develop inside the uterus, nourished by the placenta. The **gestation period** of a mammal, the time between fertilisation and birth, often reflects the metabolic rate of the organism. It is shorter in small short-lived mammals such as mice and longer in large long-lived mammals such as elephants.

Other factors complicate the length of gestation. For example, it is also limited by the size of the fetal skull (which must fit through the female's pelvis at birth) and the mobility of the mother. In species where the mother needs speed and agility to survive, the gestation period tends to be shorter. Birth is followed by a period when the mother gives close protection and suckles her young. The length of upbringing varies greatly: it depends on factors such as degree of maturity at birth and the amount of learning required to survive.

REMEMBER THIS

The need to reproduce, so passing genes on to the next generation to ensure the survival of the population, is one of the basic features of living organisms. In many organisms, the urge to reproduce can over-ride the urge to live.

REMEMBER THIS

Diploid cells contain two sets of chromosomes, haploid cells contain only one set. In animals, **somatic** (body) cells are diploid while gametes (sex cells) are haploid.

The onset of sexual maturity marks the change from **juvenile** to **adult**. Most female mammals conceive as soon as they become sexually mature and enter their first **oestrus** (season). From then on, throughout their reproductive life, they are either pregnant, feeding young or both. A female's reproductive life is usually brought to an end only by death. This has been the harsh fact for humans, too, until fairly recently. Even now, in societies where contraception is either unavailable or not used for cultural reasons, many women reproduce continuously.

THE ROLE OF HORMONES IN MAMMALIAN REPRODUCTION

For many animals the timing of reproduction is vital. If there is a severe variation in the seasons, animals must ensure that their offspring are born when food is plentiful and conditions are mild. **Hormones** that control and synchronise reproductive events are produced by three different organs: the **hypothalamus**, the **pituitary gland** and the **gonads** (sex organs).

The hypothalamus is a major point of contact between the nervous and **endocrine** (hormone) systems (see Chapter 18). Internal and external stimuli reach the hypothalamus, which responds by releasing hormones. They, in turn, control the rest of the endocrine system (Fig 20.2).

Day length is a classic example of an external stimulus. When the hours of daylight reach a threshold level, perhaps indicating the arrival of spring, the hypothalamus responds by releasing a hormone called a **gonadotrophin releasing factor** (**GnRF**). This controls the activity of the nearby pituitary gland, stimulating it to release hormones called **gonadotrophins**.

Gonadotrophins have a direct effect on the gonads (the ovaries and testes). These respond by releasing **steroid sex hormones** (Chapter 3) such as **oestrogen**, **progesterone** and **testosterone**.

REMEMBER THIS

The role of hormones in controlling the timing of reproduction in sheep is covered in detail in Chapter 6.

? TEST YOURSELF

A The hypothalamus controls the level of sex hormones via a negative feedback mechanism. Suggest how the hypothalamus would respond to a fall in oestrogen levels.

Fig 20.2 The hormonal control of reproduction. The hypothalamus can respond to external and internal nervous stimuli and is also sensitive to hormone levels in the blood. The hypothalamus controls the levels of several different circulating hormones using a negative feedback mechanism

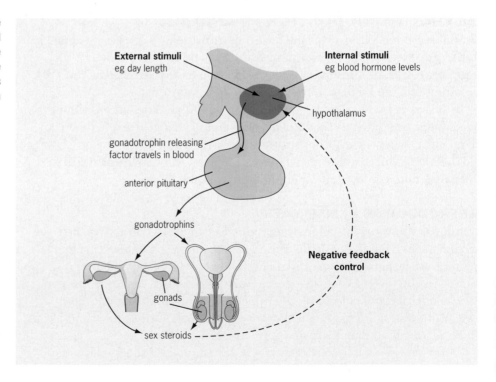

3 HUMAN SEXUAL REPRODUCTION

Sexual activity in humans has evolved to achieve more than just fertilisation. Although humans become sexually mature in their early to mid-teens, many societies have cultural and legal controls aimed at restricting sexual activity before the late teens. Although beyond the scope of this book, it is interesting to consider whether human sexual behaviour may have evolved to strengthen the emotional bond between a couple. This 'social cement' may make it more likely that couples will stay together and so provide a stable environment for raising children.

PUBERTY

Puberty is the time between childhood and adulthood: it marks the process of sexual maturing. We still don't know exactly what triggers the onset of puberty, but we know that, early on, GnRF is secreted from the hypothalamus (look back to Fig 20.2). GnRF travels the short distance to the anterior pituitary, where it stimulates the release of gonadotrophins, **follicle stimulating hormone** (**FSH**) in females and **interstitial cell stimulating hormone** (**ICSH**) in males (Fig 20.3). These two hormones are chemically identical, but have different names because they have different effects in the two sexes.

In girls, FSH targets the ovaries, which it stimulates to produce **oestrogen**. This steroid hormone is responsible for many of the female sex characteristics. In addition, oestrogen stimulates the ovaries to start producing egg cells (ovulation) and this leads to the first monthly period, or **menstruation**, an event known as the **menarche**.

To start with, periods tend to be irregular and unpredictable, and may sometimes occur without ovulation. Within about a year, hormone levels have increased to the point where they stimulate the regular development of follicles. This makes periods more regular and, unfortunately for many girls, painful. Period pain is mainly due to the hormone progesterone, which causes uterine cramps.

In boys, ICSH targets the **interstitial cells** of the testes. These are embedded in the connective tissue between the seminiferous tubules – see Fig 20.4(b). ICSH stimulates these testis cells to secrete **testosterone**, the steroid hormone that stimulates development of male sex characteristics.

THE AGE OF ONSET OF PUBERTY

The average age for the onset of puberty is 12 to 13 in girls, 13 to 15 in boys. Interestingly, this is much earlier than in previous centuries. Two hundred years ago the average age of the menarche in girls was 16, around four years later than it is now. The reason for this is almost certainly the improvement in diet. A better diet enables us to grow faster and so reach the same stage of maturity at an earlier age. In girls, the proportion of body fat appears to be important: a girl who is a dedicated athlete (a gymnast for example) and has a high muscle:fat ratio may find that the menarche is delayed. Also, girls who have started their periods but who then crash diet and lose a lot of weight may find their periods stop for a while.

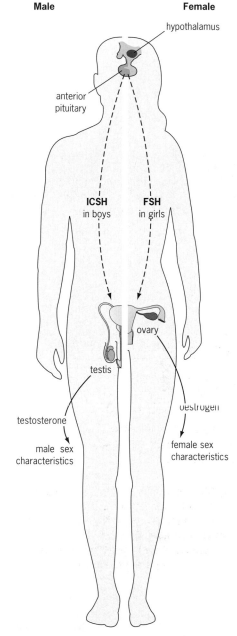

Fig 20.3 The onset of puberty is controlled by hormones.

Testosterone causes the penis, scrotum and testes to enlarge and mature. It also causes enlargement of the **larynx** (voice-box) which deepens the voice. Body hair is more extensive, with coarse hair appearing on the chest and face. Ironically, in later life, the same hormone leads to male-pattern baldness. A man's shoulders and chest tend to be large and there are often changes in facial structure, with the nose and chin becoming more prominent. The muscles tend to be better developed and generally the male has greater physical strength than the woman.

Oestrogen causes the ovaries, oviducts, uterus and vagina to mature and brings about the development of secondary sex characteristics. Breasts grow, pubic and underarm hair grows, the shape of the pelvis changes and the body fat is redistributed so that a woman tends to have wider hips and a more curved shape than a man

Fig 20.4(a) The main parts of the male reproductive system, seen from the front and the side

spermatogonia   spermatocytes   spermatids

lumen of tubule   Sertoli (nurse) cells   tails of sperm

Fig 20.4(b) The microscopic structure of the testis showing a section through a seminiferous tubule. The testis contains many seminiferous tubules. Inside each of these at any time, many thousands of sperm are maturing

THE MALE REPRODUCTIVE SYSTEM

The overall structure of the male reproductive system is shown in Fig 20.4. The male system secretes testosterone, makes and stores sperm and delivers them into the female's body.

Spermatozoa are made in the testes, a pair of organs that are held in a pouch of skin called the **scrotum**. It may seem odd that such delicate and vital organs are relatively unprotected outside the body, but there is a reason for this. The process of sperm production, **spermatogenesis**, is most efficient at around 35 °C, two degrees cooler than the core of the body. Men whose testes do not descend into the scrotum cannot produce healthy sperm.

THE PENIS

The penis introduces sperm into the female's body so that internal fertilisation can occur. Some animals do not have a penis – most birds and reptiles, for example – and their attempts at fertilisation are a little more haphazard. The male produces semen from his genital opening and simply rubs it on to the female's genital opening.

THE TESTES AND SPERMATOGENESIS

The production of sperm in human males is a continuous process, beginning at puberty and continuing well into old age: men in their nineties have fathered children. Spermatogenesis centres around the process of meiosis, and occurs at a remarkable rate: over a thousand human sperm are made every second. Look again at Fig 20.4(b) and you will see that each testis is composed of a series of **lobules**. They contain **seminiferous tubules**, the structures in which sperm production takes place. This process, shown in Fig 20.5, has three main phases:

● **Multiplication**. As large numbers of sperm are needed, cells of the **germinal epithelium** divide by mitosis to produce many **spermatogonia** (sometimes called sperm mother cells).
● **Growth**. The spermatogonia grow into **primary spermatocytes**. At this stage the cells are still diploid ($2n$).
● **Maturation**. The diploid primary spermatocytes undergo meiosis. After the first division they become **secondary spermatocytes** and when meiosis is complete they have become haploid **spermatids**. In the final part of the maturation process, spermatids differentiate into the familiar spermatozoa (sperm).

Throughout their development, sperm cells are closely associated with **Sertoli** or **nurse cells**, from which they obtain nutrients. In the lumen of the seminiferous tubule – see Fig 20.4(b) – the tails of the spermatozoa are clearly visible: their heads are attached to Sertoli cells.

Fig 20.5 The process of spermatogenesis. Sperm cells develop as they pass from the outer wall of the seminiferous tubule to the lumen

SEE QUESTIONS 1 AND 2

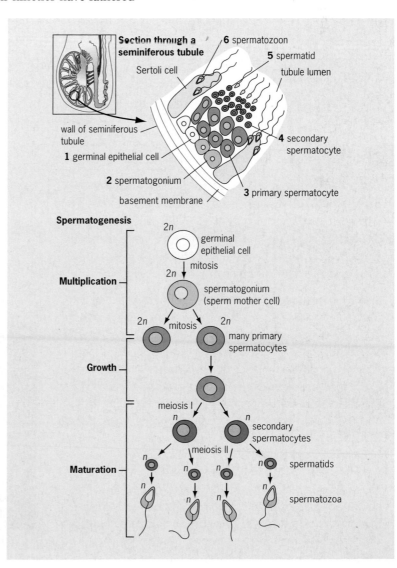

SEXUAL AROUSAL IN MALES

Men become sexually aroused by thinking about sex, as a result of physical stimulation or a combination of both. Nerve impulses from the brain pass down parasympathetic nerves (see Chapter 16) and cause arterioles leading to the penis to dilate. The penis receives more blood than can drain away, spongy **erectile tissue** in the shaft becomes filled with blood, and an **erection** results.

Flaccid (non-erect) penises vary greatly in size, largely depending on how much blood is retained in the spongy tissue. When erect, about 90 per cent are between 14 and 16 cm long. The end of the penis, the **glans**, is particularly sensitive, and continued stimulation from rhythmic thrusting eventually leads to a series of reflexes known as **ejaculation**. Stored spermatozoa are propelled along the **vas deferens** by powerful peristaltic waves. As they pass various accessory glands, different secretions are added to the sperm and the final ejaculate is a milky fluid called **semen** (Table 20.1).

Human males ejaculate, on average, about 5 cm^3 of semen, which contains between 50 to 200 million sperm. Most sperm never get anywhere near the egg, even after unprotected sex. For most of the monthly cycle, the cervix is blocked by a plug of mucus that sperm cannot penetrate. Only at around ovulation time does the mucus consistency change, allowing sperm to pass through easily.

Table 20.1 The constituents of semen

Gland	Secretion	Purpose
Seminal vesicles	fructose	energy for spermatozoa
	mucus	lubrication
	protein	forms clots, which alter consistency of semen
	prostaglandins	stimulate peristalsis in the female system
Prostate	alkaline chemicals	neutralise acid in vagina
	clotting agent	clots the protein from the prostate, forming a gelatinous mass
Cowper's gland	clear fluid	cleans urethra prior to ejaculation

---

! SCIENCE FEATURE Is the male pill in sight at last?

Until the 20th century, there were no reliable methods of contraception. Today, there are a variety of barrier and hormonal methods but women often point out that most of them are their responsibility. Only the condom really affects the male partner – the contraceptive pill, hormone injections and the cap are all used by the female. But things might soon change.

In 2001, researchers discovered a new ion channel that only occurs in the tail of sperm. It exists nowhere else in the body. Ion channels are specialised proteins that carry ions – usually sodium, potassium or calcium ions – from one side of a cell surface membrane to the other (see Chapter 4). CatSper, the ion channel in sperm, carries calcium ions and is necessary for the sperm to penetrate the outer coating of the ovum. If this channel could be blocked, the sperm would no longer be able to fertilise the egg. A small molecule that acted as a CatSper blocker would therefore be an ideal male contraceptive. It would be short acting, and so would need to be taken only when sex was likely to take place. Unlike the female contraceptive pill, it isn't hormonal, and so would not cause some of the side effects that have prevented male hormonal contraceptives from being developed. Researchers have tried

but these tend to cause lack of libido and impotence – for the perfect contraceptive that sort of misses the point.

Because CatSper is only present in sperm tails, blocking it would also be unlikely to cause side effects in other parts of the body. However, although biotech companies are currently working on such a male pill, several years of safety testing and clinical trials lie ahead.

Fig 20.6 The sperm is the target of a potential male contraceptive pill

THE FEMALE REPRODUCTIVE SYSTEM

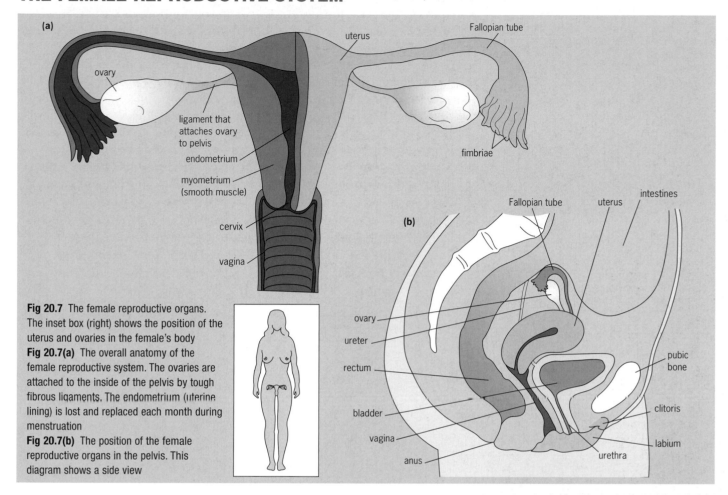

Fig 20.7 The female reproductive organs. The inset box (right) shows the position of the uterus and ovaries in the female's body

Fig 20.7(a) The overall anatomy of the female reproductive system. The ovaries are attached to the inside of the pelvis by tough fibrous ligaments. The endometrium (uterine lining) is lost and replaced each month during menstruation

Fig 20.7(b) The position of the female reproductive organs in the pelvis. This diagram shows a side view

Fig 20.7 shows the structure of the female reproductive system. This includes a pair of **ovaries**, which are the primary female sex organs. These produce egg cells, or **ova**, and also secrete the hormones **oestrogen** and **progesterone**. Each ovary is about 3 to 4 cm across and is attached to the inside of the pelvis by a **ligament** (a tough band of connective tissue). The oviducts, or **Fallopian tubes**, connect the ovaries to the **uterus** (womb). Each tube ends in finger-like **fimbriae**, which move close to the ovary at the time of ovulation.

The uterus is a compact muscular organ that nourishes, protects and ultimately expels the fetus. The human uterus is able to expand from about the size of a small orange with a capacity of about 10 cm³ to accommodate a full-term baby. This is a 500-fold increase in the capacity of the uterus. The bulk of the uterine wall is made from smooth muscle and is known as the **myometrium**. The lining of the uterus, the **endometrium**, consists of two layers. The underlying layer is a permanent **basement membrane** which produces the surface layer, the **decidua**. This layer is built up every month and shed during **menstruation**.

The **cervix**, or neck of the uterus, is a narrow muscular channel that is usually blocked by a plug of mucus. During sexual arousal the muscles of the **vagina** (a muscular tube that leads to the outside of the body) relax and glands in the vagina secrete lubricating mucus. This allows the male's penis to enter without discomfort. During childbirth, the cervix dilates to around 10 cm in diameter to allow the baby to pass through.

? TEST YOURSELF

C An ovum, the female gamete is much larger than its male counterpart, the sperm. How do the amounts of DNA in the nucleus of an ovum and sperm compare?

Fig 20.8(a) Section through an ovary showing the egg cell at different stages of development

Key labels:

8 Corpus luteum degenerates
7 Corpus luteum
ruptured follicle
1 Primary oocytes form
2 Oocyte develops within primary follicle
3 Follicle matures
4 Oocyte becomes a primary oocyte
5 Mature Graafian follicle — antrum
6 Egg is released as Graafian follicle ruptures

Fig 20.8(b) Scanning electron micrograph through a primary oocyte showing an egg surrounded by follicle cells

OOGENESIS

Throughout this section we refer to the female gamete as an **egg cell**. The egg cell is surrounded by several layers of cells and the complete unit is called a **follicle**.

The production of egg cells, **oogenesis**, takes place within the ovaries of the developing female fetus. At birth, a girl already has about 2 million **primary oocytes**. Most of these degenerate during childhood and by puberty there are only about 200 000 left. Of these, only about 450 ever mature fully – one per month throughout the female's reproductive life. As in spermatogenesis, the process of oogenesis (Figs 20.8 and 20.9) is divided into three phases:

● **Multiplication**. As the female embryo grows, **primordial germ cells** in the **epithelium** (outer layer) of the ovary go through a series of mitotic divisions to produce a population of larger cells called **oogonia**.

● **Growth**. Oogonia move towards the middle of the ovary where they grow and go through further mitotic divisions to become **primary oocytes**. Each oocyte is surrounded by a layer of follicle cells. Together they form a **primary follicle**.

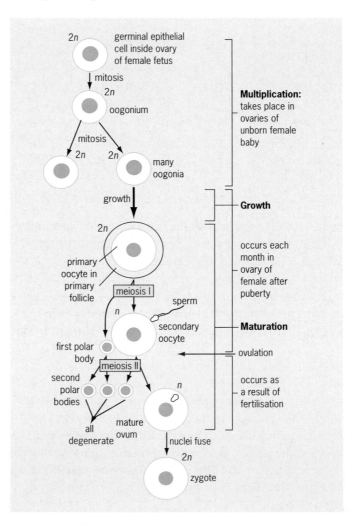

Fig 20.9 The process of oogenesis begins in the ovaries of a developing fetus but is not completed unless the egg is fertilised

● **Maturation**. From puberty onwards, a few primary follicles mature each month. Usually, only one completes its development, the rest degenerate. The remaining primary follicle grows larger, becoming an **ovarian follicle**. Its cells secrete **follicular fluid**, producing droplets which join together to form a fluid that fills the space known as the **antrum**. The mature follicle, the **Graafian follicle**, is almost 1 cm in diameter. It protrudes from the wall of the ovary just before ovulation.

Inside the developing follicle, the oocyte begins its first meiotic division. There is no need for more than one egg cell, so the second set of chromosomes formed at meiosis I is discarded, passing into a small cell (with very little cytoplasm) known as the **first polar body**. This appears to have no function, but it often completes the meiotic division, producing two similar cells; both later break down. After meiosis I, the egg cell is known as a **secondary oocyte**. It then begins the second meiotic division but gets no further than metaphase. The division is completed only if the egg cell is fertilised.

When fertilisation occurs, meiosis II is completed and the egg cell becomes the mature **ovum**. This produces another 'spare' set of chromosomes, the **second polar body**, a cell that also degenerates.

 REMEMBER THIS

Meiosis produces four daughter cells. In oogenesis, one ovum is formed; the other three cells degenerate.

SCIENCE FEATURE Contraception

The twentieth century has seen a revolution in 'family planning'. People can regulate the number and spacing of their children by using one or more of the wide range of effective methods of contraception now available (Table 20.2).

Table 20.2 The efficiency of the commoner methods of contraception

Method	Mode of action	Failure rate (pregnancies per 100 women/year)
The pill	prevents ovulation	0–3
IUD (coil)	prevents implantation	0.5–6
Condom	prevents sperm from reaching cervix	3–20
Diaphragm + spermicidal gel	prevents sperm from reaching cervix, gel kills sperm	3–25
Spermicidal gel alone	kills sperm	3–30
Coitus interruptus	penis withdrawn before ejaculation	10–40
The rhythm method	abstain from sex near to time of ovulation	15–35
Vaginal douche	washes sperm out of vagina	80
Male sterilisation (vasectomy)	vas deferens cut so that semen contains no spermatozoa	0–0.15
Female sterilisation (tubal ligation)	prevents egg from passing into uterus	0–0.05
Nothing	no contraception at all – a control for the other methods	85

'The pill'

The pill – or, to use its full name, the combined oral contraceptive pill – has been used widely since the 1960s. It contains a mixture of artificially produced oestrogen and progesterone. The pill is effective because it raises the blood hormone levels to those encountered during the secretion phase of the menstrual cycle, or those of early pregnancy. The high level of progesterone inhibits the secretion of FSH, preventing the development of new follicles and therefore egg cells. The high progesterone levels also inhibit menstruation, so the pill is usually taken for 21 days and then not taken for 7 days. This allows menstruation to take place but does not leave enough time for any follicles to develop. Strictly speaking, this is not normal menstruation but a 'withdrawal bleed' caused by the drop in progesterone.

The 'mini-pill'

This variation of the pill contains low doses of hormones, not enough to prevent ovulation, but enough to change the consistency of the mucus at the cervix. Instead of becoming permeable to sperm at ovulation, the mini-pill causes continued secretion of the thick, non-ovulatory mucus, which prevents sperm getting through.

Long-term injections

Injections that give contraception for 3 months contain progestogen, an artificial form of progesterone. Injections are intra-muscular – usually in the buttock – after which the hormone is slowly released into the bloodstream. Progestogen acts as a contraceptive by inhibiting FSH, although the cervical mucus is also thickened.

Fig 20.10(a) How levels in luteinising hormone and follicle simulating hormone change during the menstrual cycle

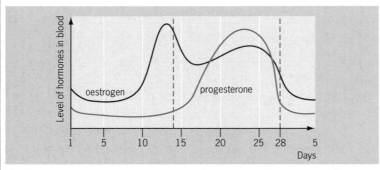

Fig 20.10(b) How oestrogen and progesterone levels change during menstruation

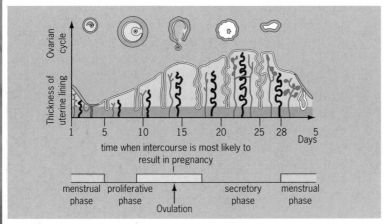

Fig 20.10(c) Stages of follicle development and changes in the thickness of the endometrium. Note that pregnancy is more likely to occur after unprotected sex between days 9 and 17, in a cycle where ovulation occurs on day 14

Fig 20.10(d) A woman's body temperature rises just after ovulation. In couples finding it difficult to conceive, taking daily temperature measurements is a good way to pinpoint ovulation

OVULATION

On around day 14 of the human female's menstrual cycle, an egg cell (a secondary oocyte) is released from the ovary in a process called **ovulation**. Pressure in the antrum builds up and ruptures the Graafian follicle, forcing the egg cell out. The egg cell is released into the body cavity, but very few get lost since the fimbriae of the Fallopian tubes hover close to the ovaries, and the ciliated lining of the tubes creates a current that gently sucks in the egg cells.

Remarkably, a woman can become pregnant even if she has lost an ovary on one side and a Fallopian tube on the other. This suggests that the functioning fimbria actively seeks out the productive ovary by moving right across to the other side of the woman's body.

THE HUMAN FEMALE MENSTRUAL CYCLE

The average length of the human menstrual cycle is 28 days, but variations from 24 to 35 days are normal, and greater variations are not uncommon. The cycle is divided into four phases, Fig 20.10(c):

- The **proliferative phase** in which the endometrium regenerates.
- The **ovulation phase** in which the ovum (egg cell) is released.
- The **secretory phase** in which the endometrium secretes nutrients in preparation for implantation.
- The **menstrual phase** in which the endometrium is shed from the body.

By convention, the first day of the period (the most obvious event) is called day 1 of the menstrual cycle. Fig 20.10 shows the timing of the phases listed above. It is important to relate the events of the menstrual cycle to the hormonal changes shown. You will need to refer to these diagrams as you read the following text.

THE PROLIFERATIVE PHASE

On day 2 of the cycle, the pituitary gland releases **follicle stimulating hormone (FSH)**. This stimulates the development of several ovarian follicles. At around day 6, one of the follicles dominates and begins to secrete oestrogen. The others degenerate. The remaining follicle develops into a **Graafian follicle** and continues to secrete oestrogen until day 14 of the cycle. Consequently, blood oestrogen levels rise. Oestrogen causes **proliferation** (growth) of the endometrium to replace the layer lost during the previous menstruation. After 14 days the repair is complete.

THE OVULATION PHASE

On day 12 to 13, the blood oestrogen level reaches a threshold level, which triggers the release of **luteinising hormone** (**LH**), from the anterior pituitary gland. The rapid increase in LH levels triggers ovulation around day 14. The ovum lives for only 24 to 36 hours. During this time it moves only a few centimetres from the ovary.

SEE QUESTION 3

THE SECRETORY PHASE

Luteinising hormone has a second effect: it causes the Graafian follicle to develop into a **corpus luteum** ('yellow body'). The name comes from the yellow appearance of the secretory cells that develop inside the 'remains' of the Graafian follicle. The corpus luteum secretes oestrogen and progesterone. Progesterone causes spiral-shaped blood vessels to grow into the endometrium. This thickened lining begins to secrete nutrients and mucus to prepare for an embryo to be implanted. During this phase the high levels of progesterone inhibit the production of FSH. As long as progesterone levels are high, the endometrium is maintained and no new follicles are stimulated. The 'contraceptive pill' takes advantage of this inhibition (see Science feature on page 349).

If the egg cell is not fertilised, the corpus luteum lasts for about 10 to 12 days and then degenerates, ceasing to secrete progesterone. This is a key event because the inhibition of FSH is lifted. The endometrium is no longer protected and the cycle can start again.

THE MENSTRUAL PHASE

The drop in progesterone and oestrogen levels causes the uterine capillaries to rupture, and the endometrium is lost from the body through the cervix, together with some blood. Renewed secretion of FSH begins around day 2, and the cycle begins again.

? TEST YOURSELF

D Many women do not have a 28 day cycle. Although the second half of the cycle (ovulation to menstruation) is usually 14 days, the first half can vary considerably. If a woman's cycle is 35 days, on which day is she likely to ovulate?

! SCIENCE FEATURE The menopause and hormone replacement therapy

The menopause is a natural event that occurs when the ovaries stop working. For most women, this happens between the ages of 45 and 54. This time is often difficult and traumatic. The lowered levels of oestrogen and progesterone are directly responsible for the unpleasant symptoms of the menopause:

● Circulatory problems such as hot flushes and night sweats.
● Psychological problems, such as depression, anxiety and insomnia.
● Skeletal problems. Oestrogen inhibits reabsorption of bone, and after the menopause, bone loss can be as great as 7 per cent per year. This condition, osteoporosis, affects the spongy bone particularly and the sufferer is more likely to break a bone.
● Oestrogen is also thought to give women some protection against some types of heart disease. Women below menopausal age are less likely than men to have heart disease, but afterwards they catch up.

Hormone replacement therapy can help to reverse many of these symptoms and effects. The basic idea behind HRT is simple: to restore hormone levels to those of the early follicular phase of the menstrual cycle using tablets, implants or transdermal patches (Fig 20.11).

Fig 20.11 One method of administering the hormones in HRT: a transdermal patch (a patch attached to the skin) that releases controlled amounts into the bloodstream. The hormones used in HRT are extracted from the urine of pregnant mares or made artificially

THE EVENTS OF PREGNANCY

If the egg cell is fertilised, the menstrual cycle is interrupted and the female's body changes in response to the events of pregnancy.

FERTILISATION

Fertilisation is a complex sequence of events that begins when the sperm reaches the egg cell (oocyte). The events of fertilisation are shown in Fig 20.12.

It may also help you to look back at Fig 20.9. The secondary oocyte is surrounded by several layers of follicle cells, the **corona radiata**, and a layer of glycoprotein, the **zona pellucida**. Before a sperm can penetrate the layers surrounding the oocyte, it undergoes a process called **capacitation**. The actual mechanism is poorly understood, but it seems to set off the **acrosome reaction**, which allows the sperm head to enter the oocyte. The **acrosome** is a bag of digestive enzymes on the tip of the sperm head (see Fig 20.12). During the acrosome reaction the bag splits, releasing the enzymes, which digest a pathway through any remaining follicle cells and the zona pellucida.

As soon as the outer membrane of the first sperm penetrates the cell surface membrane of the oocyte, a rapid reaction occurs. Many **cortical granules** fuse with the zona pellucida, forming a **fertilisation membrane**. This reaction starts at the point of entry of sperm head and spreads rapidly over the surface of the oocyte, preventing entry of other sperm. So, only one sperm enters the diploid secondary oocyte, even though many reach it at the same time. Entry of the sperm nucleus triggers the completion of meiosis II in the female nucleus, leading to the formation of the second polar body and a haploid mature ovum.

Almost immediately afterwards, a spindle forms and the paternal and maternal chromosomes come together, forming a diploid zygote. Within 12 hours, the first mitotic division takes place. Cell division is now rapid, forming a bundle of cells called a **morula** (Latin for the blackberry it resembles). As divisions continue, this becomes a **blastocyst** and moves slowly along the Fallopian tube through the action of cilia, which create a steady current of fluid towards the uterus (see Fig 20.13).

It takes around 6 or 7 days for the embryo to complete the journey down the Fallopian tube. When it arrives at the uterus on day 21 of the cycle, the lining must be in just the right condition to accept it. For a successful pregnancy there must be exact timing between the preparation of the endometrium and the development of the embryo.

IMPLANTATION

Pregnancy begins not with fertilisation but when the embryo **implants** in the wall of the uterus. This happens about a week after fertilisation and is not always successfully completed. Many women trying to conceive may have a 'near miss', when an egg is fertilised but fails to implant.

Implantation begins when the blastocyst makes contact with the endometrium (Fig 20.13), usually on the back wall. The outer layer of the blastocyst, the **trophoblast**, causes an **inflammatory-type response** (normally a response to damage) and this causes an outgrowth of the endometrium at this point. The placenta develops where the trophoblast and the endometrium interact.

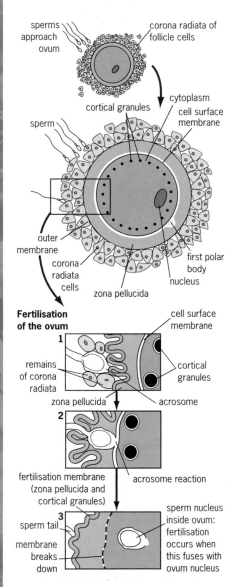

Fig 20.12 With a diameter of 100 μm, the egg cell is the largest cell in the human body. Most of the volume is taken up by inert food material that will fuel the embryo during its first few cell divisions. A spermatozoon must pass through any remaining follicle cells (the corona radiata) and the zona pellucida, before it can fertilise the egg cell. The head of the sperm enters the body of the egg cell, but the tail remains outside

If implantation is successful, the embryo begins to secrete **human chorionic gonadotrophin (hCG)**. This hormone forces the corpus luteum in the ovary to continue to secrete progesterone, thereby maintaining the endometrium and inhibiting FSH production.

The **chorion**, one of the membranes that later grows and surrounds the embryo, develops **villi** (projections) that burrow into the endometrium. These are thought to break down the mother's blood vessels, causing the chorionic villi to become bathed in maternal blood.

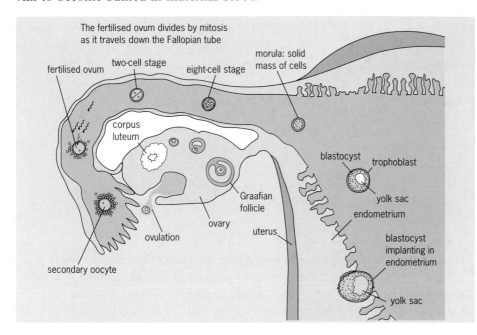

Fig 20.13 Following fertilisation, the embryo undergoes several mitotic divisions as it passes slowly down the Fallopian tube. When the embryo reaches the uterus, it implants in the endometrium

THE PLACENTA

The placenta is a temporary organ that allows the blood systems of the fetus and the mother to come into close contact, without actually mixing (Fig 20.14). The placenta allows nutrients and oxygen to pass to the fetus from the mother, and allows metabolic waste back into the mother's blood (see Table 20.3).

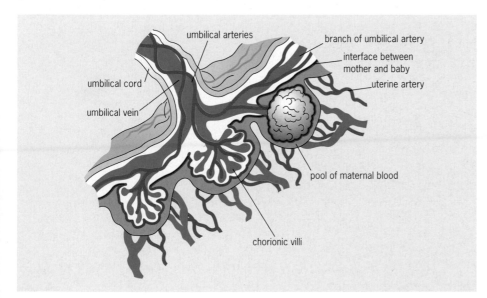

Fig 20.14 The fine structure of the placenta. The capillaries within the chorionic villi are bathed in pools (lacunae) of maternal blood. The placenta is a large disc-shaped organ that is usually attached to the back wall of the uterus. At birth, contraction of the uterine muscle causes the chorionic villi to split away from the endometrium and the placenta is delivered after the baby: hence its common name of afterbirth. Most mammals eat the placenta after the birth of their young. It is an important source of nourishment at a time of great need

The placenta is an organ adapted to maximise the exchange of materials so, as you would expect, it has a large surface area provided by the **chorionic villi**. A close look at the cells of the villi shows that the membranes are folded into microvilli and also contain many mitochondria: these two features maximise the processes of diffusion and active transport (see Chapter 4). There are many small vesicles in the cells of the villi, suggesting that substances are being absorbed by pinocytosis (see also Chapter 4).

Table 20.3 Exchange of materials across the placenta	
Mother to fetus	**Fetus to mother**
oxygen	carbon dioxide
glucose	urea
amino acids	other waste products
lipids, fatty acids and glycerol	
vitamins	
ions; Na, Cl, K, Ca, Fe	
alcohol, nicotine, many drugs	
viruses	
antibodies	

SEE QUESTIONS 4 TO 6

The placenta is an important **endocrine organ**. It secretes the hormones that maintain pregnancy, taking over from the corpus luteum at about 12 weeks. The placenta secretes progesterone (which maintains the endometrium), oestrogen (which inhibits the ovulatory cycle), human chorionic gonadotrophin (see the Science feature in Chapter 6 page 122) and **human placental lactogen** (which stimulates the development of the mammary glands).

PREGNANCY, LABOUR AND BIRTH

In humans, pregnancy lasts for an average of 40 weeks. Some of the main stages of growth of a baby are shown in Fig 20.15.

Fig 20.15 Stages in the growth of a baby in the uterus

24 days old
length 3 mm

6 weeks old
length 1.3 cm

12 weeks old
length 10 cm

24 weeks old
length 33 cm

28 weeks old
length 37 cm

32 weeks old
length 43 cm

36 weeks old
length 47 cm

40 weeks old
length 50 cm

From around the twelfth week of pregnancy, progesterone secreted by the placenta inhibits uterine contractions. The level of progesterone rises steadily until just before birth, usually around 38 weeks after conception, when it starts to fall dramatically. This lifts the inhibition of uterine contractions. The mother's anterior pituitary begins to secrete **oxytocin** and the placenta secretes prostaglandins, two hormones that actively promote contractions. Oxytocin stimulates uterine contractions at about 40 weeks. The resulting tension in the muscle and pressure on the cervix are stimuli that bring about further secretion of oxytocin, causing more powerful contractions.

Fig 20.16 Childbirth is one of the most intense human experiences. Childbirth used to be a time of great danger to both mother and child, but advances in medicine have greatly reduced the risk. Today, in the developed world, 99 out of every 100 babies born survive beyond their first birthday (200 years ago, only 54 out of every 100 lived)

There are three stages of labour:
- Stage 1. The cervix dilates (opens) to a diameter of 10 cm.
- Stage 2. The fetus is pushed out of the uterus.
- Stage 3. The placenta and umbilical cord are expelled.

LACTATION

All mammals produce milk from specialised **mammary glands**. Humans are unique in having permanent breasts but these are normally composed only of fatty tissue. Throughout pregnancy, however, oestrogen and progesterone stimulate the development of milk-producing tissue.

After birth, the first fluid released from the breasts is called **colostrum**. This contains no fat and very little sugar, but has important antibodies that 'lend' the baby immunity until it has time to develop its own. After about three to four days, normal milk is produced.

Milk is produced constantly and, although a certain amount of leaking from time to time is usual, milk flow only happens when the baby suckles (Fig 20.17). The stimulus of sucking at the nipple causes the posterior pituitary gland to secrete oxytocin. This hormone travels in the blood and causes contraction of the muscular **myo-epithelial cells** surrounding the milk glands or alveoli. This squeezes the milk out of the alveoli, through the milk ducts and into the infant's mouth. This mechanism, called the **let-down reflex**, takes several seconds to take effect but is quite powerful. If the unsuspecting infant lets go of the nipple, it can get a jet of milk in the eye.

Throughout the period of lactation the pituitary continues to secrete **prolactin**, a hormone that maintains the milk ducts and, to some extent, inhibits ovulation. This inhibition is called **lactational anoestrus** and reduces the risk of conception so soon after birth. However, it is not always effective and breast-feeding cannot be relied on as a contraceptive.

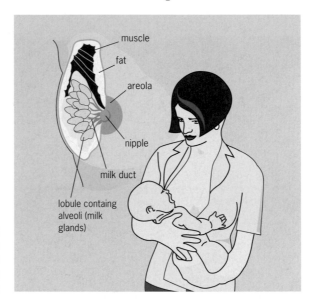

Fig 20.17 During pregnancy the milk-producing ducts grow, substantially increasing the size of the breasts. When the new baby suckles, the physical stimulation leads to a series of hormonal events that increase the amount of milk that is made

SUMMARY

After studying this chapter, you should know and understand the following:

■ **Sexual reproduction** involves two sexes which produce **haploid gametes**. Males produce **spermatozoa** by **spermatogenesis**, females make **ova** (egg cells) by **oogenesis**. Both processes involve a special type of cell division called **meiosis**.

■ At fertilisation, a new **diploid zygote** is formed. The zygote, which is genetically unique, grows and develops by a series of mitotic divisions.

■ The timing of reproductive events in humans is controlled by hormones. The **hypothalamus** controls the **pituitary gland**, which releases **gonadotrophins**. These stimulate the gonads to secrete steroids: **oestrogen**, **progesterone** and **testosterone**.

■ The human menstrual cycle lasts, on average, for 28 days. The cycle begins with **menstruation**, which occurs as the next ovum (egg cell) is

prepared inside the ovary. By day 14 a new endometrium has grown. Ovulation takes place on day 14 and the egg cell is then available to be fertilised. If the egg cell is not fertilised the endometrium breaks down and the cycle repeats.

■ In humans, the egg cell is fertilised in the Fallopian tube. The **zygote** begins mitotic divisions, becoming a **morula** and then a **blastocyst** as it moves along the tube towards the uterus. When the blastocyst has implanted in the endometrium, it is known as an **embryo**.

■ The developing embryo (which is called a **fetus** after about 8 weeks) is nourished via the placenta, a temporary organ that does the job of the fetal lungs, intestines and excretory system.

■ Birth is brought about by a series of hormonal changes that begin uterine contractions. In mammals, the infant is nourished by milk produced by the mother.

? EXAM QUESTIONS

1
a) Describe the processes involved in spermatogenesis.
b) Explain the hormonal control of puberty in boys.
c) Discuss the ethical issues of contraception.
IBO Biology Higher level Paper 2 May 1999 Q7

2 The diagram below shows some cells from a transverse section through a seminiferous tubule from a human testis.

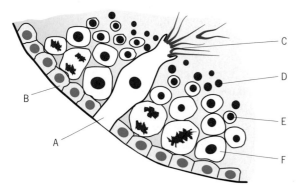

a) Give the functions of the cells labelled A and B.
b) **(i)** Give the letters of two cells which have the haploid number of chromosomes.
(ii) Give the letters of two cells which have the diploid number of chromosomes.
c) Give two ways in which the process of oogenesis in humans differs from that of spermatogenesis.
Edexcel HB6 (Synoptic) June 1999 Q3

3
a) Copy and complete the table which gives information about some of the hormones which control the oestrous cycle.

Hormone	Site of secretion	Target organ	One effect
Luteinising hormone (LH)			
Oestrogen	ovary	uterus	stimulates growth of uterine lining
Progesterone	ovary	uterus	

b) The diagram shows the way in which hormones are involved in controlling part of the oestrous cycle.
(i) Some women only produce very small amounts of FSH. Explain why these women are infertile.
(ii) Clomiphene is a drug used to treat this type of infertility. It blocks the action of oestrogen. Explain how treatment with clomiphene could be used to stimulate production of FSH.

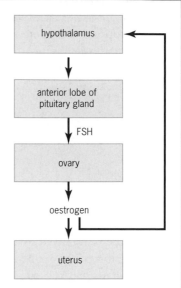

AQA (NEAB) BY01 Processes of Life June 2000 Q6

4 Give an account of the structure and functions of the placenta. [essay question]
Edexcel HB3 June 1999 Q8

5 The diagram shows a simplified diagram of a placenta and fetus.

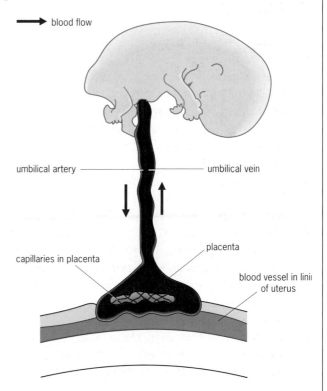

? EXAM QUESTIONS

a) Using the words, higher, lower, same, as appropriate complete the following table to compare the composition of the blood in the umbilical vein with that in the umbilical artery.

Content in umbilical vein	Compared to umbilical artery
oxygen	
urea	
glucose	
carbon dioxide	
immunoglobulins	
oestrogen	

b) On the diagram draw an arrow to show the direction of the blood flow in vessels in the lining of the uterus.

c) State four ways by which the placenta is adapted to perform its functions.

d) Suggest why the cells of the chorion, which develop into the placenta, are genetically identical to those in the embryo.

e) State how the following cells and structures are provided with nutrients.

(i) Spermatozoa in the female reproductive tract.

(ii) The developing ball of cells in the oviduct.

OCR 6918 Animal Studies March 2001 Q6

6 For each statement, write T if you consider that it is correct, and F if you consider that the statement is incorrect.

a) Both testosterone and spermatozoa are produced in the testes of mammals.

b) Fertilisation is the term used to describe the introduction of spermatozoa into the reproductive tract of the female.

c) Semen is produced by the testes.

d) Oogonia and spermatogonia are produced by mitosis.

e) Polar bodies are produced during oogenesis and during spermatogenesis.

f) Oestrogen stimulates the production of FSH.

g) LH stimulates the production of oestrogen.

h) In males FSH may be referred to as ICSH.

i) Fertilisation normally takes place in the oviducts.

j) Oxytocin is involved both in birth and in lactation.

k) A fall in oestrogen and progesterone levels in the blood causes the lining of the uterus to break down.

l) In humans the oestrous cycle is referred to as the menstrual cycle.

OCR 6918 Animal Studies March 2000 Q1

KEY SKILLS **ASSIGNMENT**

INFERTILITY TREATMENT

Nine out of ten couples who use no contraception and actively try to conceive, are successful within a year. But, one in ten couples may face the distressing problem of infertility. Many are 'sub-fertile' rather than infertile, and can be helped by the various methods described in this Assignment. However, for a few, nothing works.

In this exercise we look at the possible causes of infertility and at some of the treatments. When you have read both and worked through the questions, you should be in a position to prescribe the best course of treatment for a particular couple.

Causes of infertility

Doctors accept that a couple may need infertility treatment if they have been trying to conceive for at least a year without any success. The first objective is to establish whether either partner has an obvious problem that is preventing conception.

Female infertility may be caused by:

● Blocked Fallopian tubes.
● Altered hormone levels leading to a failure to ovulate or implant. Failure to ovulate, known as **anovulatory infertility**, is usually due to a failure to secrete the right balance of hormones.
● Cervical mucus that halts, repels or kills sperm.

1 **a)** How would a specialist find out if a woman's Fallopian tubes were blocked?
b) Why might blocked Fallopian tubes cause infertility?
c) Which hormones combine to cause ovulation?
d) What should happen to the cervical mucus around the time of ovulation?

Male infertility may be caused by:

● A low sperm count. Samples that are found to have fewer than 20 million sperm per cm^3 are said to be abnormally low.
● Production of large numbers of abnormal sperm (more than 4 per cent).
● Production of antibodies that make the sperm stick together.

2 Men who wear tight-fitting clothes, or who spend a lot of time in a hot bath, have sometimes been found to have a low sperm count. Suggest a reason for this.

Treatment methods: ovulation induction

A woman may not ovulate because the balance of hormones in her body is abnormal. To restore the hormone levels and initiate follicle development and ovulation, she can have treatment with artificial gonadotrophins or drugs that stimulate the natural secretion of gonadotrophins. One such drug, clomiphene, works by increasing FSH secretion.

After such treatment, the response of the ovaries can be followed by ultrasound, which shows how many follicles are developing in each ovary. A follicle is considered to be ready for ovulation when it reaches 17 mm in diameter. The endometrium is also checked – it should be at least 8 mm thick at this time.

When a ripe follicle is detected, ovulation can be stimulated artificially by injecting human chorionic gonadotrophin (HCG), a hormone that has a similar effect to LH. Ovulation should occur after about 36 to 48 hours, and so intercourse should be timed to coincide with this.

3 **a)** Why will increased FSH secretion help a woman who isn't ovulating regularly?
b) What is a potential problem with treatment that stimulates ovulation? Hint: this may not be a problem if the couple want a large family.

Treatment methods: *in vitro* fertilisation (IVF)

Fig 20.A1 A woman undergoing IVF treatment must endure weeks of hormone treatments, followed by the uncomfortable process of egg cell collection. Up to about 20 egg cells may be recovered. Her egg cells are fertilised by her partner's sperm '*in vitro*' and, when the tiny embryos have developed, two of these are put back into her uterus. IVF has, at best, about a 30 per cent success rate

In vitro (= 'in glass') fertilisation is what most people think of as 'test-tube baby' treatment. In IVF, the egg cells and sperm are taken from the couple and fertilised in a dish. The process of fertilisation normally takes 12 to 15 hours and after this the new embryos start to develop. Cell division is taken as a sign that fertilisation has been successful. Then, the tiny embryos (no more than balls of 8 to 16 cells at this stage – see Fig 20.A2) are placed into the uterus.

KEY SKILLS **ASSIGNMENT**

Fig 20.A2 A human embryo at the 8-cell stage. The eight large cells (red) are covered with microvilli (yellow). The smaller cells will degenerate

4 How often does an ovary normally produce an ovum?

Treatment methods: gamete intra-Fallopian transfer (GIFT)

GIFT involves stimulating the ovaries and collecting the ova in much the same way as in IVF treatment. The important difference is that in GIFT the egg cells are mixed with sperm and immediately introduced into the Fallopian tubes (Fig 20.A3), without waiting to see if fertilisation occurs. The advantage of this procedure – which would seem to be less controlled than IVF – is a 5 per cent better success rate. This is possibly because the Fallopian tube is the natural site for fertilisation: it may secrete chemicals that stimulate the process.

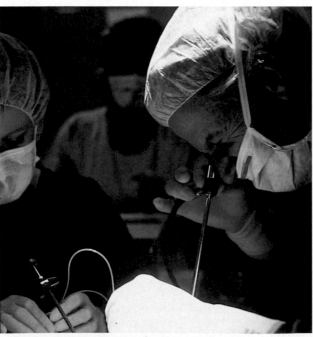

Fig 20.A3 A mixture of egg cells and sperm is being inserted into the woman's Fallopian tube through a small insertion

5 Many specialists feel that counselling couples is an important part of IVF and GIFT treatment. Think of two reasons why these treatments might cause anxiety.

Treatment methods: intra-uterine insemination (IUI)

In around 20 per cent of infertility cases, there seems to be no problem with either partner. Surprisingly, in some of these cases, intra-uterine insemination proves successful. The basic idea behind IUI is to introduce the partner's semen into the uterus.

The steps in IUI are as follows:
- Follicle development is stimulated and monitored.
- Treatment to induce ovulation is given.
- A fresh sperm sample is introduced into the uterus.

6 What advantage does IUI give over trying to conceive naturally?

After working through the Assignment so far, you should now be able to tackle the following question:

7 Copy and complete the following table, matching up the causes of infertility to suitable treatments. Remember that there may be more than one suitable treatment.

Your options are:

OI – ovulation induction

IVF – *in vitro* fertilisation

GIFT – gamete intra-Fallopian transfer

IUI – intra-uterine insemination

Problem	Possible treatments
Blocked Fallopian tubes	
Hostile cervical mucus	
Low sperm count	
Woman has antibodies against partner's sperm	
Many abnormal sperm	

8 A programme of infertility treatment can put a great strain on a relationship. Suggest reasons for this.

9 Hormone replacement therapy can be used to allow post-menopausal women to become pregnant. A donated egg can be fertilised by sperm from the woman's partner, and can lead to women as old as 60 bearing children. Discuss the advantages and disadvantages of this technology.

21

Growth and ageing

'Life is a sexually transmitted fatal condition.' As the proportion of elderly people increases in many populations in the western world, there are serious implications for our societies. Older people suffer more from degenerative diseases such as Alzheimer's and will need to be cared for, stretching the resources of health care services

In the time it takes to read this sentence, fifty million of your cells will have died. This is an old biological 'fact', but serves as a stark reminder of our mortality. The brain, like all the other organs in the body, degenerates as we get older. In most people this results in a gradual loss of memory. Many older people struggle with their short-term memory – they forget everyday things – but have very clear and lucid memories of what happened in their childhood.

Alzheimer's disease accelerates memory loss and causes a decline in many other mental abilities. Eventually, people with Alzheimer's might not remember who they are or who the people around them are, and this can lead them to behave aggressively or even violently. Few things in life can be as distressing as watching a close relative decline in this way.

The causes of Alzheimer's disease are not yet properly understood. One suggestion that might account for the symptoms is the accumulation of aluminium in the brain tissue. Post-mortem examinations show that the brains of those who have died from Alzheimer's contain significantly more aluminium than normal. Also, kidney patients who can't excrete aluminium and therefore build up higher concentrations of aluminium ions in their blood are more likely to suffer from Alzheimer's disease.

However, this evidence is circumstantial and it is not possible to say whether there is a clear causal link between aluminium and Alzheimer's disease. Aluminium deposits might just as easily be an effect caused by some other underlying problem. Further research is in progress and the evidence will need to be analysed carefully to find the answer.

1 GETTING BIGGER AND GETTING OLDER

In this chapter we look at how the body grows from birth until the late teens, and then we study some of the events that occur as the body starts to decline in old age.

WHAT IS GROWTH?

We all know what growth is, but we need a biological definition:

Growth is an increase in size brought about by the addition of more body tissue.

Growth therefore excludes temporary changes. Women who suffer from premenstrual tension can retain water in their tissues and appear to 'grow' by gaining a couple of kilograms in the few days before their period but this is not growth as such.

Human growth is **diffuse**: it occurs all over the body, not just at one growing point. A 10-year-old is not only taller than a 3-year-old but his or her organs – heart, liver, pancreas, etc. – are all larger. Not all of these organs grow at the same rate, though, and in the first part of this chapter we look at **growth patterns**. It is important to understand how internal factors such as hormones and external influences such as nutrition work together to control growth.

Fig 21.1 In the wild, mammals like the Grant's gazelle rarely reach old age. Their usual fate is to be killed and eaten by carnivores or to die of disease long before this. Modern medicine, however, has enabled humans to grow old and 'wear out'. Some of the same changes that we experience as we age are seen in the animals we protect, such as our pets

AGEING IN HUMANS

Many animals die long before they reach their full lifespan (Fig 21.1). Most are lucky to live to reproduce. A longer life, however, has its disadvantages. Organs degenerate and systems become less efficient. In the second part of this chapter we look at some of the unwelcome changes that affect the body as it gets older.

2 MEASURING GROWTH

Growth is three dimensional; people get taller, wider and the distance between their back and front increases. To give an accurate picture of overall growth, three basic criteria are commonly used:

● An increase in **body mass**.
● An increase in **height**.
● An increase in **supine** length (the distance between the top of your head and the bottom of your feet when lying flat on the floor).

BODY MASS

Body mass is quick and easy to measure. Body mass changes slightly, even from day to day and hour to hour – we weigh more when we have just eaten a large meal or when we have a full bladder, for example – but these fluctuations are minor. Larger gains in mass occur during childhood, and we use standardised charts so that children's sizes can be plotted to make sure that any potential problems in growth are spotted early (Fig 21.2).

Women gain weight as they go through pregnancy and, as people get older, they can gain weight by eating too much food and doing too little exercise. Homeostatic mechanisms try to keep our weight stable (see Chapter 11), but we can over-ride this control – all too easily, judging by the epidemic of obesity currently sweeping the developed countries (see the Science feature on page 366).

Fig 21.2 We monitor the way babies grow by weighing them regularly

HEIGHT

Like body mass, height is easy to measure. The person stands straight with heels flat on the ground and facing forwards. A horizontal bar is moved down until it touches the top of the head and the height is read off from a suitably positioned scale (Fig 21.3). Height also gives a good indication of growth over time (Fig 21.4).

Using height as the only measure of body size has its disadvantages. It cannot, for example, give information about the changes in the width of the shoulders or hips, changes that are just as important as increases in height. And it's no good for assessing the growth of babies who are too young to stand.

SUPINE LENGTH

You can only measure a person's height if they are old enough to stand. In very young children, length is measured instead. The infant is placed as flat on its back as possible. The ankles are gently pulled to stretch the child and straighten its legs. The feet are turned up vertically. The supine length can now be measured. Adults are usually between 1 and 3 centimetres taller lying down than standing up because gravity causes a slight compression of the skeleton in the upright position.

Fig 21.3 Medical checks usually involve the measurement of both body mass and height. The **body mass index** (BMI), the ratio of body mass (in kilograms) to the square of the height (in metres) gives important information about a person's health and fitness. A BMI of 20–25 is ideal, over 25 is getting a bit overweight, and over 30 is classed as clinically obese. This ratio is not reliable in very short or tall people, or in children and adolescents who are still growing

Fig 21.4 Typical height curves for girls from the ages of 2 to 18 years. The spread of values is the result of large studies that measured the growth patterns in thousands of children. The centre line is the average curve; the shaded areas show the normal range of height at each age

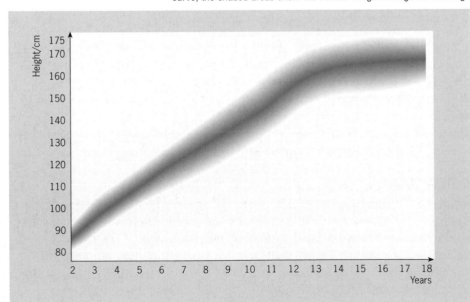

TECHNIQUES TO ASSESS AVERAGE BODY SIZE AND GROWTH

In 1759, a Frenchman, Count Philibert de Montbeillard, decided to measure the height of his new-born son. He took measurements every six months from birth until the boy was 18 years old. Collecting data like this – measuring a person or group of people many times – is called a **longitudinal study**. Studying your own child over time is easy, because they live with you, but researchers who study hundreds or thousands of people to get an idea of average growth patterns find maintaining contact with all their subjects quite difficult. People move away or forget to turn up for appointments.

An alternative approach is to use a **cross-sectional study**. This approach could be used to investigate the growth of human males between 10 and 18 years of age. The study would choose different age groups – a group of 10-year-olds, a group of 11-year-olds and so on. Each group would be large – perhaps a hundred boys – and each of them would be measured once only. By working out the average height of each age group and then plotting a graph or age against average height, this would reveal the overall pattern of growth.

The two methods can give different information and it is important to take this into account when interpreting data. Look at the graph in Fig 21.5.

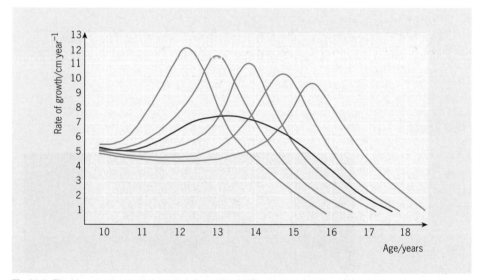

Fig 21.5 The blue curves show the rates of growth of different boys obtained from a longitudinal study. These curves differ in shape from the red curve. This curve was obtained from a cross-sectional study

The blue curves show the rate of growth in centimetres per year for five boys. This was a longitudinal study in which the height of each boy was measured at regular intervals over the full period of time. You can easily see the peak of growth that occurs at puberty. The red curve shows what happens if the average height of all the individuals in each age group is calculated and plotted – a cross-sectional study. Not really the same are they? Note how the red curve doesn't have the same 'puberty peak'.

? TEST YOURSELF

A Look at Fig 21.5. How many individuals were used in the longitudinal study?

REPRESENTING GROWTH

There are two main ways to represent growth: a **cumulative growth curve** and a curve showing **growth rate**. Fig 21.6 shows an example of each type of curve. Fig 21.6(a), a cumulative growth curve, shows the height of a child in centimetres from birth to the age of 18. Fig 21.6(b) shows growth rate. This shows the number of centimetres that the child grew each year; this is calculated easily by subtracting the child's height in the previous year, from his height in the current year.

Fig 21.6 Representing the growth of a single male child: **(a)** is the cumulative growth curve showing changes in his total height; **(b)** is a graph showing his growth rate

 REMEMBER THIS

The line in a cumulative growth curve can go up or flatten off; it never dips. Children get taller or they stay the same height, but they don't shrink. Growth rate, on the other hand, can decrease as well as increase. Fig 21.6(b) shows that a child's growth rate is higher during the first three years of life than later on, and shows the peak in growth rate that occurred in this child at puberty (see also Fig 21.5).

! **SCIENCE FEATURE** When growth becomes a health hazard

In 1995, American newspapers reported that the world's fattest man had died, weighting 465 kilograms (73 stone). The reports stated that it took a fork-lift truck to move his body from his bedroom, and that was only possible by demolishing the wall of his house.

You might think that to get as obese as this must take an enormous effort and a continuous eating binge. But, in daily terms, the excess dietary intake required to put on weight isn't really that much. If this man had started out as a 70 kg 16-year-old, he would have needed to put on only 37 grams per day to reach his final weight at the age of 45. That means eating just one chocolate bar or packet of crisps that he didn't burn off. Doesn't sound that difficult, does it?

Losing weight after you have gained it is, on the other hand, much more of a problem. The developed world, where obesity is prevalent, now has a booming industry based around people's need to diet. The medical profession now recognises the health dangers of obesity – diabetes, cardiovascular disease, gall stones are just a few of the disorders that are much more common in over-weight people. Some dieters are more successful than others and the press reports, from time to time, the 'success story' of someone who has lost half of their body weight (Fig 21.7).

Fig 21.7 The Rosemary Conley Diet and Fitness magazine Slimmer of the Year 2002 was 23-year-old Helen Jary, who lost 9 stone 9 pounds (61kg) of her original weight of 20 stone (127 kg)

3 GROWTH PATTERNS

Different organisms show differences in their patterns of growth. A crocodile is about 15 centimetres long when it hatches from its egg and can grow to a length of over 5 metres. It increases in length because all parts of it grow at more or less the same rate. As a result, an adult crocodile has very similar body proportions to a young crocodile. Its pattern of growth is described as **isometric**. Growth in a mammal such as a human follows a different pattern. The various parts of the body grow at different rates. Mammalian growth is described as **allometric**.

THE PATTERN OF HUMAN GROWTH

Humans are mammals, so it is no surprise that our growth patterns are more like those of an elephant than a crocodile (see Remember this box). Human growth is allometric and limited. Different parts of the body grow at different rates and we grow to a maximum height that is determined by both our genetic make-up and our environment. If we have two tall parents, we are more likely to be tall, and a child that has plenty of food grows taller than a child who is malnourished or suffers starvation.

Fig 21.8 shows the growth in height of a human female and a human male. Although the diagrams suggest that growth stops during adolescence, there is actually still a bit of growth later – but no more than 2 or 3 millimetres. Until the age of about 30, the length of the spine increases as small amounts of bone are added to the vertebrae. However, we usually say that in Europe and North America, growth stops at around the age of 15.5 years for the average girl and 17.5 years for the average boy.

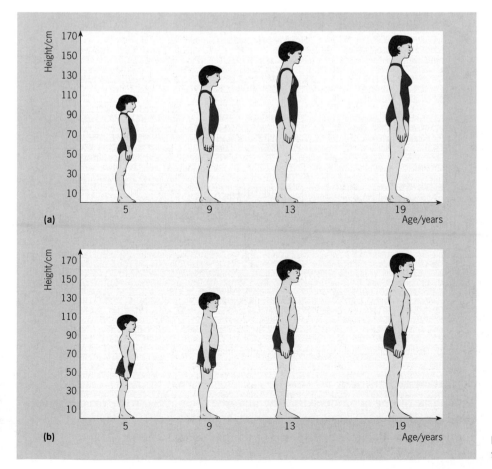

Fig 21.8 Growth in height of **(a)** a typical human female and **(b)** a typical human male

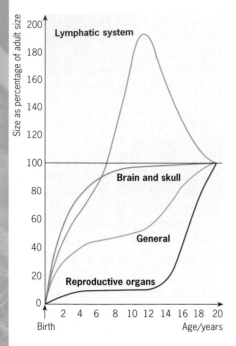

Fig 21.9 Growth curves of different parts of the human body compared with the overall growth curve

ALLOMETRIC GROWTH IN HUMANS

Allometric growth is clearly visible in Fig 21.8; the diagram shows a change in body proportions. Although organs such as the kidney and liver have a growth curve similar to that of the body as a whole, there are notable exceptions. These are shown in Fig 21.9.

The brain and skull grow very rapidly after birth, reaching 90 per cent of adult size soon after the age of 6. The lymphatic system also develops early. It has a very different growth curve from the rest of the body, reaching full size well before puberty and then decreasing. Finally, the reproductive organs grow very slowly in childhood. Maximum growth of these organs occurs during puberty.

THE CONTROL OF HUMAN GROWTH

Human growth is complex. It involves, directly or indirectly, all the systems of the body, and takes place over a long period of time, so growth obviously has to be *co-ordinated*. This is achieved by **hormones**.

Before you read this section, it would be helpful to look back at Chapter 18 and remind yourself of the ways in which hormones can affect the activities of their target cells. There are at least 12 different hormones that have a direct effect on growth at different times during a person's life. The functions of some of these are summarised in Fig 21.10.

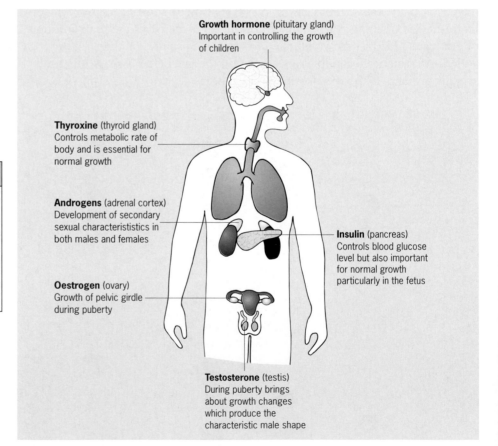

Fig 21.10 The main endocrine glands associated with human growth

GROWTH HORMONE

Growth hormone is produced by the anterior lobe of the **pituitary gland**. Human growth hormone has a rather unusual property for a mammalian hormone; it works in humans but in no other species. We say it is **species specific**.

Growth hormone is not necessary for the growth of a baby in its mother's uterus but, at birth, it takes over the control of growth. It continues to influence growth throughout childhood but then its production falls during early adulthood. Recent studies have suggested that elderly people who take growth hormone supplements are fitter and stronger, so in adults, the small amount of growth hormone that is still produced may play a role in tissue repair.

Like other hormones produced by the pituitary gland, growth hormone is secreted in pulses. For most of each day, the concentration of growth hormone in the blood is so low that it can hardly be detected, but several times each day, levels rise dramatically. Growth hormone does not itself cause growth to occur – it stimulates the release of **somatomedin**, another hormone from the liver. It is this second hormone that stimulates growth: it stimulates protein synthesis in cartilage cells at the ends of bones and in muscle cells.

THYROXINE

Thyroxine, a hormone produced by the thyroid gland, is also essential for normal growth. As well as controlling **basal metabolic rate** (see Chapter 7), and stimulating the process of growth in general, it also has a specific effect on protein synthesis and growth of the skeleton.

PUBERTY

Puberty is the period of life when the **sex organs** grow and start to produce **gametes**. **Secondary sexual characteristics** develop, such as the distribution of body hair and the differences in body shape that distinguish males from females.

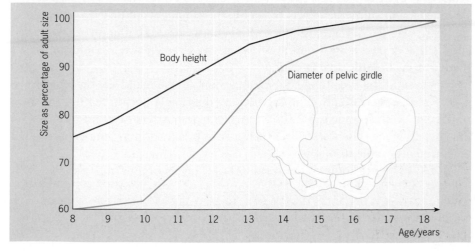

Fig 21.11(a) Being a choirboy is not a job for life: at puberty, the male voice breaks – suddenly gets deeper – as the size of the larynx increases

Fig 21.11(b) During birth, the baby passes through the mother's pelvic girdle. This graph shows that the increase in the diameter of the pelvic girdle of females occurs relatively late in puberty

THE ROLE OF HORMONES IN PUBERTY

The events of puberty are controlled by hormones, although we still do not understand the full story. To find out about the exact function of a particular hormone we would have to monitor changes in hormone concentration in the blood and then link these changes to the various aspects of puberty; but it is very difficult to measure hormone secretion accurately.

As we saw in the last section, hormones can be released into the blood in pulses rather than in a steady stream. So we would have to collect small blood samples from a large enough sample of people over a long period before we could know what was happening. Another problem is that many of the hormones involved in controlling puberty are broken down very rapidly.

The hormonal changes that control the onset of puberty are described in detail in Section 3 of Chapter 20.

4 AGEING

Fig 21.12 The proportion of old people in the populations of many less developed countries is likely to rise as we approach 2020

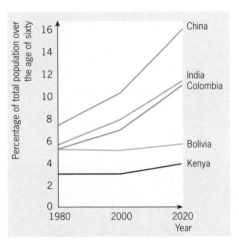

As we pointed out at the start of this chapter, few wild animals reach old age. There are many reasons for this, but predation and disease are mainly responsible. In humans, the situation is very different. Because of improvements in the standards of living in many parts of the world over the past 200 years, more people live to a ripe old age. In developed countries, we are already experiencing the effects: a 'top heavy' population. In the next few years, the same population changes will occur in the less well-developed nations. Fig 21.12 shows how, in China, India, Colombia, Bolivia and Kenya, the percentage of people over 60 is likely to increase between 1980 and 2020.

Old age affects individuals through changes in cell structure and organ function. An ageing population also affects society. An increase in the number of people living to old age places a huge strain on budgets for health and social care – one of the great challenges of the twenty-first century.

CELL STRUCTURE, ORGAN FUNCTION AND AGEING

Biologists who study the effects of ageing in humans would like to be able to define what is 'normal'. Illness and incapacity are so common among old people that complete freedom from disease is extremely rare. In addition, there is a lot of individual variation.

Let's look at the breathing system, for example: we could say that there is a 50 per cent reduction in lung capacity by the time a person reaches the age of 80. But, with so much variation, some individuals will be almost unaffected, while others will have such a severe reduction in lung capacity that they are barely able to move.

Fig 21.13 shows some of the ways old age affects organs and systems.

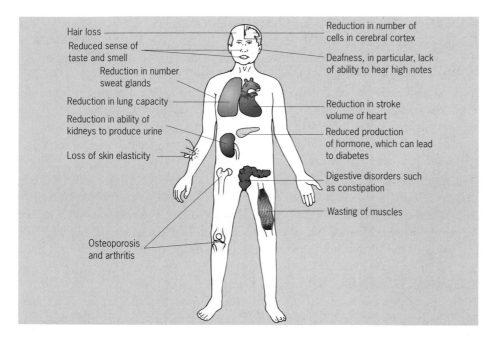

Hair loss

Reduced sense of taste and smell

Reduction in number sweat glands

Reduction in lung capacity

Reduction in ability of kidneys to produce urine

Loss of skin elasticity

Osteoporosis and arthritis

Reduction in number of cells in cerebral cortex

Deafness, in particular, lack of ability to hear high notes

Reduction in stroke volume of heart

Reduced production of hormone, which can lead to diabetes

Digestive disorders such as constipation

Wasting of muscles

Fig 21.13 Some of the many effects of ageing. It is important to remember that different individuals can be affected to different extents

---

SCIENCE FEATURE Why do we age?

Grab the skin on the back of your hand, pinch it and let it go. If you are young, it will snap back immediately. Try it (gently) with an older person and the pinched skin stays there for a while. One of the features of ageing is that our tissues lose their elasticity, causing that classic symptom of advancing years, wrinkles. Lines appear as our skin goes baggy instead of being stretched tightly over our bodies. Our hair falls out, bones soften, muscles weaken, spine curves, arteries harden, gums recede… and we may even start to enjoy gardening or making marmalade.

So what causes all this? Scientists have been pondering for years. There are no definite answers, but there are two aspects to consider. Do we age because we have a genetic clock ticking away inside us, determining how long we live – the deterministic view – or are the cumulative effects of things in our environment to blame – the non-deterministic view? The answer is almost certainly that it is a complex interaction of both. As we get older, we lose the ability to repair the damage caused by various internal and external factors.

A clue to the role played by our genes comes from the simple observation that all babies are born young. This seems obvious, but begs a fascinating question. How can it be that a baby born to a 14-year-old mother, and therefore made with a 14-year-old egg, is born at the same age as one born to a 50-year-old mother, with a much 'older' egg? It seems that there is a genetic clock inside the egg that doesn't start ticking until fertilisation.

Recent research into the ageing process of cells has focused on the ends of the chromosomes – stretches of DNA known as telomeres (Fig 21.14). When we are conceived, our telomeres are full length: the cells of a newborn baby have the potential to divide 80 or 90 times. Each time a cell divides, the telomeres

shorten (like the way that lead in a pencil wears down when you write with it). Cells seem to 'know' how old they are because they can 'count' the number of divisions they have made, and when the telomeres have been shortened to a critical length, the cell is no longer able to divide.

When cells can't divide, damaged tissues can't regenerate, and we begin to wear out. But we might in future be able to control the shortening of the telomeres, and therefore control the genetic cause of ageing. We might be able to extend life beyond 100 – which would have huge socio-economic consequences.

But how does the environment cause ageing? There are certainly external factors: exposure to sunlight, for example, accelerates the process of DNA damage. But internal factors are also very significant. Key molecules such as free radicals, reactive chemicals that react rapidly with anything around them, are capable of damaging DNA and proteins in a random fashion. One of the most damaging free radicals is the oxygen radical, O (a single oxygen atom with an unpaired electron). A small amount of all the O_2 oxygen in the air we breathe is split into free radicals. It is therefore always present in the body, and our metabolic defences cannot repair all the damage it causes. Even without external factors such as radiation, our vital molecules are damaged in time, and ageing is inevitable.

Fig 21.14 The human X-chromosome. The red areas show the telomeres

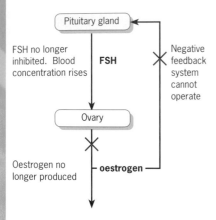

Fig 21.15 Some of the hormonal changes that accompany the menopause

AGEING AND THE REPRODUCTIVE SYSTEM

Fertility in men gradually declines with age. A man's general state of health usually determines how old he is when he loses the ability to father children. Some healthy men father children when in their 70s.

In women, however, the inability to produce children is a more marked event, usually occurring around the age of 50. This is known as the **menopause**. In the four or five years before the menopause, as the ovaries gradually fail, there is a progressive decline in the number of menstrual cycles in which an ovum is produced.

A reduction in the amount of oestrogen that the ovaries produce is the cause of many of the unpleasant symptoms of the menopause. **Hormone replacement therapy** (HRT) involves supplying extra oestrogen and can be used to control these symptoms (see Section 3 in Chapter 20). Fig 21.15 summarises the hormonal events associated with the menopause.

AGEING BONES AND JOINTS

As we get older, the skeleton suffers from a lifetime of wear and tear. The physical stresses on the body throughout life lead to two important medical conditions: **osteoarthritis** and **osteoporosis**.

OSTEOARTHRITIS

Osteoarthritis is one of the commonest conditions associated with old age (Fig 21.16). X-ray evidence suggests that nearly 80 per cent of those over 65 years have some arthritic damage to their joints. In most of them, this is not severe enough to cause painful symptoms.

Fig 21.16 Osteoarthritis is not a single disease. It describes the various conditions that result from joint failure. Joints commonly affected are **(a)** those in the hands, and **(b)** in the knees as shown by this X-ray image, and in the hips. **(c)** If the hips are severely affected, surgery might be required to replace the hip joint with an artificial one

---

! SCIENCE FEATURE Is there a gene for life expectancy?

In January 2002, researchers in the USA reported that a gene has been identified that is twice as common in babies than in people over the age of 65. We have known for some time that genetics can play a part in determining whether someone lives to old age or not (see the Assignment at the end of this chapter), but this is the first time that having a gene has been linked with dying at a young age. It is not yet known what the gene codes for but Japanese researchers, who discovered a very similar gene in mice, showed that it had a definite effect. Mice without the gene developed atherosclerosis, osteoporosis, emphysema and other conditions common in elderly humans.

The variant of the gene that seems to lead to a risk of early death is apparently found in about 3 per cent of humans – these people are homozygous for the gene (they have two copies). About 25 per cent of the populations studied carry one copy, and so are heterozygous carriers, capable of passing on the gene to their children. However, the study shows that these heterozygotes may often live longer than average, so a lot of work still needs to be done to find out if this gene really is important in preventing human ageing.

Many different factors cause osteoarthritis to develop and the way they interact is not fully understood. Experts now mostly agree that the old 'wear and tear' hypothesis is not backed up by evidence. For example, people who exercise well into old age have a much lower risk of developing osteoarthritis than their 'couch-potato' counterparts. Some researchers are now concentrating on cells in the bone and **cartilage** at joints, which send signals to each other using a huge variety of molecules, since abnormal signals may start the process of destruction in cartilage and/or bone. Whatever the actual cause, the result is loss of cartilage at the **articular surfaces** of bones that form the joints (Fig 21.17). In addition, there is an increase in bone-forming cells called **osteoblasts**. New bone may be laid down, deforming the shape of the joint.

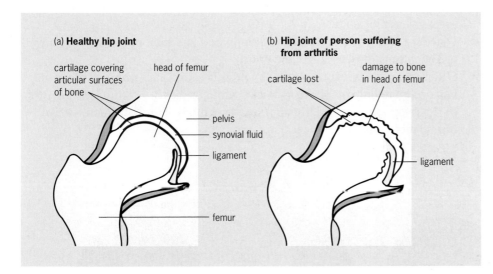

(a) **Healthy hip joint**

cartilage covering articular surfaces of bone

head of femur

pelvis
synovial fluid
ligament

femur

(b) **Hip joint of person suffering from arthritis**

cartilage lost

damage to bone in head of femur

ligament

Fig 21.17 In a healthy hip joint **(a)**, cartilage covers the articulating surfaces of the bones which form the joint. In the hip joint of a person suffering from osteoarthritis **(b)**, this cartilage has been lost

Although we cannot reverse the changes that accompany the development of osteoarthritis, we can do a lot to ease the symptoms. Keeping mobile is now recommended, and anti-inflammatory drugs and painkillers are given in the early stages. Later, a joint replacement may be necessary.

OSTEOPOROSIS

The density of bone varies with age, as Fig 21.18 shows. Young people, up to the age of about 20, show a steady increase in bone density. After 20, there is a gradual decrease in density. In men, this loss of bone is not usually a matter for concern. However, women's bone density is usually less than men's, and they suffer further reduction during the menopause. Once bone density becomes very low, the affected person is said to suffer from osteoporosis. The bones lack strength, and even very little stress can fracture them.

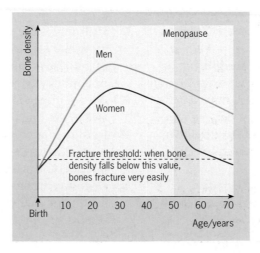

Bone density

Menopause

Men

Women

Fracture threshold: when bone density falls below this value, bones fracture very easily

Birth 10 20 30 40 50 60 70

Age/years

Fig 21.18 The effect of age on bone density

DIAGNOSIS	AGE ON ADMIT	SEX
FRACTURE OF ANKLE	17	M
FRACTURE OF NECK OF FEMUR	17	M
ACQUIRED DEFORMITIES OF TOE	17	M
INTERNAL DERANGEMENT OF KNEE	17	M

Fig 21.19 A hospital record of patients receiving X-rays

USING X-RAYS TO SEE FURTHER

We normally associate X-rays of bones with fractures, but a hospital X-ray department investigates many conditions that affect the skeleton. Fig 21.19 shows some computer records from the radiography department of a large hospital.

The pie charts in Fig 21.20 show the age on admission of males and females different conditions.

- Some conditions are the result of gender-specific behaviour. Young boys tend to be rougher in their play than girls of the same age and so they break their arms and legs more often. It is also likely that older women referred to the hospital with misshapen toes insisted on wearing tight fashionable shoes when they were younger.
- Congenital conditions are present at birth. Children with these conditions are X-rayed more often.
- Osteoarthritis is a disease of the elderly and is a major reason for X-ray investigations in the 70–79 age group.
- Older women tend to break bones often because their sense of balance has declined, but osteoporosis is also a major factor. Fractures of the neck of the femur are particularly common and many elderly women receive hip replacements.

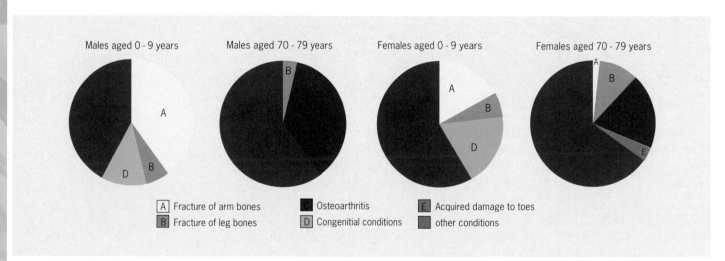

A Fracture of arm bones	■ Osteoarthritis	E Acquired damage to toes
B Fracture of leg bones	D Congenitial conditions	■ other conditions

Fig 21.20 Pie charts drawn from data like that in Fig 21.19

SUMMARY

After reading this chapter, you should know and understand the following:

- **Body mass**, **height** and **supine length** can be used to measure body size and body growth. There are advantages and disadvantages with all three criteria.
- Growth curves may be plotted using data collected from either **longitudinal** or **cross-sectional** studies. They can show **cumulative growth** or **growth rate**.
- Human growth is **allometric** and **limited**.
- Growth is controlled by hormones. These include **growth hormone** and **thyroxine**.

- **Puberty** is the period of life when the **sex organs** grow and start to produce **gametes**. The events of puberty are initiated and controlled by changes in hormone concentrations.
- Ageing brings about changes in cell structure and organ function.
- Fertility in men gradually declines with age. In women, however, the inability to produce children is associated with the menopause.
- **Osteoarthritis** and **osteoporosis** are conditions that commonly affect the skeletal system of older people.

? EXAM QUESTIONS

1 During human development, changes in height and other dimensions are not uniform. Both boys and girls experience an adolescent spurt in growth. The graph shows the mean height gain, in cm per year, for boys and girls from birth to 18 years.

a) Explain the shape of the curve for height gain per year from birth to 10 years.

b) (i) During the adolescent growth spurt, the maximum height gain per year for boys is 10.4 cm. Calculate the difference between the maximum height gain for boys and the maximum height gain for girls during their growth spurt. Show your working.

(ii) A 15-year-old boy's height was measured and found to be 176 cm. Using the graph, suggest how tall he is likely to be when he is 16 years old. Show your working.

c) Describe how increase in height is achieved.

d) During the adolescent growth spurt, changes in body mass show a pattern similar to that of height gain. Explain why there are changes in body mass in boys and girls during this growth spurt.

e) The dimensions of most, but not all, of the body organs follow the same growth curve as that of height. Name *one* body organ or body tissue which does *not* follow this pattern and explain how its growth rate differs.

Edexcel HB6 (Synoptic) January 2000 Q4

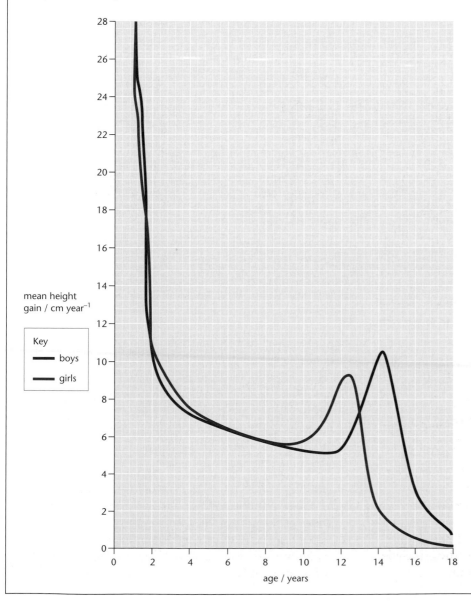

mean height gain / cm year^{-1}

Key
— boys
— girls

age / years

KEY SKILLS **ASSIGNMENT**

WHY DO WE GROW OLD?

Why do some people live longer than others? In this Assignment we look at the evidence that supports the idea that longevity is controlled by genes. Studying twins is one way of finding out more about how genes control variable characteristics. First we need to look at the ways in which twins are formed. Fig 21.A1 shows the formation of **monozygotic twins** and **dizygotic twins**.

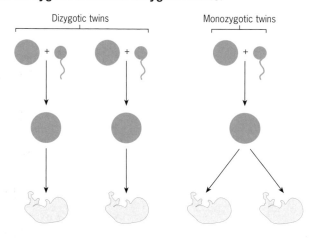

Fig 21.A1 Monozygotic twins result from the splitting of a single fertilised egg cell or zygote. Dizygotic twins develop from two separate zygotes

1 a) Use Fig 21.A1 to explain why monozygotic twins are also called identical twins.
b) Explain why it is not really accurate to describe dizygotic twins as non-identical.

In one investigation concerned with ageing, many pairs of monozygotic and dizygotic twins were studied. Records were kept of the age of the first twin when he/she died and the number of months later that the second twin died. The data were collected only on pairs of twins in which the first sibling was at least 60 years old when he/she died. The results of this investigation are summarised in Table 21.A1.

Table 21.A1 Average difference in length of life between members of twin pairs/months				
Monozygotic		**Dizygotic**		
Male	**Female**	**Same sex**		**Different sex**
		Male	**Female**	
29.4	47.6	89.1	61.3	126.6

2 a) Suggest why the data were collected only from twins in which the first sibling was at least 60 years old when he/she died.
b) Evaluate the evidence from the table that the length of a person's life is controlled by their genes.

c) Women generally live longer than men. How does this explain the figures for the difference in length of life between members of dizygotic twin pairs of the same sex and of different sex?

If individual cells are removed from an animal and cultured, most do not live forever. Similarly, most cells from different human tissues grow in a culture for only a limited amount of time. Then they degenerate and die.

In one investigation, cells were isolated from fetal skin and cultured in a suitable container with oxygen and a supply of nutrients. They grew rapidly until they had formed a layer over the inner surface of the container. At this stage they stopped dividing. The culture was then divided and placed in two new containers. The cells started growing again until, once more, they formed complete layers over the inner surfaces of the containers. The culture was divided in half repeatedly until the cells stopped dividing. The results are shown in the graph in Fig 21.A2.

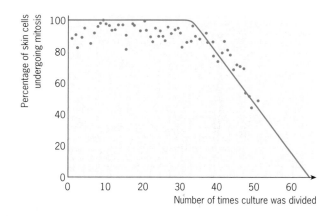

Fig 21.A2 In this graph the percentage of fetal skin cells estimated to have undergone mitosis has been plotted against the number of times the culture was divided

3 a) Look at the information about the skin in Chapter 12 and suggest the type of cell that could have been used in this investigation. Give a reason for your choice.
b) Use the information from the graph to suggest a hypothesis to explain why humans do not live forever.
c) Assuming your hypothesis is correct, predict the shape of the curve you would expect if the initial sample of skin cells had been taken from a young adult.

An inherited condition in which the process of ageing is more rapid than normal is Hutchinson–Gilford syndrome. It is controlled by a recessive allele. Sufferers show many visible signs of old age even when quite young. Fetal skin cells containing two alleles for Hutchinson–Gilford syndrome will have about 10 culture cycles before they die.

22

Genetics: the basics

If you reproduce, your DNA may well live longer than you do. This is a computer model showing the structure of DNA. The 'backbones' of the double helix (shown here as ribbons) are made from alternating phosphate and sugar groups connected to atoms of carbon in pentose molecules. Between the two backbones are the bases: adenine, guanine, cytosine and thymine

Why are we all similar? Why are we all different? These are two of the fundamental questions on which the science of genetics is based. When the sperm from your father met the egg from your mother, a unique individual was created, different from anyone who had gone before or will ever exist again.

But would it surprise you to learn that you have quite a lot of genes in common with bacteria, lettuce, cockroaches, tapeworms and ferrets? Certain essential genes – such as those that code for the enzymes of respiration – are very similar in all organisms as they are vital to this basic life process. Such genes must have been copied faithfully, generation after generation, for the last three billion years or so. It is also intriguing to realise that, although almost all organisms die young before reproducing, not a single one of our ancestors did. Their DNA must have had what it takes to be passed on. Well done Granny and Granddad, and Great Granny and Great Granddad.

1 WHAT ARE GENES?

Genes are the 'instructions' that control everything that happens inside a cell. Physically, genes are different sized stretches of DNA that code for the manufacture of proteins such as enzymes. Enzymes organise the chemistry that goes on in our bodies and are behind all the life processes, including growth, repair and reproduction. Throughout our lives, the genes remain in the nuclei, directing the activities of each of our cells.

Children grow and develop according to the genes they have inherited from their parents. It's a harsh fact, but nothing lives forever: the inevitable fate of all organisms, often sooner rather than later, is death. However, although we die, our genes live on in our offspring.

? TEST YOURSELF

A Some simple unicellular organisms such as bacteria and amoeba can reproduce by binary fission; they simply split in half. Is it correct to say that such organisms never die?

✔ REMEMBER THIS

In organisms that reproduce sexually, a set of genes from the male combines with one from the female to form a unique individual (see Chapter 20).

Fig 22.1 The film *GATTACA* presents us with a world where the privileged classes are genetically modified at birth so that all the potentially harmful genes are deleted; 'It's you, but it's the best of you.' This isn't such an unbelievable vision of the future. What do you notice about the letters used to make the title *GATTACA*?

2 STUDYING GENETICS WITH THIS BOOK

Studying genetics means learning about DNA and what it does. In this section of the book, we divide our study of this remarkable molecule into five areas:

- The **cell cycle** is the study of how cells divide. Discovering just how a single fertilised egg can develop into an organism as complex as a human is one of the central challenges of biology. When the control of cell division goes wrong, it can lead to cancer. Find out more about mitosis and meiosis, the two types of cell division, in Chapter 23.
- **Molecular biology** is the branch of science that deals with all aspects of the DNA molecule: its structure and function. How are genes used to make products? Why are some genes active when others aren't? How can DNA copy itself? How do mutations occur? This is covered in Chapter 24.
- **Genetics** is the study of inheritance: the way in which genetic information is passed on from one generation to the next. Why are only some genes passed on? Why are some characteristics always shown while others are hidden? This is covered in Chapter 25.
- **Evolution** looks at how species change and develop with time. What is the mechanism of this change? How long does the process take? Can we observe evolution in action? What do we know about the evolution of humans? This is covered in Chapter 26.
- **Genetic engineering** is the popular name for the technology that manipulates the DNA molecule. How can we change DNA? How can we transfer DNA between organisms? What are the benefits of genetic engineering? Can we cure genetic disease? Should we pursue this technology at all? This is covered in Chapter 27.

3 THE POTENTIAL OF GENETICS

Few scientific subjects make the headlines as often as advances in genetics and DNA technology. We are witnessing an accelerating revolution that is likely to have far-reaching effects. We have mapped the complete human genome and are now embarking of the daunting task of finding out where the genes are, what they do and how they interact with each other. We may have the 'book of humans', but it's huge and written in a foreign language.

The implications of this progress are enormous:

- Two out of every three people die for reasons connected to their genes. Although only a small percentage inherit a lethal genetic disease, many more inherit a *tendency* to develop a condition such as heart disease, premature senility or cancer. We call this a **genetic predisposition**. An in-depth knowledge of the human genome will make it possible for doctors to screen babies before or soon after birth for genetic abnormalities and tendencies to develop such diseases. But this raises ethical and moral questions, not just scientific ones: would you want to know just how likely you are to develop cancer in middle age? Would you want anyone else to know – life insurance companies, potential employers, friends or partners? Who would have access to your DNA files?

Fig 22.2 A complete set of human chromosomes. Each body cell contains two sets of 23 chromosomes and each chromosome contains up to 4000 genes. The total amount of DNA in a cell is called the genome. The human genome consists of about 60 000 genes

✔ REMEMBER THIS

Chromosomes condense and become visible during cell division, as we see in more detail in Chapter 23.

Normal cell division – mitosis – duplicates the DNA and then divides it accurately into two so that the two new cells formed are genetically identical.

Meiosis is the type of cell division that produces haploid cells such as eggs and sperm. This process halves the amount of genetic material that goes into each new cell and shuffles it at the same time so that no two sex cells are the same.

Extracting DNA from fossil remains such as this *Australopithecus* could give us some clues when piecing together the story of human evolution (see Chapter 26)

A human embryo 7 weeks after conception. We have developed the technology to clone animals as advanced as humans, but this area of genetics is an ethical minefield

Wheat has been bred selectively to produce a greater yield. Many new types are easier to harvest and are also resistant to disease

Fig 22.3 The applications of our knowledge of DNA

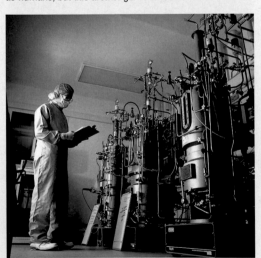

This fermenter houses genetically engineered bacteria which contain the human growth hormone gene. Inside the fermenter the bacteria multiply and produce large amounts of human growth hormone

Male and female sperm can be separated according to the amount of DNA they contain. The sperm are stored in liquid nitrogen until needed. Farm animals can be artificially inseminated to produce offspring of the required sex

- We can isolate genes for useful products, such as insulin and growth hormone, and place those genes into another organism. This is **recombinant DNA technology** (see Chapter 6). Bacteria already produce human insulin and other products on a large industrial scale, and it is also possible to transfer such genes into higher animals such as sheep so that they produce milk containing the required protein.
- We can isolate faulty genes, such as those which cause cystic fibrosis, and replace them with multiple copies of healthy genes. Although still at the experimental stage, this **gene therapy** holds much promise for the future.
- We could be able to halt some of the genetic causes of ageing, and allow more people to live even longer.

Only the bravest scientists would dare to speculate about where our genetic knowledge might eventually take us. It could be to a greatly improved world where suffering is reduced while quality of life and life expectancy are improved. Genetic engineering and agricultural techniques might give us food-producing species that can live in the harshest of conditions, so providing food in areas of most need.

But can we be this optimistic? There are still many things we don't know about living organisms and it is impossible to predict exactly the effects of interfering with an organism's DNA. Already, companies are trying to patent genetic material and its products, leading to hot debate about whether or not such natural products can be owned, like a brand name or an invention. Will genetic advances be used solely to make money, and therefore be available only to those who can pay?

The ethical implications of some very recent developments also need to be considered. The first mammals have now been cloned but the process is unreliable. This has led to a heated debate about whether or not we can, or should, clone humans.

4 WHAT GOES ON IN THE NUCLEUS?

The nucleus of every human cell contains DNA (Fig 22.4). Each nucleus contains a set number of long, elaborately coiled DNA molecules that condense into chromosomes. Genes, the individual instructions, are dotted along the chromosomes.

In an adult organism most cells are not dividing: the chromosomes are uncoiled, and active genes are producing proteins to control the activities of the cells. Not all genes are active at the same time: controlling which genes are active and which are not enables different cells to carry out different functions. For instance, all cells in the human body contain two copies of the insulin gene. Only in certain specialised cells in the pancreas, however, are the genes switched on and used to make insulin.

GLOSSARY OF GENETICS TERMS

Like many branches of science, genetics comes with its own jargon, seemingly designed to make some very straightforward ideas inaccessible to the average person. You will need to know the following terms:

Diploid: cells or organisms containing two sets of genes. Human body cells, for instance, are all diploid.

Haploid: cells or organisms containing one set of genes. Human sex cells (egg cells and sperm) are haploid.

Fig 22.4 Most of the DNA in the cell is locked away in the nucleus where, in effect, it forms a reference library. The genes, which act as instructions for all cell functions, must be passed on when the organism reproduces

Gene: a length of DNA that codes for a particular polypeptide. Some proteins consist of more than one polypeptide; these are coded for by more than one gene.

Allele: one form of a gene. There may be two or more alleles of any particular gene. Diploid organisms contain two alleles, one on each pair of chromosomes. For example, in pea plants, the gene for flower colour may have a red allele and a white allele.

Genotype: the genetic make-up of an individual.

Phenotype: the observable characteristics an organism possesses. For example, in peas the genotype Aa may result in a phenotype of red flowers. Many alleles are expressed, or used, and these contribute to the characteristics of the individual. Other genes – recessive alleles in the presence of a dominant version – are not expressed and so remain hidden.

Dominant: the allele that, if present, is expressed. If the red allele in the pea plant is dominant, the pea has a red flower, even if it has a white allele as well.

Recessive: the allele that is expressed only in the absence of the dominant. In the pea, if the white allele is recessive, the plant has white flowers only if it has two white alleles.

Homozygous: when both alleles of a particular gene are the same. In the pea example, an individual would be homozygous if it had two red alleles or two white alleles. We usually give letters to denote alleles, for example A for the dominant allele and a for the recessive allele. Individuals with genotype AA or aa are homozygous.

Heterozygous: the two alleles of a gene are different. In a pea plant, a heterozygous individual could have a red allele and a white allele, so their genotype would be shown as Aa.

5 VARIATION: WHY IS EVERYBODY DIFFERENT?

The group of people in Fig 22.5 shows what we all know: that everyone is different. Height, weight, intelligence, personality, eye colour, size of feet – the list is endless. No two people have the same combination of features. Even identical twins show subtle differences. So what makes us the way we are?

Fig 22.5 Like most species, humans show great variation between different individuals

Each human baby is born with a unique set of genes – we call this its **genotype**. However, only some of these genes affect the observable characteristics of the organism – its **phenotype**. Recessive genes, for example, remain hidden, masked by dominant versions.

The other factor that contributes to the phenotype is the environment – the circumstances in which the organism is brought up affects the way in which genes are expressed. For example, humans are born with a set of genes that give us the potential to grow to a certain height. We cannot realise this potential, however, without an adequate diet. Two hundred years ago the average height was several centimetres shorter than it is now: the difference is due to improvements in nutrition and general health care.

WHY IS VARIATION NECESSARY?

The simple answer is: survival. Variation in a population has great survival value because it greatly increases the chance that at least some individuals will adapt to changing conditions. To illustrate this, consider a **clone**, a population of genetically identical individuals. All is well when conditions are favourable, but when a serious disease comes along it is very likely that all of the organisms in the population will die. In a more varied population that has developed as a result of sexual reproduction and variation, there is a much greater chance that some of the individuals could be resistant to the disease. These individuals survive to breed, ensuring that the population, and perhaps the whole species, continues to exist.

This 'survival of the fittest' is more correctly called **natural selection**. In the long term, natural selection, acting on the variation in a population, is the main driving force behind the process of **evolution** – the subject of Chapter 26.

HOW VARIATION OCCURS

Variation arises because individual genes **mutate**. **Mutations** are changes in the genetic material of an organism, often caused by faults in the copying of the DNA (see Chapter 25). Once there is some variation to work on, the differences are maximised by the process of sexual reproduction. Using humans as an example we can see how this happens.

Firstly, gametes are produced by meiosis. This is a special type of cell division (see Chapter 23) which not only produces haploid cells, but shuffles the genes so that all eggs or sperm made by a particular individual are different. In meiosis, there are two key events that produce variation:

- **Crossover during prophase I**. Blocks of gene are swapped between chromosome pairs, bringing some new genes together and separating others.
- **Independent assortment during anaphase I**. Either chromosome from each pair can pass into the gamete. As a human cell has 23 pairs of chromosomes, 2^{23} different combinations are possible.

Secondly, following meiosis, **fertilisation** joins gametes from two separate individuals to form a genetically unique zygote from which the new diploid organism develops.

REMEMBER THIS

The phenotype of an organism = its expressed genes + effects due to the environment that it is exposed to.

REMEMBER THIS

A population is a group of interbreeding individuals of a particular species in a particular place (see Chapter 26). A species usually consists of many different populations. Some endangered species become particularly vulnerable because they consist of just one isolated population, restricted to a particular area.

CONTINUOUS AND DISCONTINUOUS VARIATION

Table 22.1 Examples of continuous and discontinuous variation	
Continuous	**Discontinuous**
Most dimensions in animals and plants such as height, weight, length of bones, number of leaves	Ability to roll tongue in humans
Tolerance to adverse conditions, such as cold, heat, dehydration	Ability to taste phenylthiourea
Human fingerprints	Flower colour in peas
	ABO blood group in humans
	Coat colour in cats

There are two main types of variation in a population: **continuous variation** and **discontinuous variation** (Table 22.1):

● Continuous variation is so-called because the factors, or variables, can be any value inside a given range. Examples of continuously varying characteristics include height and weight in humans (Fig 22.6a), and the number of leaves in plants.

● Discontinuously varying features do not display a range: they are either one thing or the other. The ABO human blood group system is a classic example of discontinuous variation: people are either A, B, AB or O. They cannot be anything in between. Height in pea plants is another example: pea plants can be either tall or dwarf (Fig 22.6b).

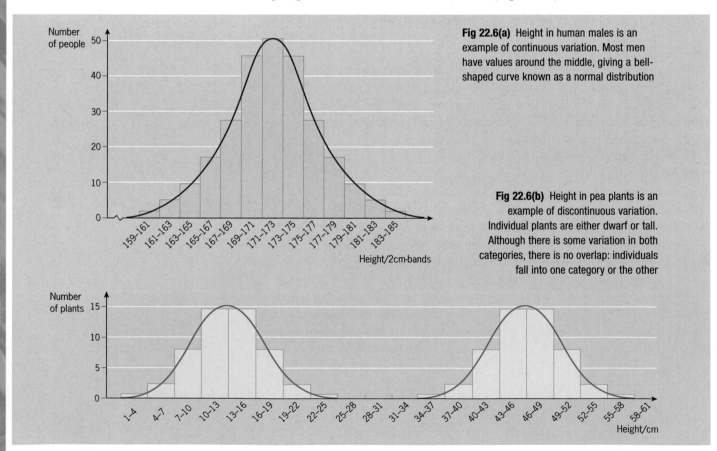

Fig 22.6(a) Height in human males is an example of continuous variation. Most men have values around the middle, giving a bell-shaped curve known as a normal distribution

Fig 22.6(b) Height in pea plants is an example of discontinuous variation. Individual plants are either dwarf or tall. Although there is some variation in both categories, there is no overlap: individuals fall into one category or the other

Often, continuous variation is **quantitative** (you can put a numerical value on it) and it is **polygenic** (controlled by several genes). Discontinuous variation is more often qualitative (you can describe it, but not measure it) and is controlled by one or only a few genes. In Chapter 25 we look at the principles of inheritance using examples of discontinuous variation.

? TEST YOURSELF

B Look at Fig 22.6(b) on the previous page. A pea plant is 34 cm in height. Is it a tall or a dwarf plant?

6 A BRIEF HISTORY OF GENETICS

The history of genetics gives an insight into the changing nature of scientific progress. Early scientific discoveries and breakthroughs were few and far between, often made by individuals who observed and studied nature and put forward radical theories to explain what they had seen. Mendel (page 423) and Darwin (page 444) are classic examples.

More recent discoveries have been made by teams of dedicated people working together, using the scientific method and latest technology to build on knowledge accumulated by others. Table 22.2 outlines some of the landmarks in this most dynamic of scientific areas.

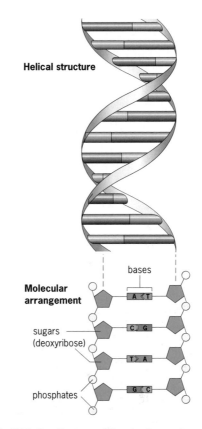

Helical structure

Molecular arrangement

bases

sugars (deoxyribose)

phosphates

Fig 22.7 The discovery of the structure and function of DNA is one of the greatest scientific discoveries of the twentieth century. The DNA molecule stores genetic information and, with the help of enzymes, makes exact copies of itself, time after time

Table 22.2 Some landmark discoveries in genetics. In many of these, the scientists mentioned by name led a much larger team of scientists who worked towards the discovery described

Year	Scientist(s)	Discovery
1858	Charles Darwin, Alfred Russel Wallace	Jointly announced their theory of evolution by natural selection: 'Survival of the fittest'
1859	Charles Darwin	Published his book *The origin of species*
1866	Gregor Mendel	Published his laws of genetics following studies on pea plants. His findings were ignored
1900	Hugo de Vries, Carl Correns, Eric von Tshermak	Discovered meiosis, providing an explanation for Mendel's laws: his work is rediscovered by all three and gains recognition
1905	Nettie Stephens, Edmund Wilson	Independently discovered the principles of sex determination: XX = female, XY = male
1910	Thomas Hunt Morgan	Proposed the theories of sex linkage, mutation, linkage and chromosome maps following work on *Drosophila* (fruit fly)
1928	Fred Griffiths	Proposed that some 'transforming principle' had changed a harmless strain of bacteria into a lethal one. The hunt for DNA began
1941	G. W. Beadle, E. L. Tatum	Irradiated the bread mould *Neurospora*, producing mutations which suggested that genes code for enzymes
1944	Oswald Avery	Purified the 'transforming principle' in Griffiths' experiment, showing it to be nucleic acid (DNA)
1950	Erwin Chargaff	Discovered that the base pairing ratios in DNA were always the same, whatever the organism (Chargaff's principles: A = T, C = G)
1951	Rosalind Franklin	Obtained high-quality X-ray diffraction studies of DNA, showing that it has a helical structure
1952	A. D. Hershey, M. Chase	Used bacteriophages to show that DNA, not protein, is the material of heredity
1953	Francis Crick, Jim Watson	Built on the work of Chargaff and Franklin to work out the 3D structure of DNA

Table 22.2 Some landmark discoveries in genetics. In many of these, the scientists mentioned by name led a much larger team of scientists who worked towards the discovery described (continued)

Year	Scientist(s)	Discovery
1958	M. Meselson, F. W. Stahl	Used radioisotopes of nitrogen to prove the semi-conservative mechanism of DNA replication
1958	Arthur Kornberg	Purified DNA polymerase from *E. coli* and used it to make DNA from nucleotides in a test tube
1961	F. Jacob, J. Monod	Propose a mechanism for switching genes 'on' or 'off': the operon hypothesis of metabolic control
1966	Marshall Nirenberg, Gobind Kharana	Cracked the genetic code: particular triplets of bases on DNA code – via mRNA – for the 20 amino acids
1970	H. O. Smith, K. W. Wilcox	Isolated the first restriction enzyme, *Hind*II, which can cut DNA
1972	Paul Berg, Herb Boyer	Produced the first recombinant DNA molecules
1973	Annie Chang, Stanley Cohen	Showed that DNA could be inserted and cloned inside a bacterium
1977	Fred Sanger	Developed a method for sequencing DNA
1977		The first genetic engineering company (Genentech) is founded, using recombinant DNA to make pharmaceuticals
1983	James Gusella	Located the gene responsible for Huntington's chorea on chromosome 4 (see Fig 24.2)
1984	Cary Mullis	Developed the polymerase chain reaction (PCR) in which minute samples of DNA can be copied, prior to analysis
1984	Alec Jeffreys	Developed DNA profiling (or 'fingerprinting')
1988		The Human Genome Project began to map the human DNA sequence
1989	Francis Collins	Identified the gene (*CFTR*) responsible for cystic fibrosis on chromosome 7
1990	French Anderson	First attempts at gene therapy: T cells of a 4-year-old girl were exposed to viruses containing working copies of her defective gene. After treatment, her immune system began working again
1994		Genetically engineered 'Flavr savr' tomatoes went on sale
1995		Transgenic sheep made to express human genes in their mammary glands, so that they produce milk containing valuable pharmaceuticals
1997	Roslin Institute, Edinburgh	First mammal cloned. Dolly the sheep fuels controversy about genetic research
1997		Dolly the cloned sheep gives birth to lambs conceived normally.
2000		Entire human genome announced
2001		First stem cells extracted from cloned human embryo Gaur, an almost extinct bovine species, cloned using a cow as a surrogate mother Pigs cloned for first time, bringing the possibility of pig organs for transplant into human patients a step closer Researchers use cow stem cells to construct a complete cow kidney in the laboratory
2001	Texas A&M University	First cat ('cc') cloned
2002		Scientists in the UK were given the go-ahead to create human embryo clones under strictly controlled conditions by a House of Lords select committee

23

Cell division

A magnetic resonance imaging (MRI) scan of a section through the brain of a 42-year-old woman. A tumour, coloured yellow, can be seen in the left hemisphere. This is a **metastatic** tumour; it has spread from another tumour elsewhere in the body

Coloured X-ray of hand showing a tumour growing on the bone on one finger

In a healthy human adult, cells divide only when they should. Some cells, such as those that line the gut, are replaced at a remarkable rate. Other cells, such as muscle cells, live longer and need to be replaced far less often. Occasionally, the systems that control cell division break down, and a cell that should be stable divides uncontrollably. Soon, a mass of tissue called a tumour forms. Tumours can be either benign or malignant.

A benign tumour is not cancerous. Cells divide within a small, confined area and the growth is often surrounded by a membrane. New cells form in the centre of the mass. This type of tumour can be dangerous if it presses on important organs, or if it grows very large. It is usually treated by surgery. Because the tumour cells are confined and do not invade other tissues, recovery and survival rates are good.

A malignant tumour is commonly known as cancer. The tumour grows at the edges, spreading out and invading the surrounding tissues (supposedly like a crab – hence the name cancer). This type of tumour is much more dangerous. Vital tissues and organs can be destroyed quickly and this can lead to death. Even if the tumour is removed from the body, actively dividing cells may have already broken off and set up secondary growths elsewhere in the body.

1 SOME BASIC PRINCIPLES

An understanding of cell division is of great importance in the study of human biology. It is the process by which we grow and develop. As adults, our body uses the main type of cell division for maintenance and repair. When control of cell division is lost, the result can be cancer (see Chapter 31). Another type of cell division halves the chromosome number in eggs and sperm (the sex cells, or **gametes**) so that when fertilisation occurs, the resulting zygote has the normal double number.

Humans, like all animals, are eukaryotic organisms: in eukaryotic cells, DNA is organised into chromosomes and is enclosed within a double nuclear membrane. Before going on to study the mechanism of cell division in eukaryotic cells, familiarise yourself with some of the basic terms and concepts by looking at Table 23.1.

AN INTRODUCTION TO CELL DIVISION

Our bodies are made from two different types of cells; body (or somatic) cells and sex cells (or **gametes**). The nucleus of each body cell contains two sets of chromosomes and so has two sets of genes. Body cells are described as diploid (from Greek, meaning 'double number') (Fig 23.1).

Mitosis is the process of normal cell division. In mitosis, the chromosomes are copied and then divided equally between the two new daughter cells. So each mitotic division produces two cells, both diploid and each with exactly the same genes as the parent cell.

Table 23.1 Some basic terms important in cell division

Term or feature	What it is or what it means
Gene	A length of DNA that codes for the production of a particular polypeptide or protein
Genome	The name given to the full set of genes in a cell. The human genome consists of between 50 000 and 100 000 genes (the exact number is not yet known)
Diploid	Diploid cells contain two versions of every gene
Haploid	Haploid cells contain one version of every gene
Chromosome	A long single molecule of DNA, organised around proteins called histones. The largest human chromosomes contain about 4000 genes. Chromosomes exist in cells all the time but they can be seen only during cell division, when they condense and separate
Homologous chromosomes	Chromosomes exist in pairs – humans have 23 pairs. Homologous chromosomes have the same genes at the same positions, but not necessarily the same versions of each gene
Chromatid	During cell division, the DNA of a cell is replicated (copied). When the chromosomes condense, they therefore appear as double structures: each unseparated chromosome within such a pair is called a chromatid
Sister chromatid	When two chromatids are genetically identical (as in mitosis), they are called sister chromatids
Bivalent	A bivalent is a pair of homologous chromosomes that line up together, as they do during meiosis
Locus	The position of a gene on a chromosome
Transcription	The process in which a molecule of mRNA is assembled on an active gene. mRNA thus becomes a mobile copy of the gene
Translation	The process of converting the code on the mRNA into a protein. This is achieved by protein synthesis: amino acids are joined in a particular order to make a protein such as an enzyme

Fig 23.1(a) A human white cell (a lymphocyte). The nucleus of this cell contains all the genes necessary to make a whole new organism, a clone of the person from whom this cell was taken. Although the technology exists to clone humans, the area is fraught with huge moral, ethical and legal problems

Fig 23.1(b) Dolly the sheep, born in 1997, was the first ever mammal to be cloned from a diploid adult body cell

Fig 23.1(c) A micropipette puts the nucleus from an adult cell into a non-fertilised egg cell that has had its own nucleus removed

Fig 23.2 This human embryo is 4 days old, but the fertilised egg has already divided mitotically four times, producing a cluster of 16 cells. The cells will soon specialise to form the tissues and organs of the body

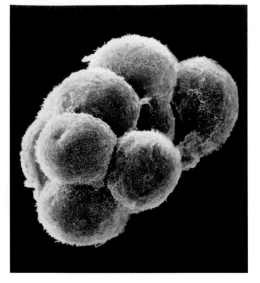

SEE QUESTION 1

Some cells in the human body are not diploid. Gametes, the sperm and eggs, contain only one copy of each gene as they have only one set of chromosomes. These cells are haploid and are produced by a special type of cell division called **meiosis**.

A male and female gamete join together at fertilisation, to form a new diploid cell called a **zygote**. This one cell divides by mitosis to produce a complete new organism (Fig 23.2). Reproduction and growth are covered in Chapters 20 and 21.

! SCIENCE FEATURE Cell turnover

As children, we grow because the cells in our bodies divide mitotically, and cell production outnumbers cell death. When we reach our adult size, our cell population stays relatively constant. The rate at which cells die and are replaced is known as **cell turnover**.

Research has shown that different tissues and organs have very different turnover rates. Brain cells, for instance, are not usually replaced, and after our 20s there is a slow but steady decline in number. This is not normally something to worry about – you won't run out. Other cells have a low turnover. Liver cells, for example, might divide only once every one to two years although, unlike the brain, they can regenerate if some cells are damaged or destroyed. Most of the high turnover cells are epithelial cells (Figs 23.3 and 23.4). These are the cells of the body that most frequently give rise to tumours. Cells of the bone marrow (Chapter 30) and cells in the testes (Fig 23.5) also divide very frequently.

Fig 23.4 Intestinal epithelial cells have a very rapid turnover: the entire gut lining is replaced every couple of days. Much of the content of these cells is digested and reabsorbed in an efficient recycling system, but even so, gut cells form a significant proportion of faecal matter

Fig 23.3 Scanning electron micrograph of human epidermis. Dead skin cells are constantly being lost and replaced by mitosis. A large proportion of household dust consists of human skin cells

Fig 23.5 The production of sperm is a remarkably intense process. The average healthy human male produces about 1000 new sperm every second. The process of sperm production, spermatogenesis, which involves meiosis, is described in Chapter 20

2 THE CELL NUCLEUS

The DNA of a cell is contained in the nucleus. In a cell that is not actively dividing, DNA exists as **chromatin** (see Fig 3.41). This has a granular, or grainy, appearance (Fig 23.6). Separate chromosomes are present, but the DNA is so spread out that we cannot tell one from another.

When a cell is about to divide, the chromosomes condense and separate. They become visible under the light microscope as dark, rod-like structures. Each pair of chromosomes is known as a **homologous pair**. This means that they both contain genes at the same positions, or **loci**. You received one chromosome of each pair from your mother and one from your father. So, although homologous pairs contain the same genes, they do not necessarily carry the same versions of each gene. This is the key to understanding why organisms vary. Alternative forms of genes are called **alleles** (see Chapters 22 and 25).

The appearance, number and arrangement of chromosomes in the nucleus are referred to as the **karyotype**. The human karyotype consists of 23 pairs of chromosomes (Fig 23.7). All diploid human cells contain 23 pairs of chromosomes and this is often given the notation $2n = 46$. This means that a diploid cell has 46 chromosomes (a haploid cell has 23). Other organisms have different numbers of chromosomes (Table 23.2).

Fig 23.6 In a non-dividing cell, the chromosomes are spread out as chromatin and appear as pale granules. In this form, the DNA is used as a template to make mRNA in the process of transcription. mRNA passes to the cytoplasm and is itself used as a template for protein production in the process of translation. In this cell, the nucleolus is also clearly visible. it is the dark sphere inside the nucleus

Fig 23.7 The 23 pairs of chromosomes in the human karyotype. Karyotypes are used to detect chromosomal abnormalities and to perform sex tests on athletes. The presence of XX confirms the athlete is female, the presence of XY shows the athlete is male

? TEST YOURSELF

B Which notation would you use to describe the number of chromosomes in human sperm?

✔ REMEMBER THIS

How can a potato have more chromosomes in its cells than humans do? It's quality, not quantity, that counts. All organisms have the same DNA but it's what the DNA codes for that makes the difference. When you translate potato DNA, you get a potato plant. When you translate human DNA, different proteins are made, and eventually a human being results.

Table 23.2 The number of chromosomes in the cells of some plants and animals, compared with humans. Note that there is no connection between chromosome number and the complexity of an organism. Some ferns have hundreds of chromosomes

Species	Number of chromosomes
Penicillin mould (*Penicillium notatum*)	5
Broad bean (*Vicia faba*)	12
Lettuce (*Lactuca sativa*)	18
Yeast (*Saccharomyces cerevisiae*)	34
Cat (*Felis cattus*)	38
Human (*Homo sapiens*)	46
Potato (*Solanum tuberosum*)	48
Chimpanzee (*Pan troglodytes*)	48
Horse (*Equus caballus*)	64
Chicken (*Gallus gallus*)	78
Dog (*Canis familiaris*)	78

3 THE STAGES OF THE CELL CYCLE

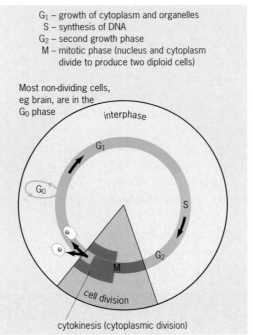

G₁ – growth of cytoplasm and organelles
S – synthesis of DNA
G₂ – second growth phase
M – mitotic phase (nucleus and cytoplasm
   divide to produce two diploid cells)

Most non-dividing cells,
eg brain, are in the
G₀ phase

interphase

G₁

G₀

S

G₂

M

cell division

cytokinesis (cytoplasmic division)

Fig 23.8 The four stages of the cell cycle: the three stages of interphase and the stage of active division in which mitosis occurs

The cell cycle is the complete sequence of events in the life of an individual diploid cell. The cycle starts and ends with cell division and consists of the stages of mitosis plus **interphase**, the interval between divisions in which the cell carries out its normal functions. Fig 23.8 shows the main stages of the life cycle of a cell.

Cells that do not divide, such as those in muscle and nerve tissue, are always in interphase: this is the normal state for a functioning cell. In interphase the DNA in the nucleus is unwound, and active genes are read to produce proteins such as enzymes.

A new cell has three options:

- It can remain stable, in interphase for many months or years. Brain and other nerve cells rarely divide, if ever.
- It can undergo mitosis within a short period of time. Skin cells and cells that line the gut all have a very rapid turnover (see the Science feature box on cell turnover on page 390).
- It can undergo meiosis. Specialised germ cells in the ovary and testes include meiosis in their cell cycle and produce egg cells or sperm.

INTERPHASE

We start looking at the cell cycle at the point when a cell has just divided. The cell is now in interphase. Interphase has three distinct stages: G_1, S and G_2.

G₁ – THE FIRST GROWTH PHASE

Just after it has been produced by division of its parent, a new cell is in the early part of the first **g**rowth phase, G_1, sometimes called G_0. Cells that do not divide remain at this point in the cell cycle because they do not need to go further: they never replicate their DNA. 'New' cells are relatively small, with a full-sized nucleus but relatively little cytoplasm.

During G_1, protein synthesis starts and the volume of cytoplasm and the number of organelles increase rapidly. The process is actually quite complicated, not least because some organelles such as mitochondria (which have their own DNA) divide independently of the cell nucleus. In later G_1, the cell takes on more 'normal' proportions: the nucleus begins to look smaller as the surrounding cytoplasm increases in volume.

THE S PHASE

The **S** phase – DNA **s**ynthesis phase – follows G_1. In this phase, the cell's DNA **replicates** (copies itself). Predictably, a cell only enters the S phase if it is going to divide. The point at which DNA replication starts is called the **restriction point**. After this, the cell becomes **restricted**, or locked into an automatic sequence that moves inevitably on to cell division.

G₂ – THE SECOND GROWTH PHASE

Before the actual mitotic cell division (see below), the cell enters a second, shorter **g**rowth phase, **G₂**, in which the proteins necessary for cell division are synthesised.

MITOSIS

Mitosis is a continuous sequence of events but, for clarity, they are divided into four distinct stages: **prophase**, **metaphase**, **anaphase** and **telophase**. The stages of mitosis are illustrated in Fig 23.9. You will need to refer to this constantly as you read the text.

Fig 23.9 The stages of mitosis

(a) Early prophase. The DNA has already replicated during interphase and the chromosomes begin to condense. Each chromosome is a double structure made from two genetically identical chromatids

— chromatin threads
— nuclear membrane
— nucleolus
— cytoplasm
— cell surface membrane
— centrioles

(b) Late prophase. The nuclear membrane has disappeared and the spindle has started to develop

— nuclear membrane
— nucleolus
— centriole
— centromere
— pair of chromatids

(c) Metaphase. The nuclear membrane has gone and the chromosomes arrange themselves on the equator (middle) of the spindle

— spindle fibres (microtubules)
— centromeres on 'equator' of spindle

(d) Anaphase. The chromatids are pulled apart and move to opposite poles

Daughter chromosomes move apart, led by their centromeres

(e) Telophase. Cytokinesis (cytoplasmic division) is beginning. The chromosomes becomes no longer individually visible

— nuclear membrane
— nucleolus
— chromatin threads
— pair of centrioles

(f) Cytokinesis. Cytoplasmic division is achieved

SEE QUESTIONS 2 TO 4

REMEMBER THIS

Exam hint – Remember the mnemonic: **IPMAT** – **I**nterphase, **P**rophase, **M**etaphase, **A**naphase, **T**elophase.

The movement of chromosomes during cell division is controlled by microtubules, which form the **spindle**. In a non-dividing cell the microtubules are found as two bundles of fibres, the **centrioles**, in an area of cytoplasm known as the **centrosome**, or **microtubule organising centre (MTOC)**. During cell division the centrioles move to opposite sides of the nucleus, from where they form the spindle.

PROPHASE

Chromatin begins to condense into chromosomes: Fig 23.9(a). At this stage, each chromosome has replicated itself and now consists of two identical **chromatids**. These are known as **sister chromatids** and are joined by a **centromere**. Unless there has been a mutation (a fault in the DNA replication), these sister chromatids are genetically identical. The subsequent stages of mitosis organise and split the pairs of chromatids, so that one chromatid of each pair goes into each daughter cell.

As the chromosomes condense, other changes occur in the cell:

● The nucleolus begins to break down.
● The centrioles move to opposite sides of the nucleus.
● The centrioles begin to assemble the spindle.
● The nucleolus disappears.
● The nuclear membrane begins to break up.

From prophase onwards, most 'normal' cell activity, such as protein synthesis and secretion, is halted until division is over.

METAPHASE

The beginning of metaphase (meta = middle) is marked by the disappearance of the nuclear membrane, which breaks down into separate vesicles, moves into the surrounding cytoplasm and joins with the endoplasmic reticulum: Fig 23.9(b).

The spindle (Fig 23.10) becomes fully developed and fills the space that was occupied by the nucleus. Then the most obvious event of metaphase happens: the chromatid pairs attach themselves to individual spindle fibres and align themselves on the equator of the spindle.

Fig 23.10 The spindle is a cradle of microtubule fibres which organise the chromosomes during cell division

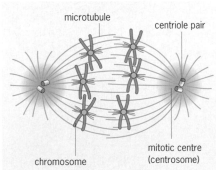

ANAPHASE

At the start of anaphase (ana = apart), as Figs 23.9(c) and 23.10 show, the chromatids are pulled apart by movements of the spindle fibres: sister chromatids are pulled to opposite poles. The newly separated chromatids are now called chromosomes, and are single structures. If you watch a film of mitosis (highly recommended for learning purposes), you will see that anaphase is the most obvious event.

TELOPHASE

In telophase (telo = final) the chromosomes reach the poles of the spindle: Fig 23.9(d). They then unravel and become indistinct, forming the familiar chromatin of interphase. The nuclear membrane re-forms and the nucleolus reappears. Soon afterwards, transcription resumes and the cell restarts protein synthesis, endocytosis and other normal cytoplasmic functions. This marks the end of mitosis.

CYTOKINESIS – DIVISION OF THE CYTOPLASM

In the events of mitosis just described, we have looked only at the splitting of the nucleus of the cell. Splitting of the cell itself is called **cytokinesis** and usually begins during anaphase. In most animal cells, microtubules form a furrow in a ring around the cell. These gradually constrict until the cells separate, as shown in Fig 23.11.

Fig 23.11 Cytokinesis in animal cells. As the events of mitosis come to an end, movement of microtubules causes a constriction around the centre of the cell. Eventually the cytoplasm divides, leaving two new daughter cells

MEIOSIS: MAKING THE GAMETES

Meiosis is the type of cell division that halves the genetic material in cells. For this reason, it is known as a **reduction division**. In humans, as in all animals, meiosis makes haploid gametes (sex cells).

Meiosis is a remarkable process that produces haploid gametes and shuffles the genes, so that each gamete produced is genetically different. This is why children born to the same parents are usually not identical: they are produced by the fusion of a genetically unique sperm and egg. Identical twins are an exception because they originate from the same fertilised egg.

THE STAGES OF MEIOSIS

Meiosis has two separate divisions. Both divisions have stages that are given the same names as in mitosis, but they have a number to denote whether they refer to the first or second division, for example anaphase I, prophase II.

It is important to remember here that a normal diploid cell contains two copies of each chromosome, one from the 'mother', and one from the 'father'. Just before meiosis, the cell's DNA becomes organised into chromosomes and replicates to form pairs of chromatids in preparation for division, just as it does before mitosis. So at the first stage of meiosis, the cell contains 92 (2 × 46) chromatids. The 23 chromosomes originally from the mother have been duplicated, making 23 identical pairs of chromatids, as have the 23 chromosomes originally from the father.

An overview of meiosis is shown in Fig 23.12. There is a basic difference between meiosis and mitosis: in meiosis, the homologous chromosomes pair up; in mitosis this does not happen. During the first part of meiosis, each homologous pair can swap sections of DNA, so that each pair becomes 'genetically mixed'.

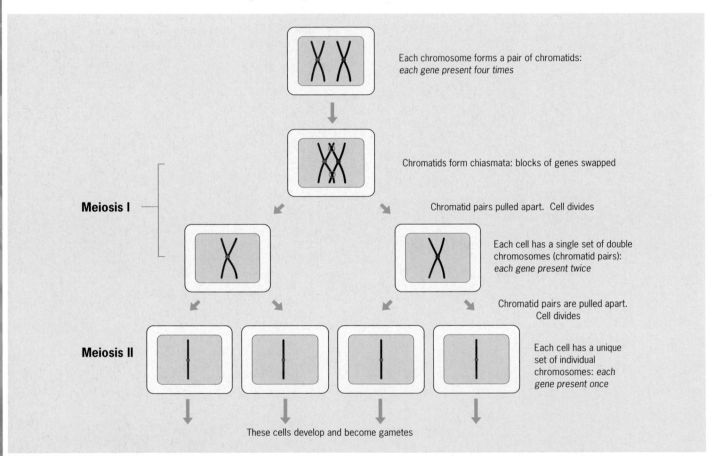

Fig 23.12 An overview of meiosis in an imaginary animal cell with just one pair of homologous chromosomes

The individual stages of meiosis are shown in Fig 23.13. In the first meiotic division (prophase I to telophase I), one of each homologous pair of 'genetically mixed' chromosomes passes into each new cell. In the second meiotic division, the individual chromatids are pulled apart so that one chromatid (now a chromosome) goes into each daughter cell. The end result of meiosis is four haploid cells, each containing a single set of chromosomes. Each cell is genetically unique.

Fig 23.13 The stages of meiosis in an imaginary animal cell showing just one of the pairs of chromosomes

Stage of meiosis First division	What is happening	What it looks like
Interphase	Just before meiosis, DNA replicates, so cells which contained two copies of each chromosome now have four. Chromosomes not yet visible	nucleolus
Early prophase I	Chromosomes become visible. Centromeres move to opposite sides of cell	centrioles — Early prophase I
Mid prophase I	Each homologous pair of chromosomes comes together to form a bivalent	Mid prophase I
Late prophase I	Each chromosome in a bivalent forms two chromatids. Genetic mixing occurs: chiasmata, the points of crossover, are visible	
Metaphase I	The bivalents arrange themselves on the equator of the spindle	spindle — Metaphase I
Anaphase I	The chromatid pairs from each homologous chromosome split apart and move to opposite poles of the cell	Anaphase I
Telophase I	Cytokinesis begins, two new cells form, each has two copies of each chromosome. These chromosomes are genetically different from those in the original cell	Telophase I
Interphase	A resting time (length varies between cell types)	
Second division		
Prophase II	A new spindle forms, at right angles to the first	
Metaphase II	Chromosomes, each of which is a pair of chromatids, align themselves on the equator of the spindle	Metaphase II (1 cell only)
Anaphase II	Chromatids are pulled apart to form two chromosomes that then move to opposite poles of the cell	Anaphase II (1 cell only)
Telophase II	Cytokinesis begins. Four haploid cells, each with only a single chromosome, have been formed. Each chromosome is genetically different	Telophase II

SEE QUESTIONS 5 AND 6

= maternal chromosome
= paternal chromosome
⊛ = position of chiasma

Fig 23.14 Chiasmata form between homologous chromosomes during late prophase I of meiosis. At these points, parts of the maternal chromosome separate and join with the paternal chromosome, and vice versa. This produces new chromosomes that are genetically different from each other, and from both the maternal and paternal chromosomes from which they are derived

PROPHASE I

Prophase I is more complex than the corresponding phase in mitosis, and can be subdivided into early, middle and late.

Early prophase I starts when the chromosomes condense and the nucleolus disappears.

In **mid-prophase I**, the homologous chromosomes, one from each parent, pair up. Each pair forms a bivalent. This does not happen in mitosis. When a chromosome pair is exactly aligned, it is said to be at **synapsis**. Next, each chromosome in a pair divides into two chromatids, giving four chromatids per bivalent.

In **late prophase I**, **recombination** – or crossover – takes place (Fig 23.14). One (or both) of the chromatids of the two homologous chromosomes breaks off at certain points and fuses with a chromatid of the other chromosome in the bivalent, forming joints called **chiasmata** (the singular, *chiasma*, means 'crosspiece'). This process ensures that blocks of genes are swapped between maternal and paternal chromosomes. The position of chiasma formation varies, even within the same species, and this produces a large variety of new gene combinations.

Prophase ends with the chromatids of each bivalent (pair of homologous chromosomes) entwined and joined by chiasmata.

METAPHASE I

In metaphase I, the nuclear membrane disappears and the spindle is fully developed. The bivalents move to the equator of the spindle in the same way as individual chromosomes do during the matching phase in mitosis.

ANAPHASE I

The chromatid pairs of a bivalent are pulled apart (Fig 23.15) because of the action of spindle fibres, a process that separates the entwined chromatids. At the end of anaphase I, the chromatid pair from one of the original homologous chromosomes is positioned at one pole of the cell and the chromatid pair from the other homologous chromosome is at the other pole. Either the maternal or the paternal chromosome can pass into either cell. This is the process that allows **independent assortment of alleles** (see Chapter 25). It is another reason why meiosis increases variation.

Occasionally, this phase is not completed successfully and a pair chromosomes fails to separate. The result is that both homologous chromosomes pass into one daughter cell, the other receiving neither. This situation can lead to conditions such as Down's syndrome.

TELOPHASE I

The spindle disappears and the nuclear envelope re-forms around the two sets of chromosomes. At the same time cytokinesis separates the cytoplasm, forming two daughter cells, ready for the second meiotic division.

INTERPHASE

The length of the resting interphase between the two meiotic divisions varies widely. It is sometimes short or even non-existent. If there is no interphase, the chromosomes remain condensed and the cell passes straight from telophase I into prophase II. In human females, however, ova may remain in interphase for decades. Basic egg cells are made in a girl before birth, but they do not complete meiosis until just before ovulation, anything up to 50 years later.

At the end of meiosis I, there are two cells, each containing two copies of each chromosome on a chromatid pair. The second meiotic division separates the chromatids, so that each daughter cell formed is haploid (has one set of single chromosomes).

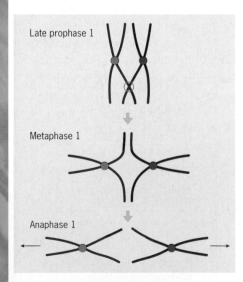

Late prophase 1

Metaphase 1

Anaphase 1

Fig 23.15 Chiasmata form during late prophase I and cross-over takes place. The newly formed chromatid pairs separate during metaphase I and then the two pairs move to opposite poles of the cell during anaphase I

THE SECOND MEIOTIC DIVISION

Prophase II

For each chromosome, the chromatid pair attaches itself to the new spindle, which forms at right angles to the first.

Metaphase II

Each chromatid pair lines up on the equator of the spindle.

Anaphase II

The chromatids are pulled apart to form two chromosomes that move to opposite poles.

Telophase II

The spindle disappears, the nuclear membrane re-forms, chromosomes unravel and cytokinesis produces two separate cells.

THE END-PRODUCT OF MEIOSIS

Meiosis produces four genetically different haploid cells, known as a **tetrad**. Genetic variation has been produced in three ways:

- The homologous chromosome pairs originate in different organisms, one maternal and one paternal, and so are genetically different.
- Blocks of genes are swapped between the chromatids of homologous chromosomes as the chiasmata form during prophase I.
- Each daughter cell can receive a copy of either chromosome from a pair, and each copy may have undergone crossover and have different genes from the other three. This is called independent assortment and is covered in more detail in Chapter 25.

> **? TEST YOURSELF**
>
> **F** How many bivalents are formed in a human cell undergoing meiosis?

> **✓ REMEMBER THIS**
>
> The importance of genetic variation should become apparent as you study evolution. Find out more in Chapter 26.

! SCIENCE FEATURE Down's syndrome

In the United Kingdom, two children in every 1000 are born with Down's syndrome (Fig 23.16a), a genetic condition that arises from a fault in meiosis. Individuals with this condition have an extra chromosome 21 (Fig 23.16b) and usually have physical and mental disabilities. They often have a small mouth with a normal-sized tongue (making eating and speech difficult), reduced resistance to disease and heart abnormalities.

The reason for the extra chromosome is usually a failure of the chromosome 21 pair to separate during anaphase I of meiosis (see Fig 23.13). Both chromosomes 21 pass into one daughter cell, and the other gets no copy at all. If a gamete with no chromosome 21 forms a zygote, the embryo fails to develop. But, if the gamete containing both chromosomes is fertilised, the resulting baby later develops Down's syndrome.

Fig 23.16(a) A child with Down's syndrome

One reason for the non-separation may be the length of time the egg spends in the ovary. In females, egg cells do not complete their meiotic division until they are 'selected' for ovulation (see Chapter 20). The longer the eggs remain in the ovary, the 'stickier' the chromosomes become and the greater the probability that they will not separate during anaphase I. This explains why the chance of having a child with Down's syndrome increases as women get older. Women below the age of 25 have less than a 1 in 2000 chance, but this rises to 1 in 50 for women over 45.

The age of the father is also important: recent studies have shown that in a significant minority of cases the extra chromosome comes from the father, that is, a normal ovum is fertilised by a sperm containing two chromosomes 21.

Fig 23.16(b) The complement of chromosomes (the karyotype) of a person with Down's syndrome, showing the extra chromosome 21

LEARNING MEIOSIS

Meiosis is a complex process that many people find difficult to learn. Here are some hints that you may find useful. First, divide your study of meiosis into three sections by looking at three questions:

What is the point of meiosis?

It makes gametes by shuffling the genes to produce haploid cells that are all genetically different from either of the parents. Each fertilisation (combination of an egg and a sperm) produces an individual that is genetically unique.

What do each of the meiotic divisions achieve?

Start by learning an overview of each division: Fig 23.13 is a good place to start. In the first meiotic division, homologous chromosomes pair up, divide into chromatids, swap blocks of genes and then separate. In the second division the individual chromatids separate. The end result is four genetically different haploid cells from one diploid original.

What are the stages of the two divisions?

Once you are confident that you can answer the first two questions, you can put some flesh on the bones. Describe one stage at a time and then draw what you have just described. Try using different colours for paternal and maternal chromosomes. Alternatively, you could make some chromosomes out of modelling clay and work through the meiotic sequence yourself. Many students find this a valuable exercise.

SUMMARY

After reading this chapter, you should know and understand the following:

- Almost all the cells in the human body are **diploid**: they contain a double set of chromosomes. The only exceptions are the gametes: sex cells – egg cells and sperm – are **haploid** and contain only one set of chromosomes.

- The cell cycle consists of four stages: G_1, in which the cell grows; **S**, in which the DNA is doubled; G_2, a second growth phase, and finally **mitosis** or **meiosis**. G_1, S and G_2 are collectively known as **interphase**.

- Mitosis is normal cell division, the process by which we grow and develop. One diploid cell divides to produce two genetically identical diploid cells.

- You can remember the stages of mitosis using the nmemonic IPMAT: Interphase, Prophase, Metaphase, Anaphase, Telophase.

- Non-dividing cells are said to be in interphase. There are no chromosomes visible and the active genes in the DNA are being used to direct the activities of the cell. If the cell is going to divide, the cell prepares itself during interphase and the DNA is replicated.

- In prophase, the chromosomes condense and become visible. The nuclear membrane disappears and the **spindle** forms. The chromosomes are double: two identical chromatids are joined at the **centromere**.

- In metaphase, the chromosomes arrange themselves in the middle of the spindle (remember, meta = middle).

- In anaphase, the double chromosomes are pulled apart (remember, ana = apart).

- In telophase, the nuclear membrane re-forms around the two new sets of chromosomes and the cytoplasm divides forming two new, genetically identical cells.

- Meiosis is a special type of cell division which produces sex cells or gametes. It is also known as a 'reduction division'. One diploid cell will divide meiotically to produce four genetically different haploid cells.

? EXAM QUESTIONS

1 The diagram shows a sperm and an egg from the same species, drawn to the same scale.

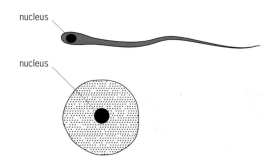

nucleus

nucleus

a) Assume that both the egg and its nucleus are spheres.
The volume of a sphere is given by $4\pi r^3/3$ where r is the radius of the sphere.
Calculate the proportion of the egg that is occupied by the nucleus.
Show your working.

b) The nucleus occupies a much smaller proportion of an egg cell than it does of a sperm.
Give **one** reason for this difference and explain its importance.

AQA (B) BYB2 Genes and Genetic Engineering June 2001 Q2

2 The diagram shows the main stages of the cell cycle. The letters A to D represent the four stages of mitosis.

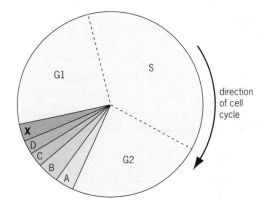

a) Identify the stage when each of the following events is taking place.
(i) DNA replication
(ii) Individual chromatids from a chromatid pair move to opposite poles of the cell.

b) What is happening during Stage **X**?

c) *Vinblastine* is an anti-cancer drug that prevents the formation of a spindle.
(i) What is the function of the spindle?
(ii) How would a drug like vinblastine help prevent the growth of a tumour?

AQA (A) BYA3 Pathogens and Disease January 2001 Q6

3 The simplified diagrams below show some of the main stages of mitosis, not in order.

a) State the name of each of the four stages shown in the diagrams.

b) Using the numbers 1, 2, 3 and 4, place the stages into the correct sequence to show the process of cell division.

c) State **two** differences between this type of cell division and meiosis.

d) The table below shows how the quantity of DNA varies with time in a cell cycle.

Time in hours	0	5	10	15	20	25	30
Quantity of DNA measured in arbitrary units	4	4	4	8	8	4	4

(i) Plot a graph of the data in the table, on graph paper.
(ii) Explain fully the changes that have happened to the cell and why the quantity of DNA varies in this way.

WJEC Module B11 June 2001 Q6

? EXAM QUESTIONS

4 The drawing indicates the appearance of a chromosome at early prophase of mitosis.

A

identical chromatids

a) With reference to the drawing,
(i) name the structure labelled **A**;
(ii) explain why the two chromatids are identical.
The diagram represents the nucleus of an animal cell ($2n = 6$) at early prophase of mitosis.

nuclear envelope

b) Copy the diagram and shade **one** pair of homologous chromosomes.
c) Draw an annotated diagram to indicate what happens in this cell at anaphase of mitosis.
d) **(i)** State the number of chromosomes which would be found in a **haploid** cell in this animal.
(ii) Explain why haploid cells need to be produced during a life cycle which includes sexual reproduction.

OCR 2801 Biology Foundation June 2001 Q6

5
a) Describe the features which would help you to recognise when a cell is:
(i) in metaphase of mitosis;
(ii) in anaphase of mitosis.

b) The diagram shows stages in the development of the female gametes in a flowering plant.

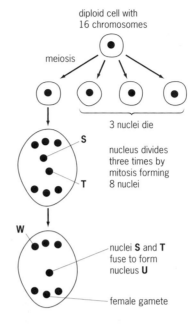

diploid cell with 16 chromosomes

meiosis

3 nuclei die

S

nucleus divides three times by mitosis forming 8 nuclei

T

W

nuclei **S** and **T** fuse to form nucleus **U**

female gamete

(i) Diploid nuclei of this species of flowering plant contain 16 chromosomes.
How many chromosomes has: nucleus **T**; nucleus **U**?
(ii) Does nucleus **W** have different alleles from nucleus **U**? Explain the reason for your answer.
(iii) During fertilisation, a haploid nucleus from a pollen grain fuses with nucleus **U**. How many chromosomes has the resulting nucleus?

AQA (B) BYB2 Genes and Genetic Engineering June 2001 Q3

6
a) Describe the behaviour of chromosomes during meiosis.
b) Explain the similarities and differences between homologous chromosomes.

OCR9264/3 Biology Linear Paper June 2000 Q8

24

DNA, genes and chromosomes

24 DNA, genes and chromosomes

Why do we age? Until a few years ago, scientists and doctors thought that our bodies simply wore out like machines. They now think that many aspects of ageing are genetically determined. Organisms develop and grow by cell division, and the same process happens in adults to repair and maintain their bodies. It seems likely that each of our cells has a genetic timer that counts how many times a cell divides: after a maximum number of divisions, the cell dies. But what sets this limit?

The latest research is focusing on the ends of chromosomes, the telomeres. Every time a cell divides, the telomeres shorten, and when they disappear completely, the cell is unable to divide further and its repair systems break down. Researchers have been intrigued to find that eggs and sperm are 'immortal' – they have an enzyme, telomerase, which ensures that the embryo's telomeres are complete. This is necessary to make sure that the genetic clock that produces a baby is set to zero. It might seem obvious, but babies need to be born with 'new' cells so that they have their full life expectancy, regardless of the age of their parents.

The discovery of telomerase opens up the exciting, if ethically difficult, possibility that we may be able to manipulate the ageing process. If we can use telomerase to repair the telomeres of normal body cells, it may be possible to extend potential life expectancy to 150 years, or more. The social and economic consequences of this are massive, and are bound to be the cause of much heated debate in years to come.

Imagine the implications of prolonging your life well past 100. How long would you work for? When would you retire? Could any government afford pensions and healthcare bills? What would happen to societies in which six, seven or even eight generations of the same family were alive together?

1 WHAT IS A GENE?

The nucleus of a eukaryotic cell has a set number of chromosomes – every human body cell has 46. Each chromosome is a single, elaborately coiled DNA molecule that has individual genes dotted along its length (Fig. 24.1). A gene is a length of DNA that codes for the synthesis of one polypeptide.

We used to think that a gene contained all the information needed to build a complete protein, such as an enzyme. However, many proteins consist of more than one polypeptide chain, so these molecules are coded for by more than one gene. For instance, haemoglobin molecules consist of four polypeptide chains, two alpha and two beta, so it takes two genes to make a complete haemoglobin molecule, one is copied twice for the alpha chains, the other twice for the beta chains.

A gene is a region within the DNA molecule

Fig. 24.1 Genes are sequences in the DNA molecule. The sequence of bases shown here represents one very short gene. In the genetic code, a group of three bases codes for one amino acid, so a protein consisting of 500 amino acids requires a gene of at least 1500 base pairs long, probably more. When unravelled, chromosomes are incredibly long – each one consists of up to 4000 genes and these make up only about 10 per cent of its total length (the rest is non-coding DNA)

GENES IN ACTION:
THE CENTRAL CONCEPT OF MOLECULAR BIOLOGY

Can we sum up the way genes work in one sentence? It is a tall order, but the underlying theme, or **central dogma**, of genetics is:

The DNA of a gene codes for the production of messenger RNA which, in turn, codes for the production of a polypeptide
(Fig 24.2).

In this chapter we look at how genes work, and at how they are copied to allow cell division, reproduction and growth.

After Crick and Watson worked out the structure of DNA in 1953, two vital questions remained:

- How does information get from the DNA in the nucleus to the site of protein synthesis in the cytoplasm?
- How does the base sequence of the DNA translate into the amino acid sequence in a polypeptide?

To answer the first question, Crick, Brener and Monod developed the **messenger hypothesis**. This states that a specific molecule, named **messenger RNA** (mRNA), is copied directly from the DNA sequence of the gene. mRNA is therefore a mobile copy of a gene. It can move from the nucleus to the cytoplasm. Here, ribosomes use mRNA as a template (pattern) to assemble amino acids in the correct order to make the required polypeptide.

Crick also answered the second question: he proposed that an 'adaptor' molecule existed which had two specific binding sites. At one side, the molecule fitted a specific DNA base sequence and at the other, a specific amino acid. This molecule was later discovered and named **transfer RNA** (tRNA) (see Fig 24.6 on page 409).

Fig 24.2 The central concept of molecular biology. The base sequence on a particular gene is copied, or transcribed, to produce a messenger RNA template. This mRNA template passes out of the nucleus and moves to the ribosomes where it is read, or translated, into a polypeptide

2 THE GENETIC CODE

Along the DNA molecule the base sequence is continuous, but it is read in blocks of three. Each three, or triplet, is called a **codon**. Each codon codes for a particular amino acid. A particular sequence of bases therefore codes for a specific amino acid sequence. When the amino acids are assembled into a polypeptide, they interact to twist and bend the chain into its final shape. When the polypeptide chain is complete, it forms all or part of a protein which performs a vital role in the organism.

You will have noticed that there are two strands of bases in the DNA molecule. One strand, called the **template strand** or **sense strand**, contains the all-important genetic code for any particular gene. The corresponding part of the other strand is simply there to stabilise the molecule. However, in the same DNA molecule, different genes are present on different sides and the enzymes that transcribe DNA can use either side as the sense strand.

> **REMEMBER THIS**
>
> The genome is defined as 'all the DNA sequences in an organism'. It is a complete set of genes, together with the non-coding DNA in between. The human genome is thought to consist of over 3 billion base pairs. Remarkably, the entire genome is present in every one of our body cells.

> **REMEMBER THIS**
>
> The position of a gene on a chromosome is known as its **locus** (plural: loci).

> **REMEMBER THIS**
>
> Retroviruses such as the Human Immunodeficiency Virus (HIV) are an exception to the central dogma. Retroviruses contain RNA from which they make a DNA copy using the enzyme **reverse transcriptase**.

> **SEE QUESTIONS 1 TO 3**

> **REMEMBER THIS**
>
> The base pairings are always the same. In DNA, A bonds with T and G with C. In RNA, T is replaced by U, so A bonds with U, and G with C (see Chapter 3).

SCIENCE FEATURE Nucleic acids — molecules with jobs

A favourite topic for biology examiners is to ask the candidate to relate structure to function. On a molecular level, the nucleic acids are classic examples. This feature builds on your knowledge of the basic structure of nucleic acids covered in Chapter 3. The differences between DNA and RNA are shown in Table 24.1.

Table 24.1 The differences between DNA and RNA			
	DNA	**mRNA**	**tRNA**
Sugar	deoxyribose	ribose	ribose
Bases	C, G, A, T	C, G, A, U	C, G, A, U
Strands	two	one	one
Size of molecule	enormous	small compared with DNA, varies with size of gene	small. constant size
Life span	long term	short term	short term
Site of action	nucleus	nucleus and cytoplasm	cytoplasm

DNA

How is the structure of DNA related to its function?

- It contains a sequence of bases that codes for a sequence of amino acids. Cells use this genetic code to make the proteins that build organisms. The base sequence is read in blocks of three, eg GAU – this is a **codon**.
- It is long, so can store a lot of information.
- It has complementary base pairing that allows replication and transcription.
- It has a stable sugar–phosphate backbone to give strength.
- It has relatively weak hydrogen bonds between the bases. This allows the two strands to come apart – again allowing replication and transcription.

Messenger RNA

mRNA is simply a mobile copy of a gene, carrying a base sequence that is complementary to – not the same as – the gene on which it was made. mRNA molecules are long, ribbon-like molecules consisting of a single strand of nucleotides. Molecules of mRNA vary in length – there are as many different ones as there are genes.

Transfer RNA

This is the adaptor molecule, connecting the genetic code to the protein. tRNA molecules have an **anticodon** on one end and an amino acid binding site at the other. The role of tRNA is to pick up a particular amino acid, bring it to the ribosome and hold it in place so it can be added to the growing polypeptide. tRNA binds to the mRNA by means of the anticodon. It could be argued that because there are 20 different amino acids, there are 20 different types of tRNA molecule. Another way of looking at it is that there are 61 different codons (64 minus 3 **stop codons** so there are 61 different types of tRNA.

Fig 24.3 Comparison of the three types of molecule

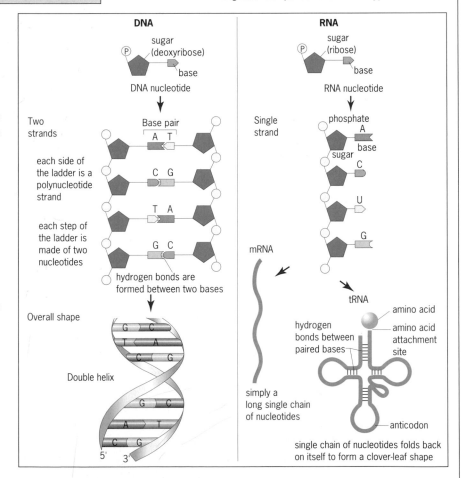

TRANSCRIPTION

DNA can be thought of as a permanent reference library: it does not leave the nucleus and so cannot be used directly for protein synthesis. Instead, the genetic code in a gene is transferred, or **transcribed**, on to a smaller RNA molecule, **messenger RNA**, which can then pass to the ribosomes to act as a template for protein synthesis. In effect, mRNA is a mobile copy of a gene.

Before transcription can begin, the DNA helix must unwind and the two halves of the molecule must come apart, so exposing the base sequence. This process begins when the DNA is 'unzipped' by specific enzymes and the enzyme **RNA polymerase** attaches to the DNA molecule at the **initiation site**. This is a base sequence at one end of the gene which effectively says: start here. Once in place, the enzyme moves along the gene (Fig 24.4), assembling the messenger RNA molecule by adding the matching nucleotides, one at a time.

Once formed, the mRNA copy begins to peel away from the DNA. When RNA polymerase has passed a particular region of DNA, the double helix rewinds. When the whole gene has been transcribed, the complete mRNA molecule leaves the nucleus. As on a production line, RNA polymerase molecules can follow each other along the gene, assembling several copies of mRNA together.

? TEST YOURSELF

A The RNA polymerase enzyme moves along a DNA sequence reading ATACGCTAT.
a) What is the corresponding sequence on the RNA molecule?
b) How many amino acids are encoded in this sequence?

SEE QUESTIONS 4 AND 5

(a) Part of the DNA molecule untwists

RNA nucleotides attach to the free bases on one strand of DNA

mRNA molecule

(b) pore in nuclear membrane

nuclear membrane

cell nucleus

DNA of chromosome (not to scale)

The mRNA moves through the nuclear membrane and attaches to the ribosomes

ribosomes

rough endoplasmic reticulum

⬜ = adenine   ⬛ = guanine   ◁ = uracil

◻ = thymine   ◼ = cytosine

TRANSLATION

Translation occurs when the base sequence on the mRNA is used to synthesise a polypeptide. The process is called translation because the information contained in the gene – and delivered by the mRNA – is translated into a polypeptide (Table 24.2 on the next page).

Protein synthesis is one of the major activities of cells, and it is highly likely that you are already familiar with many of the organelles and chemicals involved. The roles of individual organelles in the process are listed in Table 24.3. See also Chapters 1 and 3 for background information.

Translation occurs on **ribosomes**, structures that hold all the components together as an amino acid chain is made. The mRNA strand passes out of the nucleus and attaches to the ribosome. At the point of attachment, two tRNA molecules deliver the required amino acids and hold them in position so that they can be added to the growing polypeptide.

The mechanism of translation is shown in detail in Fig 24.6. The process continues until an mRNA **stop codon** – which has no tRNA molecule – moves into the ribosome's first binding site. At this point all the components separate, releasing the completed polypeptide.

Fig 24.4(a) The process of transcription. The process is controlled by enzymes that are not shown here. Starting at the initiation site, helicase enzymes separate the two sides of the DNA molecule, and then the enzyme RNA polymerase calalyses the assembly of mRNA, one nucleotide at a time, until the gene has been transcribed
(b) The sites of transcription and translation in the cell. Transcription happens in the nucleus and involves the assembly of a molecule of messenger RNA. Translation happens on the ribosomes and involves the assembly of a protein according to the code on the mRNA

? TEST YOURSELF

B How do mammals ensure that a regular supply of amino acids reaches all of their cells?

Table 24.2 The genetic code: the base sequence in each mRNA triplet code translates to a particular amino acid. This same code is used by all organisms

First base	Second base				Third base
	G	A	C	U	
G	GGG glycine	GAG glutamic acid	GCG alanine	GUG valine	G
	GGA glycine	GAA glutamic acid	GCA alanine	GUA valine	A
	GGC glycine	GAC aspartic acid	GCC alanine	GUC valine	C
	GGU glycine	GAU aspartic acid	GCU alanine	GUU valine	U
A	AGG arginine	AAG lysine	ACG threonine	AUG methionine	G
	AGA arginine	AAA lysine	ACA threonine	AUA isoleucine	A
	AGC serine	AAC asparagine	ACC threonine	AUC isoleucine	C
	AGU serine	AAU asparagine	ACU threonine	AUU isoleucine	U
C	CGG arginine	CAG glutamine	CCG proline	CUG leucine	G
	CGA arginine	CAA glutamine	CCA proline	CUA leucine	A
	CGC arginine	CAC histidine	CCC proline	CUC leucine	C
	CGU arginine	CAU histidine	CCU proline	CUU leucine	U
U	UGG tryptophan	UAG **stop**	UCG serine	UUG leucine	G
	UGA **stop**	UAA **stop**	UCA serine	UUA leucine	A
	UGC cysteine	UAC tyrosine	UCC serine	UUC phenylalanine	C
	UGU cysteine	UAU tyrosine	UCU serine	UUU phenylalanine	U

Table 24.3 The main organelles and chemicals involved in protein synthesis

Organelle/molecule	Role in protein synthesis
Nucleus	houses the DNA
Nucleolus	manufactures the ribosomes
Ribosome	site of translation – where polypeptide assembly occurs
Endoplasmic reticulum	isolates, stores and transports polypeptides
Golgi apparatus	modifies and packages polypeptides
DNA	stores the genetic information
Messenger RNA	is a mobile copy of a gene on DNA
Transfer RNA	each brings a specific amino acid to the ribosomes

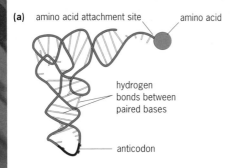

(a) amino acid attachment site — amino acid — hydrogen bonds between paired bases — anticodon

(b) amino acid attachment site — amino acid — hydrogen bonds between paired bases — anticodon

Fig 24.5 Two ways of representing the tRNA molecule. **(a)** The 3D shape of the molecule, stabilised by hydrogen bonds as the single nucleotide chain folds back on itself
(b) A schematic diagram: each type of transfer RNA transports a specific amino acid to the ribosomes, the site of protein synthesis. At one end of the molecule is the anticodon which attaches to the corresponding codon on the messenger RNA. At the other end of the molecule is the amino acid attachment site

After translation, polypeptides are processed according to their final destination:

● Those that are to be exported from the cell, such as digestive enzymes, are threaded through pores in the endoplasmic reticulum to accumulate on the inside. Here they are processed and packaged before being secreted (see pages 12 and 14).

● Polypeptides that will form membrane proteins follow the same route as those for export, but they remain on the cell surface membrane rather than being released.

● Polypeptides that will be used inside the cell, such as those that form haemoglobin, remain free in the cytoplasm.

THE ROLE OF TRANSFER RNA

Transfer RNA molecules are smaller than mRNA molecules, containing only about 75 to 80 nucleotides. Each consists of a single strand of nucleic acid folded back on itself to form a 'clover-leaf' shape (Fig 24.5). Transfer RNA molecules bring specific amino acids from the cytoplasm to the ribosome so that they attach to the growing polypeptide.

Fig 24.6(a) The ingredients needed for protein synthesis

P site: site of attachment of the growing polypeptide

A site: each new amino acid is added here

large subunit of the ribosome

small subunit of the ribosome

amino acid

tRNA

tRNA–amino acid complex

anticodon

mRNA strand

Fig 24.6(b) The process of protein synthesis

(1) The two subunits of the ribosome come together. The mRNA strand binds to the ribosome. Then an amino acid–tRNA complex with an anticodon complementary to the first codon on the mRNA, binds to the P site

(2) In the next step, an amino acid–tRNA complex with an anticodon complementary to the next codon on the mRNA strand binds to the A site

peptide bond

first tRNA leaves ribosome

tRNA joined to dipeptide

direction of mRNA movement

(3) A peptide bond forms between the two amino acids which have been brought together on the ribosome. The bond between the first amino acid and the first tRNA is broken and the tRNA molecule leaves the ribosome. The ribosome moves along the mRNA strand and the second tRNA molecule, now joined to a dipeptide, shifts across from the A site to the P site

(4) A third amino acid–tRNA complex with an anticodon that matches the third codon on the mRNA strands comes into the A site

The sequence of events from **(3)** to **(4)** is repeated until the entire length of the mRNA strand has been translated. The polypeptide is then complete

The process by which an amino acid binds to a tRNA molecule is controlled by an enzyme. The process also involves the splitting of ATP. This is important because ATP gives the tRNA–amino acid complex enough energy to form a peptide bond when the amino acid is added to the growing polypeptide.

At one end of the tRNA molecule is the **anticodon**, a three-letter base sequence that matches the codon on the mRNA molecule. At the other end is

Fig 24.7 A polysome is a cluster of ribosomes which can simultaneously translate a different section of the mRNA molecule. The situation is similar to having a message on a long ribbon, with several people reading different bits at once

a particular amino acid. So, for instance, if the codon on the mRNA reads AUG, which codes for the amino acid methionine, it needs a tRNA molecule with an anticodon of UAC which arrives at the ribosome carrying a methionine molecule.

There are 64 different codons, but three of them code for 'stop translating'. When such a codon arrives at the ribosome, no tRNA is needed. This means that tRNA molecules must match with 61 different codons. There are only 20 different amino acids, so some amino acids are translated from more than one codon. For example, the codons GGG, GGA, GGC and GGU all code for the amino acid glycine.

Like messenger RNA, transfer RNA is also made by transcription. In eukaryote DNA there are many genes whose sole function is to make tRNA. In any cell, most genes occur only once, but there are hundreds of genes that code for tRNA synthesis, to provide the vast numbers of these workhorse molecules that are needed. That tRNA genes do not code for proteins is another exception to the central dogma.

POLYSOMES AND THE RATE OF TRANSLATION

We have seen that the mRNA molecule is a long single strand of nucleotides, and that the code is translated into a polypeptide at the point of contact with a ribosome. To achieve protein synthesis at a reasonable speed, ribosomes occur in clusters, called **polysomes** or **polyribosomes**, which all translate a different bit of the mRNA at the same time (Fig 24.7).

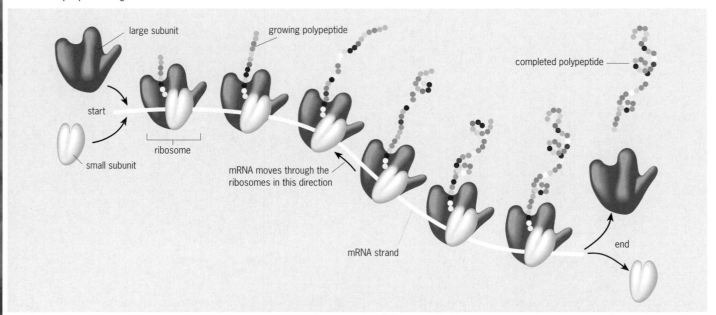

large subunit

growing polypeptide

completed polypeptide

start

ribosome

small subunit

mRNA moves through the ribosomes in this direction

mRNA strand

end

3 A JOURNEY ALONG A CHROMOSOME

If one single chromosome were unravelled, like pulling the wool out of a piece of knitting, we would be left with one long DNA strand. Although this is one single molecule, it is several *centimetres* long and contains up to four thousand genes.

How do you go about investigating such a molecule? One way is to work backwards; if you can isolate the mRNA molecules that are in the cytoplasm, you can trace their origin in the DNA.

A clever technique called **nucleic acid hybridisation** is used to pinpoint the origin of the mRNA, giving the location of the gene itself. In this technique, DNA strands are denatured: when heated to 87 °C, hydrogen bonds that hold the two strands together are broken, and the individual strands separate. If messenger RNA strands from the cytoplasm are added to the mixture, they stick to the region of DNA on which they were originally made.

Studies using this technique have unexpectedly revealed that there is some non-coding DNA in genes (Fig 24.8). Most eukaryote genes contain regions, called **introns**, which do not find their way into the mRNA molecule at all. The parts of the gene that are expressed in the mRNA are called **exons** because they are the expressed regions of the gene.

Generally, a gene consists of a DNA sequence which is present only once. The non-coding sequences, however, tend to consist of repeated or 'stuttered' sequences. Humans all have the same repeated sequence, but in different individuals it is repeated a different number of times and in different places. This is the basis of DNA profiling – see the Assignment in Chapter 3.

The function of non-coding DNA is not known. It is probably not useless: much of the repeated DNA in a chromosome occurs around the region of its centromere and may have an important role in maintaining the physical stability of the chromosome as a whole.

The DNA code can be compared to a page of writing in a book. If the letters were strung together continuously, we would find it difficult to read: the gaps between words help us to make sense of the sentences.

4 THE CONTROL OF GENE EXPRESSION

Most multicellular organisms start life as a zygote: a single fertilised cell. This cell has a complete set of genes. As the cell divides by mitosis, the DNA is copied faithfully, so each new cell also contains the same complete set of genes. How, then, do cells differentiate? How, for example, do some animal cells specialise into muscle, nerve or skin?

The answer lies in **gene expression** – how different genes are 'switched on', or **expressed**, in different cells (Fig 24.9).

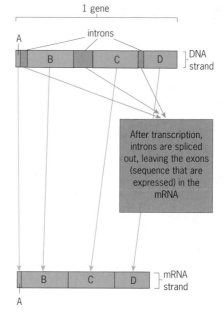

Fig 24.8 Even in a gene, there are non-coding lengths of DNA called **introns**. Only the **exons** are encoded into the mRNA molecule, and therefore expressed

Fig 24.9 A spectacular example of the power of gene expression is seen in the life cycle of some insects such as this peacock butterfly. The genes present in the fertilised egg are still present in the adult, but a different set is activated to give rise to the caterpillar and to control its metamorphosis into a pupa and then an adult – which looks like a completely different organism

At any one time, the average cell is probably using only 1 per cent of its available genes. Some of the expressed genes are used for 'housekeeping' – they code for the proteins needed by all cells, such as the enzymes involved in respiration. In contrast, other genes are expressed only in specialised cells. The genes for making insulin, for example, are expressed only in the β cells of the islets of Langerhans in the pancreas (see Chapter 11).

Selective activation of genes is the key to the development of a complex multicellular organism and, not surprisingly, it is the subject of intense research. If we could understand how and why some genes are activated while others remain dormant, we could unlock the mystery of embryonic development – how a single fertilised egg grows and differentiates into something as complex as a human being.

5 DNA REPLICATION

Whenever a cell divides, the DNA must be copied. Otherwise, there would be no reproduction and no growth. When cells divide, each new daughter cell must have a complete set of genes. **Replication**, the mechanism of DNA copying, is therefore of fundamental importance.

Just three years after Crick and Watson published their model of DNA structure in the journal *Nature*, their theory about DNA replication was confirmed by Arthur Kornberg. He put DNA into a mixture containing all four **nucleotides** (each of the four bases attached to a phosphate and a sugar), together with the enzyme DNA polymerase, and showed that the DNA could replicate without any other factors present.

In the mixture, each DNA strand unwinds and acts as a template for the construction of a new strand. The exposed strand acts as a template on which the free nucleotides arrange themselves in exactly the same sequence as the intact strand they replace. This model is called **semi-conservative replication** because each of the resulting strands of DNA contains one strand from the original DNA and one newly synthesised strand: half has been conserved and half is new.

Strong support for the semi-conservative model of DNA replication came from the work of Meselson and Stahl (Fig 24.10). By growing bacteria with bases containing the radioisotope ^{15}N (sometimes called heavy nitrogen) and then following its progress, they showed that each new DNA strand contains half the original strand.

Fig 24.10 Meselson and Stahl's experiment to support the semi-conservative replication model of DNA replication

1 To start with, many generations of *E. coli* bacteria are grown in a medium containing nucleotides labelled with the heavy nitrogen radioisotope, ^{15}N. Eventually, almost all of the bacterial DNA contains ^{15}N, and so is denser than normal DNA. A control culture of the same bacterium is grown with 'normal' ^{14}N nucleotides for comparison

2 The bacteria grown in ^{15}N are transferred to ^{14}N medium. After 0, 20 and 40 minutes, samples of the DNA are extracted, placed in a solution of caesium chloride and centrifuged. This separates out the DNA according to its density and allowed Meselson and Stahl to distinguish between DNA containing ^{15}N and ^{14}N

3 The results confirm that, initially, all the DNA contained ^{15}N. After 20 minutes the DNA had replicated, producing two strands of $^{15}N/^{14}N$ hybrid DNA. After 40 minutes, the second generation strands consisted of half $^{15}N/^{14}N$ hybrids and half all-new ^{14}N strands

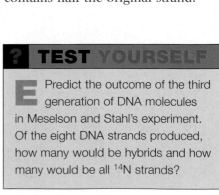

THE MECHANISM
OF SEMI-CONSERVATIVE REPLICATION

The basic mechanism of DNA replication is shown in Fig 24.11(a). DNA is copied before the cell divides (see Chapter 23) and the basic idea is simple: the helix unwinds and each strand acts as a template for the manufacture of a new, identical strand.

In practice, DNA replication is rather more complex than it appears, and is a good illustration of the importance of enzymes (Fig 24.11b). Two **helicases** use energy from ATP to separate the DNA strands. Next, **DNA binding proteins** attach to keep the strands separate. **DNA polymerase** enzymes then move along the exposed strands, synthesising the new strands, often in short segments. Finally, **DNA ligase** enzymes join the segments together, completing the new DNA strands.

Fig 24.11(a) When DNA replicates, the double helix unwinds and each exposed strand acts as a template for the synthesis of a new strand. This is a simplified diagram showing how free nucleotides are added to produce two new DNA helices

Fig 24.11(b) DNA replication is actually more complicated and several enzymes other than DNA polymerase are involved. DNA helicase unwinds the parent helix and then this is 'held open' by DNA binding proteins so that the DNA polymerase can gain access to the parent strand

DNA PROOFREADING

Mistakes in DNA replication occur regularly. The chances of the wrong base being added are between 1 in 10^8 and 1 in 10^{12}. This may seem low, but so many nucleotides are added that this rate means approximately one mistake occurs for every ten genes that are copied. Clearly, such a rate would cause the death of most organisms. This disaster is prevented by the polymerase enzymes themselves. They can detect when the wrong base has been inserted, and pause to correct the mistake. In this way 99.9 per cent of mistakes are corrected. Uncorrected mistakes give rise to **mutations**.

There are many different DNA repair mechanisms in cells to put right the genetic damage that inevitably accumulates during the life of an organism. The condition xeroderma pigmentosum illustrates the importance of these mechanisms: those affected are liable to get skin cancer because their cells cannot repair genetic damage caused by exposure to ultraviolet light (Fig 24.12).

Fig 24.12 Xeroderma pigmentosum is a rare genetic disease characterised by dry, freckled skin which is extremely sensitive to sunlight. One theory suggests that individuals with this condition, who have to stay indoors, could have given rise to the vampire legends of central Europe

SCIENCE FEATURE How do you compare the DNA of different organisms?

The technique of DNA hybridisation (Fig 24.13) can be used to compare the similarity of the DNA of different species, and is proving to be a useful tool in the search to piece together evolutionary trees.

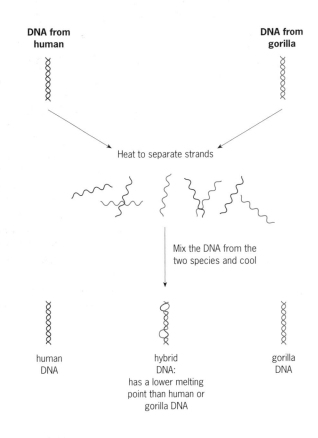

Fig 24.13 The principle of DNA hybridisation

DNA is much more stable than protein because it is held together by regular hydrogen bonds along the whole of its length. However, when DNA is heated to about 87 °C, known as the melting point of DNA, the hydrogen bonds break and the two halves of the molecule separate. When the temperature drops below 87 °C, the two strands join up: they **re-anneal**.

This is the basic process of DNA hybridisation:

DNA from two different species is cut into small fragments and heated to 87 °C so that the two strands separate. The separated strands from each species are then mixed together. As the temperature drops the DNA strands re-anneal. Some of the molecules formed will be hybrids: they contain one strand from each species.

The temperature at which the two hybrid strands re-anneal tells us about the degree of similarity between them. When two strands from the same species re-anneal, they do so at 87 °C. Closely related DNA re-anneals at temperatures slightly lower but close to 87 °C. This indicates that they have many base sequences in common and so form many hydrogen bonds along the length of the molecule. In contrast, DNA from more distantly related species must cool to much lower temperatures before they can anneal as there are fewer hydrogen bonds to hold the strands together.

For example, human DNA can be hybridised with gorilla DNA or rabbit DNA. The melting point for DNA from all three single species is 87 °C. The melting point for the human/rabbit DNA is lower than for the human/gorilla DNA. This shows that humans and gorillas have more DNA sequences in common than humans and rabbits.

POLYMORPHIC GENES

Genes that have two or more alleles are called polymorphic. Many genes have no alleles – they code for essential proteins and any mutation in them is selected against. For instance, the genes that code for the enzymes in respiration must be copied exactly time after time, or the organism dies. Some genes, however, can cope with a bit of variation. The genes that code for eye colour or hair colour are obvious examples. In humans, it has been estimated that 31.7 per cent of our genes are polymorphic.

MUTATIONS

A mutation is a spontaneous change in the genetic material of an organism (Fig 24.14b). It usually occurs when DNA is copied and cells divide. Generally, there are two types of mutation:
- a gene mutation,
- a chromosome mutation.

Both of these may occur in **somatic cells** (normal body cells) or **germ cells** (sex cells).

Fig 24.14(a) To many people, this is the image conjured up by the word mutant. But in biology, a mutation is a fault in the copying of genetic material and is usually lethal or makes no difference to the organism. Only in rare cases will mutation give rise to new variation, and it is sure to be less spectacular than science fiction writers would have us believe

Fig 24.14(b) An example of a real-life mutation. A chance mutation in a single gene of the speckled moth *Biston betularia* produced a fortunate change from the normal speckled coloration, seen in the moth on the left, to the melanic form seen on the right. In sooty areas, this new form was better camouflaged and so had a better chance of avoiding predators. More of the black moths passed on their genes to the next generation, making the melanic form dominant in industrialised areas

Somatic cell mutations are not passed on to the next generation. Generally, organisms accumulate somatic mutations as they get older, and this is thought to be one of the causes of cancer (see Chapter 31). In contrast, when a mutation arises in an egg or sperm cell which then goes on to become a zygote, the mutation is passed on to all the cells of the new individual. Germ cell mutations alter the genome of an organism.

GENE MUTATIONS

A gene mutation is a change in the base sequence of DNA. We have already seen that when DNA replicates, two identical strands are formed. Occasionally, however, a fault can occur. In a typical gene of, say, 1400 base pairs, a change in just one of the base pairs has the potential to make the whole gene useless, and can prove lethal to the organism. A change in a single base is known as a **point mutation**.

How can such a tiny change be so disastrous? It is because a change to any base will alter a codon so that it will probably code for a different amino acid. In turn, this 'wrong' amino acid can affect the way the polypeptide chain folds, so changing the shape of the whole protein molecule. It might then be unable to function. (Remember: many genes code for enzymes or parts of enzymes whose proper functioning relies on precise shape.) Table 24.4 shows the different types of gene mutation that can occur, using words instead of codons. In all cases, the original meaning of the message is lost.

SEE QUESTION 7

? TEST YOURSELF

F From the information given in Table 24.4, which type of mutation is likely to cause the fewest changes to a protein?

Table 24.4 Different types of gene mutation	
Type of gene mutation	**Effect on code (using words rather than codons)**
Normal code	THE FAT OLD CAT SAW THE DOG
Addition (frame shift)	THE EFA TOL DCA TSA WTH EDO G
Deletion (frame shift)	THE ATO LDC ATS AWT HED OG
Substitution	THE FAT OLD BAT SAW THE DOG
Duplication	THE FAT TFA TOL DCA TSA WTH EDO G
Inversion	THE TAF OLD CAT SAW THE DOG

From Table 24.4 we can see that the two least damaging types of mutation are substitutions and inversions, because these alter only one of the seven amino acids. In some cases, if the new amino acid occurs in a non-vital part of the chain, or has similar properties to the correct one, then the mutation might not actually matter at all.

In contrast, deletions and additions are potentially much more damaging because they cause **frame shifts**: only one base may be added or lost, but this causes the whole sequence to shift along, altering all the codes. In this case, the protein made is completely different from the original and is unlikely to function as it should. The Assignment at the end of this chapter deals with sickle cell anaemia, a disease caused by a fault in just one base.

HOW COMMON ARE GENE MUTATIONS?

The rate at which genes mutate varies between species and between genes at different loci. At a rough estimate, there is an average mutation rate of about 10^{-5} per locus per generation, meaning 1 in 100 000. So, in a population of one million individuals, you could expect ten to have inherited a mutation at any particular gene. The rate of gene mutation (and chromosome mutation) can be greatly increased by environmental factors, known as **mutagens**, such as ultraviolet, X-rays, alpha and beta radiation, and chemicals such as mustard gas and cigarette smoke.

CHROMOSOME MUTATIONS

During mitosis (normal cell division) chromosomes are copied, condensed and pulled apart. The end result should be that each daughter cell receives a complete set of perfect chromosomes. But faults can occur.

A chromosome mutation occurs when the structure of a whole chromosome – or set of chromosomes – is altered in some way.

Fig 24.15 shows the four basic types of chromosome mutation: deletion, duplication, inversion and reciprocal translation.

While gene mutations normally affect just one gene, chromosome mutations involve the disruption of whole blocks of genes. For this reason, many chromosomal mutations give rise to **syndromes**, which are complex sets of symptoms with a single underlying cause.

In addition to the structural mutations outlined above, another common failure is **non-disjunction**. This occurs when pairs of chromosome fail to separate during anaphase I of meiosis. Some gametes get both chromosomes, while others receive neither. Down's syndrome, for example, occurs when a gamete containing two copies of chromosome 21 joins with a normal gamete containing one, giving the affected individual three copies of chromosome 21 (see Chapter 23).

POLYPLOIDY

Occasionally, a mutation causes a whole set of chromosomes to be changed. For instance, a fault in meiosis might result in some gametes having two sets of chromosomes while others receive none. When such a diploid gamete joins with a normal one, the result is a **triploid** zygote – one with three sets of chromosomes. In another case, a fault in mitosis after

Fig 24.15 The main types of chromosome mutations
(a) Deletion occurs when a chromosome splits and a fragment is lost as the parts rejoin. In this case, a whole set of genes is lost, and so this is often fatal
(b) A duplication occurs when a section of chromosome is copied twice. This can occur when pairs of homologous chromosomes break at different places and then reconnect to the wrong partners. Thus, one of the partners would have two copies of a particular sequence, but the other would have neither. This can have serious consequences
(c) An inversion results when a broken segment is reinserted in the wrong order. Any genes contained in this region will be transcribed and translated backwards, and will therefore not make the correct proteins
(d) A reciprocal translocation occurs when two non-homologous chromosomes exchange segments. This makes the chromosome pairs of unequal size, and this can cause problems in meiosis, with daughter cells receiving the wrong chromosomes

(a) Deletion
genes

A B C D E F G → A B C F G   D E
lost

(b) Duplication and deletion (between homologous chromosomes)

A B C D E F G → A B E F G

A B C D E F G → A B C D C D E F G

(c) Inversion

A B C D E F G → A B C F E D G

this section is inverted

(d) Reciprocal translocation (between non–homologous chromosomes)

A B C D E F G → A B P Q R S

L M N O P Q R S → L M N O C D E F G

fertilisation could result in the chromosomes doubling but not separating, leaving four sets in a single **tetraploid** cell. Neither of these embryos could develop.

The possession of multiple sets of chromosomes is known as **polyploidy**. This phenomenon is very common in flowering plants and has played a vital role in their evolution. Botanists estimate that over half of the world's flowering plant species are polyploid, including some of the world's most economically important crops: sugar beet, tomatoes and tobacco.

WHY MUTATIONS ARE IMPORTANT

A mutation can have one of several effects:

- It can be lethal: an organism with a lethal mutation cannot survive and might not even develop beyond the zygote. Genes code for vital proteins such as enzymes, so many gene mutations mean that the protein is the wrong shape, and so cannot function metabolically. Chromosome mutations which involve the loss of whole blocks of genes are usually lethal.
- It can have no effect. It might occur in a non-coding part of DNA, or in a gene that is not expressed. Alternatively, it can result in a different amino acid being incorporated into the protein, but one that does not alter its ability to function.
- It can be beneficial. Occasionally, a mutation produces an improvement in phenotype. Statistically, this is a very rare event, but given the number of DNA replications and the number of individual organisms, it is bound to happen sooner or later. Beneficial mutations are hugely important because, ultimately, they are the source of all variation. Natural selection favours these 'improved' mutants (see Chapter 26).

Fig 24.14(b) on page 415 shows an example of a beneficial mutation. Following the Industrial Revolution, a mutation in the peppered moth produced a black (melanic) form that was better camouflaged against sooty backgrounds. The black moth had an advantage because it was hard for predators to see and so survived to pass on its genes to the next generation. Consequently, the mutated gene spread and the genome of the species was altered (see Chapter 26).

> **REMEMBER THIS**
>
> Meiosis is also known as reduction division. See Chapter 23 for more detailed information about this important process.

> **? TEST YOURSELF**
>
> **G** Polyploid cells can be induced experimentally by adding the chemical colchicine, a substance extracted from the crocus. Colchicine inhibits the formation of the spindle during cell division. Suggest why colchicine results in polyploid cells.

SUMMARY

- A chromosome is a large, densely staining body consisting of a single, supercoiled DNA strand. The genes along a chromosome are found at particular points called **loci**.
- A gene is a length of DNA that codes for the synthesis of a particular polypeptide or protein. Only about 10 per cent of a chromosome consists of genes; the rest is non-coding DNA.
- The genetic code is the sequence of bases contained in the DNA molecule. The code is read in groups of three bases, called **codons**. Each codon codes for a particular amino acid.
- The central dogma (principle) of molecular biology is that genes are read by **messenger RNA** which in turn moves out into the cytoplasm where it codes for the assembly of a protein.
- At the ribosome, messenger RNA attracts **transfer RNA** molecules with the right amino acids attached.

- The amino acids are held together so that peptide bonds form; in this way the protein is assembled.
- DNA replication occurs when the two strands separate and the enzyme **DNA polymerase** assembles two new strands on the originals. This is called semi-conservative replication as each old strand is conserved as half of the new DNA.
- A fault in DNA replication leads to a **gene mutation**. The base sequence is altered so that the gene no longer codes for the correct protein. Chromosome mutations can result in whole blocks of genes being lost, or translated backwards for example.
- Most mutations are harmful, some have no effect, but a few can be beneficial. Beneficial mutations might give the organism an advantage and in this way mutations give rise to variation and hence drive evolution.

? EXAM QUESTIONS

1 Give an account of the structure and replication of deoxyribonucleic (DNA) [**essay question**]

Edexcel B/HB1 June 1999 Q8

2 The diagram shows part of a DNA molecule.

a) DNA is a large molecule made up of a large number of nucleotides. Each nucleotide has three components.
(i) Name the components that make the parts of the DNA molecule labelled **A** [two components] and **B**.
(ii) Name the type of bond found at **C**.

b) 1 kilobase of DNA is a length of a DNA molecule with 1000 base pairs.
The length of 1 kilobase is 0.34 micrometres (μm).
The DNA in a human cell contains a total of 2 900 000 kilobases.
Calculate the total length of DNA in a human cell. Give your answer in millimetres (mm). Show your working.

c) Give **three** ways in which the structure of the DNA molecule enables it to carry out its functions.

AQA (B) BYB2 Genes and Genetic Engineering June 2001 Q1

3 DNA is made up of two polynucleotide strands, the sense strand and the antisense strand. Messenger RNA is transcribed from the DNA sense strand, which contains the genetic code.

a) The graph shows the number of bases found in the sense strand and the antisense strand of a short piece of DNA, and the mRNA transcribed from it.

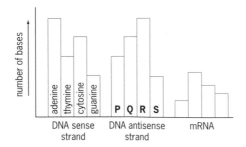

(i) Identify the base represented by each of the following letters: **P; Q; R; S**.
(ii) Explain why the total number of bases in the DNA sense strand and the total number of bases in the DNA antisense strand are the same.
(iii) Explain why the total number of bases in the DNA sense strand and the total number of bases in the mRNA are different.

b) The mRNA has a sequence of 1824 bases. How many amino acids will join to form the polypeptide chain?

c) Although DNA is double-stranded, only the sense strand determines the specific amino acid sequence of a polypeptide. Suggest a role of the antisense strand.

AQA (A) BYA3 Pathogens and Disease January 2001 Q3

4
a) Draw a labelled diagram to show the structure of an RNA nucleotide.

b) The diagram shows a molecule of an enzyme called ribonuclease. Each amino acid in the protein is indicated by a 3-letter symbol e.g. Arg = arginine.

(i) How many nucleotides are there in the mRNA molecule that codes for this enzyme?
(ii) The table gives the mRNA code for the four amino acids in the part of the enzyme labelled **X**.

? EXAM QUESTIONS

Amino acid	Symbol	mRNA code
Alanine	Ala	GCU
Glutamine	Glu	GAG
Leucine	Leu	UUA
Serine	Ser	AGU

Give the DNA code for the part of the enzyme labelled **X**.

c) **(i)** Where does translation occur in a cell?

(ii) Describe what happens during translation.

d) Explain how the structure of DNA is related to its function.

AQA (B) BYB2 Genes and Genetic Engineering January 2001 Q7

5

a) List **three** ways in which transcription differs from translation in protein synthesis.

The diagram represents a polyribosome with several translation sites.

b) Name the structures labelled **A** to **C**.

c) Name **two** molecules, in addition to the molecules shown in the figure, which are required to complete translation.

d) Describe **two** structural features which adapt tRNA to its role in translation.

OCR 9264/3 Biology Linear Paper 3 June 2000 Q4

6

a) Name the organelle where proteins are synthesised from amino acids.

b) The diagram shows a transfer RNA molecule (tRNA).

(i) During protein synthesis, which molecule is attached to the tRNA molecule at **X**?

(ii) What is an anticodon?

(iii) Give **one** way in which the structure of a tRNA molecule is different from an mRNA molecule.

AQA (B) BYB2 Genes and Genetic Engineering June 2001 Q5

7 About one in 20 000 humans has the condition albinism. This is caused by the absence of melanin, the dark brown pigment normally present in the human skin, hair, and eyes. Albinism arises from a gene mutation that causes skin cells to produce a different version of the enzyme tyrosinase. This different version of tyrosinase is faulty.

These events are shown in the diagram.

a) Explain how a gene mutation may result in a different version of tyrosinase.

b) The faulty tyrosinase does not produce melanin. Suggest an explanation for this.

AQA (B) BYB2 Genes and Genetic Engineering January 2001 Q6

KEY SKILLS **ASSIGNMENT**

SICKLE CELL ANAEMIA

In addition to the topics covered in this chapter, this Assignment assumes that you know some simple genetics (monohybrid inheritance) and evolution (natural selection).

A small minority of people, particularly those of African origin, inherit sickle cell anaemia (SCA), a condition in which their haemoglobin acts abnormally. SCA is a genetic disease: it results from a gene mutation that is inherited. Carriers of the sickle cell allele are usually perfectly healthy, and in some circumstances even have an advantage over people who possess two normal alleles. However, individuals with two sickle cell alleles have the disease.

The faulty allele makes haemoglobin S instead of the normal haemoglobin A. Problems arise when oxygen tensions are low, as they are in actively respiring tissues. At low oxygen tension, abnormal haemoglobin tends to come out of solution and crystallise, distorting the red blood cells into crescent shapes (Fig 24.A1). This has two main consequences. Firstly, sickle cells tend to form clots which can block blood vessels. Secondly, the spleen destroys abnormal sickle cells at a much greater rate than normal, causing anaemia.

Fig 24.A1 Top: Normal red blood cells.
Bottom: Abnormal red blood cells from someone with sickle cell anaemia. Haemoglobin S tends to polymerise, forming crystal-like structures that distort the red blood cells

1 What is the function of haemoglobin?

Haemoglobin molecules consist of four polypeptide chains: two α-globins and two β-globins. Research shows that sickle cell anaemia results from a fault in just one base in the β-globin gene. Here is the base sequence of the first seven codons of this gene:

Normal base sequence
GUA–CAU–UUA–ACU–CCU–GAA–GAG

Sickle cell sequence
GUA–CAU–UUA–ACU–CCU–GUA–GAG

2 **a)** How can you tell that these codons are from mRNA and not DNA?
b) Which codon has mutated?
c) What is the name given to this type of mutation?
d) Work out the amino acid sequence for both base sequences (you will need to refer to Table 24.2).
e) Explain in general terms how this slight change can produce abnormal haemoglobin.
f) Explain why other types of gene mutation can have a much greater effect on the final protein than the example seen here.

The inheritance of sickle cell anaemia

Generally, people have two copies of each gene, and this is the case with sickle cell anaemia. The two alleles involved are:

Hb^A = coding for normal haemoglobin

Hb^S = coding for sickle cell haemoglobin

So the three possible genotypes are:

Hb^AHb^A Normal

Hb^AHb^S Carriers of the disease but with few, if any, symptoms

Hb^SHb^S Have the disease

3 Sickle cell anaemia is described as an autosomal, recessive disease.
a) What do the terms **(i)** autosomal and **(ii)** recessive mean?
b) Explain why carriers are usually healthy.
c) If two carriers have children, what is the probability that the first child will have the disease?

The sickle cell gene is far commoner than you would expect for one which is selected against. The reason for this is that the carriers usually show a higher resistance to malaria than normal.

4 **a)** What is meant by the phrase 'selected against'?
b) Explain why the frequency of the sickle cell allele is higher than you would expect in some areas.

25

How genes are inherited

How genes are inherited

Woody Guthrie, the famous American folk singer, suffered from Huntington's disease. His son, Arlo, was born before his father knew he had a disease that could be passed on to his children. Luckily, Arlo is unaffected: he has not inherited his father's faulty gene.

The degeneration caused by Huntington's disease can be revealed by a brain scan. A protein called huntingtin, produced by a faulty gene that was discovered in 1993, damages nerve cells in some areas of the brain, including the cerebral cortex. The gene persists because affected people tend to reproduce before symptoms appear

About eight people in every 100 000 are born with the genetic disorder Huntington's disease. This condition, which involves premature degeneration of the brain, is caused by a single faulty gene on chromosome 4. The slow decline into dementia that develops is fatal.

Diagnosis is difficult because early symptoms are vague: sufferers can have bursts of temper, become more clumsy than usual or have difficulty remembering things. Also, the symptoms do not usually become apparent until the sufferer is in their 30s or 40s, usually after they have passed the gene on to the next generation.

Until recently, people with a family history of the disease knew that they were at risk, but they could never be sure whether they were actually carrying the gene. This led to a dreadful dilemma: should they go ahead and have children and risk passing the gene on, or should they decide to remain childless, only later to experience the anguish of finding out that they were free of Huntington's disease when it was too late to try for a baby?

Today, tests are available that can tell people whether they are carriers of the Huntington's gene. A sample of DNA can even show if an unborn baby carries the disease. This does not make the decisions involved any easier, but it does give affected families accurate information on which to base those decisions.

1 MENDELIAN INHERITANCE

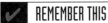

REMEMBER THIS

Some important terms and concepts relating to inheritance are covered in Chapter 22. If you are new to genetics, read that chapter before you start to study this one.

REMEMBER THIS

A gene is a length of DNA that acts as a template for the production of messenger RNA. mRNA forms a mobile copy of a gene, travelling from the cell's nucleus to the cytoplasm. Ribosomes use this template to synthesise a specific amino acid chain. Complete proteins are made from one or more of these chains.

Why don't we look exactly like our parents? Why are brothers and sisters all different (unless identical twins) when born to the same parents? As Fig 25.1 shows, there is often a marked family resemblance between parents and their children and between siblings. However, all new-born babies, except identical twins, are genetically unique. Every new-born baby has around 30 000 genes and because humans – like most animals – are diploid organisms, a baby has two versions of each gene. It inherits one version from its mother and one version from its father.

Fig 25.1 The tendency for certain features to be inherited is shown clearly in this family. However, no two people ever inherit exactly the same set of genes unless they are identical twins

It is difficult to imagine what happens when 30 000 different genes are mixed. Not surprisingly, although a vast amount of genetics research is under way, we are still a long way from completely understanding what happens when those mixed genes are passed on to the next generation. Some genes are activated while others remain hidden, some master genes switch on whole blocks of other genes.

How can we begin to make sense of it all? Amazingly, the man who first worked out the underlying rules did so without any knowledge of genes, chromosomes, DNA or cell division. Known as the father of genetics, **Gregor Mendel** (Fig 25.2) worked with pea plants. These are much simpler organisms than humans and provide a good starting point to understand some of the basic rules of **Mendelian inheritance**.

THE WORK OF GREGOR MENDEL

Gregor Mendel, a Czech monk, was the first to work out the basic laws that govern the inheritance of genes.

Fig 25.2 Gregor Mendel (1822–1884). Mendel's discoveries, like those of many scientists, were due to a combination of hard work, good scientific method, a touch of genius and a significant amount of luck

Before Mendel's work became known, it was widely thought that inheritance was due to 'blending' of different features. So, for example, a tall person married to a short person would produce children who would eventually grow to an intermediate height. Mendel showed it was wrong to assume that characteristics blended and demonstrated how individual characteristics are passed down to the next generation.

For years, Mendel bred and studied the edible pea plant (Fig 25.3). He chose these plants because they had easily observable features, they were easy to cultivate, they had rapid life cycles and their pollination could be controlled. His breeding experiments centred on the inheritance of a few features such as plant height, flower colour and pea shape/appearance.

Mendel was lucky because he chose features that were controlled by single genes. In addition, pea plants tend to fertilise themselves and produce pure breeding homozygous individuals. Had Mendel worked with more complex features or a different species, he might not have been able to work out the underlying laws of inheritance.

Mendel conducted his experiments using a good scientific method. He controlled his experimental conditions carefully, ensuring that plants were pollinated only by the pollen he transferred. He made careful observations and then repeated experiments as many times as was practicable, keeping meticulous records. Mendel was trained in mathematics and he applied statistical methods to his results in a way that was very advanced for the time.

His genius really showed in the way he interpreted his results. Mendel proposed the **particulate theory**, in which he stated that there are individual units of heredity that cannot be diluted, only hidden. Mendel came up with the idea that each pea plant has two factors for each character, one inherited from each parent. Only one unit can be present in the gamete and so be passed on to the next generation. Mendel's 'particles' are, of course, genes.

In 1866, Mendel published a detailed account of his findings, but their significance was not appreciated by the scientific community. In 1884, Mendel died without ever receiving true recognition for his work. In 1900, his work re-emerged thanks to the independent work of three other European scientists – see Table 22.2 in Chapter 22. In the time between 1866 and 1900, the process of meiosis had been discovered and this provided a mechanism that would explain Mendel's observations. All three scientists cited Mendel's original work, which had suggested the basic mechanism of inheritance.

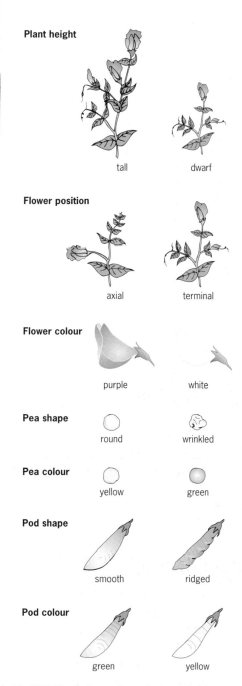

Fig 25.3 The features of pea plants studied by Mendel. Pea plants tend to self-fertilise, producing true-breeding strains. For example, white-flowered plants contain only white-flowered alleles: they are not 'hiding' any purple alleles and so always produce white flowers

MEIOSIS AND MENDEL'S FINDINGS

Mendel's findings can be summarised into two laws, both of which are explained by the process of meiosis (Fig 25.4).

Fig 25.4 An overview of meiosis. In a diploid cell, the DNA is first duplicated and then blocks of genes are swapped between homologous chromosomes. Then, in the first meiotic division, the pairs of double chromosomes are separated. Finally, in the second meiotic division, the chromatids are pulled apart and each cell receives a single chromosome

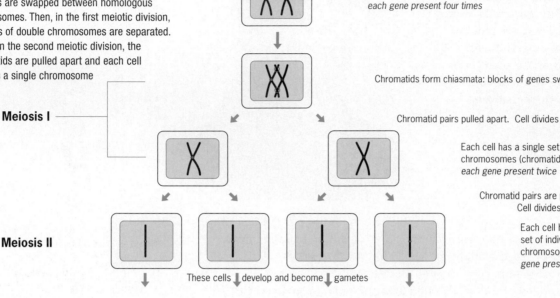

Each chromosome forms a pair of chromatids: *each gene present four times*

Chromatids form chiasmata: blocks of genes swapped

Chromatid pairs pulled apart. Cell divides

Each cell has a single set of double chromosomes (chromatid pairs): *each gene present twice*

Chromatid pairs are pulled apart. Cell divides

Each cell has a unique set of individual chromosomes: *each gene present once*

These cells develop and become gametes

Meiosis I

Meiosis II

REMEMBER THIS

Mendel's second law is perhaps easier to understand by considering an example. If you have a plant with red or white flowers and round or wrinkled seeds, the offspring can have any combination of flowers and seeds; red and round, red and wrinkled, etc. The offspring will not necessarily have the same combination as the parent.

REMEMBER THIS

In genetics, different alleles are often denoted by letters, such as Aa. Usually, the capital letter stands for the dominant allele. As a tip, if you are choosing letters to represent alleles, chose those whose capital cannot be confused with the lower case version. So, Rr is better than Ss. This is very important in exams, where your handwriting may be worse than usual.
Homozygous individuals are said to be **true breeding**. This means that they will always produce the same phenotype of offspring because they are not hiding a recessive allele.

- Mendel's **law of segregation** states:

  An organism's characteristics are determined by internal factors that occur in pairs, only one of which can be present in a single gamete.

In other words an organism has two versions of each gene, but only one of these version is passed on to the offspring. This is because meiosis separates homologous pairs of chromosomes, so that only one of each pair passes into the gamete.

- Mendel's **law of independent assortment** states:

  Each of a pair of contrasted characteristics can be combined with each of another pair.

Again, this is explained by meiosis; any one of a pair of chromosomes can pass into the gamete with any member of another pair. The significance of this is seen in dihybrid (two-gene) inheritance.

MONOHYBRID INHERITANCE: HOW SINGLE GENES ARE PASSED ON

Monohybrid (single gene) inheritance concerns the inheritance of different alleles (usually two) of a single gene. Like Mendel, we start with the pea plant. Peas have several easily observable features that are controlled by single genes (see Fig 25.3).

For example, pea plants have one gene for height. The height gene has two alleles: tall (T) and dwarf (t). Pea plants are diploid and so have two alleles. There are therefore three possible genotypes:

TT – homozygous for T (homo = same)

Tt – heterozygous (hetero = different)

tt – homozygous for t

Consider what happens when a homozygous tall plant is crossed with a homozygous dwarf plant (Fig 25.5). All the gametes from the tall plant contain a T allele and all those from the dwarf plant have a t allele. These combine at fertilisation to give offspring all with the genotype Tt. The **first generation**, known as the F_1, are all tall, because tallness is dominant. However, although they look identical to the tall parent plant, they are different in one very important respect: they are heterozygous and not homozygous.

If two of these heterozygous plants are crossed, half of the gametes from each parent are T and half are t, giving us four possible genotypes in the **second generation**, the F_2:

25% are TT

25% are tT

25% are Tt

25% are tt

The first three genotypes give tall plants (the T allele is present) but a quarter of the plants are dwarf (they are homozygous for the t allele).

It is important to realise that chance plays a very important part in genetics. If we took four F_2 pea plants we might expect to get three tall plants and one dwarf, but it is highly likely that we would get something else, such as four and none or two and two. The greater the number of offspring, the higher the chance that the numbers reflect the expected ratio. This is why Mendel had to carry out hundreds of crosses before he became convinced of the underlying ratio – see the Assignment at the end of this chapter.

THE TEST CROSS

It is often important to know whether an individual displaying a dominant trait is homozygous or heterozygous for a particular allele. For example, how can we tell if a tall pea plant is pure breeding, or if it is carrying a hidden dwarf allele? To find out we do a **test cross**. This involves crossing the tall plant with a homozygous dwarf plant (tt). If the tall plant is homozygous (TT), all the offspring produced are tall. However, if it is a heterozygote (Tt), about half the offspring are tall and half are dwarf.

MONOHYBRID INHERITANCE IN HUMANS

Clear-cut examples of monohybrid inheritance in humans are relatively rare, and often involve genetic disease where people inherit one or more faulty alleles. Genetic diseases are often recessive because faulty alleles that fail to make an important protein can be masked by normal ones that function properly. Table 25.1 shows examples of monohybrid inheritance in humans.

In contrast, some genetic disorders such as Huntington's disease (see the Opener of this chapter) are caused by dominant alleles. The alleles concerned code for a product that actively causes damage; the symptoms are not due to an allele not doing its job. Such alleles are dominant because the presence of a normal allele cannot mask the symptoms.

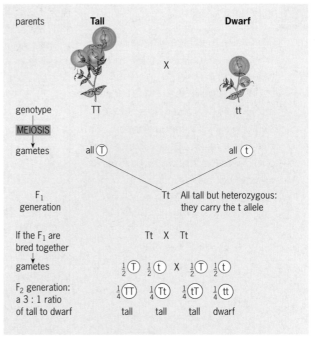

Fig 25.5 Monohybrid inheritance in pea plants. Pure breeding tall and dwarf pea plants are homozygous (TT and tt). When crossed, the plants of the first (F_1) generation are all tall but they are heterozygous (Tt), carrying the allele for dwarfness. If these heterozygous individuals are selfed (bred together) the dwarf form reappears in the next (F_2) generation in a ratio of 1:3 (ie 1 in 4). We say that the heterozygotes do not breed true

SEE QUESTION 1

TEST YOURSELF

A In pea plants, yellow seeds are dominant to green. If a heterozygous yellow-seeded plant were crossed with a green-seeded plant, what ratio would you expect in the F_1 generation?

B Why is a homozygous dwarf plant used in a test cross? Why not use a homozygous tall plant?

SEE QUESTION 2

REMEMBER THIS

The ability to roll the tongue is often cited as a simple example of inheritance in humans. However, although people seem to fall neatly into one of two categories – either you can roll your tongue or you can't – there is considerable debate about whether or not this skill can be learned, and whether it is controlled by one gene or more.

Table 25.1 Examples of monohybrid inheritance in humans

Traits	Features
Dominant traits	
Huntington's disease	See Opener
Freckles	
Dimple in chin	
Recessive traits	
Sickle cell anaemia	Haemoglobin polymerises, distorting red blood cells into a sickle shape. This leads to circulatory blockages and anaemia. See the Assignment to Chapter 24.
Attached ear lobes	
Phenylketonuria (PKU)	Inability to process the amino acid phenylalanine. Protein intake must be monitored to prevent build-up of the amino acid to harmful concentrations. Can be fatal if not treated.
Haemophilia (also sex linked – see below)	Inability to produce a critical blood-clotting factor, carried on the X chromosome. Haemophilia is now treated by supplying the affected individuals with the clotting factor.
Albinism	Inability to make the pigment melanin.
Rhesus blood group	Presence (Rh⁺) or absence (Rh⁻) of rhesus protein on red blood cells.
Cystic fibrosis	Excessive mucus production, especially in lungs and pancreas. Breathing and digestion are affected and sufferers are very susceptible to lung infections.
Galactosaemia	Inability to convert galactose to glucose in the liver. All lactose must be avoided to prevent fatal levels of galactose from accumulating.
Tay–Sachs disease	Inability to produce critical lipases, which results in fatty accumulations in the brain.
Lactose intolerance	Inability to breakdown the disaccharide lactose to glucose and galactose. Leads to vomiting, diarrhoea, flatulence. Avoiding lactose prevents symptoms.

SCIENCE FEATURE Cystic fibrosis: an example of monohybrid inheritance in humans

The disease cystic fibrosis, covered in detail in Chapter 32, is caused by the mutation of a single allele. A normal allele (C) makes a membrane protein essential for the proper functioning of certain epithelial cells. The mutated allele (c) does not code for a functional protein.

Most people have two healthy alleles: their genotype is CC. In the UK population, however, around 1 person in 25 carries a faulty allele. These carriers (Cc) are perfectly healthy because they also possess a normal allele, which makes the working protein.

However, in 1 couple in 625 (1 in 25 × 1 in 25) both partners are carriers. There is a one in four chance that any child they have could inherit both faulty alleles and therefore be a cystic fibrosis sufferer. As 1 cystic fibrosis child in 4 is born to 1 couple in 625, approximately 1 child in every 2500 has this condition.

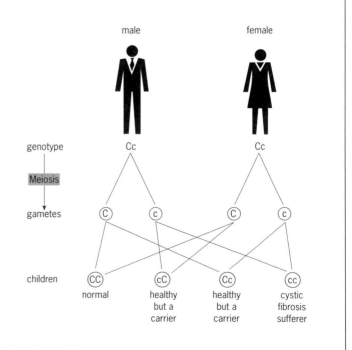

Fig 25.6 The inheritance of cystic fibrosis. If both parents are carriers, there is a 25 per cent chance that any child will be a cystic fibrosis sufferer. There is also an equal chance that a child will inherit no faulty alleles and that his/her descendants will be completely free from the disease. There is a 50 per cent chance of the child being a carrier of cystic fibrosis

INBREEDING AND OUTBREEDING

Everyone agrees that brothers and sisters or other closely related people should not have children. In most human societies such **inbreeding** is illegal for social, moral and biological reasons.

Biologically, inbreeding is undesirable because it reduces genetic variation in the offspring. Inbreeding promotes homozygosity: inbred individuals are homozygous for many more alleles than individuals produced by outbreeding. So, faulty recessive alleles are more likely to be paired up, resulting in a much higher proportion of genetic defects and disease.

In practice, inbreeding tends to happen in small, isolated human and animal populations, or as a consequence of artificial selection. Pedigree dog and cat breeding is a good example of this. To maintain the most desirable features, champion animals are often bred with near relatives. Analysis of the pedigree certificates of the champions in a particular breed will often show that the same individuals crop up again and again. The consequence of this artificial selection is a relatively high proportion of genetic defects.

Outbreeding (the breeding of unrelated individuals) is more desirable as it produces more new genetic combinations and greatly reduces the chance that faulty genes will be expressed. Many organisms have developed mechanisms which prevent inbreeding (Fig 25.7). If pedigree animals are cross-bred, rather than inbred (cross a Siamese cat with a Persian, for example), you often get a fitter, healthier animal. We say that such offspring show **hybrid vigour**, and cross-breeding is a favourite tool of plant breeders who find that hybrids often grow faster and show better disease resistance.

Fig 25.7 When woodlice lay a batch of eggs, the offspring are either all male or all female. This means that brothers and sisters cannot mate, thus reducing the chances of inbreeding

DIHYBRID INHERITANCE: HOW TWO GENES ARE PASSED ON

What happens when a tall, purple-flowered pea plant is crossed with a dwarf, white-flowered individual? Do you always get tall plants with purple flowers, or can the features re-combine, producing tall, white-flowered plants, and dwarf, purple-flowered ones? The simple answer is yes, you can get these recombinants, thanks to the process of meiosis.

The inheritance of two separate genes is called **dihybrid inheritance**. If, as is usual, the two genes are on separate chromosomes, Mendel's second law (page 424) applies: each of a pair of contrasted characteristics can be combined with each of another pair. So, an organism with a genotype of AaBb can produce gametes of AB, Ab, aB or ab. In other words, one allele from each pair passes into the gamete, and all four combinations are possible. This is due to **independent assortment** during meiosis (Fig 25.8).

? TEST YOURSELF

C There are approximately 700 000 live births every year in the UK. How many of these babies would you expect to have cystic fibrosis?

D Suggest why there is less inbreeding in most human societies today than in previous centuries.

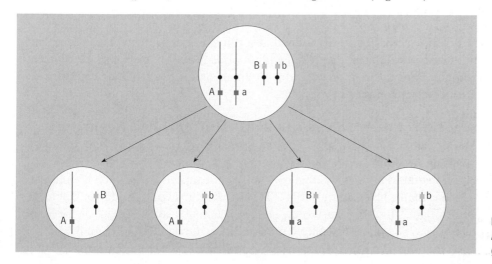

Fig 25.8 Consider a diploid cell with the genotype AaBb. Meiosis produces four different types of gametes by independent assortment

For an example, let's return to pea plants. In addition to tall and dwarf, we can consider a gene for seed texture. The allele R for round seeds is dominant to r, the allele for wrinkled seeds.

Fig 25.9 shows a cross between a pure breeding, tall, round-seeded plant (TTRR), and a pure breeding dwarf, wrinkled-seeded plant (ttrr). One allele from each pair is passed into the gametes. All of the gametes from the tall, round-seeded plant are TR and all those from the dwarf, wrinkled-seeded plant are tr. All of the F_1 generation have the genotype TtRr and are tall with round seeds.

Fig 25.9 Dihybrid inheritance in peas. The grid used to work out the phenotype ratios in the F_2 generation is called a Punnett square

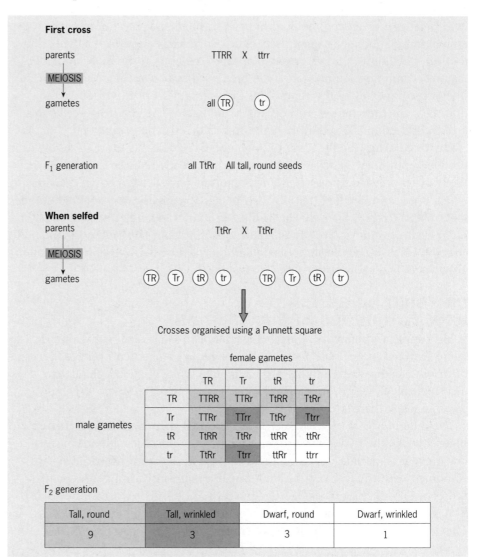

When we cross two TtRr individuals, things get more complex! Each plant produces four different gametes. Fertilisation is random, producing 16 different combinations in the F_2 generation. To organise our thoughts we can use a **Punnett square** (see Fig 25.9).

SEE QUESTIONS 3 TO 5

DIHYBRID INHERITANCE QUESTIONS IN EXAMINATIONS

Many examination questions require the candidate to work out the ratios of offspring resulting from a particular cross. If you have not practised these questions they can be very time consuming, leading to stress and silly mistakes.

The example given above (TtRr × TtRr) is the most complex dihybrid situation you can get. This is because both male and female are heterozygous for both alleles, and so produce 4 different gametes each, resulting in 16 different combinations. All other genotypes will produce fewer combinations. If you practise these combinations you can learn to work out the gametes and resulting ratios in your head without resorting to Punnett squares and counting the results.

Consider the cross TtRR × Ttrr. The first individual produces the gametes TR and tR in equal amounts. The second one produces Tr and tr in equal amounts. So, there are four possible combinations: the same number as in a monohybrid cross.

You should see that the offspring will be one quarter TTRr, two quarters TtRr and one quarter ttRr, giving three tall plants with round seeds to one dwarf plant with round seeds. You would not get any dwarf plants with wrinkled seeds.

Some exam questions give the ratios of offspring phenotypes and ask you to work backwards to the original genotypes.

If you do not get the expected ratios, consider **linkage** as an explanation (see below).

2 WHEN MENDEL'S LAWS DON'T APPLY

LINKAGE AND CROSSING OVER

So far, we have considered the inheritance of genes on different chromosomes that could be separated at meiosis. We say that genes that are on the same chromosome are **linked**. Such genes cannot obey Mendel's laws because they do not undergo independent assortment. They are inherited together unless separated by crossing over during prophase I of meiosis (Fig 25.10).

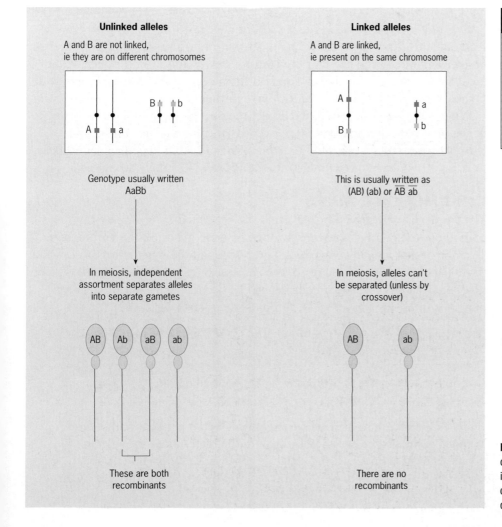

Unlinked alleles

A and B are not linked, ie they are on different chromosomes

Genotype usually written AaBb

In meiosis, independent assortment separates alleles into separate gametes

AB Ab aB ab

These are both recombinants

Linked alleles

A and B are linked, ie present on the same chromosome

This is usually written as (AB) (ab) or $\overline{AB}$ $\overline{ab}$

In meiosis, alleles can't be separated (unless by crossover)

AB   ab

There are no recombinants

Fig 25.10 When two genes are close together on the same chromosome they are almost always inherited together: they are linked. The chance of being separated by crossover depends on the distance between them

To illustrate linkage, let's consider the sweet pea plant (*Lathyrus odoratus* – a different species from the edible pea we have been looking at). In this species, the allele for purple flowers is dominant to red flowers, and the allele for elongated pollen grains is dominant to round pollen. If pure breeding plants with purple flowers and elongated pollen grains are bred with plants with red flowers and round pollen grains, the F_1 generation all have purple flowers and elongated pollen.

When the F_1 are **selfed** (bred together) you would expect a 9:3:3:1 ratio, but this does not happen. The F_2 generation are mainly like the parents (purple and elongated or red and round) but there are a few recombinants (red and elongated or purple and round). This is because the genes for flower colour and pollen shape are linked (present on the same chromosome) and so are inherited together. Any recombinants we get must therefore be due to crossing over.

Consider the set of results shown in Table 25.2.

Table 25.2 An example of a phenotypic ratio that indicates linkage		
Phenotype	Approx. expected numbers if no linkage (9:3:3:1)	Observed numbers in F_2 generation
Purple, elongated	405	336
Purple, rounded	135	31
Red, elongated	135	28
Red, rounded	45	325
Total	720	720

REMEMBER THIS

The term **locus** is used to describe the position of a gene on a chromosome.

The total number of offspring is 720, of which 59 (31 + 28) are recombinants. Eight per cent of the offspring result from crossing over and so, by convention, we can say that the **loci** for flower colour and pollen shape are 8 units apart on the chromosome. If the percentage of recombinants were higher, we would know that the loci were further apart on the chromosome.

When genes are not linked, we expect them to be separated 50 per cent of the time because there is a 50 per cent chance that alleles on separate chromosomes will be inherited together. When looking for linkage, geneticists look for a frequency of recombinants significantly less than 50 per cent.

CHROMOSOME MAPS

The frequency with which linked genes are separated by crossover can tell us a lot about the relative positions, or loci, of these genes on the chromosome. Two alleles that are close together are rarely separated by crossing over, while those located on opposite ends of the chromosome are separated almost every time. Analysis of crossover frequency can be used to work out the relative positions of alleles and so make chromosome maps (Fig 25.11).

Fig 25.11 The position of genes on chromosomes can be worked out by interpreting crossover frequencies. Consider three genes, X, Y and Z. Genes X and Z are found to have a crossover frequency of 23 per cent, so they are 23 units apart on the chromosome

Genes Y and Z have a crossover frequency of 10 per cent and so are 10 units apart. A third cross between Y and X determines the relative positions of X, Y and Z. If the order of genes is XYZ (as shown), the crossover frequency between X and Y is 13 per cent. If the order is XZY, the crossover frequency is 33 per cent.

EXAMPLE

An A-level genetics question involving linkage

This example is a modified question from an UCLES/OCR Modular Biology examination in Applications of Genetics.

The mosquito *Aedes aegypti* has a spotted and a spotless form on a grey or yellow body. The dominant allele is spotless, S, and the recessive allele is spotted, s. The allele for grey body G, is dominant to the allele for yellow body, g. A cross was made between homozygous spotless, grey-bodied mosquitoes and spotted, yellow-bodied mosquitoes. The F_1 individuals were then crossed with the double recessive strain and the numbers of the resulting phenotypes were counted. The results are shown below.

Phenotype	Number
Spotless, grey bodied	442
Spotted, yellow bodied	458
Spotless, yellow bodied	46
Spotted, grey bodied	54

Q State the genotype and the phenotype of the F_1 using the above symbols.

A The question is asking for the genotype of the first generation. You know that the parents are homozygous spotless, grey (genotype: SSGG) and spotted, yellow (genotype: ssgg). The genotype of the F_1 must be SsGg, giving a phenotype of spotless, grey bodied.

Q Using a genetic diagram, explain fully the results shown in the table.

A The table below shows the result of a cross between an F1 individual, SsGg, and a double recessive which must be ssgg.

So we have:

Parents with genotype	Spotless, grey		×	Spotted, yellow	
	SsGg			ssgg	
Gametes	SG Sg sG sg		×	all sg	
Expected ratio	25% SsGg	25% Ssgg		25% ssGg	25% ssgg
Phenotypes	Spotless, grey	Spotless, yellow		Spotted, grey	Spotted, yellow
	1 :	1 :		1 :	1

You should see that the expected results are not the same as those given in the table of results on the left, so there must be an extra complication. The key feature from the table is that there are more of the parental phenotypes than expected, and fewer recombinants. This should scream *LINKAGE* at you!

To produce a really thorough answer you should also work out the percentage of recombinants.

The total number of crosses in the table is 1000, of which 100 (46 + 54) are recombinants. This leaves you with the simple calculation to show that 10 per cent of the mosquitoes are recombinants, and so the two genes are 10 units apart on the chromosome.

3 SEX DETERMINATION

What decides whether a baby will be a boy or a girl?

In humans, sex determination depends on the inheritance of special **sex chromosomes**. Every human cell contains 23 pairs of chromosomes: 22 pairs of **autosomes** and one pair of sex chromosomes (Fig 25.12). Females have two large X chromosomes and so we say they are the **homogametic** sex. Males have one X chromosome and one smaller Y chromosome, and are therefore the **heterogametic** sex.

In the female, all eggs receive an X chromosome during meiosis (Fig 25.13). However, meiosis in the male produces sperm, half of which receive a Y chromosome and half an X. So as the sperm swim for the female's egg it's a real race: if a Y sperm fertilises the female's egg, the offspring is male. If an X sperm wins, the baby will be a female.

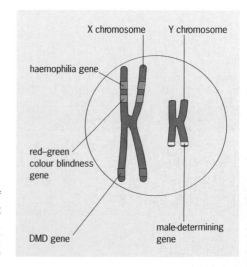

Fig 25.12 The human karyotype of 22 pairs of autosomes and 1 pair of sex chromosomes is shown in Fig 22.2. This diagram shows the sex chromosomes from a male: females have two identical X chromosomes. Note that the Y is much smaller than the X, and carries fewer genes. The Y chromosome does, however, carry the male-determining gene

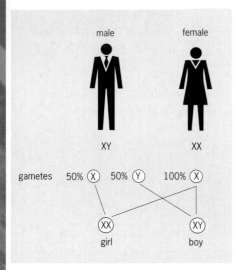

Fig 25.13 Sex determination in humans. Half of the sperm carry an X chromosome, half carry a Y

In humans, about 114 boys are conceived for every 100 girls. We do not yet fully understand the reason for this. Interestingly, however, more male embryos die and so at birth the proportions are down to 106 to 100. By puberty the numbers are equal and in old age the females outnumber males by two to one.

In the 1990s, a group of genetic researchers demonstrated that the Y chromosome carries just 60 functional genes, including a male-determining gene called **SRY**. All embryos are female unless the active SRY imposes maleness upon it. When a male and female embryo share the same blood supply, it is possible that the SRY can produce hormones that can force an XX embryo to develop as a male. This is an extremely rare situation in humans but it is common in cattle.

Androgen Insensitivity Syndrome (AIS) is condition that can cause a genetically male child to develop outwardly as a female. The receptor proteins for the male hormones (androgens) don't work properly, so the embryo fails to read the signals that should turn it into a male. This syndrome may go undetected until the teens, when puberty fails to happen. Genetic analysis will reveal that the 'girl' is in fact a male. Although she will look and act like a girl, she will be unable to have children and will not menstruate.

4 SEX-LINKED INHERITANCE

Genes carried on the sex chromosomes are said to be **sex-linked**. Human females have two X chromosomes, which, like the autosomes, carry two alleles of every gene. Females therefore have two sets of sex-linked alleles. In males, however, the Y chromosome is smaller and cannot mirror all of the genes on the X chromosome, so males have only one set of most sex-linked alleles. This is why males suffer from the effects of X-linked genetic diseases more often than females. There are no known Y-linked diseases or conditions, probably because the Y chromosome carries so few genes.

A classic example of a sex-linked trait transmitted by the X chromosome is **haemophilia**. People suffering from this disease do not make Factor VIII, an essential component in the complex chain reaction of blood clotting (see Chapter 30). In addition to the problems caused if they injure themselves, haemophiliacs suffer from internal bleeding as a result of normal activity. Bleeding at the joints during even light exercise is a particular problem. However, haemophiliacs can usually live a full and active life by having regular injections of Factor VIII.

Haemophilia is caused by a sex-linked recessive allele. Females have a pair of alleles but males possess only one. So, if a male inherits the haemophilia allele, he has the disease since he cannot possess another healthy allele to mask its effect.

Using the following notation:

X^h = the haemophilia allele on the X chromosome
X^H = the healthy allele on the X chromosome
Y = the Y chromosome which does not carry either allele
we can see that the possible genotypes are:

X^HX^H = healthy female
X^HX^h = healthy, carrier female
X^HY = healthy male
X^hY = haemophiliac male

SEE QUESTION 6

? TEST YOURSELF

E What would the notation be for a female haemophiliac? Suggest why they are much rarer than male haemophiliacs.

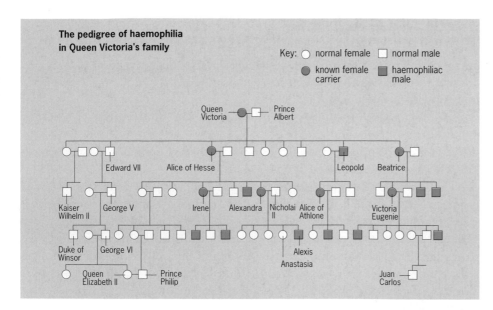

The pedigree of haemophilia in Queen Victoria's family

Key:
- ◯ normal female
- ⬜ normal male
- ⬤ known female carrier
- ▣ haemophiliac male

Fig 25.14 Queen Victoria (1819–1901) had nine children and passed the haemophilia allele on to two daughters and one son. Her eldest son, Edward VII, did not inherit the faulty allele and so the disease is not carried by his descendants, the present-day British Royal Family

Fig 25.15 Duchenne muscular dystrophy affects mainly boys. The body muscles are gradually replaced by fibrous tissue and become weak, usually confining sufferers to a wheelchair by the age of 10. Few live beyond 20. One male child in 4000 is born with this disease, sometimes caused by a spontaneous mutation, so the mother may not even be a carrier. Gene therapy holds new hope for the future

? TEST YOURSELF

F If a female carrier of the haemophilia allele marries a normal male (as was the case of Queen Victoria and Prince Albert) and they have a son, what is the probability that he will have haemophilia?

Fig 25.16 A person with red–green colour blindness cannot distinguish between red, green, orange and yellow. On these Ishihara test charts, a person with normal vision would see numbers but a person with red-green colour blindness would see only a random pattern of dots. This condition is caused by a defective allele that codes for the one of the three groups of light-sensitive cone cells in the retina. Eight per cent of males but only 0.4 per cent of females suffer from colour blindness

We can study the inheritance of haemophilia using a **pedigree diagram**. The word pedigree means 'of known ancestry' and applies as much to humans as it does to domestic animals. The pedigree of the Royal Family can be traced back to 1066 and shows that Queen Victoria was a carrier of the haemophilia allele and passed it on to four of her nine children (Fig 25.14). Other sex-linked traits in humans include Duchenne muscular dystrophy (DMD, Figs 25.12 and 25.15) and red–green colour blindness (Fig 25.16).

5 MULTIPLE ALLELES

Sometimes there are more than two alleles for a particular gene. Such genes are said to be **polymorphic.** It has been estimated that 31.7 per cent of human genes are polymorphic.

A classic example of a polymorphic gene is the ABO blood group system in humans. The entire human population can be classified into A, B, AB or O depending on the proteins they carry on the membrane of their red blood cells. The ABO system is controlled by one gene (I) with three alleles, I^A, I^B and I^O. The I^A allele codes for A proteins, I^B for B proteins and IO for no relevant proteins. I^A and I^B are codominant over I^O. Each individual inherits two alleles which combine to produce the blood group as shown in Table 25.3.

Blood group testing can sometimes be used to disprove (but not prove) parentage. Table 25.4 shows, for instance, that a man of blood group AB must pass on either an I^A or an I^B allele, and so cannot possibly be the father of a child with blood group O. DNA fingerprinting, of course, provides a much more accurate (if more expensive) test which can say with a great deal of certainty who *is* the father. The Assignment in Chapter 3 looks at the process of DNA fingerprinting.

Table 25.3 The genetics of the ABO blood group system

Genotype	Phenotype (blood group)
$I^A I^A$ or $I^A I^O$	A
$I^B I^B$ or $I^B I^O$	B
$I^A I^B$	AB
$I^O I^O$	O

? TEST YOURSELF

G If a child has the blood group B and his mother is blood group O, can his father be a man with blood group A? Say why.

Table 25.4 Possible blood groups of children born to parents of particular blood groups

			Paternal blood group			
			A	B	AB	O
		Genotypes	$I^A I^A$ or $I^A I^O$	$I^B I^B$ or $I^B I^O$	$I^A I^B$	$I^O I^O$
Maternal blood group	A	$I^A I^A$ or $I^A I^O$	A, O	A, B, AB, O	A, B, AB	A, O
	B	$I^B I^B$ or $I^B I^O$	A, B, AB, O	B, O	A, B, AB	B, O
	AB	$I^A I^B$	A, B, AB	A, B, AB	A, B, AB	A, B
	O	$I^O I^O$	A, O	B, O	A, B	O

6 POLYGENIC INHERITANCE

So far in this chapter, we have been concerned with **discontinuous inheritance**: all-or-nothing features controlled by single genes. Many features, however, are **continuous** and are controlled by several genes that act together. We describe the inheritance of such characteristics as **polygenic inheritance**.

Skin colour in humans is an example of polygenic inheritance. The depth of skin colour depends on the amount of melanin present. This is determined by at least three genes. As an example of this type of inheritance at work, imagine that the alleles A, B and C contribute cumulatively to the overall amount of melanin but the alleles a, b and c do not. So, someone with the genotype AABBCC would have the darkest skin, while the genotype aabbcc would confer a very pale skin colour.

As Fig 25.17 shows, a cross between two purely homozygous individuals would result in an intermediate (heterozygous) F_1 generation, but the F_2 generation would show a wide variation in skin colour depth. The graph shows a **normal distribution**: most individuals have skin colours in the midrange. This range is one of the characteristic features of continuous variation (see Chapter 22).

✔ REMEMBER THIS

The inheritance of height in humans is polygenic.

? TEST YOURSELF

H Non-identical twins are born to a couple who both have the genotype AaBbCc. Is it possible for one baby to be born with white skin and the other to have black skin? Explain your answer.

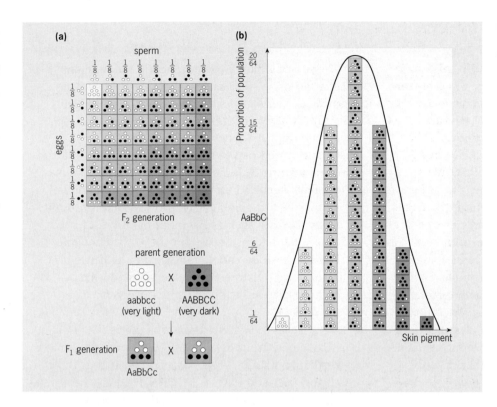

Fig 25.17 Human skin colour is an example of polygenic inheritance which shows a normal distribution in the F$_2$ generation

7 EPISTASIS (WHEN GENES INTERACT)

Up to now, we have discussed the inheritance and effect of genes that act on their own. In practice, however, many genes interact with each other: the presence of one allele often affects the expression of others. We call this interaction **epistasis**. A classic example of epistatic gene interaction is coat colour in mice. The natural coloration of wild mice is called **agouti** and is produced from banded hairs (Fig 25.18).

Two genes are involved, each with a dominant and a recessive allele. We will represent them by the notation Aa and Bb, where A and B are dominant over a and b. The allele A codes for the ability to produce hair pigment: AA and Aa mice have pigmented hairs but all aa individuals are albinos. The B allele codes for the ability to make hair with graduated coloration: BB and Bb mice have graduated hair, bb mice have hair that is all one colour.

Only mice that have both dominant alleles A and B show the agouti coloration. These can have the genotype AABB, AaBB, AABb or AaBb. Mice that are aa cannot make pigment, so, whether they are BB, Bb or bb, no bands are visible. Mice that can make pigments but not banded hair (genotypes AAbb or Aabb) are plain black.

? TEST YOURSELF

I Two agouti mice, genotypes AaBb, are bred together. What phenotypic ratio would you expect in the next generation?

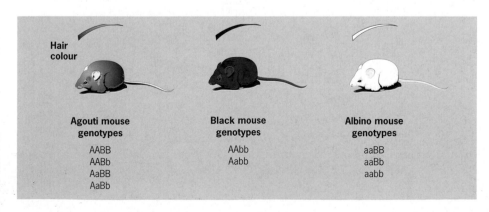

Hair colour

Agouti mouse genotypes	Black mouse genotypes	Albino mouse genotypes
AABB	AAbb	aaBB
AABb	Aabb	aaBb
AaBB		aabb
AaBb		

Fig 25.18 The base of the hair of a mouse with agouti coloration is dark brown/black fading to lighter brown towards the tip

SUMMARY

- Humans, like most organisms, are **diploid**; each cell carries two versions, or **alleles**, of each **gene**. If both alleles are the same, the organism is **homozygous** for that allele. If the pair are different, the organism is **heterozygous**.

- During **meiosis**, only one of the alleles passes into the **gamete**. This is Mendel's first law.

- **Monohybrid inheritance** is the study of one gene. A cross of AA × aa gives all Aa in the **first generation (F_1)**. A cross of Aa × Aa gives a 3:1 phenotypic ratio in the **second generation (F_2)**.

- **Dihybrid inheritance** involves two separate genes, usually on separate chromosomes. Homozygous individuals AABB × aabb produce offspring that are all AaBb in the F_1 generation. However, owing to **independent assortment** during meiosis, AaBb individuals produce gametes of AB, Ab, aB and ab. If bred together, AaBb individuals produce 16 different genotypes, and 4 different phenotypes with a ratio of 9:3:3:1.

- Genes present on the same chromosome are said to be **linked**. Such genes are inherited together unless separated by crossover in prophase I of meiosis. The greater the distance between the genes on the chromosome, the more often they are inherited separately.

- Humans have 23 pairs of chromosomes: 22 pairs of **autosomes** and one pair of **sex chromosomes**. The X chromosome is larger than the Y and carries more genes.

- The sex of a baby is determined at the moment of conception. Because gametes are produced by meiosis, half of the sperm carry a Y chromosome and half carry an X. If a Y sperm fertilises the egg, a male results; X gives a female.

- Genes carried on the sex chromosomes are said to be **sex-linked**. Females have two complete sets of sex-linked genes, but males have only one copy of all but the few genes carried on the Y chromosome.

- Some diseases such as haemophilia are sex-linked; caused by defective genes on the X chromosome. Females have two alleles and so can be carriers. Males have only one allele and so cannot be carriers: they are either healthy or they have the disease.

- Some genes have more than two alleles. The human ABO blood group is an example of such a **multiple allele**. In this case, one gene has three alleles, A, B and O.

- Many characteristics are controlled by many different genes. We describe such characteristics as **polygenic**; they often produce a range of values. Human height is a good example of a polygenic characteristic.

- The presence of one allele can affect the expression of an allele of a different gene. This type of gene interaction is called **epistasis**. For example, one allele may give an organism the ability to make pigment, while another allele determines the colours and/or the distribution of that pigment.

? EXAM QUESTIONS

1 In rabbits, the fat beneath the skin can be either white or yellow. The F_1 generation from a cross between a white-fat rabbit and a yellow-fat rabbit all had white fat. When these offspring were interbred, some of the F_2 generation had white fat and some had yellow fat.

a) What type of inheritance does this character show?

b) Choosing appropriate symbols for the two characters, state the **genotypes** of the parent rabbits.

c) (i) Draw a genetic cross diagram to show how the genotypes of the F_1 generation are produced.
(ii) State the likely proportion of each **phenotype** in the F_2 generation.

d) Explain how the genotypes of the white-fat rabbits in the F_2 generation could be determined. Use diagrams where appropriate.

e) What conclusion would you draw if all the F_2 generation yellow-fat rabbits were male?

The yellow colour in the fat is due to xanthophyll, a plant pigment absorbed from the food.

f) Suggest why this pigment is absent from the fat of white-fat rabbits.

OCR 6912 Inheritance Part 1 March 2000 Q1

2 Cystic fibrosis is a common inherited disorder amongst Europeans. It is caused by a single gene. The diagram below is part of a family pedigree showing the inheritance of this disorder.

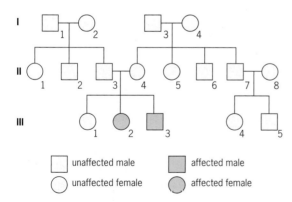

unaffected male affected male
unaffected female affected female

a) Describe *one* piece of evidence from the information given on the diagram which suggests that the allele causing this inherited disorder is recessive, and *one* piece of evidence that it is autosomal.

b) If the percentage of children born with cystic fibrosis is known, then it is possible to calculate the frequency of the recessive allele in the population. This information can be used to estimate the percentage of people in the population who are heterozygous for this gene, using the equation below:

$$p^2 + 2pq + q^2 = 1$$

where **p** is the frequency of the normal (dominant) allele
q is the frequency of the recessive allele for cystic fibrosis
p² is the frequency of people homozygous for the normal allele
q² is the frequency of people homozygous for the recessive allele
2pq is the frequency of heterozygous people (carriers)

(i) If **p** is 0.978 and **q** is 0.022, calculate the frequency of the heterozygous genotype in the population.

(ii) Explain the significance of the frequency of the heterozygous genotype of this genetic disorder in the population.

c) A person whose family includes people with cystic fibrosis may wish to know if they or their partner are carriers of this inherited disease, before deciding to have children. Suggest how an individual with the heterozygous genotype for this disorder could be identified.

d) Suggest *two* reasons why chorionic villus sampling is more acceptable than amniocentesis in detecting disorders in a fetus.

Edexcel HB3 June 1999 Q6

3 Eye colour in humans is controlled by a pair of alleles of a gene where the allele giving brown eyes is dominant to the allele giving blue eyes. Both parents of a blue-eyed man, John, were brown-eyed. He married a brown-eyed woman, Sara, whose father had brown eyes and mother blue eyes. Sara had a blue-eyed sister. John and Sara had a brown-eyed child.

a) (i) Devise suitable symbols for the alleles for brown eyes and for blue eyes.
(ii) Fill in the boxes and circles on the pedigree (family tree) below to show the genotype of each individual.

? EXAM QUESTIONS

b) Is it likely that the gene is carried on the X chromosome? Explain your answer.

c) For animals such as mice it is possible to carry out a cross to find out whether an individual showing the dominant allele is heterozygous or homozygous.

(i) Use a genetic cross diagram to show how this might be done.

(ii) In humans such an experiment is not ethically permissible. Explain briefly how it might be possible to determine whether a human was heterozygous or homozygous for a dominant allele.

OCR 6912 Inheritance Part 1 March 2001 Q1

4 Events in meiosis may give rise to genetic variation in daughter cells.

a) Using **diagrams**, show how **(i)** independent assortment and **(ii)** crossing over, may result in such variation for two heterozygous pairs of alleles, P and p, Q and q.

b) Name, and give an example of **one** other source of genetic variation, other than meiosis.

OCR 6912 Inheritance Part 1 November 2001 Q2

5 In *Mirabilis,* the four o'clock plant, one gene with two alleles T and t gives rise to tall or short plants with the tall allele being dominant to short.
Another gene with two alleles R and r gives rise to red, white or pink flowers, with neither allele showing dominance to the other.
A pure breeding tall, red-flowered plant was crossed with a short, white-flowered plant and all the F_1 offspring were found to be tall with pink flowers. On intercrossing (selfing) the F_1 an F_2 generation was obtained with six different phenotypes as follows:

tall with red flowers   short with red flowers
tall with pink flowers   short with pink flowers
tall with white flowers   short with white flowers

a) **(i)** Copy and complete the genetic diagram showing clearly the genotypes of the F_1 and F_2 generations together with the appropriate gametes in each case.

parents	TTRR	×	ttrr
gametes	.....		.....
F_1		.....	

gametes	TR	.....	tR	.....
....	.....	.....	.....	.....
Tr	.....	.....	.....	
....	.....	.....	.....	
tr	.....	.....	.....	

(ii) Indicate the **proportions of each phenotype** you would expect to find in the F_2.

tall with red flowers   short with red flowers
tall with pink flowers   short with pink flowers
tall with white flowers   short with white flowers

b) **(i)** What name is given to the type of inheritance shown by the **flower colour** gene?

(ii) Suggest how this gene may operate to give red, white or pink flowers.

OCR 6913 Inheritance Part 2 March 2001 Q2

6 In the fruit fly, *Drosophila melanogaster,* the allele **r** for white eyes is sex-linked and is recessive to the allele **R** for the normal red eye colour. Sex-linked genes are carried on the X chromosome. In fruit flies, the sex chromosomes are XX in females and XY in males.

a) When a white-eyed female fly is crossed with a red-eyed male, all of the male offspring have white eyes.

(i) Draw diagrams to represent the chromosomes, and alleles present on them, of the parents in this cross.

(ii) State the genotype and phenotype of the female offspring.

b) A gene which affects the size of the wings in fruit flies is not sex-linked (autosomal). The allele for short (dumpy) wings, **d**, is recessive to the allele, **D**, for normal wings.
A female fly was heterozygous for the genes for eye colour and for wing size. This female was crossed with a white-eyed, short-winged male.

(i) Use a genetic diagram to show the expected results of this cross.

(ii) List the genotypes and full phenotypes of the male offspring produced.

Edexcel B/HB1 June 1999 Q7

KEY SKILLS **ASSIGNMENT**

The chi-squared test

The chi-squared test is a simple statistical method that scientists use to tell if their results are significant or due to chance. For instance, if you were to roll a die, you would have a one in six chance of getting any particular number. If you rolled the die six times, you would probably not get each number once; you may get three twos or some other combination. However, the more times you roll the die, the greater the probability that all six numbers will be represented evenly.

If you are playing a game with someone who gets a higher proportion of sixes than normal, you might suspect them of playing with a loaded die. Without investigating the die itself, how could you decide whether that person was just lucky or that the difference was due to cheating? You could use the chi-squared (χ^2) test.

The nature of science and the null hypothesis

Most people accept that there is a link between smoking and lung cancer but, in science, it is virtually impossible to prove anything absolutely. What you can do is gather support for your hypothesis.

You could do some research into lung cancer victims and divide them into smokers and non-smokers. You could then use the χ^2 test to decide if there was a significant difference in the number of smokers who developed lung cancer compared with the non-smokers.

To use this statistical test you must approach the problem from a specific angle, and this involves developing a null hypothesis.

In the lung cancer example, you would use the hypothesis: 'There is no difference in the incidence of lung cancer between smokers and non-smokers'. You could then perform the χ^2 test (or some more sophisticated statistics) and come up with the conclusion that 'there is a very low probability that these results are due to chance' in which case you can reject the null hypothesis and support the idea that there is a link between smoking and lung cancer.

1 Write a null hypothesis that you might use to investigate the following questions:

a) What is the effect of sunbathing on the incidence of skin cancer?

b) What is the effect of cooking on the vitamin C content of vegetables?

The test itself

The basic idea behind the χ^2 test is that you compare the observed results with those you would expect. The extent to which the results vary is sometimes described as 'goodness of fit'. This means do the actual results fit the results that we had expected?

The formula for the χ^2 test is:

$$\chi^2 = \sum \frac{(O - E)^2}{E}$$

In this formula

O = the observed result,

E = the expected result and

Σ = the sum of

In practice it is best to use a table like the one below to progressively work out all the steps needed by the formula. In this worked example, a geneticist performed one of Mendel's classic pea experiments and got the following results.

Yellow, round	318
Yellow, wrinkled	103
Green, round	106
Green, wrinkled	33
Total	560

The geneticist wanted to know whether these results did actually reflect the expected 9:3:3:1 ratio, or whether they deviated so significantly that other forces were at work, and that the 9:3:3:1 hypothesis was in doubt. She did the following calculations:

Category	Hypothesis	Observed	Expected	$O - E$	$(O - E)^2$	$\frac{(O - E)^2}{E}$
Yellow, round	9	318	315	3	9	0.028
Yellow, wrinkled	3	103	105	−2	4	0.038
Green, round	3	106	105	1	1	0.009
Green, wrinkled	1	33	35	−2	4	0.114
Total		560				= 0.189

So the value of χ^2 is 0.189. What next? The hard work is now done and we only need one more piece of information; the number of degrees of freedom, which is a measure of the spread of the data. The number of degrees of freedom is always one less than the number of categories of information. So in this case, it is four categories minus one, which equals three degrees of freedom.

KEY SKILLS **ASSIGNMENT**

We now look up the value of 0.189 with three degrees of freedom on a table of χ^2 values. Table 25.A1 is an extract from the χ^2 tables.

From Table 25.A1 we can see that our value of

Table 25.A1 Extract from χ^2 tables

	Probability (eg 0.5 = 50%)									
	0.99	0.98	0.95	0.9	0.7	0.5	0.3	0.2	0.1	0.02
1 D of F	0.00016	0.0063	0.0039	0.0158	0.148	0.455	1.07	1.64	2.71	5.41
2 D of F	0.0201	0.0404	0.103	0.211	0.713	1.39	2.41	3.22	4.60	7.82
3 D of F	0.115	0.185	0.352	0.584	1.42	2.37	3.66	4.64	6.25	9.84

0.189 lies between the 0.98 and 0.95 column. From this we can say that the probability that our results are significant is between 95 and 98 per cent. Put another way, it tells us that the probability that these results are due to chance is between 2 and 5 per cent. By convention, scientists take the cut-off point as 5 per cent. If the test shows that there is a less than 5 per cent probability that your results are due to chance, you can reject your null hypothesis.

2 Use the χ^2 formula to work out the following:
a) A student tosses a coin 100 times. She gets 64 tails and 36 heads. Is the coin biased or are the results due to chance?
b) A geneticist managed to get his hands on a breeding pair of incredibly rare Matabili dung beetles. The male had green eyes and a blue abdomen. The female had pink eyes and an orange abdomen. When he crossed them he got the following results:

Table 25.A2

Phenotype	Numbers of offspring in F_1
Green eyes, blue abdomen	563
Pink eyes, blue abdomen	190
Green eyes, orange abdomen	183
Pink eyes, orange abdomen	64
Total	1000

Do these results conform to the expected 9:3:3:1 ratio?
c) Apart from their diet, these beetles look like being ideal organisms for the study of genetics. What features make an organism suitable for the study of inheritance in the laboratory?
d) Much progress in the field of genetics has been made using the fruit fly *Drosophila*. Find out some of the features that can be studied using *Drosophila*.

Chi-squared in examinations

If you have worked your way through the above examples, you may have concluded that chi-squared involves a lot of arithmetic and is time consuming. The examiners have come to the same conclusion and therefore it is unlikely that you will be set a full investigation to work through from scratch: they want to test your biology, not your maths. However, this test is a very useful tool and examiners will want you to appreciate its uses, and possibly to do some simple calculations.

26

Evolution

Adam and Eve: were they really the start of the human race?

Humans are a very strange species. We stand on two legs, naked apart from hair on the top of our heads and at a few strategic places. We are covered in sweat glands, we have a flat face and a long neck. Males have very large genitals compared to body size, while females have permanently swollen breasts. An alien zoologist looking at humans for the first time might be surprised that we dominate the Earth.

In the absence of any survival equipment, such as sharp claws or teeth, great speed or protective armour, why have we been so successful? The answer lies, of course, in our brains. Given our body size, we have the largest brain of any animal. Coupled with dextrous hands, our brains allow us to use language, co-operate and make tools to a far greater extent than any other species.

But where did humans come from? Until fairly recently, the answer was clear. 'And the Lord God formed man of the dust of the ground, and breathed into his nostrils the breath of life; and man became a living soul' (Genesis 2:7). Women followed soon after: 'And the rib, which the Lord God had taken from man, made he a woman, and brought her unto the man' (Genesis 2:22).

Until relatively recently most people believed this, or another explanation based on religious beliefs. Some still do, but Charles Darwin's theories on evolution changed fundamental scientific thinking. It is clear that humans evolved from an ape-like ancestor.

1 WHERE DO LIVING THINGS COME FROM?

To most human eyes, the living world seems fairly static. Like produces like: elephants give birth to baby elephants, frogs make frogs. It seems that no new species appear. There are literally millions of different species on Earth. Where did they come from?

THEORIES ON THE ORIGIN OF SPECIES

Ideas about the origins of species – including ourselves – have been central themes in philosophical and religious thought for centuries. Until the nineteenth century, virtually everybody in the Christian world looked to the Bible. The book of Genesis provided the answers: all animals and plants were created at the same time, by the great Creator. Even the eminent taxonomist Linnaeus, who in 1742 published his system for classifying all living plant species known at the time, said 'there are as many species as God created in the beginning'. Other religions had different versions of the same creationist idea.

The idea that species were not fixed, and that the infinite variety of living things had developed through a slow process of evolution, seems to have arisen first in early Greek philosophy and reappeared from time to time throughout the following centuries. It failed to gain lasting acceptance because no explanation could be found to show how the process might have come about. In addition, in some cultures, any 'non-religious' thought was strongly discouraged, even on penalty of death. A catalyst in the development of the theory of evolution was the discovery of fossils (Fig 26.1). These were clearly the remains of extinct creatures. Why were they no longer alive?

Fig 26.1 The discovery in the nineteenth century of fossils such as this *Triceratops* fuelled the idea that the living world was not static. Different layers of rock held different sets of organisms and this led to the acceptance that the living world was not fixed: plants and animals could change with time

One Christian explanation said that extinct species were the victims of a great global catastrophe, and that a new set of organisms had been created to replace them. It was even suggested that only the passengers aboard Noah's Ark survived a great global flood. When it was pointed out that different rocks contained different types of plants and animals, it seemed that there would have had to be many such floods, and so the **catastrophe theory** gradually lost credibility.

An early attempt to explain the mechanism of evolution came from the French naturalist Lamarck, who had made extensive studies of different life forms. He was convinced of the process of evolution, but the mechanism he proposed in 1809 to explain it was wrong. Lamarck supposed that organisms adapt to their environment by developing new structures and losing old ones. Such differences, acquired during the animal's life, were then passed on to the next generation. As an example he cited the webbed feet of water birds. Lamarck suggested that, in an effort to swim, the birds extended their toes and so stretched the skin between them. The stretched condition was then inherited and this process was repeated until it produced a fully webbed foot. With no knowledge of genes, chromosomes or DNA, this idea of the **inheritance of acquired characteristics** was perfectly reasonable for the time.

? TEST YOURSELF

A What was wrong with Lamarck's theory of evolution?

! SCIENCE FEATURE How old is the Earth?

Accurately estimating the age of the Earth is central to the theory of evolution. In 1650 James Usher (then the Archbishop of Armagh) announced that his study of Hebrew literature had led him to the conclusion that the date of creation was the year 4004 BC. Dr Lightfoot, vicar of the University church at Cambridge, took it a step further and proclaimed that the date of creation was 9 a.m. on 23 October, 4004 BC. Baron George Cuvier later stated that the Earth must be much older, more like 70 000 years.

Faced with the large array of different fossils, nineteenth-century geologists decided that the Earth was much older still. In 1830, Charles Lyell suggested that the true age of the Earth would turn out to be millions, rather than thousands of years. He was the first to propose that old rocks could be uncovered or brought to the Earth's surface (by volcanic activity, for example), so revealing the remains of long-dead species.

Ultra-sensitive dating techniques that measure the decay of radioactive isotopes have now put the age of the Earth at about 4 600 000 000 (4.6 billion) years. This is an important breakthrough because, although our brains cannot comprehend such a time scale, it fits in well with the idea of progressive evolution. The Assignment at the end of this chapter should help you to put this geological time scale into perspective.

✔ REMEMBER THIS

A working definition of evolution is 'a change in the genetic composition of a population over time'. It has also been defined as 'a gradual process in which new species develop from pre-existing ones'. But as we shall see in this chapter, populations can evolve without changing into new species.

✔ REMEMBER THIS

The cornerstone of Darwin's theory is that the mechanism of evolution is natural selection – this is known as **Darwinism**. The application of twentieth-century knowledge about chromosomes, genes and DNA to Darwin's theories is called **neo-Darwinism**.

2 CHARLES DARWIN AND THE THEORY OF EVOLUTION

In a book with the shortened title of *The Origin of Species*, Charles Darwin included these four ideas in his theory of evolution:

● **The living world is changing**, not static.

● **The evolutionary process is usually gradual**, not a series of jumps such as catastrophes and re-creations.

● **The common ancestor theory**. This suggested that closely related species evolved from one basic ancestor. Darwin said that man and apes evolved from a common ancestor, but the popular press took this to mean that humans had recently evolved from apes, leading to cartoons such as the one in Fig 26.2.

● **The theory of natural selection** as a mechanism to explain evolution.

Fig 26.2 This cartoon first appeared in *Punch* in 1861. It shows how poorly people of the time understood Darwin's theories

---

⚠ SCIENCE FEATURE Charles Darwin

Charles Darwin (Fig 26.3) was the son of a doctor. After studying medicine at Edinburgh he changed his mind and prepared for a career in the Church. However, despite going to Cambridge to study theology (religion), Darwin retained his interest in biology and geology.

Fig 26.3 Charles Darwin, 1809 to 1892

Darwin was offered the post of naturalist on the HMS *Beagle*, embarking on a scientific survey of South American waters. It was observations he made during this trip that first stimulated the young Darwin into contemplating the origins of species. As well as becoming convinced of the process of evolution, he also developed an idea about how it might happen. This trip was not only the turning point in Darwin's life; it was also the start of the greatest ever revolution in our perception of the living world, and of our place within it.

On returning home, Darwin buried himself in his studies, determined to prove or disprove his ideas about the evolution of species. Like Lamarck, Darwin developed his theories with no knowledge of genes, chromosomes or DNA. Mendel's work with peas (see Chapter 24) was going on during Darwin's lifetime but it remained hidden in an obscure journal.

Even though he managed to build up an overwhelming case for evolution, Darwin was reluctant to publish his theories, knowing the conflict they would cause with the Church. For more than 20 years Darwin carried on gathering evidence to support his theories. He was finally pushed into publication after exchanging letters with **Alfred Russel Wallace**: he too had come to the same conclusion about evolution. They agreed to make a joint announcement, and the Darwin–Wallace paper was presented to the Linnaean Society of London in July 1858.

In November 1859, Darwin published his revolutionary work, *The Origin of Species by Means of Natural Selection*. Few books before or since have caused quite such a controversy. Although Church leaders largely remained quiet, many people bitterly attacked it because it questioned the theory of creation. Darwin was called 'the most dangerous man in England'.

NATURAL SELECTION: 'SURVIVAL OF THE FITTEST'

Darwin's observations of the living world led to three basic conclusions:

● Generally, organisms produce far more offspring than can possibly survive.

● Living things are locked in a struggle for survival. They compete for food, space, mates, etc. In short, they are trying to eat and not be eaten, so that they can reproduce.

● Individuals of the same species are rarely identical: they show **variation**.

There was nothing particularly original about these observations, or in the idea that organisms evolved. What Darwin did, however, was to put all these factors together and suggest **natural selection** as the mechanism for evolutionary change.

Natural selection is commonly simplified to 'survival of the fittest'. Biologists define fitness as the ability of an organism to survive and reproduce. The fittest organisms survive to produce more offspring than less fit ones. In this way, the characteristic of a population can gradually change from generation to generation.

To illustrate the principle, imagine a population of blackbirds in a woodland. Like most organisms, they reproduce sexually and there is variation in the population. Some have better reflexes than others, forage for food more successfully, others have a better immune system. Normally, there is competition for resources such as food, mates and nesting sites. In this situation, the fittest birds survive and produce more offspring than others. The fitter birds may gather more food and so rear larger families than less fit individuals (Fig 26.4). Or they may rear two clutches of eggs in the time it takes others to produce one. The vital point here is that the *fittest individuals pass their genes on to more of the next generation.*

The severity of an organism's circumstances is called the **selection pressure**. The greater the selection pressure, the faster the evolutionary process. When organisms have a short life cycle, the process can be very swift indeed. In the 1950s farmers in the UK and Australia took the drastic step of introducing a deadly disease, myxomatosis, into the ever-expanding rabbit population (Fig 26.5). This virus killed over 99 per cent of the rabbits: this is selection pressure at a massive level. Only the rabbits whose genes gave them resistance to myxomatosis were able to survive and pass on their genes to the next generation.

> ✔ **REMEMBER THIS**
>
> Fitness in biology is a measure of reproductive success. So, a fit flowering plant produces and disperses more seeds than its competitors. Fit fungi are more successful because they produce more spores.

Fig 26.4 Blackbirds that can gather the most food tend to raise the greatest number of offspring, and so pass on their genes to more of the next generation

Fig 26.5 When the European rabbit is introduced into countries where it is not endemic, it tends to out-compete the native herbivores. With few predators around, the rabbit population can reach plague proportions. When this has happened in the UK and Australia, farmers faced ruin. The deadly myxoma virus was introduced into the rabbit population, killing 99 per cent. The remaining 1 per cent were naturally resistant to the virus and are the ancestors of today's, largely resistant, rabbit population

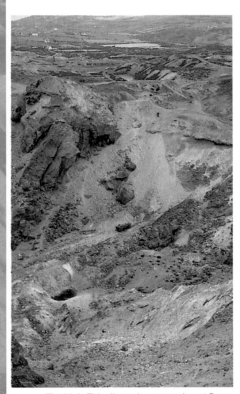

Fig 26.6 This disused copper mine at Parys Mountain in Anglesey (above) is devoid of life. Copper is a metabolic poison that kills virtually all plants. The grass *Agrostis tenuis* (right) is one of the few that has evolved copper resistance

NATURAL SELECTION IN ACTION

Many people tend to think that evolution is something that happened in the past and that the world of today is the finished product. But this is far from the truth: evolutionary forces are still at work and natural selection is continually changing the characteristics of populations, often with great speed. Consider the following examples:

● The widespread use of antibiotics such as penicillin rapidly led to the development of resistant strains of bacteria. Penicillin works by inhibiting one of the enzymes that the bacteria need in order to manufacture cell walls. However, many resistant strains can make an enzyme, penicillinase, that breaks down the antibiotic.

● The use of warfarin as a rat poison has led to the development of warfarin-resistant rats. Warfarin is an anti-coagulant that kills rats by inducing **haemorrhage** (internal bleeding). Resistant rats have a modified enzyme in their blood-clotting system that allows their blood to clot in the presence of warfarin.

● Copper tolerance in the grass *Agrostis tenuis*. Copper is a metabolic poison that prevents many plants from growing on copper-polluted land (Fig 26.6). However, resistant plants have evolved. These can transport copper out of their cells, so that it accumulates in the cell wall and does not interfere with the plant's metabolism. (In non-polluted areas, the normal type of grass out-competes the copper-resistant strain.)

The organisms in these examples have a short life cycle and high reproductive capacity, and so can respond quickly to the selection pressure placed on them. The effects of selection pressure on organisms that live longer and have a lower reproductive capacity take much longer to become apparent.

DIFFERENT TYPES OF NATURAL SELECTION

From what you have learned about natural selection, it would be tempting to assume that it is just a mechanism for change. However, this is not always the case: it can be a means of keeping things static. Many organisms, crocodiles and sharks for example, have not changed significantly for millions of years.

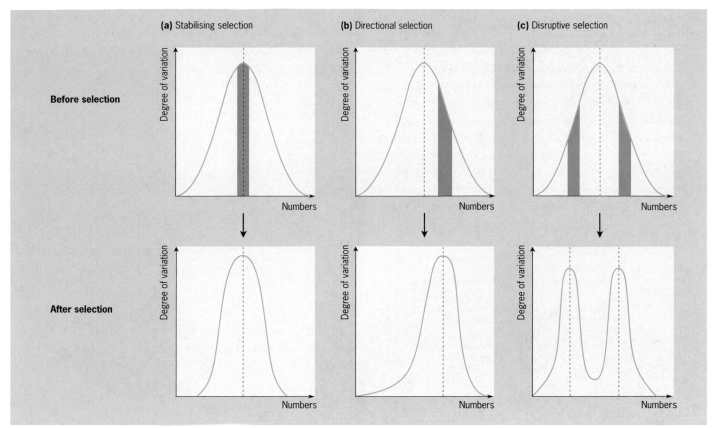

Fig 26.7 The different types of natural selection. Individuals in the shaded areas in the top row of graphs have a selective advantage

There are basically three different types of natural selection:

● **Stabilising selection** (Fig 26.7a). In a favourable and unchanging environment, stabilising selection can lead to a 'standardising' of organisms by selecting against the extremes. An example of this is human birth weight. Studies show that particularly large or small babies have a higher mortality than those in the mid-range.

● **Directional selection** (Fig 26.7b). This mechanism tends to occur following some kind of environmental change that causes selection pressure. The organisms with a particular extreme of phenotype may have an advantage. At some stage, directional selection must have operated to produce the long neck of the giraffe. Probably, in times of food shortage, only the tallest individuals could reach enough food to survive and pass on their genes to the next generation.

● **Disruptive selection** (Fig 26.7c). This type of selection acts at both ends of the distribution, favouring individuals with extreme rather than average characteristics. Rarer than the other two types, disruptive selection is important in the formation of new species, where natural selection acts differently on different populations. An example of this is seen in Darwin's finches (see page 449) where disruptive selection has acted on beak size. In some situations, those with longer, thinner beaks would have succeeded in exploiting insects as a food source, while those with shorter, stronger beaks would have succeeded in cracking seeds. Those with average beaks were adapted for nothing in particular and would have reproduced less successfully.

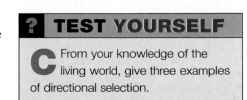

? TEST YOURSELF

C From your knowledge of the living world, give three examples of directional selection.

! SCIENCE FEATURE Artificial selection

Humans have practised artificial selection, or selective breeding, for thousands of years. We haven't needed a scientific understanding of genetics to do this: animals or plants with the most desirable features are simply crossed or mated. For example, if cows with the highest milk yield are impregnated by the healthiest and fastest-growing bulls, many of the next generation are big, healthy cows that produce large quantities of milk. Those that do not have the desired features are not used for breeding. In this way, artificial selection rather than natural selection is used to bring about great changes in the phenotype in a short time: animals and plants can be 'domesticated' in just a few generations.

Selective breeding is still important in many different areas of food production:

● Increasing the growth rate and the meat, milk or egg production of livestock.
● Increasing the disease resistance of crop plants (increasing the disease resistance of animals by selective breeding is not as reliable).
● Changing the muscle to fat ratio of livestock, so that more lean meat is produced.
● Increasing the yield and nutritive value of crop plants such as wheat.
● Increasing the tolerance of plants and animals to drought, heat or pollutants.

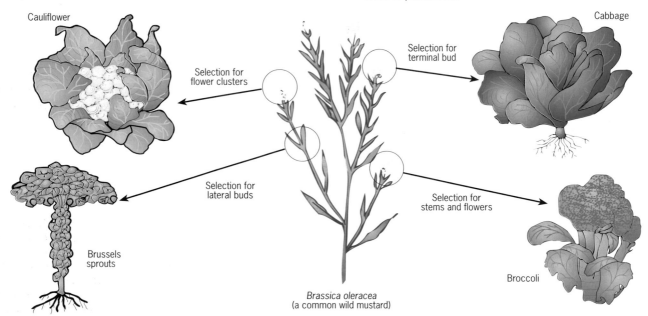

Fig 26.8 Many familiar vegetable types have arisen from the artificial selection of the wild mustard, *Brassica oleracea*. In the original wild population, plants show variation in the size and shape of leaves, stems, terminal buds, lateral buds and flower clusters. Selection for individual features rapidly produces the familiar crops you see here

Fig 26.9 Like all pedigree animals, this champion Abyssinian cat is the product of many generations of artificial selection. Breeders consider the wedge-shaped face a desirable characteristic and so have concentrated on this feature. Breeding animals in this way can lead to problems, and some people strongly disagree with breeding animals for superficial 'perfection'

3 THE EVIDENCE FOR EVOLUTION

Today there are few scientists who seriously doubt the theory of evolution, or that it happens by natural selection. Here we consider some lines of evidence.

FOSSILS

Palaeontology, the study of fossils, supports the theory of evolution. Most fossils are found in sedimentary rocks, formed by layers of silt that preserve the bodies of organisms before they have a chance to decay in the normal way. In many places (see Fig 26.10) it is clear that different layers, or **strata**, represent different time periods. By studying the fossils in these strata we find obvious evidence for the gradual evolution of some species.

GEOGRAPHICAL DISTRIBUTION

A study of the distribution of species presents strong evidence that organisms evolved 'into' their ecosystem rather than being placed there during a single act of creation.

During his time in South America, Darwin encountered a classic piece of evidence for evolution. On the South American mainland, finches have short, straight beaks to crush seeds. Other species have taken advantage of the other food sources. Woodpecker finches, for instance, have long beaks for extracting insects from holes in trees. On the Galapagos Islands, however, things are different. For one reason or another, the finches managed to get to the islands, while many other species of birds did not. In the absence of competition, the finches have evolved to take advantage of the range of food sources on offer (Fig 26.11).

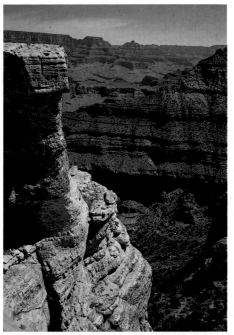

Fig 26.10 The Grand Canyon in Arizona. The horizontal rock strata can be clearly seen here and span some 500 million years of geological history

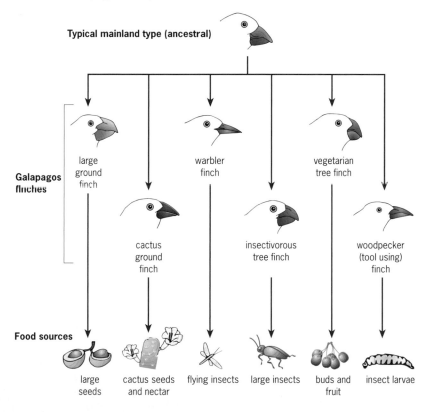

Fig 26.11 Darwin's finches on the Galapagos Islands. In a process called adaptive radiation, one common ancestor has evolved into many different species of finch, each exploiting a particular food source, such as seeds, insects and buds. Interestingly, Darwin did not immediately recognise the significance of all of these different species; he simply collected a large number of specimens in a bag and presented them to a bird specialist on his return home

! SCIENCE FEATURE Evolution on islands

As Darwin found out on his voyage, islands are particularly interesting places to study evolution. Young, volcanic islands such as the Galapagos (Fig 26.12) are often colonised by plants whose seeds have been blown there by the wind, and by flying animals which have been driven off course by storms. In later years, of course, colonisation by non-flying animals (such as rats and cats) is likely to be a result of human activity.

The evolution of birds on islands can be very spectacular. Free from predators, birds often lose the power of flight and sometimes grow very large. Before the Maori settlers arrived from Polynesia, New Zealand was a bird-dominated ecosystem. Huge flightless birds, **moas**, some as tall as 3.5 metres, evolved to fill the niches normally occupied by mammals. They were easy prey for the Maoris, however, who hunted them to extinction within a few centuries. Ironically, the only survivor of the moa family, the kiwi, has become the country's national emblem.

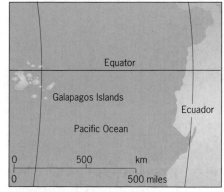

Fig 26.12 The Galapagos Islands are an archipelago situated near the Equator in the Pacific Ocean, about 600 miles west of Ecuador

COMPARATIVE ANATOMY

The anatomy (body plan) of plants and animals can reveal a lot about evolutionary relationships. Fig 26.13 shows one of the classic examples of comparative anatomy, the **pentadactyl limb** (five-fingered limb). This structure appears in various forms throughout the vertebrates: the feet of reptiles and amphibians, the human hand, the bat's wing, the seal's flipper, for example.

Fig 26.13 The vertebrate pentadactyl limb provides strong evidence for evolution. In all these examples, the basic bone layout is the same (in the diagram, the same type of bone is given the same colour in each animal) but the limbs have become modified in different ways according to the way the animal moves

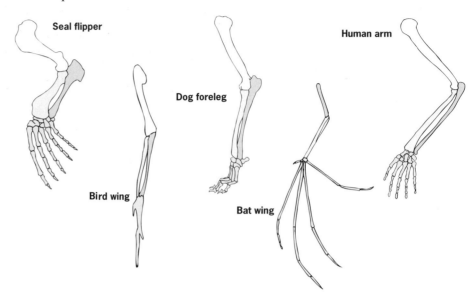

This basic five-fingered plan has been modified in a variety of ways to suit the animal's mode of locomotion. Structures with a common origin but with a different function are said to be **homologous**. So, the wing of a bat is homologous to a human hand. The possession of the pentadactyl limb in so many different organisms is persuasive evidence that these animals have evolved from a common ancestor.

In contrast, totally unrelated structures can become adapted for the same function. The wings of insects and the wings of birds, for example, have a totally different anatomy and origin in the embryo, but are similar in shape because they perform the same job. Such structures are said to be **analogous**.

? TEST YOURSELF

D Are the following pairs of structures analogous or homologous?
a) Trout's fin and dolphin's flipper.
b) Horse's leg and bat's wing.

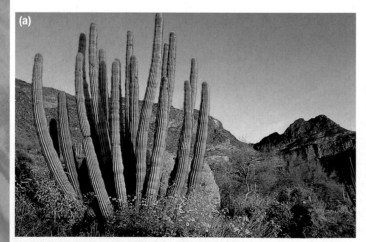

Fig 26.14 Convergent evolution occurs when unrelated species solve the same evolutionary problem in a similar way
(a) Cacti, which are native only to the Americas, have evolved to cope with desert living. The leaves have become spines which lose no water and protect the plant. The fleshy stem expands and stores water and has taken over the job of photosynthesis

(b) You might guess that this euphorbia from Africa was a cactus, but you would be wrong. The two plants have evolved in different continents and are unrelated; they look very similar because they have developed the same features to survive desert conditions

The process by which unrelated species evolve to resemble each other is called **convergent evolution** (Fig 26.14). A classic example of this is seen in seals (mammals), penguins (birds) and fish. Because of their need to swim efficiently, all have evolved the same streamlined shape. The forelimbs of the seal and the wings of the penguin have both evolved to resemble fins.

Fig 26.14(c) (left) The flipper of the whale and the fin of the shark (right) are a classic example of convergent evolution. The two animals are completely unrelated and the internal structure of forelimbs is totally different. However, they are both the same shape because of the job they have to perform

EMBRYOLOGY

Studying the development of an embryo reveals some interesting evolutionary secrets. The scientist Ernst Haeckl said 'ontogeny recapitulates phylogeny' which means that the evolutionary development of a species (its **phylogeny**) is mirrored by the changes of each individual from birth to maturity (its **ontogeny**).

A number of vertebrate embryos follow a very similar path from fertilisation through several stages of early development – so similar that it is often difficult to tell them apart (Fig 26.15). For example, all vertebrate embryos develop branchial arches; these are relics of the gills possessed by aquatic ancestors. However, although the study of the developing embryo provides further evidence for evolution, it does not give us much information about the actual evolutionary process.

Fig 26.15 Four developmental stages in embryos of fish, turtles, chicken and humans. You can see that the early stages are very similar in all four. For example, in the drawings in the second column, all four embryos have branchial arches, the 'folded' areas just under the head at the front. Only in the later stages do differences become obvious

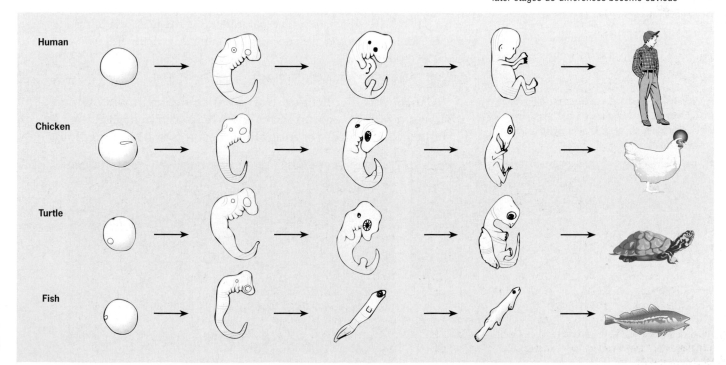

COMPARATIVE BIOCHEMISTRY

The biochemistry of different species shows some striking similarities. Many chemicals such as nucleic acids, ATP and cytochromes, organelles such as ribosomes, and pathways such as respiration, are almost universal. This is strong evidence that all living things have a common ancestry.

4 HOW A NEW SPECIES EVOLVES

We have looked at natural selection and seen that it can lead to stability or change. In this section we examine how natural selection can lead to the formation of a new species.

WHAT IS A SPECIES?

We instinctively think that we know what the word 'species' means, but an accurate definition is difficult, if not impossible. This is a common definition:

> **A species is a population or group of populations of similar organisms that are able to interbreed and produce fertile offspring.**

This would seem to be reasonable; a horse and a donkey can mate to produce a mule (Fig 26.16), which is sterile, so the horse and the donkey can be said to belong to different species. However, this definition runs into trouble. There are many closely related species, wolves, coyotes and domestic dogs, for example, which can interbreed to produce perfectly fertile offspring. The only reason why the wolves and coyotes remain distinct is that the hybrids do not compete as successfully as the pure strains.

The difficulty we have in defining the term species stems from the very nature of the process of evolution. New species evolve from pre-existing ones, and this is usually a slow transition, which may take many generations and thousands of years. There are many living species that are still in the process of separating from each other. The wolf and the coyote almost certainly evolved from a common ancestor and for all practical purposes are separate species (Fig 26.17). If left to evolve naturally over the next few thousand years, the two species may well accumulate enough genetic differences to prevent interbreeding.

So, we can modify the definition to:

> **A species is a collection of recognisable organisms which shares a unique evolutionary history and is held together by cohesive forces of reproduction, development and ecology.**

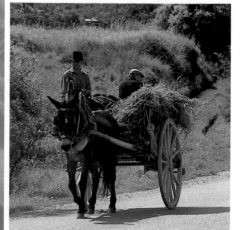

Fig 26.16 'And nothing to look forward to either.' The mule is the product of a mating between a donkey and a horse, two species that have different numbers of chromosomes. It cannot reproduce because it is sterile. Because of its stamina, the mule has been used as a beast of burden for thousands of years

Fig 26.17 The coyote (*Canis latrans* – left) and the wolf (*Canis lupus* – right) are classed as different species, but they are capable of interbreeding and producing fertile offspring

Although more of a mouthful, it is a more satisfactory description because it recognises that all members of a species are usually (but not always) recognisable, and that they share the same 'biology', occupying the same niche in the ecosystem and having particular social structures, courtship, mating habits, etc.

THE DEVELOPMENT OF NEW SPECIES

Speciation, the development of new species, happens when different populations of the same species evolve along different lines. **Populations** are groups of interbreeding individuals of the same species occupying a particular geographical region. Examples of populations may be the water fleas in a pond, all the oaks in a forest or the elephants in a game reserve.

It is thought that speciation usually happens in the following way:

- Part of a population becomes isolated in some way that prevents them from breeding with the rest of the population. Very often, the breakaway population will, by chance, have a different genetic constitution from the original (see genetic drift – page 456).
- The two populations experience different environmental conditions, so that natural selection acts in different ways, causing the genetic make-up of the two populations to differ.
- Eventually, genetic differences accumulate to a point where individuals from separate populations that come together again can no longer interbreed.

It is worth mentioning that new species can also arise from a hybridisation between two different species. This particularly applies to plants, and plant breeders take advantage of this ability to produce new crop varieties.

There are two basic methods of speciation:

- **Allopatric speciation** (*allo* = different, *patris* = country), where the two populations are physically separated.
- **Sympatric speciation**, where the two populations are reproductively isolated in the same environment.

HOW POPULATIONS BECOME ISOLATED

Usually, species evolve when populations become isolated by physical barriers such as stretches of water or mountain ranges. This is known as **geographical isolation** (Fig 26.18).

SEE QUESTIONS 1 AND 2

Fig 26.18 The principle of geographical isolation
(a) The original population of mice has a 50:50 distribution of the alleles A and a, ie A is as common as a. A river changes its course and cuts off a few individuals. In the new, isolated population the allele frequency, by pure chance, is 4:10. This change in allele frequency is called the founder effect. The fewer the individuals in the breakaway founder population, the more dramatic the change in allele frequency is likely to be
(b) The founder population starts to grow, but they are in different conditions from the original population. The alleles A and a were neutral up to now, but then a water-borne disease spreads through the new population, which are confined to marsh land. The aa genotype confers resistance on those who carry it. By natural selection, the A allele is selected against, and so the a allele becomes even more frequent. In this way, the genetic make-up of the two populations diverges

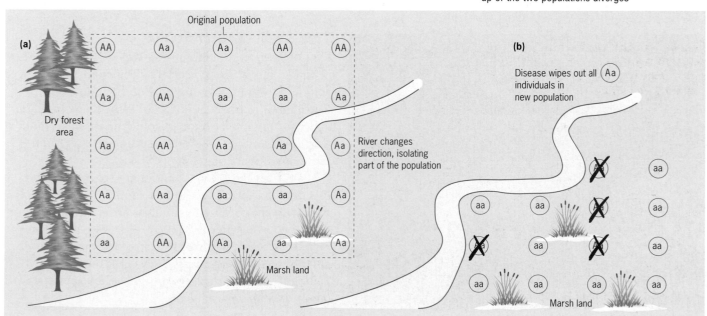

Evidence suggests that geographical isolation is the most common way in which populations become reproductively separated. However, speciation can happen even when two populations overlap. For instance, two populations of frogs may have a different mating call, and the females may respond only to calls of males in 'their' population. Table 26.1 summarises the different isolating mechanisms.

Table 26.1 A summary of isolating mechanisms	
Mechanism	**How it works**
Prezygotic mechanisms, ie gametes don't come into contact	
Geographic	Two populations do not come into contact due to some physical barrier, eg mountain range, water.
Temporal	Different timing, eg flowering times or mating seasons do not coincide.
Behavioural	Individuals of other populations are not acceptable as mates, eg wrong courtship, mating call or pheromones.
Mechanical	Structural differences prevent mating, eg the shapes of the genitals in insects.
Postzygotic mechanisms, ie even if mating takes place	
Hybrid non-viability	Hybrid zygotes fail to develop to sexual maturity.
Hybrid sterility	Hybrids do not produce functional gametes.
Hybrid weakness	Hybrids have reduced survival or reproductive rates.

5 EVOLUTION IN ACTION

The process of evolution is generally slow, and it is no surprise that for centuries people thought that the living world was static. However, there are situations where you can see the process in action. Organisms with short life spans and high reproductive capacities, such as bacteria or insects, can evolve in a very short time indeed – see page 446.

Also of great interest to evolutionary biologists is a **polymorphism**, a situation where two or more different forms, **morphs**, of a species exist in a population. Each morph confers a different survival advantage in different situations. Polymorphisms such as that of wing colour in the peppered moth (below) allow us to see how natural selection leads to changes in allele frequency, and so we can observe evolution in action.

THE PEPPERED MOTH

Fig 26.19 shows the original speckled form of the peppered moth *Biston betularia*. This moth was common in the UK at the time of the Industrial Revolution. In areas of heavy industry, many buildings and trees became covered in soot. The speckled moth lost its camouflage in these areas and so was eaten more often by its predators.

Luckily for the moth, a mutation occurred around this time that produced a black moth known as the **melanic form**. In industrialised areas the melanic moth had a selective advantage. However, in 'cleaner' areas the reverse was true: the speckled moth thrived and the melanic form was rarely seen. Both morphs belong to the same species, but their survival depends largely on the environment.

Fig 26.19 The two forms of the peppered moth, normal (speckled) and melanic (black). Against the background of the bark of a tree shown in this photograph, the speckled moth on the right is almost invisible while the melanic form on the left is far more likely to be seen and caught by a predator. However, it becomes invisible on a black surface such as a sooty wall, and so is more likely to survive in an industrial environment

HOW CAN WE STUDY EVOLUTION?

If evolution is so slow, how can we observe it happening? An important principle here is that evolution involves a *change in allele frequency*. For instance, when myxomatosis was introduced into the rabbit population, the alleles for resistance were selected for, and so they increased in frequency. In any particular situation, if we can show that allele frequencies are changing, we know that evolutionary forces are at work.

Evolutionary studies are usually based on interbreeding populations, which are known as **demes** or **Mendelian populations**. The sum total of all the genes circulating within such a population is called the **gene pool**.

POPULATIONS AND ALLELE FREQUENCY

When the gene pool remains more or less the same, a population is not evolving. Natural selection may be acting, but it is often a force for stability rather than change. However, if the gene pool of a population is changing, natural selection is acting as a force for change and this is evolution in action.

To show how we measure allele frequency, imagine a population of 500 rats. In the population are two alleles for coat colour: black (B) is dominant over brown (b). The rats are diploid – each individual has two alleles – so the population contains 1000 coat colour alleles. If 600 of the alleles are B, then we can say that the frequency of B is 60 per cent, or 0.6 (statisticians prefer decimals). As there are only two alleles, it follows that the frequency of b must be 40 per cent or 0.4. (Remember that the sum of the allele frequencies must be 100 per cent, or 1.)

If one gene is dominant to the other – as in this case – the symbol p is used for the frequency of the dominant allele B, and q is used for the frequency of b. It is clear that $p + q$ must equal 1.

The values of p and q change when a population evolves. To find out if this is happening we use the **Hardy–Weinberg principle**.

THE HARDY–WEINBERG PRINCIPLE

The principle, named after the British mathematician G.H. Hardy and the German biologist W. Weinberg, states that there will be no change in the frequency of alleles in a population so long as the following conditions are met:

- The population is very large, diploid and reproduces sexually.
- Mating is totally random – there is no tendency for certain genotypes to mate together.
- All genotypes are equally fertile, so there is no natural selection.
- There is no emigration or immigration.
- No mutations occur.

We can use the Hardy–Weinberg rule to estimate allele frequencies using just the information about the frequency of a particular genotype. Consider the example of albinism in humans. One in 10 000 humans are born albinos – they have two copies of a recessive gene which means they cannot make the pigment which gives colour to hair, skin and eyes (Fig 26.20). The frequency of the allele for albinism in the population is difficult to estimate because carriers of the allele are not albinos. So, how can we find the frequency of the albinism allele?

REMEMBER THIS

$p + q = 1$
Frequency of dominant allele + frequency of recessive allele = total number of alleles of that gene.

Fig 26.20 This person is an albino. The white hair, pink eyes and pale skin are the result of an inability to make the pigment melanin

If we call the normal allele A and the albinism allele a, most of the population are AA, some are carriers: Aa or aA, and a small number are albinos: aa. Using the Hardy–Weinberg formula we can estimate the percentage of the population who carry the a allele. Remember that p and q represent the frequency of the A and a alleles respectively. The Hardy–Weinberg equation:

$$p^2 + 2pq + q^2 = 1$$

Put into words, it says:

The frequency of AA individuals, added to the frequencies of Aa, aA and aa, must equal 1.

We also know that $p + q = 1$, and we can use these two equations to work out the proportion of heterozygotes in the population, and any other frequencies involved:

1 in 10 000 is a frequency of 0.01%, or 0.0001 expressed as a decimal. 1 in 10 000 has the genotype aa, so the frequency of aa; qq or q^2, has the value of 0.0001.

So $q = \sqrt{0.0001} = 0.01$.

If $q = 0.01$, then p must be $1 - q$, which is 0.99.

So, in other words, 99 per cent or 0.99 of the alleles in the human population are A.

From this we can calculate that the number of carriers of the albino allele $2pq$ is:

$$2pq = 2 \times 0.99 \times 0.01 = 0.0198 \approx 0.02$$

So, almost 2 per cent of the population carry the albinism allele.

WHEN ALLELE FREQUENCIES CHANGE

The Hardy–Weinberg principle states that allele frequencies remain constant as long as certain conditions apply. It follows that when these conditions are not met, the allele frequencies can change and the population can evolve. Consider what happens when the conditions listed on page 455 are not met.

● The **population is small**. If the population is small, chance plays a large part in determining which alleles pass to the next generation. The smaller the population, the greater the probability that the genes of one generation will not be accurately represented in the next. The name given to a change in allele frequency caused by chance is **genetic drift**. These are two important causes of genetic drift: **bottlenecks**, when a population declines and then recovers; and the **founder effect**, when a population starts from just a few individuals (see below). As Figs 26.18 and 26.21 show, this can result in a drastic change in the allele frequency.

Original population

New population

50:50 allele frequency

Population passes through a bottleneck: only a few survive. By chance there is 70:30 allele frequency

70:30 allele frequency

Fig 26.21 An illustration of genetic drift. In the original 'population' there are 1000 beads; half red, half blue. If ten beads are extracted at random, there is a high probability that there won't be five of each. If these ten 'organisms' then become the founders of a new population, the proportion of colour (the allele frequency) will be significantly changed, in this example to 7:3

- **Mutations occur**. Mutations are ultimately the origin of all genetic variation (see Chapter 24). They are rare, and most are harmful or neutral. Occasionally, however, a mutation gives an organism a survival advantage, or the environment changes so that previously harmful/neutral mutations become advantageous. For example, a mutation in a bacterial gene that leads to the production of a modified enzyme could make the bacterium resistant to an antibiotic. That individual would survive, even in the presence of the antibiotic, and pass its resistance gene on to the next generation.

- **Individuals move in or out of populations**, that is, immigration or emigration. If the new individuals breed, this movement may lead to gene flow. When a few individuals move into a new environment and start a population, we see the founder effect. Like the beads in Fig 26.21, the individuals in the founder population are probably not representative of the original population. A group of six human survivors on a desert island, for example, might have two red-haired people and two blondes. A population founded by these people would have a much higher percentage of red-heads and blondes than the population in their country of origin. In the most extreme case, a new population can be founded by one individual, such as a self-fertilising plant or a pregnant female animal.

- **Mating is non-random**. In many cases, individuals choose mates with a particular genotype. In this case, mating is not random, it is **assortive**. Such assortive mating can change allele frequency.

- **Natural selection occurs**. When there is competition for resources – and this is usually the case in an ecosystem – natural selection acts and only the most successful individuals pass on their genes.

6 EVOLUTION: OUR PLACE IN THE GREAT SCHEME OF THINGS

There are millions of different species on Earth, and we have given all the ones we know about a scientific name. The gorilla is simply *Gorilla gorilla*, the chimpanzee is *Pan troglodytes* while we have given ourselves the name *Homo sapiens*, which means 'wise man'. A scientific name tells us both the **genus** and the **species**. The generic name has a capital latter while the specific name does not. When written, scientific names should always be in italics or underlined.

We classify organisms according to the similarities between them. Understanding relationships between species allows us to piece together the story of life on Earth. Classification is therefore a bit like a huge family tree over a massive time scale. Just like an address has a house number, a road, a town, etc, we classify organisms into progressively narrower groups. Table 26.2 shows this 'address', which we more correctly call a **taxonomic hierarchy**. Taxonomy is the science of naming and classifying organisms and the term hierarchy reflects the layered structure of the family tree of living organisms that consists of groups within groups.

HUMANS AND THEIR CLOSE RELATIVES

Humans vary in size, shape, colour, blood groups and in thousands of other ways, but we are all one species, *Homo sapiens*. *Homo* is a genus, a set of species that are grouped together because of their close similarity. There are no living species considered to be close enough to humans to be put in our

REMEMBER THIS

Scientific names are important. If you call an organism by its scientific name, any biologist in the world, whatever language they speak, will know what you are talking about. The bird known as the sparrow in America is a completely different species from the one we are familiar with in Britain, but confusion can be avoided as the two birds have completely different scientific names. Most scientific names have Latin or Greek origins.

SEE QUESTIONS 3 AND 4

REMEMBER THIS

Organisms do not evolve according to a set pattern, just to make the taxomist's life easier. A detailed study of a particular organism might show that it falls neatly into Kingdom, Phylum, Class, Order, Family, Genus and Species. Or it might not. This is why you sometimes find extra groups such as sub-phyla, super-families or sub-species.

✔ REMEMBER THIS

Taxonomic groups do not overlap. For instance, there are no animals that are both reptiles and amphibians. An animal must belong to one class or the other; for example, a toad is an amphibian and an iguana is a reptile. Neither animal can be part reptile and part amphibian.

genus, so we are the only species described as *Homo*. However, there have been human ancestors that belonged to the genus *Homo*. Humans are also the only species in the family Hominidae. However, the basic body plan shared by humans and the great apes is so similar that biologists have put us all in the super-family of Hominoidae. The word **hominoid** describes any ape-like or human-like creature.

Table 26.2 The taxonomic hierarchy: how we classify human beings

Taxonomic group	Humans	Explanation
Kingdom	Animalia	There are five kingdoms, animals, plants, fungi, protoctista and bacteria.
Phylum	Vertebrata	We are animals with backbones.
Class	Mammalia	There are five classes, fish, amphibians, reptiles, birds and mammals. The defining features of mammals are fur/hair and milk production.
Order	Primates	Primates are adapted for life in the trees. Key primate features are covered in the main text.
Family	Hominidae	Human-like creatures – includes several extinct species.
Genus	*Homo*	No other living species is sufficiently like humans to be included in this genus.
Species	*sapiens*	We are the thinking species, modern man.

HUMAN EVOLUTION

So how did humans evolve? It is widely accepted that humans evolved in Africa, from the same ancestors that gave rise to the great apes; chimps (Fig 26.22), gorillas and orang-utans. We did not, however, evolve from these species, as is often thought. Piecing together a complete evolutionary picture is difficult. If you took all the remains that have ever been found of ancestral humans, they would all fit into an average-sized living room. However, we do have some vital fossil evidence that has allowed us to piece together a likely **lineage**, or line of development (see the Science feature on the next page).

Fig 26.22 Chimpanzees show the classic ape characteristics; a sloping face with protruding jaw and heavy brow ridges. It is debatable whether the chimp or the gorilla is man's closest relative – the evidence is conflicting

THE FIRST PRIMATES

The first primates to evolve appeared about 70 million years ago. We owe a lot of our human features to our primate ancestry. Primates have the following key features:

● Flexible limbs with gripping hands. Hands and feet have **opposable digits** – thumbs that can oppose all the other digits, giving an excellent grip.
● Finger nails rather than claws. This allows some protection, but doesn't interfere with the sensitive fingertips. Primate hands are highly dextrous. This makes it easier to obtain food and it allows more effective grooming. It has also helped us develop the ability to make tools.
● Flat faces with forward-facing eyes. This allows stereoscopic vision, giving judgement of distance, an essential skill for a life in the trees. Primates also have colour vision, which is believed to help them find ripe fruit.
● Large brains. Primate brains have large areas that process information from the eyes, ears, mouth and skin.
● Socially, primates tend to live in family groups, giving birth to single offspring that are dependent on their parents for a long time.

⚠ SCIENCE FEATURE Making sense of fossils

This photograph shows the fossil remains of Lucy, a young girl who lived approximately 3.5 million years ago. Looking at it can illustrate quite well how fossils are able to tell us a lot more than they appear to at first glance.

How can we tell this fossil is a 12-year-old girl?

You can sex a human, or humanoid, skeleton from the shape of the pelvis. Females have a wider pelvis, with a larger gap to allow a baby's head to pass through. Size and bone development indicates age.

How do we know the age of these remains?

Fossils are formed in sedimentary rocks. When this girl died, her skeleton was covered with layers of silt and and sand in conditions that protected it from the normal processes of decay. The soft tissues such as the skin, muscles and internal organs decay, but the bones undergo a chemical reaction that turns them into rock. In time, the sediments bury the fossils deeper and deeper. We can estimate the age of the skeleton by comparing this layer with other strata, but for a more accurate measurement we can use potassium/argon dating – see the Assignment at the end of this chapter.

What did she look like when she was alive?

From the face we can see some obvious features: the sloping face, protruding jaw and prominent eyebrow ridges; she was obviously not a modern human. We can reconstruct the soft tissues in the same way that forensic artists do with the skull

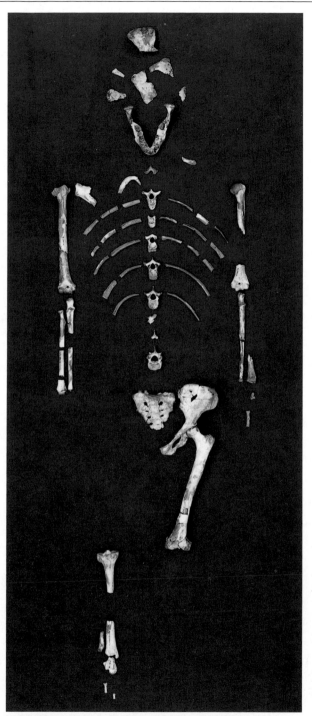

Fig 26.23 This is Lucy, the most complete specimen of *Australopithecus afarensis* ever found. She was discovered in 1974 in what is now Ethiopia, and named after the Beatles song 'Lucy in the sky with diamonds' that the archaeologists Don Johanson and Tom Gray were playing at the time of the discovery.
Overall, the great skill in piecing together fossil evidence comes from knowing where to look, and in recognising hominid remains when you see them. There are incredibly skilled people who can look at a pile of rubble and spot a tooth, jaw fragment or vertebrae of a primitive human, without mistaking it for modern ape or human remains. Who knows what remains are yet to be discovered?

of a murder victim. If you know the muscle structure and the average depth of the soft tissues at strategic points on the skull, you can use clay, and a lot of skill, to reconstruct the face.

How large was her brain?

This is easy to do – you just look at the space left inside the skull. You can estimate the volume by filling the cranium with water or sand and then emptying into a measuring cylinder. You can even make a latex cast of the brain cavity and measure its volume more accurately by displacement.

What did she eat?

There are several clues about diet. Generally, the more vegetation in the diet, the stronger the muscles which pull the jaw from side to side, and the flatter the molars. This girl has teeth very similar to ours, so she probably had a varied, omnivorous diet. Sometimes microscopic examination of the scratches on the enamel can tell us a lot about the diet of a fossilised human.

How do we know that she walked upright?

Firstly, the hole where the spinal cord goes into the skull (the foramen magnum: Latin for 'big hole') is under the skull so that the head pivots on the spinal column. Also, the leg bones are thicker than those of the arm, and the pelvic girdle is stronger than the shoulder girdle. This implies that the weight was placed mainly on the legs rather than evenly distributed.

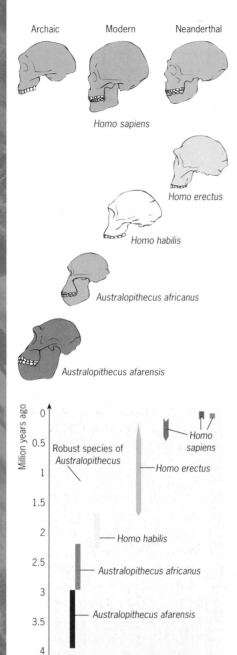

Fig 26.24 The major evolutionary stages of the hominids.
Above: The development of skull shapes from *Australopithecus afarensis* to *Homo sapiens sapiens*
Below: The place in time and cranial capacity of the hominids

Table 26.3 The history of the hominids

Time period	What was going on?
8–4.4 million years ago	This period is sometimes called the 'fossil void', which is frustrating because, at some time in this period, the hominids, the family that gave rise to man, appeared.
4.4 million years ago	*Australopithecus ramidus* appeared. Discovered in Ethiopia in 1992, this is the earliest known hominid. Evidence suggests that *A. ramidus* was bipedal (walked on two legs) and had a small brain but we know little about his lifestyle.
4–3 million years ago	The time of *Australopithecus afarensis*. The first, almost complete skeleton, Lucy (Fig 26.23) was discovered in 1974. Lucy was like a modern human from the waist down, suggesting bipedalism, but her upper half was distinctly ape-like. Lucy and her relatives evolved in a time of climatic change, when the great forests gave way to extensive grasslands. The largely forest-dwelling ape-men and women were forced to survive out in the open. In this situation natural selection would favour individuals who were more intelligent, showed greater social co-operation and were better at using tools. Some experts think that this is what drove evolution to produce modern humans.
3–2.2 million years ago	*Australopithecus africanus* existed. A specimen found in 1924 in South Africa shows a distinctly human jaw and dentition, but the cranial capacity was still small: more ape-like than human.
2.3–1.7 million years ago	*Homo habilis*, 'handy man', appeared. This is the earliest creature to be assigned to the genus *Homo*. *Homo* is distinguished from *Australopithecus* by progressive changes: a larger brain, a higher skull, a flatter face and a more rounded jawbone. It is possible that *H. habilis* made primitive tools, such as sharp-edged stones that could cut up animal carcasses. *H. habilis* was replaced by, or evolved into, *Homo erectus*.
1.8–0.5 million years ago	*Homo erectus* was a larger species with a larger brain. He was almost as big as modern man, although his skull still had the primitive, heavy-browed ape-like form. This species existed for over a million years.
0.4–million years ago	*Homo sapiens* appeared in Europe and Africa. Two sub-species are H*omo sapiens neanderthalensis* (Neanderthal man) and *Homo sapiens sapiens* (ourselves). Neanderthals had slightly larger brains than *H. sapiens sapiens*; they were large, muscular, barrel-chested people with long, globe-like skulls and long noses. They made tools and there is evidence to suggest that they buried their dead. The most recent Neanderthal fossils are only 32 000 years old. Studies of DNA suggest that *Homo sapiens sapiens* first appeared in Africa about 150 000 years ago: just a moment ago in the geological time scale (see the Assignment). They began to migrate from Africa about 100 000 years ago. Why Neanderthal man died out remains a mystery.

PRIMATE FOSSILS

The oldest primate fossils are about 70 million years old. These looked more like modern tree-shrews, but had certain primate features. Over the next 30 million years various primate species evolved and became extinct, until the first anthropoids appeared about 40 million yeas ago. These were like modern monkeys, living in trees and with a well-developed social structure.

In 1924 the fossilised skull of a child was found in a quarry at Taung in South Africa. After much speculation, this specimen was named *Austrolopithecus africanus* which means 'Southern ape from Africa' (see Fig 26.24). Since then, several hundred austrolopithecines have been discovered, mainly in Africa. It seems highly likely that humans evolved from this type of 'subhuman'. The outline of human evolution that is shown in Table 26.3 and in Fig 26.24 is highly speculative, and can change with the discovery of one new fossil. A comparison of the skeletons and dentition of a gorilla, and early hominid (Lucy) and a modern man, is shown in Fig 26.25.

Fig 26.25 A comparison of the skeletons and jaws of a gorilla, Lucy and a modern man

SUMMARY

When you have finished this chapter, you should know and understand the following:

- **Evolution** is defined as a change in the genetic composition of a population over time. New species develop from pre-existing species in this way.

- The basic mechanism of evolution is **natural selection**. This acts on the genetic variation that occurs in a population. Individuals whose genotype gives them an advantage over others are more likely to pass their genes on to the next generation.

- A **population** is a group of interbreeding individuals of the same species in the same geographical area.

- New species are formed when populations become isolated in some way so that they cannot interbreed. Natural selection then acts on the different populations, changing the frequency of **alleles**. With time, the differences between the two populations can become so great that an individual from one cannot interbreed with an individual from the other.

- Humans can speed up the evolutionary process by **artificial selection**. Breeding organisms with desirable features, such as cattle with a high milk yield, changes the allele frequency rapidly. Within only a few generations, organisms are produced which are very different from the original wild stock.

- The evolutionary process is fastest in species which have a large population size, a large reproductive capacity (fecundity) and a rapid life cycle. Many disease-causing and pest organisms fall into this category.

- Many endangered species have a low reproductive capacity and a long life cycle.

- Evolution happens when there is a change in the allele frequency. The **Hardy–Weinberg principle** states that, in a large population, the allele frequency remains constant unless some outside force acts to change it. These forces are chance (the smaller the population, the greater the effects of chance), migration of organisms, selective mating and natural selection.

- When the conditions for the Hardy–Weinberg equilibrium are not met, the allele frequency changes and the population evolves.

- Humans evolved in Africa from ape-like ancestors. Key stages include bipedalism (walking on two legs), an increase in brain size, the use of tools, the development of language, social co-operation and the evolution of agriculture.

? EXAM QUESTIONS

1 Read the passage and then answer the questions that follow.

Myxoma virus causes a disease of rabbits called myxomatosis. In the early 1950s, this virus was used to control rabbit populations in Australia. The virus is widespread amongst Brazilian rabbits which suffer only a slight skin disorder. In European rabbits, the virus rapidly leads to a fatal disease. Following infection, death occurs within approximately two weeks. In Australia, the disease caused a 99.5% mortality rate in the rabbit population when it was first introduced in the years 1950–51, but this declined to a 30% mortality rate by 1957–58. By 1959, the rabbit population was increasing again. A study of the infection was carried out to determine the reason for the decreasing effectiveness of the disease as a means of biological control. A single strain of the virus was used throughout the whole study. Each year, numbers of uninfected rabbits were caught and injected with the virus strain. The symptoms of the disease and mortality rates were monitored and some of the results are presented in the table.

Year of capture	Percentage of rabbits with symptoms		
	severe & fatal	moderate	mild
1953	95	5	0
1954	93	5	2
1955	61	26	13
1956	75	14	11

a) What evidence is there that the geographical isolation of rabbits leads to variation?

b) **(i)** One hypothesis for the change in mortality rate was that the rabbits were developing an inheritable resistance to the disease. What evidence is there in the passage to support this hypothesis?
(ii) Suggest another hypothesis to account for the decreasing effect of the disease in the wild rabbit population.
In Australia, rabbits are now controlled by the application of a single pesticide chemical.

c) **(i)** Other than elimination of all rabbits, what might be the long term effect of this on the rabbit population?
(ii) In what way could the use of the pesticide be regarded as an example of artificial selection?

OCR 6912 Inheritance Part 1 Mar 2000 Q3

2 Read the following passage and answer the questions below, from R. Dawkins, *Unweaving the Rainbow*.

The standard neo-Darwinian view of the evolution of diversity is that a species splits in two when two populations become sufficiently unalike that they can no longer interbreed. Often the populations begin diverging when they chance to be geographically separated. The separation means that they no longer mix their genes sexually and this permits them to evolve in different directions. The divergent evolution might be driven by natural selection (which is likely to push in different directions because of different conditions in the two geographical areas). Or it might consist of

random evolutionary drift (since the two populations are not held together genetically by sexual mixing, there is nothing to stop them drifting apart). In either case, when they have evolved sufficiently far apart that they could no longer interbreed even if they were geographically united again, they are defined as belonging to separate species.

a) **(i)** According to the passage, what commonly starts the divergence of two populations?
(ii) From your knowledge of isolating mechanisms suggest **two** other events which may prevent breeding between two populations.
(iii) When might it be said that populations should be defined as belonging to different species?
Further evolutionary divergence leads to the formation of larger taxonomic groups. The table shows part of the classification of a rat.

b) **(i)** Complete the names of the principal taxonomic groups in the correct order in the following table.
(ii) Copy and complete the classification of a mammal, **the rat.**

Taxonomic group	Classification of rat
kingdom	
	Rodentia
family	Muridae
	Rattus
	norvegicus

c) Explain briefly why Lamarck's theory of inheritance of acquired characteristics is not acceptable to Darwinist and neo-Darwinist biologists.

OCR 6912 Inheritance Part 1 Mar 2001 Q4

3 The figure below shows the way in which four species of monkey are classified.

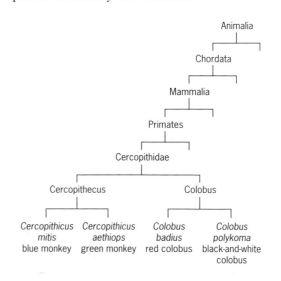

? EXAM QUESTIONS

a) This system of classification is described as hierarchical. Explain what is meant by a hierarchical classification.

b) **(i)** To which genus does the green monkey belong?
(ii) To which family does the red colobus belong?

c) What does the information in the diagram suggest about the similarities and differences in the genes of these four species of monkey?

NEAB Biology Continuity of Life Molecule Test March 1998 Q3

4 The diagrams below show the skull of a Neanderthal man and that of a modern human, *Homo sapiens.*

a) State *two* differences, excluding overall size, between these two skulls.

b) Describe *one* way in which fossil remains, such as those of Neanderthal man, can be dated.

Neanderthal man 5cm   **modern man** 5cm

Edexcel HB2 June 1999 Q2

KEY SKILLS **ASSIGNMENT**

THE HISTORY OF THE WORLD

Rocks, and the fossils they contain, can now be dated by radiometric dating techniques, which measure levels of certain isotopes. We can use this technique to construct a geological timescale – a biological history of the world. This Assignment outlines the process of dating using radioactive isotopes and puts the history of life on Earth into perspective.

Dating rocks and fossils

The principle behind radiometric dating is simple: some elements exist as radioactive isotopes that decay at a predictable rate. Measurement of this decay allows us to estimate the age of the rock.

For example, when volcanoes erupt, the lava contains radioactive potassium, ^{40}K. This decays steadily into argon (^{40}Ar). The half-life of ^{40}K is 1.3 billion (1.3×10^9) years. So, in 1.3 billion years, half of the potassium decays. Half of the remaining potassium will decay in the next 1.3 billion years, and so on. By measuring the ratio of radioactive potassium to argon, scientists can estimate the age of the rock strata to the nearest 50 000 years.

In many places, layers of volcanic lava alternate with sedimentary rocks that contain fossils. If we can date the lava, we can also estimate the age of the fossils. Some fossils of primitive bacterium-like organisms are thought to be around 3 500 000 000 years old.

1 **a)** What does half-life mean?
b) What is the half-life of ^{40}K?
c) What fraction of the original ^{40}K will be left after 3.9 billion years?

Carbon dating

This process can be used to date samples that still contain some organic material. Cosmic radiation is constantly bombarding the atmosphere, turning $^{12}CO_2$ into $^{14}CO_2$. This becomes fixed into organic molecules by the process of photosynthesis and then passes up the food chain. When an organism dies, the ^{14}C decays into ^{14}N, with a half-life of 5730 years. Carbon dating can be used to measure the age of any carbon-containing material up to 50 000 years old.

2 A human skull is found to have only one-sixteenth of the ^{14}C it would have has when 'new'. Estimate its age.
The Earth is currently estimated to be about 4 600 000 000 years old. This timescale is impossible to comprehend, but if we condense this huge span into one year we can begin to put it into perspective (see the Earth calendar on the next page).

3 **a)** In our one-year scale, how many seconds represent 1000 years?
b) How long in actual time is represented by one minute, one hour and one day?
c) Use your answer to (a) to estimate how long the dinosaurs lived for.

KEY SKILLS **ASSIGNMENT**

The Earth calendar

Date	Event
1 Jan	Earth forms
26 Feb	oldest rocks – nothing was permanent before this time
23 March	first life – bacteria-like organisms obtain energy from organic molecules (food) by the process of glycolysis
April–May	organic food runs out; photosynthesis evolves, makes food and releases oxygen as a by-product
30 June	oxygen atmosphere fully developed; aerobic respiration evolves some time after this
17 Sept	first eukaryotes
19 October	first co-ordinated multicellular colonies (like sponges)
5 Nov	first invertebrates (worms and jellyfish)
15 Nov	rapid increase in invertebrate diversity
20 Nov	first fish
21 Nov	first coral reefs
29 Nov	first colonisation of land – vascular plants appear
4 Dec	first amphibians
6 Dec	first trees
7 Dec	first insects
8 Dec	first reptiles
17 Dec	first mammals
19 Dec	first birds and dinosaurs
21 Dec	first flowering plants
25 Dec	last of the dinosaurs
26 Dec	first primates
31 Dec	
14.00	first upright bipedal hominids
21.00	*Homo erectus*
22.45	recent Ice Age starts
23.00	evidence that primitive humans can cook food
23.48	Neanderthal man appears
23.54	*Homo sapiens* appears
23.55	Australian aborigines settle
23.58.49	agriculture invented
23.59.28	beginning of recorded history
23.59.29	pyramids built
23.59.41	Buddha born
23.59.45	Christ born
23.59.53	Battle of Hastings
23.59.57	start of the scientific method
23.59.58	Australia 'discovered' by Cook; the Industrial Revolution
23.59.59	most of the knowledge contained in this book is discovered

4 The rise of humans as the dominant species has been meteoric.

a) What features of humans account for our success?

b) What dangers does this success bring?

5 Write down what you feel to be the four most significant advances made by humans. These could be inventions or social and cultural advances. If you are working in a group, use your own list to provide a discussion.

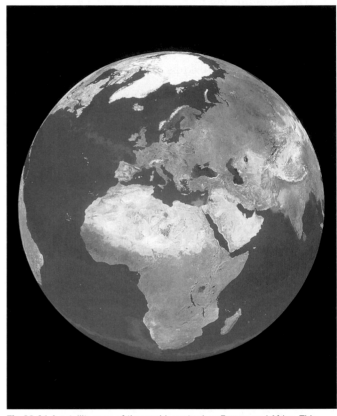

Fig 26.A1 A satellite map of the world, centred on Europe and Africa. This image is a combination of hundreds of images acquired by NOAA weather satellites from orbits some 820 km above Earth. It didn't always look like this. The shape of the land masses has been subject to constant change, shaping patterns of evolution. When land masses separate, two halves of a population can also be separated and proceed to evolve in different directions

27

DNA technology

This is 'Rainbow', cc's genetic donor. Note the differences in coat colour between cc and Rainbow

cc, the first-ever cloned cat shown here at seven weeks old, was born on 22 December 2001, at the College of Veterinary Medicine, Texas A&M University. The announcement of the successful cat cloning was delayed until the animal had completed its shot series and its immune system was fully developed

Allie and cc. Allie is cc's surrogate mother

In February 2002, researchers announced that a cat had been cloned for the first time. The kitten, called cc (the typist's old abbreviation for carbon copy), was born two months earlier, but had to be checked carefully to make sure she was normal and healthy, before the announcement was made.

The cloning of cc highlights some of the technological and ethical issues surrounding cloning technology in this first decade of the twenty-first century. cc was cloned by transplanting DNA from Rainbow, a female three-coloured (tortoiseshell) cat into an egg cell whose nucleus had been removed, and then implanting this embryo into Allie, the surrogate mother. cc is genetically identical to her genetic donor, Rainbow, but she isn't an exact physical copy, underlining the delicate balance between genetics and environment. Coat colour in cats is determined partly genetically, and partly by environmental factors, such as conditions in the womb. In the past couple of years, newspaper reports have hinted that a cult in America, who have developed cloning technology, intend to help a couple whose 10-month-old baby died tragically. The idea is to clone the dead child from frozen cells to 'replace' the child. However, cloning cannot bring the dead back to life; this is a seriously misguided experiment that has been widely condemned by other cloning scientists all over the world.

cc also demonstrates just how difficult and 'wasteful' cloning is. In their first attempt, researchers obtained the cells used to make the clone from the skin cells of a donor cat. Eggs from other cats were used for the next step. Their chromosomes were removed and replaced with the DNA from the skin cells, creating cloned embryos that were then transplanted into surrogate mothers. In total, 188 such nuclear-transfer procedures were performed, which resulted in 82 cloned embryos that were transferred into seven recipient females. Only one cat became pregnant, with a single embryo. But this pregnancy miscarried and the embryo was surgically removed after 44 days. In the next attempt the scientists tried using cells from ovarian tissue to receive the DNA from the cat to be cloned. Five cloned embryos made in this way were implanted into a single surrogate mother. Pregnancy was confirmed by ultrasound after 22 days and a kitten was born on 22 December 2001, 66 days after the embryo was transferred. So, out of 87 implanted cloned embryos, cc is the only one to survive. This is comparable to the success rate in sheep, mice, cows, goats and pigs, but would be unacceptably low for people.

The group in America says that it has dozens of women who have volunteered to be surrogate mothers. But have they been told of the high risk of losing the baby that they are being asked to carry? Are the parents really willing to let so many women go through such heartache, in order to produce a baby that will never replace the one they lost?

Most scientists agree that human cloning for reproductive purposes should not be attempted, but animal cloning looks as though it will expand further in the next few years. The company in Texas that part-funded the research that produced cc hopes to use the technology to provide commercial cloning of companion animals for pet owners. Less controversially, the cloning of prize cattle, and perhaps successful racehorses, might become big business. But perhaps the most acceptable face of animal cloning will be the application of the technology to preserve endangered species.

1 THE NEW GENETICS

Genetics seeks to explain the mechanisms of inheritance and evolution and, like many other 'pure' sciences, it can be applied to solve practical problems. In the 'new genetics' that has developed rapidly in recent years, humans are taking a much more active role in determining how genes pass between individual organisms. Not only can we study DNA, we can manipulate it and use it to our advantage. This branch of genetics is called genetic engineering, or DNA technology. This chapter considers the basic ideas behind this new technology.

Transferring genes between organisms offers many exciting prospects. In agriculture, for example, genes from nitrogen-fixing bacteria could be put into plants which would then fix their own nitrogen and so make some of their own fertiliser. This could greatly improve the efficiency of food production. Crop plants given genes that code for disease resistance or tolerance to pollutants could be grown on previously inhospitable land, such as industrial land or desert margins. DNA technology also has many potential applications in medicine.

The genome of an organism, its entire collection of DNA sequences, can now be explored and mapped. Theoretically, any individual gene can be located, cut out, replaced or inserted into another organism. Of course, this sort of research is regulated very carefully, and the ethical, moral and social consequences need to be taken into account (Fig 27.1).

WHAT IS ALREADY POSSIBLE

We can sequence the genes in whole organisms; several bacteria, some plants, animals and the human genome are now fully sequenced (see the Science feature on page 471).

We can take individual genes and make multiple copies: this is **gene cloning**.

We can extract genes that code for useful products and insert them into other organisms: this is **recombinant DNA technology**. Organisms produced in this way are called **transgenic organisms**. More genes that code for human products could be transferred into microorganisms for mass production. Products already made in this way include insulin, somatostatin, alpha-1-antitrypsin, human growth hormone and blood clotting substances such as Factor VIII. See Chapter 6 for more details.

We are just starting to be able to replace defective genes with healthy ones: this is **gene therapy**. Some successful gene therapy trials have been carried out, but this area is fraught with technical and ethical problems (see Chapter 32).

We can clone parts of organisms, using **stem cells**. These are undifferentiated cells, found in embryos and in adults, that can be persuaded to turn into particular tissues and organs. Scientists have recently cloned kidneys using stem cells in cows (see the Science feature on page 478).

We can clone whole animals. Many farm animals – goats, sheep, pigs and cows – have been cloned and other animals such as the cat, and endangered species of animals have either just been cloned, or will shortly be cloned.

In this chapter we focus on the techniques involved, and we look briefly at some of the possibilities for the future and at some of the ethical implications of genetic technology.

REMEMBER THIS

Genetic engineers manipulate DNA to allow individual genes to be transferred from one organism to another. Often the two organisms are from completely different species.

Fig 27.1 'You were so keen to see what you could do, you didn't stop to think about whether or not you should.' This is a scene and a quote from the film *Jurassic Park*. It is quite unlikely that we will ever be able to recreate dinosaurs, but the quote reflects an important point: all genetic research has the potential for abuse and we should think very carefully about its consequences

2 GENE HUNTING

✔ REMEMBER THIS

DNA is common to all organisms, and it always has the same basic structure. This allows us to combine genes from organisms as diverse as humans and bacteria.

? TEST YOURSELF

A Suggest what the enzyme reverse transcriptase does.

Fig 27.2 The technique called FISH (fluorescent *in situ* hybridisation) was used here – a genetic probe containing a fluorescent marker seeks out the gene. This micrograph shows a deletion on chromosome 17. On the normal chromosome on the left the normal gene shows up as two green dots. These are absent on the chromosome on the right

The first problem that genetic engineers tackle is finding the gene they are interested in. Every human body cell contains 46 chromosomes, each having several thousand genes. So how, for example, do you go about finding the insulin gene that you want to insert into a bacterium? Let's look at three different approaches.

● **Isolate the gene from the rest of the genome**. This has been compared with looking for a needle in a haystack, when the needle is also made from straw. However, there are several sophisticated techniques that can be used to map chromosomes. For instance, if you know part of the base sequence in the required gene you can make a genetic probe. This is a single-stranded piece of matching DNA labelled with a radioactive or fluorescent marker. When mixed with the genomic DNA (DNA from the genome) – in the right conditions – the probe seeks out the complementary base sequence and binds to it, showing you exactly where the gene is (Fig 27.2).

These techniques can now be combined with the vast database that has been produced by the Human Genome Project (see Section 4 of this chapter). This database of all the genes in the human genome is freely available to all scientists, and they can look up likely genes and gain predictions of function, using computer programs that construct the most likely protein from the DNA sequence. The possibilities are extending so fast, it is impossible to give more than a brief overview in this book.

● **Use messenger RNA**. If you can isolate the mRNA molecule that is a copy of the gene that you are trying to find, you can use it as a genetic probe as described above. You can also use it to make a DNA copy using the enzyme reverse transcriptase. This means essentially that you make an 'artificial' gene. For example, you could find the mRNA that codes for insulin in the cells where the insulin gene is expressed – the cells in the islets of Langerhans. DNA made from mRNA is called complementary DNA, or cDNA. Artificial genes are shorter than those in the genome because they contain no introns (Chapter 24).

● **Work backwards from the protein**. If you work out the amino acid sequence of the desired protein – a relatively easy process – you can make a piece of DNA that codes for it. It would obviously be a tedious process to make a large gene, but now there is equipment that can synthesise artificial DNA quickly and easily. The end-product of this process is also a short, artificial gene made of cDNA.

3 MANIPULATING DNA USING ENZYMES

Once you have located the gene you want, the next step is to use it. Enzymes are the tools of the trade here. Some of the enzymes used in genetic engineering are shown in Table 27.1. See also Chapter 6.

Table 27.1 Some enzymes commonly used in genetic engineering in addition to those featured in the main text

Enzyme	Function
Exonucleases	Remove terminal base from DNA – useful in analysis of base sequences.
DNase	Makes random cuts in DNA: chopping into varying-sized fragments.
Kinases	Add groups such as phosphates to DNA: this helps with labelling and analysis, particularly when phosphates are made with ^{32}P.
Polymerases	Synthesise nucleic acid molecules from nucleotides. This is essential for the polymerase chain reaction (PCR) – see Science feature on page 472. Reverse transcriptase is a polymerase that makes DNA from RNA. It occurs naturally in retroviruses such as the human immunodeficiency virus, HIV.

RESTRICTION ENZYMES

Restriction enzymes can be thought of as 'molecular scissors' – they cut DNA strands at specific points (Table 27.2). More properly called **restriction endonucleases**, they are made by bacteria in response to attack by viruses called bacteriophages. Their name reflects their function: they restrict damage by chopping bacteriophage DNA into smaller, non-infectious fragments, and they make their cuts inside the nucleic acid molecules.

There are many different endonuclease enzymes, produced originally by different species of bacteria. Each enzyme cuts DNA at a different base sequence, a point known as the recognition site. The enzymes make staggered cuts in the DNA (Fig 27.3), commonly called sticky ends. These can combine with complementary sticky ends, and this allows lengths of DNA that have been removed from one organism to be spliced – inserted – into the DNA of another. *Eco*RI, for example, is a common endonuclease isolated from the bacterium *Escherichia coli*. This is a very popular enzyme with genetic engineers because it is readily available. It is also reliable: it cuts DNA at its recognition site with precise accuracy, showing very little **star activity** (cutting DNA at sites other than the recognition site).

Table 27.2 The target sites of some restriction enzymes		
Enzyme	**Bacterial origin**	**Recognition site**
*Eco*RI	*E. coli*	G|AATTC CTTAA|G
*Hind*III	*H. influenzae*	A|AGCTT TTCGA|A
*Bam*HI	*B. amyloliquefaciens*	G|GATCC CCTAG|G

SEE QUESTION 1

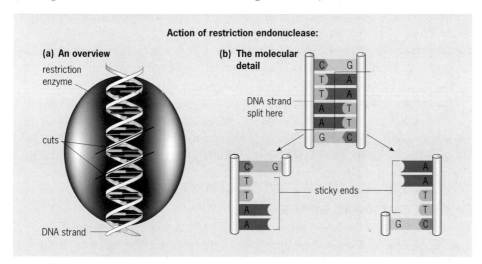

Action of restriction endonuclease:

(a) An overview
restriction enzyme
cuts
DNA strand

(b) The molecular detail
DNA strand split here
sticky ends

Fig 27.3 Endonuclease enzymes are wonderful tools for the genetic engineer because not only do they cut DNA, but they make staggered cuts, leaving 'sticky' ends which make re-joining DNA strands much easier. Endonucleases cut at the sugar–phosphate bonds in the DNA molecule, leaving only the hydrogen bonds intact. At room temperature and above, these bonds are not enough to hold the molecule together, and so it splits, producing a 'sticky end'

You might be asking, 'If restriction enzymes cut DNA, why is the bacterium's own DNA not damaged?' The answer lies in a process called **methylation**: the bacteria add a methyl ($-CH_3$) group to the sequences in their own DNA that could act as recognition sites. This prevents the restriction enzymes doing any damage, presumably because the methylated region no longer fits into the active site of the enzyme.

DNA LIGASES

The sticky ends of the DNA strands produced by the action of restriction enzymes are joined together by enzymes called DNA ligases (ligate means to join/bind). The normal function of a DNA ligase enzyme is to join strands of DNA during replication (Chapter 24). They are also involved in DNA repair. Genetic engineers use them as 'molecular glue' and they are essential partners for restriction enzymes. An example is T4 DNA ligase, which is made by the bacterium *E. coli*.

✔ **REMEMBER THIS**

The prefix *endo-* means inside; the prefix *exo-* means outside. An endonuclease cuts up a DNA strand by making cuts along the whole length of the molecule. In contrast, an exonuclease cuts DNA by chopping off the nucleotides at the ends of the molecule.

? **TEST YOURSELF**

B Different restriction enzymes have differently sized recognition sites. Those in Table 27.2 are described as 'six cutters' because their recognition sites are six bases long, but four or five cutters are also common. Why do you think four cutters produce a larger number of fragments than five cutters?

4 THE HUMAN GENOME PROJECT

The 1990s saw the beginning of the most ambitious project ever to be undertaken in molecular biology: the Human Genome Project (HGP). The aim of this project was simple: to map out the sequence of DNA that makes up all the human genes, but it was a phenomenal task. Although it was estimated that the project would take until 2005 to complete, it was actually published ahead of schedule, in the summer of 2001. Technology advanced very rapidly, enabling much of the sequencing work to be done much faster than had been anticipated.

THE MAIN RESULTS:
WHAT THE HUMAN GENOME SEQUENCE IS

The human genome is divided into 46 chromosomes. Each of these chromosomes has now been completely sequenced; the complete gene sequence from the DNA, pooled from several human individuals, has been read to show the order of the 3.2 million base pairs it contains (Figs 27.4 and 27.5). Obviously, we are all different, but when the overall size of the genome is taken into account those differences are less significant than you might think. Between humans, DNA differs by only 0.2 per cent, or 1 in 500 bases. These differences are being mapped and can reveal some interesting information about disease risk, but the overall sequence is still sufficiently similar in all humans to make the entire sequence very useful.

Surprisingly, there are fewer genes that had been expected – only about 30 000, compared with the 100 000 genes that had been estimated about 10 years ago. A nematode worm has 18 000 genes, a fruit fly has 13 000, so it's quality, not quantity, that obviously counts!

Fig 27.4 Computer analysis of DNA sequencing autoradiograms, used to decode the base sequence of lengths of DNA

WHAT ARE WE USING THE SEQUENCE FOR?
FINDING OUT WHAT GENES DO

The application of the knowledge that has been gained by decoding the human genome is the most exciting part of this project. We are now at the very start of doing this: new technologies are now being developed to monitor the expression of different genes in disease. For example, researchers looking at colon cancer can already monitor how different genes are turned on at each different stage of cancer development, and have plotted out lots of pathways that lead to cancer. Discovering the function of many of the genes in the human genome will produce exciting new information that should help us understand more about how the body works, and how it develops disease.

GENETIC SCREENING

In the short term, the genome is also enabling scientists to set up more and better genetic tests that give advance warning to people of their disease risk. Hopefully, this will allow people to choose a lifestyle to help prevent the disease, and to start preventive drug therapy where necessary. People with a known genetic tendency to develop high cholesterol, for example, could choose to eat a low-fat diet, and perhaps take drugs such as statins, to lower their cholesterol in middle age. This would prevent them succumbing to atherosclerosis, and the forms of heart disease that follow (see Chapter 31).

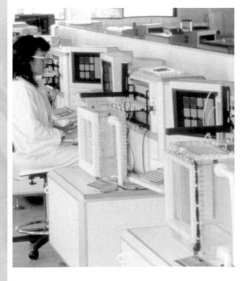

Fig 27.5 Automated electrophoresis equipment used to study the structure of DNA

CHOOSING AND DEVELOPING DRUGS MORE EFFECTIVELY

In the medium term – the next 15 years or so – genome sequence information is likely to become important in prescribing drugs. At the moment, if you have a particular disease, there is a range of drugs that you can be given, but this is often fairly narrow – and they are not tailored to your individual genetic background. The result is adverse reactions that kill hundreds of thousands of people worldwide each year. The new data about the human genome will make it possible to identify which people will react badly to which drugs, so that doctors will be able to avoid them and try something else. It may also become possible to design drugs that are better suited to people with a particular genetic background. Knowing all the genes in the human genome will give researchers new targets to aim for.

GENE THERAPY

Gene therapy, using the genes themselves as medicines, is in many ways the most obvious application of the human genome data. But it is also the most controversial, with a number of deaths linked to experimental treatments.

Gene therapy is currently being developed for single gene disorders such as cystic fibrosis, Huntington's disease, sickle cell anaemia and muscular dystrophy. The basic concept is that the abnormal gene can be cut out and replaced by a healthy version, delivered by a tamed virus. Other multifactorial diseases, such as heart disease may also be treatable to some extent when the therapy is developed further.

GERMLINE ENGINEERING

The most long-term application of the human genome sequence may be germline engineering. This is really highly theoretical at the moment, and is likely to require much thought and ethical debate between scientists, governments and the public before it becomes a reality. The idea here is to identify an abnormal gene and then to correct it in the sex cells – the eggs and sperm – that are used to pass genetic information to offspring. No subsequent generation would then be afflicted by their ancestors' gene defect.

Fig 27.6 The US geneticist Dr Craig Venter, founder of The Institute fo Genomic Research (TIGR) in Maryland, USA. He is standing in the DNA sequencing laboratory at TIGR

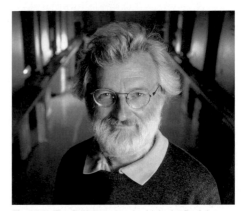

Fig 27.7 The British molecular biologist Dr John Sulston, director of the Sanger Centre in Cambridge, UK. He is standing in the main sequencing room

---

! SCIENCE FEATURE Human conflict and the human genome map

The image of a scientist as an unfeeling and slightly weird boffin is a common stereotype, but scientists are very human, as the background to this enormously successful scientific project demonstrates. The draft human genome, the sequence of most of the human chromosomes, was announced jointly in June 2000 by the Human Genome Project, a publicly funded programme that included scientists from all over the world, and Celera Genomics, a US private company that used a different technique to produce essentially the same data. Simultaneous publication of the research behind the announcement followed in February 2001. This public show of togetherness masked a bitter dispute that rumbled on during the year before the announcement, and which was still rumbling on when this chapter was written (in March 2002).

The two groups used a slightly different technique to read the 3.2 billion base pairs of DNA that make up the human genome.

The HGP method broke the genome up into manageable segments that were mapped before sequencing took place. Celera took a more sledgehammer approach: it broke the genome into tiny pieces and then used massive computing power to reassemble the read fragments. The arguments originally were about the quality of the work and the time it took. The HGP camp were put on the defensive because Celera managed to produce the sequence in a fraction of the 10 years it took them, and for all the public money that they spent. Celera were criticised for the quality of their work, and for 'cribbing' essential information from the publicly available HGP work, which was updated as work progressed. Many scientists hoped that the joint announcement of the results would be the end of the rows, but a paper in a leading medical journal contained an article from leading HGP scientists, again criticising the quality of the Celera sequence.

! SCIENCE FEATURE The polymerase chain reaction (PCR)

The first recombinant DNA molecules were produced in the early 1970s when bacteria were made to copy foreign DNA along with their own. However, in 1983 an American called Cary Mullis cloned DNA without bacteria using the relevant enzymes. This process, the polymerase chain reaction (PCR), is gene cloning in a test-tube. It allows DNA to be amplified (copied many times).

The idea behind PCR is very simple (Fig 27.8). You mix together your original piece of DNA with the enzyme DNA polymerase in a solution of nucleotides. Next, add some primers, short pieces of DNA which act as signals to the enzymes, effectively saying 'start copying here'. You then heat the DNA so that it denatures into two strands. The thermostable DNA polymerase gets to work and produces two identical strands of DNA (see Chapter 24 for the detailed mechanism of DNA replication).

In a second cycle of reactions the two strands become four, and so on. Typically, the cycle is repeated about 20 times and within an hour you have millions of copies of the original piece of DNA.

This technique has many applications. Forensic scientists use it to amplify tiny DNA samples from spots of blood or hair roots, to obtain enough material for forensic analysis by DNA profiling. PCR also allows archaeologists to study small samples of DNA from historical material such as the preserved bodies found in peat bogs or ice graves.

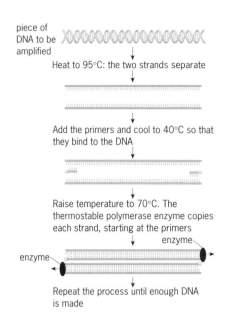

Fig 27.8 The polymerase chain reaction

SEE QUESTIONS 2 AND 3

5 CLONING GENES

Armed with restriction enzymes and ligases, genetic engineers can create recombinant DNA molecules. However, once a recombinant DNA fragment has been made *in vitro* ('in glass'), it usually needs to be copied repeatedly so that enough molecules are available for analysis or for commercial use.

Although DNA fragments can be multiplied in a test-tube (see the Science feature above on PCR), it is often much easier to put the DNA into organisms such as bacteria which will then adopt the new DNA as their own, and copy it for you. This is called gene cloning. The basic procedure is outlined in Fig 27.9. As an added bonus, the bacteria can also express the gene, making large amounts of the product.

SEE QUESTION 4

✔ REMEMBER THIS

The word *protocol* is often used in genetic engineering. It means 'standard procedure' and so a cloning protocol would be a standard method for cloning a gene.

✔ REMEMBER THIS

Gene therapy is an important part of DNA technology; see Chapter 32 for a full explanation and discussion of gene therapy.

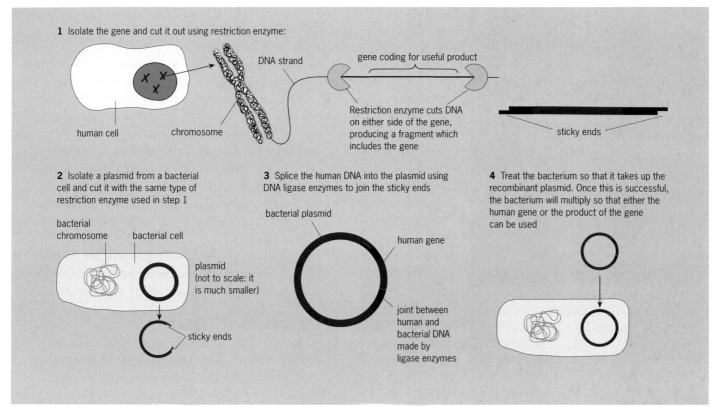

Fig 27.9 The principles of recombinant DNA technology. A human gene that codes for a useful product, such as insulin, is transferred into a bacterium which then multiplies to form a large population. The bacteria all express the human gene, and large amounts of the gene product can be recovered

USING VECTORS TO TRANSFER GENES

Human genes are often transferred into bacteria or yeast because these microorganisms multiply quickly to form a huge population that expresses the gene and makes large quantities of the gene product. This can then be extracted and purified. The microorganisms are grown in a fermenter, a giant vat with ideal conditions for growth (Fig 27.10).

PUTTING GENES INTO BACTERIA

Putting a human gene into another organism is not an easy task. The first step is to attach the gene to a vector, or carrier. One such vector is a plasmid (Fig 27.11), a tiny circular piece of DNA that occurs in bacteria (see Chapter 1). Like restriction enzymes and ligases, plasmids can be bought commercially.

The plasmid is cut using the same endonuclease enzyme used to cut the human DNA to obtain the gene. Using the same endonuclease is important – the DNA must be cut at the same base sequence to produce complementary sticky ends that will join up. The gene and the plasmid are mixed together and then a DNA ligase is added to join up the sticky ends. The DNA molecule produced is circular, like the original plasmid.

In the second step, the genetic engineer must induce the bacterium to accept the plasmid. One technique involves soaking the bacteria in ice-cold calcium chloride and then incubating them at 42 °C for 2 minutes. Nobody seems to know exactly why this works, but it does. Bacteria which have accepted the plasmid now contain recombinant DNA, and are, by definition, transgenic organisms.

The conversion process is not very reliable: for every bacterium that takes up the recombinant plasmid, about 40 000 do not. So how do you tell which ones are recombinant? A clever trick here is to insert two genes into the

Fig 27.10 Transgenic bacteria or yeasts are cultured in large cylindrical sterile fermenters, although the process is not strictly speaking fermentation. These containers were originally used for making wine and the name has stuck

Fig 27.11 False-colour transmission electron micrograph of plasmids of bacterial DNA from the bacterium *Escherichia coli* (*E. coli*). This plasmid, called pBR322, is the one most commonly used in genetic engineering work

---

 REMEMBER THIS

To make exact copies of colonies of bacteria growing on a Petri dish, the process of **replica plating** can be used. This involves pressing a sheet of velvet or other suitable material on the original colony and then pressing it down on several sterile agar plates. This is commonly used to identify strains of recombinant bacteria – one or more replica plates will contain a particular antibiotic, and only the resistant strains will survive and grow.

---

plasmid: the one you want to use and one that makes the recombinant bacteria easy to detect. For instance, you can add a gene that confers antibiotic resistance and then culture the bacteria on agar plates containing the antibiotic. Only the bacteria with the modified plasmid will survive and grow.

An alternative is to add a second gene that codes for an enzyme that metabolises a coloured substrate. When bacteria are grown on agar plates made with the substrate, colonies that have taken up the plasmid are a different colour from colonies of non-recombinant bacteria.

Once you have identified colonies of recombinant bacteria, you can start pure cultures. Transgenic bacteria are often described as 'sick' because they multiply more slowly than normal bacteria: the population doubles every 30 minutes instead of every 20 minutes. Within ten hours, one transgenic bacterium can produce over a million copies of itself, each one containing a working clone of the original gene.

Modified bacteriophages also make good vectors. They reproduce by inserting their own DNA into the DNA of a host bacterium. Bacteriophages can transfer larger amounts of DNA than plasmids but, at present, their use is limited.

PUTTING GENES INTO EUKARYOTIC CELLS

Usually, a gene is transferred into a different organism so that it can be expressed, making greater volumes of product. This happens only when the gene is able to use the mRNA, ribosomes and Golgi apparatus, etc, for protein synthesis. Some eukaryotic genes are not expressed effectively by prokaryotic cells, and so must be transferred into another eukaryotic cell. Yeast, a single-celled fungus, is often used.

Producing recombinant eukaryotic cells is more difficult than producing recombinant bacteria. The cell walls of fungal cells are a major barrier. But they can be digested away with suitable enzymes to form a protoplast, a cell without a cell wall. In this 'naked' state, the yeast cell will accept plasmids.

Other methods of transferring DNA into eukaryotic cells include the following techniques.

- **Microprojectiles**. It is possible to shoot DNA into host cells. Tiny pellets of metal are coated with DNA and fired at high speed at the target cells. Remarkably, some of the cells recover and accept the foreign DNA.
- **Electroporation**. This involves exposing the host cells to rapid, brief bursts of electricity to create temporary gaps in the cell surface membrane, through which foreign DNA can enter.
- **Liposomes**. In this more subtle method, DNA is inserted into liposomes – spheres formed from a lipid bilayer. The liposomes fuse with the surface membrane of the host cell and introduce the DNA into the cytoplasm.
- **Calcium phosphate precipitation**. Plasmids are mixed with calcium phosphate. As this precipitates, the grains that form contain DNA. These grains can enter cells by endocytosis (see Chapter 4).
- **DNA injection**. DNA can be inserted directly into a cell using a very fine pipette called a micromanipulator that avoids the inevitable hand tremor (see Fig 27.12). Tracey, a transgenic sheep, was created when the gene for alpha-1-antitrypsin was manually injected into the cells of a sheep embryo. This method leaves much to chance: many embryos have to be treated before one takes up the foreign DNA.

Fig 27.12 Injecting DNA directly into a sheep cell. The cell is held steady by gentle suction from the pipette on the left, while the even finer pipette on the right penetrates the cell surface membrane. This technique is used to produce transgenic and cloned animals

6 STEM CELLS AND CLONING

Cloning is a difficult topic for anyone; it is not only complex biology, but it is also a tricky subject ethically. What you feel about cloning may not just depend on your knowledge of science – it may also depend on your religious, moral or other views. We don't have enough space in this book to do more than a brief introduction. This field is moving so fast, it will also be out of date within months. To keep up, and to find information to help you make your own mind up about cloning, refer to good journals, such as *Scientific American* and *New Scientist* (both are available online), and look at the BBC online Science and Technology news site.

Fig. 27.13 Coloured SEM of stem cells taken from umbilical cord blood

CLONING WHOLE ORGANISMS

In 1997, when scientists succeeded in cloning a sheep, there was a lot of fuss in the media. Was it ethically and morally right to clone animals as advanced as mammals? Would this lead to scientists attempting to clone humans?

The definition of a clone is 'a genetically identical copy' and the term can apply to individual strands of DNA, individual cells or whole organisms. Cloning is, in fact, a common natural process: all organisms made by asexual reproduction are clones of their parent, unless there is a mutation. Even humans make clones – they are more commonly called identical twins.

There is much commercial interest in the cloning process because it can produce exact copies of organisms with desirable characteristics. If, for example, you have a 'perfect' apple tree that produces large amounts of delicious, healthy, blemish-free apples, you can clone it rather than risk losing its valuable genotype in the lottery of sexual reproduction.

Cloning plants is easy, and can be done with relatively little equipment in schools and colleges. Cloning mammals, however, is a lot more difficult but technically feasible. There are two approaches to cloning whole organisms, embryo splitting and nuclear transfer.

SEE QUESTIONS 5 TO 7

EMBRYO SPLITTING

Eggs and sperm or fertilised embryos are collected from prize specimens of cattle, for example. Cells can be pulled apart when the embryo is at the 8 or 16 cell stage and each fragment can then develop into a complete embryo. Each embryo is a clone of the original. The process can be repeated many times and then the cloned embryos can be introduced into the uteruses of normal cows. After the normal gestation period, several very average animals give birth to a herd of prize specimens (see Chapter 6 for more details).

NUCLEAR TRANSFER

Cloning is based on nuclear transfer, the same technique scientists have used for some years to copy animals from embryonic cells. Nuclear transfer involves the use of two cells. The recipient cell is normally an unfertilised egg taken from an animal soon after ovulation. Such eggs are poised to begin developing once they are appropriately stimulated. The donor cell is the one to be copied. A researcher using a high-power microscope holds the recipient egg cell by suction on the end of a fine pipette and uses an extremely fine micropipette to suck out the chromosomes, sausage-shaped bodies that incorporate the cell's DNA (see Fig 27.12). (At this stage, chromosomes are not enclosed in a distinct nucleus.) Then, typically, the donor cell, complete with its nucleus, is fused with the recipient egg. Some fused cells start to develop like a normal embryo and produce offspring if implanted into the uterus of a surrogate mother.

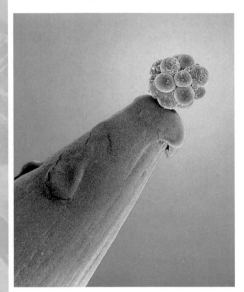

Fig 27.14 A coloured scanning electron micrograph (SEM) of a human embryo at the 10-cell stage on the tip of a pin. The ball of cells (orange) of the embryo is known as a *morula,* a cluster of almost identical, rounded cells, each containing a central nucleus. This 10-cell embryo is about three days old. It is at the early stage of transformation from a single cell to a human composed of millions of cells

---

! SCIENCE FEATURE Will the cloning bubble burst?

Apart from the success rates and embryo wastage that occur in the production of an animal clone, scientists are becoming concerned about the long-term effects of cloning. Dolly the sheep, the first sheep to be cloned, was touted as a perfectly healthy, normal animal. She has had several normal lambs and all seemed to be well. However, in January 2002, it was announced that Dolly has developed premature arthritis. She is only 6, and arthritis is very rare in sheep of this age. This report has been followed by others; in February 2002, a paper reported that cloned mice live significantly shorter lives than mice of the same genetic background produced by normal mating. Researchers are now beginning to think that the trauma to the cells caused during the cloning procedure, where a nucleus is removed from a normal body cell and then transferred into a zygote, could cause

problems. The cloned mice all had problems with their immune systems and it may be that we don't understand enough about the delicate process of development at the early embryo stage to find out what has gone wrong.

Steps are now being taken to introduce more gentle cloning methods, which allow two cells to fuse, but it will still be a shock for a cell that is making all the proteins that keep a body cell going one minute to suddenly switch to become a zygote, ready to develop into an embryo. Some leading scientists have called for researchers to be very honest about their cloning experiments – some, particularly those in companies, might not reveal 'failed' cloning experiments because of the damage it could do to the company, but these failures might help to solve some of the puzzles.

TECHNICAL AND ETHICAL PROBLEMS WITH CLONING

Dolly the sheep and cc the cat were created by the process of nuclear transfer, a process that many people feel will give rise to cloned humans at some stage in the future. As we saw in the opener to this chapter, the technique sounds simple, but we can't do it reliably. It took over two hundred attempts to get Dolly and almost 100 to get cc, the cloned cat. Many laboratories report that only 1 to 2 per cent of cloned embryos survive to become live offspring. Even some clones that survive through birth die shortly afterwards. The recently cloned gaur, an endangered species of wild ox from Asia, lived only a few days before dying from a common dysentery.

This unreliability has two main implications for human cloning that intends to produce living babies:

- Each attempt requires an individual egg. Sheep don't tend to complain when you stimulate their ovaries and harvest the eggs, but humans are a different matter. Very few women will give up eggs for any reason other than their own pregnancy, so finding 200 ova for each attempt is always going to be a problem. The women in the USA may have been persuaded to do this by the cult aiming to clone a human (see opener), but whether they will actually do so remains to be seen.
- In humans, failure brings great distress. Miscarriages and malformed offspring always cause great distress to all concerned in the process.

In 2001, the UK government passed a law that allowed scientists to produce human embryo clones solely for the purpose of making replacement tissues that could be used in transplant procedures. This was challenged by the pro-life lobby, but in February 2002, a House of Lords Select Committee upheld this decision, and the Human Fertilisation and Embryology Authority, which regulates such research is now likely to give researchers licences to experiment on early stage human embryos. Any human cloning that intends to produce a full-term pregnancy is banned in the UK, Europe and the USA and many other countries.

STEM CELLS:
THE REASON BEHIND CLONING HUMAN EMBRYOS?

If the objective is not to make a human baby, then why clone? The answer comes down to stem cells. These cells are capable of differentiating into any type of adult body cell. Theoretically, you could say that every cell in the body has the capacity to do this, as they all contain the same genome. However, once a cell has gone through a pathway of differentiation, and has become, say, a brain cell, there is little you can do to shift it back again. Although the adult body does contain some stem cells, they are few and far between and have proved difficult to isolate. A much better source, in terms of numbers of cells and ease of extraction, is to use early stage embryos.

As you can see from the Science feature on the next page, stem cells may be useful for making transplantable organs in culture. Stem cells might be used to treat a variety of serious diseases caused by damage to cells, perhaps including AIDS as well as Parkinson's, muscular dystrophy and diabetes. They may be useful for repairing the damage to spinal cords and nerves that are caused in the many road accidents each year, or for replacing the damaged cells in people with Alzheimer's disease. The possibilities are, apparently, endless.

 REMEMBER THIS

The American cult that says it intends to clone a human baby is based in the USA, but it cannot do its experiments there. Some countries in the world do not have strict regulations on cloning, and scientists from the organisation have set up their operation where they can do so legally, for the moment. Global appreciation of the problem is increasing, and this 'loophole' may be closed during the next few years.

? TEST YOURSELF

C There is a common misconception that people can have their cells frozen so that they can be cloned after their death to 'live again'. How would you explain to someone that this is utter rubbish?

 REMEMBER THIS

Stem cells are the cells in the embryo that have the potential to become all other tissue types. For this reason they are called **totipotent**. There are stem cells in adult bodies, notably in the bone marrow, but these can give rise to only a few tissue types: for this reason they are called **multipotent**. Research on embryonic stem cells therefore has greater potential to repair and replace damaged tissues. Embryonic stem cells have the added advantage that they have not accumulated the mutations that stem cells obtained from adults will have.

! SCIENCE FEATURE The first cloned kidneys

It is only at the beginning of this century that the idea of 'growing' organs for transplant has moved from the realms of science fiction and fantasy into reality. In January 2002, researchers revealed that they had produced a cow kidney using a single skin cell from the ear of an adult cow. This was fused with a donated cow egg that had been stripped of its own genetic material. A small electrical current was then used to stimulate the fused cell to become an embryo. The embryo was rich in stem cells that have the potential to become a wide range of body tissues.

Some of the stem cells were removed from these early embryos and treated with chemicals and growth factors to push them towards development pathways that made them into fully mature kidney cells. They were grown on a biodegradable kidney-shaped scaffold designed by a team from Harvard Medical School. The scientists produced several miniature kidneys each a few centimetres long. These were transplanted back into the adult animal, alongside its existing organs, where they started to produce urine.

Obviously, this study is at a very early stage; it will be some time before fully functioning kidneys are produced that can replace a diseased kidney completely. However, it does provide hope for the future; donation of organs such as kidneys and hearts has always been a problem, and people needing transplants still face long waits to see if a suitable donated organ will become available. One day, in the not too distant future, it may become possible to grow a custom-built kidney, exactly compatible with a particular patient.

SUMMARY

- Genetic engineers isolate, cut out and transfer genes between organisms. Genetic engineering technology has many applications, including making human products such as insulin on a large scale.
- The Human Genome Project has now produced the complete sequence of all 30 000 genes in the human genome. Knowing this sequence will provide a big boost for research in many areas of biology and medicine.
- Strands of DNA are cut using **restriction endonuclease** enzymes, which always cut DNA at a particular base sequence. Strands of DNA can be joined by using DNA **ligase enzymes**.
- In order to transfer DNA into a living cell such as a bacterium, a **vector** (carrier) is needed. Common vectors are **plasmids** (circles of DNA found in bacteria) or viruses.
- Once the foreign DNA is inside the host cell, it is adopted by the host. Genes in the foreign DNA are expressed using the host cell's synthetic machinery (ribosomes, mRNA, etc). In addition, the foreign DNA is **cloned** (copied) every time the host cell divides.
- **Gene therapy** is a technique that replaces faulty genes with healthy ones. It provides great hope for sufferers of genetic diseases such as cystic fibrosis.
- The **polymerase chain reaction** (PCR) allows samples of DNA to be cloned in a test-tube using the necessary enzymes, instead of inside a host cell.
- Many recent developments in this field have made cloning whole animals and organs possible. However, although the technical problems have been overcome, the ethical difficulties still pose a great problem for this area of research.

? EXAM QUESTIONS

1 The diagram outlines the way in which the gene for human factor VIII, a protein which is necessary for the clotting of blood, is incorporated into bacterial DNA and inserted into a bacterium.

a) With reference to the diagram, state:
(i) the type of enzyme used to cut the factor VIII gene from human DNA;
(ii) the name of the small circular ring of DNA;
(iii) the term used to describe the regions labelled **A**;
(iv) why these regions have been added to the factor VIII gene;
(v) the word used to describe the bacterial DNA which now contains the human factor VIII gene.
b) Suggest why it is considered better to use genetic engineering as a source of human factor VIII rather than material obtained from human blood.
c) Describe how DNA replicates. (You may use annotated diagrams to assist your explanation if you wish.)

OCR 2801 Biology Foundation January 2001 Q6

2 The flow chart shows one method of finding the sequence of nucleotides in DNA.

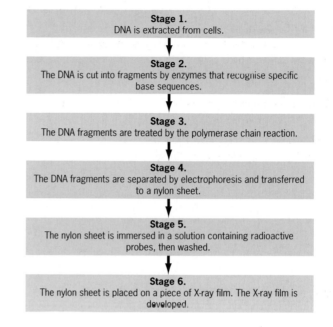

Stage 1.
DNA is extracted from cells.

Stage 2.
The DNA is cut into fragments by enzymes that recognise specific base sequences.

Stage 3.
The DNA fragments are treated by the polymerase chain reaction.

Stage 4.
The DNA fragments are separated by electrophoresis and transferred to a nylon sheet.

Stage 5.
The nylon sheet is immersed in a solution containing radioactive probes, then washed.

Stage 6.
The nylon sheet is placed on a piece of X-ray film. The X-ray film is developed.

a) Name the type of enzyme which is used in **Stage 2.**
b) Why is the polymerase chain reaction used in this procedure (**Stage 3**)?
c) What is meant by electrophoresis (**Stage 4**)?
d) **(i)** In Stage 5, the nylon sheet is immersed in a solution containing radioactive probes. Explain the function of these radioactive probes (**Stage 5**).
(ii) Explain why only some nucleotide sequences show up when the X-ray film is developed.

AQA (B) BYB2 Genes and Genetic Engineering June 2001 Q6

? EXAM QUESTIONS

3 The diagram illustrates the polymerase chain reaction.

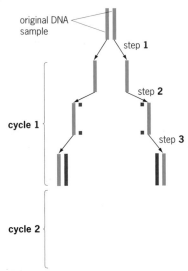

a) **(i)** What method is used to split the original DNA sample into two strands during Step **1**?
(ii) Describe what happens during Steps **2** and **3** in order to produce a new DNA strand.
b) Draw the DNA molecules that would be present at the end of **Cycle 2**. Use the same method of shading to distinguish original DNA strands and the new DNA strands.
c) Describe **one** example of the use of the polymerase chain reaction.

AQA (B) BYB2 Genes and Genetic Engineering January 2001 Q3

4
a) **(i)** Describe how a gene may be taken from a mammalian cell and inserted into bacterial cells.
(ii) Describe how bacteria containing the transferred gene can be cultured on a large scale.
b) Mutation of the gene responsible for production of the enzyme galactosidase causes a liver disorder. The liver is unable to metabolise certain carbohydrates, and toxic products accumulate.
(i) Describe **one** way in which the structure of the DNA of a gene may be changed as a result of a mutation.
(ii) The disorder can be treated by introducing the gene for galactosidase into a harmless virus, then injecting the transformed virus into the patient. A liver cell containing DNA from the transformed virus produces galactosidase. Describe how a cell synthesises a protein such as galactosidase.
(iii) The transformed virus enters liver cells but does not usually enter other cells in the body. Suggest **one** explanation for this.

AQA (B) BY82 Genes and Genetic Engineering June 2001 Q7

5 In gene therapy healthy genes are cloned and used to replace defective genes.
a) Cystic fibrosis is caused by a mutant gene that produces defective CFTR.
(i) Give **one** difference in structure between normal CFTR and defective CFTR.
(ii) Explain why defective CFTR does not transfer chloride ions across cell membranes.
Haemophilia is a genetic disease in which blood fails to clot. Factor IX is a protein used to treat haemophilia.
b) Factor IX can be made by cloned mammalian cells. Explain what is meant by a clone.
c) The DNA sequence that codes for factor IX protein is extracted from human DNA. This human DNA sequence is then introduced into a fertilised egg from a sheep.
Describe **one** way, other than using enzymes, in which a DNA sequence could be introduced into fertilised eggs.
d) **(i)** Another technique is to treat a culture of human body cells chemically so that the cells will take up a DNA sequence.
Explain **one** advantage of using a culture of human body cells rather than fertilised eggs from sheep.
(ii) An antibiotic resistance gene is added to the cultured human body cells along with the gene for factor IX protein. A dose of antibiotic concentrated enough to kill human cells is then added to the cell structure.
Suggest **one** explanation for adding the resistance gene and using the antibiotic.
(iii) Some scientists wish to use this technique to clone cells from human embryos. The cloned cells could then be implanted into humans to cure diseases such as muscular dystrophy or Parkinson's disease.
Use your knowledge and understanding of genetic engineering to explain the arguments for and against using cells from human embryos in this way.

AQA (B) BYB2 Genes and Genetic Engineering June 2001 Q8

? EXAM QUESTIONS

6 A new variety of tomato has been produced by genetic engineering. This variety contains a synthetic gene that blocks the action of a natural gene that would make the fruit soften rapidly once ripe. It also contains a marker gene.

The marker gene added by the scientists makes this variety of tomato resistant to the antibiotic, kanamycin. It is possible that this gene could be taken up by disease-producing bacteria in the human gut. In humans, kanamycin is used to treat certain types of gut infections.

Using information from the passage, explain the advantages and disadvantages of putting this new variety of tomato on the market.

AQA (B) BYB 2 Genes and Genetic Engineering January 2001 Q4

7 The diagram shows how insulin can be made using genetically modified bacteria.

a) **(i)** The human insulin gene is obtained from mRNA, rather than DNA. Suggest why.

(ii) Name the enzyme used to make a single-stranded DNA copy of the mRNA coding for insulin.

(iii) The table shows a sequence of bases from the mRNA coding for insulin. Copy and complete the table to show the sequence of bases you would expect in a single-stranded DNA copy.

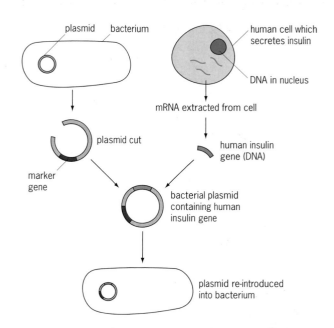

b) What is the role of DNA ligase in producing genetically-modified bacteria?

c) The plasmid contains a marker gene coding for antibiotic resistance. Explain the importance of this marker gene.

AQA (A) BYA2 Making Use of Biology January 2001 Q6

mRNA base sequence	U	C	A	A	C	C
DNA base sequence						

KEY SKILLS **ASSIGNMENT**

CYSTIC FIBROSIS AND GENE THERAPY

Cystic fibrosis (CF) is a genetic disease that is due to one defective allele (see Chapter 25). About one person in 25 carries the defective allele, but they also carry the normal allele and do not suffer from the disease.

1 Why don't heterozygotes suffer from the disease?

2 A couple are both carriers for cystic fibrosis.
a) What is the probability that their first child will have cystic fibrosis?
b) What is the probability that their second child will have cystic fibrosis?

The normal allele codes for a protein called cystic fibrosis transmembrane regulator (CFTR). This essential membrane protein in epithelial cells transports chloride ions out of the cells and into mucus. Normally, when chloride ions are secreted, sodium ions follow, and this decreases the water potential of the epithelial mucus. Water follows outwards by osmosis, making normal, watery mucus, which can be moved by the cilia (tiny hairs) lining the airways.

Cystic fibrosis sufferers make a protein that differs in just one of its 1480 amino acids. Although slight, this fault prevents CFTR from functioning normally. Chloride ions and sodium ions cannot be secreted, and so the mucus becomes much thicker than normal. Dead epithelial cells also accumulate in the mucus, adding to the general congestion of the airways.

3 Explain how a fault in just one amino acid can have such a drastic effect.

Sticky mucus is also a real problem in the pancreas. The mucus blocks the pancreatic duct, preventing the secretion of pancreatic juice.

4 Suggest why cystic fibrosis sufferers have to take tablets containing digestive enzymes.

Treatment of CF includes regular physiotherapy in which the chest is patted to dislodge the mucus. Even so, infections are common and most CF sufferers have to take a variety of antibiotics according to the infection they have at the time.

Gene therapy for cystic fibrosis

Now, one exciting possibility is that the faulty genes can be replaced by healthy ones. In 1989 the cystic fibrosis allele was located on chromosome 7. The base sequence of the healthy gene was then compared to the defective allele, and the nature of the fault was narrowed down at the molecular level. This opened up the exciting possibility that if healthy genes could somehow be introduced into the epithelial cells, they might be expressed and so make the correct membrane protein, solving the problem for a time.

The basic steps are as follows:

1. The CFTR gene is isolated, and cut out.
2. The gene is cloned many times.
3. The genes are encapsulated, either by putting them into liposomes (spheres made from lipid) or viruses.
4. The gene particles are inhaled, so that they can pass into the epithelial cells of the lung to be incorporated into the DNA of the cells.

Once in place, if all goes well, the healthy genes are transcribed, making the correct protein.

5 **a)** What is used to cut the CFTR gene out of chromosome 7?
b) Name two different methods of cloning the CFTR gene.
c) Suggest why the pancreatic genes would be more difficult to replace.
d) Suggest why repeat treatments at regular intervals would be necessary.

Germline gene therapy

This area of research is an ethical minefield. So far in this Assignment we have looked at body cell gene therapy – the replacement of the faulty allele in certain cells of the body. However, the eggs or sperm of the sufferer would still carry the defective alleles. In germline gene therapy, the original alleles in the zygote would be changed. The individual would grow and develop with healthy alleles in all cells and all of the individual's children would inherit the healthy allele.

Germline gene therapy sounds attractive but there are problems, so much so that this area of research is banned in the UK, and in many other countries. Ethically, there is the familiar 'where do you stop?' problem. A new treatment for cystic fibrosis or haemophilia would be desirable, but having the technology might allow for all sorts of temptations. Our knowledge of the human genome could reveal, for example, genes that code for intelligence, height or skin colour. Tampering with these characteristics would lead to obvious problems.

Biologically, germline gene therapy is potentially dangerous because we know very little about how genes function in the embryo. Tampering with genes in the zygote could have effects that might become apparent only in later life.

6 In discussion groups, prepare a case which you could present either for, or against, germline gene therapy.

28

Health and disease: an introduction

Living to a grand old age is now a more attainable goal for more people than it was a hundred years ago

Currently, a man living in the United Kingdom can expect to live until he is 84; a woman until she is 87. Just a hundred years ago things were very different: in 1900, average life expectancy was between 45 and 55 in this country. So what accounts for this massive change?

If you ever take a look around cemeteries or churchyards, you will notice that some people from the 1800s did live to their 70s or 80s or beyond. The maximum age to which a human being can live does not seem to have extended very much in the past 100 years. What has changed is the likelihood of dying when young. Infant mortality, the number of children under 5 that did not survive, was very high in the 1900s. Then, about 150 children in every 1000 died at birth and perhaps another 200 did not make it to their 5th birthday. Today, less than 20 children in every 1000 die before they are 5. The average life expectancy has increased because of factors such as better living conditions, better medicine and better diet.

Today we know that these sorts of environmental factors combine with our genes to determine how long we live. Researchers have been studying very long-lived people to see if they can identify which genes are important. Recently, a cluster of genes on chromosome 4 was found to be more common in people who reached the age of 90, but finding out exactly what these genes do is likely to take time.

1 DEFINING HEALTH AND DISEASE

Until this century, the commonest causes of death were the 'catchable' diseases – those carried by bacteria, viruses and other organisms too small to detect. Then the microscope showed that diseases such as dysentery and tuberculosis were caused by **infectious agents** whose spread could be worked out and then controlled. Later came antibiotics, vaccines and other sophisticated medicines.

Today, the commonest causes of death in the western world are all **non-communicable**, caused in part by our lifestyle and in part by the fact that modern medicine prevents most of us from dying from diseases that are **communicable**. In this section of the book, we look first at infectious diseases, and then at the ways in which the body defends itself against these potentially lethal invaders. We progress to the biggest killers, the lifestyle diseases, before finishing with genetic diseases and some thoughts for the future.

WHAT DO WE MEAN BY HEALTHY?

Health is difficult to define accurately; we can say that it isn't being ill or having a disease, but laying down a set of characteristics that describes being healthy is more of a problem. Health can be thought of as a dynamic state in which a person is able to cope well with their internal and external environments in order to maintain a state of wellness. The internal environment includes our genes and our psychological approach to life, as well as our exposure to microorganisms that cause infectious diseases. The external environment includes our physical living conditions, the relationships we have with other people and whether we have enough money to eat a healthy diet and have a healthy lifestyle.

The World Health Organisation defines health like this, recognising that mental as well as physical health is important:

Health is a state of complete physical, mental and social well being.

However, few of us attain this sort of health for long periods of time. What if you lose an arm in an accident? When you have had time to heal the injury, you cannot grow another arm and be complete in a physical sense, but you are not really ill. What about someone with a genetic disorder such as Down's syndrome? A definition of health must be flexible enough to include such special cases.

WHAT DO WE MEAN BY DISEASE?

Defining disease is more straightforward. Diseases arise when the body's functions are disturbed or when its structures are altered. Diseases can be defined as communicable if they are caused by infectious agents and can be passed from one person to another. Non-communicable diseases are caused by a broad range of environmental and genetic factors. It is important to remember that the development of a particular disease and the way it responds to treatment varies from person to person.

2 THE MAIN CAUSES OF DISEASE

COMMUNICABLE DISEASE

Communicable or infectious diseases are those that can pass from one person to another. These diseases are caused by microorganisms such as bacteria and viruses (Figs 28.1, 28.2 and 28.3). Although common in many developing countries, infectious disease is much rarer in modern Europe because of effective treatments such as vaccines and antibiotics.

NON-COMMUNICABLE DISEASE

Non-communicable diseases seem to be connected with our lifestyle. Cancer, for example, is a disease that can affect many different parts of the body. This disease clearly does not have a single cause. Smoking tobacco is the major cause of lung cancer, and too much sunbathing can lead to skin cancer. People who inherit certain genes can be at an increased risk from some cancers, particularly those that affect the colon or the breast. Viruses seem to be involved in the development of cervical cancer.

HOW DISEASE-CAUSING FACTORS CAN INFLUENCE EACH OTHER

Cystic fibrosis is an **inherited disease**. It is due to a fault in a single gene. The normal allele of the gene makes a protein responsible for transporting chloride ions across cell membranes. If a person inherits one faulty allele, they are fine because they have a normal allele that still makes the protein. People develop the disease only if they inherit two faulty alleles, which makes cystic fibrosis a **recessive characteristic**.

People with no normal allele cannot make any functional transport protein and, as a result, produce extremely thick mucus. This affects organs such as the lungs and the pancreas. So cystic fibrosis has a genetic cause. At the same time, people who inherit the condition are likely to suffer from chest infections caused by microorganisms that become trapped in the mucus in the lungs.

? TEST YOURSELF

A Suggest why communicable diseases are not as common in Europe as they were two hundred years ago.

Malaria is a severe disease that affects many people living in the tropics. It is caused by *Plasmodium*, a single-celled parasite that is transmitted from the blood of one person to that of another by *Anopheles* mosquitoes. People who can afford preventive medicines, mosquito nets over beds and insect repellents are less likely to be bitten by mosquitoes than the poorer members of the community who cannot pay for such protection.

In addition, people who live in areas where malaria is common often have genes that cause them to make slightly different sorts of haemoglobin. This changes the red blood cells so that the malaria parasite cannot infect them as easily and they have some resistance to developing the symptoms of malaria. However, the change in the red blood cells causes problems; the red cells lose their biconcave shape and start to look like crescents, or sickle blades. Such people have the disease sickle cell anaemia, which can itself be life-threatening (see Chapter 24).

So, with malaria, there is a link between lifestyle, infection and genetics. Whether or not a particular person catches malaria and then dies from it depends on all three factors.

> ✔ **REMEMBER THIS**
>
> Find out more about *Plasmodium*, malaria and the *Anopheles* mosquito in Chapter 29, Section 4.

⚠ SCIENCE FEATURE Communicable diseases: infections

Many communicable diseases are caused by microorganisms.

Fig 28.1 Smallpox has now been totally eradicated. In 1967, when the World Health Organisation began its campaign to eliminate this dreadful viral disease, there were ten million cases spread through thirty different countries. The last case was recorded in 1977 in Somalia

FATHER THAMES INTRODUCING HIS OFFSPRING TO THE FAIR CITY OF LONDON.
(*A Design for a Fresco in the New Houses of Parliament.*)

Fig 28.2 This Punch cartoon shows three diseases caused by bacteria: diphtheria, scrofula (a form of tuberculosis) and cholera. Between them, they affected large numbers of people in London and other parts of the United Kingdom in the mid-nineteenth century

Fig 28.3 *Plasmodium*, the malarial parasite, is a small one-celled organism belonging to the Kingdom Protoctista. It is carried by the *Anopheles* mosquito and invades human red blood cells. Worldwide, malaria kills about 2.7 million people every year

! SCIENCE FEATURE Non-communicable disease

Lifestyle diseases

Lifestyle diseases are those that result from the conditions in which we live.

Genetic diseases

It has been estimated that approximately 2 per cent of the UK population have some form of genetic disease.

Degenerative diseases

Age brings about gradual decline in the efficiency of many of the systems in the human body.

Fig 28.4 Heart disease is the greatest single cause of death in England and Wales. It accounts for nearly one in three of all deaths. Many environmental factors are linked to a greater risk of heart disease. These include stress, smoking, a high-fat diet and lack of exercise

Fig 28.5 Cystic fibrosis is caused by a single gene. It is a recessive characteristic. Individuals who inherit two defective alleles cannot make a protein responsible for transporting chloride ions across cell membranes. As a result, they produce extremely thick mucus, which affects the functioning of organs such as the lungs. Physiotherapy helps to disperse this mucus

Fig 28.6 Osteoarthritis is a painful condition that affects the joints of many older people. The cartilage at joints is lost and this may result in damage to the underlying bone, so that the sufferer has difficulty in moving the joints

3 PATTERNS OF DISEASE

The earliest people were hunter–gatherers who lived in small isolated groups. What did they die of?

Even if they had a reliable food supply, they still had to search for food whatever the weather. Some of the very old and the very young might have died of the effects of extreme cold or heat, and fighting between rival groups could have accounted for a proportion of deaths. However, the most likely cause of death was from infectious disease: not the massive epidemics of measles, plague or smallpox that killed so many later in history, but chronic (long-lasting) diseases such as tuberculosis, malaria and leprosy.

Why was this? The simple answer is that there weren't enough people to sustain an epidemic of a short-lived infection such as measles. Today, there are lots more people and infectious diseases can spread more easily. In Africa, for example, measles is **endemic** – on any one day, someone, somewhere in Africa has measles. If the people who are infected have contact with many other people who have no immunity to measles, a large proportion of the population can become infected all at the same time. This is an **epidemic**.

The typical hunter–gatherer society probably consisted of no more than a couple of hundred people and they probably had little contact with outsiders. Diseases such as measles would not have been able to spread easily, and would have never become a problem (Fig 28.7).

 REMEMBER THIS

An **endemic** disease is one that is always present in the population. An **epidemic** is an outbreak of disease that spreads rapidly through a population, affecting a large number of people.

Fig 28.7 Measles is always present in very large populations, but it is often absent from smaller populations. Such evidence suggests that diseases such as measles were unknown in early hunter–gatherer populations

Chronic diseases such as tuberculosis, malaria and leprosy were much more of a problem. These conditions affect the body for a long period of time and infection could have been passed on even when contact between different groups of people did not happen very often.

When people began to domesticate animals and cultivate crops, human societies changed enormously. The land could support far more people and communities grew larger. As settled agriculture developed, towns and cities were built, and trade brought increasing contact between different groups of people. From that time, we have been literally plagued by infectious diseases.

? TEST YOURSELF

B Use the graph in Fig 28.7 to estimate the minimum size of the population in which cases of measles are always present.

! SCIENCE FEATURE Measles and the controversy over vaccination

If you are fit and well-nourished, measles is an unpleasant childhood infection that makes you ill for two weeks, and then you get better. However, it affects some people much more seriously, particularly if they are weakened by malnutrition or other diseases. In many developing countries, measles is a big killer. Fortunately, a measles vaccine has been available for some time and children in the United Kingdom and other countries in the developed world don't usually catch this disease.

In 1998, a study of children who had received the combined MMR vaccine – for measles, mumps and rubella (commonly known as German measles) – suggested that some children developed bowel disease after the vaccination, and then went on to develop autism. This is a condition in which children, and adults, have difficulty relating to other people, and show unusual behaviour. But was the vaccine responsible?

A great controversy has raged since, with some parents believing passionately that the vaccine is to blame for their children's problems. Another study that was already going on at the time has now come up with the closest we will probably ever get to the 'truth' about how safe the MMR vaccine is. In January 2001, researchers announced the results of following nearly 2 million children in Finland over a period of 14 years. They found no link between the vaccine and the incidence of either bowel disease or autism. This number of subjects in a study is enormous, and leaves very little to chance, so most experts now conclude that the vaccine is not a cause of autism.

Fig 28.8 A child with measles

Agriculture influenced disease in two other ways:
- The cultivation of crops led to changes to the environment. For example, simple irrigation systems that made more water available for crops provided the perfect breeding grounds for the larval forms of the parasites that cause malaria and schistosomiasis (see Chapter 29).
- Closer contacts between humans and their domestic animals allowed new diseases to emerge, such as smallpox, plague and influenza.
- Trading links allowed diseases to spread rapidly.

INFLUENZA, AN EXAMPLE OF A COMMUNICABLE DISEASE

Outbreaks of influenza or 'flu occur every year, with huge epidemics, or pandemics spreading throughout the world every 30 years or so. The disease is caused by a virus that affects the epithelial cells that line the nose and throat (Fig 28.9). After a short period of incubation, 'flu causes headache, fever and aching muscles and joints. In most years, a bit of rest and drinking plenty of fluids allows most people to get better within about a week. However, the old and the very young can be very ill and thousands of people with die from 'flu every winter.

'Flu passes between people rapidly because it is airborne. When we breathe out, sneeze or cough, tiny droplets of mucus and saliva fly into the air. If you have 'flu, these droplets contain 'flu viruses and this is how they are passed on to other people.

Influenza virus

RNA polymerase

RNA strand

protein coat

Proteins which help virus invade new cells and escape from infected ones

lipid coat

Fig 28.9 The electron micrograph shows influenza viruses leaving an infected cell. The drawing shows the main features of an influenza virus

A GLOBAL DISEASE

The influenza virus has different forms. Two of these, **Type A** and **Type B**, are associated with major epidemics.

Influenza epidemics are most often caused by Type A viruses. Type B is not as common but outbreaks of Type B influenza occur on a global scale. Such large-scale epidemics are known as **pandemics**.

One pandemic broke out in Europe in 1556 and lasted until 1560. Historians estimate that during this period almost 20 per cent of the population of the United Kingdom died of influenza. The worst 'flu pandemic that is recorded in recent history occurred in 1918–1919, just at the end of the First World War. Experts estimate that this very virulent form of 'flu killed between 40 and 60 million people worldwide. It affected just as many people who were in their 20s as the elderly and the young, spreading rapidly in the troubled conditions of the Western Front (Fig 28.10).

Fig 28.10 A major 'flu pandemic killed about 50 million people in the period immediately after the First World War. There was little doubt that the speed at which the disease spread from country to country and continent to continent was partly due to troops returning home

Fig 28.11 Vaccination against influenza is offered to the most vulnerable members of the community

REMEMBER THIS

Symptoms are indications of a disease that are noticed by the patient.
Signs are indications of a disease that are picked up by a doctor but not necessarily by the patient.

PROTECTION AGAINST INFLUENZA

Why is it that you can have a single attack of measles and then be immune for life, but you can get influenza many times?

A person who has had influenza becomes immune to the strain of virus that caused it. However, the influenza virus has a very high rate of mutation. This means that different strains arise, each with slight differences in its outer protein coat. The body might be immune to the strain responsible for an outbreak of influenza one year, but this immunity does not protect against other strains that cause later outbreaks of 'flu.

Influenza vaccines are available (Fig 28.11). They are usually given to those at risk such as elderly people and those with asthma and heart conditions. The actual vaccine produced is effective only against the two or three strains of the virus common that year. It is necessary to have another injection with a different vaccine the following year since new strains will probably have evolved.

4 IDENTIFYING A DISEASE

LOOKING FOR THE OBVIOUS

Before a doctor can treat a disease effectively, he or she must identify the condition that is making the patient ill. Someone usually goes to the doctor with a collection of symptoms. For example, a woman finds that she gets tired very easily. Her husband thinks she looks rather pale and she notices that she gets out of breath when she exerts herself. The doctor observes the patient and listens to her describing these symptoms, and suspects that she is suffering from anaemia. Anaemia is not a disease itself, but it has a cause. It could be due to:

- a shortage of iron in the diet – iron is an important part of haemoglobin molecules;
- loss of blood – this might result from heavy periods or perhaps an ulcer that bleeds constantly;
- destruction of red blood cells caused by sickle-cell anaemia, an inherited condition. An affected person has an abnormal form of haemoglobin that transports oxygen with reduced efficiency. The red blood cells of someone with this condition have a shorter life span than normal red blood cells.

The doctor looks for clinical signs that might show which one of these is most likely. Treatment can begin only when an accurate diagnosis has been made.

IDENTIFYING PANCREATITIS

The pancreas is an important organ located just below the stomach (Fig 28.12). It has two main functions. It produces the hormones insulin and glucagon that regulate blood glucose concentration, and it secretes digestive enzymes. The enzymes include trypsin which breaks down proteins, lipase which digests fats and amylase which hydrolyses starch to maltose.

Pancreatitis is a severe inflammation of the pancreas, caused when the enzymes that normally break down food in the intestine start to break down and destroy the tissue of the pancreas itself. It can be acute (when it comes on suddenly) or chronic (long term).

Pancreatitis can be difficult to identify. The pancreas is in the first loop of the small intestine and is close to the liver and stomach, so pain in this area could be caused by any of pancreatitis, liver disease, stomach ulcers or gallstones. A correct diagnosis is vital, and this is where biochemical tests come in handy.

If the pancreas tissues are damaged rapidly, the enzymes in the cells escape into the blood. Tests involve checking whether lipase and amylase are in a blood sample. If they are, it is a good indication that the patient has acute pancreatitis.

Fig 28.12 The pancreas is situated in the first loop of the small intestine and near to a number of other important organs. Diseases that affect any of these organs could give similar symptoms

(Diagram labels: liver: hepatitis; stomach: gastric ulcer; gall bladder: gallstones; duodenum: ulcer; pancreas: pancreatitis)

> **? TEST YOURSELF**
>
> **E** Trypsin is an enzyme secreted by the pancreas. Explain why:
> **a)** you would expect to find trypsin in the faeces of a healthy person;
> **b)** the amount of trypsin in the faeces will decrease in a person with chronic pancreatitis.

If the condition is long term, however, the pancreas tissue is more slowly and progressively destroyed, and the ability to secrete enzymes declines gradually. This is chronic pancreatitis and is identified by measuring the amounts of enzymes in the patient's faeces. Recently, researchers have found that chronic pancreatitis is common in people who have a mutation in the gene that is faulty in cystic fibrosis. Although their mutation causes a less severe set of symptoms, it might soon be possible to develop a genetic screening technique to identify people at risk from developing pancreatitis later in life.

PHENYLKETONURIA: AN INHERITED ENZYME DEFICIENCY

Phenylketonuria is an inherited condition that affects approximately 1 in every 10 000 people. All new-born babies in the United Kingdom are tested for phenylketonuria. In the heel or **Guthrie** test, the midwife takes a blood sample from the heel of the baby and tests the sample for the presence of phenylalanine, an amino acid.

In a healthy baby, phenylalanine is converted into tyrosine by the enzyme phenylalanine hydroxylase (Fig 28.13). This enzyme is missing in a baby with phenylketonuria, and phenylalanine and other metabolic substances made from it build up in the baby's blood. These substances interfere with brain development, so it is very important that the condition is identified as soon as possible after birth.

phenylpyruvate → phenylalanine → [enzyme: phenylalanine hydroxylase] → tyrosine → proteins

Fig 28.13 Part of the biochemical pathway by which phenylalanine is converted to other substances in the body. Each step in this biochemical pathway is controlled by a different enzyme

Suppose the enzyme phenylalanine hydroxylase were missing. Let's look at the possible effects on the biochemical pathway, and how we could use these effects to show that the patient has phenylketonuria.

- First, there would be no enzyme. So we could do a direct test for the enzyme and show that it was absent from the patient's blood.
- If phenylalanine hydroxylase were missing, its substrate, phenylalanine, would not be converted into tyrosine, so the amount of phenylalanine should be higher than normal, and the amount of tyrosine should be lower or absent. We could test for raised phenylalanine levels and reduced tyrosine levels.
- Finally, the biochemical pathway from phenylalanine to phenylpyruvate might be affected. There is more phenylalanine present and, since it cannot be converted to tyrosine, it might be converted to phenylpyruvate instead. We could test for an increase in phenylpyruvate.

> **? TEST YOURSELF**
>
> **F** Suggest how phenylketonuria might be treated by altering the amounts of phenylalanine and tyrosine in the diet.

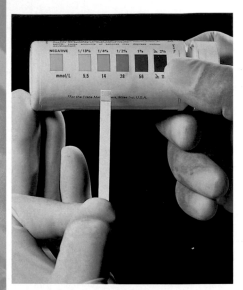

Fig 28.14 The Clinistix™ test

? **TEST YOURSELF**

G Suggest two advantages of people with diabetes being able to test a urine sample for glucose with Clinistix™ rather than with Benedict's solution.

Fig 28.15 An X-ray of a dislocated elbow joint

Fig 28.16 An ultrasound image of a human fetus

? **TEST YOURSELF**

H Explain which of the methods for looking inside the body would be most suitable for:
a) examining the ovaries of a woman undergoing fertility treatment in order to see the developing follicles;
b) examining the inside of a knee joint before performing a cartilage operation.

DIABETES AND TESTING FOR GLUCOSE

People with diabetes are unable to control the concentration of glucose in their blood (see Chapter 11). The blood glucose concentration rises and glucose appears in the urine. A hundred years ago, the condition was always fatal. Now, as long as it is correctly diagnosed, people can be treated successfully and lead normal lives.

Monitoring the amount of glucose in the blood and urine is an essential part of this treatment. Early methods relied on chemical tests such as Benedict's test in which glucose is used as a reducing agent. More modern methods make use of enzymes. Enzymes are highly specific and sensitive. This makes them ideal for analysing small samples. Today, people with diabetes use Clinistix™ to test for the presence of glucose in urine (see Fig 28.14 and page 119 in Chapter 6). Enzymes are also involved in the Glucowatch® Biographer, which measures the amount of glucose in the blood (see Chapter 6).

LOOKING INSIDE THE BODY

Sometimes, the only way to make an accurate diagnosis is to look inside the body. Cutting it open is one way, but there are always risks with surgery. The risks are rare, but it is safer to use **non-invasive** methods where possible. Several techniques now produce images of what is going on inside the body. The images sometimes look very strange to the untrained eye, but a skilled technician or doctor can interpret them.

X-RAYS

X-rays have a much shorter wavelength than visible light. They penetrate different tissues in the body to different extents. Although they are most often used to look at bones and teeth (Fig 28.15), X-rays can show other organs if a suitable contrast medium is used. For example, if someone drinks a **barium meal** (a solution containing salts of the heavy metal barium), the barium absorbs the X-rays and details of the stomach and small intestine show up on the final image of X-rays taken shortly afterwards.

The disadvantage of X-rays is that they damage DNA. For this reason, X-rays are not used to examine the reproductive organs or to scan the developing baby inside a pregnant woman, and the people who use X-ray equipment have to take safety precautions.

ULTRASOUND

Ultrasound is high-frequency sound, so high that the human ear cannot detect it. An ultrasound beam passes into the body and bounces off dense tissue. The reflected waves produce an image that shows internal organs in great detail. Unlike X-rays, it does not damage DNA and so can be used to check the development of a baby in the uterus (Fig 28.16).

ENDOSCOPY

An endoscope is a long tube with a light at one end and a means of displaying an image at the other. It can be used to examine the inside of hollow body parts such as the oesophagus and stomach (see Figs 8.2–8.6, pages 152–154) or the trachea and bronchioles. Many modern endoscopes rely on fibre optics. This means that they are small in diameter and flexible. Endoscopes are used for a variety of purposes such as inspecting the external ear and eardrum for signs of damage or infection, or for looking at the inside of the gut.

SUMMARY

After reading this chapter you should know and understand the following:

■ **Health** is difficult to define, but it is more than just an absence of disease.

■ **Communicable** or infectious diseases may be spread from person to person and are caused by **microorganisms**.

■ **Non-communicable** diseases have many causes. They can result from our **lifestyle** or from our **genes**, or they can occur as we age as a consequence of the process of **degeneration**.

■ Infectious diseases became increasingly important as agriculture developed and humans began to live in large settled communities.

■ Diseases are characterised by distinct **signs** and **symptoms**.

■ Diagnosis may be confirmed by a range of **biochemical tests**. Many of these rely on **enzymes**.

■ **X-rays**, **ultrasound** and **endoscopy** allow us to investigate the internal structure and functioning of the body.

KEY SKILLS **ASSIGNMENT**

THE CHILDREN IN THE CEMETERY

Today, Lyme Regis is a small, pleasant seaside town on the Dorset coast. It is a popular tourist spot set in a bay in a region noted for its attractive scenery.

In Victorian times, however, it was very different. There was considerable poverty and overcrowding in the lower part of the town along local riverbanks. Just over a hundred years ago, the medical officer of health wrote that it was common for dwellings with just one living room and one bedroom to house a family of ten. In these conditions, communicable diseases spread very rapidly.

Fig 28.A1 shows the seasonal pattern of deaths from communicable diseases typical of towns such as Lyme Regis in the middle of the nineteenth century.

1 Suggest an explanation for:
a) the pattern of deaths from respiratory diseases;
b) the pattern of deaths from diseases affecting the gut.

Fig 28.A2 shows the results of an investigation of the burials in Lyme Regis Cemetery in the period 1856 to 1900. The upper line on the graph shows the total number of burials per year; and the lower line shows the total number of burials of children aged under 16 years.

2 **a)** Explain what is meant by a disease epidemic.
b) What is the evidence from the graph that epidemics were responsible for many deaths in the period shown on the graph?

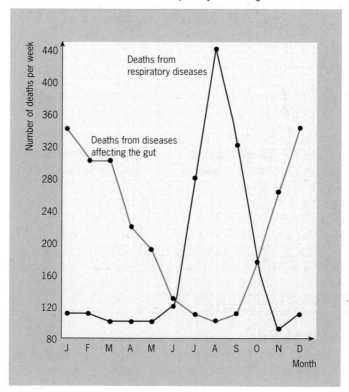

Fig 28.A1 Graph showing the seasonal pattern of deaths from communicable diseases of the respiratory tract and gut

KEY SKILLS **ASSIGNMENT**

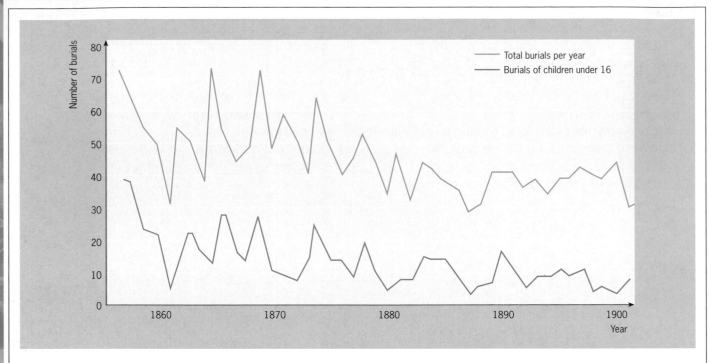

Fig 28.A2 Burials in Lyme Regis Cemetery during 1856–1900

Important diseases that produced epidemics in Lyme Regis during the nineteenth century included whooping cough, measles and smallpox.

3 Suggest why these diseases are no longer a major cause of death in the United Kingdom.

4 **a)** Copy and complete the table below to show the percentage of the total number of burials that were children under 16.

Decade	Number of burials		Percentage of total burials
	Total	Under 16	that were under 16
1866	539	224	
1876	519	145	
1886	402	105	
1896	373	75	
1916	313	39	
1936	308	21	
1966	301	7	

b) Plot the data in the table as a suitable graph to show how the percentage of total burials that were children under the age of 16 years changed over the period shown.

c) Use the information in this Assignment to suggest why there was a change in percentage of total burials that were children under the age of 16 over the period shown.

Cholera is a bacterial disease that produces severe vomiting and diarrhoea. Unless treated promptly, the patient becomes severely dehydrated and death can occur within 24 hours of the symptoms first appearing. Epidemics of cholera caused large numbers of deaths in nineteenth-century Britain. After an outbreak in Portsmouth, a local doctor wrote:

'At present the part of Portsmouth called Portsea was one large cesspool, for there could not be less than 16 000 cesspools daily permitting 30 000 gallons of urine to penetrate into the soil. Just reflect on the character of the well-water of a district which became mixed every year with 365 times 30 000 gallons of urine, to say nothing of a host of other abominations!'

5 **a)** Use this information to suggest how the epidemic of cholera spread through Portsmouth.
b) Suggest why cholera is no longer a major cause of death in the United Kingdom.

29

Infectious disease

An electron micrograph showing an ebola virus particle. Ebola virus particularly targets liver cells and macrophages: massive destruction of the liver is a hallmark feature of infection

We are used to hearing about new treatments, new vaccines and new medical advances on an almost daily basis. But it comes as something of a shock when a new disease appears. Everyone knows about AIDS, but other new diseases have appeared within the past 30 years. Ebola virus, for example, first emerged in Zaire (now Democratic Republic of Congo) and Sudan in 1976. Over 500 cases were reported and infection proved very dangerous: almost 88 per cent of victims died in Zaire and 53 per cent perished in Sudan. Ebola haemorrhagic fever occurred again in 1978, in virtually the same place in Sudan, but only three other cases have been reported since.

The 'shock, horror, gasp' factor of Ebola fever has driven journalists into a frenzy. Reports, articles and even films predicted the mayhem that would follow if Ebola virus hit the United States, even though the number of people who have ever died from Ebola is tiny. But important questions need to be answered: Where has this virus been hiding? Why did it suddenly emerge to cause disease? Could there be a large epidemic?

The animal reservoir for Ebola virus seems to be a species of monkey that lives in areas of dense forest that are rarely visited by people. Yes, a few places like this still exist, and as people start to go into them, they come into contact with such monkeys. The virus is transmitted and what was a harmless organism in the monkey is deadly in its new host.

In Zaire and Sudan, Ebola virus spread quickly between people because of close contact. The centre of the epidemic in Zaire involved a missionary hospital where needles and syringes were re-used without sterilisation. Most of the staff of that hospital died and there were a few cases involving people taking care of the sick or preparing bodies for burial, but the epidemic was self-limiting. Because of its relatively short incubation period and the severe symptoms it causes, it is unlikely ever to cause a worldwide epidemic.

1 MICROORGANISMS AND DISEASE

Disease-causing microorganisms have been responsible for enormous numbers of human deaths. They have frequently altered the course of history. The species that probably had the greatest effect on human populations was *Yersinia pestis*, the bacterium responsible for the great outbreaks of plague such as the Black Death and the Great Plague. The Black Death swept through Europe in the middle of the fourteenth century, killing almost a quarter of the entire population. The Great Plague of 1665 had a devastating effect on the population of London and other British cities and towns (Fig 29.1).

Plague is not a disease confined to the rat-infested slums of the Middle Ages. It still exists today. Every year there are about 15 cases in the western United States, and the disease is still able to terrify. An outbreak of plague in India in 1994 brought headline fears in many newspapers of a new epidemic.

But it didn't happen: only a handful of people died. A disease that once killed millions of people can now be readily controlled using **antibiotics** and **antiseptics** and by preventing the conditions that lead to its **transmission**.

In this chapter we look at the different organisms that cause infectious disease. In Chapter 30 we look at the ways in which the immune system deals with infectious disease.

WHAT MAKES MICROORGANISMS DANGEROUS?

Microorganisms are very common in and on the human body. There are more bacteria living on your skin than there are people in the world. Most bacteria cause you no harm; the few that do are called **pathogens**. It is important to find out which microorganisms are responsible for which diseases for, until the infection has been identified, it is difficult to give effective treatment.

The nineteenth-century biologist Robert Koch put forward a set of basic principles for identifying a pathogen: these principles are known as **Koch's postulates**.

● The microorganism must always be present when the disease is present. It should not be present if the disease is absent.
● It should be possible to isolate the microorganism from an infected host and grow it in culture.
● When cultured microorganisms are introduced into a healthy host, the disease should develop.
● It should be possible to isolate the microorganism from the new host.

By following these principles, researchers have identified the microorganisms responsible for many diseases. But there are problems in applying Koch's ideas, especially when humans are involved. For example, it would be unethical to introduce a suspected pathogen to humans to see whether or not they developed a disease. In addition, some people fail to show disease symptoms after being infected. They might be resistant to that disease or they might be **carriers** that can pass on the disease to others, even when they appear disease-free themselves (Fig 29.2).

Fig 29.1 Part of the stained glass window in the church at Eyam in Derbyshire. It tells the story of how plague was brought to the village in a bundle of cloth from London. This bundle was opened by the tailor George Viccars who became the first of the 267 out of the total of 350 people in the village to die from plague

Fig 29.2 Typhoid fever is a severe disease of the digestive system caused by the bacterium *Salmonella typhimurium*. In carriers, people who are infected but show no signs of the disease, the bacteria living in the gut pass out of the body in the faeces. Typhoid Mary, who lived in New York in the twentieth century, was a carrier of *Salmonella typhimurium*. She gained a bit of a bad name for herself when, unwittingly, she shared her infection with many of the people she came into contact with. While she remained the picture of health, they sickened and died

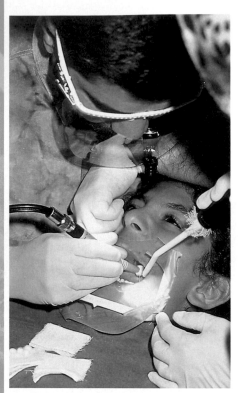

Fig 29.3 During dental treatment, the patient is at risk from infection (through the wounds in the gums), but so is the dentist. The patient can pass on airborne viral infections and there is a slight risk of blood-to-blood transmission

? TEST YOURSELF

B Study the photograph of the dentist at work in Fig 29.3. What precautions are being taken to avoid transmission of any infection:
a) from the dentist to the patient?
b) from the patient to the dentist?

Fig 29.4 The power of a sneeze. When someone sneezes, the droplets that explode from the mouth and nose start their journey at over 150 kilometres per hour

? TEST YOURSELF

C Everyone gets colds, but measles, mumps and diphtheria are now very rare in the UK. Can you guess why this is? (Clue on page 526 in Chapter 30.)

HOW DO MICROORGANISMS GET INTO THEIR HOST?

Your body provides an ideal environment for pathogens to multiply rapidly. It's warm and wet and there are plenty of nutrients. Fortunately for you, your skin prevents most of them taking advantage of the ideal conditions inside. The outer layer of skin consists of dead, dry cells made of an indigestible protein called **keratin**. Microorganisms find it difficult to penetrate this layer to enter the body. They are more likely to use the slightly easier entry route offered by the **mucous membranes** of the breathing system (Fig 29.3), the gut, the urinary system or the reproductive system. But the body has various strategies to make things as difficult as possible. The mechanisms involved are explained in Chapter 30.

In order to cause disease, pathogens must be transmitted from one person to another. Many of them cannot survive for very long outside the body. The main **transmission routes** are:
● through the air in droplets of **mucus** or **saliva**,
● in contaminated food or water,
● by sexual contact with another person,
● through breaks in the skin.

AIRBORNE INFECTIONS

Many microorganisms are transmitted through the air. These include influenza, colds and measles, which are all caused by viruses; and the bacterial diseases tuberculosis, diphtheria and meningitis. When an infected person coughs, sneezes or even just breathes out, tiny droplets of mucus and saliva are expelled (Fig 29.4). These droplets carry microorganisms with them. The larger ones fall to the ground rapidly, but the smaller ones can remain suspended in the air for long periods of time. In poorly ventilated, crowded conditions, diseases such as influenza can spread very rapidly.

CONTAMINATED FOOD AND WATER

Every year over 5 million people die from infections that cause diarrhoea. Many victims are small children. In Bangladesh three out of five children under 5 die from infant diarrhoea. The microorganisms responsible spread in water or food contaminated by human faeces.

DIRECT CONTACT

Direct contact between the skin of one person and that of another is unlikely to produce infection. Sexual contact is another matter, however. Microorganisms that spread like this include the human immunodeficiency virus (HIV), the virus that causes genital herpes, the bacterium responsible for gonorrhoea, the spirochaete that causes syphilis and *Candida*, the yeast that causes thrush (Fig 29.5).

Fig 29.5 Coloured scanning electron micrograph of the fungus that causes thrush, *Candida albicans*. This yeast-like fungus is common in the mouth, vagina and gut

THROUGH THE SKIN

The outer layer of the skin consists of tough cells that protect the body very effectively against infection. But this layer can be broken, and cuts and grazes form an entry point for bacteria. For example, tetanus is a serious disease that affects the nervous system. The bacteria that cause tetanus are common in soil and can enter the body through puncture wounds and animal bites.

Mosquitoes, ticks and lice all feed by puncturing the skin. In doing this they transmit microorganisms directly into the blood. One of the most important insect-borne diseases is malaria, which is transmitted by *Anopheles* mosquitoes (Fig 29.6).

Fig 29.6 An *Anopheles* mosquito feeding on human blood. Over 1.5 million people per year die from malaria

THE DIFFERENCE BETWEEN INFECTION AND DISEASE

Infection is not the same as having a disease. Before the disease develops, there is an **incubation period** in which the pathogen multiplies in the cells and tissues of the host. Think about what happens when you have been in a room with someone who is suffering from a cold. The infected person might sneeze and expel tiny droplets of mucus and saliva carrying the cold viruses on them. You inhale these particles and they get into your throat. There they attach to the epithelial cells in the mucous membrane. Like all cells in the body, these cells have cell surface membranes that contain many different protein molecules. The type of proteins present depends on the cell type, and on the genetics of the individual. The virus attaches only if the right protein is present in the cell surface membrane.

Fig 29.7 Coloured transmission electron micrograph of human immunodeficiency virus (HIV) budding from an infected T-lymphocyte human blood cell. Four viruses are shown at different stages of budding

The virus then enters the cell and begins a cycle of reproduction and colonisation of new cells (Fig 29.7). But infection is not entirely one-sided. The body has a complex immune system (see Chapter 30). This might prevent the symptoms of the disease from ever developing. Even if some illness does follow, the immune system works to allow us to recover quickly.

THE EFFECTS OF DISEASE

Infectious disease has two main effects. Firstly, the pathogen can damage the cells of the body directly. The human immunodeficiency virus, HIV, gives rise to the symptoms of AIDS mainly as a result of destroying large numbers of a type of white blood cell. As a result, the immune system stops working effectively and this allows other infections or certain types of tumour to develop. Similarly, some of the symptoms of malaria result from a loss of red blood cells or the blocking of capillaries that supply particular organs in the body.

Alternatively, the pathogen can cause the release of substances that injure the body. Many bacteria, for example, produce **toxins**. **Exotoxins** are released by bacteria as they grow. **Endotoxins**, which form part of the bacterial cell wall, escape into the body when the bacterium dies and the cell wall breaks down.

Diphtheria is a good example of a bacterium that causes disease because it produces a toxin. It was a dangerous bacterial infection until about 40 years ago, when it used to kill many children. Since then, antibiotics and an effective vaccine have virtually eliminated it.

Diphtheria has an incubation period of 2 to 6 days, and then a sore throat and fever develop. A membrane forms across the throat and makes it difficult to breathe (Fig 29.8). As the bacteria multiply, they release a toxin into the blood. This toxin inhibits protein synthesis and results in damage to many parts of the body, particularly to the heart and the nerves. Interestingly, this toxin is produced only in bacteria of the species *Corynebacterium diphtheriae* that are themselves infected by a **bacteriophage**. The toxin is actually produced by one of the bacteriophage genes.

Fig 29.8 The infected throat of a child with diptheria

2 INFECTIONS CAUSED BY BACTERIA

Bacteria are **prokaryotes** (see Chapter 1). Each bacterial cell is a mass of cytoplasm enclosed by a cell surface membrane and a cell wall. Bacterial cells have no nucleus and their DNA is in the form of a loop in the cytoplasm. There are none of the membrane-bounded organelles such as mitochondria, endoplasmic reticulum and lysosomes that are found in eukaryotic cells. Fig 29.9 shows a typical bacterium, and summarises how different bacteria can cause disease.

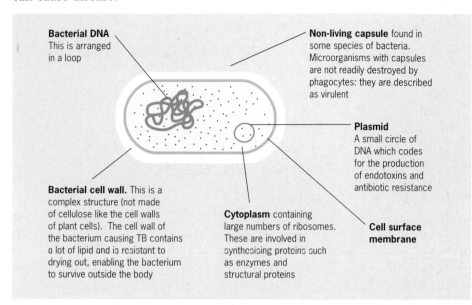

Bacterial DNA This is arranged in a loop

Non-living capsule found in some species of bacteria. Microorganisms with capsules are not readily destroyed by phagocytes: they are described as virulent

Plasmid A small circle of DNA which codes for the production of endotoxins and antibiotic resistance

Bacterial cell wall. This is a complex structure (not made of cellulose like the cell walls of plant cells). The cell wall of the bacterium causing TB contains a lot of lipid and is resistant to drying out, enabling the bacterium to survive outside the body

Cytoplasm containing large numbers of ribosomes. These are involved in synthesising proteins such as enzymes and structural proteins

Cell surface membrane

Fig 29.9 Some of the strategies that bacteria use to invade the human body and cause disease

Bacteria don't need much to grow and reproduce. Give a single bacterial cell carbohydrate or some other source of carbon, a supply of nitrogen and a few mineral salts, and it can divide to produce a population of millions. It is easy to mix all these substances together in the laboratory to produce an artificial medium on which to culture bacteria. It's even easier to provide them with some leftover food in the home. In both cases, the bacteria start to increase rapidly.

A typical population curve is shown in Fig 29.10. The actual rate of growth depends on environmental conditions. Low temperatures or a shortage of nutrients, for example, limit the rate of growth.

Fig 29.10 A population growth curve for the bacterium *Escherichia coli* growing on an artificial medium. Bacterial food poisoning results from eating foods that have been contaminated with certain types of bacteria or with the toxins that they have produced. Food that gives rise to this type of food poisoning has often been stored under conditions in which bacteria can multiply rapidly

Lag phase	When introduced to a medium, bacteria take time before they start to increase in number. They need to produce the enzymes necessary to digest the nutrients in the medium. This requires activation of genes and the synthesis of proteins.
Exponential phase	Bacteria reproduce by binary fission. Once the cell has grown and reached a certain critical size, it divides in two. In this way, the population grows very rapidly. In optimum conditions, it may double in as little as 20 minutes. During the exponential stage, there are plenty of nutrients available, and toxic waste products have not accumulated to the extent that they slow down bacterial growth.
Stationary phase	Population growth begins to slow down and finally stops. There is a balance between the number of new bacterial cells being added to the population and the number dying. Nutrients are beginning to be in short supply and there may be a build-up of waste products which will inhibit the growth and reproduction of the bacteria.
Decline phase	Conditions in the culture become increasingly limiting. Nutrient supplies continue to fall and waste products to build up. Bacterial cells are unable to survive in these conditions and die in increasing numbers.

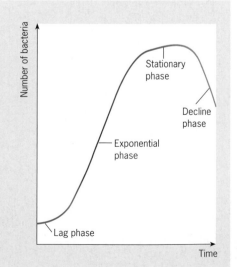

TUBERCULOSIS

Tuberculosis (TB) is a disease caused by an infection with the bacterium *Mycobacterium tuberculosis*. Globally, each year, there are 3.8 million cases of TB, and 2.6 million of them prove fatal.

In the United Kingdom and other more developed countries, the incidence of tuberculosis fell dramatically during the twentieth century. Many factors contributed to this decline. There was a general improvement in living standards during this time, and advances in medicine such as the discovery of effective antibiotics and vaccines have made tuberculosis treatable and preventable. However, there are now signs that the number of cases is rising again (Fig 29.11). What has gone wrong?

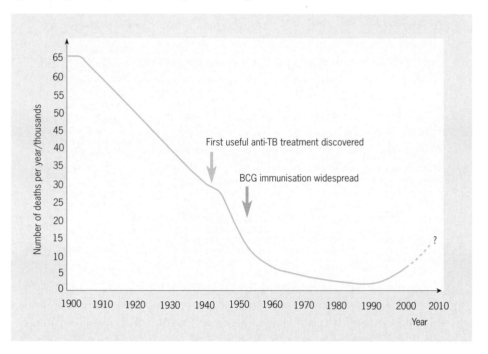

Fig 29.11 The graph shows the change in incidence of tuberculosis during the twentieth century and predicts what might happen during the next decade. Six countries account for over 50 per cent of the current TB epidemic. According to WHO estimates, 4.5 million of the 8 million new cases that occur each year are in India, China, Bangladesh, Pakistan, Indonesia and the Philippines. The problem in these countries is largely due to three factors: multidrug-resistant TB (MDR-TB), HIV and economic problems

Part of the problem is that **drug-resistant** strains of mycobacteria have evolved. Surveys in selected sites in Asia show high levels of TB that is resistant to multiple drugs and so cannot be treated with the most powerful anti-TB drugs. The number of people being immunised has decreased and large numbers of people have become more susceptible because they have AIDS and the decreased immunity that goes with it. Experts predict a further increase in the number of TB cases because of higher rates of TB transmission to HIV-positive people.

Table 29.1 shows some of the groups of people in the UK who are at greater risk of tuberculosis.

Table 29.1 Tuberculosis risk groups in the 1990s in the UK	
Risk group	**Reasons for greater likelihood of developing tuberculosis**
AIDS patients	AIDS patients have immune systems that are less effective at combating infection. Tuberculosis is more common in AIDS patients.
Recent immigrants	Recent UK immigrants, especially from the Indian subcontinent, are less likely to have been immunised against tuberculosis. They are also more likely to have had an earlier infection.
Vagrants	Misuse of alcohol and other drugs is associated with an increased risk of tuberculosis. Once diagnosed, it is difficult to ensure the necessary long-term treatment.

SIGNS AND SYMPTOMS OF TB

Tuberculosis can affect many of the organs in the body. In almost 80 per cent of cases it affects the lungs. Someone with primary tuberculosis has a fever, tends to lose weight and often has a persistent cough. They might cough up blood because the bacteria destroy the lung tissue: a chest X-ray shows this damage as a dark shadow (Fig 29.12). The condition can be confirmed by identifying the bacteria in a sputum sample or in a small piece of lung tissue removed with the help of an endoscope.

TREATMENT AND CONTROL OF TB

There is no completely effective vaccine against TB of the lungs. **BCG vaccine** (bacille Calmette–Guérin) was developed in 1921. It is very useful in preventing certain types of TB including the more severe types that can affect babies during the first year of life. The vaccine, which is a live but **attenuated**, or altered, strain of *Mycobacterium tuberculosis*, is given at or soon after birth to 85 per cent of babies worldwide. The vaccine provides cross-protection against leprosy, but is unlikely to protect adults against TB, and some people question why children in the UK are still given the BCG vaccine at the age of 13.

The disease itself can be treated successfully with antibiotics. Combinations of drugs are used to discourage the development of strains that are resistant to particular antibiotics. Since tuberculosis is infectious, health workers trace all the people who have been in close contact with patients. These contacts can then be screened for the disease and treated if necessary.

In 1998, a group of researchers announced that they had decoded the entire genome of the tuberculosis bacterium. Knowing the sequence marks a new phase in the battle against one of the most successful pathogens that affects humans. The sequence contains 4 411 529 base pairs, which combine to form over 4000 individual genes. The genome is already yielding information that could prove invaluable to future research on anti-TB drugs and vaccines.

Fig 29.12 In this chest X-ray, the fluffy yellow areas are where the bacteria that cause tuberculosis have damaged the lung tissue

SALMONELLOSIS: SIGNS AND SYMPTOMS

There are many different species of bacteria that belong to the genus *Salmonella*. Only about ten of them cause food poisoning, known as **salmonellosis**. When food contaminated with the bacteria is eaten, the

Fig 29.13 Some ways in which *Salmonella* food poisoning can spread

WC

Contaminated water used on crops and for drinking

Sewage discharged at sea contaminates shellfish

Infected faeces enter water supply

Contaminated food or water ingested

Flies carry bacteria from sewage to food

Diarrhoea follows infection

Contaminated hands carry bacteria to food

✔ **REMEMBER THIS**

If you want to avoid salmonellosis or other food poisoning, avoid buying and eating food after the recommended date. Store perishable food in a cool place or, preferably, in a refrigerator. Do not re-freeze food that has already been frozen and thawed once. Avoid the risk of contamination by maintaining good personal hygiene and by keeping raw and cooked foods separate from each other. Make sure that cooking and re-heating is thorough at sufficiently high temperatures.

SEE QUESTION 1

bacteria pass into the intestine. Other bacteria that cause food poisoning do not multiply once they are inside the body – they cause illness as they release toxins. Salmonella species actually enter the cells lining the small intestine and multiply actively. As the population increases, some bacteria die and release endotoxin. This is the toxin that causes the typical symptoms of food poisoning or **gastroenteritis**. Bad personal hygiene and poor food preparation skills can spread *Salmonella* food poisoning (Fig 29.13).

The endotoxin released as the bacteria die irritates the lining of the intestine, and causes the unpleasant and typical symptoms of gastroenteritis, nausea, vomiting, diarrhoea and abdominal pain.

TYPHOID FEVER

Typhoid fever is caused by another species of *Salmonella*. It is a much more severe disease than salmonellosis. The patient often has a long-lasting fever, becomes delirious, and the spleen becomes inflamed. The wall of the intestine can be so badly damaged that it bleeds.

TREATMENT AND CONTROL OF SALMONELLOSIS AND TYPHOID FEVER

Most cases of food poisoning are left to run their course. Sometimes, loss of body fluid may be severe, particularly in infants. Then, applying **oral rehydration therapy**, ORT (see Chapter 13), or using an **intravenous drip** will replace body fluids. Antibiotics may be used to treat typhoid fever.

The best way of dealing with *Salmonella* or any other organism that causes food poisoning is to avoid getting it in the first place: careful attention to basic food hygiene removes much of the risk.

3 INFECTIONS CAUSED BY VIRUSES

It is difficult to decide whether viruses are living organisms at all. They do not show several of the characteristics that define a living organism; for example, they don't respire or feed. But they are more than just a collection of very complex chemicals. Their origins are also rather uncertain. Some biologists think they are stray bits of DNA that have 'escaped' from the genomes of higher plants and animals.

Viruses vary considerably in size but they are all very small (Fig 29.14). The virus responsible for polio, for example, is one of the smallest and is only 20 to 30 nm in diameter. The herpes virus is much larger; its diameter is about 250 nm. In addition to their small size, viruses have a very simple structure. They consist of a piece of genetic material, DNA or RNA, surrounded by a protein coat and sometimes also by a membrane.

In order to multiply, a virus must take over the host cell. It uses the cell's organelles and biochemical processes to make more virus particles. Viruses therefore have a great capacity to harm the cells and tissues of their host, and they cause a huge range of diseases, including influenza and AIDS.

AIDS

AIDS, short for **acquired immunodeficiency syndrome**, is characterised by the destruction of vital cells in the immune system. It is caused by the human immunodeficiency virus (HIV) (see page 534). The disease was first recognised in the early 1980s and the virus itself was identified in 1986. It has spread dramatically (Fig 29.15). As of the end of

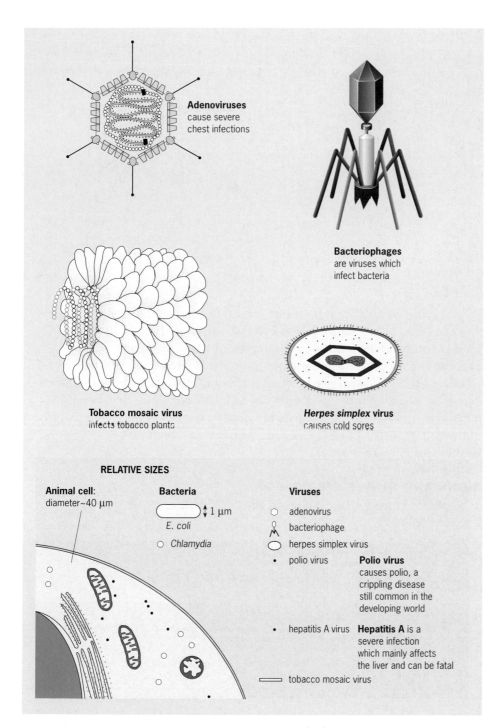

Adenoviruses cause severe chest infections

Bacteriophages are viruses which infect bacteria

Tobacco mosaic virus infects tobacco plants

Herpes simplex virus causes cold sores

RELATIVE SIZES

Animal cell: diameter~40 μm

Bacteria
↕ 1 μm
E. coli
○ *Chlamydia*

Viruses
○ adenovirus
bacteriophage
○ herpes simplex virus
• polio virus
• hepatitis A virus
tobacco mosaic virus

Polio virus causes polio, a crippling disease still common in the developing world

Hepatitis A is a severe infection which mainly affects the liver and can be fatal

Fig 29.14 In the blue box, six different viruses have been drawn to the same scale as the two bacterial cells *E. coli* and *Chlamydia*, and an animal cell. Viruses are generally much smaller than bacteria, but there is some overlap. *Herpes simplex*, one of the larger viruses, is similar in size to *Chlamydia*, one of the smaller bacteria. The magnified views of four of the viruses show the appearance of individual virus particles

REMEMBER THIS

Facts about AIDS

Worldwide, approximately one in every 100 adults aged 15 to 49 is HIV-infected. In sub-Saharan Africa, about 8.4 per cent of all adults in this age group are HIV-infected. In 16 African countries, the prevalence of HIV infection among adults aged 15 to 49 exceeds 10 per cent.

An estimated 5 million new HIV infections occurred worldwide during 2001; that's about 14 000 infections each day. More than 95 per cent of these new infections occurred in developing countries. In 2001 alone, HIV/AIDS-associated illnesses caused the deaths of approximately 3 million people worldwide, including an estimated 580 000 children younger than 15 years.

Worldwide, more than 80 per cent of all adult HIV infections have resulted from heterosexual intercourse.

(Data from the National Institutes of Health, USA.)

2001, an estimated 40 million people worldwide – 37.2 million adults and 2.7 million children younger than 15 years – were living with HIV/AIDS. More than 70 per cent of these people (28.1 million) live in sub-Saharan Africa; another 15 per cent (6.1 million) live in South and Southeast Asia.

HIV is a **retrovirus**: it carries the enzyme **reverse transcriptase**. Once the virus infects a cell, often a T lymphocyte, this enzyme allows the cell to make virus DNA from virus RNA. The virus DNA is then inserted into the cell's own DNA. Here it acts as a gene. It might do nothing for a long time but eventually it will cause the production of more virus RNA. This will result in thousands of new viruses that burst out through the cell surface membrane and infect other cells. This life cycle is summarised in Fig 29.16.

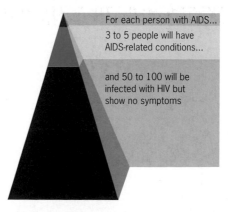

For each person with AIDS... 3 to 5 people will have AIDS-related conditions...

and 50 to 100 will be infected with HIV but show no symptoms

Fig 29.15 The HIV pyramid

Fig 29.16 HIV infects vital cells in the immune system, notably the helper T lymphocytes. This diagram shows how a single virus can infect a host cell and multiply to produce many thousands of new viruses

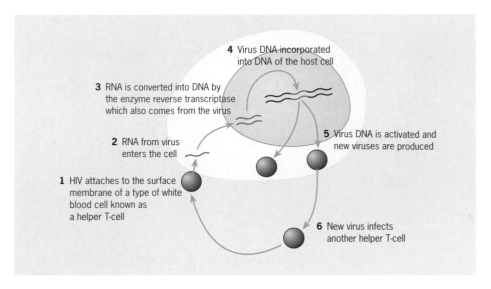

4 Virus DNA incorporated into DNA of the host cell

3 RNA is converted into DNA by the enzyme reverse transcriptase which also comes from the virus

2 RNA from virus enters the cell

5 Virus DNA is activated and new viruses are produced

1 HIV attaches to the surface membrane of a type of white blood cell known as a helper T-cell

6 New virus infects another helper T-cell

SEE QUESTION 2

HIV is often transmitted from one person to another during sex, but the virus can also spread when drug users share dirty needles, when infected blood is transfused, or when contaminated blood is used to make blood products. When HIV enters the body, it can be many years – maybe 10 or 15 – before AIDS develops. A very small number of people are known to have been HIV positive since the mid-1980s and have not so far developed the syndrome. Perhaps they never will, and AIDS researchers are trying to find out why. This might provide really important clues about how AIDS can be held in check in other people.

SIGNS AND SYMPTOMS OF AIDS

Table 29.2 shows the main stages in the development of AIDS from first infection with HIV.

Fig 29.17 The black swelling shows that this person has Karposi's sarcoma, a form of cancer, which affects blood vessels in the skin. Karposi's sarcoma is much more common in people with AIDS than it is in the rest of the population

Table 29.2 The stages in the development of AIDS		
Stage	**Number of helper T lymphocytes per mm³ of blood**	**Main features of stage**
A	over 500	No symptoms but patient has antibodies to HIV in the blood.
B	200–499	Patient usually has fever or diarrhoea which lasts for a month or more. He or she might have infections such as thrush, caused by *Candida*, a type of yeast.
C	less than 200	Likely to have one or more of the following diseases known to be related to AIDS: • Tuberculosis • Karposi's sarcoma, a form of cancer in which large black or brown tumours appear on the skin (Fig 29.17) • Pneumonia

TREATMENT AND CONTROL OF AIDS

Drugs are available for the treatment of AIDS, and they are getting more effective all the time. However, they are expensive and can only slow the progress of the disease – there is currently no cure. An effective vaccine is thought to be a long way in the future because HIV can change its surface proteins and evade host immune responses. Gene therapy might provide some hope but, until it is developed, the current priority is to limit further spread by advising people how to reduce their risk of exposure to HIV.

HIV is capable of infecting another person only when it is passed inside lymphocytes or macrophages (see Chapter 30). Apart from blood itself, semen and vaginal secretions contain large numbers of these cell types, making body fluids high-risk sources of infection.

Practising safe sex – using condoms during every sexual encounter – is very important. Using a barrier method of contraception not only prevents pregnancy, it cuts down the chance that HIV in one person's semen or vaginal secretions will get into his or her partner's blood. Drug users who share dirty needles are also at risk. Programmes that provide free, sterile needles on demand are designed to cut down HIV transmission by this route.

4 PARASITES AND DISEASE

Bacteria and viruses are **microorganisms**; they are too small to be seen with the naked eye. However, some larger organisms also cause disease. These are the **protoctists** and the **parasitic worms**. In Europe, they are relatively rare, but on a global scale they have an enormous impact. Some examples of these larger parasites are shown in Fig 29.18. The roundworm *Ascaris lumbricoides* is the subject of the Assignment at the end of this chapter.

It is all too easy to concentrate on the horrors of parasitic infections. As biologists we should not forget that parasites as a group provide one of the best examples of **adaptation**.

A useful definition of a parasite is as follows:

> **A parasite is an organism that lives in or on a host organism. It gains a nutritional advantage from this relationship while the host suffers a disadvantage.**

Fig 29.18 Some common human parasites. Over half the world's population is infected by at least one of them
(a) Roundworm
(b) Tapeworm: head end
(c) Hookworm
(d) Bilharzia: male (fat) and female (thin) flukes
(e) Malaria: thread-like sporozoites in blood

(a)

(b)

(c)

(d)

(e)

It might seem that this is a very easy way of making a living, but think about the problems a parasitic worm faces, living in your intestine. There are obvious advantages in lying in a soup of pre-digested food. There is no need to produce digestive enzymes, a complex nervous system is unnecessary because food does not need to be searched for, and a locomotory system is almost a disadvantage since the worm might move to less favourable areas of the gut.

On the other hand, there are disadvantages. Parasites are made of the same complex macromolecules as the food that you eat, so they must avoid being digested and must try to stay put despite the constant contractions of the gut muscles. It's also difficult to move from the intestine of one person into the intestine of another. Because of this, life cycles can be very complex and many involve **intermediate hosts**. Since the chances of completing the life cycle and reaching a new host are small, many parasites produce vast numbers of offspring.

MALARIA

Malaria is caused by four species of the protoctist *Plasmodium* – *P. vivax*, *P. falciparum*, *P. ovale* and *P. malariae*. They are transmitted by the bite of the *Anopheles* mosquito. The life cycle of *Plasmodium* and an introduction to malaria is on page XXX in Chapter 36.

During the 1960s, malaria was successfully eradicated from the warmer areas of southern Europe and the United States, and its incidence was reduced significantly in many countries in the tropics. But the disease is now on the increase again. The WHO estimates that each year 270 million new malaria infections occur worldwide, along with 110 million cases of illness and 1.5 million deaths. Malaria is endemic in at least 100 countries of the tropical and subtropical regions of the world, there are an astounding two billion people at risk of malaria infection, and malaria is the cause of a quarter of all childhood deaths in Africa.

There are several reasons why the disease is so widespread: the cost of maintaining control programmes is high, insecticide resistance has become more common in the mosquito that transmits malaria, and the malarial parasite itself has developed drug resistance.

SIGNS AND SYMPTOMS OF MALARIA

One of the most characteristic features of malaria is the pattern of fever (Fig 29.19), with temperature peaks at regular intervals. These correspond to a new generation of parasites bursting out of the red blood cells, destroying them and causing anaemia. The parasites can block capillaries, limiting the blood supply to organs in the body: the brain, digestive system and kidneys can all be affected.

Fig 29.19 A temperature chart for a patient with malaria. Note the peaks that occur at regular intervals. This cyclic pattern of fever is characteristic of malaria

TREATMENT AND CONTROL OF MALARIA

One way to combat malaria is to break the cycle of infection. This can be done by killing the mosquitoes with insecticides, clearing the pools of stagnant water in which the larvae live and preventing the insects from biting humans. Where possible, the risk can be reduced by sleeping under mosquito nets (Fig 29.20), wearing long-sleeved shirts and trousers after dark when mosquitoes are most active, and using mosquito repellents. But all these measures are too expensive for many local people. Also, they are not certain to work.

ANTI-MALARIA DRUGS

The anti-fever properties of the bitter bark of the tree *Cinchona ledgeriana* were known in Peru before the fifteenth century. **Quinine**, the active ingredient in the bark, was first isolated in 1820. **Chloroquine** is now the most commonly used anti-malaria drug in Africa. It lowers fever and reduces the number of parasites in the body. Resistance to chloroquine, however, is widespread and has been rising steadily since the early 1980s (Fig 29.21). For this reason, it is used in combination with other drugs. Research continues into resistance-reversing drugs, but so far no useful drugs have emerged.

Drugs based on **artemisinin** are important treatments for severe malaria. These are derived from the Chinese plant *Artemesia annua* (or quinghaosu). They are effective, fast-acting, but relatively expensive drugs.

Artemether and quinine are the anti-malarial drugs most suitable for the treatment of severe malaria in children.

Fig 29.21 The map shows that malaria is confined to tropical and subtropical regions. Chloroquine has been used very successfully in the control of malaria but resistance to it is now a major problem

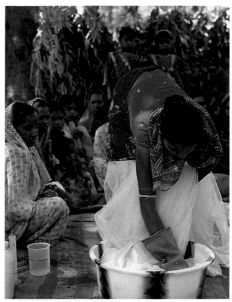

Fig 29.20 The use of mosquito nets soaked in insecticide can be very effective in controlling malaria. Unfortunately, the cost may be beyond the resources of people living in malarial areas

✔ REMEMBER THIS

An infusion of quinghaosu (*Artemesia annua*) has been used for at least the past 2000 years in China. Its active ingredient artemisinin has only recently been scientifically identified.

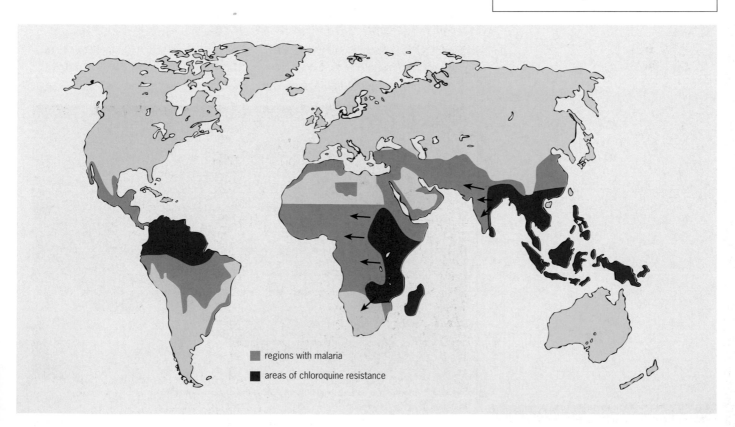

regions with malaria

areas of chloroquine resistance

⚠ SCIENCE FEATURE Will there ever be a vaccine for malaria?

Despite decades of research effort, no one has yet managed to develop a vaccine that is effective against a protoctist such as *Plasmodium*. Several potential malarial vaccines have been developed, but many have run into problems. The vaccine developed by Manuel Patarroyo in Colombia has been tested in people and seemed to be the answer everyone was waiting for However, further trials showed that the vaccine was not protective, and the arguments about it have raged eve since.

Currently, molecular biologists are using two main tactics to find a suitable vaccine. Some are trying to identify antigens that appear on the parasite's surface at various stages of its life cycle. The idea is to put several *Plasmodium* genes into the harmless **Vaccinia virus**, and then introduce it as a recombinant vaccine into the body. Once inside, it divides a few times and sets up a good immune response to *Plasmodium*. Some of these recombinant vaccines look promising and may be tested in humans very soon.

The other approach is to develop a DNA vaccine for malaria. The aim is to isolate the *Plasmodium* genes that provoke an immune response during infection, and to put them into a plasmid. This is a ring of DNA obtained from bacteria. The DNA itself is introduced into the body where it uses the human cell machinery to make the proteins coded for in the genes. The person's immune system detects these proteins, and makes an immune response. It is this response that could combat the real infection if the person is later exposed to *Plasmodium*.

Another new approach, announced late in 2001, is to express proteins that can stimulate immunity to *Plasmodium* parasites in goat's milk. Researchers showed that it was possible to produce transgenic mice that could express proteins from *Plasmodium* in their milk and experiments were then started to try to produce goats to do the same thing. The advantage of this strategy is that a herd of goats could probably produce enough vaccine for the whole of sub-Saharan Africa.

Trials of various types of malarial vaccine are in progress and new approaches are being investigated. Even so, it could be at least five years before we have a real hope of a useable vaccine for humans.

Fig 29.22 An effective malarial vaccine would cut the misery of malaria infection for millions of people

SEE QUESTIONS 3 AND 4

SCHISTOSOMIASIS

Schistosomiasis, also known as **bilharzia**, results from infection by one of three species of a blood fluke called **Schistosoma**. The three species are: *Schistosoma mansoni* (Africa and South America), *Schistosoma haematobium* (Africa and the Middle East) and *Schistosoma japonicum* (Asia). Like many other parasitic worms, these flukes have a very complex life cycle (Fig 29.23).

Schistosomiasis is generally believed to rank second after malaria as the most important parasitic disease in the world. A total of 500–600 million people are at risk of infection and more than 200 million people in 74 countries succumb to the disease. Of these, 800 000 die each year and 120 million others have severe symptoms.

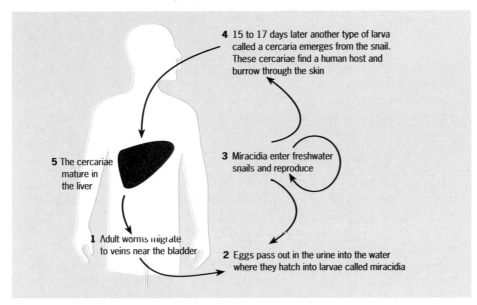

4 15 to 17 days later another type of larva called a cercaria emerges from the snail. These cercariae find a human host and burrow through the skin

5 The cercariae mature in the liver

3 Miracidia enter freshwater snails and reproduce

1 Adult worms migrate to veins near the bladder

2 Eggs pass out in the urine into the water where they hatch into larvae called miracidia

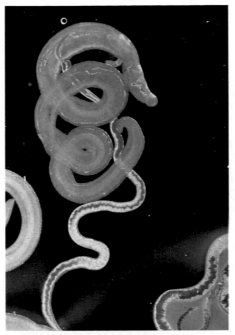

Fig 29.23 Schistosomiasis and its cycle of infection. Part of the life cycle takes place in humans and part in water snails. Unfortunately the snails are very difficult to kill. In the photo the male schistosome is blue and the female is white.

SIGNS AND SYMPTOMS OF SCHISTOSOMIASIS

The body responds to an infection with schistosomes by trying to 'wall off' the eggs that are produced by the worms within the veins. This scar tissue can cause the liver to enlarge or can lead to blockages in the bladder and kidney. The severity of the damage is directly related to the numbers of parasites in the body. Since the parasite does not reproduce inside the human host, clinical signs do not appear until the number of worms in the body reaches a critical level.

TREATMENT AND CONTROL OF SCHISTOSOMIASIS

The main drug used to treat schistosomiasis is highly effective and has few side-effects. However, it does not prevent reinfection – only continued, repeated treatments will keep the disease under control – and it is expensive. This creates an impossible situation because the people worst affected by the disease are the ones least able to afford treatment. Increased use of drugs also increases the likelihood that drug-resistant strains will evolve.

Tackling the snail that carries the schistosome is also difficult. The chemicals used to kill the snails are thought to be harmful to the environment. Furthermore, the increased use of irrigation worldwide has aided the spread of schistosomiasis by expanding the habitat of the freshwater snail (Fig 29.24).

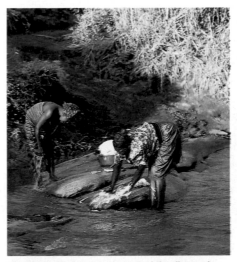

Fig 29.24 The best way to control the disease is therefore for people to avoid the freshwater areas that form the snail's breeding ground. This is good in theory, but if all supplies of freshwater are affected, where can the local people bathe and wash clothes? Until the poverty and poor living conditions of many of the areas where schistosomiasis is rife are tackled, people have no choice but to expose themselves to repeated infection

? TEST YOURSELF

The cercariae emerge from the water snails during the daytime. How is this behaviour an advantage to the parasite?

? TEST YOURSELF

J What is the advantage to the malarial parasite in living in a medium that has the same water potential as its own cytoplasm?

HOW PARASITES ARE ADAPTED FOR THEIR LIFESTYLE

The parasites that cause malaria and schistosomiasis illustrate how specialised parasites become as they adapt to living in the blood system of their human host. The two parasites and their adaptations are compared in Table 29.3.

Table 29.3 Adaptations to a parasitic lifestyle: *Plasmodium* and *Schistosoma* compared

Adaptation	Plasmodium	Schistosoma
Attaching to the host	The malarial parasite spends much of its adult life inside the cells of its host. Attachment is not a problem.	Adult worms live in the veins around the bladder. Because they are small they could easily be washed away by the blood. Suckers help them attach to the walls of the veins.
Resisting host defence mechanisms	The blood is a dangerous place for any parasite because of the white blood cells of the immune system (see Chapter 30). Within an hour of entering the body, malarial parasites are inside liver cells. Here they hide from the immune system. The antigens on the surface of a malarial parasite change frequently and the immune system cannot keep up.	Adult worms have a remarkable way of avoiding the host's immune system: they disguise themselves by covering up with molecules from the host's red blood cells.
Reproduction and reproductive organs	There are many reproductive stages in the life cycle of the malarial parasite. The parasites reproduce in the mosquito, but this insect is small compared to a human. Once in their main host, malaria parasites also reproduce in the liver and in the blood cells, causing a rapid increase in parasitic load.	The life cycle of this worm depends very heavily on chance. A person must urinate into fresh water; the miracidia larvae that hatch from the eggs must find a snail host and the cercariae must encounter a human shortly after emerging from the snail. The production of large numbers of offspring is therefore the best way of ensuring survival. Each egg laid by an adult worm can produce up to 200 000 cercariae.
The life cycle	Usually, blood from one person does not mix with that of someone else. So malaria parasites cannot depend on person-to-person contact for transmission. A second host is used. In malaria, the female *Anopheles* mosquito is the organism. When she bites someone with malaria, she takes in infected red blood cells. These parasites complete their life cycle in her body, finally migrating to her salivary glands. When she bites someone else, she injects a small amount of infected saliva, and malaria is passed on.	Part of the life cycle of *Schistosoma* is spent inside its human host; part is spent inside a freshwater snail. In many parts of the tropics where schistosomiasis is endemic, humans spend much of their time around areas of fresh water. They use it for washing clothes, irrigating crops, watering animals and bathing. This increases the chances of the parasite being passed from one person to another.
Reduction of body systems	The malaria parasite spends most of its life inside one of its host's red blood cells. It does not have to move to find food, and the medium in which it lives has the same water potential as its own cytoplasm. It has therefore lost the capacity to locomote and to regulate cell water content – features of many free-living single-celled organisms.	Schistosomes do not have a complex nervous system because conditions they find in the blood are remarkably constant. They cannot move either: a locomotory system is also unnecessary since the parasites' survival depends on it being anchored in the host.

SUMMARY

After reading this chapter you should know and understand the following:

■ Robert Koch produced a series of principles that enabled biologists to identify the microorganism that causes a specific disease. These principles are known as **Koch's postulates**.

■ In order to cause disease, a pathogen must gain access to the body. This involves penetrating the body's defences.

■ **Transmission** of pathogens can take place through the air, in contaminated food or water, by sexual contact or through the skin.

■ For infection to cause disease, the pathogen must multiply inside the host. Disease can result from **direct damage** to the cells and tissues of the host or from indirect damage caused by **toxins** produced by the infecting microorganism.

■ Some species of bacteria and viruses are pathogenic. So are some species of larger organisms such as protoctists and worms. These are **parasites** and they show many **adaptations** to their way of life.

? EXAM QUESTIONS

1 The diagram shows the structure of the influenza virus.

haemoglutinin protein helps virus penetrate cells

neuraminidase protein helps virus escape from cells

lipid envelope

nucleic acid

polymerase enzyme helps synthesise RNA

nucleoprotein

protein coat

a) Which **two** biochemical components, indicated in the diagram, are common to all viruses?

b) The virus shown can be described as a *non-cellular pathogen*. Explain the meaning of this phrase.

c) Give **three** differences between viruses and bacteria.

OCR 6911 Cells and Structures March 2000 Q7

2

a) **(i)** Name the specific type of white blood cells invaded by HIV.

(ii) Describe precisely what happens when these white blood cells are destroyed.

b) The graph shows the numbers of AIDS cases diagnosed in the United Kingdom from 1983 to 1997.

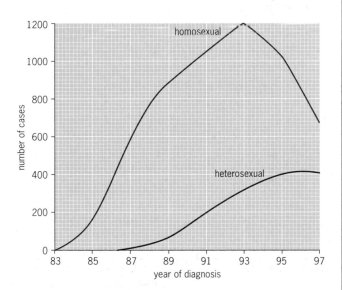

homosexual

heterosexual

year of diagnosis

number of cases

(i) Suggest an explanation for the trend in the incidence of AIDS in homosexuals from 1993 to 1997.

(ii) Calculate, for the period 1991 to 1995, the mean rate of increase in AIDS cases transmitted by heterosexual contact. Show your working.

c) Give **two** methods of transmission of HIV other than sexual contact.

AQA BY08 Health and Disease June 2000 Q4

? EXAM QUESTIONS

3 Malaria is a disease caused by the parasite *Plasmodium*.

a) Explain why *Plasmodium* is described as a parasite.

b) Most of the life cycle of *Plasmodium* is spent inside a red blood cell of its main host.
Suggest how each of the following is an adaptation to its way of life.
(i) *Plasmodium* has no locomotory structures.
(ii) *Plasmodium* has no mechanism for regulating its water content.

c) *Plasmodium* survives for long periods in the blood of its main host. Explain how.

AQA (A) BYA3 Pathogens and Disease January 2001 Q2

4 Over 40% of the world's population live in areas where there is a risk of malaria. The disease causes widespread suffering and death especially among children.

a) Name the organism that causes malaria.

b) State **two** geographical areas of the world where there is a high risk of malaria.
Malaria is transmitted by a vector, the mosquito *Anopheles*. The diagram shows the life cycle of *Anopheles*.

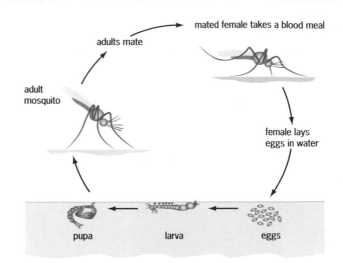

c) With reference to the diagram.
(i) describe the role of *Anopheles* in the transmission of malaria;
(ii) outline the biological factors that make malaria a difficult disease to control.

OCR 2802 Health and Disease June 2001 Q5

KEY SKILLS **ASSIGNMENT**

ROUNDWORMS

Ascaris lumbricoides is a large roundworm that lives as a parasite in the human intestine. A few roundworms in the gut of an otherwise healthy person have relatively little effect. But if that person is already suffering from malnutrition, it's a different story. The parasite absorbs most of the little food that is eaten, making the malnutrition much worse.

Roundworms also cause problems when they are present in large numbers. The mass of worms in the intestine can block the passage of waste through the colon, causing severe pain. If the worm infection is not treated, the intestine can rupture and the affected person is then seriously ill and can die without emergency medical attention.

Female worms lay eggs that can contaminate food or that can be spread by hand–mouth contact. Adult male worms have an average length of 15 cm, while females can be half as long again.

Fig 29.A1 shows an adult female roundworm and a cross section through its body.

Adult female

Cross section

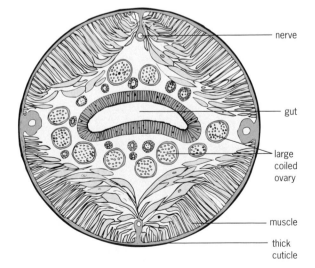

nerve

gut

large coiled ovary

muscle

thick cuticle

Fig 29.A1 *Ascaris lumbricoides*

Fig 29.A2 When a roundworm infection is treated, many worms can be expelled. This photograph shows the number of worms present in a person with a relatively light infestation

1 Look at Figs 29.A1 and 29.A2 and suggest two features of *Ascaris lumbricoides* that represent adaptations to its life as a parasite in the human intestine.

Ascaris lumbricoides is very common in Nigeria, a large country with a population of over 150 million. Many people live in rural communities.

We cannot estimate exactly what proportion of the population is infected with roundworms: different studies have produced wildly different figures that range from 10.8 per cent to 96.3 per cent.

2 Suggest why it is difficult to estimate accurately the proportion of the population affected by roundworms.

The studies that have been carried out suggest that the proportion of people affected differs according to the group of people that you are studying. Table 29.A1 shows the results of five different surveys.

Table 29.A1		
Place	**Urban/rural community**	**Percentage of sample infected with roundworms**
Ibadan	urban	48.1
Ewekoro	urban	52
Benin City	urban	19.5
Oyo State villages	rural	71.4
Lagos State villages	rural	74.2

KEY SKILLS **ASSIGNMENT**

3 **a)** Why is it not possible to compare the urban and rural results shown in Table 29.A1 by taking averages for the urban and rural studies?

b) Suggest why the percentages of infected people in the urban and the rural areas are different.

Table 29.A2 shows the relationship between the father's occupation and the percentage of children infected with roundworms.

Table 29.A2

Father's occupation	Percentage of children infected with roundworms
professional	9.5
clerical officer	26.7
semi-skilled worker	20.2
trader	32.1
unemployed	51.4

4 Explain the relationship between the fathers' occupations and the percentage of children who are infected with roundworms.

A study was carried out to see how effective treatment for roundworm infections was in primary schoolchildren. Each child was treated three times and, after each treatment, the total number of worms passed out in the faeces was counted (Fig 29.A3).

5 **a)** Describe the effects of treatment on the children in the 5–6 age group.

b) How does age affect the success of the treatment?

Elephantiasis

The effects of the swelling disease elephantiasis is shown in Fig 29.A4. The condition is caused by the nematode worm *Wuchereria bancrofti*. The worm blocks lymph vessels. Infection usually sets in and fluid accumulates until the affected limb is several times its normal size.

Fig 29.A4 The swelling disease elephantiasis is caused by an accumulation of fluid due to an infection of the nematode worm *Wuchereria bancrofti*. The worm blocks lymph vessels and so stops lymph draining away

6 **a)** Explain why blockages in the lymph vessels of the legs would cause fluid to build up.

b) How do you think the worm that causes elephantiasis gets into its human host?

For research and discussion

7 Several other worms cause disease in the developing world. Schistosomes and liver flukes are just two others. Find out how many people in the world are affected by these four parasites. Why are they common in the developing world but virtually unheard of in a temperate developed country such as the UK? Are there parasitic worms that cause problems in more affluent areas of the world?

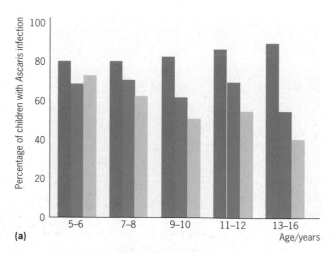

(a)

Fig 29.A3 The percentage of children with *Ascaris lumbricoides*

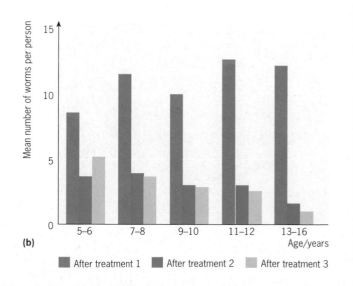

(b)

■ After treatment 1 ■ After treatment 2 ■ After treatment 3

30

Defence against disease

Defence against disease

A false-colour electron micrograph of a mast cell, a type of white cell involved in allergic reactions. The small red vesicles contain histamine, the chemical that produces many of the symptoms of allergy. Many people take antihistamines to reduce their allergy to such things as pollen, house dust or animals

A few years ago, a major medical journal reported the tragic case of a teenage girl who died within 20 minutes of taking a health supplement containing 'royal jelly', the substance made by queen bees. Sudden death in young people usually makes headlines, particularly when it seems to be related to a trivial event such as taking a normally harmless food supplement. But similar deaths have also occurred as a result of a bee sting.

The victims all have in common a very rare and severe allergy. Just as some of us suffer from hay fever in the summer because we over-react to pollen in the air, some people over-react to the 'foreign' proteins produced by bees.

Normally, when someone is stung by a bee, the result is a painful red swelling that disappears again in a couple of days. In someone who is allergic to bee protein, the effects can be catastrophic. As soon as the poison gets into their bloodstream, many of their white cells release large amounts of histamine and other chemicals that cause severe inflammation. Fluid builds up in the tissues and the smooth muscles all contract. The whole body is affected and both blood volume and blood pressure quickly plummet. Sometimes, the airways in the lungs narrow so much that the affected person can no longer breathe. These symptoms, collectively known as anaphylactic shock, can happen very quickly and can be fatal unless immediate treatment with adrenaline and antihistamines is available.

1 INTRODUCING IMMUNITY

We are all surrounded by bacteria, viruses, fungi and other organisms that are capable of invading our bodies and causing disease (Fig 30.1). We are able to overcome infections by these **pathogens** (disease-causing organisms)

Fig 30.1 Spots!

This baby boy has chicken pox, caused by the varicella-zoster virus

This girl has impetigo, a highly infectious bacterial infection

because we have an **immune system**. This is a complex system involving many different cells and tissues that allows us to develop **immunity** – resistance to infections.

Common pathogens include bacteria, fungi, viruses and protoctists. This last group includes microscopic parasites such as Plasmodium, which causes malaria, and larger parasitic animals such as tapeworms. A pathogenic organism is able to:

● break through the physical barriers of the body and enter tissues or cells;
● resist the efforts of the immune system to destroy it, long enough to multiply inside the host's body;
● get out of one host and into another;
● damage the host's tissues – either directly, or indirectly by means of **toxins** (poisons) that it releases. Some bacterial **exotoxins**, such as the one produced by E. coli O157, which has caused epidemics of gastroenteritis in the past few years – see page 2 – are very powerful and can be fatal.

The easiest way to start understanding the immune system is to look at its overall functions, rather than concentrating on the individual parts. Fig 30.2 summarises the main lines of defence that an organism comes up against when it tries to infect a healthy person. We look at each of these in more detail in the sections that follow.

Our study of the immune system covers:

● The ability of the immune system to distinguish between invading organisms and the tissues of its own body. This concept of **self** and **non-self** is important in determining whether the immune system keeps the body healthy, or whether it is overcome by infection.
● Problems of the immune system. In the Science Feature box on page 523, we see how the body can react against its own tissues, causing **autoimmune disease**.
● How medicine today can manipulate the immune system to provide life-saving treatments such as vaccines, blood transfusions, organ and tissue transplants and specific cancer therapy.

? TEST YOURSELF

A What conditions inside our bodies make it ideal for the growth of microorganisms?

✓ REMEMBER THIS

Pathogens are disease-causing organisms. Such diseases are communicable or infectious: they can pass from one person to another. Other diseases, such as diabetes and cancer are non-communicable.

Fig 30.2 An overview of the body's defences: non-specific responses are general responses to damage. They include inflammation and phagocytosis of debris. Specific responses are targeted against individual types of microorganism

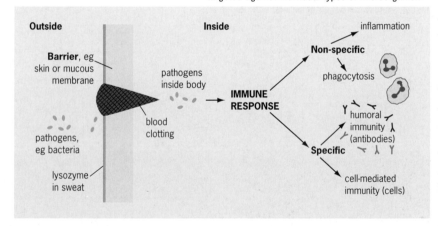

2 THE BODY'S BARRIERS TO INFECTION

One of the most obvious ways to avoid infection is to stop potential pathogens getting into the body in the first place. The four main strategies that the body uses are summarised below:

● **Mechanical** defence. **Nasal hairs** filter the air that is drawn into the upper airways. **Cilia**, which line the airways, sweep bacteria and other particles away from the lungs.
● **Physical** defence. The skin, made from **stratified squamous epithelium** (see Chapter 2), forms a tough, impermeable barrier that normally keeps out bacteria and viruses. The **mucous membranes** that line the entry points to the body such as the nose, eyes, mouth, airways, genital openings and anus produce fluids and/or sticky mucus. These fluids trap microorganisms and stop them attacking the cells underneath.

SEE QUESTIONS 1 AND 2

? TEST YOURSELF

B Smoking leads to paralysis of the cilia that line the airways. Suggest why this causes problems.

- **Chemical** defence. Fluids such as sweat, saliva and tears contain chemicals that create harsh environments for microorganisms. Sweat contains **lactic acid** and the enzyme **lysozyme**, both of which slow down bacterial growth. Stomach acid kills many microorganisms that manage to get that far. When we are injured, blood clots at the injury site, sealing the breach to prevent entry of bacteria (see the next section).
- **Biological** defence. Normally, a vast number of non-pathogenic bacteria live on the skin and mucous membranes. These do not harm the body but they out-compete pathogenic bacteria, preventing them from gaining a foothold from which to launch a full-scale infection.

3 BLOOD CELLS AND DEFENCE AGAINST DISEASE

BLOOD CLOTTING

Whenever blood vessels break, blood leaks out and **clots** (Figs 30.3 and 30.4). We can see this happening when we cut ourselves, but blood can also clot deep inside the body. Clotting enables the body to avoid blood loss and, at the surface, to prevent infection. It is important that blood clots only when it should, because when a blood clot, a **thrombus**, blocks a vital blood vessel, it can cause a fatal heart attack or stroke.

Fig 30.3 Blood clotting is a cascade reaction: a complex set of chemical reactions that occur one after the other

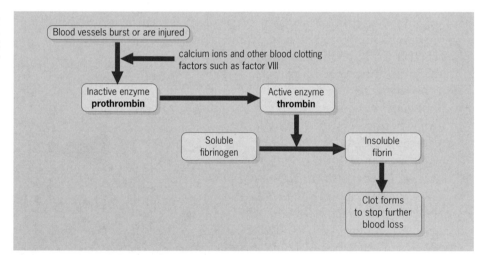

Fig 30.3 outlines the major steps involved in the control of blood clotting. When blood vessels are injured or burst, a **cascade reaction** is initiated. The activation of one molecule leads to the activation of many more. In the final steps of the process, an inactive enzyme, **prothrombin**, is converted to active **thrombin**. This form of the enzyme converts soluble **fibrinogen** into insoluble **fibrin**. The fibrin fibres form a mesh that traps red blood cells and forms a clot (Fig 30.4).

Fig 30.4 A scanning electron micrograph of a blood clot. You can see red blood cells trapped in a fine mesh of fibrin. Blood clotting is a complex process that is set off when blood vessel walls are damaged and finishes when soluble fibrinogen is converted into an insoluble fibrin mesh

If any of the factors in the cascade are missing, the blood cannot clot. This occurs in the condition **haemophilia**, a sex-linked genetic disease in which the sufferer (usually a male) cannot make factor VIII. The genetics of haemophilia are discussed in Chapter 25.

CLASSIFICATION OF WHITE CELLS

Fig 30.5 is a blood smear showing some white cells – **leucocytes** – along with many red blood cells. (The red cells outnumber the white by about 700 to 1.) White cells are made in the bone marrow and are found throughout the body. They can move and are able to squeeze between cells, passing freely in and out of the circulation. Individual types of white cell are classified according to their appearance, origin or function (Table 30.1).

Fig 30.5 In this blood smear, the white cells stand out because their nuclei are stained. The numerous red cells, which have no nuclei, do not take up the purple stain

Table 30.1 Different types of white cell. Neutrophils, eosinophils and basophils have a granular cytoplasm and are therefore called granulocytes. The other types are called agranulocytes

Cell type	Diagram	How to recognise them	Relative abundance %	Function
Neutrophils		lobed nucleus	57	phagocytosis
Lymphocytes		large round nucleus; little cytoplasm	33	specific immunity; B cells make antibodies. T cells involved in cell-mediated immunity
Monocytes		large, kidney-shaped nucleus	6	phagocytosis; monocytes develop into macrophages: general 'rubbish collecting' cells
Eosinophils		stain red with eosin	3.5	associated with allergy
Basophils		stain with basic dye	0.5	release chemicals such as histamine that are responsible for inflammation

REMEMBER THIS

The process of thrombosis causes blood to clot inside vessels. The clot itself – a thrombus – can be fatal if it prevents blood from reaching vital tissues such as heart muscle. A coronary thrombosis is a common cause of heart attack while a thrombosis in the brain can cause a stroke.

? TEST YOURSELF

C The saliva of several species of leech contains an anticoagulant. What are anticoagulants and why does the leech need them? Could 'leech spit' have any medical uses?

SEE QUESTIONS 3 AND 4

? TEST YOURSELF

D Use Table 30.1 to identify the white cells shown in Fig 30.5.

REMEMBER THIS

White cells spend only about 10 per cent of their time in the blood, and so cannot really be called white blood cells. Leucocyte is the general name for all white cells.

4 THE NON-SPECIFIC IMMUNE RESPONSE

When the body is damaged by cuts, scratches or burns, or is attacked by a pathogenic organism that manages to breach its defences, it produces a **non-specific immune response**. It is called a non-specific response because it occurs in response to tissue damage itself, not to the cause of the damage.

INFLAMMATION

Inflammation is a rapid reaction to tissue damage. Whether it is in response to a cut, insect bite or a heavy blow such as a sport injury, the classic signs of inflammation are always the same:

- **Redness**: blood vessels dilate, increasing blood flow to the area.
- **Heat**: also caused by the extra blood flow.
- **Swelling**: the extra blood forces more tissue fluid into damaged tissues.
- **Pain**: swollen tissues press on receptors and nerves. Also, chemicals produced by cells in the area stimulate the nerves.

Inflammation (Fig 30.6) is triggered by damaged cells. Ruptured cells and some white cells (**mast cells** and **basophils**) release 'alarm' chemicals such as **histamine**. These substances dilate blood vessels and the increased blood flow leads to the classic signs of inflammation. The 'alarm' chemicals also attract white cells that remove bacteria and debris by **phagocytosis**.

WHY INFLAMMATION IS USEFUL

Inflammation prevents the spread of infection and speeds up the healing process. It also provides a way of telling the rest of the immune system what is going on. When microorganisms are phagocytosed, fragments of their cells (particularly molecules that were originally on their surface) are processed by the phagocytes. Some of these surface molecules, which we call **antigens**, allow the specific immune system to recognise and remember the type of microorganism that has tried to invade the body.

Antigens stimulate the specific immune system to produce cells and chemicals that bind specifically to that antigen, and to no others. Find out more about specific immunity in the next section of this chapter.

Fig 30.6 The familiar effects of inflammation: reddened skin, swollen tissues and tenderness

✔ REMEMBER THIS

An antigen is a molecule or part of a molecule, for example a protein, that is detected by the specific immune system as 'foreign': not part of the host's body. The immune system responds to an antigen by producing a very specific protein, an antibody, which reacts with that antigen, and that antigen only.

Fig 30.7 This phagocyte (a neutrophil) is engulfing a speck of dust. These white cells occur in the bloodstream and throughout the tissues. Each phagocyte (which is spherical when not active) can ingest between 5 and 25 bacteria before the toxic products of breakdown kill the cell. At sites of infection there may be many dead, liquefied white cells that, together with dead bacteria and other debris, form pus

> ### SCIENCE FEATURE Autoimmune disease: when the body attacks its own tissues
>
> In people who have an autoimmune disease, the mechanism that enables the immune system to tell what is self and what is non-self breaks down. T and B cells (types of white cell) begin to attack the body's own cells and tissues. This is the underlying cause of **multiple sclerosis, insulin-dependent diabetes, myasthenia gravis** and **rheumatoid arthritis**.
>
> In multiple sclerosis, T cells attack the myelin sheath around nerves. This severely limits nerve function, resulting in loss of movement and sometimes in blindness.
>
> In insulin-dependent diabetes, the body makes antibodies that destroy the β cells in the islets of Langerhans in the pancreas. This means that the pancreas becomes unable to produce insulin and the affected person can no longer control their blood glucose.
>
> In myasthenia gravis, the body makes antibodies that attack the motor end plates, the specialised synapses that connect motor nerves to muscles. If the motor end plates become damaged, the muscles cannot contract. The first symptom of this disease, which affects one in 30 000 people (mainly female), is rapid tiring during exertion. The muscles become progressively more unresponsive and the affected person can have difficulty breathing.
>
> People with rheumatoid arthritis suffer from swollen and deformed joints (Fig 30.8).
>
>
>
> **Fig 30.8** This person suffers from rheumatoid arthritis. The cartilage at the joints has been attacked by the immune system, causing swollen and painful joints

PHAGOCYTOSIS

Neutrophils are the commonest type of white cell. Together with monocytes, they are known as **phagocytes** because of their ability to 'eat' pathogens by phagocytosis (Fig 30.7). In this process, the white cell engulfs the pathogen, takes it into a vacuole inside its cytoplasm and then digests it with **lytic** enzymes (see Chapter 4).

5 SPECIFIC IMMUNITY

The specific immune system protects the body from 'invasion' by microorganisms and parasites and also makes sure that the body's defences do not turn on its own tissues. The specific immune response is made up from two different systems that co-operate closely:

- **Humoral immunity**, also called **antibody-mediated immunity**, involves only chemicals: no cells are directly involved. The chemicals, called **antibodies**, attack bacteria and viruses before they get inside body cells. They also react with toxins and other soluble 'foreign' proteins. Antibodies are produced by white cells called **B lymphocytes**, or **B cells**.
- **Cell-mediated immunity**, as the name suggests, involves cells that attack 'foreign' organisms directly. Activated **T lymphocytes**, or **T cells**, kill some microorganisms, but they mostly attack infected body cells. The body uses cell-mediated immunity to deal with multicellular parasites, fungi, cancer cells and, rather unhelpfully, tissue transplants.

HUMORAL AND CELL-MEDIATED IMMUNITY

Fig 30.9 summarises the differences between humoral and cell-mediated immunity and Fig 30.10 shows how antigens from pathogens stimulate antibody production by B cells.

Fig 30.9 Lymphocytes are made in the bone marrow by stem cells. Two distinct types of lymphocytes develop, T lymphocytes and B lymphocytes (T cells and B cells). Both T and B cells migrate from the bone marrow to the lymph glands, but the T cells go via the <u>T</u>hymus while the B cells go straight from the <u>B</u>one marrow

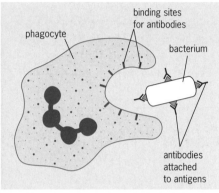

Fig 30.10 Pathogens such as bacteria are covered in antigens, which stimulate B cells to produce antibodies. Many antibodies coat the pathogen, labelling it as foreign to stimulate attack by phagocytes

B CELLS AND HUMORAL IMMUNITY

B cells are produced in the **bone marrow** and are distributed throughout the body in the **lymph nodes**. B cells respond to the 'foreign' antigens of a pathogen by producing specific antibodies. Antibodies are complex proteins that are released into the blood and carried to the site of infection. B cells do not fight pathogens directly.

An antibody, or **immunoglobulin**, is a Y-shaped protein molecule that is made by a B lymphocyte in response to a particular antigen. Fig 30.11 shows the overall structure of an antibody molecule. Antibodies interact with the antigen and render it harmless.

When a pathogen tries to invade the body for the first time, each of its antigens activates one B cell, which divides rapidly to produce a large population of cells. All the new cells are identical (we say they are **clones**) and they all secrete antibodies specific for the invading pathogen. When the infection is over, most of the newly made B cells die: their job is done. We describe this sequence of events as a **primary immune response**.

✔ **REMEMBER THIS**

A clone is a set of genetically identical individuals. In immunology, a clone refers to a population of B cells, all of which can produce the same antibody because they are genetically identical.

So that the body can respond more quickly next time, some of the activated B cells persist in the body for several years. These **memory cells** 'remember' what the pathogen is like and, if it tries to invade again, they divide rapidly to produce an even greater number of active B cells, all capable of secreting specific antibody. This response is called a **secondary immune response** and is very much quicker and more effective than the primary response (Fig 30.12).

This ability of the immune system is central to **vaccination**. A vaccine stimulates the body to produce a primary immune response to a particular pathogen, without becoming infected by it. A subsequent booster produces a secondary response. Later, if the pathogen tries to invade, the body can mount a very fast response and the person does not become ill.

Fig 30.11 (a) Antibodies are Y-shaped molecules consisting of four polypeptide chains, two heavy and two light. Much of the molecule is constant but the tips of the Y are variable and match precisely part of a particular antigen molecule

Fig 30.11 (b) A computer-generated 3-D image of an antibody molecule (green) bound to an antigen (red)

SEE QUESTION 5

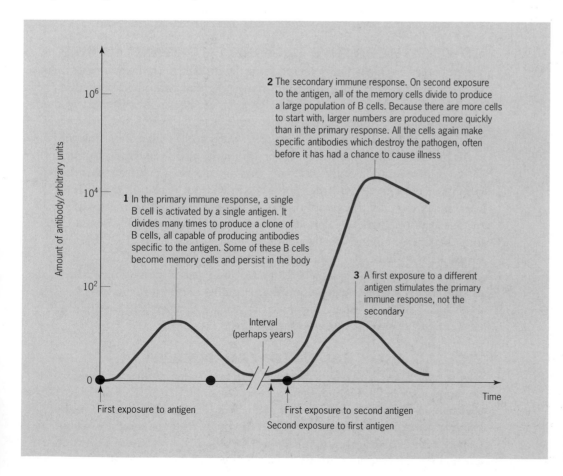

Fig 30.12 The immune system can remember antigens

Fig 30.13 Vaccinations are an effective way of stimulating the body's own defences so that we need not suffer the infectious diseases, such as measles, mumps and whooping cough, that used to be a common feature of childhood

VACCINATIONS

Several infectious diseases overwhelm the normal primary immune response and so can be fatal on first exposure. Thankfully we are able to speed up the specific immune response by giving vaccines against the pathogens that cause them (Fig 30.13). The basic idea behind a vaccine is that it contains some form of the pathogen, so that it stimulates memory cells to develop, ready to destroy the real pathogen should it be encountered. Obviously, the vaccine can't simply be the pathogen itself, or the toxins it makes. Somehow, the vaccine must be made less **virulent** – less able to produce disease. Examples of this are shown in Table 30.2.

Table 30.2 Recommended vaccination schedule in the UK		
Vaccination	**Type of vaccine**	**Age due (UK vaccination schedule)**
Diphtheria	killed organism	2, 3 and 4 months, 3–5 years
Tetanus	modified toxin	2, 3 and 4 months, 3–5 years
Whooping cough	killed organism	2, 3 and 4 months, 3–5 years
Polio	live, non-virulent	2, 3 and 4 months, 3–5 years
Haemophilus influenzae type B (Hib) for meningitis	purified bacterial capsule	2, 3 and 4 months, 3–5 years
Measles	live, non-virulent	12–18 months, 3–5 years
Mumps	live, non-virulent	12–18 months, 3–5 years
Rubella	live non-virulent	12–18 months, 3–5 years
BCG (bacille Calmette–Guérin for tuberculosis)	live, non-virulent	10–14 years
Hepatitis B	genetically engineered antigens	for people at risk, eg health professionals
Meningococcal group C conjugate vaccine	purified bacterial components of *Neisseria meningitidis* group C	2, 3 and 4 months, but also for all young adults (under 25)

HOW MOTHERS GIVE IMMUNITY TO THEIR BABIES

When babies are born, they emerge from the protective environment of their mother's uterus. They are exposed to many potential pathogens. Healthy babies have a fully functional immune system and can mount primary immune responses to many different antigens straight away. However, babies also get a bit of help from their mother, who gives them some of her own immunity.

Antibodies pass across the placenta and, after birth, the supply continues through breast milk. Babies have very porous intestines that can absorb these large proteins directly into the bloodstream without digesting them. These large pores close by the age of one year. We call this kind of immunity – passed from one person to another – **passive immunity**. It does not last long, because the antibodies are broken down within a few days, but it can help a baby to fight off common pathogens.

Passive immunity is also used to treat some types of poisoning, such as snakebites. **Antiserum**, blood that contains antibodies specific to a particular snake venom, is produced in horses, purified and then given to people who have been bitten by a snake.

T CELLS AND CELL-MEDIATED IMMUNITY

Like B cells, T cells respond to specific antigens. When a pathogen first infects the body, each individual antigen stimulates a single T cell. This divides to form a **clone**, in the same way that B cells do. Some of the **activated T cells** become **memory cells** and persist in the body, ready to mount a secondary response if the pathogen attacks again. The others, however, do not produce antibodies. They develop further to become one of three types of T cell:

- **Helper T cells** are so called because they help with, or rather control, the rest of the specific immune response. They tell B cells to divide and then to produce antibodies, they activate the two other sorts of T cell (see below), and they activate macrophages, telling them to get ready to phagocytose pathogens and debris.
- **Killer T cells** attack infected body cells and the cells of some larger pathogens (eg parasites) directly. The two cells face each other, membrane-to-membrane, and the killer T cell punches holes in its opponent. The infected cell or parasite loses cytoplasm and dies.
- **Suppressor T cells** are a sort of safety cut-out mechanism. When the immune response becomes excessive, or when the infection has been dealt with successfully, these T cells damp down the immune response. Obviously, this is a good idea: if the body continued to make antibody and stimulate more and more T and B cells to divide, even when there was no need, this could damage the body and would be, at best, a waste of resources.

ALLERGIES

An **allergy** or **hypersensitivity** is an over-reaction to the presence of a normally harmless substance called an **allergen**. Some common allergens are pollen, house dust mites, animal fur and feathers, fungal spores, insect bites and penicillin.

The commonest symptoms of an allergy are sore eyes, runny nose, sneezing and asthma. Many of these symptoms result from an inflammation of the mucous membranes, caused by **mast cells**, which release chemicals such as histamine. Many anti-allergy treatments suppress mast cells or neutralise histamine – chemists sell many **antihistamines** in the pollen season. For more information about asthma see the Assignment in Chapter 9.

6 BLOOD TRANSFUSIONS AND ORGAN TRANSPLANTS

In the past 100 years, advances in modern medicine have led to the technology and knowledge that allow doctors to give one person's cells, tissues and organs to another person. One of the main objectives in developing these life-saving treatments has been to overcome the natural reaction of the immune system to destroy transplanted cells and tissue, which it 'sees' as non-self.

BLOOD GROUPING AND TRANSFUSIONS

Towards the end of the nineteenth century, medical scientists realised that accidents were often fatal simply due to blood loss. Losing large volumes of blood sent victims into shock and they died, even though their injuries were otherwise not too severe. Many women, in particular, often died when they lost blood when giving birth. Different doctors tried transfusing blood from a healthy person into the injured person, to try to restore their blood volume.

Sometimes this worked and sometimes it didn't. When it failed, the results were disastrous. We now know that if the **blood groups** of the **donor** (the person giving the blood) and the **recipient** (the person receiving it) are different, the recipient's immune system reacts against the donated blood, producing a massive and deadly immune response.

In the early 1900s, an Austrian scientist, Karl Landsteiner, discovered that red blood cells from different people had different sets of antigens on their surface. The entire human population can be placed into one of four main blood groups: **A**, **B**, **AB** and **O**, according to the antigens on their red blood cells.

Tetracycline is bacteriostatic. It binds to bacterial ribosomes, preventing RNA attachment

Chloramphenicol is bacteriostatic. It prevents transfer of amino acids to ribosomes

amino acids

DNA → RNA ✗ ribosomes

proteins

Erythromycin is bacteriostatic. It binds to bacterial ribosomes and prevents translation

Penicillin is bactericidal. It stops the formation of cross-bridges in peptidoglycan, an important molecule in bacterial cell walls. This weakens the cell wall. The bacteria absorb water and burst

HOW ANTIBIOTICS WORK

Antibiotics target processes such as protein synthesis that are subtly different in microorganisms compared with the cells of humans and other animals. Different antibiotics have different modes of action (Fig 30.16).

Fig 30.16 Some antibiotics kill the microorganisms they target. These are said to be bactericidal. Bacteriostatic antibiotics do not kill microorganisms; they prevent them multiplying and give the immune system a better chance to overcome the infection

---

⚠ SCIENCE FEATURE Producing penicillin on a large scale

Penicillin is produced commercially using a strain of fungus called *Penicillium chrysogenum*. This is an aerobic organism that needs a high level of oxygen to grow in culture. For this reason, it is grown commercially in fermenters with a relatively small volume – 40–200 dm³. These are easier to

Fig 30.17 Commercial production of penicillin. The process involves several stages; scaling up of the original fungal culture, fermentation, batch fill and draw, filtration, and then three steps of downstream processing. The final product is 99.5 per cent pure

aerate effectively than much larger fermentation tanks. (For general information about commercial fermentation, see Chapter 6.)

The commercial production of penicillin is shown in Fig 30.17. The fungus produces most penicillin at temperatures between 25 and 27 °C. During the first 40 hours or so, the fungus is growing in the culture and increasing in biomass. Once it has reached a certain concentration in the culture, it starts to produce large amounts of penicillin, but this continues

for only a short time before the fungus starts to run out of nutrients and begins to die. To get the most out of each fermentation, new nutrients are added after 40 hours to stimulate a further increase in biomass. 20–40 per cent of the culture is removed, to be processed so the penicillin can be purified, and new culture medium replaces it. This process, called **batch fill and draw**, can be carried out up to ten times before the fermentation tank needs to be cleaned out completely.

It has been said that the most dangerous place for you if you are sick is in hospital! This, of course, is a great overstatement, but many bacteria that exist in hospitals are resistant to common antibiotics, and the problem is getting worse.

Antibiotics were undoubtedly one of the major medical advances of the twentieth century. They saved the lives of millions of people. But nothing is perfect and there is no such thing as a wonder drug. Since the 1940s when antibiotics were first used widely, resistant strains of bacteria have arisen that can cause infections that are very difficult to treat. How did this happen?

Bacteria, like all living organisms, vary. In any population of bacteria, the vast majority can be killed by an antibiotic such as penicillin. However, there might be one that has the gene that allows it to produce the enzyme penicillinase. This enzyme breaks up penicillin molecules and bacteria that produce it are resistant to the action of penicillin. If such a penicillin-resistant bug can evade the body's defences long enough to reproduce, in a day or two – bang – a new infection that is even more difficult to treat.

Penicillin resistance has become common in the past few years and bacteria have also become resistant to several other antibiotics. We now recognise that over-prescribing has played a role, and doctors are now much more reluctant to give antibiotics for every cough and cold. Medicine now faces the challenge of staying one step ahead to develop new antibiotics faster than bacteria evolve resistance.

Fig 30.18 Cells of this highly antibiotic-resistant strain of the bacterium *Staphylococcus aureus* are seen dividing. In hospitals, it infects the wounds of patients, causes internal abscesses and produces boils by entering the skin through hair follicles

SUMMARY

After studying this chapter, you should know and understand the following:

- We are surrounded by potential **pathogens** (disease-causing organisms) such as bacteria and viruses which can cause disease if allowed to enter and multiply inside the body.

- The body's defence system consists of barriers to keep microorganisms out, and mechanisms to detect and destroy those that do enter. Together, these mechanisms are the **immune response**.

- The immune system is capable of non-specific responses and specific responses. Non-specific mechanisms (**inflammation** and **phagocytosis**) occur in response to any invading microorganism. Specific mechanisms allow the body to recognise and fight individual types of microorganism. There are two specific responses: **cell-mediated immunity** and **humoral immunity**.

- **Lymphocytes** (white cells) are responsible for the specific immune response. **T cells** mature in the thymus gland and **B cells** come directly from bone marrow. Both are able to recognise foreign antigens that come from pathogens.

- B cells are responsible for humoral immunity: they secrete specific proteins called **antibodies**.

- T cells are responsible for cell-mediated immunity and help to control the overall specific response. **Killer T cells** attack pathogens or infected cells. **Helper T cells** activate B cells, telling them to secrete antibodies. **Suppressor T cells** damp down the immune response when the infection is over.

- **Auto-immune diseases** occur when the body's immune system attacks its own tissues. Examples include rheumatoid arthritis, multiple sclerosis and myasthenia gravis. **Allergies** occur when the body 'over-reacts' to a normally harmless substance such as pollen.

- When cells or tissues are taken from one individual and given to another, they are often recognised as 'foreign' and destroyed. This process is often called **graft rejection**. To minimise the chance of rejection in blood transfusions and organ transplants, it is important to match blood and tissue types.

- **Antibiotics** are important drugs for treating infections caused by bacteria. In the time since antibiotics were first developed and used in the 1940s, bacteria have evolved to become resistant to them. Antibiotic-resistant strains of many bacteria are becoming common and represent a problem for doctors and researchers.

❓ EXAM QUESTIONS

1 Outline the role of each of the following in body defence against disease.

a) skin

b) mucous membranes

c) neutrophils

d) T cells and B cells

e) proteins

OCR 6915 Health and Disease March 2001 Q2

2

a) Explain how the defence mechanisms of the body reduce the chance of entry by a pathogen.

b) Explain how the body responds both generally and specifically to pathogens that enter the blood.

AQA (NEAB) BY08 Health and Disease June 2000 Q7

3 The diagram shows one way in which white blood cells protect the body against disease.

a) Describe what is happening during **Stage 1.**

b) **(i)** Name Organelle **A**.

(ii) Describe the role of Organelle **A** in the defence against disease.

AQA (A) BYA3 Pathogens and Disease January 2001 Q1

4 B lymphocytes are found throughout the body in the blood and in the lymphatic system. When an antigen enters the body, some of these B lymphocytes respond and change into plasma cells that secrete antibodies.

a) State the part of the body where B lymphocytes originate.

b) Explain the difference between an antigen and an antibody.

The drawing shows a plasma cell and a B lymphocyte.

7 μm

plasma cell

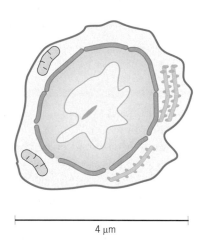

4 μm

B lymphocyte

c) With reference to the drawing,

(i) state **three** ways in which the structure of the plasma cell differs from the B lymphocyte;

(ii) explain the reasons for the differences you have described.

d) Explain why the production of antibodies is much faster when the same antigen enters the body on a second occasion.

OCR 2802 Health and Disease January 2001 Q2

? EXAM QUESTIONS

5 MMR is the combined vaccine used against measles, mumps and rubella. It contains attenuated microorganisms.

a) What is an attenuated microorganism?

b) Vaccines protect against disease by stimulating the production of memory cells. Describe how memory cells protect the body from disease.

c) The graph shows the number of reported cases of whooping cough during the period 1950 to 1975.

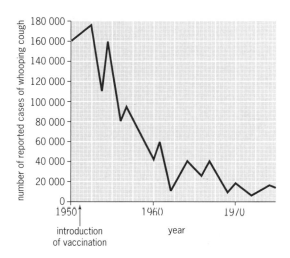

Describe and explain what the graph shows about the number of reported cases of whooping cough during the period 1952 to 1960.

d) The number of reported cases of whooping cough increased during the 1980s. Suggest **one** reason why.

AQA (A) BYA3 Pathogens and Disease January 2001 Q8

6 Three patients were tested to determine their blood groups. Samples of their blood were mixed individually with anti-A and anti-B antibodies. The results are shown in the diagram.

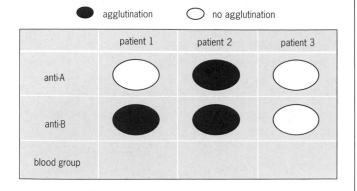

a) Complete the bottom row of the table to show the blood group of each patient.

b) **(i)** When blood from **Patient 1** was mixed with anti-B antibodies, agglutination occurred. Explain why this happened.

(ii) **Patient 1** required a blood transfusion. Which of the other two patients could safely give him blood? Explain your answer.

c) Genetic fingerprinting is more useful than ABO blood groups in identifying individuals from blood left at the scene of a crime. Explain why.

AQA (A) BYA2 Making Use of Biology January 2001 Q2

7

a) Define what is meant by the term antibiotic.

b) Give **two** examples of antibiotics, and for each, suggest its mode of action.

c) Suggest **three** problems caused by the use of antibiotics.

d) What contribution did Fleming, Florey and Cheyne (Chain) make to the development and use of antibiotics?

OCR 6915 Health and Disease March 2001 Q3

KEY SKILLS **ASSIGNMENT**

AIDS AND THE HIV VIRUS

What is AIDS?

Since its discovery in the early 1980s, few scientific topics have been as controversial as AIDS. Short for Acquired Immune Deficiency Syndrome, AIDS arises when the immune system ceases to function effectively. It is caused by the human immunodeficiency virus (HIV), shown in Fig 30.A1, which infects and destroys vital cells of the immune system, notably the helper T lymphocytes.

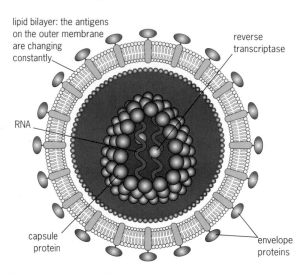

Fig 30.A1 The structure of the human immunodeficiency virus

1 HIV and AIDS are the same thing. True or false? Explain.

When the HIV enters the body, it can take many years before AIDS develops. Some people take well over 10 years to develop full-blown AIDS and some people have been infected since the mid-1980s and have not as yet developed the syndrome. Perhaps they never will.

Telltale signs that the immune system is no longer effective are the appearance of diseases that would normally cause no trouble. These include Kaposi's sarcoma (a type of skin cancer), fungal infections (eg thrush), bacterial infections (eg tuberculosis) and viral infections (eg herpes).

2 Outline briefly how viruses replicate.

HIV is a retrovirus, and contains the enzyme reverse transcriptase. Once the virus infects a cell, this enzyme allows the cell to make viral DNA from viral RNA. The foreign DNA is then inserted into the cell's own DNA where it acts as a gene, causing the manufacture of more viral RNA and, subsequently, many thousands of new viruses which eventually burst out of the cell and infect others.

3 Outline the essential difference between retroviruses and other viruses.

HIVs primarily affect a group of lymphocytes called helper T cells. These cells normally activate other lymphocytes, including B cells.

4 **a)** What is the function of B cells?
b) What will happen if B cells cannot become activated?

The transmission of HIV

Accurate knowledge of how the virus is transmitted has become a matter of basic survival for many people. Infection occurs when the virus is transferred from person to person, via body fluids. This usually occurs during sexual intercourse. The risk of someone becoming infected through sexual activity depends on the number of partners they have, the proportion of infected partners and the nature of their sexual activity.

5 What is the easiest and most usual way that a sexually active person can avoid HIV infection?

Prevention or cure?

Treating HIV infection is difficult. The viral DNA finds its way into cells in the brain, intestine and retina as well as white cells, and every time the cells divide, the viral DNA is copied. Much research is under way to develop an AIDS vaccine, but efforts have been thwarted so far by the virus's remarkable ability to vary the shape of the chemicals in its outer coat.

6 Suggest why it is difficult to make a vaccine for HIV.

Scientists agree that the most hopeful line of attack is to control the replication of HIV in the body and therefore delay the development of AIDS, perhaps indefinitely. However, this is not yet possible. The best we can do at the moment is to delay the onset of full-blown AIDS by a combination of treatments.

7 Discuss or write down details of any AIDS awareness campaigns you have seen. Which approaches do you feel are the most effective? What approach would you take if you were designing a campaign for young people?

Fig 30.A2 An AIDS information poster in Kenya

31

Lifestyle diseases

Lifestyle diseases

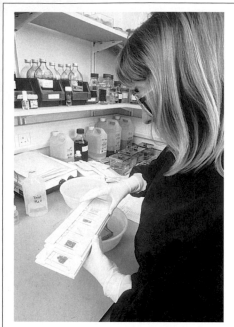

A technician preparing microscope slides of human brain slices to be checked for new variant CJD. CJD is caused by something called a prion – this is a self-replicating molecule that is a protein rather than a nucleic acid

In 1996, a handful of deaths were linked with a type of dementia called Creutzfeldt–Jacob disease (CJD). CJD had been known for a long time, but, with an average age of 27, these victims were much younger than the people who usually developed CJD. All the signs pointed to a new variant form.

When the UK government announced there was a chance that new variant CJD had been passed to people in the beef they ate, there was an incredible furore. Bovine spongiform encephalopathy (BSE), the bovine form of CJD, was known to be widespread in the British herd. Was everyone who ate beef regularly at risk of new variant CJD? The resulting debate demonstrates how difficult it is to pinpoint a lifestyle risk factor for a disease that takes several years to develop.

BSE was first identified in cattle in 1986, and scientists concluded that this bovine form had developed from a similar disease in sheep, called scrapie. Why the scrapie agent jumped species isn't really understood. But the practice of recycling slaughtered sheep and, later, infected cows, to make protein supplements for cows, helped to establish a massive infection of BSE in British cattle. Many were culled to stem the epidemic and to lessen the possibility of infected meat entering the human food chain.

By the end of 1996, scientists had analysed the chemistry of the agent that causes new variant CJD and showed that it was closely related to BSE in cattle. Research since then has demonstrated that some of the affected people definitely contracted new variant CJD from eating beef. New regulations have been introduced to prevent such transmission happening again but the long incubation times of CJD make it impossible to be sure that more people are not infected. In the next couple of years it will be possible to test someone's blood to see if they are carrying new variant CJD and a decision may be made to test all blood donated in the UK. Until then, however, we just don't know if we are still at risk of a large-scale CJD epidemic.

1 WHAT ARE LIFESTYLE DISEASES?

In this chapter, we look at a group of diseases influenced by the sort of life we lead. Such **non-transmissible diseases** or **lifestyle diseases** are the major cause of death in the more developed countries (Fig 31.1).

Cardiovascular disease, which affects the heart and circulatory system, many types of **cancer** and some diseases of the respiratory system can all affect people as a result of things they eat, drink or smoke, and the way they live. Getting enough exercise at all times of life, for example, can help to reduce the risk of developing cardiovascular disease.

In this chapter, our study of lifestyle diseases includes:

- The aspects of lifestyle that are linked with disease: smoking, alcohol, poor diet and lack of exercise. These factors affect a person's chance of developing a particular condition, so they are often called **risk factors**.
- Cardiovascular diseases. In their most severe form, heart attacks and strokes can be fatal. Less severe forms of cardiovascular disease don't kill but they disable people to such an extent that their **quality of life** is very poor.
- Cancers. A cancer is a mass of cells that multiply in an uncontrolled way, invading surrounding healthy tissue. Often cancers form tumours, or swellings.

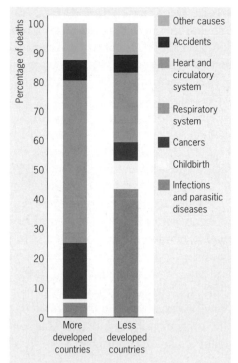

Fig 31.1 This bar chart shows the percentage of deaths from a number of different causes in more developed and less developed countries

TEST YOURSELF

A Suggest why a higher percentage of people die from cancers in more developed countries.

RISK FACTORS FOR LIFESTYLE DISEASES

Some bacteria are responsible for specific forms of heart disease and some viruses are associated with particular types of cancer, but the main causes of non-transmissible disease are features of our lifestyle. A risk factor is an aspect of our lifestyle that increases our chances of developing a disease. Table 31.1 shows some of the most important ones.

Table 31.1 Some examples of risk factors	
Factor	**Level of risk**
Smoking	There is good evidence that cigarette smoking is directly responsible for about 90 per cent of all cases of lung cancer. Smoking also makes us more likely to succumb to diseases such as bronchitis and emphysema and it increases the risk of developing cancers in other parts of the body.
Drinking alcohol	Regular drinking of small amounts of alcohol could have a beneficial effect on the circulatory system. However, there is plenty of evidence that shows that regular heavy drinking is linked to an increased risk of developing cancer of the mouth, throat, oesophagus and liver. Between 2 and 4 per cent of all cancers are thought to be due directly to alcohol consumption.
Exposure to ultraviolet light	Sunbathing is dangerous! Over the past 25 years, the incidence of skin cancers has doubled. Most of the increase is thought to be due to greater exposure to ultraviolet rays in sunlight (Fig 31.2). Pale-skinned people can burn easily and a handful of bad episodes of sunburn in childhood can double your chances of getting skin cancer later in life. Also, experts now believe that the protective ozone layer in the upper atmosphere, which acts as a barrier to the dangerous UV rays in sunlight, has been seriously eroded by pollution.
What we eat	The links between diet and heart disease and between diet and cancer are more complex. Some foods are associated with greater risk of developing some diseases. A high fat diet, for example, seems to increase risk of developing cardiovascular disease, but other environmental, lifestyle and genetic factors are also important, and teasing out which factors have the greatest effect is very difficult.

Fig 31.2 The 'pale and interesting' look went out of favour in the early part of the twentieth century and tanned skin was considered healthy. However, in the last 25 years of the twentieth century, cases of skin cancer rose dramatically and the tanned body became less fashionable. Beach holidays are still popular as we move through the first decade of the twenty-first century, but there are signs that this sort of over-tanned look is losing favour. Perhaps 'pale and interesting' is on its way back

TEST YOURSELF

B Suggest two reasons that could explain why skin cancer is more common today than it was in 1950.

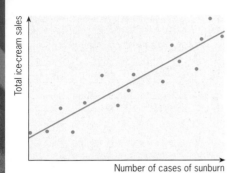

Fig 31.3 Ice-cream sales plotted against number of cases of sunburn shows a correlation. But does it prove that ice-cream causes sunburn?

INVESTIGATING RISK FACTORS

Proving that a risk factor causes a specific disease is not easy. Sometimes two things seem to be linked or **correlated**, but this does not mean that one causes the other. Fig 31.3 shows what happens if we use data from a British seaside town in summer and plot the number of cases of sunburn against ice-cream sales.

The number of cases of sunburn seems to increase with the number of ice-creams sold: there is a clear positive correlation between the two sets of data. However, we all know that eating ice-cream does not cause sunburn. A third factor, the hot weather, is involved. When it's hot, people are more likely to lie in the sun and get sunburnt, and they are also more likely to buy ice-cream.

Medical researchers try to identify risk factors so that people have reliable information to know what to avoid (whether they choose to or not is another matter). Pinning down risk factors to particular diseases is very difficult because we are continually altering our lifestyles. We might eat a good diet and take regular exercise in our teens and then fall into bad habits in our 20s. Our exposure to environmental factors changes, often beyond our control. New preservatives could be added to food we eat, or there can be changes in the levels of pollution that contaminate the air we breathe. Any of these changes can alter our chances of developing a particular disease.

2 ATHEROSCLEROSIS AND CARDIOVASCULAR DISEASE

Cardiovascular disease is a major lifestyle disease in most developed countries. It claims about 175 000 lives per year in the UK and almost a million lives per year in the USA. As its name suggests it involves the **cardiovascular system** – the heart and blood vessels that transport blood throughout the body (see Chapter 10).

Cardiovascular disease is more common in people over 50, but it can happen in younger people. It shows itself in many ways but it has a common underlying physiological cause: the narrowing and clogging up of arteries as a result of **atherosclerosis**. We look at this process in detail later in the chapter, but first we look at the different forms of cardiovascular disease.

TYPES OF CARDIOVASCULAR DISEASE

Cardiovascular disease is a general term. The symptoms and effects of different forms depend on which blood vessel is affected. The severity of the condition depends on the extent to which the vessel is blocked. Generally, partial blockages in small arteries or veins are less life-threatening than complete blockage of a major vessel. A summary of different forms of cardiovascular disease is given below and illustrated in Fig 31.4.

A HEART ATTACK

A heart attack, or **myocardial infarct**, happens when the blood supply to the heart muscle is interrupted. The cause of the interruption is usually a blockage of one of the coronary arteries. This can be due to a blood clot that has travelled to the heart from elsewhere in the body, or it can be due to atherosclerosis. When blood cannot get oxygen to the heart muscle, the affected part of the heart dies. A heart attack causes severe pain in the chest and left arm. Mild heart attacks are dangerous but do not kill; careful treatment using drugs to 'thin' the blood to prevent another blockage, followed by a change in lifestyle can help to prevent another attack.

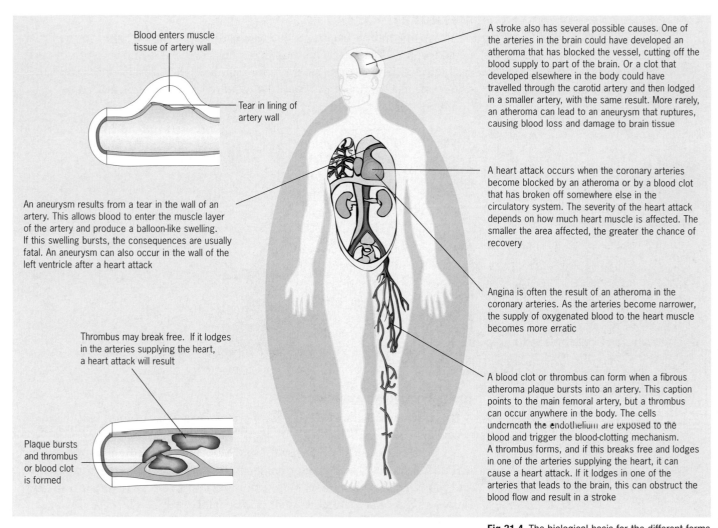

Blood enters muscle tissue of artery wall

Tear in lining of artery wall

A stroke also has several possible causes. One of the arteries in the brain could have developed an atheroma that has blocked the vessel, cutting off the blood supply to part of the brain. Or a clot that developed elsewhere in the body could have travelled through the carotid artery and then lodged in a smaller artery, with the same result. More rarely, an atheroma can lead to an aneurysm that ruptures, causing blood loss and damage to brain tissue

An aneurysm results from a tear in the wall of an artery. This allows blood to enter the muscle layer of the artery and produce a balloon-like swelling. If this swelling bursts, the consequences are usually fatal. An aneurysm can also occur in the wall of the left ventricle after a heart attack

A heart attack occurs when the coronary arteries become blocked by an atheroma or by a blood clot that has broken off somewhere else in the circulatory system. The severity of the heart attack depends on how much heart muscle is affected. The smaller the area affected, the greater the chance of recovery

Angina is often the result of an atheroma in the coronary arteries. As the arteries become narrower, the supply of oxygenated blood to the heart muscle becomes more erratic

Thrombus may break free. If it lodges in the arteries supplying the heart, a heart attack will result

A blood clot or thrombus can form when a fibrous atheroma plaque bursts into an artery. This caption points to the main femoral artery, but a thrombus can occur anywhere in the body. The cells underneath the endothelium are exposed to the blood and trigger the blood-clotting mechanism. A thrombus forms, and if this breaks free and lodges in one of the arteries supplying the heart, it can cause a heart attack. If it lodges in one of the arteries that leads to the brain, this can obstruct the blood flow and result in a stroke

Plaque bursts and thrombus or blood clot is formed

Fig 31.4 The biological basis for the different forms of cardiovascular disease

If the heart attack is severe, and a large part of the heart muscle is affected, the heart can experience **ventricular fibrillation**: instead of co-ordinated beating, the individual muscle fibres contract in an unsynchronised way. The affected part, usually the left ventricle, sometimes becomes unable to pump blood and the heart stops. People who have had severe heart attacks can be **resuscitated** using electric shocks delivered to the chest (Fig 31.5); this restarts the heart and enables it to re-establish a normal beating rhythm.

Fig 31.5 Paramedics resuscitating a heart-attack victim. When someone has a major heart-attack, prompt emergency treatment is necessary, first to save their life and, secondly, to to give them a good chance of recovery.

A STROKE

A stroke, or cerebral haemorrhage, results if the supply of blood to the brain is interrupted. The symptoms vary according to the size and location of the area affected. There might be no more than a tingling sensation or weakness down one side of the body, but at its most severe, a stroke can lead to permanent paralysis (Fig 31.6) or death.

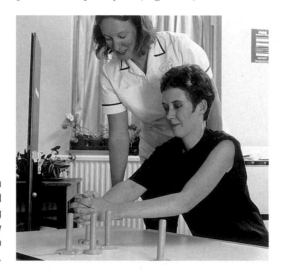

Fig 31.6 One-sided paralysis can result from a stroke. Here a 29-year-old woman, who had a stroke in the right side of her brain, is being encouraged to pick up and move each peg by gripping it between both hands. This is to help her regain some use of her left arm.

OTHER CONDITIONS

A **thrombosis** is the blockage of an artery caused by a blood clot. It can occur anywhere in the body.

An **embolism** occurs when a mobile blood clot (called an embolus, hence embolism) moves through the circulatory system and then lodges in a vein and blocks it. If a pulmonary vein (from lungs to heart) becomes blocked, this causes a pulmonary embolism that can be fatal.

Angina pectoris is a severe pain in the chest that occurs when the coronary arteries are unable to supply sufficient oxygen to the heart muscles.

Thrombophlebitis occurs when a vein is damaged and develops a blood clot that blocks it. Blood clots in the leg veins can occur in patients who are forced to remain in bed for long periods.

THE LINK BETWEEN ATHEROSCLEROSIS AND CARDIOVASCULAR DISEASE

To understand how and why cardiovascular disease develops, we need to look at three things:

● How does an atheroma form?
● How does atherosclerosis lead to cardiovascular disease?
● What are the risk factors associated with atherosclerosis and how do they increase the chance of atheroma formation?

THE FORMATION OF AN ATHEROMA

Atheromas, areas or **plaques** of fatty material, develop on the inside of arterial walls. Remind yourself of the structure of an artery wall by looking at Fig 31.7. As plaques build up, they can affect the circulatory system by:

● Blocking the flow of blood through the artery.
● Initiating the formation of blood clots. Blood clots that form inside coronary arteries are particularly dangerous. They can break free and lodge in the smaller coronary arteries, causing some of the heart muscle to die. This is what initiates a heart attack. If blood clots interfere with the blood supply to the brain, the result is a stroke.

tunica adventitia containing collagen fibres

tunica media containing smooth muscle and elastic fibres

endothelium

lumen

Fig 31.7 Cross section through a large artery. The space through which the blood flows is called the **lumen**. Surrounding this is a layer of smooth flat cells called the **endothelium**. Outside this lining is the **middle layer** or **tunica media**, which contains a high proportion of smooth muscle and elastic fibres. Finally, there is an **outer layer** or **tunica adventitia** containing fibres of the structural protein collagen

---

! SCIENCE FEATURE Heart failure

The term heart failure suggests a condition in which the heart suddenly comes to a complete stop. This doesn't happen – heart failure usually develops slowly, often over years, as the heart gradually loses its pumping ability and works less efficiently. The underlying cause is usually coronary artery disease.

How serious the condition is depends on how much pumping capacity the heart has lost. Nearly everyone loses some pumping capacity as they get older but the loss is significantly more in heart failure.

There are two main types of heart failure:

● Systolic heart failure occurs when the heart's ability to contract decreases. The heart cannot pump with the force required to push enough blood out through the aorta. Blood coming into the heart from the lungs can back up and cause fluid to leak into the lungs, a condition known as **pulmonary congestion**.

Fig 31.8 This person has a bad case of oedema

● Diastolic heart failure occurs when the heart has a problem relaxing. The heart can't fill with blood because the muscles of the atria and ventricles have become stiff. People with this form of heart failure accumulate fluid, especially in the legs, ankles and feet. Their condition is called **oedema** (Fig 31.8).

Tiredness, breathlessness and excess fluid retention are common symptoms but because heart failure usually develops slowly, these obvious signs don't often appear until the damage to the heart is serious. This is partly because the heart adjusts to cope with the effects of heart failure by getting bigger, developing thicker muscle and by contracting more often.

Heart failure is treated by a combination of lifestyle changes and drug treatment. Patients are advised to stop smoking, lose weight if necessary, avoid drinking alcohol, and change their diet to reduce the amount of salt and fat they eat. Regular, moderate exercise is also helpful, but this needs to be done only with a doctor's approval.

However, heart failure is always a serious condition because it poses an increased risk of sudden death, through cardiac arrest. Drugs are usually used to control the symptoms and to try to prevent them getting worse.

● Diuretics help reduce the amount of fluid in the body and are useful for patients with fluid retention and high blood pressure.

● The drug digitalis increases the force of the heart's contractions, helping to improve circulation.

● Recent studies have shown that drugs known as angiotensin-converting enzyme (ACE) inhibitors can be useful. ACE inhibitors improve survival among heart failure patients and can slow, or perhaps even prevent, the loss of heart-pumping activity.

Arterial blood supplies all the body's organs with nutrients, including triglycerides and cholesterol. These are lipids and have several important functions in the body: they form cell membranes and they act as energy stores. Lipids combine with proteins to form **lipoproteins** before they can be transported in the blood.

Lipoproteins that contain cholesterol are engulfed by white cells called **phagocytes** that occur in the artery wall, just under the endothelium. If the blood contains large amounts of cholesterol-rich lipoproteins, large numbers of stuffed phagocytes accumulate. Their rich 'filling' gives them a foamy appearance and they are, in fact, often referred to as **foam cells**. When patches or streaks of foam cells start to appear on the inside of artery walls, this is recognised as the first sign of atherosclerosis. At this stage, there is little effect on the flow of blood through the artery and no discernible symptoms.

Fig 31.9 Areas of fatty material called atheromas build up in the walls of arteries. This is one of the first signs of cardiovascular disease. Eventually it can almost block the artery

SEE QUESTIONS 1 AND 2

? TEST YOURSELF

D Explain why an exercising muscle needs more glucose.

HOW DOES ATHEROMA LEAD TO CARDIOVASCULAR DISEASE?

Gradually, usually over many years, muscle cells and fibres begin to grow over the affected patches, giving rise to a **fibrous plaque**. This bulges into the lumen of the artery and starts to obstruct the flow of blood (Fig 31.9). This can have one of several effects:

● The artery can become very narrow. Depending on the position of the narrowed artery, the oxygen supply to vital parts of the body can be reduced. If, for example, the arteries feeding the heart muscle are affected, the result can be **angina**.

● The blood vessel can become completely blocked. Blockage of an artery that takes oxygenated blood to a vital organ can lead to many different problems. If the coronary arteries become blocked, the result can be a heart attack. If arteries in the brain are affected, this can lead to a stroke.

● The blood vessel can burst. A burst artery, or **aneurysm**, inside the body is very serious. If one of the main arteries near the heart burst, this is usually fatal.

● The plaque can burst into the blood vessel, damaging the endothelium. When the cells underneath come into contact with blood, this triggers the blood-clotting mechanism. A blood clot or thrombus forms in the vessel. If this breaks free, it can travel around the body and block an artery or a vein, causing an embolism or a thrombus.

RISK FACTORS ASSOCIATED WITH ATHEROSCLEROSIS AND CARDIOVASCULAR DISEASE

As we have seen, the underlying process of atherosclerosis can have many different effects, depending on which part of the body is involved. The risk factors for **cardiovascular diseases** (CVDs) are also complex because many different factors affect the way this whole body system functions. We shall look at the three main ones that have been identified.

HYPERTENSION

Blood vessels are more than a series of pipes that take blood to and from different organs. They react to the changing needs of the body supplying, for example, more blood to the muscles during exercise. When we make our muscles work hard, they need more oxygen and more glucose and they also produce more carbon dioxide. The graph in Fig 31.10 shows how the supply of blood to different organs changes as we do different activities.

Think about an exercising muscle. How can it get more blood? Although blood vessels are flexible, they don't just expand and let more blood through. If this were to happen, the blood would just flow more slowly and no extra oxygen would actually reach the muscle. An increase in blood pressure is needed. This speeds

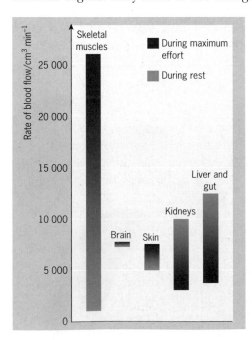

Fig 31.10 The blood flow to different organs in the body changes as we exercise. The flow to muscles and the skin increases. The flow to the brain stays the same. The blood supply to organs like the kidneys and the gut decreases

up the blood flow so that more blood reaches the muscle in a given time, supplying more oxygen and glucose to muscle cells and removing waste products more rapidly. Contraction of the smooth muscle in the walls of the body's arteries helps to bring about this increase in pressure.

High blood pressure is perfectly normal as long as it results from short-term fluctuations in blood flow. Problems start when the blood pressure stays higher than normal while someone is at rest. This person is suffering from high blood pressure, or **hypertension**.

Hypertension sets up a sequence of events that becomes a vicious circle. This is summarised in Fig 31.11. Smooth muscle in the artery walls responds to an increase in blood pressure in the same way that skeletal muscle responds to extra exercise – it increases in size. This results in a thickening of the artery wall and a narrowing of the lumen. The volume of blood passing through the arteries remains the same, causing the blood pressure to rise further. At higher pressure, the blood also exerts a greater force on the endothelial cells that line the artery and these cells are more likely to become damaged. Small flaws in the lining of the artery increase the risk that an atheroma will form. This, in turn, makes it more likely that a thrombosis will develop.

Two things increase the risk of hypertension. A high salt intake is one. If you increase the salt concentration of your blood, its water potential falls. A lower water potential means that blood volume increases as water is drawn from the cells of the body into the plasma. The higher the volume of the blood, the greater its pressure. The other is smoking.

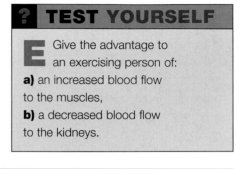

? TEST YOURSELF

E Give the advantage to an exercising person of:
a) an increased blood flow to the muscles,
b) a decreased blood flow to the kidneys.

Fig 31.11 This flow chart shows some of the ways in which hypertension can affect the body

THE LINK BETWEEN SMOKING AND ATHEROSCLEROSIS

If you smoke, you are much more likely to develop atherosclerosis. However, there is no straightforward explanation of how smoking actually causes the formation of atheroma. Several mechanisms could be involved:

- Cigarette smoke contains nicotine. Nicotine is a **vasoconstrictor**: it narrows the blood vessels and leads to an increase in blood pressure. High blood pressure can damage the artery walls and so increase the likelihood of atheroma formation.
- Smoking increases the level of cholesterol-rich lipoproteins in the blood. This accelerates the formation of the fatty streaks that are the first sign of atherosclerosis.
- Smoking increases the amount of fibrinogen circulating in the blood. This increases the chance of a blood clot forming if an atheroma breaks up in the artery.

THE LINK BETWEEN DIET AND ATHEROSCLEROSIS

Although newspapers and magazines can give the impression that all cholesterol is bad, it is an essential component of cell surface membranes. Cholesterol is used by the body to synthesise hormones such as oestrogen and testosterone. It becomes a danger only when it is present in high concentrations in the plasma.

Cholesterol is a lipid and, like other lipids, it is insoluble in water. In order to be transported in the blood it is combined with a protein, forming a complex molecule called a **lipoprotein**. There are different sorts of lipoprotein, which have varying amounts of cholesterol and other lipids. **Low-density lipoprotein** (**LDL**) is commonly called 'bad' cholesterol. This type of lipoprotein contains high amounts of cholesterol and can contribute to the formation of atheromas. **High-density lipoprotein** (**HDL**) is commonly called 'good' cholesterol because, not only does it not contribute to atherosclerosis, it actually helps to scavenge some of the cholesterol from LDL particles, making them less dangerous.

It is best to have a total blood cholesterol level of 200 milligrams per cubic centimetre of blood. Of this total, the amount of LDL should be low – between 100 and 130 mg cm^{-3}. HDL should make up the rest. If someone's total blood cholesterol is higher than 200 mg cm^{-3} or if LDL is over 130 mg cm^{-3}, or if HDL is lower than 35 mg cm^{-3}, they are at increased risk of developing atheromas.

? TEST YOURSELF

F LDL receptors are proteins. Use this information to suggest why some people may be genetically more susceptible to heart attacks than others.

The sequence of events by which cholesterol contributes to the formation of an atheroma is summarised in Fig 31.12. LDL particles that are excess to the requirements of the liver and other organs leave the blood and penetrate the walls of blood vessels, particularly in places where the wall has been damaged. Phagocytes then engulf them but are unable to break down the cholesterol that they contain. This accumulates in the cytoplasm of the phagocytes which then become foam cells (see page 541). The appearance of foam cells is the first recognisable sign of atherosclerosis.

Receptor on cell surface membrane

LDL particle in blood

Liver cells with receptors on the cell surface membrane into which LDLs fit

Surplus LDLs pass through endothelium where they are engulfed by phagocytes to form foam cells

Fig 31.12 How LDL causes atheromas

! SCIENCE FEATURE Jim's story

Jim is 53. When he was younger, he used to play a lot of football but he hasn't taken much exercise recently. He just doesn't have time. He has had a lot of stress at work in the last few years, trying to hang on to his job, and he smokes and has put on a bit of weight. He thinks that's normal in a man his age and he thinks he is fit enough. He can't remember when he last had to take a day off because he was ill.

26 March

'I was mowing the lawn today when I suddenly got a really sharp pain in my chest. I felt quite funny for a bit and I had to sit down. I'll have to get some more indigestion tablets for when we have a good fried breakfast.' Later in the day, the pain came

back every time Jim got up. By the evening he felt as if his chest was being crushed. His wife called the doctor and he was rushed into hospital with a suspected heart attack.

27 March, lunchtime

Jim is resting on the ward. The immediate danger has passed, but he has been having tests all morning and he still feels rough. A diagnosis must be confirmed before treatment can begin.

31 March

Like Jim's, most heart attacks are not fatal. With careful treatment and good advice, the patient usually makes a full recovery and can lead a normal life. One of the most important aspects of treatment is advice on lifestyle changes that can reduce the risk of a further heart attack.

! SCIENCE FEATURE Jim's story continued

The most obvious first step for someone like Jim is to stop smoking. The link between this habit and heart disease has been established by a great weight of evidence. A middle-aged man who smokes is three times more likely to suffer a heart attack than a non-smoker.

'I've also got to change what I eat. Far less of the fried breakfasts from now on.'

Most blood cholesterol doesn't come directly from food. It's produced in the liver from saturated triglycerides that come from the animal fats in our diet. So it is sensible to reduce the intake of lipid and animal fats so that they make up no more than 25 to 30 per cent of the total energy intake.

1 September

Jim is now fully recovered from his heart attack, but there have been a lot of changes in the past six months.

'The thought of almost dying gave me a real scare. It's been difficult, but I've managed to stop smoking altogether and now eat much less fatty foods. I followed the diet that the hospital gave me and joined an exercise programme. I'm now two stones lighter and I actually feel better now than before the heart attack.'

Blood tests

Table 31.2 shows the results of some of Jim's blood tests. Jim has also had a chest X-ray (Fig 31.13) and an **electrocardiogram** (ECG). An ECG records the wave of electrical activity that travels through the heart during a cardiac cycle (see page 188 in Chapter 10).

Fig 31.13 A chest X-ray shows how Jim's heart and lungs **(a)**, differ in size and shape from those of a healthy person **(b)**

Life after a heart attack

There is strong evidence that strenuous exercise benefits the heart. Jim joined a study that looked at the effect of exercise on people who had recently had a heart attack. It showed that, after exercising for three months, he was 20 per cent less likely to die than someone who hadn't joined an exercise programme as part of their recovery (Fig 31.14).

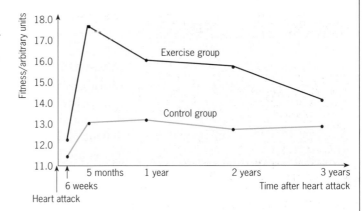

Fig 31.14 This graph shows how a programme of exercise affects fitness. The group that followed the programme suffered fewer deaths than the control group

Table 31.2			
Compound	**Reason for test**	**Concentration in blood/arbitrary units**	**Normal range in healthy people/arbitrary unit**
Cholesterol	A high blood cholesterol concentration is often associated with an increased risk of heart disease.	8.2	3.6–6.7
Glucose	High blood glucose concentration is also a risk factor. This test showed that Jim had diabetes although he hadn't know it. People with diabetes are at a slightly higher risk of heart disease.	13.6	less than 7
Aspartate transaminase	Aspartate transaminase and lactate dehydrogenase are enzymes normally found inside heart muscle. During a heart attack the muscle is damaged and some of the enzymes are released into the blood. Jim's test shows an elevated level of both enzymes, confirming his heart attack.	235	5–40
Lactate dehydrogenase		2266	300–650

TREATING CARDIOVASCULAR DISEASE

Although emergency resuscitation is the first treatment needed to save the life of someone who has just suffered a major heart attack or stroke, there are many different treatments for underlying CVD.

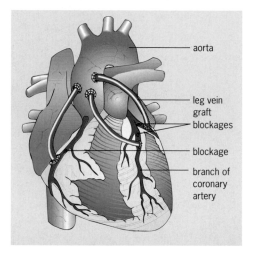

Fig 31.15 The principle of a triple heart bypass

HEART BYPASS SURGERY

A **coronary artery bypass graft operation**, sometimes called CABG or 'cabbage', is done to reroute blood around clogged coronary arteries to improve the supply of blood and oxygen to the heart. The surgeon sometimes takes pieces of vein from the leg of the patient, sews one end into the aorta and positions the other end to a coronary artery below the blocked area (Fig 31.15). In this way, the blockage is bypassed and normal blood supply is restored. It is also possible to detach one of the arteries that lead into the chest wall and then to graft it into the coronary artery. The number of grafts made is indicated by the way the surgeon describes the operation – a double heart bypass involves two grafts, triple and quadruple bypass operations use three and four.

ANGIOPLASTY

Angioplasty is used to treat angina or to reduce the risk of a heart attack. In **balloon angioplasty**, a thin tube with a deflated balloon at the end is introduced through the skin and fed into the coronary artery that has become narrowed by atherosclerosis (Fig 31.16). The surgeon checks its progress using a real-time high-speed X-ray image and, once the end of the tube is in the right place, the balloon is inflated gently. This pushes through the obstruction, making the lumen of the artery bigger.

Sometimes in balloon angioplasty, a short length of rigid tubing called a stent is left in the artery. This keeps the vessel open and prevents further narrowing at that point. Although angioplasty is most often done in coronary arteries, it can be carried out in any blood vessel in the body.

Fig 31.16 Balloon angioplasty and stent

COMMON DRUG THERAPIES

Hundreds of drugs are used in the treatment of heart disease. Some have already been discussed in more detail in this chapter, but Table 31.3 provides an overview of some of the main classes.

Table 31.3

Drug type	Drug	Used to treat	Effect on the body
Coronary vasodilator	Isosorbide dinitrate and other nitroglycerins	Angina.	Relaxes veins in the body. The amount of blood that returns to the heart is reduced and so the heart has to work less hard.
Beta-blockers	Inderal	Angina; to prevent heart attacks in people with coronary artery disease; to reduce death rate in people who have suffered a heart attack.	Blocks receptors on the heart that respond to adrenaline and related hormones, preventing the heart from working harder when these hormones are released.
Diuretics	Bemtanide	High blood pressure; congestive heart failure.	Causes kidneys to retain less water, reducing excess fluid in the body.
Calcium channel blockers	Cardizem	Angina; coronary artery disease; high blood pressure.	Relax the muscular walls of the arteries. In coronary arteries, this enables better blood supply to heart muscle. In arteries in the rest of the body, it reduces blood pressure.
Cholesterol-lowering drugs	Fluvastatin	High blood cholesterol and high LDL.	Reduces total blood cholesterol, targeting LDL specifically by inhibiting an enzyme that is involved in LDL production. Used in combination with a low-fat, low-cholesterol diet.
Anti-arrhythmics	Digoxin	Heart rhythm problems; angina; high blood pressure.	Different problems have different effects. Digoxin increases the strength of contraction of the heart muscle as well as helping to prevent abnormal heart rhythms.
Anticoagulants	Warfarin, aspirin	Post-heart attack or stroke treatment to prevent further attacks.	Makes blood less likely to clot, reducing the danger of thrombosis and thrombophlebitis.

Fig 31.21 A micrograph of cells taken from the cervix of a woman during a routine smear test. The pathologist then examines the tissue for signs of abnormal mitosis. Here, the cancerous cells are clearly seen because they stain red and have relatively large nuclei

Fig 31.22 Tumours arise when DNA in specific genes is damaged

WHY IS CANCER SO DANGEROUS?

Tumours interfere with the activity of the cells in the tissues of organs that surround them. Benign tumours can compress tissues, preventing normal blood flow or nerve function. Malignant tumours do even more damage, invading surrounding tissues and killing normal cells in the process. Cancer can also spread to other parts of the body by the process of **metastasis** (Fig 31.20).

Many tumours form from epithelial cells and, in order to break away, they must escape from the proteins that bind them to the underlying basement membrane that is characteristic of epithelial tissue (see Chapter 2). They do this by secreting protein-digesting enzymes. They also produce reduced amounts of the adhesion molecules that usually hold cells together.

THE BIOLOGICAL BASIS OF CANCER

Although there are many different types of cancer, we now know that all cancers are the result of DNA damage in cells. If this damage occurs in specific genes, control of cell division in that cell can be lost and it can begin to divide continuously (Fig 31.22).

Cell division is described in detail in Chapter 23. You might want to revise this before going on to look at how control of the process can break down.

GENES AND THE CONTROL OF CELL DIVISION

In Chapter 23 we saw that different types of cell in the body divide at different rates. Some, such as the neurons in the brain, do not divide at all after we reach the age of about 2. Others, such as liver cells, divide only rarely. Others, notably epithelial cells, divide as often as once every couple of days. Even so, all cell division is tightly controlled at the genetic level.

Some genes in the cell code for proteins that activate other genes to produce growth factors, chemicals that start a sequence of events that leads to cell division. Other genes code for proteins that inhibit cell division. Genes are turned on and off according to the cell type and its position in the body. Needless to say, the interplay of all the different genes involved is very complex. However, scientists have identified a few genes that, when they suffer a mutation, can cause the cell to lose control of its division pattern. These cells have been named **proto-oncogenes**, or just **oncogenes**.

The human genome contains many oncogenes. One, the *ras* oncogene, is on chromosome 11. It codes for a chemical called G-protein, which basically acts as an on-switch for cell division. Normal G-protein is usually inactivated very quickly by cellular enzymes. This means that the G-protein on-switch for cell division is only temporary and is switched off most of the time.

If the DNA in the *ras* oncogene is damaged, however, it codes for a G-protein that cannot be inactivated. In a cell with a defective *ras* gene, the switch for cell division always says 'on'. Cells that lose control of cell division can become cancerous, and *ras* mutations are associated with 20 to 30 per cent of all human cancers.

However, mutations in oncogenes are much more common than the cancer rate would suggest. This is because the cell has a second, back-up control. It also has genes called **tumour suppressor genes**. These prevent cells dividing too quickly, until the body's immune system can either kill the rogue cells, or until the cell's enzymes manage to repair the damage in the DNA (Fig 31.23).

The tumour suppressor genes do an excellent job, but their DNA too can be damaged and they can suffer mutations. If these genes fail to work, this back-up system is lost and any cells that start dividing uncontrollably are more likely to go on to develop into cancers. A gene called *p53*, on chromosome 17, has been well studied, and about half of all human cancers have been linked with a mutation in this tumour suppressor gene.

Oncogenes and tumour suppressor genes can both mutate because of DNA damage caused during life. Occasionally, though, faulty forms are passed on to the next generation. People who are born with a gene defect in one of these genes are said to have a genetic predisposition to cancer. They are more likely to develop cancer and often do so at a relatively early age (see page 553 and Table 31.5).

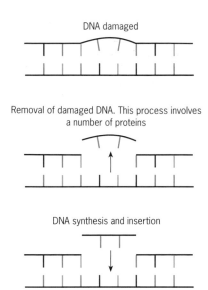

DNA damaged

Removal of damaged DNA. This process involves a number of proteins

DNA synthesis and insertion

Fig 31.23 Cells have many different ways of repairing DNA. This process is also controlled by genes. Mutations affect genes that repair other, damaged genes, as well as genes that control cell division

SEE QUESTIONS 5 AND 6

RISK FACTORS AND CANCER

So, two different classes of gene in the cell can fail and can result in cancer. But what does the damage to the DNA of these genes? Many things in our environment are carcinogenic. Some of the risk factors for cancer are connected with our lifestyle, others are present in our environment. In both cases, we can usually modify our cancer risk by changing our behaviour. This chapter looks at just a few of the major lifestyle and environmental factors that affect cancer risk.

SMOKING

Smoking is now one of the major causes of lung cancer in the developed world. Each year, lung cancer due to smoking leads to around 40 000 deaths from cancer in the UK (a large sports stadium holds about 40 000 people) and to around 175 000 deaths in the USA.

The link between smoking and lung cancer has been recognised for many years but it is only recently that scientists have identified a compound in tobacco smoke that binds to and damages the *p53* tumour suppressor gene. The longer a person smokes and the greater the number of cigarettes they have a day, the greater the chance that the *p53* gene will be damaged.

Fig 31.24 If you smoke, your lungs will eventually become clogged up with tar, and will look like this

SCIENCE FEATURE Eat those greens

There is a great weight of evidence that eating lots of fresh fruit and vegetables decreases the risk of developing cancer – so much evidence, in fact, that governments and health organisations in Europe, the USA and Australia all advise that we each eat five portions of fresh fruit or vegetables every day.

But why are fruit and veg so protective? One theory is that the high levels of vitamins they contain, particularly vitamins A and the B group, act as anti-oxidants. This type of chemical mops up free radicals that form in the body all the time, and that can damage DNA. The problem is that several large trials have shown that although taking vitamin supplements does no harm, it doesn't actually lower cancer risk.

Other chemicals in fruit and vegetables are responsible. Three that have already been isolated are:

● Chemicals called dithiol-thiones that occur in vegetables such as broccoli, cauliflower and cabbage seem to have several effects on the body. They activate liver enzymes that deal with carcinogens and they inhibit tumour growth.

● Sulphorane, another chemical isolated from broccoli, also activates helpful liver enzymes.

● Genistein, a compound found in soya beans, prevents the formation of the blood vessels that tumours need to support their rapid growth.

Many studies are now planned, and scientists expect to develop new drugs based on natural plant compounds that can cut cancer deaths by prevention.

Fig 31.25 These vegetables are known to help reduce the likelihood of cancer

DIET

There is strong evidence that some of the things we eat and drink contain substances that are carcinogenic:

● Drinking alcohol is linked to 3 per cent of cancers in the western world, causing higher rates of mouth, throat and oesophageal cancer. Smokers who drink large amounts of alcohol are particularly at risk.

● Heavily salted and smoked foods also increase the risk of developing mouth, throat and oesophageal cancer. About 1 per cent of all cancers are due to eating salted and smoked foods regularly.

● Eating a diet very high in over-cooked meats is thought to increase the risk of stomach cancer, because charring red meat (such as blackening it on a barbecue) produces breakdown products that are carcinogenic.

● Food contaminated with aflatoxins, substances produced by the fungus *Aspergillus flavus*, have been linked to specific cases of liver cancer. In some parts of Africa, groundnuts (peanuts) contaminated with this fungus are a staple food and liver cancer is common. Aflatoxins are chemically modified by liver enzymes. This modified aflatoxin affects the base guanine, which is part of the DNA molecule.

Apart from these specific examples, the cancer risk of eating specific foods is quite difficult to assess. People's diets are extremely varied and many studies involving hundreds of thousands of people are just beginning to uncover some definite patterns. However, the American Cancer Society states that, on the basis of current scientific evidence, one-third of the 564 800 cancer deaths that occurred in the USA in 1996 were related to diet.

At the moment, the generally accepted view is that the diet to avoid is one high in red meat, particularly meat that is charred, high in animal fats, and low in fibre, fresh fruits and vegetables (see the Science feature above).

EXPOSURE TO RADIATION

We are all exposed to a common source of radiation – the Sun. Exposure to sunlight is linked with **melanoma**, a form of skin cancer that accounts for about 2 per cent of all cancer deaths. It occurs most often in fair-skinned people and is caused by the ultraviolet light in sunlight. Particularly dangerous are the higher-frequency UVB rays that can damage DNA: UVB radiation causes 90 per cent of all skin cancers (Fig 31.26).

The cancer risk posed by other forms of radiation seems to be low. Excessive radiation such as the fallout from the atomic bomb blasts at Hiroshima and Nagasaki led to cancers in the survivors, but the low levels emitted by nuclear power plants do not seem to increase cancer risk. There is also little evidence to support the theories that objects such as mobile phones, microwaves and overhead power lines increase the risk of cancer.

Fig 31.26 Skin cancers are more likely to develop on parts of the body that are exposed to a lot of sunshine

EXPOSURE TO CHEMICAL CARCINOGENS

Substances such as asbestos, benzene, methanal (formaldehyde) and diesel exhaust are carcinogenic. We know this because of instances in the past where people have been exposed to high levels – usually through their work.

EXPOSURE TO MICROORGANISMS

Specific microorganisms are linked to the development of some cancers:

- The Epstein–Barr virus, which causes glandular fever, has been linked to some cancers of the lymph gland in the western world and with nose and throat cancer in Asia.
- HIV leads indirectly to an increased cancer risk: people with AIDS have a much higher rate of a skin cancer called Kaposi's sarcoma because their immune system cannot weed out rogue cancer cells.
- Infection with genital wart viruses is associated with higher rates of cancer of the cervix. Women who have many sexual partners are more likely to be exposed to wart viruses and so run a higher risk of developing cervical cancer.
- Infection with the bacterium *Helicobacter pylori*, recently found to cause stomach ulcers and gastritis, also causes higher rates of stomach cancer.

> ✓ **REMEMBER THIS**
>
> Today there is much greater control of work environments, but many work-related cancers are expected in the developing world as countries start industrialisation programmes.

CANCER AND HEREDITY

About 5 to 10 per cent of cancers arise because of faulty genes that can pass from one generation to the next. In the past 10 years, several such genes have been identified. However, before the genes were recognised, it was clear that some cancers were hereditary:

- Some families have many members all with the same type of cancer.
- Cancer tends to develop early – in people under 50.

Study of the DNA from members of such families has provided a lot of information about genes and cancer. Table 31.5 shows some of the genes that, when they are faulty, confer a very high cancer risk.

Table 31.5 Genes and cancer risk

	Gene	Tumour type	Gene class
Breast cancer	BRCA1	breast, ovary	tumour suppressor
	BRCA2	breast (both sexes)	tumour suppressor
	p53	breast, sarcoma	tumour suppressor
Colon cancer	MSH2	colon, endometrium, other	mismatch repair
	MLH1	colon, endometrium, other	mismatch repair
	PMS1,2	colon, other	mismatch repair
	APC	colon	tumour suppressor
Melanoma	MTS1 (CDKN2)	skin, pancreas	tumour suppressor
	CDK4	skin	tumour suppressor
Neuroendocrine cancer	NF1	brain, other	tumour suppressor
	NF2	brain, other	tumour suppressor
	RET	thyroid, other	oncogene
Kidney cancer	WT1	Milms' tumour	tumour suppressor
	VHL	kidney, other	tumour suppressor
Retinoblastoma	RB	retinoblastoma, sarcoma, other	tumour suppressor

CANCER – THE PATTERNS OF DISEASE

In the nineteenth century, cancer caused the deaths of fewer people than it does today. We must be careful how we interpret this statement. A hundred and fifty years ago infectious diseases were much more important as causes of death than they are now. Because of this, the average age of death was lower than it is today, and fewer people lived to the ages where cancers become increasingly common. In addition, medical treatment was far more basic and methods of diagnosis meant that the disease could be recognised only when it had reached an advanced stage.

If you were to compare data that shows the deaths from different types of cancer in one particular year, with the corresponding figures from 10 years earlier, you would notice differences. Between 1986 and 1996, for example, stomach cancer declined because of the discovery that various gastric problems, including cancer, were caused by infection by *Helicobacter pylori*. Treating people for this infection cut stomach cancer cases significantly during this time.

You might also notice an increase in the incidence of some cancers. Breast cancer cases, for example, increased between 1986 and 1996 because of advances in methods of screening and diagnosis. Much of the apparent increase was probably due to doctors being able to detect the condition earlier than they would have before 1986.

However, often it is not as easy to account for differences. Most of the time, statistics on cancer death rates and cancer incidence are difficult to interpret.

It might be better to ask some questions: Are the differences actually statistically significant? An apparent increase in cases might be only a reflection of an underlying trend, or the differences could be just the result of chance. Has there been a change in exposure to particular risk factors? Could we explain any changes in the incidence of lung and bladder cancer in terms of changes in smoking habits? Is the picture incomplete, and does this confuse the issue? Why are some cancers more common in one sex than in the other?

Answers to questions such as these are important if we are to understand the factors affecting the trends shown by such data. The Assignment at the end of this chapter provides some more information about how patterns in the incidence of a particular type of cancer can shed light on the factors that cause it.

THE BATTLE AGAINST CANCER

CANCER PREVENTION

Although drugs and food supplements to prevent cancer are a possibility, they are likely to take time to develop. In the meantime, changing your lifestyle could help you reduce the risk of cancer in later life. Four steps seem to be the most important:

- Never smoke. If you smoke, stop.
- Eat a well-balanced diet in which animal fat makes up less than 25 per cent of the calories. Include at least five portions of fresh fruit or vegetables every day, and try to obtain most of your protein from plant sources (beans, peas, grains, etc). Avoid very salty or smoked foods and charred meats, particularly red meats.
- Take regular exercise and maintain a sensible weight.
- Drink alcohol in moderation.

SCREENING FOR CANCER

Successful cancer treatment often depends on getting an early diagnosis. The longer a malignant tumour is left, the larger it grows and the more likely it is to spread and affect other organs. Surgery can be used to remove a localised

? TEST YOURSELF

Look back at Section 1 in this chapter, and you will notice that the lifestyle recommendations that help to prevent cancer are much the same as those that help prevent heart disease. Why do you think large numbers of people in the UK and the USA ignore them most of the time?

tumour but it is of little use if the condition has spread throughout the body.

Skin cancer is one of the commonest forms of cancer in younger people, but at least it is easily noticed. Surgical removal is very effective and results in 97 per cent of people surviving five years after the operation. Cancer of the colon is a very different matter. There are no visible signs and early symptoms are vague. Most people don't take them seriously and end up visiting their doctors for the first time when the cancer is already quite advanced. Not surprisingly, the five-year survival rate for this form of cancer is much lower, at only 37 per cent.

Regular screening can help to detect cancers early, but the advantages have to be weighed against its high financial cost. The NHS has screening programmes for two of the most common female cancers, breast cancer and cervical cancer. Breast cancer screening is carried out every three years by mammography (Fig 31.27). It targets women over 50 years old since breast cancer occurs most often in women over this age and because breast tissue in younger women is too dense for mammography to be effective. The breast cancer programme has probably caused a 20 per cent fall in death rate, simply because cancers have been picked up and treated early enough.

The national cervical screening programme was launched in 1988 and has cut deaths from cervical cancer by about 30 per cent. The programme recommends that a smear is taken every three years. Cells are stained and examined under a microscope, and those likely to become cancerous can be identified (look back to Fig 31.21). Early treatment is then possible.

Screening programmes for other cancers might not be so effective. Prostate cancer is common, particularly in older men, and there are methods of identifying it at an early stage. These include using ultrasound or testing the blood for high concentrations of prostate-specific antigen or PSA, an enzyme secreted by cells in the prostate. An enlarged prostate gland produces more PSA than a healthy one.

But is screening cost effective? Can it be used to prevent cancer deaths? The answer to some extent depends on where you live and what kind of healthcare system you have locally. In the USA, where private medical insurance bears the cost, PSA screening is commonplace, even though there are some doubts about its reliability. Early treatment of prostate cancer is carried out routinely. In the UK, the cost of a general screening programme would be an enormous burden on the NHS and would save relatively few lives: most men with fully developed prostate cancer are very elderly and are usually at far higher risk of dying from some other condition.

Fig 31.27(a) A patient is undergoing routine mammography. This enables tumours to be identified early, and increases the chance of treatment being successful

Fig 31.27(b) This mammogram has been enhanced with the aid of a computer. The denser tissue of the tumour shows up as a white area

TREATING CANCER

Surgery, chemotherapy and radiotherapy are important cancer treatments (Fig 31.28). They are all constantly being improved by research and have varying success rates, depending on the type of cancer and its site in the body. Drug treatment for leukaemia in children, for example, now cures 9 out of 10 children. Lung cancer can be much less easy to treat: the lungs cannot be completely removed by surgery, and the tissue is delicate and easy to damage.

Fig 31.28 This person has a tumour in the left eye and is undergoing proton radiation therapy

New strategies are approaching cancer treatment from several different angles. Table 31.6 describes some of them.

Table 31.6 New strategies for treating cancer

Treatment	What it does	Current status
Conformal radiation therapy	New technology allows radiation beams to be concentrated only on the tumour, avoiding damage to surrounding healthy tissue. The exact 3D position of the tumour is found using CT scanning or MRI scanning, and the positioning of the treatment beams is determined by computer.	In relatively widespread use: new types of tumours are treated every year.
Neutron and proton radiation therapy	X-rays and gamma rays are commonly used in radiotherapies; protons and neutrons can also be used.	Recent studies show that protons can treat small tumours of the spine that lie near vital organs, and that neutrons can treat salivary gland tumours.
Immunotherapy	New drugs to treat cancer are being developed all the time, but some of the most promising new drugs recognise substances found on the surface of tumour cells, and attack them while ignoring healthy cells.	Clinical trials are in progress. Some drugs are in use.
Differentiating agents	Drugs that do not kill tumour cells but force them to differentiate into cells with a specific function. When this happens, the cells stop dividing and tumour growth stops.	Several new drugs are in early clinical trials.
Angiogenesis inhibitors	Chemicals that stop the fast growth of new blood vessels into a tumour. Without a good blood supply, the tumour expands more slowly and can be easier to treat by radiotherapy or more traditional forms of chemotherapy.	Several angiogenesis inhibitors are undergoing clinical trials in patients, and other compounds that show potential are in an earlier stage of research.
Oncogene blockers	Scientists have been looking at how mutated oncogenes lead to cancer. They are trying to develop drugs to block the product of the damaged gene, so that the cell does not continue to divide uncontrolled.	Any cancer that arises because of a mutation in an oncogene.
Gene therapy	If a tumour suppressor gene becomes damaged, it no longer inhibits cancers from forming. In addition to treating the cancers that result, researchers are looking at the possibility of replacing the damaged gene in tumour cells.	Experimental studies using cells in culture have been quite successful. Clinical trials could soon begin.

SUMMARY

When you have studied this chapter, you should know and understand the following.

- **Cardiovascular disease** (CVD) and **cancer** are the two main lifestyle diseases.
- Lifestyle diseases are **non-transmissible**. They are associated with different risk factors such as smoking and diet.
- The underlying cause of heart disease is the process of **atherosclerosis**. This is commonly called hardening of the arteries.

- The major risk factors for CVD are hypertension (high blood pressure), smoking and diet.
- Although many cases of CVD are preventable and new treatments are being developed all the time, this disease is a major killer in developed countries.
- Cancer occurs when cells start to divide out of control. Many factors – genetic and environmental – cause this to happen.
- The major risk factors for cancer are smoking and diet. Many cases of cancer are preventable.

? EXAM QUESTIONS

1

a) Explain why atheroma may result in cardiovascular disease.

b) Three of the risk factors associated with the development of cardiovascular disease are eating a diet high in saturated fat, eating excessive quantities of salt and smoking.
Explain why these factors increase the risk of developing cardiovascular disease.

AQA(NEAB) BY08 Health and Disease June 2000 QB

2 Smoking is a risk factor associated with coronary heart disease. Smoking is known to raise the concentration of fibrinogen in the blood, promote the aggregation of platelets and reduce the ability of arteries to dilate. Use this information to:

a) explain the effect of smoking on blood pressure;

b) explain how smoking might lead to the formation of a blood clot.

AQA (A) BYA3 Pathogens and Disease January 2001 Q7

3 The diagram shows the heart of a person who has had a coronary artery by-pass operation. The person had suffered from coronary heart disease as a result of atherosclerosis.

a) (i) Describe what had happened in the coronary artery to make this operation necessary.
(ii) Describe the likely effects on the heart muscle if the operation had not been carried out.
Governments promote measures to reduce the number of cases of coronary heart disease.

b) State **three** of these measures.

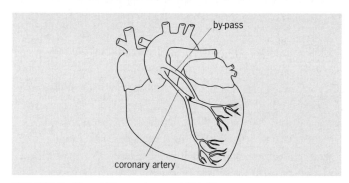

OCR 2802 Health and Disease January 2000 Q6

4

a) The diagram shows how a tumour is produced from a single normal body cell.
(i) What type of cell division occurs when a cancerous cell develops into a tumour?

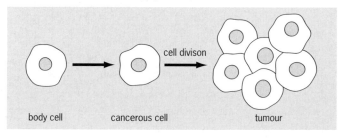

(ii) Give **two** characteristics that would indicate that a group of cells is a tumour.

b) (i) Explain how ultraviolet light may cause skin cancer.
(ii) Suggest why the incidence of skin cancer is higher in Northern Australia than in Southern Australia.

? EXAM QUESTIONS

c) Explain how the method of reproduction of the fungus responsible for athlete's foot allows the disease to be transmitted from person to person.

AQA(NEAB) BY08 Health and Disease June 2001 Q2

5 The line drawings are of transverse sections of lung tissue. They are drawn to the same magnification. Normal lung is shown on the left; the drawing on the right shows lung tissue from a person suffering from a disease. Structure **X** is not a blood vessel.

a) Name: **(i)** the structures labelled **A**;
(ii) the structure labelled **X**;
(iii) the disease illustrated in the diagram on the right.

b) Compare the following features of the normal and diseased sections, and explain how each of the differences you describe reduces the ability of a person suffering from this disease to take up oxygen.
(i) The surface area for gaseous exchange.
(ii) The relationship of the blood vessels to the surface of **A**.

WJEC Biology (Modular) B12 June 2001 Q6

6 Tobacco smoke contains over 4000 different chemicals. The table shows the effects of three components of tobacco smoke.

Component	Effects on the body
Tar	deposited in the gaseous exchange system; irritates the lining of the trachea, bronchus and bronchioles
Nicotine	absorbed in the blood; stimulates central nervous system to increase heart rate and blood pressure
Carbon monoxide	absorbed in the blood; binds to haemoglobin to form carboxyhaemoglobin

a) With reference to the table, explain how long term smoking may lead to
(i) heart disease;
(ii) chronic bronchitis.
In an experiment to investigate the effect of tar on the gaseous exchange system, tar collected from tobacco smoke was painted on to the skin of some laboratory animals. These animals later developed cancerous growths on their skin.

b) Outline the effect that tar had on the skin cells of these animals.

OCR 2802 Human Health and Disease June 2001 Q6

KEY SKILLS **ASSIGNMENT**

HEART DISEASE

Generally, the underlying cause of heart disease is one of the following:

● structural problems;
● inadequate coronary blood supply;
● faulty electrical conduction.

Structural problems

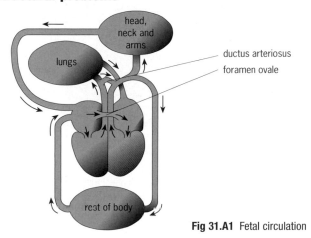

Fig 31.A1 Fetal circulation

Structural problems are defects in the anatomy of the heart, often present from birth. The diagram shows the differences in the fetal circulation before birth. Two important features are the presence of the **foramen ovale** and the **ductus arteriosus.**

The foramen ovale is a hole between the right and left atria. A valve in the vena cava directs blood through this hole so that a large proportion of venous blood passes right into the left side of the heart. The ductus arteriosus is a vessel that connects the pulmonary artery to the aorta. Its purpose is to by-pass the lungs, directing most of the blood into the aorta. The combined effect of the foramen ovale and the ductus arteriosus is to by-pass the lungs.

1 Why is it advantageous to by-pass the fetal lungs?

2 From which organ does the fetus obtain its oxygen **a)** before birth and **b)** after birth

At birth, rapid changes in the circulation are necessary. The normal sequence of events is as follows: at birth the lungs inflate with air and the pulmonary capillaries fill with blood, greatly increasing the blood flow in the pulmonary vein. As a result, the blood flow through the foramen ovale is reversed; it starts to go from left to right. This causes a small valve in the left side of the foramen ovale to close, preventing further left to right flow. In most people, the difference in pressure between the left and right side keeps the valve permanently shut, giving the tissue a chance to bind permanently.

The ductus arteriosus closes soon after birth. The reason for this appears to be a change in the oxygen content of the blood which causes a constriction of the muscular wall. This drastically reduces the blood flow and, within a few days, blood flow ceases completely.

If the foramen ovale fails to close, the baby will have a 'hole in the heart'. The severity of this depends on the size of the hole. Large holes need surgery, but many people, including athletes, have led a normal life with a small hole in the heart.

3 What problems will be caused by a hole in the heart?

Valve problems

There are two main valve problems:

● a leaky valve due to damage to one or more of the cusps;
● a narrowing – or **stenosis** – of the valve opening.

4 What would be the likely outcome of damage to one or more of the cusps?

5 Why would a narrowing of the valve entrance cause problems?

A faulty valve can be detected by the sound it makes – known a **heart murmur**. With leaky valves there is a lub 'swish' instead of the lub-dup, caused when some blood squirts back through the valve. A stenosed valve makes a different type of murmur caused by the narrowing of the opening. In both cases the best treatment is usually replacement with an artificial valve.

Inadequate coronary blood supply

This is the commonest cause of heart disease, and also the most preventable. There is a very extensive network of vessels which supplies blood to the myocardium - the heart muscle itself. The build up of fatty deposits inside the arterial walls – known as **atherosclerosis** – will reduce and even block the blood flow through the coronary artery. This problem, known as **coronary heart disease (CHD)** or **ischaemic heart disease,** has varying degrees of severity, from mild chest pain to a full heart attack. Research has shown that the people who develop coronary heart disease always have one or more of the following factors:

● high blood cholesterol
● high blood pressure
● cigarette smoking
● obesity
● lack of exercise
● diabetes mellitus
● genetic predisposition

KEY SKILLS **ASSIGNMENT**

Two common consequences of CHD are:

- **Angina pectoris**. When blood flow is reduced the heart muscle is unable to meet increase performance when the need arises. This leads to a pain called angina pectoris (commonly just 'angina'). Any extra strain on the heart – even walking upstairs – causes severe chest pains that makes the sufferer rest before they can continue.

- **Myocardial infarction** – a heart attack. When the blood flow to a particular part of the muscle is blocked, that part of the muscle will die. Infarction can cause death in several different ways. The heart may fibrillate – contract rapidly and uncontrollably. Alternatively, the cardiac output can drop below the minimum requirement. In some cases, the dead area of muscle may rupture, causing the heart to literally burst. The body can recover from a mild infarction, but extensive ones are nearly always fatal.

6 Imagine you are a GP faced with an overweight couple in their late forties. How would you persuade them to change their lifestyle?

7 Discuss the effectiveness of campaigns you have seen aimed at reducing the risk of developing heart disease. What approach would you suggest?

Faults in the conduction system

Contrary to popular belief, the heart doesn't have to stop to cause death – it just has to stop pumping blood effectively. Many heart attacks are caused by **arrhythmias** – a disruption of the normal rhythm of the heart. If the atria and or the ventricles don't contract at the right time they won't even be able to supply the myocardium with blood.

The ECG trace is a very useful tool in detecting cardiac arrhythmias.

A common arrhythmia is a **heart block**. This results from a blockage in the conducting system – often in the atrio-ventricular node – and may result from a myocardial infarction as described above.

8 What will happen if the impulse created by the sino-atrial node doesn't reach the ventricles?

Coronary heart disease and other traumas such as electrocution can cause **ventricular fibrillation** – the rapid and uncontrolled contraction of the ventricles. If this is not corrected rapidly, death is inevitable. The usual way to stop ventricular fibrillation is to use a **defibrillator** – a machine that applies a very strong current across the heart for a short period of time. This works because the electric current depolarises all of the heart cells at the same time. All activity stops for 3–5 seconds, during which time the normal electrical activity has a chance to re-establish itself.

Fig 31.A2 An artificial pacemaker is basically a small box that houses small batteries and complex electronics. The device is fitted under a muscle in the upper thorax, and a wire leads down a vein into the heart. The wire ends in an electrode, seen at the base of the heart. The electrode touches the heart muscle, initiating (setting off) the cardiac cycle. New, sophisticated pacemakers can sense changes in breathing, movement and body temperature and can adjust heart rate accordingly

A patient with a permanent heart block can be fitted with an artificial **pacemaker**. These devices vary according to the nature of the fault; some will stimulate the atria, others just the ventricles. New, more sophisticated pacemakers can sense changes in breathing, body temperature and vibration, adjusting the rate accordingly.

The actual device consists of a small box that houses the lithium batteries and the complex electronics that create the impulse. The electrical impulse is delivered to the heart by means of a lead passed down the subclavian vein and the superior vena cava (Fig 31.A2).

Replacement hearts

When a patient's heart becomes permanently damaged, the prospects for a long and healthy life are not good. Despite the fact that the heart has an apparently simple mechanical function, attempts to make an artificial one have defeated many of the world's best engineers. Until an artificial heart is made, the only source of an artificial heart is from a transplant.

9 What features would be needed by a perfect artificial heart?

32

Genetics, disease and the future

Genetics, disease and the future

Joseph Merrick, the Elephant Man, in 1889

Joseph Cary Merrick, better known as the Elephant Man, became well known when the story of his life was depicted in the film released in 1980. Merrick's physical problems were severe – his head was swollen and distorted and had many bony outgrowths. His skeleton was abnormal and he had difficulty walking because one of his legs was much shorter than the other. People at the time assumed he was also retarded and treated him as a curiosity and a freak, but he showed no signs of mental impairment.

Merrick died in 1890 and his remains have been studied by several generations of medical scientists. Many have assumed that Merrick suffered from neurofibromatosis, commonly called Elephant Man disease. People with mild forms of this genetic disorder, which occurs in about one in every 4000 births, sometimes have brownish spots on their faces or bodies. Merrick, however, had no skin markings, and over the years other inconsistencies have been noticed. Recently, studies were made of the skeletal remains of Merrick using the most up-to-date medical technology to see if the mystery could be solved.

The X-ray studies and CT scans showed that Merrick's skeleton was heavily disfigured and had large numbers of bony tumours. The pattern of bony growth suggests that Merrick actually suffered from the far more rare Proteus syndrome, of which only 120 cases have been reported worldwide. This disease, first identified 20 years ago, is triggered by dividing cells in bone that cause abnormal growths in the skull and other body tissues. No one yet knows which genes are involved in Proteus syndrome and there is no effective treatment.

1 GENETIC DISEASES

A genetic disease is one that results from a problem with our genes. Such diseases used to be described as 'running in the family' or 'in the blood'. Now we know that conditions can be passed on from one generation to another because they are caused by a faulty gene.

Some genetic diseases, such as cystic fibrosis (Fig 32.1), are caused by a single gene defect. Others are much more complicated. Not only are several genes involved; whether we develop that disease also depends on whether we are exposed to factors in our environment and in our lifestyle.

In this final chapter, we summarise the topics relevant to genetic diseases that we have covered elsewhere in the book (Table 32.1) and we look in detail at two other genetic conditions: achondroplasia (dwarfism), and Tay–Sachs disease. In the second section, the prospects that gene therapy offer for the treatment of genetic disease are explored, and in the final section of this chapter we highlight some of the ethical issues that are raised, not only by genetic disease, but also by the new treatments themselves.

SEE QUESTION 1

Fig 32.1 Cystic fibrosis is a disease that results from a defect in a single gene. The picture on the left is a coloured scanning electron micrograph of the mucus (red) in the lung of a person with cystic fibrosis. Percussion treatment shown on the right helps to clear the mucus. Learn more about cystic fibrosis on page 487 and Chapter 27

Table 32.1 Topics that you might find it useful to revise while reading this chapter	
Topic	**Chapter**
Important concepts in genetics	22
Ethics of genetic engineering and gene cloning	22
	27
Genetic manipulation and cloning	27
Genes and cancer	31
The human genome project	27
Monohybrid inheritance in humans	25
Cystic fibrosis	25
	27
	28
Down's syndrome	23
	24
Sex-linked inheritance	25
Haemophilia	25
	30
Duchenne muscular dystrophy	25
Red–green colour blindness	25
Xeroderma pigmentosum	24
Sickle cell anaemia	24
Huntington's disease	25
Albinism	26
Phenylketonuria	28

ACHONDROPLASIA

Many conditions cause a lack of height. In some, the bones do not grow and develop normally. More than 100 specific skeletal dysplasias, as they are called, have been identified, and achondroplasia is the most common. It occurs about once in every 25 000 births in all races and with equal frequency in males and females.

Achondroplasiacs, people with disproportionately short stature, often refer to themselves as dwarfs, little people or short-statured persons. Their intelligence and mental development are generally normal. Affected men average 131.6 cm in height, while women average 123.4 cm. There seems to be little or no relationship between the height of achondroplasiac parents and the adult height of their children. It is possible for two dwarf parents, for example, to have a child that grows to average adult height.

Someone who has achondroplasia (Fig 32.2) is disproportionately short. Their head is large and their arms and legs are short compared to their trunk. This shortness is particularly noticeable in the upper arms and thighs. Achondroplasiacs also tend to have a prominent forehead, a flat or even depressed area between their eyes, a protruding jaw and crowded teeth. The typical signs of achondroplasia can be seen at birth, and the condition can be diagnosed early.

Fig 32.2 This teenager with achondroplasia is testing a modified car

WHAT CAUSES ACHONDROPLASIA?

This is a genetic condition that is due to a mutation of a single gene. The mutated gene is dominant. The condition can be passed from one generation to the next or it can result from a new mutation in a gene from average-sized parents. Nine out of ten children born with achondroplasia have average-sized parents, and no other family member is affected. The condition is all or nothing – a person with this mutation has achondroplasia, there are no mild forms.

In 1994, researchers discovered that a mutation of the fibroblast growth factor receptor-3 gene (*FGFR3*) on human chromosome 4 causes achondroplasia. Exactly how the mutations bring about the condition is not yet known.

THE INHERITANCE OF ACHONDROPLASIA

A couple in which one partner is achondroplasiac and the other is of average height has a 1 in 2 chance of having a child with achondroplasia. If the child does not inherit a mutated copy of the gene for achondroplasia, growth during childhood and adult height are usually within the normal range.

Average-sized children of parents with achondroplasia have no increased risk of having achrondroplasiac children themselves. If both parents have achondroplasia, there is a one in four chance that they will have a baby who inherits two altered copies of the gene that leads to achondroplasia. These children are said to have homozygous achondroplasia and this condition is almost always lethal in the first year of life.

TREATMENT OPTIONS

At present there is no specific treatment to promote growth in achondroplasia. Growth hormone treatment seems to increase the rate of growth during the first year of treatment, but does not increase adult height. Surgery to lengthen the legs and arms of people with achondroplasia is being done experimentally in the USA, but achondroplasiacs themselves have mixed views on whether this is a step in the right direction.

! SCIENCE FEATURE Genetic screening and diagnosis

The opener in Chapter 25 describes how the singer Woody Guthrie, along with other sufferers, only found that he had Huntington's disease when the symptoms appeared. By that time he was about 40, and had already had a son to whom he could have passed the disease. Now it is possible to screen new-born babies for an increasing list of genetic abnormalities.

All babies in the UK are tested routinely for phenylketonuria. This condition and the heel test that detects it (Fig 32.3) are described in Chapter 28. Cystic fibrosis, sickle cell anaemia (Fig 32.4) and Tay–Sachs disease can also be detected by screening techniques.

A simple blood test can tell Tay–Sachs carriers from non-carriers. Blood samples can be analysed by either enzyme assay or DNA studies. The enzyme assay is a biochemical test that measures the level of Hex-A (a vital enzyme called hexosaminidase A – see the text overleaf) in a person's blood. Carriers have less Hex-A in their body fluid and cells than non-carriers.

Fig 32.4 Prospective parents can find out if they are likely to have children with sickle-cell anaemia. Electrophoresis is the technique used to screen their DNA for the sickle-cell gene

Fig 32.3 Soon after birth, babies have the Guthrie test which screens for phenylketonuria. This genetic disease can cause brain damage if untreated, but a special diet, if followed carefully, can prevent any problems

DNA-based testing to test carriers of Tay–Sachs (people who show no symptoms but who are able to pass the recessive mutated gene on to their children) looks for specific mutations, or changes, in the gene that codes for Hex-A.

Since 1985, when the Hex-A gene was isolated, over 50 different mutations in this gene have been identified.

The limitation of DNA-based carrier testing for Tay–Sachs disease is that not all known mutations in the Hex-A gene are detected by the test, and others have yet to be identified. Since some carriers cannot be identified by DNA analysis alone, a blood test is always performed as well.

In fact, DNA testing can give more information when done in conjunction with the blood test, especially in cases where both members of a couple are carriers. The two tests complement each other and can reveal what type of Tay–Sachs disease is most likely to occur in the couple's child, if they decide to go ahead with a pregnancy after receiving genetic counselling. It also makes it easier to correctly diagnose Tay–Sachs in the developing fetus at an early stage of pregnancy.

TAY–SACHS DISEASE

Tay–Sachs disease (TSD) is a fatal genetic disorder in children that causes progressive destruction of the central nervous system. The disease was named after Warren Tay (1843–1927), a British doctor who identified the first patient with TSD in 1881, and Bernard Sachs (1858–1944), an American neurologist who first described the cellular changes in Tay–Sachs disease.

Tay–Sachs disease is caused by the absence of a vital enzyme called hexosaminidase A (Hex-A). Without Hex-A, a fatty substance builds up to abnormally high levels in cells, especially in the neurones of the brain. This causes progressive damage to the cells. The destructive process begins in the fetus early in pregnancy, but the signs of the disease are not apparent until the child is several months old. By the time a child with Tay–Sachs disease is 3 or 4 years old, the nervous system is very badly damaged and all children with classical TSD die early in childhood, usually by the age of 5.

There is no cure or effective treatment for Tay–Sachs disease. However, active research is being done in many laboratories around the world. The use of enzyme replacement therapy has been explored but its effectiveness has been very limited. Because the disease affects brain cells that are protected by the blood–brain barrier, the replacement enzyme cannot get to the cells that need it.

THE GENETICS OF TAY–SACHS DISEASE

There are two genes that code for the enzyme Hex-A; both are found on chromosome 15. If either or both Hex-A genes are active, the brain produces enough of the enzyme to prevent the abnormal build-up of the fatty deposits in its cells. Carriers of Tay–Sachs disease – people who have one mutated copy of gene as well as one copy of the active gene – are healthy. They do not have Tay–Sachs disease.

A carrier has a 50 per cent chance of passing the mutated gene on to his or her children; any child who inherits one mutated gene is a Tay–Sachs carrier like the parent. If both parents are carriers and their child inherits two copies of the mutated gene, that child will have Tay–Sachs disease since he or she cannot produce any functional Hex-A.

When both parents are carriers of the Tay–Sachs gene, there is a 1 in 4 chance with each pregnancy that their child will have Tay–Sachs disease, and a 3 in 4 chance (75 per cent) that their child will be healthy. Of their unaffected children, there is a 2 in 3 chance that each child will be a carrier, like the parents. This pattern of inheritance is called monohybrid inheritance (Fig 32.5 and see Chapter 25).

SEE QUESTION 2

IDENTIFYING CARRIERS

Tay–Sachs often occurs in families with no history of the disease. The Tay–Sachs gene can be carried without being expressed through many generations. Before 1970, the only way to learn if you were a Tay–Sachs carrier was to be the parent of a baby with Tay–Sachs. Now, safe and reliable carrier testing is available to identify Tay–Sachs carriers. Most important, testing can identify carrier couples who are at risk for bearing a child with Tay–Sachs – before a tragedy occurs (see the Science feature on page 565).

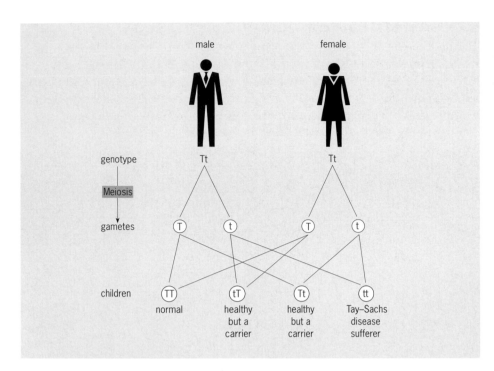

Fig 32.5 The inheritance of Tay–Sachs disease. If both parents are carriers, there is a 25 per cent chance that any child will be a sufferer. There is also an equal chance that a child will inherit no faulty alleles and that his/her descendants will be completely free from the disease. There is a 50 per cent chance of the child being a carrier

2 GENE THERAPY

The new technology of gene therapy promises to revolutionise medicine in the twenty-first century (Fig 32.6). The idea that it can be used to treat genetic diseases such as cystic fibrosis and Tay–Sachs disease has been around for several years but it is taking longer than expected to develop gene therapy into an acceptable and safe treatment.

Recently, scientists have also begun to look beyond just the diseases caused by a defect in a single gene and are considering gene therapy as a potential treatment for all sorts of problems, from cancer and heart disease to AIDS.

Fig 32.6 Cause and treatment of cystic fibrosis. Chromosome pair 7 is highlighted in pink, with the location of the cystic fibrosis gene mutation arrowed

Electrophoresis of the gene sequence shows that base sequence TTC is present in an unaffected person (left), and absent in someone with cystic fibrosis.

The healthy gene is isolated before being inserted into a virus. Cystic fibrosis patients use an inhaler with the virus that delivers the healthy gene to lung cells

GENE THERAPY FOR GENETIC DISEASES

The first person to be given gene therapy was Ashanti DeSilva. She became the first patient with severe combined immunodeficiency (SCID) (Fig 32.7) to be treated in September 1990, when she was 4 years old. A team in the USA led by French Anderson gave Ashanti four infusions of cells containing the working gene that she lacked. Over the four months of the treatment, her condition improved.

SEE QUESTION 3

Fig 32.7(a) A SCID baby lives in a sterile compartment and is handled by medical staff wearing full body gowns to avoid passing on any infections

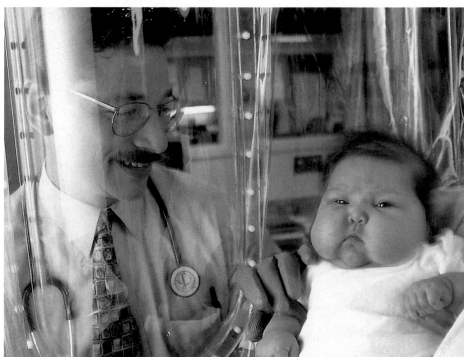

Fig 32.7(b) This Apache baby is having gene therapy: some of his stem bone marrow cells are removed, the gene for the enzyme he lacks is inserted into them and they are returned to his body. Then he can produce normal immune cells

With the help of follow-up treatments she has now been transformed from a small girl who was constantly ill and could not leave the house, to a normal, healthy and lively teenager. She continues to do well and, apart from needing regular injections of one of the enzymes she cannot make naturally, she lives a normal life.

Early in 2000, 10 years after the first successful gene therapy trial to treat SCID, researchers in Paris managed to treat two babies, aged 8 and 11 months, with gene therapy. The babies both had SCID, but a slightly different form from that of Ashanti DeSilva. Their treatment involved taking out some bone marrow and selecting a set of blood stem cells. Stem cells are the cells that differentiate into all sorts of blood cell types, including the white blood cells that make up the immune system. By introducing the gene that the babies were missing into these stem cells, the researchers managed to get the correct gene into all of their immune cells. Both babies developed a fully functional immune system within a few weeks and can now live normal lives. A similar process was used in the summer of 2001 on 18-month-old Rhys Evans at Great Ormond Street Hospital in London.

HOW GENE THERAPY WORKS

One of the main problems that make gene therapy difficult is finding out how to transfer genetic material containing corrected, healthy genes into cells that contain the faulty gene. Scientists have now developed several methods.

Modified viruses are often used as carriers, or vectors. Viruses are useful because they are able to penetrate cells and to introduce their genetic material into the host cell. Before researchers can use viruses in therapy, however, they must remove the genes that code for proteins that viruses use to reproduce themselves. When those genes are replaced with corrective genes, the vector virus becomes a delivery unit that is identical to the original virus on the outside, and can transport useful genes into cells, but which cannot cause illness.

Three main approaches to gene therapy are currently possible.

EX VIVO GENE THERAPY

The most commonly used technique in gene therapy is known as *ex vivo* because the gene transfer is done outside the body. Cells with defective genes are removed from the patient and given normal copies of the affected DNA before being returned to the body. This therapy has generally targeted blood cells because many genetic defects alter the function of one or other of the blood cells. However, the problem with this approach is that blood cells have limited life spans. Corrected cells can be introduced into the body and they can work well, but after a few weeks they die and the patient's problems return, making repeat treatments necessary to keep the person healthy.

A better approach is to target the stem cells of the bone marrow. Stem cells survive as long as the patient does, so if these immature cells that give rise to the full array of blood cells in the circulation can be given the correct gene, the treatment should last indefinitely.

Although scientists have been able to obtain stem cells from human bone marrow, it has proved difficult to get genes into these cells. But advances have been made recently. In 1993 a group in Los Angeles treated three new-born babies with SCID by inserting genes into their stem cells. Their blood cells began producing the critical enzyme that they lacked and they grew into healthy children.

REMEMBER THIS

In vitro means literally 'in glass'. This term usually refers to research done on cells isolated from the body and grown in the laboratory. *In vivo* means literally 'in life'. This terms usually refers to research done on living animals or people. *Ex vivo* means 'outside life'. It is really only another way of describing work done on living cells in the laboratory, and means the same as '*in vitro*'.

! **SCIENCE FEATURE** Setbacks plague gene therapy development

Although a handful of young children have received successful gene therapy for severe immunodeficiency syndrome (SCID), few other gene therapy trials have worked very well. A low point for this new technology came in September 1999 when a teenager, Jesse Gelsinger, died during a gene therapy trial. There was a full investigation and there were indications that many research rules had been broken. The investigation was widened to other centres, mainly in the USA, and there were strong suggestions that many poor results in trials were being covered up because of the financial damage they could cause to the companies funding the trials. Many gene therapy programmes in the USA were put on hold. These have now started again, but the rules governing trials have been made stricter.

French Anderson, the pioneer of gene therapy, commented after the recent gene therapy success in April 2000 (see text) that although 340 gene therapy trials were on-going in the USA, only a handful are showing promising results. There are still many problems to overcome and some of them might need unusual solutions. Recent research suggests that lentiviruses may be better gene vectors than other viruses. They can get a gene into cells in the body with higher efficiency and can be modified so that the gene they carry is expressed only in some types of cell. However, a major problem is that lentiviruses are retroviruses and one of the most useful of these is HIV-1 – the virus that causes AIDS. Although the new vectors are extensively modified, they will need to be tested very well to make sure there is no chance of them reverting and becoming able to cause an immunodeficiency in the patients they are used to treat.

US halts gene tests after youth dies

Effects of experimental treatment 'not known'

Duncan Campbell
in Los Angeles

The controversial use of an experimental gene therapy treatment has been suspended throughout the United States after the death of a teenager.

Some medical experts are suggesting that experimental forms of treatment are being put into operation too quickly, before all possible tests have been carried out.

Jesse Gelsinger, 18, from Arizona, died on September 17 in the University of Pennsylvania hospital in Philadelphia. He had been suffering from a serious, potentially fatal, metabolic disease.

Since his death, all such treatment has been halted by federal officials until the precise cause of death is established.

Thousands of patients in the US have been treated with variations of gene therapy.

The doctors treating Gelsinger injected a genetically engineered virus into his liver at the highest dosage allowed by the federal regulations governing such treatment.

Now federal officials plan to ask 100 other researchers throughout the United States who have been involved in such treatment to inform them if there have been any serious side-effects of their experiments.

"This was a tragic and unexpected event," James Wilson, director of the university's institute of gene therapy, told the Washington Post after Gelsinger's death.

"I hope in a month we'll have looked at every angle so we can share what we have learned form this."

Another 17 patients who have been undergoing the same treatment as Gelsinger — although not necessarily at such high dosages — have reportedly suffered no side effects.

The experiment involves

> 'This was a tragic and unexpected event. We'll share what we have learned from this'

using a genetically modified adenovirus, one of the viruses which cause the common cold, to correct a defective gene in the patient. It has so far only been used on the very ill who have not appeared to respond to any form of more conventional treatment.

The father of the dead youth, Paul Gelsinger, said he hoped that his son's death would aid the fight to discover a cure for the condition.

"I lost a hero," he said of his son, who had fought a long and debilitating battle against the disease.

The death coincides with a serious debate among doctors in the US about the speed with which genes are being used in different forms of treatment when little is still known about their effects.

There have been suggestions that some methods are being employed before the full results of the potential of gene treatment is known from laboratory experiments.

There is also a continuing argument, driven by the religious right, about the ethics of using live genes.

Fig 32.8 Jesse Gelsinger died during a gene therapy trial

IN SITU GENE THERAPY

In *in situ* treatment, vectors carrying the corrective genes are introduced directly into the tissue where the genes are needed. This approach makes most sense for problems that are localised, but it cannot correct disorders that affect several systems of the body.

In situ gene therapy is being explored for several diseases. For cystic fibrosis, workers have introduced gene vectors containing healthy copies of the cystic fibrosis gene into the lining of the bronchial tubes. As a first step towards treating muscular dystrophy, other researchers have injected a gene directly into muscle tissue in animals to investigate the possibility of re-engineering the body to make normal muscle proteins. Several teams have also inserted 'suicide' vectors into tumours. These carriers contain a gene that is intended to make cancer cells self-destruct when treated with certain chemotherapy drugs.

IN VIVO GENE THERAPY

This third approach is still in the experimental stages but it is likely to be the gene therapy of the future. Gene carriers will simply be injected into the body, in much the same way that drugs are given now. Once in the body, the carriers will find their target cells, ignore other cell types, and transfer their genetic information efficiently and safely.

! SCIENCE FEATURE Gene therapy before birth

The US scientist French Anderson is currently working towards carrying out gene therapy on a fetus that is affected by a genetic disease that is usually fatal before birth.

Because of the enormity of the ethical issues involved, he has raised the possibility of *in utero* gene therapy for public discussion, even before the techniques that would enable it to go ahead have been developed. He argues that if society decides that this sort of treatment should not be allowed, there is little point in spending the time and the money to work on it.

In the next few years, Anderson hopes that it will be possible to take some of the bone marrow stem cells from a fetus with untreatable SCID or who has α-thalassaemia, transform the cells using a corrective gene, and then reintroduce the cells into the fetus at about 15 weeks. Fig 32.9 shows a 15-week-old fetus.

If the therapy works, the child would be born healthy and would never suffer from the symptoms of the disease it has been treated against. However, the problem is that it might not work – if the baby dies before birth, you might say that nothing has been lost. However, the possibility exists that some sort of partial cure could result and the baby could be born, only to suffer a short life of severe illness and disability.

Fig 32.9 The CT scan of a 15-week-old fetus

3 PROBLEMS AND ETHICS

The people who have been trying to develop gene therapy techniques have had to overcome many technical problems. *In situ* therapy, for example, is still hampered by a lack of safe ways to implant corrected genes into various organs. *Ex vivo* therapies do not have this problem, but the genes that are transferred do not always yield good quantities of the encoded proteins. And, in both forms of treatment, once the genes enter the cells, they can insert themselves randomly into the DNA of the chromosomes. Sometimes this is harmless, sometimes it is not. If the gene sent in as therapy manages to disrupt a tumour suppressor gene (see page 551) that normally protects the body against cancer, a tumour could result.

Development of *in vivo* gene therapy has been slow because researchers have found it difficult to create delivery units that will allow the genes to find their way into the correct cell and that can avoid the immune system of the body.

Cost is also a potential hurdle to gene therapy. Because we are still at an early stage of gene therapy research, it is still quite expensive to do and is carried out only in the major medical centres of the developed world. However, French Anderson, the pioneer of the first gene therapy procedure, thinks that, ultimately, gene therapy will become relatively inexpensive and perhaps in 50 years, may be used regularly and routinely to cure many different medical conditions.

However, the financial and technical challenges of gene therapy are nothing compared to the ethical dilemmas that crop up. So many different questions need to be raised and considered. We end this chapter with just a few for you to mull over:

● Should gene therapy be used to prevent disease?
● All the gene therapy carried out so far has affected only the somatic cells, not the germline cells – the eggs or the sperm. Do you think germline gene therapy could ever be justified?
● Where should the line be drawn for genetic therapy? How will it be possible to avoid people trying to use gene therapy to practise eugenics (altering racial characteristics)?
● In 1998, scientists managed to grow human stem cells from embryos in culture for the first time. This raises the possibility of engineering an entire human being. Do you agree that this type of research should continue?

? EXAM QUESTIONS

1   The photograph below shows a karyotype of a boy with Down's syndrome.

a) State the number of chromosomes shown in this karyotype.
b) On the karyotype, label the Y chromosome.
c) Down's syndrome is an example of polysomy, and may arise as a result of non-disjunction during the formation of a female gamete.
(i) State what is meant by the term *polysomy*.
(ii) Explain how fertilisation by normal sperm of a female gamete formed by non-disjunction may result in Down's syndrome.
Edexcel B/HB1 June 1999 Q5

2   Cystic fibrosis is a genetic disorder caused by the inheritance of two recessive alleles of a single gene. The gene codes for a protein in the cell surface membrane.
a) (i) What is a gene?
(ii) What structural features of a gene allow it to code for a protein?
b) A healthy couple have a son with cystic fibrosis. What is the probability of their next child being a daughter with cystic fibrosis?
Show your working.
AQA(NEAB) BY08 Health and Disease June 2000 Q3

3   Early detection of hereditary disorders may be carried out using the techniques of amniocentesis or chorionic villus sampling (CVS).
a) Describe briefly how a sample obtained by **one** of these techniques is processed for a diagnosis to be made.
b) Name **one** inherited disorder which may be detected using these techniques.
c) Outline **two ethical** issues concerning the use of these techniques.
OCR 6912 Inheritance Part 1 March 2000 Q2

33

Ecology: the basics

Ecology: the basics

| Competition for a mate | Rainfall | Parasites | | Number of other species in ecosystem | Seasonal changes in climate | Plant producers, eg grass | Competition for food |

Fig 33.1 Ecology is the study of whole organisms: we learn how an organism interacts with individuals of its own species and other species, and with the physical environment that surrounds it. The jigsaw pieces show just a few of the factors that influence the lives of these zebras in the savannah grasslands of Africa

The jigsaw pieces labels read: Competition for a mate, Rainfall, Parasites, Number of other species in ecosystem, Seasonal changes in climate, Plant producers, eg grass, Competition for food, Number of animals in population, How much water is available, Hours of sunlight, Temperature, Altitude of environment, Catastrophes – drought, hurricanes, fire, etc, General climate, Interaction with mate, Action of decomposers on body after death, Concentration of minerals in the soil, Time needed to rear young, Competition for space and territory.

REMEMBER THIS

The word ecology is loosely used to mean environmental concern or environmental conservation. These topics are important, and they depend heavily on knowledge gained through ecology, but they are not the same thing. We define ecology as: **'the scientific investigation of living organisms in their natural surroundings'**.

1 THE SCIENCE OF ECOLOGY

Ecology is a branch of biology. The word ecology was first used by Ernst Haeckel in the 1860s. It comes from two Greek words: *oikos* meaning home and *logos* meaning understanding. His definition of ecology was 'the knowledge of the sum of the relations of organisms to the surrounding outer world, to organic and inorganic conditions of existence'. Put more simply, ecology is the study of organisms in their natural surroundings. It looks at how they are adapted to their environment and how they interact with both the living and non-living world around them.

Each organism and each physical feature of an environment is separate, but they all interact and interlock, forming a complex system that is a bit like a three-dimensional jigsaw puzzle: ecologists study different parts of the puzzle and then try to work out how it is put together (Fig 33.1).

ECOLOGY AND OTHER BRANCHES OF SCIENCE

Ecology is closely related to two other branches of science: **environmental biology** and **environmental science**. The first of these looks more generally at living things in the environment. The second uses the biology, geology and the physical sciences to help us to explore our environment.

It is also impossible to study ecology without bearing in mind the genetics, evolution and behaviour of the organisms at the centre of a study. Ecological genetics and behavioural ecology are beyond the scope of this book but it is important to remember that each organism you study in the context of its environment has developed to fit that environment through the process of natural selection (see Chapter 26).

PRACTICAL ECOLOGY

As in other branches of science, progress in ecology depends heavily on scientific investigation. Looking back at our definition of ecology, it should come as no surprise that many of these studies are done outside. Field studies (Fig 33.2) involve a great deal of observation and can include experimental studies, though these are more often carried out in the laboratory where they can be more strictly controlled. Ecologists use these basic approaches together to gain a complete picture of the organism or the environment that is being investigated.

Fig 33.2 Safety and field trips. Ecologists conduct studies outside and must consider safety when working near water or in isolated areas. Here, a student is identifying organisms from a pond, and two students are measuring the rate of water flow in a stream

The exciting thing about ecology is that it is still a relatively young science. In just over 100 years of ecological study, we have barely begun to even scratch the surface. Many environments of the world remain unexplored and there are many organisms that we know of, but whose habits and living conditions are known only sketchily, or are a complete mystery. Unlike biochemistry or physiology, ecology offers an A-level student the chance to do original practical work, and perhaps to make an important scientific discovery (Fig 33.3).

Fig 33.3 Habitats change frequently. Porlock used to have a mile-wide strip of farmland between it and the sea. In 1996, the shingle barrier that formed the sea defences was breached and a salt marsh with saltwater lakes has developed rapidly. This is a new habitat that now supports birds such as egrets. Coastal areas are subject to many changes like this, creating new areas for observing the processes of colonisation and adaptation

---

> ### ⚠ SCIENCE FEATURE Why do human biologists need to study ecology?
>
> You might be wondering what a section on ecology is doing in a human biology textbook. Ecology is sometimes thought of as a study of nature – lots of field trips and wading around in mud looking at insects. But ecology is much more wide-ranging.
>
> The quality of our environment and the way we interact with the outside world are major factors that determine our health. Substances that contaminate our environment in water, food, air and soil contribute to the untimely death of millions of people every year.
>
> The problem is worst in the developing world where 4 million babies and young children die from diarrhoea every year because of dirty water and bad food. Over 1 million people die from malaria each year and over 267 million others suffer recurrent symptoms. Treating these illnesses with medicines is only of limited use: to really tackle them effectively, it is crucial to find out what is causing them and take steps to reduce people's risk. That doesn't mean just finding out what bacteria, virus or parasite is causing the disease directly. It means identifying the environmental factors that are allowing the disease to spread.
>
> Many third world countries, for example, have seen huge increases in cases of malaria with the more widespread availability of plastic 'disposable' containers. When these are dumped, they provide the perfect trap for rainwater and then form a wonderful stagnant pool in which the Anopheles mosquito can lay its eggs. Recognising this as a problem has helped to prevent many cases of malaria.
>
>
>
> **Fig 33.4** We have a two-way relationship with our environment, and our actions can have unexpected consequences. These people in the Philippines live by recycling rubbish. The plastic bags around them look harmless enough but they could be lethal

---

Fig 33.5 The biosphere is the part of the Earth's surface that supports life. It extends from the bottom of the deepest oceans into the part of the Earth's atmosphere that contains breathable air: in total a vertical distance of between 30 and 40 kilometres. It covers most of the hydrosphere, the surface layers of the lithosphere and some of the atmosphere. The hydrosphere is the part of the Earth's surface that is covered by water. The lithosphere is the layer of soil and rock that forms the Earth's crust. The atmosphere is the blanket of gases that envelop the planet, maintaining a mean temperature of about 7 °C. This temperature is crucial to life because it allows most of the water on Earth to exist as a liquid

2 TERMS AND CONCEPTS IN ECOLOGY

The relationships between an organism and the physical features of its environment, and between all the other organisms that live with it, are incredibly complex. Any description of them uses terms that you have probably not come across before. The next section gives an overview of some basic terms and concepts.

THE BIOSPHERE

All the living organisms that we currently know of live on Earth. But not all parts of the Earth support life. The part that does forms a sort of 'skin', the **biosphere**, which is shown in Fig 33.5.

THE BIOSPHERE IS DIVIDED INTO BIOMES

The biosphere is made up of lots of different areas that have very different environmental conditions. We call each of these fairly broad areas **biomes**. The major biomes of the world are listed in Table 33.1 and some of these are shown in Fig 33.6. Many of the best-studied biomes are the **terrestrial biomes** (those that exist on land) but there are also **aquatic biomes** (those that exist in water).

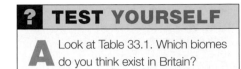

? TEST YOURSELF

A Look at Table 33.1. Which biomes do you think exist in Britain?

Table 33.1 The major biomes of the world

Biome	Type	Major features
Savannah	terrestrial	Tropical grassland with few trees. Example: the Serengeti in Africa.
Temperate grassland	terrestrial	Grassland with hot summers and cold winters. Some broad-leaved plants. Example: the prairies in the USA.
Desert	terrestrial	Hot and dry. A few highly specialised plants and animals. Example: the Kalahari desert in Africa.
Tundra	terrestrial	Cold for most of the year: very short growing season. No trees: some specialised plants. Example: Northern Canada.
Tropical rainforest	terrestrial	Tropical climate. Lush vegetation. Great diversity of animal and plant life. Example: forests of equatorial Africa.
Temperate rainforest	terrestrial	Cool and wet. Tallest trees in the world. Example: redwood forests of North America.
Temperate deciduous forest	terrestrial	Hot summers and cold winters. Dominated by broad-leaved deciduous trees. Example: oak woodlands of central and western Europe.
Lakes and ponds	aquatic: freshwater	Large bodies of standing freshwater. Example: Great Lakes of North America.
Streams and rivers	aquatic: freshwater	Flowing freshwater. Example: the Amazon.
Marine rocky shore	aquatic: sea water	The border between the land and the sea. Most rocky shores show **zonation**: there are characteristic bands of different environmental conditions that lead to colonisation by very different plants and animals.
Coral reef	aquatic: sea water	The tropical rainforests of the seas. Reefs form in warm, shallow waters and form a biome that contains an incredible diversity of living organisms.

Fig 33.6 The biomes of the world show great extremes of environmental conditions

(a) Arctic tundra

(b) Desert

(d) Rainforest

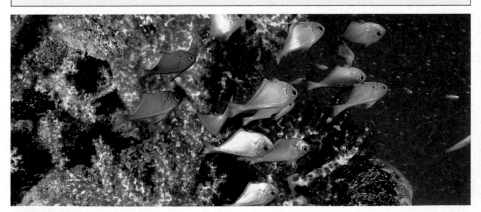

(c) Coral reef

THE ECOSYSTEM CONCEPT

The **ecosystem** is the basic functional unit of ecology. It is a single working unit that consists of a group of interrelated organisms and their physical environment (Fig 33.7). We can divide any one of the biomes described above into many different ecosystems.

When we study a particular organism in its ecosystem, we look at the physical features that might affect it, such as rainfall, soil type, temperature and so on. These physical or **abiotic** features help to determine what range and type and numbers of living organisms live inside the ecosystem. Of course, we also look at the living or **biotic** part of the ecosystem, to see how our chosen organism relates to the other organisms present. This is particularly important if we are looking at the impact of human activities.

On pages 580 to 582, we look at the abiotic and biotic features that affect the distribution of organisms in an ecosystem in much more detail.

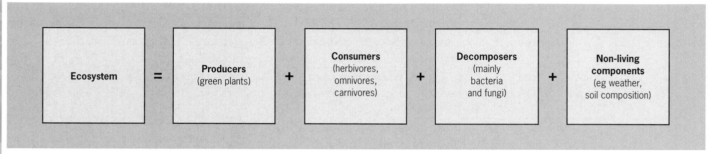

Fig 33.7 A schematic diagram showing the essential features of an ecosystem

Ecosystem = Producers (green plants) + Consumers (herbivores, omnivores, carnivores) + Decomposers (mainly bacteria and fungi) + Non-living components (eg weather, soil composition)

LEVELS OF ORGANISATION IN AN ECOSYSTEM

Individual living organisms may look randomly arranged in an ecosystem, but they are organised into recognisable units. Fig 33.8 provides an overview of organisation in an ecosystem.

Fig 33.8 An ecosystem has two basic parts: the abiotic part and the biotic part. Within both of these, there are further levels of organisation

FROM ECOSYSTEM TO INDIVIDUAL

We can study the human body and narrow our investigation from a body system to an important organ, to the level of the individual cell. In a similar way, we can focus in on the components of an ecosystem. Each one contains a **community** of organisms, which consist of many different **populations**. Each population is made up of many **individuals**. It is important to define these basic features of an ecosystem:

- A community is a collection of groups of organisms from different species that live in close association in the same ecosystem. Organisms in a community interrelate (see Chapter 36).
- A population is a group of individuals of the same species that live in a particular area at any one moment in time. We can look at a population of rabbits or a population of beech trees. Different populations interact in an ecosystem, forming a community.
- An individual is a single organism within a population. Some populations contain organisms that are genetically identical. Most populations of bacteria, for example, are identical because they have developed as a result of asexual reproduction. But many populations contain individuals that are the result of sexual reproduction: these individuals show genetic variation (Fig 33.9 and see Chapter 26).

ENVIRONMENT, HABITAT AND ECOLOGICAL NICHE

The divisions listed above relate primarily to the *living* components of an ecosystem. We can also sub-divide an ecosystem with reference to its *non-living* components. The term **environment** describes the overall physical surroundings that occur in a biome or an ecosystem (although it is often used more generally: people speak of the environment of the Earth).

Within the environment of a single ecosystem are **habitats**, individual areas in which particular groups or individual organisms live. The habitat of any organism is its normal home. Some organisms have only one habitat, for example, the giant panda lives only in bamboo forest. Others, such as humans, live in many different ecosystems and biomes and so have many different habitats.

Another important term is **ecological niche**. To describe an organism's niche, we need to show how that organism relates to the physical and biological components of its surroundings. It is not just where it lives, it is also how it lives and what it does.

Fig 33.9 This individual stoat is part of a larger population. Like other members of the group, she is genetically unique. She has evolved to become well adapted to her woodland environment and, if her particular genes allow her to live a long life, she will pass her genes on to several new members of the next generation

? TEST YOURSELF

C What is the difference between an organism's habitat and its ecological niche?

3 THE BASIC FEATURES OF AN ECOSYSTEM

Let's consider an example of a real ecosystem. Fig 33.10 shows the main features of a familiar aquatic ecosystem, the garden pond. This is simplified (a real pond contains more organisms), but you can see that an ecosystem is made up of physical or abiotic components such as water, mineral nutrients, soil and rock, and living or biotic components such as fish, snails, insects and water plants.

Fig 33.10 A schematic diagram showing a garden pond as an ecosystem. Every living organism in the pond is affected by abiotic factors such as temperature, light levels and mineral availability, and by biotic components – other living organisms

Think about one of the insects in the pond. It can only survive and live there if the physical conditions in the pond suit it. So the pH and the temperature of the water must be within a suitable range. The insect also needs to interact with other organisms: it needs to eat smaller organisms or to feed on the sap of plants; and it needs to be with other insects of the same species so that it can find a mate and reproduce. And there is a 'grey area': some of its physical needs, dissolved oxygen for instance, are only met if the insect lives with organisms that release oxygen as a product of photosynthesis.

THE ABIOTIC COMPONENTS OF AN ECOSYSTEM

The abiotic components of an ecosystem give it a physical form, providing places for organisms to live. Physical conditions help to determine where each organism survives best: plants adapted to dry, almost desert conditions will not thrive if a river changes its course and the environment becomes boggy and wet. The nature of the physical environment in any ecosystem depends on:
- the rock and soil types present;
- the type of landscape: mountain range, flood plain, etc;
- the position on Earth: altitude, latitude, etc;
- the climate and weather, including light, temperature, water availability, wind and water movements;
- the potential for catastrophe: fire, flood, etc.

✔ **REMEMBER THIS**

In a real ecosystem the abiotic and biotic factors all act together to determine the environmental conditions for each organism. Both interact to constantly change the conditions in the ecosystem. How this happens depends on the ecosystem and the physical conditions. The Science feature on the next page gives a thought-provoking example.

Some abiotic factors, such as the position of an ecosystem on Earth, remain unchanged as time passes. Others, such as the landscape and soil type, change very slowly. Although the basic climate of an ecosystem is usually pretty constant, you can get daily and often hourly changes in weather. Catastrophes such as fires and floods are sudden and short-lived, but they can produce severe, long-lasting effects (Fig 33.11).

Fig 33.11 In some ecosystems, catastrophic events are necessary. Some species of plant, such as *Banksia* from Western Australia (left), reproduce only after a serious fire (below). *Banksia* seeds are protected by closed capsules that can be broken open only by the heat of the flames: without exposure to fire, the seeds cannot germinate

---

! **SCIENCE FEATURE A modern-day plague**

The abiotic and biotic parts of any ecosystem interact in many different ways, some subtle and some not so subtle. Australia provides one of the most dramatic examples of how disrupting the delicate balance of an ecosystem can devastate both the environment and the populations of living organisms that depend on it.

Two hundred years ago, Australia was hot, dry and covered predominantly by grasslands, with some areas of forest. When European settlers arrived they started to clear the forests and brought in two alien species of mammal: sheep and rabbits. The rabbits in particular multiplied extremely rapidly in conditions that suited them very well, and were soon eating the grass faster than it could grow.

Ten years after just 24 rabbits were introduced, the wild population had grown to millions and started destroying crops and wild vegetation as well as any grass that was left. They out-competed native mammal species, driving many of them to extinction. This plague continued until the late 1950s, when farmers introduced the myxoma virus. This killed over 99 per cent of the rabbits, with a few surviving because they were naturally resistant. Today, Australia is in danger of being overrun by another rabbit plague.

Fig 33.12 A rabbit suffering from the symptoms of myxomatosis, the deadly viral disease that wiped out most of the rabbit population. Most rabbits today are resistant to the virus

Fig 33.13 Meercat sentries keep a look out for predators while other members of the group search for food and look after the young

Fig 33.14 Black-chinned humming birds pollinate the penstemon as they drink its nectar

BIOTIC COMPONENTS OF AN ECOSYSTEM

The biotic environment that surrounds an organism in an ecosystem results from the activities of all the other organisms living there. Complex relationships called intraspecific interactions occur between organisms of the same species. Interspecific interactions happen between organisms from different species (Table 33.2).

Table 33.2 Types of interactions that occur between organisms in an ecosystem

Activity	Type of interaction	What happens
Reproduction	intraspecific	Location of, selection of and competition for a mate.
Caring for young	intraspecific	Both parents, one parent or, more rarely, older siblings feed, protect and shelter the young.
Social behaviour	intraspecific	Animals co-operate to find food and to defend themselves against competitors or predators (Fig 33.13).
Competition	intraspecific	Organisms compete for resources such as food, space, light, water and mineral nutrients.
Reproduction	interspecific	Animal vectors pollinate some species of plant and help to disperse the fruits and seeds of others (Fig 33.14).
Caring for young	interspecific	Rare, but some species rear the young of another species.
Mutualism	interspecific	Two organisms both benefit from a long-term association.
Parasitism	interspecific	Individuals from one species use another species to provide food and shelter while giving nothing in return. Many hosts actually suffer from the interaction: many parasites cause disease.
Predation	interspecific	All animals obtain food by eating another organism.
Protection	interspecific	Many species make use of other species to hide from predators. Some copy the markings of other species to mimic warning signals that predators avoid.
Competition	interspecific	Organisms in a community compete for resources such as food, space, light, water and mineral nutrients.
Defence	interspecific	Organisms have developed various strategies to defend themselves against predators, parasites and competitors.

4 AN OVERVIEW OF THIS SECTION OF THE BOOK

In this section of the book, we provide an introduction to ecology at advanced level, with an emphasis on what is important for human biologists to know. In the next chapter there is a thorough overview of the importance of energy and a description of the biochemical processes of respiration and photosynthesis. In Chapter 35 we look at recycling in the natural world, with detail on the carbon cycle and nitrogen cycle. Chapter 36 is a study of populations and, finally, in the last chapter in this section, we look at how humans affect their environment and look at some of the big issues that seem likely to threaten our environment in the future.

34

Energy in living systems

34 Energy in living systems

An understanding of energy is of fundamental importance in biology, from the study of individual molecules and reactions to global issues such as the use of fossil fuels, pollution, damage to the ozone layer and the greenhouse effect

All living organisms depend on energy. Most of them live by the energy that comes from sunlight. Studying energy and how it flows through organisms and ecosystems is valuable for many topics in human biology.

Part I of this chapter gives an overview of some of the key concepts that you will need to understand in order to study energy flow in living systems. Part II gives a fairly detailed account of photosynthesis. This topic is crucial to understanding how energy flows through ecosystems and it is a core topic on most human biology syllabuses. Part III concludes the chapter with the process of cell respiration. Again we have included details of the biochemical pathways involved in this vital life process, but the information is presented in a step-wise format, so that you can build up your understanding.

With all three topics, remember that you should concentrate on understanding the relevance and overall importance of the process, not the details of individual biochemical reactions and pathways.

PART I: OVERVIEW OF KEY CONCEPTS

1 SOLAR ENERGY, PHOTOSYNTHESIS AND RESPIRATION

This planet, and all the organisms on it, must have energy and almost all of it is supplied by our nearest star, the Sun. A huge amount of sunlight reaches the surface of the Earth but only about 2 per cent of it is actually absorbed by plants. The rest of the energy heats up the land, air and water, preventing the planet from freezing.

In **photosynthesis**, plants use radiant energy from the Sun to convert simple inorganic substances (mainly carbon dioxide and water) into larger organic molecules (glucose, starch, lipids and proteins). The plant tissues built from the products of photosynthesis form food for organisms that cannot make their own, such as animals, fungi and most bacteria. Some of these feed on other organisms. Ultimately, therefore, the products of photosynthesis provide the energy which most living things need to carry out the processes of life. This energy is made available to organisms through the process of **cell respiration**.

We look at the biochemistry of photosynthesis and cell respiration later in this chapter, but first we revise some key concepts.

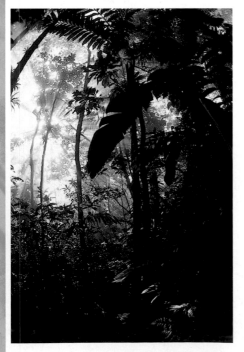

Fig 34.1 Photosynthesis by blue-green bacteria was responsible for the appearance of oxygen in the Earth's atmosphere 2 billion years ago, and photosynthesis by plants maintains the oxygen level at 20 per cent of the atmosphere today

2 WHY DO WE NEED ENERGY?

Humans need a constant supply of energy to maintain their life processes:

● **Growth and repair** of cells and tissues. Energy is required for the biochemical reactions that build large organic molecules from simpler ones. For example, energy is needed to build proteins from amino acids.

● **Active transport**. Energy is required to move some substances in or out of cells. The transport of amino acids from the small intestine into the blood is achieved by active transport. Active transport often takes place against a diffusion gradient (see Chapter 4) and it allows the body to control its internal environment more efficiently.

● **Movement**. All movement requires energy, and human movement occurs on several levels:

inside cells, eg chromosomes separating

whole cells, eg sperm swimming (Fig 34.2)

tissues, eg muscles contracting

whole organs, eg a heart beating

parts or whole organisms, eg talking, walking.

● **Temperature control**. Humans are warm-blooded animals, **homoiotherms**, and we use around 70 per cent of the energy from respiration to maintain the body at a constant 37 °C.

Fig 34.2 Movement requires energy. To swim far enough up into the Fallopian tube to reach an egg, a sperm cell needs enough energy to enable it to swim over 7500 times its own length – the equivalent of a 10 kilometre swim for an adult human

3 HOW DO LIVING ORGANISMS GET THE ENERGY THEY NEED?

In order to respire and so release energy, organisms need a supply of food. Food can be obtained in two ways:

● Some organisms can make their own food. These organisms are called **autotrophs** (meaning 'self-feeders') and they use energy from the surroundings to make the materials they need. There are two kinds of autotroph, the **photoautotroph** and the **chemoautotroph**. Photoautotrophs, such as green plants, photosynthesise, using sunlight as an energy source. Chemoautotrophs, all of which are bacteria, use energy made available from chemical reactions other than those involved in photosynthesis. Some bacteria involved in nitrogen recycling are chemoautotrophs (see Chapter 35).

● Some organisms need to obtain ready-made food because they cannot make their own. Such organisms are called **heterotrophs**. Heterotrophs must obtain their food, either directly or indirectly, from organisms that carry out photosynthesis. They either eat their food (ingesting it into a gut, as most animals do) or digest the food first and then absorb the nutrients, as decomposers, mainly fungi and bacteria, do (Fig 34.3).

Human nutrition is covered in more detail in Chapter 7.

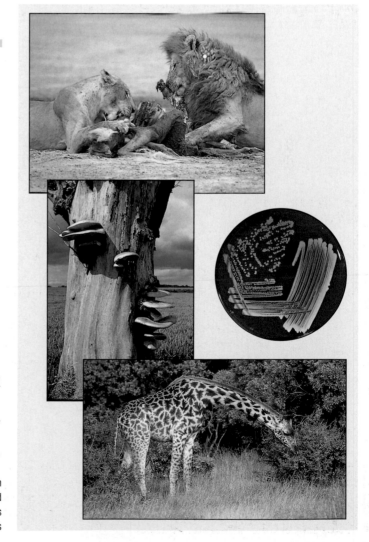

Fig 34.3 All of these organisms have something in common – they are heterotrophs. They cannot make their own food and must take it, ready made, from an outside source. As you can see, there are several different ways of doing this

4 ENERGY FLOW: WHERE DOES ENERGY COME FROM, WHERE DOES IT GO?

As Figs 34.4 and 34.5 show, as little energy is transferred at each point in a food chain, there is relatively little energy available for the final consumers, usually large carnivores. For this reason, such animals are quite rare.

Think of your own diet. You consume large amounts of food but very little of it is incorporated into your body. Compare your own body mass with an estimate of the mass of all the food you have ever eaten: you can see that a lot of energy must have been lost as heat. This heat is used to keep your body temperature at a constant 37 °C. So much is needed because you are constantly losing heat to the air, your clothes and the objects you touch.

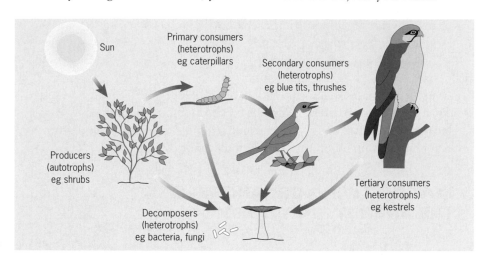

Fig 34.4 Energy flow through a food web

Of the light energy from the Sun that reaches the Earth, only about 42 per cent actually gets through to the surface of the planet. The rest is either reflected by the atmosphere, or is absorbed by it, a process that warms the atmosphere.

Of the light energy reaching the Earth's surface, about 2 per cent is absorbed by plants. The rest goes to produce the heat that warms up the land and the oceans.

Of the energy absorbed by plants, only 0.1–1.6 per cent is incorporated into plant tissue – and even some of this is lost as heat because the plant itself must respire.

Of the plant tissue eaten by herbivores, about 20 per cent (at most) will become incorporated into the tissues of the animal. Most of the rest is lost as heat as the animal respires.

Of the animal tissue eaten by a carnivore a maximum of 30 per cent becomes incorporated into the body of the carnivore. Most of the rest is lost as heat.

Fig 34.5 The basic idea of energy flow. At the start of any food chain there must be an organism capable of making food – usually a green plant. From there, the energy is passed along the chain within food molecules. Some of this energy is eventually used to make the cells, tissues and organs of heterotrophic organisms

5 ENERGY TRANSFER IN ECOSYSTEMS

In photosynthesis, green plants and some algae and bacteria use the energy from sunlight to power chemical reactions that combine simple inorganic molecules to form complex organic compounds. We say that the autotrophs in an ecosystem are **producers**: the products that they make form the food that ultimately feeds all other organisms in the ecosystem.

The heterotrophs in an ecosystem meet their need for energy by feeding on other organisms. We say that they are the **consumers**. Generally, **primary consumers** are **herbivores**: they obtain their energy by eating producers. **Secondary consumers** are **carnivores** or **scavengers**: they eat primary consumers. **Tertiary consumers** are **carnivores** that prey on other meat-eaters. This chain of dependence is usually called a **food chain** (Fig 34.6). The distinct levels of each chain are called **trophic levels**.

Fig 34.6 A simple flow chart summarising the overall structure of a food chain that has four trophic levels

As one organism eats another, there is a transfer of energy from the bodies of the producers (or consumers) into the bodies of consumers through the trophic levels of a food chain. Energy transfer occurs in one direction only: it is not recycled, and much of the energy escapes from the ecosystem as heat (see Fig 34.13).

Another major group of heterotrophs in an ecosystem are the **decomposers**. These are mainly bacteria and fungi. They break down the bodies of dead animals and plants, usually by extracellular digestion (Fig 34.7) and absorb the soluble products. Transfer of energy from producers and consumers to decomposers is also one-way, with a high proportion eventually lost as heat. The process of decomposition allows mineral nutrients to be cycled in the ecosystem (see page 629).

FOOD CHAINS AND FOOD WEBS

An example of a food chain is shown in Fig 34.8. Most animals have a varied diet. A food chain is therefore only part of a much larger feeding picture: the **food web**. Take our top freshwater carnivore, the pike. It does not feed on stickleback alone but also on roach and insects such as dragonfly nymphs and pond skaters.

Most rivers and lakes include a wide variety of producers, ranging from the microscopic plankton to the larger pondweeds, rushes and flowering plants.

Fig 34.7 Fungi grow into their food (in this case, human skin), secreting enzymes which digest the food externally. The soluble products of digestion are then absorbed through the walls of the hyphae

Freshwater food chain

phytoplankton ⟶ zooplankton ⟶ water fleas (*Daphnia*) ⟶ stickleback ⟶ pike

Fig 34.8 An example of a freshwater food chain. Each chain has five trophic levels (most food chains have fewer trophic levels than this)

⚠ SCIENCE FEATURE Food webs and pesticides

In developing a new pesticide, scientists look for a chemical that kills only the pest that we want to get rid of, harms no other species, breaks down into something harmless after a short time, does not lead to the development of resistance, and is cheap. This is a pretty tall order: no pesticide yet developed is perfect, but modern pesticides are much safer than some of those developed and used in the 1950s and 1960s.

DDT (dichlorodiphenyltrifluoroethane) was used as a pesticide in many countries after World War II. It breaks down slowly, remaining in the environment for between two and five years.

During the 1950s and 1960s it became obvious that DDT was affecting whole food webs. Fig 34.9 shows what happened when DDT was transferred up the trophic levels of a food chain in an ecosystem in the USA. DDT is not excreted and concentrates in the fatty tissues of organisms. So, although the concentration of DDT in zooplankton, the primary consumers in the chain, was only 0.04 parts per million, its concentration in the tissues of a top carnivore such as the osprey or bald eagle was 25 parts per million. The bodies of the large birds of prey converted the DDT into a substance that made their eggshells very fragile. As a result, very few birds managed to breed and their numbers fell quickly.

In 1972, DDT was banned and since then, the numbers of the large predatory birds have recovered. However, there are still problems because of the illegal use of DDT.

Fig 34.9 Bioamplification of DDT through an aquatic food chain. DDT accumulates in the fatty tissues and cannot be excreted. The fat of the osprey contains over 8 million times more DDT than the water at the bottom of the food chain in which the producers live

PYRAMIDS OF NUMBERS

Food chains and food webs tell us a great deal about the feeding relationships that occur between organisms in an ecosystem, but they are **qualitative** rather than **quantitative**. That means that we know which organisms are part of the chain or web, but we have no idea of the exact numbers of organisms involved at each trophic level.

Clearly, numbers are very important. There are many more individual producers than primary consumers, and, as you carry on along a food chain, the numbers of organisms at each trophic level continue to decrease. At the same time, the body size of individual organisms usually increases. It is difficult, if not impossible, to find out exactly how many producers are eaten by a primary consumer and so on, but we use a pyramid of numbers to show the general idea (Fig 34.10).

Two odd things can happen when you use **pyramids of numbers**. Sometimes the base of the pyramid is very narrow, indicating that there are fewer producers than primary consumers. This seems not to make sense, but these examples arise when a few large plants such as trees produce food for thousands of tiny plant-feeders, such as aphids (Fig 34.11a). In pyramids of numbers that represent food chains that have a population of parasites at the top, the upper level can be much larger than the level below it. This is because many parasites can feed on one host (Fig 34.11b). We say that these pyramids are **inverted pyramids**.

Fig 34.10 A simple pyramid of numbers. The width of each box represents the relative number of organisms present in each trophic level at any one particular time. In this example, many grass plants are needed to sustain all the antelope that one individual lion needs to eat to survive

Fig 34.11 Two examples of inverted pyramids of numbers. Neither is completely inverted; they simply do not conform to the expected pyramidal shape because one of the levels is distorted
(a) One sycamore tree can feed many thousands of aphids, making the first trophic level of this pyramid of numbers much smaller than the second
(b) This pyramid is complicated by the top trophic level, which involves a parasite. Many parasites can infect one host. *Wuchereria bancrofti* is a nematode worm that causes the disease elephantiasis in humans (see Chapter 29)

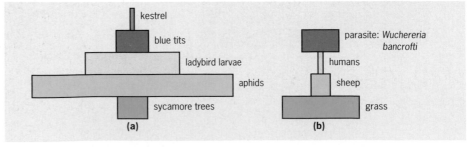

PYRAMIDS OF BIOMASS

Ecologists often use **pyramids of biomass** (Fig 34.12a), to avoid the problem of inverted pyramids of numbers. Pyramids of biomass show the mass of all the organisms at each trophic level – dry mass is the most useful measure. This is obviously extremely difficult: obtaining the dry mass of organisms means killing them, drying them and measuring the remains. Not surprisingly, very few pyramids of biomass have ever been determined in this way.

However, using samples of organisms and their wet mass allows ecologists to make estimates and devise models. But, even this type of pyramid can be inverted, as Fig 34.12(b) shows.

To obtain an even better model of an ecosystem, one that avoids inverted pyramids altogether, we must look at the way energy is transferred between the different trophic levels.

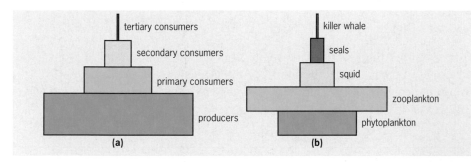

Fig 34.12(a) A generalised pyramid of biomass for a terrestrial ecosystem
(b) In the waters of the Antarctic, the mass of zooplankton is five times that of the phytoplankton. If we draw up a pyramid of biomass for one of the food chains, we get an inverted pyramid. The zooplankton eat the phytoplankton so quickly that the latter never get the chance to attain a large biomass. Instead, phytoplankton have a high turn-over rate, reproducing very quickly. So the biomass of organisms produced each year is large, although the number alive at any one time is often less than the number of consumers

PYRAMIDS OF ENERGY

Fig 34.13 shows a typical pyramid of energy. It shows the amount of energy that is transferred along the food chain and how much is lost as heat at each level. The more levels in the chain, the less energy there is available at the top. Consequently, very few food chains have more than four or five levels. It also explains why the top carnivores such as the big cats and large whales are the first to become endangered when their ecosystems come under pressure from humans.

EFFICIENCY OF ENERGY TRANSFER BETWEEN DIFFERENT TROPHIC LEVELS

As we saw earlier in this chapter, the energy transfer between organisms at different trophic levels of a food chain is never anything like 100 per cent efficient. In fact, less than 30 per cent of the energy available ever reaches the organisms in the next level. There are three main reasons for this:

● Not all the organisms in one trophic level are eaten by organisms from the next level. Many die and provide food for decomposers.
● Not all food that is eaten is digested and not all digested food is absorbed – some is lost as faeces.
● Most of the energy gained from the food that is absorbed is lost through the processes of respiration and excretion (production of metabolic waste).

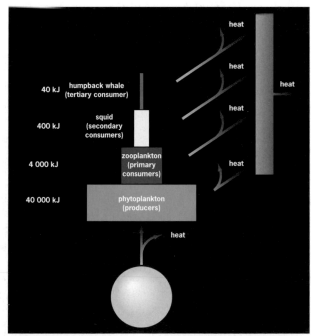

Fig 34.13 A pyramid of energy for an Antarctic food chain similar to the one shown as a pyramid of biomass in Fig 34.12(b). As you go up the pyramid, there is a 90 per cent loss of energy at each trophic level. This energy, of course, does not disappear: it is 'lost' as heat. For a more detailed investigation of energy transfer, see the Assignment for this chapter

SEE QUESTION 2

? TEST YOURSELF

D List the foods you ate at your last meal. Sketch out the food chains that produced them and, for each food chain, say which trophic level you, the consumer, represent.

PART II: PHOTOSYNTHESIS

6 THE IMPORTANCE OF PHOTOSYNTHESIS

It is difficult to overstate the importance of photosynthesis – the process that feeds the world and replenishes the atmosphere with oxygen. Organisms that photosynthesise are able to capture the energy of the sun and use it to make carbohydrate molecules such as sugars and starches, and, indirectly, other essential compounds such lipids, proteins and nucleic acids. It is no exaggeration to say that photosynthesis turns thin air into food.

Photosynthesis is a complex process, but it can be summarised as:

$$\text{Carbon dioxide} \quad \text{and} \quad \text{water} \xrightarrow[\text{chlorophyll}]{\text{light}} \text{glucose} \quad \text{and} \quad \text{oxygen}$$

$$6CO_2 \quad + \quad 6H_2O \quad \rightarrow \quad C_6H_{12}O_6 \quad + \quad 6O_2$$

You shouldn't think of this as a chemical equation, even though it looks a lot like one. In reality, the carbon dioxide molecules never react directly with water molecules, so this is more a summary of photosynthesis, showing what goes in and what comes out.

! SCIENCE FEATURE Artificial photosynthesis

Plants are able to convert sunlight into a high-energy fuel in a complex yet very efficient process that scientists have so far been unable to duplicate. In recent years, however, progress has been made towards creating a type of artificial photosynthesis. The benefits in the long term could be huge; cheap, clean energy sources.

A team at the Massachusetts Institute of Technology in the USA has developed a protein that captures energy from a light source and stores it in the form of hydrogen gas. Creating a molecule to replace a leaf – 'photosynthesis in a beaker' – could help revive interest in the Sun as a source of energy.

One possible spin-off of this technology is that in future Armed Forces operations in the field could rely on biological methods to produce electricity through photosynthesis, rather than relying on fossil fuels and bulky, heavy batteries.

Over millions of years of natural selection plants have evolved to make optimum use of the different wavelengths of light, and plants convert 98 per cent of the sunlight they receive into energy. Conversely, current artificial solar energy systems are only 10–15 per cent efficient. Using plant proteins to produce electricity – an area of research known as biological photovoltaics – could make the process much more efficient; perhaps by 40–50 per cent.

It is possible that in the future soldiers could have photosynthetic coatings on their Kevlar helmets that could produce enough energy for their electronic equipment. Other equipment and vehicles could also be covered with these solar converters. An added benefit is that the protein coatings would make whatever they coat more difficult to detect by electronic means since they would mimic the natural environment.

Fig 34.14 The light infantry? In future, helmets may by able to photosynthesise and produce enough energy to power a soldier's equipment

7 PHOTOSYNTHESIS FACTS

- Organisms that can photosynthesise include higher plants (Fig. 34.15), algae (including the seaweeds) and photosynthetic bacteria.
- Organisms that can photosynthesise are called **producers**, or **autotrophs** ('self-feeders').
- The process involves absorbing the light energy of the sun and converting it into chemical energy. The compounds made, such as sugars, starch, lipids and protein, contain this energy in their chemical bonds.
- Globally, photosynthesis makes billions of tonnes of organic compounds every year; one estimate puts global production at 70×10^{12} kg per year.
- Up to 50 per cent of all photosynthesis takes place in the sea, in the tiny algae known as phytoplankton (Fig 34.16).
- Virtually all ecosystems are supported by photosynthesis – it is the most important way energy gets into an ecosystem.
- Photosynthesis needs light, chlorophyll, carbon dioxide, water and a suitable temperature. The rate of photosynthesis is limited by the factor in shortest supply.
- Photosynthesis is a complex step-by-step process – each step is catalysed by a particular enzyme.

Fig 34.15 Where did this tree come from? We know that the material to make our bodies comes from the food we eat, but what about plants? This tree may weigh upwards of 3000 tonnes, but the vast majority of the organic compounds in the wood didn't come from the soil – they were made by photosynthesis using carbon dioxide from the atmosphere. A huge amount of water was also needed from the roots, but most of the dry mass comes, quite literally, from thin air

Fig 34.16 When you think of photosynthesis on a global scale you probably think of rainforests or fields of crops, but the tiny algae in the oceans produce just as much food as land plants. These are diatoms, one type of single-celled algae that are responsible for producing huge amounts of food in the world's oceans. As one of the many species that makes up phytoplankton (plant plankton) they support vast marine and freshwater ecosystems

8 PHOTOSYNTHESIS AND RESPIRATION

If you look at the overall summary for respiration, where:

glucose and oxygen → carbon dioxide and water + energy,

you can see that photosynthesis and respiration are effectively the reverse of each other. Photosynthesis traps energy in organic molecules, and respiration releases it. Overall, photosynthesis is the reduction of carbon dioxide to form organic molecules, while respiration is the oxidation of organic molecules to form carbon dioxide. It is important to remember that plants also respire day and night, but in order for them to accumulate organic molecules and to grow, their overall rate of photosynthesis must exceed their rate of respiration.

9 WHAT HAPPENS IN A LEAF?

In most plants the leaf is an organ of photosynthesis. As Fig 34.17 shows, the leaf is perfectly adapted to maximise the process. The most actively photosynthesising cells in the leaf are the **palisade cells**, and the whole structure of the leaf is geared towards providing these cells with what they need and taking away their waste products.

Fig 34.17 Journey to the centre of a leaf. The chlorophyll pigments are housed on discs called thylakoids that are piled into grana inside chloroplasts. Most chloroplasts are found in the palisade and spongy mesophyll cells in the centre of leaves

section through leaf
- upper epidermis
- palisade mesophyll
- spongy mesophyll
- lower epidermis
- many chloroplasts in each spongy mesophyll cell

section through chloroplast
- outer membrane
- inner membrane
- stroma (the 'dark reaction' of photosynthesis occurs here)
- granum (the 'light reaction' of photosynthesis occurs here)

two grana
- each granum is made up of a stack of thylakoid discs
- extensions of the thylakoid disc connect separate grana

section through granum
- a granum
- thylakoid membrane (a phospholipid bilayer containing chlorophyll and proteins)
- thylakoid space
- chlorophyll with associated proteins for light absorption
- a thylakoid disc

antenna complex in thylakoid membrane
- light energy
- pigment molecules
- energy transfer
- reaction centre chlorophyll molecules

Adaptations of the leaf shown in Fig 34.17:
- It is thin, so that gases can diffuse in and out quickly.
- It has a large surface area to maximise light absorption.
- It has a thin layer of wax – the **cuticle** – to reduce evaporation from the upper surface.
- It has a rich network of **xylem** fibres to deliver water and minerals, and **phloem** fibres to remove the product of photosynthesis (mainly as sucrose). This vascular tissue also gives the leaf more rigidity.
- The cells of the upper epidermis have no chloroplasts, so that more light gets through to the palisade cells.
- The palisade cells are deep and tightly packed. The chloroplasts within them can move so that all have a chance of obtaining all the light they need.
- Guard cells surround pores called **stomata** (singular = stoma; 'hole') opening in the day to allow carbon dioxide in, then closing at night to reduce water loss.
- Loosely packed **spongy mesophyll** cells create air spaces, making gas exchange more efficient between the atmosphere and the palisade cells. The air spaces also trap a little carbon dioxide, allowing some photosynthesis even when the stomata are closed.

? TEST YOURSELF

E In order to make photosynthesis in the palisade cell as efficient as possible, what must be supplied and what must be taken away? Use the basic photosynthesis equations to help you.

CHLOROPLASTS

Chloroplasts are the organelles of photosynthesis. In the leaves of higher plants there can be as many as 100 or more chloroplasts in each cell, but more commonly 20–30. Their detailed structure is shown in Fig 34.18.

Fig 34.18 False colour transmission electron micrograph of chloroplast. The grana are connected by thin membranes; intergrana lamellae. Note the starch grains and lipid globules

lipid droplet

stroma: the site of the light-independent reactions

The stroma is a fluid containing enzymes that use ATP generated in the light-dependent reactions of photosynthesis to fix carbon dioxide into sugar

thylakoid: the light-dependent reactions happen here

The thylakoid membrane is a phospholipid bilayer that has many chlorophyll molecules embedded in it. These absorb light energy and then transfer it to other protein molecules. This stage of photosynthesis generates ATP

outer membrane

The outer membrane is a phospholipid bilayer that controls movement of molecules in and out of the chloroplast

starch grain

Excess carbohydrate made during photosynthesis is temporarily stored as starch grains

stroma thylakoid

membrane-bound ribosomes

free ribosomes

inner membrane

granum

The thylakoids are arranged in stacks, each called a **granum**. This greatly increases the efficiency of the light-dependent reactions by capturing most of the light energy that enters the cell

CHLOROPLAST FACTS

- They are usually shaped like biconvex discs, about 5 µm across and 2 µm deep. The flattened shape increases the rate of the exchange of materials.
- They have a double outer membrane.
- There is an elaborate system of membranes inside the chloroplast that provides a large surface area for housing the chlorophyll pigments. The disc-shaped membranes that house the chlorophyll are called **thylakoids**, which are stacked into **grana** (singular = granum). Usually, the thylakoids are at right angles to the incoming light.
- The fluid surrounding the grana is called the **stroma**.
- Chloroplasts have their own DNA and ribosomes so they can synthesise their own proteins.
- It is thought that chloroplasts may have originated as free-living algae that became incorporated into a larger cell – see the endosymbiont theory on page 22.

CHLOROPHYLL AND LIGHT

In order to understand the role of chlorophyll in photosynthesis it helps to know something about the nature of light. Fig 34.19(a) shows the electromagnetic spectrum – the range of different types of radiation. Most of the radiation that reaches the Earth from the Sun falls in the region of 400–700 nm wavelength, so not surprisingly this is what our eyes have evolved to see; it is the **visible spectrum**. When stimulated by the whole 400–700 nm range we perceive white light but, as Fig 34.19(b) shows, this can be divided up into the familiar seven colours. Chlorophyll has also evolved to make use of this available light.

What we think of as chlorophyll is actually a mixture of compounds that fall into two basic types; the **chlorophylls** and the **carotenoids** (Fig 34.20). The two main chlorophylls, *a* and *b*, are virtually identical molecules that absorb light mainly in the red and blue areas of the spectrum. The carotenoids, such as **β-carotene**, harvest light from different wavelengths, thus allowing plants to make much better use of the available light. Energy absorbed by the carotenoids is eventually passed on to the chlorophyll molecules before being used in photosynthesis. The carotenoids also act as a safety mechanism, protecting the leaf from the damaging effects of too much light.

(a)

| gamma rays | X rays | UV | | infrared | micro wave | radio wave |

(b) blue missing / red missing

chlorophyll solution

Fig 34.19(a) The electromagnetic spectrum ranges from the long wavelength (and lowest energy) radio waves to the shortest wavelength (high-energy) gamma rays. Visible light forms just a small part of the spectrum, between ultraviolet and infared
Fig 34.19(b) White light splits into its component colours when passed through a prism. When white light is first passed through chlorophyll the red and blue light is absorbed, which is why plants appear green

Fig 34.21 The structure of chlorophylls *a* and *b*. The molecule consists of a head to absorb light and a tail to anchor it to the thylakoid membrane. The head is a complex *chlorin* ring structure with a magnesium ion at the centre. Different chlorophylls have slight variations on the side groups of the head, which allow different wavelengths of light to be absorbed

Key

→ co-ordinate linkage

X is CH_3 in chlorophyll *a*
is CHO in chlorophyll *b*

Fig 34.20 If you grind up a leaf with a little sand (to break up the cell walls) and a suitable solvent such as propanone, you can run a paper or thin layer chromatogram (see page 42 in Chapter 3) to separate out the different pigments that make up chlorophyll

The structure of **chorophylls *a*** and ***b*** is shown in Fig 34.21. The ring structure surrounding the magnesium ion contains five smaller rings consisting of alternating single and double bonds. In reality each bond is partly double; a **conjugated system**. These bonds absorb light, so the more of them there are, the greater the light absorbance. Chlorophyll *a* absorbs strongly in the 660–700 nm range; red light. It also absorbs to a lesser extent in the blue range.

The wavelengths of light that stimulate photosynthesis are shown by an **action spectrum**, obtained by varying the wavelength of light a plant receives and then measuring the rate of photosynthesis. When the chlorophyll pigments are extracted, and their absorption at the different wavelengths is measured, we get an **absorption spectrum**. Fig 34.22 shows that the two graphs are very similar, showing that the chlorophylls are the pigments responsible for photosynthesis.

SEE QUESTIONS 3 AND 4

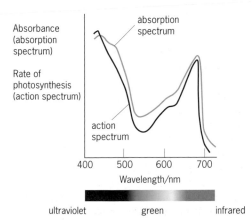

Fig 34.22 The action spectrum shows the wavelengths of light that stimulate photosynthesis, while the absorption spectrum shows the wavelengths of light absorbed by isolated chlorophyll pigments

10 AN OVERVIEW: THE FIRST STEP TO UNDERSTANDING PHOTOSYNTHESIS

Photosynthesis is often seen as a difficult topic, but approaching it in the right way will help you make sense of it. Firstly, you need an overview or framework that you can fit the details on to later. Photosynthesis can be divided into two easy steps as summarised below and in Fig 34.23:

● **Step 1** The light-dependent reaction. Light hits chlorophyll, which emits two high-energy electrons. These electrons pass through a series of electron transfer reactions that make ATP and NADPH. The electrons lost by the chlorophyll are replaced when water is split, a process that also produces oxygen as a by-product.
● **Step 2** The light-independent reaction. ATP and NADPH are used to power a series of reactions known as the **Calvin cycle**. Overall, these reactions reduce carbon dioxide into the all-important carbohydrate.

For some A-level specifications, all the detail you need (and more) has been covered in the chapter so far and the summary at the end. Other specifications require the more in-depth account that follows.

Fig 34.23 An overview of photosynthesis

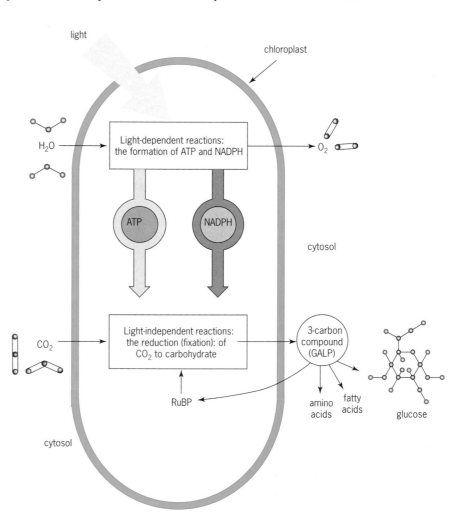

11 THE LIGHT-DEPENDENT REACTION

In order to understand the light-dependent reaction, you need to be familiar with the fine structure of the chloroplast, and in particular the arrangement of the chlorophyll molecules on the thylakoid membranes. As Fig 34.17 shows, chlorophyll molecules are arranged in funnel shapes, known as **antennae complexes**. The mouth of the funnel is occupied by several different types of pigment, allowing light to be harvested across more of the visible spectrum.

When a photon of light hits any of the molecules of chlorophyll at the mouth of the complex, the electrons they contain absorb the energy and are raised from a low-energy **ground state** to an **excited state**. As the energy moves from molecule to molecule down the antenna complex, electrons are first excited, then fall back to the ground state as they pass their energy on. The energy passes on down the antenna complex until it eventually reaches two chlorophyll *a* molecules at the bottom, known as the **special pair**.

The stages in the light-dependent reaction (Fig 34.24) are as follows:

- **Stage 1** A photon of light hits photosystem II and is channelled down to the reaction centre, which emits high-energy electrons.
- **Stage 2** The electrons are picked up by electron acceptors. In doing so the electron acceptor becomes reduced while the chlorophyll is oxidised. The electrons are then transferred along a series of electron carriers in the thylakoid membrane. The energy in the electrons is used to make ATP – this is **photophosphorylation**.
- **Stage 3** The electrons, having fallen to a lower energy level, are absorbed by the chlorophyll molecules in photosystem I.
- **Stage 4** A photon of light hits the reaction centre in photosystem I, which also emits two high-energy electrons. Again, the electrons are channelled down a series of electron carriers but this time the electrons combine with H^+ ions to reduce NADP into NADPH.

✔ REMEMBER THIS

The light-dependent reaction takes place on the thylakoid membranes, while the light-independent reaction takes place in the stroma of the chloroplasts

? TEST YOURSELF

F What is the difference between oxidation and reduction?

G Which one involves a molecule gaining energy?

SEE QUESTION 5

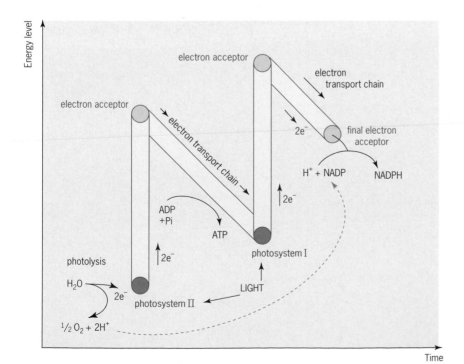

Fig 34.24 Schematic diagram of the light-dependent reaction

CYCLIC AND NON-CYCLIC PHOTOPHOSPHORYLATION

Photophosphorylation refers to the production of ATP using the energy from light. There are two methods of doing this: cyclic and non-cyclic. ATP production as described above is called **non-cyclic phosphorylation**, because the electrons are not recycled. Electrons that start off in water molecules are passed through the process until they end up in NADPH, which is then used to make carbohydrate.

However, ATP can also be made by **cyclic phosphorylation**. In this case some of the electrons from photosystem I are channelled back on to the first electron transport chain, and their energy is used to make ATP before returning to photosystem II. This process therefore makes ATP but not NADPH. No water is needed to provide electrons, and so no oxygen is produced.

12 THE LIGHT-INDEPENDENT REACTION

The light-independent reaction has in the past been referred to as the *dark reaction* but this misleading name has been dropped because it does not happen in the dark. The light-independent reaction happens in the light because it needs the ATP and NADPH that the light-dependent reaction produces.

The light-independent reaction takes place in the stroma (fluid part) of the chloroplast. The process is also known as the **Calvin cycle** after Melvin Calvin, the American scientist who first worked out the details in the 1950s.

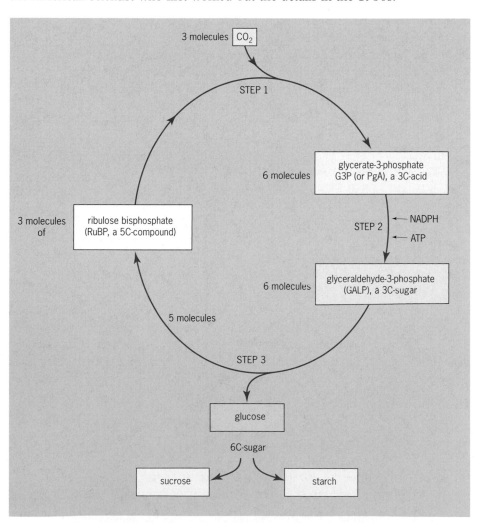

Fig 34.25 The light-independent stage of photosynthesis. See text for an explanation of each step

The steps in the Calvin cycle are shown in Fig 34.25.

- **Step 1** Carbon dioxide diffuses into the stroma of the chloroplast and combines with a 5-carbon sugar called **ribulose bisphosphate** (RuBP). This forms an unstable 6-carbon compound that immediately splits into two 3-carbon compounds called **glycerate 3-phosphate** (G3P). The enzyme that catalyses this reaction, **ribulose bisphosphate carboxylase** (mercifully shortened to **RUBISCO**), is the most abundant enzyme on Earth, accounting for up to 60 per cent of the protein in many leaves.
- **Step 2** The energy in ATP and the reducing power of NADPH are used to convert G3P into glyceraldehyde phosphate (GALP).
- **Step 3** GALP is used in two ways: five out of every six molecules are used – via a series of reactions – to make more RuBP to continue the cycle. One molecule in every six is used to make glucose.

Note that the carbon dioxide puts in one single carbon atom for every turn of the cycle. Thus the cycle must turn six times to make one glucose molecule. If the glucose were allowed to accumulate, it would lower the water potential of the cell, so it is converted into starch almost as soon as it is made, and this accounts for the large starch grains that can be seen in actively photosynthesising chloroplasts.

Plants move carbohydrate around in the form of sucrose. The plant's main storage compound, starch, is a large molecule that cannot pass across the cell membrane, so it must be converted into a smaller compound before it can pass into the phloem tissues of the plant.

PHOTORESPIRATION

It is thought that photosynthesis evolved in an atmosphere that contained much less oxygen and more carbon dioxide than it does today. The enzyme that catalyses the reaction between RuBP and carbon dioxide, RUBISCO, can also catalyse a reaction between RuBP and oxygen – so oxygen and carbon dioxide compete for the same active site. When oxygen levels are high and carbon dioxide levels low, the oxygen reaction is favoured.

$$O_2 + RuBP \xrightarrow{\text{RUBISCO}} \text{glycolate (a 2-carbon compound)} + G3P$$

Compare this to the usual reaction of the Calvin cycle

$$CO_2 + RuBP \xrightarrow{\text{RUBISCO}} 2 \times G3P \text{ (2} \times \text{3-carbon compound)}$$

You can see that one molecule of G3P is made instead of the usual two in the Calvin cycle. The glycolate cannot be fixed into carbohydrate, and instead passes into the mitochondria where it is respired.

Owing to its dependence on light, and the fact that it takes in oxygen and releases carbon dioxide, this process is called **photorespiration**. This process is wasteful to a plant because it means that some of the carbon dioxide that could be fixed into carbohydrate is lost. It has been estimated that photorespiration makes photosynthesis 30–40 per cent less efficient. However, some plants have overcome this problem – see the section on **C4 plants** below.

13 VARIATIONS ON THE THEME OF PHOTOSYNTHESIS

The mechanism of photosynthesis described above is known as the **C3 pathway** because the first stable product of the Calvin cycle is a 3-carbon compound, glycerate 3-phosphate. Most familiar plant species are C3 plants but there are two important variations on the theme: **CAM plants** and **C4 plants**.

CAM PLANTS DO IT IN THE DARK

C3 plants, the plants we have discussed so far, would not survive long in the desert. Their open stomata would prove to be a liability, they would rapidly lose water, dry out and die. In drier habitats some plants have overcome this problem by opening their stomata to take in carbon dioxide only at night. This drastically reduces water loss as the plants then keep their stomata tightly closed during the hours of heat and daylight. But how do they store their carbon dioxide to make it available for the light-dependent phase of photosynthesis? The key is that they use carbon dioxide to make an organic acid, which they store during the night and then break down in the day to release the essential carbon dioxide.

Plants that photosynthesise in this way are called **CAM (crassulacean acid metabolism)** plants. The carbon dioxide that enters the leaves at night is fixed into the organic acid **malate** by the enzyme phosphoenolpyruvate (PEP) carboxylase, and is then stored in the vacuoles until the daytime.

When the sun rises, the stomata of CAM plants close, preventing the loss of precious water. The malate is broken down to provide carbon dioxide according to the simple reaction

$$\text{malate} \rightarrow \text{pyruvate and } CO_2$$
$$(C4) \quad \rightarrow \quad (C3) \quad\quad (C1)$$

During daylight, C4 plants make their carbohydrate in the same way as 'standard' C3 plants. CAM plants are well suited to climates where there are high day and low night temperatures, especially where there is occasional drought. They tend to be native to semi-arid and tropical environments (Fig 34.26).

Overall, CAM plants overcome the problem of water loss by
● opening their stomata at night and closing them during the day;
● having fewer stomata than temperate plants;
● having a low surface area to volume ratio, which reduces evaporation.
These measures result in CAM plants losing as little as 50–100 g of water for every gram of dry mass produced, compared with 400–900 g for C3 plants.

Fig 34.26 The prickly pear, *Opuntia,* is a CAM plant. This species thrives in warm, dry climates. When introduced into Australia in the 1920, this species out-competed the native plants and took over large areas of farmland until cactus moth caterpillars were introduced to control it

C4 PLANTS LIKE IT HOT

In the tropics many species, including some commercially important plants such as sugar cane, maize (Fig 34.27) and millet, exhibit remarkably rapid growth compared with C3 plants. They have achieved this mainly by overcoming the problem of photorespiration (see above). How have they managed to do this?

In 1966 two Australian scientists, Hatch and Slack, discovered that C4 plants were much more efficient at taking up carbon dioxide.

The key to C4 photosynthesis comes from the modified structure of the leaf. Surrounding the vascular bundles (the xylem and phloem tissue) is a layer of **bundle sheath cells** that have different chloroplasts.

The stages in C4 photosynthesis as shown in Fig 34.28 are as follows:
● **Stage 1** Carbon dioxide is fixed in the mesophyll cells. Instead of having RuBP and the enzyme RUBISCO, C4 plants use PEP (phosphoenolpyruvate) and the enzyme PEP carboxylase. This enzyme has a higher affinity for carbon dioxide than does RUBISCO, and it does not react with oxygen. The big advantage is that all the available carbon dioxide is fixed; none is lost on photorespiration. The 4-carbon compound oxaloacetate gives C4 plants their name.

$$PEP + CO_2 \xrightarrow[\text{PEP carboxylase}]{} \text{oxaloacetate (4C)}$$

● **Stage 2** Oxaloacetate is converted into malate (or a similar compound), which is shunted across into the chloroplasts of bundle sheath cells.

Fig 34.27 C4 plants such as maize show some of the fastest growth in the plant kingdom

- **Stage 3** The malate is used to make carbon dioxide, pyruvate (a 3-carbon chemical) and hydrogen, which is used to make NADPH.

- **Stage 4** The pyruvate is moved back into the mesophyll cells and used to make more PEP, while the carbon dioxide is now fixed into carbohydrate in the same was as in C3 plants, ie using RuBP and the enzyme RUBISCO. Note that in C4 plants the RUBISCO is always surrounded by a high concentration of carbon dioxide so that it is not inhibited by oxygen.

14 WHAT LIMITS PHOTOSYNTHESIS?

This is a very important question, because if commercial plant growers can increase the rate of photosynthesis, they can make plants grow faster, feed more people and/or make more money. The law of limiting factors states that the *rate of a process is limited by the factor in shortest supply*. For example, if carbon dioxide is in short supply, providing the plant with more carbon dioxide will continue to increase the rate of photosynthesis until some other factor – light intensity for example – becomes limiting. In this section we look at each limiting factor. Table 34.1 gives a brief overview of the way in which farmers can increase the growth of crops.

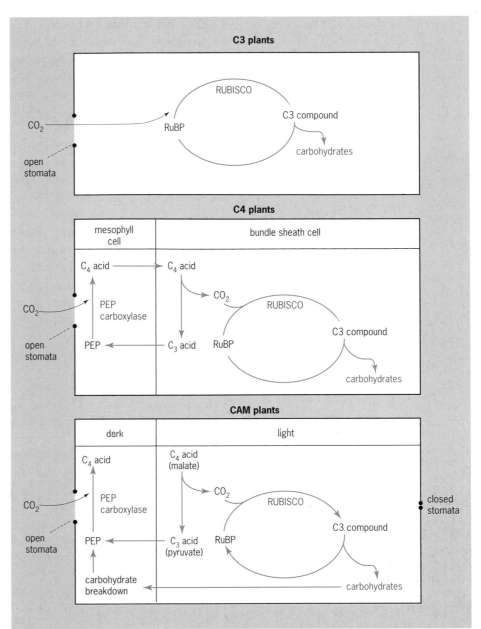

Fig 34.28 Comparisons of C3, C4 and CAM hotosynthesis

Table 34.1 How to increase the supply of limiting factors			
Factor	**How supply can be increased**	**Is it economically viable...**	
		In a glasshouse?	**In a field?**
Temperature	burn a fossil fuel, eg paraffin	usually	no
Carbon dioxide	burn a fossil fuel, eg paraffin	usually	no
Mineral ions	add fertiliser	usually	usually
Light	artificial light	sometimes, with a valuable crop	rarely
Water	irrigate or spray	essential	rarely in UK, essential in some climates

LIGHT INTENSITY

Light intensity is defined as the number of photons falling on a given area in a given time. Geographically, light intensity varies according to latitude – the closer to the equator, the greater the light intensity. Light intensity also varies according to the time of day, the time of year, the amount of cloud cover and whether the plant is shaded by other plants or objects. Generally, as light intensity increases so does the rate of photosynthesis until another factor – often carbon dioxide – becomes limiting. Plants have evolved to cope with different light intensities; for example plants on the forest floor have to photosynthesise in much lower light intensities than the trees above. How do they manage?

Plants respire all the time, but only photosynthesise when there is light. Photosynthesis will begin after dawn and the rate will increase until it equals the rate of respiration – this is the **compensation point**. After the compensation point photosynthesis will exceed respiration and so the plant will accumulate organic compounds. Plants that can photosynthesise in low light intensities are shade-tolerant and have a low compensation point. In contrast, shade-intolerant species have a high compensation point and so can only grow in bright environments. They cannot tolerate shading.

WATER SUPPLY

Although water is essential for photosynthesis, this represents only a tiny proportion of the water that flows through the plant in the transpiration stream. A lack of water will normally cause a plant to wilt and affects all other metabolic processes long before the process of photosynthesis is affected directly.

CARBON DIOXIDE

Carbon dioxide is the commonest limiting factor when light intensity is sufficient. The atmosphere contains only about 0.035 per cent carbon dioxide, which is about 35 parts per 10 000. When the carbon dioxide levels are low and oxygen levels are high the enzyme RUBISCO can be inhibited – see the section on photorespiration and C4 plants above.

TEMPERATURE

Photosynthesis, like all metabolic processes, is temperature-dependent. As the temperature increases the kinetic energy of the molecules increases, they 'bounce around' more, so that diffusion of gases in and out of the leaf is faster, and there are more collisions between enzymes and substrates. The optimum temperature for C3 plants is about 25 °C, while for C4 plants it is about 35 °C. Above these temperatures a metabolic imbalance develops so that photosynthesis becomes less efficient, while extremes of temperature will of course denature the enzymes.

INORGANIC IONS

Plants need a variety of mineral ions from the soil in order to remain healthy. Of particular relevance to photosynthesis are nitrogen and magnesium; both are constituents of the chlorophyll molecule. Without these ions the plant develops yellow leaves, a condition known as chlorosis.

PART III: CELL RESPIRATION

15 CELL RESPIRATION: A VITAL LIFE PROCESS

Understanding cell respiration can be a demanding task. You will probably find it useful to read Part I of this chapter first: it looks at key concepts that you will need to understand before you tackle the details given in this chapter.

All living things obtain the energy they need from **respiration**, a chemical process that breaks down simple food molecules such as glucose. Animals, including humans, digest food to produce these molecules, which are absorbed into the blood and then transported round the body.

Multicellular organisms mainly respire **aerobically** (using oxygen). Oxygen is absorbed by lungs, gills or the body surface and is usually distributed around the body by the **circulatory system**. Only when food molecules and oxygen are together inside cells can the complex process of aerobic cell respiration begin.

SOME IMPORTANT DEFINITIONS

The terms 'respiration' and 'breathing' seem to be used interchangeably by many people, including scientists, and this can lead to confusion. As a basic guide, we use the following definitions.

Respiration is the process that releases the energy in organic molecules such as sugars and lipids. Respiration is a complex, multi-stage process that takes place in *all* cells of *all* organisms, *all* of the time (unless the organism is dormant). You may remember that respiration is one of the seven signs of life. To avoid confusion with breathing, it is often called tissue respiration or cell/cellular respiration. In this book we call it **cell respiration**.

> **!** SCIENCE FEATURE Sources of energy

In humans, as in all organisms, obtaining energy from food is a priority that cannot be ignored. Tragic scenes from parts of the world devastated by famine show the terrible consequences of malnutrition and starvation.

We obtain most of our energy by respiring the food that we have eaten recently. Carbohydrate is the primary source of energy, but we also respire small amounts of fat and protein. When the energy provided by recent meals runs out, as it does when we are asleep, for example, our bodies start to respire molecules stored in our tissues. These are replaced when we next eat, usually first thing in the morning when we break our fast.

During prolonged starvation, the body's stored carbohydrate runs out after about a day, and then the body must obtain its energy by respiring fat. An average 70 kg man will have about 10 to 15 kg of stored fat in his body. This is a large energy store, but one that is not readily available (fat can only be respired when some carbohydrate is still available). Increasingly the body is forced to turn to protein, the very fabric of the cells, for the energy it desperately needs.

The body can withstand the loss of about half of its protein only. Complete starvation, where no food at all is available, leads to death within weeks. The exact cause of death varies. It can be an infection such as pneumonia (starvation weakens the immune system) or circulatory problems due to a much reduced blood volume.

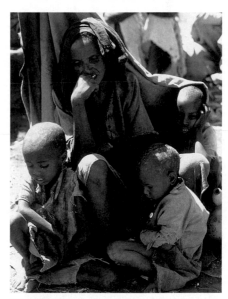

Fig 34.29 Famine is still a fact of life for people in many parts of the world

Fig 34.30 Living organisms have developed various organs to increase their ability to exchange gases. In humans, this gas exchange occurs at the lung surface

Organisms that carry out aerobic cell respiration need to take in oxygen, and they produce carbon dioxide, a waste product that must be removed. Larger organisms have had to develop specialised organs with large surface areas to increase their capacity for gas exchange (Fig 34.30). (See also Chapter 9.)

Breathing is the mechanical process that supplies oxygen to the body to drive respiration and that removes the carbon dioxide produced. Breathing *ventilates* the gas exchange surfaces.

Aerobic respiration is respiration that requires oxygen.

Most organisms respire aerobically: aerobic respiration releases a relatively large amount of energy and gives them a survival advantage over anaerobic organisms.

Anaerobic respiration is respiration without oxygen.

Some organisms, mainly bacteria, can only respire anaerobically. A few more, yeast for example, can turn to anaerobic respiration when there is no oxygen. Some animal tissues (eg muscle during strenuous exercise) and plant tissues (eg plant roots in waterlogged soil) are able to respire anaerobically if circumstances demand it.

16 AEROBIC CELL RESPIRATION

'Aerobic cell respiration' describes the cell processes that require oxygen to release energy from all types of organic molecules. In many organisms, cell respiration involves the breakdown of a variety of molecules from their food. Humans respire mainly sugars with a small percentage of amino acids and fatty acids but, when the need arises, such as during starvation, the balance can change.

✔ **REMEMBER THIS**

Although our muscles can work anaerobically for a short time, the brain must have a continuous supply of oxygen. A lack of oxygen for just 3 minutes is enough to cause brain damage.

? **TEST YOURSELF**

H Why is it important for organs involved in gas exchange to have a large surface area? What other features would you expect these organs to have?

AEROBIC CELL RESPIRATION OF GLUCOSE

To make a complex topic easier to study, we shall concentrate first on the aerobic respiration of only one type of food molecule: glucose.

For any chemical reaction to take place, energy is required to break existing molecular bonds. The process of forming new bonds may either require or release energy. For there to be a *net release of free energy* (the whole point of cell respiration), the products of cell respiration must be at a lower energy level than the reactants. Chemists say the products are more **thermodynamically stable** than the reactants. The basic equation we use to sum up glucose respiration is:

$$C_6H_{12}O_6 + 6O_2 \rightarrow 6CO_2 + 6H_2O + ENERGY$$

The total energy contained in the reactants (the molecule of glucose and the six molecules of oxygen on the left) is greater than the total energy contained in the products (the six molecules of carbon dioxide and the six molecules of water on the right). The difference is the energy released by cell respiration.

In reality, glucose respiration is a sequence of many different reactions. These can produce up to 36 molecules of a chemical called **ATP** per molecule of glucose. This large amount of energy can do useful work in the cell:

$$C_6H_{12}O_6 + 6O_2 \rightarrow 6CO_2 + 6H_2O + 36ATP$$

The steps involved in the production of ATP depend heavily on redox reactions. These are reactions that involve oxidation of one reactant and reduction of another. Before going on to look at the detailed biochemistry of glucose respiration, take a look at the box on **redox reactions** on page 607, to make sure that you are comfortable with this important concept.

RESPIRATION OR COMBUSTION?

It is often said that we 'burn up' food in respiration but this is not really true. The energy contained in glucose *could* be released by setting fire to it (Fig 34.31). The glucose would be oxidised to carbon dioxide and water, but the reaction would be rapid and uncontrolled.

During the process of cell respiration, glucose and other organic molecules are broken down in small stages, some of which release energy. This step-by-step breakdown (Fig 34.32) is more gradual than combustion. The whole process of respiration is controlled by enzymes that transfer energy in food molecules to ATP, a substance that is able to power cell processes by supplying on-the-spot, instant and usable energy in controlled amounts.

THE AEROBIC RESPIRATION OF OTHER ORGANIC MOLECULES AND ANAEROBIC RESPIRATION ARE DEALT WITH ON 614 AND 619.

? TEST YOURSELF

How is cell respiration similar to the working of a petrol engine?

Fig 34.31 The energy in food can be released by combustion, but almost all of it is lost as heat, raising the temperature of the surroundings to levels that could not be tolerated inside any living cell

Energy in molecules

glucose and oxygen

Combustion has one step and occurs very rapidly

energy lost as heat

CO₂ + H₂O

Time

Energy in molecules

glucose and oxygen

Respiration occurs in small steps and releases energy in controlled amounts. Less energy is lost as heat

CO₂ + H₂O

Time

Fig 34.32 The difference between combustion and respiration. Although the difference in energy between the reactants (glucose and oxygen) and the products (CO₂ and water) is the same in both cases, the way in which the energy is released is totally different

ALL ABOUT ATP

ATP stands for **adenosine triphosphate**. ATP is a relatively small, soluble organic molecule that consists of a base (adenine), a sugar (ribose) and three inorganic phosphate (P_i) groups (Fig 34.33). ATP has a *high **free energy of hydrolysis***, which means that when ATP is **hydrolysed** (Fig 34.34), a relatively large amount of energy is released. In hydrolysis, ATP reacts with water, producing ADP and inorganic phosphate. Because of its solubility and small size (relative molecular mass 507), ATP can diffuse rapidly around cells and so can supply energy where it is needed. Metabolically active cells, such as those in muscle and in secretory tissue, synthesise and break down many ATP molecules.

SEE QUESTION 6

Fig 34.33 The structure of ATP. In the computer graphics image, adenosine is blue, pentose is white and the phosphate groups are red. This molecule is present in all cells and is used to provide energy. The purpose of cell respiration is to re-synthesise ATP at the same rate as it is used up

Fig 34.34 The hydrolysis (splitting) of ATP. Under the control of the enzyme ATPase, the terminal phosphate group of ATP is removed and combined with water. This reaction releases free energy which can be used to drive energy-requiring reactions such as muscular contraction

HOW DOES ATP RELEASE ENERGY?

When ATP loses a phosphate group to become ADP, adenosine *di*phosphate, the reaction is **exergonic** (it *releases* energy). This can be used to do useful work within the cell. The same amount of energy is released when ADP loses another phosphate group to become AMP (adenosine *mono*phosphate). Less energy is released when the last phosphate group is lost.

Many of the reactions within cells are **endergonic**: they *require* energy. This energy is supplied by *coupling* them with reactions that involve ATP breakdown.

Endergonic reaction: amino acid A + amino acid B → dipeptide

Exergonic reaction: ATP → ADP + P_i + energy

It is often said that ATP is a 'high-energy molecule' or that is contains 'high-energy phosphate bonds'. Neither statement is really true. Many molecules have more energy than ATP, and many others contain exactly the same phosphate-to-phosphate bonds. ATP is an important molecule in living systems because it can lose its terminal phosphate group *readily*, releasing just enough energy to power biological processes without producing excess heat.

ATP → ADP + P_i + energy (30.6 kJ mol^{-1})

During strenuous exercise, when the muscles quickly use up all their ATP, energy is supplied by a back-up chemical called **creatine phosphate** (CP). Since CP has a higher free energy of hydrolysis than ATP, its breakdown releases enough energy to make more ATP instantly and so allows exercise to continue.

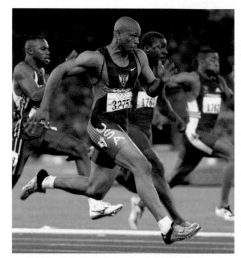

Fig 34.35 For explosive bursts of effort, such as a short sprint, most of the energy comes from ATP already present in the muscles

---

! SCIENCE FEATURE Redox reactions

Redox is short for *red*uction and *ox*idation, two chemical processes that often occur together in the same reaction.

Oxidation reactions may involve the addition of oxygen or the removal of hydrogen from molecules, but the important underlying rule is that a molecule that is oxidised *loses* electrons.

Conversely, reduction reactions involve a *gain* of electrons. As electrons carry energy, a molecule that has been reduced, and that therefore has gained one or more electrons, will usually carry more energy than the oxidised form of the same molecule.

Reduction and oxidation reactions usually occur together, because if one chemical loses electrons another must gain them (Fig 34.36).

The concept of redox reactions is a vital one in biology. The molecules that make up living things are produced, directly or indirectly, by photosynthesis. This process reduces carbon dioxide to form organic molecules (such as sugars) that contain energy. These can be later oxidised in the process of cell respiration to release energy. Lipids contain more hydrogen than carbohydrates: they have more C—H bonds than C—O bonds.

For this reason, lipids are said to be more highly reduced chemicals than carbohydrates and more energy can be released by the oxidation of their many C—H bonds. Because they contain more energy per gram than other substances, lipids are ideal storage compounds (see Chapter 3).

Cell respiration is the release of energy from food by progressive oxidation.

In cell respiration, glucose is oxidised to form carbon dioxide and oxygen is reduced to form water. Most of the ATP created during this process comes from the final stage: a series of redox reactions that occur on the mitochondrial membrane.

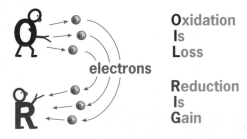

Oxidation
Is
Loss

Reduction
Is
Gain

Fig 34.36 Remember – OIL RIG: oxidation is loss, reduction is gain

Fig 34.37 The overall process of cell respiration, showing the order of the four main stages

THE FOUR MAIN STAGES IN GLUCOSE RESPIRATION

The complete process of aerobic respiration of glucose can be divided into four distinct processes:

● **Glycolysis**; glucose is split into pyruvate.
● **Pyruvate oxidation** (the 'link' reaction); pyruvate is oxidised into acetate.
● The **Krebs cycle**; electrons are stripped off the acetate.
● The **electron transport chain**; the energy in the electrons is used to make ATP.

The way in which the four processes relate to each other is shown in Fig 34.37, and Table 34.2 gives the overall function of each stage. When learning respiration, the two compounds to look out for are ATP and NADH. The production of ATP, as we have seen, is the whole point of the process, but what is NADH?

Overall, respiration involves the removal of electrons from the glucose molecules, and then using the energy in the electrons to make ATP. NADH is an electron carrier.

$$NAD^+ \quad + \quad e^- \quad \rightarrow \quad NADH$$

$$\text{Coenzyme} + \text{electron} \rightarrow \text{reduced coenzyme}$$

NAD^+ is a coenzyme that picks up electrons from the processes that release them (glycolysis, link reaction, Krebs cycle) and delivers them to the process that converts them into ATP; the electron transport chain. You might like to think of ATP as cash and NADH as a credit note that can be turned into cash later.

Table 34.2 The overall process of glucose breakdown. Stages 2, 3 and 4 can be thought of as the aerobic (oxygen-requiring) reactions and take place in the mitochondria

Stage	Site within cell	Overall process	Number of ATP molecules produced*
1 Glycolysis	cytosol	glucose is split into 2 molecules of pyruvate	2 per glucose
2 Pyruvate oxidation	matrix (inner fluid) of mitochondria	pyruvate is converted into acetyl Co-A	none
3 Krebs cycle	matrix (inner fluid) of mitochondria	acetyl Co-A drives a cycle of reactions which produces hydrogen	2 per turn, so 4 per glucose
4 Electron transport chain	inner membrane of mitochondria	hydrogen drives a series of redox reactions which release enough energy to make ATP	up to 32 per glucose

*This table indicates that, for each molecule of glucose, 38 molecules of ATP are produced. The actual figure is nearer 36 molecules. See the Example on page 614.

MAKING ATP BY PHOSPHORYLATION

In the cell, ATP is made by adding a phosphate group to a molecule of ADP. This process is called phosphorylation and it is carried out in two different ways.

Substrate-level phosphorylation is the process in which a phosphate group from a substrate molecule (a molecule other than ATP, ADP or AMP) is transferred to a molecule of ADP, giving a new molecule of ATP. This form of phosphorylation occurs in glycolysis and in the Krebs cycle.

In **oxidative phosphorylation**, ATP is synthesised using free phosphate groups. The energy required is obtained from a series of redox reactions in the electron transport chain. So whenever you see 'oxidative phosphorylation', think 'electron transport chain'.

GLYCOLYSIS

In this first stage a glucose molecule is converted, by a series of enzyme-controlled reactions, into two molecules of pyruvate, a 3-carbon compound. The process, shown in detail in Fig 34.38, yields relatively little energy (only two molecules of ATP) but this stage does not require oxygen and takes relatively little time to complete, so it can provide immediate energy.

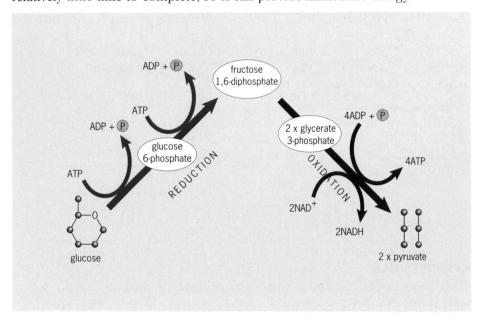

Fig 34.38 Glycolysis is a two-stage process. The 'uphill' part involves raising glucose to a higher energy level by using ATP. In the 'downhill' part, the products are oxidised, yielding two molecules of pyruvate, two molecules of reduced coenzyme and a net gain in ATP

Glycolysis takes place in the **cytosol**, the fluid part of the cytoplasm, so this basic process takes place all over the cell. The overall process consists of ten different reactions, each catalysed by a specific enzyme.

The first half of glycolysis actually uses ATP to raise the energy level of the original glucose molecules by adding phosphate groups to form **fructose 1,6-diphosphate**. This must be done if glucose is to have enough energy to complete the second half of the process.

Fructose 1,6-diphosphate is then split into two molecules of a 3-carbon compound, **glyceraldehyde 3-phosphate**. These are subsequently oxidised to pyruvate by a series of five reactions that also produce four molecules of ATP and two molecules of reduced coenzyme (NADH).

So, overall, per glucose molecule, glycolysis produces:

- two molecules of ATP (four are created but two are used up in glycolysis);
- two molecules of **NADH** (**reduced coenzyme** which later feeds electrons into the electron transport chain);
- two molecules of pyruvate (which enters the 'link' reaction if oxygen is available).

PYRUVATE OXIDATION: THE 'LINK' REACTION

This reaction links glycolysis with the Krebs cycle. It is often thought of as part of the Krebs cycle but we shall consider it to be a separate reaction.

In the presence of oxygen, pyruvate (the 3C molecule produced by glycolysis) moves from the cytosol to the mitochondrial matrix where it is oxidised into **acetate** (a 2C molecule), producing carbon dioxide as a by-product. This reaction also produces two molecules of NADH. The acetate is picked up by a carrier molecule, **coenzyme A**, and **acetyl coenzyme A** is formed.

Overall, the reaction is:

$$2 \text{ pyruvate} + 2NAD^+ + 2H_2O \rightarrow 2 \text{ acetate} + 2NADH + 2H^+ + 2H_2O$$

These terms are used interchangeably but they refer to the same substance. Fig 34.39 shows the difference: pyruvic acid in solution becomes ionised and, like all acids in solution, it loses a hydrogen ion. In the ionised state it is known as pyruvate. The same applies to other organic acids such as lactic acid and acetic acid, which become lactate and acetate respectively.

Fig 34.39 Pyruvate, lactate and acetate and their corresponding acids

M In a human, what happens to the carbon dioxide made by the reactions in the Krebs cycle?

N Why does the Krebs cycle turn twice for every glucose molecule?

KREBS CYCLE

The **Krebs cycle** is a series of reactions named after Sir Hans Krebs, the biochemist who first worked out the details. It is also known as the **citric acid cycle**, the **tricarboxylic acid cycle** or simply the **TCA cycle**. The main purpose of the Krebs cycle is to provide a continuous supply of electrons to feed into the electron transport chain.

The details of the Krebs cycle, which takes place in the matrix of the mitochondria, can be seen in Fig 34.40 on the next page. Overall, per turn, the Krebs cycle produces:

● 3 molecules of NADH;
● 1 molecule of $FADH_2$;
● 1 molecule of ATP (by substrate-level phosphorylation);
● 2 molecules of CO_2;
● 1 molecule of oxaloacetate (to allow the cycle to continue).

The Krebs cycle will 'turn' twice for every glucose molecule that enters it, so, per glucose molecule, six molecules of NADH and two molecules of $FADH_2$ are produced. NADH and $FADH_2$ are particularly important because they carry the electrons which power the next stage of glucose respiration.

NAD (**nicotinamide adenine dinucleotide**) and FAD (**flavine adenine dinucleotide**) are **coenzymes**, organic compounds that are catalysts for reactions. As their name implies, coenzymes work closely with enzymes. Unlike enzymes, coenzymes are not proteins. NAD and FAD carry electrons from the **electron donors** in glycolysis, pyruvate oxidation and the Krebs cycle to **electron acceptors** in the electron transport chain in the mitochondrial membrane.

When NAD and FAD accept electrons, they are reduced, becoming NADH and $FADH_2$. In this form they are called

reduced coenzymes. Reduction of a coenzyme always requires association with a **dehydrogenase enzyme**, an enzyme that removes hydrogen from other molecules. The hydrogen removed by the dehydrogenase enzyme is split into an electron and a hydrogen ion. NAD (or FAD) accepts the electrons produced and the hydrogen ions play an important part in the electron transport chain.

In the human body, NAD is synthesised from vitamin B_3 (nicotinic acid) and FAD is made from vitamin B_2 (riboflavin): this explains why both vitamins are essential components of our diet.

Fig 34.40 The process of cell respiration. Most A-level students do not learn all the intermediate stages, but it is a good idea to look at the biochemical pathways involved to understand several important points about cell respiration:

- This is an example of a metabolic pathway. It looks complex but, like a street map, it is easy to follow once you know where you are and where you would like to go. Each individual reaction is a relatively simple step.
- Each stage is catalysed by a specific enzyme.
- You can see how the respiration of different foods (glucose, fatty acids and amino acids) is closely interconnected. The central pathway, in black, shows the complete breakdown of a molecule of glucose. The path taken by fatty acids is shown in green and the path taken by amino acids is shown in red.

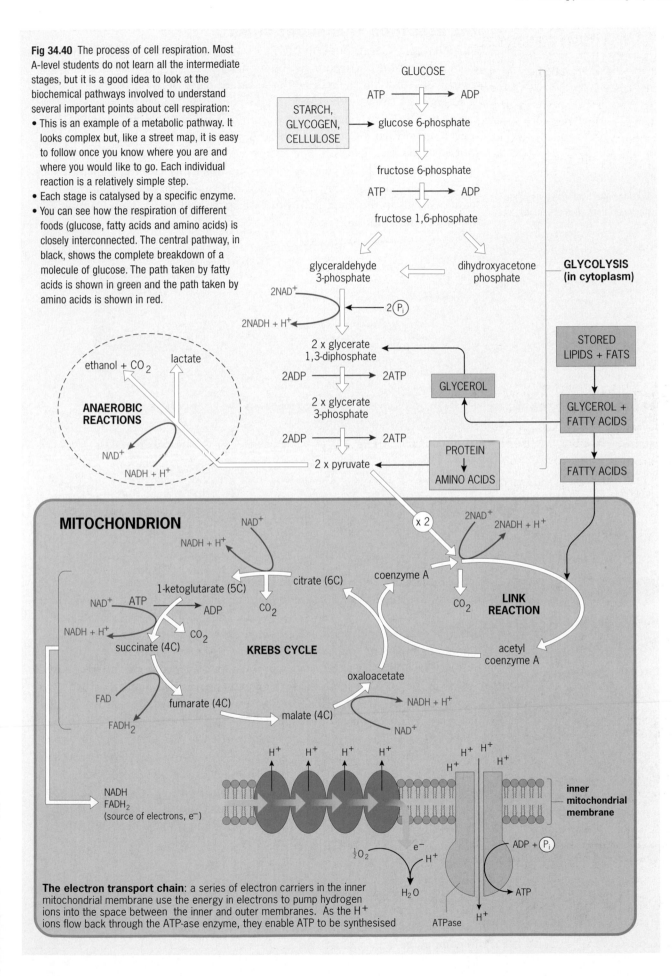

The electron transport chain: a series of electron carriers in the inner mitochondrial membrane use the energy in electrons to pump hydrogen ions into the space between the inner and outer membranes. As the H$^+$ ions flow back through the ATP-ase enzyme, they enable ATP to be synthesised

THE ELECTRON TRANSPORT CHAIN

This final phase of respiration produces the largest number of ATP molecules. The electron transport chain can be thought of as the 'pay day' for all the 'hard work' that has been done in the earlier parts of the process.

As Fig 34.41 shows, the electron transport chain consists of a series of carrier molecules, which will first accept an electron (thereby becoming reduced) and then lose it again (so becoming oxidised). At each of these transfers the electrons lose some energy, which can be used to power the active transport of hydrogen ions across the inner mitochondrial membrane (see Fig 34.42 on the next page). This results in a high concentration of hydrogen ions in the outer mitochondrial space and a low concentration in the inner mitochondrial space.

Fig 34.41 The electrons and hydrogen ions made during the first stages of respiration are finally used to synthesise ATP

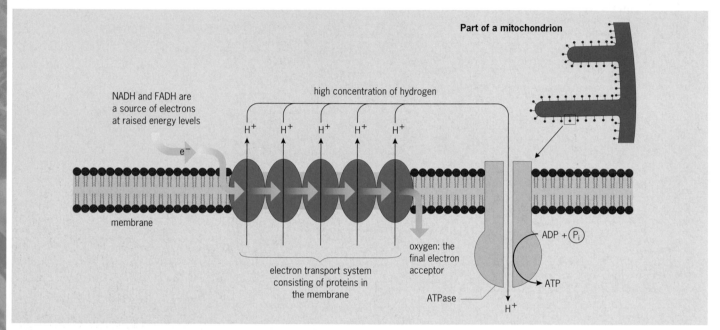

Part of a mitochondrion

NADH and FADH are a source of electrons at raised energy levels

high concentration of hydrogen

e^-

membrane

electron transport system consisting of proteins in the membrane

oxygen: the final electron acceptor

$ADP + P_i$

ATPase

ATP

H^+

TEST YOURSELF

O Muscle cells and cells that manufacture and secrete large amounts of hormones have more mitochondria than, say, a skin cell. Can you say why?

P The function of a kidney tubule cell is to move substances by active transport. Suggest why these cells are packed with mitochondria.

SEE QUESTIONS 7 AND 8

Because of the difference in concentration, hydrogen ions leak back into the inner compartment. The only route that they have is through the middle of the stalked granules – ATPase enzymes. As the stream of hydrogen ions flows down the concentration gradient, enough energy is released to allow free inorganic phosphate molecules to be added to ADP, forming new molecules of ATP. The flow of H$^+$ ions is sometimes known as a **proton motive force**.

This model of ATP synthesis by **oxidative phosphorylation** is called the **chemiosmotic hypothesis** and was first proposed by Peter Mitchell in 1961. Mitchell received the Nobel Prize for Chemistry in 1978.

The end-products of the electron transport chain are low-energy electrons and hydrogen ions that combine with oxygen to form water. Although this reaction is at the end of the process, it is a key one: it is why most organisms need oxygen. If there is no oxygen to mop up the electrons and hydrogen ions from the electron transport chain, the pathway cannot be completed. This is a disaster for the cell as the intermediate compounds of glucose respiration build up and the cell loses its ability to make most of the ATP it needs.

Looking at the separate processes involved in the breakdown of one molecule of glucose makes it easy to lose track of the overall purpose of glucose respiration: ATP production. The following example gives you the chance to analyse the contribution to ATP production made by each of the different stages.

Fig 34.42 Mitochondria are the organelles that house the enzymes, substrates and other chemicals associated with aerobic respiration. Pyruvate oxidation and the Krebs cycle take place in the matrix (fluid) centre of the organelle, while the electron transport chain takes place in the inner membrane itself

(a) Micrograph of a pancreatic cell: the large numbers of mitochondria suggest that this is a metabolically active cell. In this case ATP is needed to synthesise the digestive enzymes (proteins)

Fig 34.42(b) Micrograph of a single mitochondrion. Generally, the more metabolically active the cell, the more folds (cristae) there are on the inner membrane

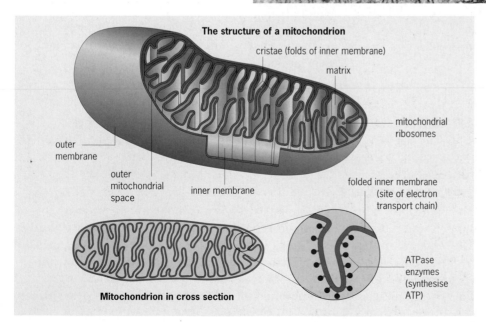

The structure of a mitochondrion

cristae (folds of inner membrane)

matrix

mitochondrial ribosomes

outer membrane

outer mitochondrial space

inner membrane

folded inner membrane (site of electron transport chain)

ATPase enzymes (synthesise ATP)

Mitochondrion in cross section

Fig 34.42(c) 3D illustration of a mitochondrion

● EXAMPLE

Q How many ATP molecules are produced, per molecule of glucose, in aerobic cell respiration?

A ATP can be made in one of two ways, by **substrate-level phosphorylation** and by **oxidative phosphorylation** (see page 608). We shall take each process in turn.

Substrate-level phosphorylation

Glycolysis and the Krebs cycle produce ATP in this way. We can complete the following table:

Stage	ATP produced
Glycolysis	2 (2 used but 4 made)
Pyruvate oxidation	0
The Krebs cycle	2 (1 per turn)
Total	4

Oxidative phosphorylation

This is the production of ATP by the electron transport chain. It relies on a continuous supply of 'high-energy' electrons from the preceding processes. Before we can calculate the final ATP total, we must calculate the number of electrons being provided by NADH and $FADH_2$.

We can draw up a second table, as follows. Remember, pyruvate oxidation and the Krebs cycle occur twice for every glucose molecule put into the system.

Stage	NADH produced	$FADH_2$ produced
Glycolysis	2	0
Pyruvate oxidation	2	0
The Krebs cycle	6 (3 per turn)	2 (1 per turn)
Total per glucose	10	2

We know that:

3 ATP molecules are produced for every molecule of NADH fed into the chain,

and that:

2 ATP molecules are produced for every molecule of $FADH_2$ fed into the chain.

We can now estimate the final ATP total:

10 NAD molecules produce 3 ATPs, giving a total of	**30**
2 FADH molecules produce 2 ATPs, giving a total of	**4**
Therefore, the total number of ATP molecules produced by oxidative phosphorylation is	**34**
The total number of ATP molecules produced by substrate-level phosphorylation is	**4**

The grand total per glucose molecule is therefore 38 molecules of ATP,

most of which are produced by the electron transport chain.

In practice, the figure is usually 36 molecules of ATP. Because the mitochondrial membrane is impermeable to NADH, the 2 molecules of NADH made during glycolysis cannot carry their electrons directly into the electron transport chain.

To get over this, most cells have a shuttle system in which the electrons are released by the NADH and passed across the mitochondrial membrane where they are picked up by $FADH_2$. As we saw earlier, $FADH_2$ produces only 2 ATPs compared with the 3 made by NADH, thus reducing the final total by 2.

This is the ATP total under *ideal* conditions. Several other factors can also reduce ATP production even further, but these are beyond the scope of this book.

AEROBIC CELL RESPIRATION OF OTHER FUELS

So far, we have studied the respiration of glucose, but many organic molecules can be fed in the same central pathway to create ATP.

LIPID BREAKDOWN

The lipid stores in the body can be used when carbohydrate is in short supply. Stored triglycerides are broken down into **glycerol** and **fatty acids**. The latter are further broken down to give several 2-carbon acetyl fragments by a process called **beta-oxidation**. These fragments are combined with coenzyme A and so enter the Krebs cycle and, eventually, the electron transport chain. Glycerol is converted into **dihydroxyacetone phosphate** so that it can be used as a fuel in glycolysis.

PROTEIN BREAKDOWN

Adult humans need up to 60 g of protein per day. We cannot store protein, so excess **amino acids**, the building blocks of proteins, are degraded in the liver by the process of **deamination**. Enzymes in the liver separate the amine group from the rest of the molecule, leaving an **organic acid**. The amine group is released as free **ammonia**, which enters the **ornithine cycle** (see Chapter 11) to be incorporated into **urea** and excreted from the body in the urine. The organic acids are fed into the Krebs cycle and are respired (see Fig 34.40).

17 THE RESPIRATORY QUOTIENT (RQ)

The **respiratory quotient**, or **RQ**, is a ratio of gas exchange, worked out by comparing oxygen uptake with carbon dioxide production. The RQ can be used to determine the type of food being respired. The basic equation for the respiration of glucose:

$$C_6H_{12}O_6 + 6O_2 \rightarrow 6CO_2 + 6H_2O + \text{energy}$$

shows that, for every glucose molecule respired, six molecules of oxygen are required and six molecules of CO_2 are produced. Equal numbers of gas molecules occupy the same volume and so the volume of oxygen taken up by an organism is equal to the volume of CO_2 produced.

> **REMEMBER THIS**
>
> Lipids can fuel aerobic exercise but not glycolysis. Aerobic exercise is therefore a good way to burn excess fat.

> ### EXTENSION The origins of respiration: some speculation

Even the simplest life form needs to release the energy locked in organic molecules and it is thought that glycolysis was one of the first metabolic pathways to evolve.

Since there was very little free oxygen at the time, aerobic respiration was impossible. It seems reasonable to assume that the process of glycolysis, as practised by early, bacteria-like organisms (Fig 34.43), used up much of the available food, giving the organisms a major problem: that of energy supply. Solving this problem proved to be a massive evolutionary step: the development of photosynthesis. Not only did photosynthesis provide a new supply, but it also released free oxygen as a by-product. The oxygen in the atmosphere is due to photosynthesis.

The build-up of oxygen in the atmosphere kick-started another evolutionary leap: the appearance of organisms that could respire aerobically. This probably happened fairly quickly because of the massive survival advantage gained by organisms that respire aerobically to produce the much larger amounts of ATP.

It is at this stage that prokaryotic organisms, which could respire aerobically, may have developed a close association with large cells, so forming primitive eukaryotic cells. Find out more about the **endosymbiont theory** on page 22.

Fig 34.43 *Clostridium botulinum*, the anaerobic bacterium that causes botulism (serious food poisoning), is found in food such as meat and fish that have been badly canned. Anaerobic bacteria such as these are believed to have been the only organisms that could have survived in the oxygen-poor environment 3500 million years ago

WORKING OUT THE RESPIRATORY QUOTIENT

The RQ is worked out using the simple formula:

$$RQ = \frac{\text{volume of } CO_2 \text{ given out}}{\text{volume of } O_2 \text{ taken in}}$$

For the glucose equation, the RQ is: $6 \div 6 = 1$.

USING THE RESPIRATORY QUOTIENT

If we measure the volumes of oxygen and carbon dioxide that are exchanged by an organism and find that the RQ is 1, we can infer that the organism is respiring mainly carbohydrate. Different substrates will produce different RQ values:

- Carbohydrate respiration gives an RQ of 1.
- Protein respiration gives an RQ of 0.8–0.9.
- Lipid respiration gives an RQ of 0.7.
- Fermentation gives an RQ of infinity.

The problem with trying to work out what substrate is being respired by looking at RQ values is that many organisms, including humans, respire more than one substrate. In humans, carbohydrate is the main fuel for respiration, but a small amount of protein is also respired. The RQ figure obtained is therefore variable.

USING A RESPIROMETER

The rate of aerobic respiration can be estimated by measuring how much oxygen is taken up or how much heat is produced. Gas exchange, including oxygen uptake, is measured using a sealed chamber called a **respirometer**.

A SIMPLE RESPIROMETER

A simple respirometer is shown in Fig 34.44. Organisms placed inside exchange gases and so alter the composition of the air around them.

Fig 34.44 A simple respirometer

To measure gas exchange due to a respiring organism, we could remove the air after a fixed time and analyse it. A simpler method is to place a carbon dioxide absorber such as sodium hydroxide in the respirometer chamber. Over a set time, a volume of oxygen is used up. Carbon dioxide is produced but is absorbed by the sodium hydroxide. So, overall, the volume of gas falls by the volume of oxygen used. The gas pressure inside the chamber falls and coloured liquid is drawn along the tube towards the chamber. The volume of oxygen used up equals the volume of the liquid moved in the fixed time. The following example shows how the respiration rate is calculated from readings taken from the tube.

EXAMPLE

Q a) When measuring the gas exchange of an organism with a simple respirometer, how do you calculate the volume of oxygen used?

b) In what units do you express the rate of respiration?

A a) The position of the liquid along the tube is recorded at the start and end of the set time and the distance the liquid moves is calculated. The volume of liquid (equal to the change in oxygen volume) is a cylinder inside the tube.

Therefore:

$$\text{volume of oxygen used up (in cm}^3) = \pi \times r^2 \times h$$

where r = radius or diameter/2 (in cm) and h = distance moved by the liquid (in cm).

b) The rate of respiration is given as the volume of oxygen used per organism (or per unit mass, eg per gram) per unit time; for example, as cubic centimetres of oxygen per gram per hour. So:

$$\text{rate of respiration} = x \text{ cm}^3 \text{ O}_2 \text{ g}^{-1} \text{ h}^{-1}$$

A MORE COMPLEX RESPIROMETER

The simple respirometer shows the principles of respirometry, but has these limitations:

- Changes in temperature or atmospheric pressure make gases expand or contract. These changes affect the distance the fluid moves, so the measured distance is not just due to the change in oxygen volume.
- The chemicals used might alter the composition of the gases in the chamber.
- It is difficult to restart the experiment without taking the apparatus apart.
- The volume change calculated using the diameter of the tube may be inaccurate.

To avoid these problems, a more sophisticated respirometer has been developed. This is shown in Fig 34.45.

screw clip

1 cm³ syringe

three-way tap

small organisms (eg maggots)

gauze

potassium hydroxide solution

potassium hydroxide solution

capillary U-tube containing coloured water

Fig 34.45 A more complex respirometer. The left tube contains the organisms while the right tube acts as a control. In this way, any changes in fluid level that are not due to the organisms, such as changes in room temperature and pressure, are balanced out by the control tube. The experiment can be set up again using the syringe, which will also accurately measure the changes in volume due to the respiring organisms

This apparatus works on the same principle as the simple respirometer, but the syringe allows the volume change to be measured directly. After a set time, the syringe can be pulled up until the dye goes back to its starting point. This is a neat way to find the exact volume of oxygen used without having to work out the volume of a cylinder, and it allows the experimenter to move the experiment back to the start at the same time.

● EXAMPLE

Using the respirometer

Q The respirometer in Fig 34.45 was set up with 50 g of blowfly (bluebottle) larvae and left for 1 hour. The fluid moved up the left side-arm of the U-tube by 5 cm. It was moved back to its original level by drawing 2 cm³ of air into the syringe.

a) What is the purpose of the syringe and why does it give this respirometer an advantage over the simple respirometer?

b) If there were 10 larvae, what would be the respiratory rate per organism?

A a) The syringe allows movement in the tube to be measured directly as a volume. This removes the need to calculate the volume of the cylinder to find out how much liquid is in the tube.

b) The volume of oxygen used up = 2 cm³. Per larva:
respiratory rate = 2/10 = 0.2 cm³ O_2 h⁻¹

Calculating the respiratory quotient

Q When the experiment was repeated using water instead of KOH, the fluid moved up 1 cm, not 5 cm.

a) What does this tell us about the volume of carbon dioxide produced?

b) What is the respiratory quotient for the larvae?

A a) Carbon dioxide was not being absorbed, and its volume is one unit less than the oxygen volume. So:

$$CO_2 \text{ out} = 4 \text{ units}$$
$$O_2 \text{ in} = 5 \text{ units}$$

b) $\dfrac{CO_2 \text{ out}}{O_2 \text{ in}} = \dfrac{4}{5} = 0.8$

Blowfly diet

Q a) Assuming the larvae are only respiring one substrate, suggest which one.

b) From what you know about the diet of blowfly larvae, does this seem reasonable?

A a) Proteins

b) Yes: they consume a large amount of protein in rotting meat.

Respirometer design

Q How does the addition of the tube on the right (known as a thermobar) help to overcome the problem of changes in atmospheric pressure and temperature?

A Any pressure changes in the tube on the left are mirrored and cancelled out by those in the right tube.

❗ SCIENCE FEATURE Cyanide!

Cyanide is a famous poison, often featured in crime fiction and films (Fig 34.46). It is a very effective toxin because it inhibits one of the enzymes of the electron transport chain. This stops the flow of electrons and effectively prevents aerobic respiration, the source of most of the body's ATP. Without ATP, the muscles go into spasm and the body experiences convulsions. Spasm of the muscles that control breathing usually leads to death by asphyxiation (lack of oxygen to the brain).

Fig 34.46 'A faint smell of bitter almonds and the victim lay dead, the back arched, the eyes staring.'

RESPIRATION, ENZYMES AND TEMPERATURE

Like all metabolic reactions, respiration is controlled by enzymes. The Q10 law (page 93) states that the rate of an enzyme-catalysed reaction will approximately double with every 10 °C rise in temperature between 4 and 40 °C. So, with a temperature increase of 10 °C, much more ATP becomes available to power other metabolic processes.

A dramatic example is seen in **ectothermic** animals such as reptiles, which may have wide variations in body temperature during a 24-hour period (Fig 34.47). Their rate of respiration, and therefore their metabolic rate, is roughly twice as fast at 30 °C than it is at 20 °C. Such organisms are only active when their body temperature allows them to react and move quickly.

18 ANAEROBIC RESPIRATION

Anaerobic respiration does not require oxygen, so it can take place in oxygen-poor environments such as stagnant water or deep soil. It can even occur in parts of an aerobic organism that are starved of oxygen, such as muscles during strenuous exercise.

In anaerobic respiration, glucose is broken down into pyruvate but, because no oxygen is available, the pyruvate cannot be broken down any further. Instead of entering the link reaction and the Krebs cycle, the pyruvate is converted to a waste product. The nature of the waste product depends on the organism. Fig 34.48 shows the waste products of anaerobic respiration in different organisms.

Fig 34.47 Reptiles are ectotherms: they cannot control their body temperature in the way in which mammals and birds do. This lizard must absorb sunlight in order to increase its body temperature to the level that allows its enzymes to function at their optimum efficiency

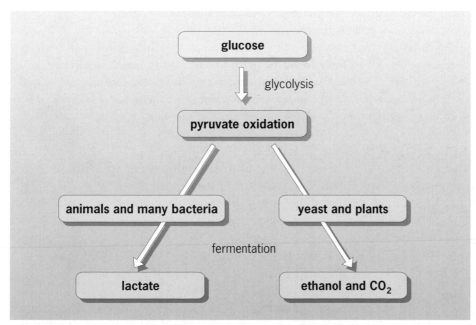

Fig 34.48 The main problem with anaerobic respiration is the build-up of NADH and the shortage of NAD+. To solve this, H+ is added to pyruvate to make lactate or ethanol and carbon dioxide. Fermentation is glycolysis plus the reduction of pyruvate to either lactate or ethanol and carbon dioxide

Generally speaking, animals and some bacteria convert pyruvate into **lactate** by a simple reduction reaction. In contrast, plants and fungi such as yeast convert pyruvate to **ethanol** (alcohol) and carbon dioxide.

In animals, vigorous exercise leads to anaerobic respiration in the muscles and the resulting build-up of lactate causes **fatigue**.

Fig 34.49 Fermentation in different microorganisms can be used to make a variety of useful products

Fig 34.50 Yeast is a single-celled fungus that usually respires aerobically. When deprived of oxygen, it switches to anaerobic respiration: sugars are not completely broken down and the by-products are ethanol (alcohol) and carbon dioxide

? **TEST YOURSELF**

T What is the difference between fermentation and glycolysis?

Fig 34.51 A baby with muscle stiffness caused by tetanus. The toxins produced by the anaerobic bacterium, *Clostridium tetani*, are causing the baby's muscles to go into spasm

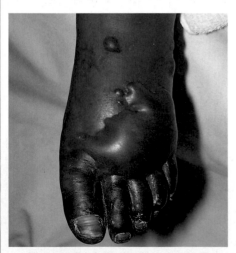

Fig 34.52 This is the dreaded gangrene. Tissue death, caused by a poor blood supply, can lead to infection by the anaerobic bacterium *Clostridium welchii*. This organism causes gas gangrene and the toxins produced can be lethal if not treated quickly

Anaerobic respiration in microorganisms is generally known as **fermentation**. As Fig 34.49 shows, this process can give rise to many useful products. Bacteria that produce lactic acid are used in the manufacture of dairy products such as yoghurt (see page 111). The tangy taste is due to the high concentration of this organic acid. In addition to lactic acid and ethanol, other microorganisms can make solvents such as acetone and butanol.

Anaerobic respiration in yeast is also known as **alcoholic fermentation** (see Fig 34.50 and Chapter 6). This process was discovered by accident and alcoholic drinks such as wine and beer were made for many centuries before the science behind the process was known. For fermentation, yeast simply needs a source of carbohydrate, anaerobic conditions and a suitable temperature. So wine can be made from many organic materials, such as peaches, elderberries or even potato peelings, as well as grapes.

Since alcohol is an intermediate product in the breakdown of sugar, it contains energy that can be released by respiration. Many drinks, particularly beer, are also rich in carbohydrates that have not been turned to alcohol. Heavy drinkers become overweight because they drink large amounts as well as eating meals that, by themselves, provide enough calories to satisfy the body's energy demands.

ANAEROBES AND DISEASE

Many common bacteria can only thrive in anaerobic conditions. Two such bacteria are *Clostridium tetani* and *Clostridium welchii*.

C. tetani usually thrives deep in soil, where there is little oxygen, but is present in small numbers almost everywhere. The bacteria can enter the human body through a deep cut or wound. Once inside, they multiply and produce a toxin that makes muscles go into spasm (Fig 34.51). This is **tetanus**, or lockjaw (named because of its effect on the jaw muscles), which can be fatal if it affects the muscles used for breathing.

C. welchii can cause **gas gangrene**. If it infects areas of dead tissue, such as frost-bitten toes or injured limbs that have a poor blood supply, it turns the tissues black (Fig 34.52). In such 'ideal' conditions, the bacteria multiply rapidly, producing a foul-smelling gas and several lethal toxins. Antibiotics help if the condition is treated early. If not, the affected limb may need to be amputated to prevent the toxins spreading further into the body.

As *C. tetani* and *C. welchii* can *only* respire anaerobically, they are known as **obligate anaerobes**. Organisms such as yeast can respire by either pathway and are known as **facultative anaerobes**. Organisms that thrive only when they are respiring aerobically are called **obligate aerobes**.

SUMMARY

By the end of this chapter, you should know and understand the following:

PART I: OVERVIEW OF KEY CONCEPTS

- Energy transfer through organisms in an ecosystem is one-way.
- Green plants are **autotrophs**, organisms that can make sugars from carbon dioxide using energy. Organisms that cannot make their own food are called **heterotrophs**. Not all autotrophs use light as a source of energy, water as a source of hydrogen or carbon dioxide as a source of carbon.
- In a **food chain**, energy is passed from **producers** to **primary consumers** to **secondary** and **tertiary consumers**. Each level of the food chain is called a **trophic level**. In most ecosystems, many food chains are linked together in a **food web**.
- There are three main types of **ecological pyramid**: **pyramids of numbers**, **pyramids of biomass** and **pyramids of energy**. Pyramids of energy show that over 90 per cent of the energy transferred to the next trophic level is lost as heat.

PART II: PHOTOSYNTHESIS

- **Photosynthesis** is a process that captures the light energy of the Sun and uses it to make organic molecules. Initially, glucose is made but plants use this basic carbohydrate to make the other essential organic molecules such as lipids, proteins and nucleic acids. Photosynthesis happens in chloroplasts; specialised organelles that house the chlorophyll pigments and all of the other compounds essential to the process. Photosynthesis is essentially a two-stage process; the light-dependent reaction and the light-independent reaction.
- The **light-dependent reaction** happens on the **thylakoids**, membranes that contain the chlorophyll molecules.
- When a photon of light hits a molecule of chlorophyll, the pigment becomes excited and emits two high-energy electrons.
- These electrons pass down an electron transport system that synthesises ATP – this is called **photophosphorylation**.
- Energy from the excited electrons is also used to split water; this is **photolysis**.
- Photolysis produces electrons to replace those lost by the chlorophyll and protons (hydrogen ions) and oxygen are made as by-products.
- Electrons are used to make NADPH (also called reduced NADP).

- ATP and NADPH are essential for the light-independent reaction, and oxygen is released into the atmosphere.
- The **light-independent reaction** happens in the **stroma**, the fluid in the centre of the chloroplast.
- It involves the reduction of carbon dioxide by a series of reactions known as the **Calvin cycle**.
- ATP and NADPH are used to reduce the carbon dioxide. Supplies of these compounds depend on light, so, when it gets dark, the light-independent reaction finishes soon after the light-dependent reaction.
- The essential stages of the Calvin cycle are:
  - Stage 1 Carbon dioxide combines with the 5-carbon compound **ribulose bisphosphate (RuBP).** This reaction is catalysed by the enzyme **RUBISCO** (ribulose bisphosphate carboxylase).
  - Stage 2 This produces a highly unstable 6-carbon compound that immediately splits into two molecules of **glycerate 3-phosphate (GP)**.
  - Stage 3 ATP and NADPH are used to reduce the GP into **glyceraldehyde-3-phosphate (GALP)**.
  - Stage 4 Some of the glyceraldehyde-3-phosphate is used to make carbohydrate, but most of it is used to make more RuBP to continue the cycle. For every glucose molecule produced, five molecules are RuBP are re-synthesised.
- CAM plants save water by opening their stomata at night, storing carbon dioxide in an organic acid before releasing it for photosynthesis in the daytime.
- C4 plants avoid photorespiration, allowing them to make use of all the available carbon dioxide and therefore to grow faster.
- The rate of photosynthesis is determined by the factor in shortest supply; this could be light availability, carbon dioxide levels or temperature.

PART III: CELL RESPIRATION

- **Respiration** is the release of energy from food. The energy released is transferred to ATP, a chemical that can provide the cell with energy in small, instant, controllable amounts.
- Respiration consists of many reactions. Each reaction is controlled by a different enzyme. Some of the reactions occur in the cytoplasm and some occur inside the **mitochondrion**.
- Complete **aerobic respiration** of glucose has four stages: **glycolysis**, **pyruvate oxidation**, the reactions of the **Krebs cycle** and the reactions of the **electron transport chain**.

SUMMARY

- Glycolysis: glucose is broken down into two molecules of **pyruvate**. There is a net gain of two molecules of ATP. This happens in the cytoplasm of cells.
- **Pyruvate oxidation** (the 'link' reaction): in the presence of oxygen, pyruvate enters the mitochondrion and is converted into acetyl coenzyme A.
- The Krebs cycle: a series of reactions, fuelled by acetyl coenzyme A, which produces two molecules of ATP and many electrons and hydrogen ions.
- The electron transport chain: a series of redox reactions, fuelled by the electrons and protons from the preceding three processes, as supplied by the coenzymes NAD and FAD. Electron transport releases enough energy to synthesise most of the ATP produced by aerobic respiration.

- Most food chemicals – carbohydrates, lipids and proteins – can be **respired** to provide energy.
- The rate of respiration can be measured as a respiratory quotient, the ratio of carbon dioxide evolved to oxygen absorbed. It can be calculated from measurements made using a **respirometer**.
- Most organisms respire **aerobically** (in the presence of oxygen). The substrate (eg glucose) is broken down completely to produce carbon dioxide, water and a lot of ATP.
- Some organisms and some tissues respire **anaerobically**. In the absence of oxygen, substrates are only partially broken down, leading to waste products such as lactate (in animals and bacteria) or ethanol and carbon dioxide (in plants and yeast).

? EXAM QUESTIONS

1 The diagram shows part of a food web found in a woodland community.

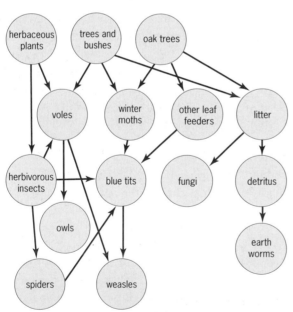

a) Name **one** example, shown in the web, of **each** of the following:
(i) a decomposer;
(ii) a tertiary consumer;
(iii) an omnivore;
(iv) a producer;
(v) a predator which is an invertebrate.

b) For the food chain:
oak tree → winter moths → blue tits → weasels, construct labelled diagrams to show:
(i) a pyramid of numbers:
(ii) a pyramid of biomass.

c) (i) Not all of the sunlight which strikes the leaves of the oak tree is trapped in the products of photosynthesis. Suggest why this is so.
(ii) Explain the difference between *net* production and *gross* production.
(iii) In this community only about 3.5% of the energy available to herbivores is actually taken in them by them. Explain what happens to the bulk of the net production.

d) Studies of the community have shown that the population size of blue tits remains remarkably constant over several years.
(i) Name any **one** density-independent factor which could lead to a sudden fall in the population size.
(ii) Suggest **two** density-dependent factors which might operate to restore population size following such a fall.

WJEC B12 June 2001 Q2

? EXAM QUESTIONS

2 The chart below shows the energy flow through a temperate forest. The energy flow is shown per square metre per year (kJ m^{-2} year^{-1}).

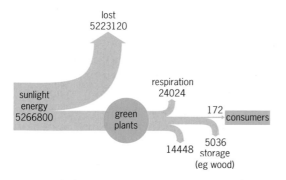

[Source J.R. Grosz *et al*, *Scientific American* (1978), *238*, No 3, page 92]

a) **(i)** Calculate the net production of the plants from the data given.
  (ii) Calculate the gross production of the plants from the data given.

b) The chart shows that 99.17% of the sunlight energy in the temperate forest is lost. Predict with reasons whether a greater or lesser percentage of sunlight energy would be lost in a desert.

c) Only a small part of the net production of plants in the temperate forest passes to herbivores. Explain the reasons for this.

IBO Biology Standard Level Paper 3 November 1999 QG1

3

a) Draw a curve of the action spectrum for photosynthesis on the axis below.

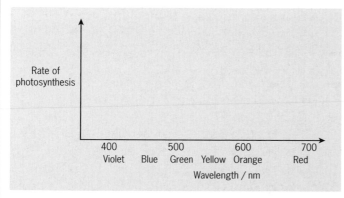

b) Explain the relationship between the action spectrum and the absorption spectra of photosynthetic pigments.

IBO Standard level Paper 3 November 1999 QC3

4 The results of an investigation of photosynthetic pigments in ivy leaves (*Hedera helix*) is shown below.

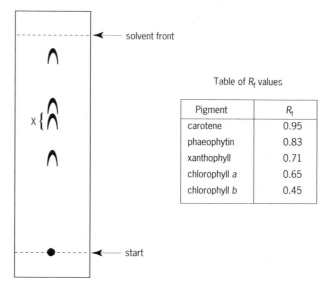

Table of R_f values

Pigment	R_f
carotene	0.95
phaeophytin	0.83
xanthophyll	0.71
chlorophyll *a*	0.65
chlorophyll *b*	0.45

a) Measure the distance travelled by spot X.

b) **(i)** The solvent front travelled 100 mm. Calculate R$_f$ for spot X where
  R$_f$ – distance moved by spot/distance moved by solvent.
  (ii) Identify the pigment in spot X using the table of R$_f$ values above.
  (iii) State **one** way, apart from R$_f$ values, that could be used to identify the spot X.

c) Deduce which of the photosynthetic pigments listed in the table of R$_f$ values was not present in *Hedera helix*.

d) Suggest **one** advantage to the plant of having more than one photosynthetic pigment.

IBO Standard level Paper 3 May 1999 QC1

5 The diagram below is a summary of the light-dependent reaction of photosynthesis.

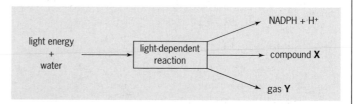

a) **(i)** Name the compound **X**.
  (ii) Name gas **Y** and describe how it is produced.
  (iii) State where in a chloroplast the light-dependent reaction takes place.

b) Name *one* pigment which is responsible for the capture of light energy during the light-dependent reaction of photosynthesis.

Edexcel B2 June 1999 Q4

? EXAM QUESTIONS

6 Adenosine triphosphate, ATP, provides energy within the cell.

a) Excluding respiration, state **three** different cellular activities which use ATP.

b) When ATP is used, the reaction may be summarised as:

$$ATP \rightarrow ADP + Pi + 30.6 \text{ kJ mol}^{-1}$$

Consider this reaction and the statements below, then write either T or F after each question to indicate whether the statement given is true (T) or false (F).

The reaction is exergonic.
The ADP molecule now contains two nucleotides.
The reaction is anabolic.
The ATP contains a purine base, a hexose (6C) sugar and three phosphate groups.
The ADP contains a purine base and a pentose (5C) sugar.

c) Explain precisely, **where** in the metabolic pathway, and **why**, ATP is **used** in the process of respiration.

d) State **where**, precisely, most ATP is synthesised in an animal cell.

OCR Molecules and Life March 2001 Q4

7 The diagram represents some of the biochemical reactions of aerobic respiration. Study the diagram, then answer the questions which follow.

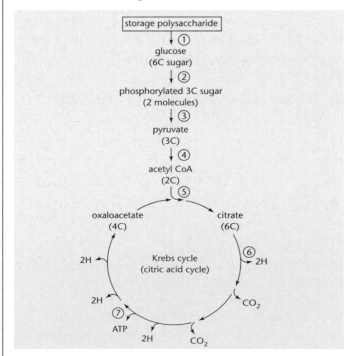

a) Although oxygen is not mentioned, state how the diagram indicates that aerobic respiration is occurring, rather than anaerobic respiration.

b) Which reaction, represented by a numbered arrow, indicates that a molecule of carbon dioxide is released at that point?

c) **(i)** At what numbered stage is ATP used?
(ii) Why is ATP needed at this stage?

d) Which numbered stage represents a coenzyme being released for recycling?

e) Which numbered stage represents the point at which a respiratory substrate enters the mitochondrion?

f) What type of enzyme would be necessary at stage 6?

g) What happens **immediately** to the removed 2H?

h) Where **precisely** would you expect to find acetyl CoA in a cell?

OCR Molecules and Life June 2000 Q7

8 The diagram below summarises the electron-transport chain, which is a stage of cell respiration. A, B and C represent electron carriers

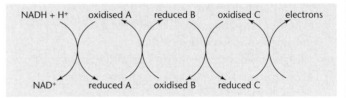

a) **(i)** Describe what happens to the electrons and hydrogen ions at the end of this stage.
(ii) State where in a mitochondrion the electron transfer chain is situated.

b) Name *one* process which produces $NADH + H^+$.

c) ATP is synthesised as a result of this stage of respiration. State *two* uses of ATP in a cell.

Edexcel B/HB1 June 1999 Q4

KEY SKILLS **ASSIGNMENT**

ARE PLANTS GOOD AT PHOTOSYNTHESIS?

One way to measure photosynthetic efficiency is to look at the proportion of light that plants can convert into stored energy in food.

Let us start with the energy in a glucose molecule. We can base our calculations on a mole of glucose molecules. A mole is the amount of a substance equal to its relative molecular mass in grams. A mole contains 6×10^{23} particles. When a mole of glucose is burned it releases about 2820 kJ.

1 How much does a mole of glucose weigh?

In the first half of this century, scientists believed that 24 photons of light were needed to synthesise one mole of glucose. Let us see if they were right. You have met the following summary reaction for photosynthesis:

$$6CO_2 + 6H_2O \rightarrow C_6H_{12}O_6 + 6O_2$$

Look back at Fig 34.24 and the text that describes it.

2 **a) (i)** How many electrons are needed to release one molecule of oxygen at PSII?

(ii) How many electrons are needed to release 6 molecules of oxygen?

b) If one photon of light is needed for each electron excited, then how many photons would be needed to generate 6 molecules of oxygen at PSII?

But there is a complication to consider. The chloroplast has two photosystems, which do not act independently of one another. Each electron released from PSII ends up at PSI, where another photon is absorbed.

3 Taking into account both photosystems, what is the minimum number of photons required to produce one molecule of glucose?

The latest research shows that the figure is closer to 54 photons per glucose molecule, but perhaps the difference you find may be caused by problems in making accurate measurements.

Scientists have compared the amount of energy that would be trapped in a glucose molecule with the energy in the photons needed to produce it. The actual efficiency, the yield, for most crop plants is below 1 per cent.

4 List as many factors as you can which could affect the efficiency of photosynthesis.

Crop researchers have estimated the energy in joules of light falling on a crop and compared it with the yield observed at harvest time. The UK receives sunlight at 100 joules per metre squared per second ($100 \ J \ m^{-2} \ s^{-1}$) in a 24-hour period.

5 **a)** Calculate the average energy input per hectare per year. (1 hectare = 10 000 m^2)

b) Assume the energy value of the crop of potatoes shown in Fig 34.A1 is 17.85 kilojoules per gram dry mass (17.85 kJ g^{-1}). If light were converted to chemical energy with an efficiency of 5 per cent, then what is the maximum crop (dry mass) you could expect in tonnes per hectare? (1 tonne = 1000 kg)

Fig 34.A1 Potato crop in Norfolk

6 Copy and complete the following table:

Yield	% conversion of light to chemical energy	Total dry mass/tonnes hectare^{-1}
Possible maximum	5	
Record UK crop		22
Average		3

7 The values of the record and average conversion efficiencies are very low. Find out the strategies that scientists and farmers are using to improve on them.

KEY SKILLS **ASSIGNMENT**

WHY ARE BIG FIERCE ANIMALS SO RARE?

We normally think of large predators like lions or killer whales as being the most successful organism in their ecosystem because they can take the food that they need without worrying about predators themselves. But if you look at their numbers relative to the numbers of other animals further down the food chain, it is clear that these top carnivores are really quite rare. In this Assignment, we see why.

1 Look at Fig 34.A2. This shows an energy flow diagram for a food chain in the Serengeti.

a) What is process Z?

b) What are organisms X?

c) How many trophic levels are there in the food chain?

d) Calculate the percentage efficiency with which light energy is converted into food energy in the producers.

e) What percentage of the light energy absorbed by plants is eventually available for secondary productivity in the lion?

Fig 34.A2 In this schematic diagram, the producers are the grass plants which make up most of the vegetation. The primary consumers are deer and the secondary consumers are lions

2 Look at Fig 34.A3.

Fig 34.A3 A comparison of the proportion of energy lost as heat and in faeces and urine in a deer and a lion which eat 3000 kJ of food

a) For every 3000 kJ of food eaten, what amount of energy is available to the deer for growth and repair?

b) What amount is available to the lion for this purpose?

c) Try to suggest why the deer loses a higher proportion of energy in faeces and urine than the lion.

d) Why should the lion use up more energy in respiration than the deer?

e) Suggest why the proportion of energy available for secondary productivity is about three times greater in the lion than in the deer.

3 From the information you have gained so far, why are large carnivores so rare?

Discussion questions

4 Do you think that the fact that humans are omnivores has anything to do with their evolutionary success?

5 Some people hold the view that, if humans were all vegetarian, the world food shortage would be largely solved. Do you agree or disagree, and why?

6 Several species of fish, such as trout and turbot, can be farmed. These fish are carnivores. Do you think this is an efficient way of producing food? Discuss the reasons for your answer.

35

Recycling in living systems

THE CARBON CYCLE

Life on Earth is based on carbon compounds (see Chapter 3). Carbon circulates between the abiotic and biotic parts of the environment, as shown in Fig 35.2. Autotrophs capture carbon during photosynthesis, and all living organisms return it to the atmosphere as they respire and also when they die and decompose.

Fig 35.2 The carbon cycle. Photosynthetic organisms fix about 80 billion tonnes of carbon each year. But this represents less than 1 per cent of the total carbon in the biosphere. Carbon can be locked up for long periods of time in limestone rocks, inorganic solutions (containing soluble carbonates) and fossil fuels such as coal: these reservoirs of carbon are often called carbon sinks. When volcanoes erupt, they return carbon dioxide to the carbon cycle

THE ROLE OF MICROORGANISMS IN THE CARBON CYCLE

By looking at Fig 35.2 or Fig 35.3 you should be able to see that carbon would still cycle through the biosphere for some time if decomposers were removed. However, because of their general role in decomposition, decomposers play a key role in the carbon cycle. Bacteria and fungi feed by secreting digestive enzymes on to organic material to digest it. Some of the 'freed' mineral nutrients are absorbed by these microorganisms. Others escape into the soil.

SEE QUESTION 1

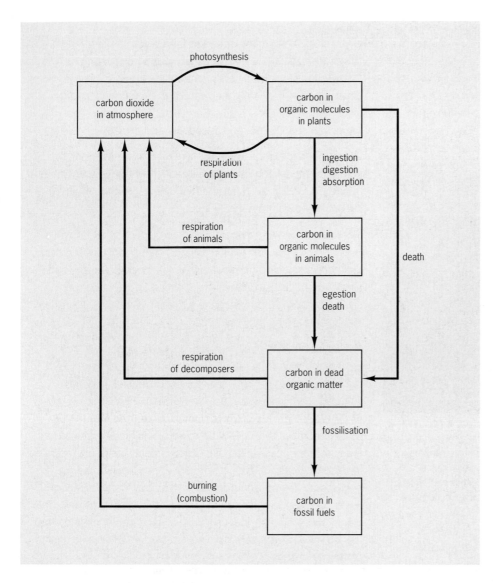

Fig 35.3 An exam/revision version of the carbon cycle. You should find this diagram easy to draw. Learning to draw this, perhaps with your own illustrations and notes, is a good revision exercise

HOW HUMANS INFLUENCE THE CARBON CYCLE

Humans influence the carbon cycle by two activities not practised by other species. Firstly, humans burn fossil fuels such as coal, oil and gas. This has been going on in a big way since the start of the Industrial Revolution in Europe. Burning such fuels is the only way that their 'locked' carbon is released back into the atmosphere.

The second activity is cutting down forests, particularly rainforests in recent times, to make room for farms, fields and homes. This increases the amount of carbon dioxide in the atmosphere because it reduces the total number of photosynthetic organisms, the 'trappers' of CO_2. The possible effects of these activities are discussed in the Science feature box on the next page.

! SCIENCE FEATURE Carbon dioxide, the greenhouse effect and global warming

Carbon dioxide is one of the main greenhouse gases. Together with other gases, it reduces the heat that can escape from the atmosphere (Fig 35.4). This effect is often called the **greenhouse effect**. Although pollution could be increasing the greenhouse effect, this phenomenon is not caused by pollution: it is a natural effect due to the presence of an atmosphere. In fact, having an atmosphere is one of the reasons that life developed on Earth: it allows surface temperatures to be much more stable than they would otherwise be.

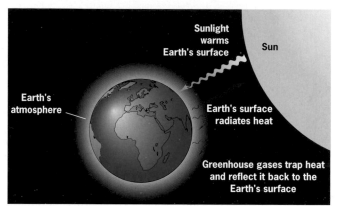

Fig 35.4 The greenhouse effect. Energy from the Sun is either transmitted, reflected or trapped by the atmosphere. Energy radiated back into space is at longer wavelengths (in the infrared region) than the energy arriving from the Sun, and the consequence is that more heat enters than can escape. Greenhouse gases (CO_2, CH_4, nitrous oxides, CFCs), together with water vapour in clouds, contribute to the 'blanketing' effect of the atmosphere

Accurate measurements of atmospheric carbon dioxide concentrations began in the mid-1950s, in Hawaii and at the South Pole. These sites were chosen because they are far away from any major sources of pollution. The curve in Fig 35.5 clearly shows an annual cycle of peaks and troughs corresponding to summer and winter. Superimposed on this is a small but steady rise in overall CO_2 levels. Levels of other greenhouse gases such as methane, nitrogen oxides and chlorofluorocarbons (CFCs) have also increased during the same period.

What are the likely effects of global warming?

Many scientists think that an increase in carbon dioxide in the atmosphere will cause global warming. They believe that the average temperatures on the Earth's surface will increase, causing widespread disturbances in world climate.

Using all the evidence and data available at the moment, a computer model has been developed to predict what might happen in the future. This projection shows that the mean surface temperature might increase.

It is important to remember that any projection is an informed guess: we cannot say exactly what will happen. Nevertheless, the world's scientific community has reached consensus that if greenhouse gas concentrations continue to rise at current levels, global mean temperature will rise by 2 °C above pre-industrial levels by the year 2030.

The effect of any global warming that does occur will not be felt evenly around the globe. There will be greater warming at the high latitudes (at the poles, for example) and at middle latitudes than at the equator. The Northern Hemisphere will warm more than the Southern Hemisphere, because it has more land mass (land heats up more quickly than water). More areas of the world would have extreme heat waves and more forest and grassland fires. The average sea level would rise, perhaps by 2–4 centimetres every 10 years. Many low-lying areas could become flooded and maybe even submerged. This would cause huge problems for developing countries and small islands, which are much more vulnerable to the adverse effects of climate change.

What is being done about the problem of global warming?

Everyone agrees that cutting emissions is a good idea. This means reducing the amount of fossil fuels that we burn: we could make a big difference by improving the efficiency of heating systems and by reusing 'waste' heat from industry. Wherever possible, we should use renewable energy sources such as wind, water and solar power. Although many people object on safety grounds, making more use of nuclear energy would also help to reduce emissions of greenhouse gases.

In addition, we should limit the destruction of forests and use methods of agriculture that do not irreversibly damage the land. Of course, all of this is cancelled out if the human population continues to increase, so efforts should also be made to stabilise the world's population, to prevent further pressure on energy resources.

Fig 35.5 The rising level of carbon dioxide in the atmosphere. Data were collected at the Mauna Loa observatory in Hawaii. There is good evidence to suggest that changes in the average temperature at the surface of the Earth are closely linked to changes in the levels of greenhouse gases

! **SCIENCE FEATURE** Carbon dioxide, the greenhouse effect and global warming continued

What are the problems?

The United Nations Framework Convention on climate change agreed that industrialised countries would reduce CO_2 emissions first, providing developing countries space for development. The Kyoto Protocol (1997) set out that industrialised countries would reduce their emissions by 5.2 per cent compared with 1990 levels by 2012. However, not all industrialised countries have put their full backing behind this policy and other strategies are becoming necessary.

Industrialised countries have put forward the idea of creating carbon sinks by planting new forests that will act as a sink for carbon dioxide. Whether this would work or not remains to be seen, but the commitment to reduce emission of greenhouse gases during 2008–2012 is being put in jeopardy by the USA and Japan, who propose to use carbon sinks to meet their reduction commitments without curbing emissions from fossil fuels.

THE NITROGEN CYCLE

Like carbon, nitrogen is vital to organisms for the formation of proteins, nucleic acids and their products. Although nitrogen is all around us (it makes up 79 per cent of the air we breathe), most living organisms cannot access this supply. Nitrogen gas is unreactive because the two atoms that make up the N_2 molecule are bound together by a strong triple bond. Nitrogen can, however, be fixed by microorganisms, and then it cycles through ecosystems, as shown in Fig 35.6. Fig 35.7 shows an exam-friendly version.

SEE QUESTIONS 2 AND 3

Fig 35.6 The nitrogen cycle. Bacteria play an important role in the movement of nitrogen in the biosphere

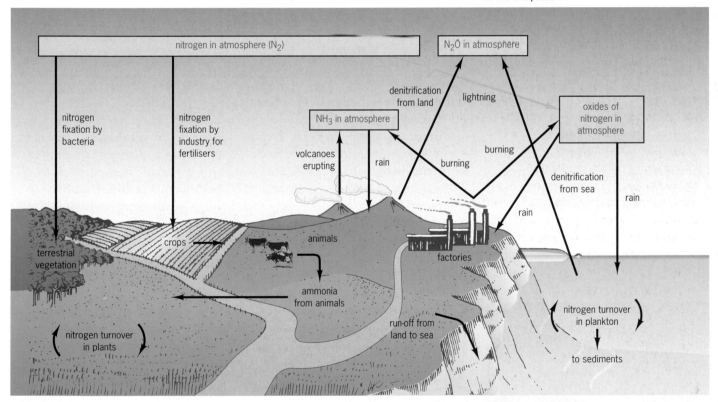

Nitrogen-rich waste from animals (compounds of ammonia, for example) is converted to nitrates by nitrifying bacteria. These microorganisms also fix nitrogen by converting atmospheric nitrogen into useful products. In terrestrial ecosystems, nitrogen is fixed by free-living nitrifying bacteria in soil and by symbiotic bacteria that live in nodules, the swellings on the roots of legumes (plants like peas and beans). In aquatic ecosystems, some blue-green bacteria fix nitrogen. Nitrogen returns to the atmosphere through the action of denitrifying bacteria.

✔ **REMEMBER THIS**

Decomposers stop organic material accumulating in an ecosystem. They also ensure that essential nutrients, such as nitrates, sulphates and phosphates, are released back into the soil.

Fig 35.7 An exam/revision version of the nitrogen cycle. You should find this diagram easy to draw. Learning to draw this, perhaps with your own illustrations and notes, is a good revision exercise

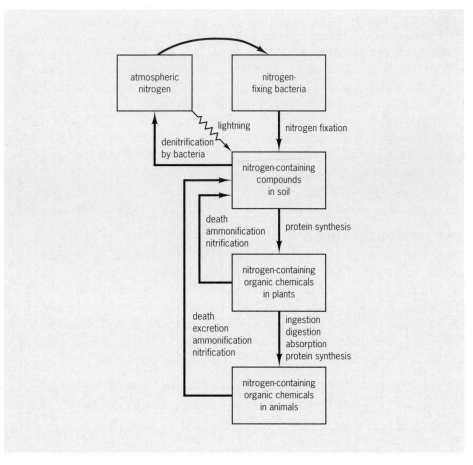

A Traditionally, farmers used to rotate crops with clover and grass every fourth year. Suggest how this could improve the yield of their crops. Would it have any positive effect on the environment?

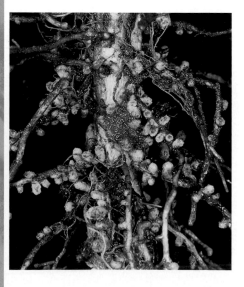

These processes, in more detail, are:

● **Nitrogen fixation**. This is carried out mainly by soil bacteria such as Azotobacter. Some bacteria, such as Rhizobium are **symbiotic**: they live in the root nodules of some plants such as peas and beans (Fig 35.8).

● **Ammonification**. This is the breakdown of organic nitrogen such as proteins and urea into ammonia. It is also known as **saprotrophic decay** and is usually thought of as 'rotting', or **decomposition**. Bacteria and fungi carry out most of the ammonification in the nitrogen cycle.

● **Nitrification**. Nitrifying bacteria such as Nitrosomonas oxidise ammonia to nitrite (NO_2^-). Other bacteria, such as Nitrobacter oxidise nitrite to nitrate (NO_3^-). These oxidation reactions release energy that the bacteria use to make food (they are chemoautotrophs).

● **Denitrification**. Bacteria such as Pseudomonas denitrificans obtain energy by using the nitrite or nitrate ion as an electron acceptor for the oxidation of organic compounds. They live in conditions with low oxygen levels such as waterlogged soils, and can be important in reducing nitrate levels further.

Fig 35.8 Legumes such as peas and beans have nodules on their roots. These house nitrogen-fixing species of the bacterium *Rhizobium*. The bacteria convert nitrogen gas from the atmosphere into ammonia. This passes out of the root nodules, dissolves in soil water and is then taken up by plant roots. The relationship between the bacteria and the plants is therefore symbiotic because both benefit: the plants get a steady supply of nitrogen and the bacteria get a home and a constant supply of sugars from the plant

SUMMARY

After reading this chapter you should know and understand the following:

- Energy transfer through organisms in an ecosystem is one-way. Elements such as carbon and nitrogen cycle between living organisms and the abiotic environment.
- **Decomposers** break down the dead bodies of producers and consumers and play an important role in mineral cycling. Most decomposers are bacteria and fungi.
- The **carbon** and **nitrogen cycles** show how these elements circulate between the biotic and abiotic parts of an ecosystem. You should appreciate the overall cycle and be able to draw a clear summary diagram.
- Carbon dioxide is a **greenhouse gas**: together with other gases in the atmosphere, it acts as a heat insulator. The presence of carbon dioxide in the atmosphere helps to maintain temperatures on Earth that are warm enough to sustain life.

- Human activities have led to an increased concentration of carbon dioxide in the atmosphere. Many scientists believe that this is causing a general increase in average temperatures throughout the world, a phenomenon commonly called **global warming**.
- Microorganisms play an important role in the nitrogen cycle. Different species of bacteria carry out the processes of **nitrogen fixation** (the conversion of atmospheric nitrogen into useful products), **ammonification** (the breakdown of organic nitrogen such as proteins and urea into ammonia), **nitrification** (the oxidation of ammonia to nitrite and the conversion of nitrite to nitrate) and **denitrification** (use of the nitrite or nitrate ion as an electron acceptor for the oxidation of organic compounds).

? EXAM QUESTIONS

1 The diagram represents several important aspects of the carbon cycle. Study the diagram and then answer the questions that follow.

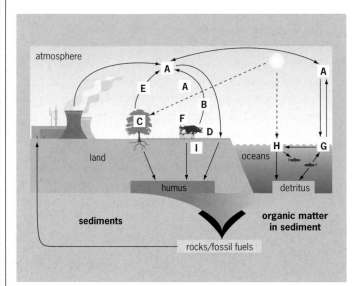

a) Explain briefly (in biological terms) why the chemical compound **A**, around tree **C**, has been shown to vary between 0.03% and 0.04% over a 24 hour period.

b) What biological process is represented by **B**?

c) Name the general class of storage biochemical represented by **D**.

d) When might process **E** be most easily detected?

e) Name a storage compound of carbon found in the liver of animal **F**.

f) Organic compounds of dead organisms are represented by **I**. State the term given to the microorganisms of the carbon cycle which utilise these compounds.

g) Suggest what chemical compound **G** represents.

h) **H** represents phytoplankton. Suggest a reason why this could reduce the concentration of **A** on a long-term basis.

i) **A** and **G** are said to be 'in dynamic equilibrium'. Suggest what this means.
OCR 6910 Molecules and Life June 2000 Q6

2 Give an account of the ways in which nitrogen is cycled in ecosystems. **[essay question]**
Edexcel B2 June 1999 Q8

3 Organisms, both plants and animals, are able to exist together in the same environment. They interact, both with members of their own species and with other species.

a) State the word or phrase which best describes each of the following.
(i) The place where an organism lives.
(ii) A number of **different** species interacting in a particular place.
(iii) All members of the **same** species in a particular place.
(iv) The ecological role of an organism.
(v) The first trophic level in **all** food chains.
(vi) A natural unit of living and non-living parts, interacting to produce a stable system.
Some farmers spread waste from sewage works on their fields. This waste contains nitrogen compounds such as protein, urea and ammonia.

b) **(i)** Describe how the nitrogen compounds present in the sewage waste are converted naturally into a form which can be taken up by the plants in the fields.
(ii) Suggest a possible **disadvantage** of spreading this waste on the fields.
OCR 2801 June 2001 Q2

36

Populations

Gatherings of people like this are common today but could not even have been imagined a hundred thousand years ago

For hundreds of thousands of years, the population of human beings on Earth grew at a steady rate. But in the past 200 years, the world's population has increased from 1 billion to over 6 billion people. World population is currently growing by 80 to 90 million people each year. At 20.02.02 GMT on 20.02.2002 (an interesting time point), the world population was estimated at 6 206 852 159 people. It will probably reach 6.5 billion in 2005, and experts estimate that it will pass the 7 billion mark at the end of the year 2012.

This rate of growth is projected to continue, leading to a global population of around 9 billion people by 2030. The rate of increase and the sheer massive numbers of human beings that this population explosion will generate are unprecedented in the history of the planet. Even with our constantly developing technology and ability to adapt, making the Earth's finite resources stretch to accommodate everyone will be a huge challenge.

Trying to predict what will happen to the population further in the future is difficult because it depends on so many factors. If the population carries on increasing at the current rate, a century from now there could be as many as 15 billion people.

But this is the worst case scenario: United Nations projections assume that family planning will become widespread all over the world and that birth rates everywhere will fall. Even taking into account that life expectancy would continue to increase as medicine continues to advance, the UN estimates that population growth will slow down during the early part of this century and will stabilise around the year 2100.

1 HOW POPULATIONS GROW

REMEMBER THIS

The **biotic potential** is the maximum rate at which a population can increase in size when there are no limits on its growth.

The **environmental resistance** which affects a population is made up of all the factors that act together to limit its size.

When the biotic potential and the environmental resistance reach a balance, the size of the population levels off: it reaches its **carrying capacity**.

Populations are changing constantly. Imagine a new population starting in a particular area with a handful of individuals and ideal environmental conditions. A good example would be a pair of healthy rabbits introduced on to a small island that had plenty of edible vegetation and no predators. The population would grow slowly at first: this is phase A in the graph shown in Fig 36.1.

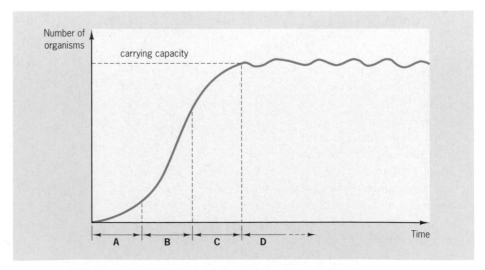

Fig 36.1 This graph shows an idealised curve of population growth. As the curve A to C looks like a flattened S, it is called an S-shaped or sigmoid curve. The text describes phases A to D in detail

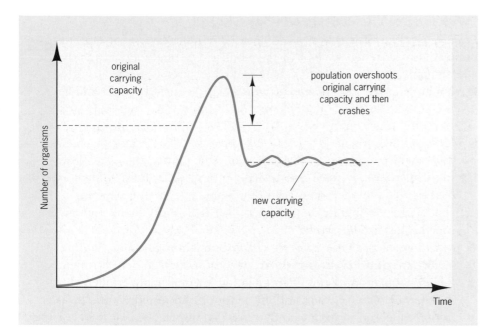

original carrying capacity

population overshoots original carrying capacity and then crashes

new carrying capacity

Number of organisms

Time

Fig 36.2 A population crash follows a sudden increase in numbers which exceeds the carrying capacity. If the environment is damaged by the temporary overshoot, there may be a new, reduced carrying capacity

Fig 36.3 Stable, irruptive and cyclic population curves

Stable

Number of organisms

Time

Irruptive

Number of organisms

Time

Cyclic

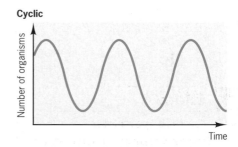

Number of organisms

Time

After a few months, the population would start to grow very rapidly and would double at regular intervals. This type of growth is called **exponential growth**, and is illustrated as phase B in the same graph. As more rabbits were born, they would then breed and there would be a huge rabbit population in a couple of years.

At some point, the population would be prevented from increasing further by various **limiting factors** in the environment – phase C. When food, space and water supply are stretched to their limit, more rabbits would die and they might breed less successfully. (If possible, they would leave the environment.) Because there is less food to go around, they would eventually eat the plants faster than the plants could grow.

Most populations stabilise at some point and the growth curve becomes flat. The number of individuals in the population at this point is the **carrying capacity** of the environment – phase D. Sometimes there is a sudden increase in numbers and a population overshoots its carrying capacity (Fig 36.2). The environment cannot provide the resources for this number of organisms and a **population crash** usually follows. The population dies back below the level of the carrying capacity, and then takes time to stabilise. If the overshoot has damaged the environment, the population might stabilise at a different level.

The two growth curves described in Figs 36.1 and 36.2 occur in laboratory situations but you will not find them in real ecosystems. Fig 36.3 shows the three main types of growth curve that occur in natural conditions:

- **Stable**: the population size remains at roughly the same level over a long time. A study of a population of trees in a well-established woodland would produce this sort of curve.
- **Irruptive**: the population size is basically stable, with the occasional dramatic increase followed by a population crash. This type of population growth occurs in algae. A population is mostly stable but, in hot weather, when water is contaminated by phosphate or nitrate fertilisers, there is an **algal bloom** (Fig 36.4).
- **Cyclic**: the population increases and decreases in a regular cycle of growth and die-back.

Fig 36.4 An algal bloom on a dyke

WHAT LIMITS POPULATION SIZE?

In an ecosystem, **abiotic** (physical) factors in the environment can limit population size. When we look at the whole range of abiotic factors that affect population size, we can classify them into three general categories:

- Limiting factors that are *always present*. These include space and light. A cliff that supports a population of kittiwakes, for example, is a fixed size and can accommodate a fixed number of birds (Fig 36.5).
- Limiting factors that *vary predictably over time*. These include changes in temperature and weather patterns in different seasons which lead to population movement or a change in behaviour among different organisms. As winter approaches, birds that live in an ecosystem in the summer might migrate to a warmer place (Fig 36.6). Small mammals such as dormice hibernate.
- Limiting factors that are *unpredictable*. Some unpredictable limiting factors are **density dependent**: they cause a greater effect if there is a high population than if the population is quite small. Sudden changes in the availability of water and nutrients, or a localised natural disaster, for example, can devastate an overcrowded population but leave a widely dispersed population relatively untouched.
- Other factors are **density independent**: they can devastate a population, whatever its size. Fires, floods, severe frosts and other catastrophes, for example, might destroy part or the whole of an ecosystem without warning (Fig 36.7).

SEE QUESTION 1

Fig 36.5 The size of the ledges is an unchanging limiting factor. The kittiwakes can get nesting sites only by displacing other birds

Fig 36.6 Arctic terns are the world's champion travellers. In their search for food and the right breeding conditions, they range from deep inside the Arctic Circle, over the Atlantic to the pack ice of Antarctica

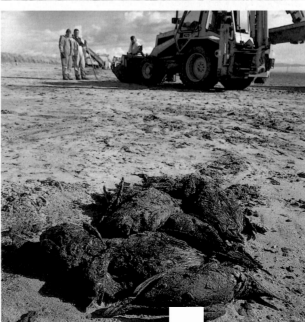

Fig 36.7 Catastrophes, including those caused by human activity such as oil spills, have a density-independent effect on populations

2 INTERACTIONS BETWEEN ORGANISMS

Every living organism in an ecosystem affects the others. Some of these interactions are **intraspecific**: they take place between organisms of the same species. Others are **interspecific**: they take place between organisms of different species.

⚠ SCIENCE FEATURE Growth and bacterial populations

Bacteria are among the fastest growing organisms on Earth. A population of *Escherichia coli*, for example, can double in size in only 20 minutes, if the conditions are very favourable. This doubling time is called the **growth rate** of the bacterial population. The doubling time is a fixed characteristic of each type of bacteria, and it can be used to identify unknown cultures.

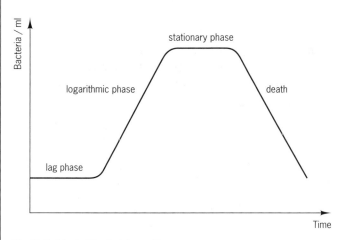

Fig 36.8 A typical bacterial growth curve

A typical growth curve for a population of bacteria is shown in Fig 36.8. There are four basic stages of growth:

The **lag phase** occurs just after some bacteria have been placed in a new container of culture medium. During this period, the bacteria are adjusting to their new surroundings, new food source and new temperature. Their doubling time is relatively slow.

When the **exponential** or **log phase** begins, the bacteria really get going and multiply very quickly. They have plenty of food and waste has not built up to dangerous levels. They make as many new bacterial cells as they can.

After a period of rapid growth, the population then enters the **stationary phase**. The bacteria run out of room or space, or they have produced toxins that limit their growth. The bacteria are still dividing, but the growth rate is equal to the death rate, and no increase in numbers occurs.

In the **death phase**, the balance shifts again, as more bacteria die than are produced by cell division. For the first time since the bacteria were placed in the new culture, their population shrinks.

Interspecific interactions in a bacterial population

The curve shown in Fig 36.9 describes the growth of a single population of bacteria. When more than one type of bacteria is present in the culture fluid, the two species compete. This is an example of an interspecific interaction.

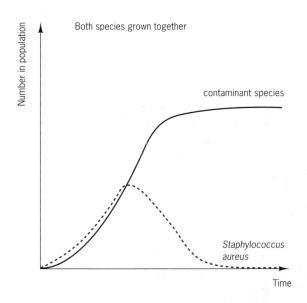

Fig 36.9 Interspecific competition between two bacterial populations

In microbiology laboratories, it is common for cultures to become accidentally contaminated. If a researcher is trying to grow up a culture of *Staphylococcus aureus* to isolate polysaccharides to study their potential as vaccines, it can be disastrous for a second species of bacteria to gain access to the culture. Fig 36.9 shows what can happen when the contaminant is a bacterial species with a faster doubling time. Both species grow equally well at first, but the contaminant species enters the exponential growth phase first, and its numbers increase at a faster rate than the *S. aureus*. The contaminant species may also be more comfortable metabolising different sugars in the medium, and can take more advantage of the food available. After a time, the *S. aureus* population enters a short stationary phase, followed quickly by a decline in population. By the end of the culture time, the researcher returns to the incubator to find a pure culture, not of the *S. aureus* that was expected, but of the rogue contaminant.

INTRASPECIFIC INTERACTIONS

Relationships between organisms of the same species in an ecosystem fall into three main categories:

● Reproduction and care of the young.
● Social behaviour.
● Competition for resources.

REPRODUCTION AND CARE OF THE YOUNG

Organisms of the same species interact at many different stages of sexual reproduction. In some species the relationship is brief and minimal: a wind-pollinated plant simply releases pollen into the air and accepts pollen from another plant of the same species. In others, the association is complex and long term. For example, swans, like humans, find a mate and stay together for many years, sometimes for life (Fig 36.10). Both parents feed and care for the young and, to some extent, provide companionship and help for each other.

In both examples, the level of interaction achieves the aim of reproduction: producing enough surviving offspring to maintain the population level in the ecosystem.

In many animal species, reproduction involves interactions with more than just one other member of the same species. Many animals compete with each other for a suitable mate: for example, male red deer fight fiercely for the control of groups of females. Animals that are more social, such as termites and meercats, provide joint care for each other's young. Some adults in the group, not necessarily just the parents, guard, transport and protect the young, while the others go off to search for food.

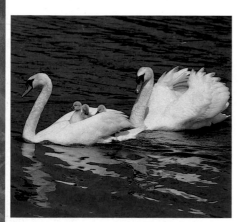

Fig 36.10 The female swan carries her young, while the male displays to deter intruders

SOCIAL BEHAVIOUR

Many animals are social: they live in groups rather than as a collection of individuals that just happen to live in the same place. Group living has its advantages and disadvantages. Animals that live in groups are usually less likely to be eaten by predators than solitary animals. There are several reasons for this:

● A large number of potential prey in one place can confuse a predator – the predator cannot decide which one to attack.
● Different individuals in a large group of animals can perform different tasks: some can look for food, others care for the young and some can act as look-outs for predators. Large groups therefore spot approaching predators much sooner than animals on their own and so can take more successful evading action.
● A large number of animals are more successful than just a few at defending against an attacking predator. If a predator threatens, elephants group together, forming a circle to protect their young, which are herded into the centre of the group.

Animals that live in groups also stand more chance of hunting or finding food successfully (Fig 36.11, and the Science feature on the next page). Hyenas, for example, hunt in packs and can bring down a deer that is much larger than one hyena would be prepared to tackle. Other group associations can help to protect animals from the cold as they huddle together.

Fig 36.11 There is evidence that bees can communicate the position of good pollen sources by elaborate dancing displays to other bees in the hive. Here, a worker bee with full pollen sacs is dancing to indicate the amount, distance and direction of its source of pollen

SCIENCE FEATURE Costs and benefits of social behaviour

Prairie dogs are rabbit-sized rodents with small ears and short legs (Fig 36.12). They live in large groups in the plains of North America digging a complex of tunnels up to 30 metres long. There are two species of prairie dog, the black-tailed and white-tailed forms, and each community or 'town' may contain up to a thousand animals.

Prairie dogs graze above ground during the day when there are many predators around such as coyotes, badgers and hawks. They have 'sentries' sitting upright outside the burrows keeping a watchful eye for predators. The appearance of a predator provokes a series of whistling barks from the sentries.

A cost–benefit analysis has been carried out on the social behaviour of the prairie dog. An anti-predator response was tested by dragging a stuffed badger through a colony. In the larger groups of prairie dog, the predator was detected sooner. Black-tailed prairie dogs live in larger groups than their white-tailed cousins. Consequently the black-tails can afford to spend significantly less time scanning their surroundings for predators (35 per cent of their average daily time budget compared with 45 per cent in the white-tails). Less time being vigilant means more time to feed.

So why don't white-tailed prairie dogs live in large groups as well? One answer is that living in large groups has costs as well as benefits. Within either species, individuals in large groups fight more often than those in small groups. Also disease-spreading fleas are more common in large groups (3.3 fleas per black-tailed burrow, 0.5 fleas per white-tailed burrow).

The evolution of group behaviour therefore depends on balancing advantages against disadvantages. Other factors may also be involved – black-tailed prairie dogs live in open plains, for example, and, since there are many predators around, it is safer for them to live in larger groups.

Fig 36.12 A prairie dog keeping guard as sentry

COMPETITION FOR RESOURCES

Most of the disadvantages of group living arise because animals have to compete with each other for vital resources. Hyenas may make a bigger kill when they hunt in a pack, but the meat has to fill more stomachs. If food becomes scarce, rivalry within the pack becomes intense and some animals may need to find a new habitat to survive. Animals of the same species also compete for mates, shelter, water and social position.

INTERSPECIFIC INTERACTIONS

A full list of the interactions that can occur between individuals of different species in an ecosystem is given in Chapter 33, Table 33.2. In this chapter we will look in detail at four of these:
- Predator–prey relationships.
- Parasitism and disease.
- Mutualism and commensalism.
- Competing for resources.

PREDATOR–PREY RELATIONSHIPS

A basic definition of a predator, which includes animals that feed on plants, is:

An organism that eats another organism to obtain energy and nutrients.

The relationship between predator and prey is based on a constant battle. The actual 'catching and eating' part of the battle is only a small part of the story. The more serious conflict involves evolution.

REMEMBER THIS

Symbiosis comes from a Greek word that means 'to live together'. We use it to describe a close physical interaction between two organisms from different species. There are three types of symbiosis:
Parasitism: one of the partners benefits, the other is harmed.
Commensalism: one partner benefits and the other is unaffected.
Mutualism: both partners benefit from the association.

Fig 36.13 Bats have evolved sophisticated sonar systems to detect flying insects with great accuracy. Some insects, however, have evolved the ability to detect bat sonar signals and respond by closing their wings and dropping out of the sky. This makes it much harder for the bat to catch them and they gain an important survival advantage over other insect species

All species evolve by the processes of genetic mutation and natural selection (see Chapter 26): prey species are constantly evolving and becoming better able to cope with predators; predator species are constantly evolving to out-manoeuvre their prey (Fig 36.13). Sometimes, either the predator or prey loses the long-term battle. If the prey evolves to completely avoid a predator, and that predator has no alternative food supply, the predator species can be driven to extinction. If a predator out-evolves its prey, it will do well in the short term. But, if the prey species becomes extinct, the predator then needs to find an alternative food supply to survive.

In general, predators thrive when they develop adaptations that allow them to eat well and make the most of their food. Herbivores, animals that eat plants, often have specialised mouthparts and digestive system to gain the most nourishment from the plants available to them. Many herbivores are ruminants: their digestive system has four stomachs and they chew their food twice, once after eating it and once after regurgitating it.

Prey species continue to survive because they have evolved defence mechanisms against predators. In animals, these include **camouflage**, **spines** and **barbs** and **chemical deterrents** (Fig 36.14). Many, particularly insects, produce foul tasting or poisonous chemicals and usually advertise this by the colours and patterns on their bodies and wings.

Although plants cannot run away from grazing animals, they are not totally defenceless. Their defences include thorns, tough tissues and thick waxy cuticles. Many plants have also developed **secondary compounds**, chemicals that have no metabolic role but which put off potential grazers. Examples include tannins in oak leaves and cardiac glycosides in foxglove plants.

Fig 36.14 Animals use many different techniques to avoid capture by predators

When under threat, a porcupine runs backwards at the enemy with quills raised. They can fall off, and an embedded quill can cause a septic wound that would kill a lion

The camouflage of this iguana matches the colours of the trunk and lichens in a Peruvian rainforest

The deadly sting of a scorpion deters predators as well as being useful for hunting food

PARASITISM AND DISEASE

Predators usually kill their prey: parasites have a different strategy because they depend on their prey, the **host** (a different species), staying alive. They live and reproduce using the resources of the host organism that is damaged in the process. This is why many parasites cause disease in humans and other animals (Table 36.1). The more successful parasites cause less damage to their host. The tapeworm, for example, can survive in the human digestive system for many years, causing relatively minor damage in the affected person.

Table 36.1 Some major parasitic diseases and the parasites that cause them		
Disease	**Parasite**	**Symptoms and effects**
Malaria	*Plasmodium* (4 different species of this protoctist parasite cause malaria)	Regular cycles of fever: disease transmitted to people by the *Anopheles* mosquito (see Science feature on page 647).
Sleeping sickness	*Trypanosoma* species (also a protoctist parasite)	Transmitted to people by insects such as the tsetse fly which bite infected animals.
Leishmaniasis (also called kala-azar and oriental sore)	*Leishmania* species (another protoctist parasite)	Transmitted to humans via sandflies.
Schistosomiasis (also called bilharzia)	*Schistosoma* (blood fluke)	Flukes live in water (their intermediate host is a water snail) and infect humans who bathe.
Tapeworm infection	*Taenia solium* (tapeworm)	Parasite lives in intestine: passed to humans in infected and undercooked meat.
Toxoplasmosis	*Toxoplasma* species (protoctist parasite)	Causes weakness and fever passed from mother to baby across placenta. Can be transmitted by domestic cats.

PARASITIC ADAPTATIONS

Parasites show very specialised features that fit them to their lifestyle (Fig 36.15). All parasites show some general adaptations:

- They have ways of getting into the body of the host. Fungi that parasitise plants produce the enzyme **cellulase**, which breaks down plant cell walls. The schistosome (blood fluke), which causes bilharzia, can bore through human skin (see Chapter 29).
- They have structures that enable them to attach to their host. Tapeworms have a ring of hooks called a **scolex** that they insert into the wall of the host's intestine.
- They have lost the organ systems that they no longer need. Tapeworms and other gut parasites have, for example, lost their own digestive system because they can absorb already-digested food across their body surface. We call this loss of structure and function **parasitic degeneration**.
- They protect themselves against the internal defences of the host. Trypanosomes, which cause sleeping sickness in humans, produce an outer layer which varies constantly in its structure. Just as the host's body starts to make antibodies (see Chapter 30) to one type of coating, the trypanosome produces a different coat which the antibodies cannot recognise.
- They often have complex life cycles that allow them to infect new hosts. See the Science feature on malaria on page 647.
- They show great **fecundity**, or capacity for reproduction. Many tapeworm segments break off every day and pass out of the body in the faeces. Each segment can contain hundreds or even thousands of eggs.

? TEST YOURSELF

A Organisms can interact intraspecifically or interspecifically. What do these two terms mean?

Fig 36.15 Right: The beef tapeworm. The head is at the thin end, attached to many hundreds of segments. They break off and pass out of the host (cattle), to infect another victim. The ruler is 30 cm long. Above: At the head end (the scolex) are four suckers and a ring of hooks that attach the tapeworm to the gut wall of its host

Fig 36.16 There are 26 species of clownfish which all associate with sea anemones. This colourful fish hides from predators among the venom-producing tentacles of the sea anemone

 ### REMEMBER THIS

Legumes, with their root nodules filled with bacteria (see Chapter 35), are an example of a mutualistic association.

MUTUALISM AND COMMENSALISM

The term **mutualism** is used to describe a relationship between two organisms from different species that receive mutual benefit from the association. Neither is harmed and both do better by being together than they would alone. The African bird, the oxpecker, for example, cleans the skin of the rhinoceros, removing all the parasites such as ticks and mites. The bird eats the insects, gaining the advantage of the nutrients.

Two organisms that are **commensal** live together and, while one of the partners benefits, the other does not. The relationship between clownfish and sea anemones is a good example of commensalism (Fig 36.16). The fish hide from predators in the mass of the anemone's stinging tentacles. Although the anemone is unharmed, it does not seem to benefit in any way.

COMPETING FOR RESOURCES

Species can compete in different ways. Some organisms stop the organisms of other species from getting near a particular resource. Wrens, for example, defend food plants in their territory. In other situations, two species have access to the resource, but one is better at exploiting it.

Some species have evolved to **coexist**, sharing the resources that are available in an ecosystem. This is termed **symbiosis**. Some plants, for example, have a mesh of roots close to the surface of the soil to trap water from short showers of rain. Other neighbouring plants have long roots that take advantage of deeper groundwater. Both species are using the same resource, but they are sharing it in a way that allows both to benefit. This is called **resource partitioning** and it also occurs between animal species.

SCIENCE FEATURE The fight against malaria

Malaria is a disease caused by a parasite. Four species of protoctist of the genus *Plasmodium* are responsible for 200 to 300 million cases of malaria every year, worldwide. The disease is common in the tropical and sub-tropical regions of the world – the home of 40 per cent of the world's population. The death rate is appalling, somewhere between one million and five million people each year, many of them children under 5.

Malaria is extremely difficult to control for several reasons. Firstly, it has a complex life cycle (Fig 36.17). Any four of the *Plasmodium* species that cause it can be passed on to human hosts by any one of 60 species of the *Anopheles* mosquito. Draining swamps and marshes and using insecticides to spray mosquito-breeding areas was temporarily successful and reduced the spread of malaria during the 1950s and 1960s. However, not every species of carrier mosquito was wiped out. Many of them have developed resistance to the early insecticides and are now much more difficult to control.

Secondly, the parasite itself is a tricky customer. It produces symptoms that get worse and then better in cycles. Anti-malarial drugs target some but not all stages in the life cycle (Fig 36.18). The sufferer therefore needs to take the drugs over a long period of time, or the parasite 'hides' from their effects. The common anti-malarial drugs have been used so widely that they have put selection pressure on the parasite and it has now developed resistance to many of them.

Although new drugs are being developed all the time (some of the most recent are based on extracts of rainforest plants), workers now concentrate on prevention. Any stagnant water that can be eliminated is removed and mosquito nets in houses are dipped in a safe insecticide such as permethrin. Biological methods of control of the mosquito larvae are being used – fish such as guppies are being bred to eat them. Vegetation around houses is cleared, and trees are being planted in marsh areas to destroy the wet breeding grounds of the mosquito.

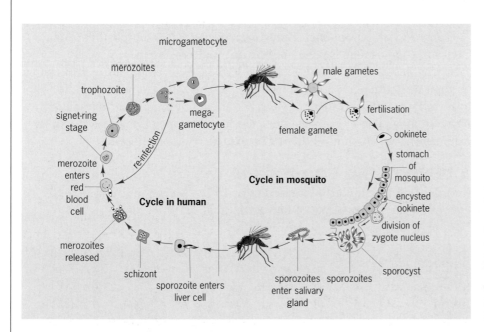

Fig 36.17 The life cycle of the malarial parasite, *Plasmodium vivax*. The *Anopheles* mosquito is the first host; humans are the second host. People with malaria get regular cycles of illness because merozoites are continually released. When the parasite is 'hiding' in the red blood cells, symptoms are not too bad: it is the sudden rush of foreign proteins in the bloodstream that leads to the bouts of fever that are typical of malaria

Fig 36.18 The merozoite stage (left) and signet ring stage (right) of *Plasmodium vivax* in human red blood cells

3 A WIDER LOOK AT COMMUNITIES IN AN ECOSYSTEM

We have seen that ecosystems are dynamic systems and that the physical parts of an ecosystem, such as the weather, affect and interact with the organisms that live there. In this section we look at how an ecosystem forms.

Imagine a completely physical environment such as the bare rock and ash left behind after a volcano has erupted and then become quiet. After the ground cools and water re-enters the environment, some plants start to grow there. This process is called **colonisation**. After a time, other species of plant establish themselves and eventually, probably after many years, woodland forms. This process is called **succession**.

Succession also describes the more minor changes in the vegetation of an ecosystem that occur after a less catastrophic environmental change. For example, after a small patch of trees in woodland die after a series of hot, dry summers, the land becomes repopulated by the same sorts of trees very quickly because their seeds are already present in fertile ground.

To distinguish between these two types of situation, we say that the process that results in new vegetation on a bare site is **primary succession**. This takes time, and the final appearance of the community depends on which plant spores and seeds come into the area from neighbouring ecosystems. The process that repopulates a previously well-vegetated area after a minor environmental change is called **secondary succession**.

Strictly, succession is used only to describe the development of and changes in plant communities. In any ecosystem, however, the plant communities help to form the conditions that enable other classes of organism (animals, bacteria, fungi, etc) to colonise the environment. Once a stable community has formed, we say that a **climax community** is established.

SEE QUESTION 2

A DEVELOPING ECOSYSTEM

In order to understand the stages that occur in succession, think of an environment consisting just of bare rock (Fig 36.19).

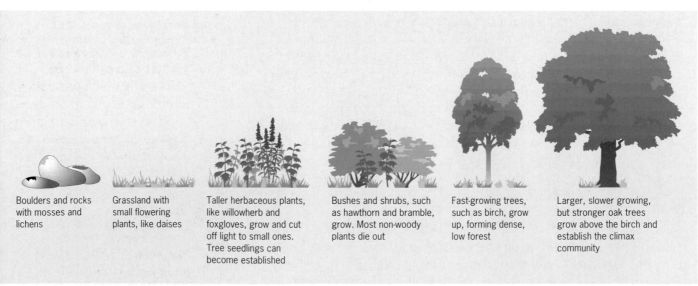

| Boulders and rocks with mosses and lichens | Grassland with small flowering plants, like daises | Taller herbaceous plants, like willowherb and foxgloves, grow and cut off light to small ones. Tree seedlings can become established | Bushes and shrubs, such as hawthorn and bramble, grow. Most non-woody plants die out | Fast-growing trees, such as birch, grow up, forming dense, low forest | Larger, slower growing, but stronger oak trees grow above the birch and establish the climax community |

Fig 36.19 Succession in a bare site to woodland

SCIENCE FEATURE Studying succession

The process of succession normally takes decades or centuries, but there are two common ways of studying the process:

● Clear a patch of ground and watch what happens to the bare soil.

● Study a sand dune system near a beach and observe the changes that occur as you move inland.

1 Clearing a patch of ground

Remove all the plants, fence it off from grazing animals, and observe the changes. The problem is that it takes at least 50 years, but eventually the forest is re-established. There are two conditions that need to be met if the ecosystem is to develop as normal:

● The soil must initially have relatively low humus content.

● There must be no grazing animals that can get on to the plot.

The climax community that develops depends on the climate. The process outlined in the text on this page will not happen, for example, on exposed hillsides where the soil is too thin or the wind too harsh. Other examples of climax vegetation around the world include rainforest, cloud forest, tundra, grassland and coniferous forest.

2 Studying a sand dune system

Sand dunes are useful areas to study because you can observe colonisation and succession without having to wait 50 years. You can see some of the changes associated with the development of ecosystems as you simply walk inland (Fig 36.20). Near the sea, the dunes are at their youngest and wind tends to pile up the sand. As a result, the profile changes from year to year. Sand is a very difficult medium for plants; water and nutrients drain straight through, and the constant shifting makes it very difficult for the roots to anchor the plant. However some pioneer species, notably marram grass, have a dense root system that binds the sand together. This holds water and humus particles and makes the whole dune more permanent. Once the marram grass has made the environment less hostile, other plants such as ragwort, willow and grasses can take over. As you move inland the sand gets darker because the humus content increases, as does the species diversity.

Fig 36.20 Sand dune succession

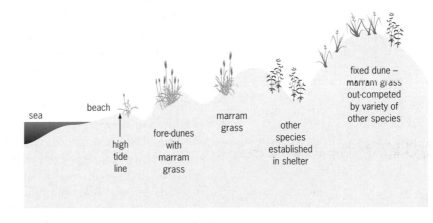

Colonisation occurs in many different stages called **seres**. The rocks first provide a home for spiders that feed on insects in the area. Succession begins as mosses and lichens start to grow on the bare rocks. These first plant species are called **pioneer species**. Over many years, perhaps thousands of years, the environment is battered by rain and wind and other climatic events, and the surface of the rock is broken down into smaller fragments. The mosses and lichens help soil to form in cracks and crevices, and eventually a few small broad-leaved plants start to grow. The first ones are what we might think of as weeds – herbaceous plants that can grow in relatively nutrient-poor soil. They have a rapid life cycle, then die back and increase the humus content of the soil. Typical coloniser species are grasses, daisies, dandelions and clover.

Succession proper is then under way. Succession occurs because the colonisers change the habitat. Once the colonisers have improved the quality of the soil, more species can grow. The **diversity** of herbaceous plants increases greatly as conditions become more favourable. Grasses and then

✓ REMEMBER THIS

Lichens consist of algae growing inside a fungus. The fungus provides anchorage and protection from drying out. In turn the algae can photosynthesise and give the fungus organic compounds that would otherwise be unavailable from the bare rock.

✓ REMEMBER THIS

Herbaceous plants have no woody tissue, in contrast to shrubs and trees.

REMEMBER THIS

It is a good idea to learn one example of succession in detail, and make sure that you can name some plant *species* at each stage, rather then talking vaguely about 'trees and shrubs', etc.

SEE QUESTIONS 3 AND 4

plants with root nodules that have nitrogen-fixing bacteria start to grow. Taller plants cut off the light and so out-compete the shorter ones. The greater diversity of plants attracts more insects that in turn attract birds and small mammals. At this stage grazing can have a marked effect and prevent any further succession. Many plants, including tree saplings, have their growing points at the top of the stem, and herbivores such as rabbits and sheep prevent any further growth. In contrast grasses grow from the base of the stem, so they thrive despite constant grazing.

ESTABLISHMENT OF THE CLIMAX COMMUNITY

In this phase, small woody plants – shrubs such as hawthorn and bramble – begin to dominate. In turn these are out-competed for light by fast-growing tree species such as birch, which form a low, dense forest. Eventually the large but slow-growing trees – notably the oak – begin to dominate until the **climax community** is established and there is *no further succession*.

4 BIODIVERSITY

Biodiversity, a word that is the popular contraction of the term **biological diversity**, is used generally to refer to the variety of life on Earth. More specifically, we use it to describe the numbers of species present in a particular habitat or to show the range of types of organism found there (Fig 36.21). Take the sea bed, for example. It has a large number of species but it is largely made up of a few basic types: annelid worms, molluscs and echinoderms. A rocky shore may have the same number of species but there is often a greater variety of animals and plants: arthropods, annelid worms, molluscs, echinoderms, cnidarians, fish, birds and seaweeds. We say that the shore has a greater **species diversity**: it has a greater **species richness** (number of different species) and a greater **species disparity** (range of different species types) than the sea bed. One of the most diverse ecosystems in the world is the tropical rainforest (Fig 36.22).

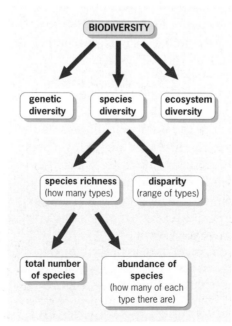

Fig 36.21 What biodiversity means

Fig 36.22 The tropical rainforests of the world encompass a massive diversity of all forms of life

Diversity can be used to indicate the 'biological health' of a particular habitat. A slow increase in plant species diversity is normal for stable habitats such as hedgerows. If a habitat suddenly begins to lose its animal or plant types, ecologists become worried and search for causes (a pollution incident, for example).

You might also have heard of the term **genetic diversity**. Different species are obviously genetically different from each other, but there is also genetic variation within a species (no two people look identical, for example).

A third form is **ecosystem diversity**. The environment is an important factor that determines the number and types of species present in an ecosystem. At its simplest, ecosystem diversity is described as the number of different types of habitat in a given area.

WHY BIODIVERSITY IS IMPORTANT

When a species becomes extinct, many other species can be affected: no type of living organism exists independently. No one can really predict the long-term effects of repeated extinctions. Organisms have, of course, always become extinct, because of evolutionary pressures. What concerns us today is the rate at which species are disappearing because of the impact of human populations.

Although the diverse world of animals and plants provides us with a massive resource of food, we actually use only 30 species or so to provide 90 per cent of all the food we eat worldwide. Plants have already given us many useful medicines (Fig 36.23 and see Science feature below) and there are undoubtedly more compounds still to be found. In addition, many different animals and plants stabilise soil and so minimise erosion, support fisheries, provide income through tourism and provide beauty and interest.

Fig 36.23 Aspirin, the most commonly used drug in the world, was originally extracted from the bark of the willow

---

SCIENCE FEATURE Medicines from the rainforest

The weather conditions in the tropical rainforests around the Equator have encouraged the evolution of areas of high plant and animal diversity. The indigenous human populations who live and work in the rainforests need to make a living, and working for developers who sell timber and other products to the developed world is an obvious means of doing this. The consequences for the environment are serious. Some scientists estimate that three species of living things are being made extinct every hour, because the rainforest is being destroyed.

But what difference can a few plants make? Possibly the difference between life and death for many people. Screening plants from the rainforest has shown that they contain many compounds that protect them against predators and parasites. These compounds might also be useful in human medicine. Since 1990, at least four compounds have been identified that might provide drugs for use against the virus that causes AIDS (Acquired Immune Deficiency Syndrome).

The work of developing medicines needs sophisticated chemical and pharmaceutical techniques, but it cannot be done unless plants that are collected for screening can be

identified accurately and consistently. Speed is also important. If destruction continues at its present rate, the rainforests will have been completely wiped out by the end of the century.

Fig 36.24 Scientists who travel to remote areas of rainforest to search for and identify plants which may be useful in medicine are called ethnobotanists. They work with local traditional healers to narrow down the choice of plants to study. Samples are sent back to laboratories for testing. Research programmes are now being set up to share the profits from successful compounds, so that indigenous people can benefit from the discovery, instead of suffering exploitation

MEASURING BIODIVERSITY

Estimates for the total number of species that still exist range from 1 million to 50 million. We base our estimates on the knowledge we already have of numbers and types of species found in particular habitats and on the rate at which our knowledge is increasing. Between 1978 and 1987, 367 new vertebrates, 173 new annelids and 7222 new species of insect were identified.

WHY DO WE NEED TO MEASURE BIODIVERSITY?

There is more to measuring biodiversity than just being able to impress your friends by knowing how many species of living things there are on Earth. We need to measure the diversity:

● To compare the diversity of the same habitat over time, to assess, for example, whether pollution is damaging that habitat.
● To compare two different habitats at the same time, to find out which is the more diverse.
● To check whether a new habitat is being colonised by the number of species that it should. This would be important in the case of a polluted river or lake that had been 'cleaned up' and then repopulated with living organisms.

PUTTING A NUMBER ON BIODIVERSITY – THE SIMPSON DIVERSITY INDEX

Being able to put a number on the diversity of a habitat makes the life of an ecologist much easier. A **diversity index** allows us to estimate the variety of living things in a particular area. However, we must make sure that the value we use accurately reflects all aspects of diversity. We could, for example, simply count the number of different species present:

Stream 1 (100 animals)	16 species
Stream 2 (100 animals)	10 species

The conclusion would be that stream 1 was more diverse. But there is a problem: this method gives an idea of species richness, but it does not reflect how many of each species are present. In stream 1, there may be 85 animals of one species and just 1 of each of the others. In stream 2, there may be 10 animals of each species. Ecologists would say that stream 2 was the more diverse habitat, and so our first conclusion would be wrong.

To avoid this problem, we can use the Simpson index. Its formula takes into account that diversity depends on the number of individuals of each species present and the species richness (the number of different species present). The higher the value of the index, the greater the variety of living organisms found in the area.

$$D = \frac{N(N - 1)}{\Sigma n(n - 1)}$$

where D is the Simpson diversity index, N is the total number of individuals of all species found, n is the total number of organisms of a particular species found, and Σ means 'the sum of'.

? TEST YOURSELF

B The Simpson index was used to calculate the diversity of an oak wood and a conifer plantation. The values obtained were 19.6 for the oak wood and 12.04 for the conifer plantation. Which habitat is the more diverse?

⬤ EXAMPLE

Q Assume that you have studied a particular habitat and you have recorded the types of 20 animals you have found. The different types are represented by letters of the alphabet, and the record shows five species:

A A A A B B A C C B B A B B A D C C D E

What is the diversity of this habitat?

A Putting the figures into the equation for the Simpson index:

$$D = \frac{20 \times 19}{(7 \times 6) + (6 \times 5) + (4 \times 3) + (2 \times 1) + (1 \times 0)} = \frac{380}{86} = 4.42$$

SCIENCE FEATURE How old is my hedge?

As you next walk down a country lane, take a good look at the hedges that you pass. Many of these structures date from the Middle Ages. They were planted by medieval farmers as boundaries to mark off their land and to keep in livestock. They probably planted shrubs such as hawthorn a few feet apart. The gaps at the base of these plants then allowed waves of colonisers.

First of all, fast growing, non-woody annual plants such as typical garden weeds would have thrived: chickweed, groundsel and shepherd's purse for example. Then, non-woody perennials such as bluebells, primroses, stinging nettles, dandelions and thistles may have become established. These plants can produce food reserves to survive the winter. Eventually, woody perennials, shrubs and small trees such as elder, holly, and even some of the larger forest trees could have invaded the hedge.

Fig 36.25 In theory, it is possible to date a hedge by looking at the range, or diversity of plants that are in it. The more diverse the community of the hedge, the older it is

SUMMARY

By the end of this chapter you should know and understand the following:

■ A **population** grows **exponentially** in ideal conditions. Growth slows down because of **limiting factors** in the environment. This produces an **S-shaped** or **sigmoid growth curve**.

■ When the size of the population becomes stable, we say that this is the **carrying capacity** of the environment. Sometimes, population growth exceeds the carrying capacity and a **population crash** follows. The population may then settle at a new carrying capacity.

■ Three main types of growth curve occur in nature: **stable**, **irruptive** and **cyclic**.

■ Some factors that limit population growth are **density dependent**: their effects are worse in a large, closely packed population. Others are **density independent**: their effects do not depend on the population distribution or size.

■ **Intraspecific interactions** occur between organisms of the same species; **interspecific interactions** occur between organisms of different species. You should know some examples of each type.

■ The living part of an ecosystem develops by the process of **succession**. If the area has never been **colonised** before, a newly formed volcanic island, for example, the process is called **primary succession**. If the area was previously covered with vegetation, a forest site that suffers a severe fire, for example, **secondary succession** occurs.

■ There are three types of **biological diversity**: **species diversity**, **genetic diversity** and **ecosystem diversity**.

■ Measuring biodiversity is important because it allows us to monitor a habitat over time, to compare two different habitats and to find out if a new habitat is being colonised by the correct number and type of species. This is vital if we are to assess the impact of our 'clean-up' efforts after environmental tragedies such as oil spills.

❓ EXAM QUESTIONS

1 A population of brown trout in a river was studied over a period of ten years. Each year the size of the population was measured in early summer, when it included the newly hatched young, and again at the end of the summer.

The results are shown in the graph.

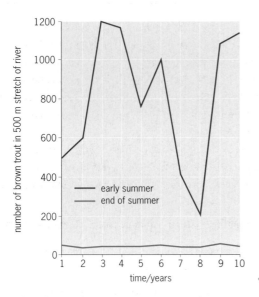

It is suggested that the change in population size between early and late summer is due mainly to one or more density dependent factors.

(i) Explain what is meant by a *density dependent factor.*

(ii) Describe and explain the evidence from the graph which suggests that the change is mainly due to the density dependent factors.

b) Brown trout only grow well in water which is above pH 5.5 Explain **one** way in which the pH of the water might become more acidic than this, other than by pollution or acid rain.

AQA(NEAB) BY05 Ecology June 2000 Q4

2
a) Give the ecological term that best describes each of the following.

(i) all the living organisms in a particular habitat

(ii) the particular set of conditions in a habitat to which a species is adapted

(iii) two species using the same resource in the same habitat

(iv) movement of a species between two habitats to make use of their resources at different times.

b) Explain what is meant by *succession* in a habitat.

AQA (NEAB) BY05 Ecology June 2000 Q1

3 The distribution of two species of plant, greater plantain and white deadnettle, on a roadside verge was investigated. The results are shown in the bar graph.

a) **(i)** Describe how the data on distribution could have been obtained.

(ii) Describe what the data show about the distribution of greater plantain.

b) The drawings show the two species of plant.

greater plantain

10cm

white deadnettle

Use evidence from the drawings to suggest a hypothesis that explains the distribution of greater plantain.

c) A diversity index for all of the plants growing on this roadside verge was calculated at three-year intervals. The table shows the results.

Year	Diversity index
1990	6.1
1993	4.5
1996	3.1

Explain what these results indicate about the plant community on this verge.

AQA(NEAB) BY05 Ecology June 2000 Q3

? EXAM QUESTIONS

4 The dog whelk, *Nucella lapillus*, is a marine snail found on rocky shores. It feeds on barnacles and molluscs by drilling holes in their shells or prising them open.

An investigation was carried out into the choice of prey species by the dog whelk. Sixty dog whelks were collected from a rocky shore where they had been feeding on barnacles. The dog whelks were given a mixture of two species of barnacle, *Semibalanus balanoides* and *Elminius modestus*. After 10 days, the number of each species of barnacle which had been eaten and the number remaining uneaten were recorded.

The observed results are shown in the table below.

Species	Number eaten	Number remaining	Total
S. balanoides	137	146	283
E. modestus	166	249	415
Total	303	395	698

Data from Barnett BE (1979)

A chi-squared (χ^2) test was then carried out on these results to test the null hypothesis that dog whelks eat equal percentages of each of the two barnacle species.

The expected results for this test are shown in the table below.

Species	Number eaten	Number remaining	Total
S. balanoides	122.9	160.1	283
E. modestus	180.1	234.9	415
Total	303.0	395.0	698

The formula for calculating chi-squared (χ^2) is given below.

$$\chi^2 = \sum \frac{(O-E)^2}{E}$$

where O = observed values, E = expected values.

a) **(i)** Use the formula to calculate the value of χ^2. Show your working.

(ii) The table below gives some values for χ^2 with one degree of freedom.

P(%)	20	10	5	2	1
χ^2	1.64	2.70	3.84	5.41	6.63

Does your calculated value of χ^2 enable you to accept or to reject the null hypothesis? Give a reason for your answer.

b) The dog whelks ate 48% of the *S. balanoides* and 40% of the *E. modestus*. Suggest possible long-term effects this could have on the populations of the two species.

c) The investigation was repeated using dog whelks which had not been fed for a period of time. The results for this second experiment are summarised in the table below.

Species	Number available	Number eaten	Percentage eaten
E. balanoides	296	161	54
E. modestus	365	191	

(i) Calculate the percentage of the available *E. modestus* which was eaten. Write your answer in the space in the table.

(ii) Describe the effects of starvation on the feeding behaviour of dog whelks.

d) The barnacles feed by filtering plankton from sea water. The plankton contains microscopic plants (phytoplankton) and microscopic animals (zooplankton) which feed on these plants.

(i) Construct a food web to show these feeding relationships.

(ii) Name the trophic levels occupied by the barnacles in this food web.

Edexcel B5/HB5 June 1999 Q2

KEY SKILLS **ASSIGNMENT**

AN UNCERTAIN FUTURE

In Chapter 26 we looked at how humans evolved from their primate ancestors. In this Assignment, we investigate what has happened to the human population over the past 12 000 years and we raise some questions about what will happen in the next 12 000.

We start our story at the start of the 'Agricultural Revolution' in about 10 000 BC, when there were around 10 million people on Earth. People had changed from being hunter–gatherers to being farmers. With this new settled lifestyle, there was a dramatic increase in the birth rate and a decline in the death rate.

1 **a)** Suggest why nomadic travellers might have had fewer children than farmers living in permanent settlements.
b) Why would life expectancy be lower in hunter–gatherer societies?

By the time of the First Egyptian Dynasty (3000 BC), the global human population had increased from about 10 million people to 100 million. By the height of the Roman Empire and the birth of Jesus Christ, there were 300 million. Civilisations flourished in three main regions of the world: Southwest Asia, China and the American continent.

2 What properties do you think these areas might have had that enabled them to support large numbers of people?

Table 36.A1 shows how population continued to increase until 1800, when it reached the 1 billion milestone.

Table 36.A1 Population trends to the first billion humans							
Year	0	1000	1250	1500	1750	1800
Population (in billions)	0.30	0.31	0.40	0.50	0.79	1.00

3 Plot the figures given in Table 36.A1 as a line graph. Extend the time axis to the year 2100 AD.
a) Estimate how many people would have been alive in the year 2000 if the rate of growth had remained constant after 1800 AD.
b) Current estimates put the world population at 7 billion in the year 2012. Why is your estimate less than this?
c) If population trends continue like this, in which year will the number of people on Earth reach 11.6 billion, the estimated maximum population that the planet's known resources can support?

Staggeringly, adding yet another billion people after 1800 took only 123 years, and another billion was added 33 years after that. Since 1960 we have been in the middle of an incredible population explosion with another billion people being added every 12 years.

4 Experts think that population growth will slow down during this century and stabilise at around 11 billion people. What factors are likely to be responsible?

Most of the Earth's land surface is now inhabited by humans. Table 36.A2 shows the 15 most populated countries today and gives an estimate of how their individual populations will grow in the next 50 years.

Table 36.A2 Human population, in millions, of the 15 most populated countries in the world with estimates of their population in the year 2050 AD			
2000		**2050**	
Country	**Population**	**Country**	**Population**
China	1232	India	1530
India	945	China	1514
USA	269	Pakistan	357
Indonesia	200	USA	348
Brazil	161	Nigeria	339
Russia	148	Indonesia	318
Pakistan	140	Brazil	243
Japan	125	Bangladesh	218
Bangladesh	120	Ethiopia	213
Nigeria	115	Iran	170

5 Explain the different reasons why China, Indonesia and the USA all have a lower place in 2050 than in 2000.

6 Imagine that it is the year 2100. You have been asked to write a book telling the history of the previous century. The population of Earth stands at 3 billion people. Explain in about one thousand words what happened.

Fig 36.A1 The suffering caused by starvation, famine and overcrowding seems likely to continue as the human population continues to increase exponentially

37

Human activity and the environment

Human activity and the environment

Many problems in the environment happen when people use a resource for today, hoping that everything will be OK tomorrow. In the Philippines sea-horse fishing is very important to the poor fishermen and their families: they can sell sea-horses to traders who then sell them on to the lucrative Chinese medicine market in Hong Kong and elsewhere. Without the sea-horses, many of the people would not be able to feed their families.

As fishing continues, the sea-horses have started to disappear. Fewer sea-horses means that the fishermen catch fewer and earn less and so start to catch smaller and younger sea-horses to supplement their income. The population is further depleted and a vicious circle is set up.

In a ground-breaking environmental project, scientists have studied the life cycle and behaviour of sea-horses and are working with local fishermen to try to change fishing habits, so that the sea-horse population recovers.

Rather than take a standard environmentalist hard-line view and try to persuade the villagers to stop fishing altogether to protect all the sea-horses, the team recognises the conflict between human need and the needs of the environment. They have shown the fishermen that an enforced 'safe area' will allow sea-horses to breed without threat, and have taught them to avoid catching pregnant male sea-horses. In this way, the scientists have proved to the local community that they can help to restore sea-horse numbers. Not only will the sea-horse population be saved for the future, the fishermen will get back their source of income and will be free from the worries of starvation.

Sea-horses are unique in the animal kingdom: the male not the female gets pregnant and gives birth

1 WHY DO HUMANS AFFECT THEIR WORLD SO MUCH?

Fig 37.1 We think of the British countryside as 'natural' but, in fact, it is almost entirely artificially made. Only a few areas in Scotland, Wales and Northern England remain in their 'original' form – all farmland and most forests are the result of human influence on the environment

We humans are a unique species: we live in most of the ecosystems of the world, adapting our habitats to suit us (Fig 37.1), rather than having to adapt to the environment as most other species do. And, unlike other species, our behaviour affects the environment on a global as well as a local level.

The main human activity that affects the world environment is farming. Close behind are industry and mining for fuels such as coal, oil and gas. The more technologically advanced the society, the greater the overall environmental impact of each individual person. So, for example, a child born in Britain into a fairly well-off family may live for 80 years or more. In that time the amount of resources they use is enormous: think of the furniture, electrical appliances, cars, fuel, food, clothing, etc used up by someone like this. At the other extreme, a child born in a poor family in India or South America might live only 20 years. They could live in poorly built housing, or be homeless and forced to live on the streets, probably never get quite enough to eat, and would certainly never use a car or any significant amount of fuel.

In this chapter, we take a brief look at some of the complex issues involved in controlling the effect people have on the ecology of the Earth. There is room only to give an overview of some of the problems and potential solutions.

2 FARMING AND ITS EFFECTS ON THE ENVIRONMENT

It is very common to hear environmentalists complain that farming is destroying hedgerows and wildlife habitats, that woodlands and forests are being cut down to make new farmland and that it should all be stopped. In addition, farmers are often portrayed as hard-line businesspeople who put profit before anything else. There is a clear conflict here, but the issue is not as simple as these points of view suggest.

We must consider five basic facts:
- People need to have food to eat: staple foods need to be cheap.
- There are more people on Earth to feed than ever before.
- Farmers need to make a living.
- Farming needs land.
- Some farming techniques are damaging the environment so badly that the capacity of the land to produce food is being reduced.

The key to solving the problem of food production is to develop farming methods that are efficient enough to provide the amount of food that we need in the short term, but that protect the environment to ensure that food production levels can be maintained, or even increased, in the future.

FOOD PRODUCTION ON A GLOBAL SCALE

On a global scale, food production has increased during the past 50 years (Fig 37.2). Until the early 1980s, this increase matched the increase in world population, so that, overall, more people had enough to eat than ever before. There were still, of course, countries and areas where food supply did not meet demand and many millions of people were malnourished or starving.

Since 1984, population growth has increased faster than food production. Part of the reason for this has been the damaging effects of farming practices which produce food in the short term, but lead to long-term environmental damage, particularly soil erosion and desertification (see page 668). Very poor countries where food production has fallen since the mid-1980s have tried to compensate by importing food from Europe and the USA where more food is produced than is needed by the local population. However, this only increases the debt of the poor countries, leading to more poverty and, in the long term, more malnutrition and starvation.

Many scientists believe that soil erosion, water shortages and pollution from excessive use of pesticides and fertilisers will lead to a crisis in food production during the twenty-first century. Only a change in farming techniques might prevent this.

THE CONCEPT OF SUSTAINABLE AGRICULTURE

The problem of providing food for an increasing world population without reducing the capacity of the land to produce food is a difficult one, but it has some solutions. By using farming techniques that are sustainable, rather than successful in the short term only, many areas of the world should be able to continue food production at the right level for the foreseeable future.

Moves towards sustainable agriculture would include:
- Using new varieties of crop plants, perhaps produced by genetic engineering, which can survive and give good yields in poor soils, dry conditions and without the need for expensive pesticides and fertilisers (see Chapter 27).

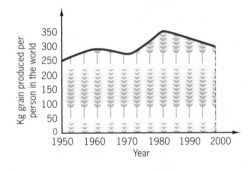

Fig 37.2 The amount of grain produced per person in the world between 1950 and 1999

SEE QUESTION 1

- Making use of non-traditional food sources – more plants that have nitrogen-fixing bacteria in their roots (these need no fertiliser), insects (many cultures of the world eat insects and their larvae as protein-rich delicacies) and unusual animals. In Europe, for example, ostrich farming is becoming more widespread.
- Reducing the amount of fish caught in the seas to levels at which the sizes of the populations do not continue to decrease. At the same time, the number of fish farms could increase.
- Encouraging and rewarding farming methods that do not lead to soil erosion and other environmental problems.
- Using biological pest control (Fig 37.3), rather than artificial pesticides which can persist in the environment.
- Encouraging farmers in poor countries to grow a selection of crops that can feed local people, rather than cash crops such as tobacco and coffee for export.
- Increasing financial and technological aid to poor countries to allow them to set up more sustainable methods of farming and food production.

FARMING IN THE TWENTY-FIRST CENTURY

MONOCULTURE

Until relatively recently, farms were diverse businesses where the farmers would grow several different crops, keep a variety of animals and perhaps have an orchard or two. The produce would be sold in local markets. But in today's supermarket-dominated food industry, farmers are encouraged to grow a single crop such as oilseed rape or wheat, so that the supermarkets have fewer growers to deal with. The word **monoculture** refers to the growing of the same crop in large fields, often year after year. A particular crop will rapidly deplete the soil of particular nutrients at particular depths, and often lead to the accumulation of pests. Without humus to bind it together, soil becomes more powdery and is likely to be eroded by wind and rain, leading to dust storms and the accumulation of silt in rivers, lakes and oceans.

REMOVAL OF HEDGEROWS

Hedgerows can be hundreds of years old, and represent an important habitat for many species. Farmers remove hedgerows for a variety of reasons:

- So that they can use large machinery more efficiently.
- To increase the area for growing crops.
- To avoid the need for maintenance (cutting, etc).
- To prevent them shading the crops.
- To remove what is often seen as a reservoir of crop pests.

The loss of hedgerows has several disadvantages. Hedges, as well as being aesthetically desirable, shelter crops from wind and therefore minimise soil erosion. Without hedgerows, the species diversity of the countryside is lowered as many species of insect, plant, bird and small mammals lose their habitat.

Fig 37.3 Until recently the only biological control for whitefly, a common plague on greenhouse-grown tomatoes (top) was *Encarsia formosa* (centre), a parasitic wasp which eats only adult flies, leaving larvae to develop and so cause a recurring problem. Gardeners and commercial growers can now obtain the beetle *Delphastus pusillus* (bottom) which kills larvae as well as adult flies, so providing much more effective and long-lasting control

USE OF FERTILISERS

One of the key differences between agriculture and a natural ecosystem is that harvesting removes nutrients from the soil. Fertilisers are used to replace these nutrients. There are two types of fertilisers:

- Organic – faeces and urine of animals and humans (sewage sludge or farmyard manure).
- Inorganic – liquid or (usually) pellets containing mineral ions, mainly nitrate, phosphate and potassium.

   Both types are added to crops with the same basic aim: to increase crop yield. Both provide mineral ions, but with manure the release of ions is gradual as the manure is decomposed by microbial action. The advantages and disadvantages are given in Table 37.1.

Table 37.1 Organic v inorganic fertiliser		
Type of fertiliser	**Advantages**	**Disadvantages**
Organic	cheap – farms often generate their own not easily lost by leaching improves soil – better humus levels, water retention, aeration and texture	variable (usually low) nutrient content slow release of nutrients difficult/expensive to store and handle may contain plant or animal pathogens – causing disease may contain metal residues that are passed into food chain requires heavy machinery that can compact soil
Inorganic	exact composition known – soil balance can be controlled easy to store and handle can be applied with light machinery – avoids soil compaction	expensive – it's a commercial product most components soluble – rapid leaching into rivers applied in concentrated form can cause osmotic damage to plants can cause acidification of soil

BALANCE OF FOOD PRODUCTION AND CONSERVATION

For economic reasons, farmers are under great pressure to:

- Concentrate on a small number of crops and grow these in large areas of monoculture.
- Remove hedgerows to make more growing space and to make the use of heavy machinery easier.
- Use large amounts of chemical fertilisers to increase yield from the depleted soil.
- Drain marshy areas and remove woodland.
- Use a wide variety of pesticides.

The two activities of producing food and conservation would seem to be exclusive, but there is some room for compromise. There are several strategies that farmers can adopt to minimise damage to the environment.

SOME METHODS OF CREATING MORE SUSTAINABLE AGRICULTURE

- Use more organic manure, which improves soil structure by providing more humus so retaining more water. As the humus decays the nutrients are released slowly so there is less chance of them being leached away.
- Delay the application of chemical fertilisers until the main growing season so that more is absorbed before it is washed away.
- Leave crop stubble over the winter (ie plough later) so that there is less bare soil to blow away.
- Rotate crops – growing a different crop year on year makes better use of minerals available at each soil depth. It also prevents the accumulation of crop-specific pests.
- Set aside – leave areas of wilderness to develop.
- Stop destroying hedgerows.

3 HUMAN ACTIVITIES AND POLLUTION

Pollution occurs when substances are released into the environment in harmful amounts, usually as a direct result of human activity (Fig 37.4). Most pollution results from excessively high concentrations of substances, but there are exceptions.

Some pollutants are harmless substances already found in the natural environment. They cause problems when they are produced in large quantities by human activity; examples include carbon dioxide and organic waste (sewage). Other pollutants produced by human activity are compounds that are known to be toxic, such as pesticides. These often remain in the environment for a long time because there is no living organism to break them down. A third type of pollutant is a chemical that passes all the safety tests, but then has totally unexpected effects, as chlorofluorocarbons have done. Pollutants can affect the air, the sea, waterways and the land.

Fig 37.4 Examples of how we humans pollute our environment.

Fig 37.4(a) The outlet pipe in the background is discharging raw sewage which is being washed up onto this beach

Fig 37.4(b) Effluent from the chemical works is polluting the river which supplies it with vital water

Fig 37.4(c) This landfill site will eventually be levelled and should look more pleasant, but methane production under the surface might still cause problems

Fig 37.4(d) Pollutants in smoke from industrial plants add to the acid content of rain

AIR POLLUTION: ACID RAIN

To many people, walking through a high Swiss meadow in summer during a shower of rain sounds idyllic. Peace and quiet, fresh mountain air and refreshing rain. But what if the rain has the pH of vinegar or lemon juice (Fig 37.5)?

In Switzerland, Norway, Sweden and other parts of Scandinavia, acid rain falls on some of the most unspoilt natural landscape in the world. It starts off, as far as 1000 kilometres away, as emissions of sulphur and nitrogen oxides released from the more highly industrial areas of Europe. Although there are some natural sources of these gases, by far the largest amount is generated by burning fossil fuels: the petrol used to power road vehicles and the oil, coal and gas used in power stations (Fig 37.6).

Fig 37.5 Unpolluted rain has a pH of about 5.6 because of the CO_2 that occurs naturally in the atmosphere: rainwater in most industrialised countries has a pH of between 4 and 4.5. Some cities in the USA are bathed in acid fog with a pH as low as 2.3 which is 1000 times as acidic as normal rain, and the same acidity as lemon juice

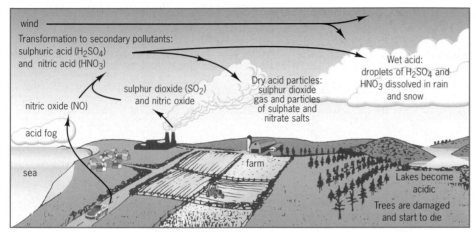

Fig 37.6 As they are carried in the air, oxides of sulphur and nitrogen form secondary pollutants such as nitric acid vapour, droplets of sulphuric acid and particles of nitrates and sulphates. These chemicals reach the surface of the Earth in two forms: wet, as acid rain, snow or cloud vapour; or dry as acidic particles

Normally, rainwater is mildly acidic because of the presence of dissolved carbon dioxide. The pH is normally no lower than 5.6. Any precipitation (rain, hail, sleet or snow) below this value is defined as acid rain. The cause of acid rain is a reaction between certain gases and the water that makes up clouds. The following gases are mainly responsible:

- Sulphur dioxide, which reacts with water to make sulphuric acid (H_2SO_4). Sulphur dioxide is emitted naturally from volcanoes but the vast majority comes from burning fossil fuels, notably coal, that contain a lot of sulphur. Power stations and metal smelting industries are the major sources of sulphur dioxide, not petrol or diesel engines.
- The oxides of nitrogen, nitrous oxide (N_2O), nitric oxide (NO) and nitrogen dioxide (NO_2). Collectively known as NO_x, these gases combine with water to make nitric acid (HNO_3). NO_x gases are made in engines when nitrogen in the air combines with oxygen during combustion.

Perhaps the most extreme example of acid rain was a 'smog' (smoke/fog) that descended on London for a whole week in 1953. Its pH was measured at 1.6 and was responsible for the deaths of 4000 people who suffered from breathing problems (asthma, bronchitis and emphysema).

The effects of acid rain on the environment are still causing arguments. It is quite difficult to get good evidence to demonstrate a definite link between the occurrence of acid rain and the effects that ecologists and environmental scientists have noticed. For example, many European and North American forests have been damaged during since the 1970s, especially at high altitudes and at the edges of some forests, which have large areas of dying trees (Fig 37.7). In Britain the pattern of loss of lichen species seems to follow the

Fig 37.7 These trees in the Czech Republic have been killed by acid rain that has resulted from emissions of sulphur dioxide and oxides of nitrogen given off by heavy industry all over Europe

distribution of sulphur dioxide pollution. In fact, lichens are so sensitive to this sort of pollution that they are used as pollution indicators. Other possible effects of acid rain are listed in Table 37.2.

Many countries, including Britain, have now recognised acid rain as a serious problem and have been reducing their emissions of sulphur dioxide since the mid-1970s.

Table 37.2 Possible effects of acid rain	
Possible effect of acid rain	**How it might happen**
Fish deaths in Scandinavian lakes and elsewhere	Indirect mercury poisoning: more acidic water is thought to convert inorganic mercury compounds in lake sediments into compounds which are soluble in the fatty tissues of fish.
	Indirect aluminium poisoning: in acidic conditions, more aluminium ions are released from the soil and washed into lakes. Fish respond by making excess mucus which clogs their gills, with fatal results.
Damage to statues and buildings	Chemical reactions between components of acid rain and rock.
Overstimulation of plant growth	Acid rain falling on the ground means excess nitrogen. The plants then use other nutrients faster, reducing soil fertility.
Thought to make human respiratory problems, such as asthma, worse	Irritation of surface membranes in lungs and bronchi.

REMEMBER THIS

Ozone (O_3) is present in the atmosphere as a whole in very small amounts, a few parts per million at the most. However, it is not evenly spread. It occurs as a definite layer in the stratosphere, 15–50 kilometres above the Earth's surface. Low-level ozone occurs at ground level when bright sunlight acts on pollutants produced by heavy traffic – we see this as photochemical smog in large cities in high summer.

Fig 37.8 In the mid-1980s, scientists discovered something unexpected: sunlight on the Antarctic in spring causing immense destruction of ozone and a huge hole appearing in that part of the ozone layer. This 'ozone hole' has been noticed every year since, and appears to be growing

REMEMBER THIS

Brands of products that carry the phrase 'CFC-free' may do so as a marketing ploy to persuade the consumer that they are more 'environmentally worthy' than other products. In fact, by law, no aerosols sold in Britain contain CFCs.

AIR POLLUTION: THE OZONE LAYER

The effects of acid rain occur hundreds of kilometres away from the source of the pollutants that cause it. This is bad enough, but other pollutants affect the whole planet. Examples include the effect of greenhouse gases on global warming (see Chapter 35) and ozone depletion.

Ozone, O_3, is a form of oxygen: its molecules contain three oxygen atoms, rather than the two that occur in molecules of atmospheric oxygen. Both gases have been a major feature of the Earth's atmosphere for the past 450 million years. Oxygen is vital to life: it is the fuel for aerobic cell respiration. The role of ozone is less direct: it forms a sort of sunscreen high up in the atmosphere, shielding the biosphere from the harmful effects of ultraviolet light that beams down from the Sun.

Chlorine- and bromine-containing compounds, mainly chlorofluorocarbons (CFCs), that were released into the atmosphere during the twentieth century seem to be causing a measurable thinning of the ozone layer (Fig 37.8). This could cause serious problems in the future.

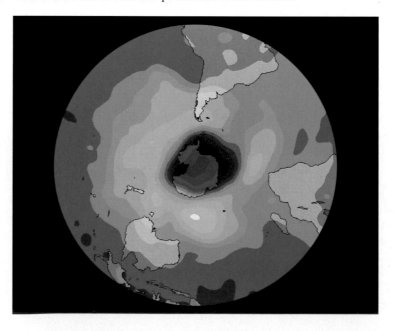

Originally, CFCs were thought to be 'ideal' chemicals: they were non-flammable, non-toxic, non-corrosive, cheap to make and could be used as coolants in fridges, propellants in aerosol spray cans, and in the manufacture of the packing material Styrofoam. However, in the mid-1970s, scientists showed that CFCs rise slowly into the stratosphere and are then converted by the action of ultraviolet light into chlorine atoms. These accelerate the breakdown of ozone into O_2 and O. CFC molecules can stay in the stratosphere for about 100 years, and each one can destroy hundreds of thousands of molecules of ozone.

CFCs have now been phased out in most countries of the world, but their effects continue. Even if no more CFCs were put into the air from tomorrow, the ozone layer would still take about 100 years to recover from the worst of the damage. Less ozone in the stratosphere allows more of the harmful ultraviolet radiation from sunlight to reach the Earth's surface. Scientists predict that this will lead to a huge increase in skin cancers and cataracts (Fig 37.9).

Fig 37.9 In countries such as Australia, New Zealand, South Africa and Chile, which are protected only by a very thin layer of ozone for many months of the year, there have already been many more cases of skin cancer than ever before. Children in Australian schools are required by law to wear hats and blocking creams whenever they go outside

WATER POLLUTION: GETTING INTO HOT WATER

When you think of water pollution, you probably think of chemical effluent (Fig 37.10), outflows of raw sewage or huge oil spills. While these undoubtedly are examples of pollution, probably the most overlooked cause of damage to inland aquatic ecosystems is the warm water released as a by-product of many industrial processes. Fish and water-living invertebrates are killed by warm water, not because they are scalded directly, but because warm water can carry much less dissolved oxygen than cool water and they 'suffocate' (Fig 37.11).

Fig 37.10 The chemical outflow is on its way to the river Mersey

Fig 37.11 As well as causing the death of fish directly, by depriving them of oxygen, warm water can also encourage the growth of parasites which would otherwise not be a problem

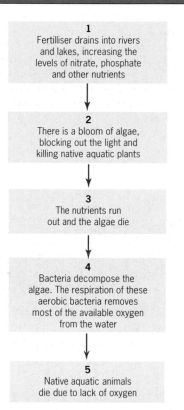

SEE QUESTIONS 2 AND 3

1
Fertiliser drains into rivers and lakes, increasing the levels of nitrate, phosphate and other nutrients

2
There is a bloom of algae, blocking out the light and killing native aquatic plants

3
The nutrients run out and the algae die

4
Bacteria decompose the algae. The respiration of these aerobic bacteria removes most of the available oxygen from the water

5
Native aquatic animals die due to lack of oxygen

Fig 37.12 A summary of eutrophication

? **TEST YOURSELF**

B Describe how you would measure BOD.

WATER POLLUTION AND SEWAGE

Sewage is good stuff. It's a natural product that is an essential part of most ecosystems. Call it what you will, but faeces and urine from any source are a mixture of undigested food, cells from the gut lining, urea, salt and lots of living bacteria and protoctists. When acted upon by decomposers (or saprophytes), sewage releases vital nutrients such as nitrate that are essential for renewed plant (or algal) growth.

Problems with sewage start when there's too much of it to break down. If the entire sewage production of a village or town is pumped into one stream or river, the natural balance is upset.

Contamination of rivers, lakes and seas by sewage causes two main problems. Firstly, it can introduce potentially dangerous pathogens (organisms that can cause disease) into the environment (see Fig 37.4a). This is a particular problem if the water is drunk by animals or people. The second problem is that sewage and other organic wastes are decomposed by the action of aerobic bacteria. These use large quantities of oxygen, leaving less for other organisms living in the water. Sewage contamination can lead to **eutrophication** (Fig 37.12).

The term **eutrophic** means over-fertile. A eutrophic river or lake contains more nitrates and phosphates than normal, and abnormal growth, particularly of algae, can occur. Note that the algae themselves do not reduce the oxygen content. They photosynthesise, so actually increase the oxygen levels for a short while. The opposite of eutrophic is **oligotrophic**: this means under-fertile. An oligotrophic river cannot sustain much life.

BIOLOGICAL OXYGEN DEMAND

It is easy to find out the oxygen requirements of water from a particular site by taking a sample of the water and measuring its oxygen content, usually with an electronic meter. After keeping it sealed and in the dark for 5 days at 20 °C, the oxygen content is measured again. The rate at which the oxygen has been used up is the **biological oxygen demand** (**BOD**). The BOD of unpolluted river water is about 3 mg O_2 per litre (dm^3) of water, per day. Raw sewage uses over 100 times more oxygen in the same time (Table 37.3).

Although pollution by sewage is a major cause of eutrophication, there are others. If large amounts of nitrate- and phosphate-rich fertilisers are put on to farmland, the excess can run off and contaminate local rivers and lakes. Industrial processes which pollute local waterways with large amounts of concentrated sugars or other organic waste products can also cause the BOD of the water to increase significantly.

Find out more about eutrophication in the Assignment at the end of this chapter.

Table 37.3 Typical BOD values per day of different types of water

Water type	BOD value per day
Clean river water	3 mg dm^{-3}
Water from polluted stream	10 mg dm^{-3}
Domestic sewage, untreated	250–350 mg dm^{-3}

CONTAMINATION OF WATER BY OIL AND DETERGENTS

The names *Exxon Valdez*, *Braer* and *Sea Empress* may sound familiar, even if you can't quite remember that they are oil tankers that have famously run aground causing huge oil spills (Fig 37.13). These pollution disasters make headline news but, in 1993, a study by Friends of the Earth estimated that over 1000 times the amount of oil carried by the *Exxon Valdez* (which polluted the coast of Alaska in 1989) is lost in the USA each year – not through national catastrophes, but through the normal operations of washing tankers and releasing the oily water, from pipeline and storage tank leaks and from accidental loss from offshore oil wells.

The effects of oil spills on ocean ecosystems depend on:

- The amount of oil released.
- How far away from the shore the spill happens.
- The weather conditions, water temperature and speed of ocean currents.
- The type of oil released: an area affected by a spill of refined oil takes over three times longer to recover than a similar area that has a crude oil spill.

In the clean-up operation following a large spill, oil is dispersed by detergents. The aim is to spread the oil molecules to limit damage to the local environment. Unfortunately, many detergents are extremely toxic and lead to many sea-bird deaths.

SPOILING THE LAND

People can pollute areas of land directly by dumping toxic waste, by industrial processes such as mining and by simple neglect: leaving non-biodegradable rubbish behind after a picnic at a local beauty spot, for example. However, human activity, particularly farming, is indirectly causing two main problems: **deforestation** and **desertification**. Both have far-reaching global effects.

DEFORESTATION

Deforestation has been practised by humans for centuries. As human population and technology have increased, so has the need to use wood. (The discovery of fire has had an enormous impact.) Many European forests were cleared by various different settlers from about the fourteenth century onwards. Today, the human population is larger than ever before and the process of deforestation has now spread to all parts of the world, particularly the tropical rainforests.

Deforestation is a process that results from the conflict between immediate human need and the necessity to protect the environment from serious and long-term damage. Tropical hardwoods are prized as building materials throughout the world, and other types of wood are used to make the thousands of tonnes of paper that we use every day. Many people in the poor countries of the world depend on wood or charcoal for fuel, and wood, twigs, crop residues and grass are used for fuel in many of the 'better off' countries such as China, Brazil and Egypt. As well as cutting down forests to use the wood they provide, deforestation is also carried on to clear land for farming. People have cleared large areas of forest for plantations of other trees (rubber and palm oil), and for cattle ranches and enormous wheat fields, in order to fulfil the needs of the human population for more raw materials and more food.

37.13 This cormorant was caught in the oil spill that happened at Manobier beach, Wales, in February 1996 after the tanker *Sea Empress* ran aground. Birds covered in oil cannot swim very easily and sometimes tire and drown. Those which end up on the beach try desperately to clean their feathers, ingesting and inhaling large amounts of oil which damages their digestive system and often causes pneumonia

✔ REMEMBER THIS

Many countries worldwide are now working together to try to reduce deforestation: people are being encouraged to manage forests **sustainably**. This means using different techniques to make sure we have the wood we need without destroying large areas of forest. People are also being persuaded to use **paper substitutes**.

There is growing investment in **rainforest communities** to find new medicines from plants and **eco-tourism** is being promoted. If people pay to see wildlife in its natural surroundings, this contributes to the income of an area and makes it less likely that habitats are destroyed.

The consequences of deforestation are potentially serious. Locally, clearing the land of trees and plants leads to increased soil erosion (see below). Globally, reducing the biomass of productive trees may contribute to the greenhouse effect (see Chapter 35) as less carbon dioxide is used in photosynthesis. Also, loss of rainforest, one of the most diverse ecosystems of the world, is likely to affect humans and other organisms in ways we cannot predict, although we already know that many of the plants found in rainforests are useful sources of powerful drugs (see Chapter 36). Balancing the needs of people with those of the environment is difficult.

DESERTIFICATION

Soil is an important abiotic factor that determines the distribution of organisms in an ecosystem: few plants can grow without the anchorage of soil or the water and minerals it supplies to them. When the soil is lost, a process called **soil erosion** occurs. This is a worldwide problem.

Soil erosion is a natural process: wind and water movements move surface debris and topsoil from one place to another. In ecosystems that are undisturbed by human activity, healthy plant roots 'bind' the soil, and soil is lost and made at about the same rate. Problems occur when the soil is removed faster than it is formed: this can happen because of farming, deforestation, building, over-grazing by animals, physical damage by off-road vehicles or fire and some kinds of mining.

Farming is probably the worst culprit. Topsoil is eroding on about a third of the world's cropland. Each year, the land must feed 90 million more people with about 25 billion tonnes less topsoil. In Africa, soil erosion is now 20 times faster than it was 30 years ago. The cattle-producing areas are now severely affected by desertification (Fig 37.14). But, there are solutions to soil erosion and desertification, and many of these are starting to be put into practice.

Fig 37.14 Areas of severe desertification occur all over the world. Desertification is defined as the process that reduces the productive potential of hot dry areas by 10 per cent or more. Moderate desertification is a 10–25 per cent drop in productivity: severe desertification is a 25–50 per cent drop. At its most severe, desertification actually creates deserts

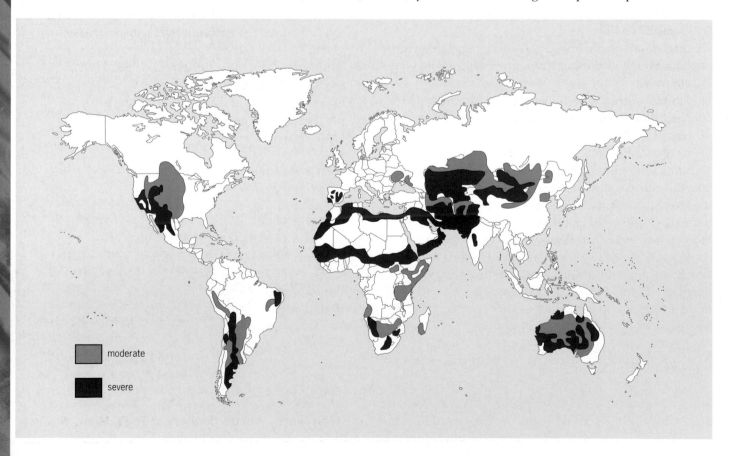

moderate

severe

SUMMARY

After reading this chapter you should know and understand the following:

- Humans influence the environment of the Earth on a *global* as well as a *local* level.
- Farming is the major human activity that causes environmental damage.
- People need to farm to feed the increasing human population but many farming techniques cause long-term damage to the land, lowering its capacity to grow food in the future.
- To avoid this problem it is important that farmers start to use **sustainable** methods of agriculture.
- Many human activities cause **pollution**. Pollutants can affect the air, the sea and rivers and lakes, and the land.

- Acid rain, use of CFCs and other substances that destroy the ozone layer, and increased production of greenhouse gases all contribute to air pollution.
- Water can be contaminated by organic pollution. This can lead to a sequence of events called **eutrophication**. Heat, oil and other substances can also pollute waterways.
- When forests are cleared to make new farmland, fewer trees are available to photosynthesise, we lose species diversity and the land becomes vulnerable to **erosion**.
- Severe soil erosion can lead to **desertification**. Many sustainable methods of agriculture try to minimise the effects of soil erosion.

? EXAM QUESTIONS

1 Read the article, about afforestation, and answer the questions that follow.

1   The places where monocultures of non-native conifers are being planted are amongst the last areas of wild land in Britain.
    Upland peat bogs are prepared for conifers by deep
5   ploughing and cutting drainage ditches into streams and rivers.
    When the young trees are planted out, their growth is assisted by the application of a phosphate fertiliser.
10 Rapidly growing trees become infested with pests and these must be controlled. Chemicals are often used.
    Airborne pollutants, such as sulphur dioxide and nitrogen oxides, cause acid rain which runs off
15 the pine trees into drainage ditches.

a) Name the habitat which is threatened by planting monocultures of conifers.

b) Suggest how the following processes, described at lines 4 to 15, might affect the quality of the water in streams and rivers which flow through the plantations.
  (i) Cutting ditches (line 5).
  (ii) Applying phosphate fertiliser (lines 8–9).
  (iii) Pest control (lines 10–11).
  (iv) Trapping of airborne pollutants by pine trees (lines 13–15).

c) The Forestry Commission publishes guidelines to help to reduce the environmental impact of forestry.

Timber growers are advised to stop drainage ditches short of streams and rivers so that instead of emptying directly into the water course, the blind end of ditches fill up and overflow. The water then trickles over natural land vegetation before entering the stream or river.
Suggest how this might reduce the impact on water quality of the application of phosphate fertiliser [lines 7–15].

d) (i) What alternative to chemicals [lines 10–12] might be used to control the pests?
  (ii) Give **one** disadvantage of the method you suggest in part (i).

WJEC B12 June 2001 Q5

2 The figure represents a simplified version of the nitrogen cycle.

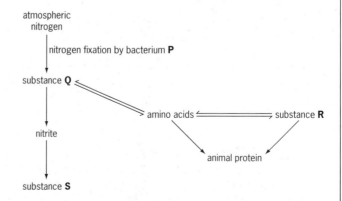

a) Give a suitable name for each of **P**, **Q**, **R** and **S**.

? EXAM QUESTIONS

b) State **precisely** why substance **S** is important for the growth of crops.

Large quantities of substance **S** are applied to agricultural land as a component of chemical fertilisers. There is concern about its effect on the environment because much of **S** that is applied to the land eventually leaches out of the soil and into ground water.

c) If substance **S** passes into drinking water, suggest **one** harmful effect it can have on the human body.

If **S** and other residues of chemical fertilisers accumulate in slow-flowing water or lakes, rapid growth of algae may occur (an algal 'bloom').

d) (i) State the name given to the process by which this happens.

(ii) Explain briefly how this process could destroy most of the community of a lake.

e) Apart from the loss of **S** due to leaching, why is it necessary for farmers to add chemical fertilisers to their fields every year?

OCR 6910 Molecules & Life March 2000 Q3

3 Read the following passage, which is extracted from an article written by an employee of the Environment Agency, then answer the questions that follow.

At the heart of most of our work is the simple principle that the organisms which inhabit an environment provide the best, and the only direct measure of whether that environment is in good order. An examination of the diversity and abundance of organisms is often sufficient to give a broad view of environmental quality or the impact of specific inputs or disturbances.

This has formed the basis of much of our work for many years, with the main emphasis on the use of freshwater macro-invertebrates to assess river quality. These include worms, snails and the larval stages of insects such as mayflies and caddis flies, which inhabit the bottom sediments. They are used because they are found in virtually all rivers throughout the year, and there are many different types with a wide range of pollution sensitivities. Sampling and identification are relatively straightforward, and we have considerable experience at interpreting the resulting data. The communities can be described by the Biological Monitoring Working Party (BMWP) Score, which is the sum of the scores allocated to each group of invertebrates found in the sample,

with the most sensitive species scoring highest. Since the latest 1970s, staff at the Institute of Freshwater Ecology have developed a computer package which has made biological assessments of river quality as reliable as those using the more traditional water chemistry techniques and have firmly established the use of macro-invertebrates in the Agency.

A simple impact assessment may be carried out by collecting samples upstream and downstream of the discharge point. A standardised technique involving a willing biologist with waders and a pond net is used to collect these samples. The communities found below the input are compared with those at the upstream 'control' site. A simple visual examination of the species list may be sufficient when, for example, an absence of sensitive organisms such as mayflies, and an abundance of tolerant groups such as oligochaete worms give a classic indication of sewage pollution.

The table shows how indices can vary around a sewage discharge to a river and indicates the type of fauna which may be found.

Site	BMWP score	Notes
50 m upstream of sewage discharge	154	A wide range of sensitive species with no group being numerically dominant.
50 m downstream of sewage discharge	79	A restricted fauna with no sensitive species and large numbers of individuals of a few tolerant groups.
200 m downstream of sewage discharge	103	A few sensitive species have returned and numbers of individuals from tolerant groups are falling.
500m downstream of sewage discharge	139	Most sensitive species have returned and numbers of individuals from tolerant groups are low.

a) Use the data in the table to plot a graph of BMWP score against distance along the stream and use the graph to estimate the distance from the sewage outlet where the stream is likely to be completely recovered.

b) Describe the 'standardised technique' that the 'willing biologist with waders and a pond net 'is likely to be using.

c) Name **two** substances dissolved in the water, that could be usefully measured to determine pollution using 'traditional water chemistry techniques'.

OCR 6916 Environmental Biology March 2001 Q5

KEY SKILLS **ASSIGNMENT**

JOURNEY ALONG A RIVER

As Fig 37.A1 shows, rivers tend to be relatively unpolluted near to their source. As they flow on towards the sea they are contaminated with organic pollution and industrial effluent, creating increased levels of pollution and eutrophication.

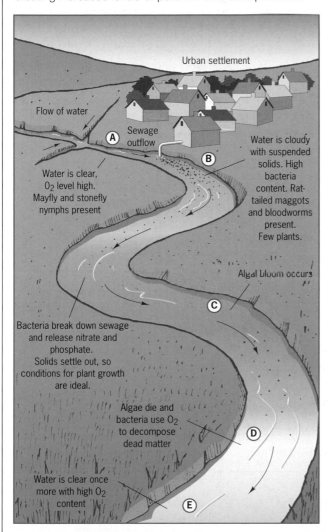

Fig 37.A1 This diagram shows the different conditions found along a river

It is often difficult to tell exactly how polluted the water is at any point, just by looking at it. We need to sample the water and sample the organisms that live there to find out whether the water is safe.

There are various methods which allow us to find out which organisms live in a particular part of the river. One method is a kick sample, where stones or gravel are kicked into the river and the disturbed organisms caught with a net placed downstream. The number and diversity of organisms found are used to give the stretch of waterway a diversity index (see page 652).

1 What sort of diversity index would you expect to find in:
a) a stretch of unpolluted river, such as point A in Fig 37.A1?
b) a stretch of polluted river, such as point B in Fig 37.A1?

2 **a)** Suggest which abiotic factor is likely to be the most important in determining whether water can support a large or a small number of species?
b) What method would you use to measure this factor?

3 Initially, the water at point B is turbid (cloudy and murky). From your knowledge of the nature of sewage, explain why few animals or plants can grow in this zone of the river.

4 What do you think the BOD value may be for the water in this part of the river?

5 A student who was feeling particularly brave did a kick sample in this area and found very few species. They were the classic indicator species of heavy organic pollution: rat-tailed maggots, bloodworms and tubifex worms – see Fig 37.A2.

Fig 37.A2 The larva of a rat-tailed maggot, some bloodworms and tubifex worms. The tails of the tubifex worms protrude from tubes that the worms make using mucus and debris

a) What is an indicator species?
b) Explain why each of the three species shown in Fig 37.A2 is able to tolerate low oxygen levels.
c) When water is tested in the laboratory, the bacterium *E. coli* is sometimes found. What conclusion can be drawn from the presence of this organism?

KEY SKILLS **ASSIGNMENT**

More sophisticated investigations of the water further below the sewage outlet produced the data shown in Fig 37.A3.

Research and discussion questions

Find out about the basic principles behind sewage treatment. What major health problems does effective sewage treatment prevent?

Find out what the current laws are in the UK that control the discharge of sewage into rivers and into the sea. Do you think they are strict enough?

6 Look at Fig 37.A3(b) and (c)
a) Use your knowledge of the nitrogen cycle to explain why there is first a peak in the levels of dissolved NH_4^+, followed by a peak in NO_3^-.
b) The population of sewage fungus falls to 0, just before point C. Suggest what this tells us about the type of respiration of the sewage fungus.

7 At point C, the water begins to look like pea soup – conditions have become perfect for the growth of algae. Contrary to what you might first think, the problem here is that the water is too fertile – it has become eutrophic.
a) Use the graphs shown in Fig 37.A3(a) and (b) to list three factors that are responsible for the algal bloom.
b) Explain why the algal bloom is harmful to the normal clean-water organisms.
c) Why do the populations of the different invertebrates present in the river water peak at different times?

8 Look at Fig 37.A3(d).
How do the detritivores help to change the water conditions between points C and E? Find out the names of some common detritivores that might be found in this sort of river.

Once the algae have absorbed the available nutrients, they start to die. Once again, bacterial activity becomes a problem, as decomposers rot the dead algae and use up large amounts of oxygen. After this, if there is no more organic discharge, the water further downstream begins to recover. Eventually, the water becomes clean once more

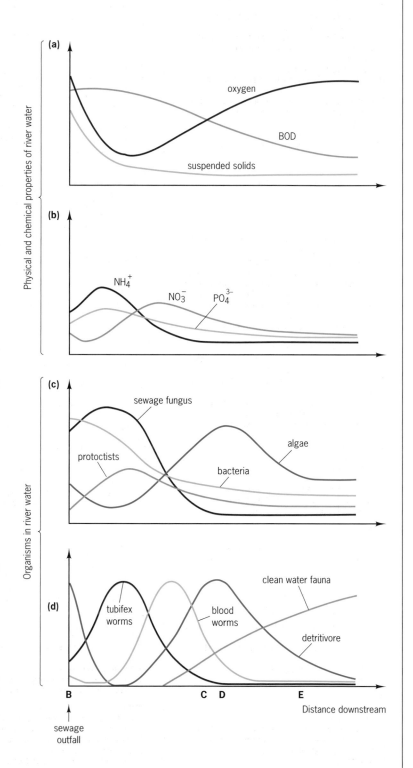

Fig 37.A3 Changes in the river after a sewage outfall

ANSWERS TO TEST YOURSELF QUESTIONS

CHAPTER 1
A **(a)** The full stop is about 0.3 mm = 300 μm. **(b)** 1 million.
B Dispersal of oil pollution, breakdown of waste plastic.
C 1000.
D They are all secretory. You would recognise them by their prominent ER and Golgi bodies.
E Because the ER is a network of thin membranes which would not show up under a light microscope.
F All its tissues would be destroyed.
G We could destroy tumours.
H Because they could digest the organelle/cell that made them.
I To respire and make ATP/energy for swimming; sperm swim incredible distances and use a vast amount of energy.
J **(a)** 500, **(b)** 200, **(c)** 5000.
K The root system, because it receives no light.

CHAPTER 2
A Ciliated epithelium.
B Smooth muscle because its contraction is slowest.
C It fatigues too quickly.
D The reproductive system.

CHAPTER 3
A 3. No, because there is no dot at the right position for C.
B Q=25/50=0.5, R=15/50=0.3. R is the least soluble.
C Sucrose consists of glucose and fructose, both reducing sugars. In the sucrose molecule, the reactive groups of the two component sugars point inwards and so become 'hidden' in the 3D structure. They only become available to react with Benedict's solution when the sucrose has been split apart.
D The glycosidic link, which is between carbon 1 and carbon 4.
E Links: sucrose 1,2; lactose 1,4.
F It would cause an osmotic imbalance – the cytoplasm would become too concentrated.
G Amylose because it has more ends to add or remove glucose units from.
H Because it is broken down/built up even quicker – suiting the metabolic needs of animals.
I Because few organisms have the enzymes that can degrade cellulose.
J Oleic and linoleic acid.
K Oleic and linoleic. They are both unsaturated.
L 46 x 5 cm = 2.3 metres.
M TAGCAATGG.
N Bases (RNA has U instead of T), sugars (RNA has ribose, DNA has deoxyribose), strands (DNA has 2, RNA has 1), life-span (DNA is long, RNA is short).

CHAPTER 4
A 1 mm.
B Water-loving, water-hating. In a phospholipid, the phosphorus end is attracted to water, the lipid end repels water.
C Because the atoms/molecules in liquids and gases are free to move, but those in solids aren't free to move.
D To maintain the diffusion gradient.
E Lung, intestine, leaf, kidney, placenta, root.
F Because water passes out of the bacteria by osmosis, dehydrating and killing them.
G You must have a solution to compare it to: hypertonic to what?
H Cyanide blocks aerobic respiration, so virtually no ATP is available.

CHAPTER 5
A Protein synthesis.
B Protein, lipid, nucleic acids.
C Because they all contain the same food type: starch.
D Tears, sweat, in the lysosomes of some white cells.
E It stimulates glycogen production.
F Peptide bonds.
G Proteins are denatured. This is irreversible.
H Because that is the optimum temperature for enzymes.
I It would be denatured by the acid.
J There would be no activity at all: the plot would go along the x-axis.

CHAPTER 6
A The castrated ram has no testes and does not produce testosterone or pheromones; he has nothing with which to turn on the ewes, and they would just ignore him.
B Discussion question, no answer provided.
C Fermentation is a vague word, and in biotechnology is taken to mean the metabolics/respiration of microbes. Fermentation can be either aerobic or anaerobic. Glycolysis is the first, anaerobic, process in respiration.

CHAPTER 7
A Small organisms have a large surface area:volume ratio and so lose heat faster than larger organisms. They compensate by using most of the energy produced by respiration to maintain their body temperature.
B Benedict's test: add blue Benedict's and heat; will give an orange-red precipitate if sugar is present.
C Biuret test: will change from blue to lilac if protein is present.
D The symptoms are those of vitamin A poisoning.
E The meal is reasonably high in energy, complex carbohydrate and fat. It would be a good basis for a lunchtime meal for an adult man. The brown bread provides some protein, fibre and most of the B vitamins; the cheese and butter provide mainly fat, and vitamin A. It lacks vitamin C, iron and other minerals; it could be improved by the addition of a side salad, or just by adding some salad – tomatoes, green salad leaves and cucumber to the sandwich, or by following it up with a few pieces of fruit.
F Exercise increases basal metabolic rate and the muscle to fat ratio; both will use up more energy, so a better diet will prevent more fat being added, while exercise will help take existing fat off.

CHAPTER 8
A They would have to eat smaller, more frequent meals as they would have no stomach to store large meals.
B About 2–3.
C Endopeptidase cuts within the chain, exopeptidase removes the end amino acid.
D **(a)** You would expect rennin production to decrease with age. It does. Young mammals on milk diets need it but older animals do not. **(b)** No, since only mammals produce milk.
E So they do not self-digest the organs that produced them.
F The small intestine is a very absorptive region and can take in many of the nutrients produced by bacteria; only a small amount of absorption takes place in the colon, so nutrients produced here will just be lost in the faeces.

CHAPTER 9
A 7.5 litres min⁻¹.
B One five thousandth.
C Dead space would be enormous and diver would breathe the same air in and out. Also, the chest muscles are not strong enough to keep expanding the lungs against the water pressure at that depth.

CHAPTER 10
A **(a)** 6 and **(b)** 15 litres per minute.
B **(a)** Increase. **(b)** Decrease.
C 120 mmHg = systolic pressure, 70 mmHg = diastolic pressure.
D **(a)** Increase. **(b)** Decrease.
E Hb 'steals' oxygen from an oxygen-rich environment and delivers it to tissues that are oxygen-poor.

CHAPTER 11
A Because they detect a change in internal conditions and then adjust it back to the normal level.
B The urine of diabetic dogs has high concentrations of sugar. This is very attractive to flies.
C They are metabolically active and are involved in a rapid exchange of large amounts of materials.
D Any excess of the daily need would be deaminated and excreted.

CHAPTER 12
A Humidity changes our ability to lose heat by sweating. Low humidity means a higher upper lethal temperature

because we can sweat and lose heat by evaporation more effectively.
B They must be below, otherwise they would disappear as old skin was shed and replaced by new skin.
C Piloerection because we have so few hairs.
D To replace the sodium and chloride ions that they lose constantly by sweating.

CHAPTER 13
A 7.5 litres.
B 14.4 times per hour.
C Blood pressure would drop, filtration would become ineffective, kidneys would suffer damage and fail.
D The loop of Henle is shaped like a hairpin, and it behaves in the same way as a countercurrent multiplier.
E Dark yellow urine indicates dehydration. Brown = trouble!
F Much more lost in faeces – dehydration.
G Proximal and distal tubules, ascending limb of loop of Henle.
H It makes the blood more concentrated, either by adding salt or removing water or both.

CHAPTER 14
A We are using the lactic anaerobic system. This means that respiration can't be completed and so the intermediate compound (lactate) builds up.

CHAPTER 15
A A neurone is a single specialised nerve cell, a nerve is a bundle of neurones in connective tissue.
B **(a)** Active transport mechanism needs specific membrane protein and ATP. **(b)** It moves particles against a diffusion gradient rather than down a diffusion gradient.
C **(a)** To ensure that a neurone responds to a specific minimum level of stimulation. **(b)** Continuous depolarisation and fatigue of neurone.
D Myelinated nerves conduct impulses more quickly.
E Narrow to make diffusion of the neurotransmitter as fast as possible and high resistance to ensure that the action potential can't jump across.
F Because the transmitter is only made on one side.
G No action potential.
H So they don't tear and can be easily stitched up.

CHAPTER 16
A Discussion question, no answer provided.
B Right motor area.

CHAPTER 17
A They have to support more weight.
B The spinal cord.
C Ligaments connect bones together at joints; ligaments need to be flexible to allow the joined bones to move.
D A tendon joins a muscle to a bone. As the muscle contracts it exerts a force on the bone, moving it. If the tendon were elastic, the bone would move less than the change in muscle length.
E **(a)** Smooth, **(b)** smooth, **(c)** skeletal, **(d)** cardiac, **(e)** skeletal.

CHAPTER 18
A The digestive juices undiluted by food would attack the stomach lining. This could cause an ulcer.

CHAPTER 19
A Chemoreceptor; exteroreceptor.
B Exteroreceptors since they respond to stimuli from outside the body.
C The light-sensitive film.
D **(a)** The ciliary muscles relax.
(b) The lens becomes thinner due to tension on the suspensory ligaments. **(c)** The pupil dilates.
E Night sight (ie vision in dim light) needs rods that work properly. Vitamin A is necessary for the formation of retinal in rhodopsin.
F **(a)** Blue, green and red cones. **(b)** Blue cones only.

CHAPTER 20
A It would secrete more GnRF, thus stimulating more gonadotrophin release and therefore more oestrogen release.

B Less energy and resources are wasted; the changes of successful fertilisation are increased.

C They both contain the same amount of DNA. Both gametes are haploid – they contain only one copy of each chromosome.

D Day 21.

CHAPTER 21

A 5.

B No. Growth occurs evenly in the crocodile's body, so the ratio stays the same.

C (a) The brain is small at birth to allow the baby's head to pass through its mother's pelvis. It then develops rapidly to enable the child to become independent and have a better chance of survival. (b) The reproductive organs do not mature until we are large enough and emotionally mature enough to have children. In our society, cultural pressures delay childbearing until well past the age when females are physiologically ready to have babies.

D Between the ages of 2 and 3 years.

E Human tissue.

F (i) The person might have a condition that prevents them from making it. (ii) The hormone is produced in pulses and the sample was taken between pulses. (iii) The hormone is quickly converted into another form or is broken down as soon as it is released.

G The number of younger people in the population is growing at an equivalent rate.

CHAPTER 22

A No: the species certainly goes on but individuals frequently perish.

B Tall.

CHAPTER 23

A DNA contains the genetic code, RNA is built up on it, then RNA is used as the instructions for making proteins.

B $n = 23$.

C G_1, S and G_2.

D The amount of DNA would be halved at each division.

E (a) In ovaries and testes respectively. (b) Because it reduces the chromosome number (by half).

F In a human cell there are 23 bivalents.

CHAPTER 24

A (a) UAUGCGAUA. (b) 3.

B They eat and digest protein, absorb amino acids, transport them in the circulation from the gut to cells.

C Anabolic.

D 61 different tRNAs carry only 20 different amino acids.

E Six: all ^{14}N strands, two hybrids.

F Substitution.

G With no spindle, the chromosomes double but cannot separate, so a cell with twice the normal number of chromosomes results.

CHAPTER 25

A 50:50 (half Yy and half yy).

B If you use a plant that is TT, all the offspring of a cross with a Tt or a TT plant will be tall and you wouldn't gain any information.

C 700 000/2500 = 280.

D In previous centuries, villagers married people from the next village. Because we now have more transport, people move around the world more easily and so people meet and mate with people to whom they are unlikely to be related.

E XhXh. Female haemophiliacs must be a double recessive whose mother is a female carrier and whose father is a male sufferer. The chance of two such people meeting and marrying is unlikely and male haemophiliacs often don't live long enough to reproduce.

F One in two chance.

G No. If the mother is O and the father is A, the child could be only A or O.

H Yes, one could be aabbcc (white) and the other could be AABBCC (very dark).

I 9 agouti, 3 black, 4 albino.

CHAPTER 26

A Acquired characteristics don't affect an organism's DNA and so cannot be passed on to the next generation.

B Insects have short life-spans and high reproductive capacity; whales and rhinos have long life-spans and only produce a handful of offspring during their life time.

C Peppered moth, *Biston betularia*: selection of melanic form following industrial revolution. Hawks: selection for great visual acuity. Cheetah: selection for high-speed running.

D (a) Analogous. (b) Homologous.

CHAPTER 27

A It performs transcription in reverse: it makes DNA from RNA.

B Because the smaller recognition site is more common in any DNA sequence.

C An individual has a brain with memories – these memories are not contained in the cell used for cloning. In cloning you may create a new brain but not the memories.

CHAPTER 28

A We now have vaccines and antibiotics, a better standard of living, less poverty and better sanitation.

B 2500.

C Knowing what to avoid would help you avoid it.

D Larger population, more crowded cities.

E (a) Most protein digestion occurs in the small intestine, before the pancreatic duct. (b) Pancreas does not produce as much trypsin when it starts to fail because of pancreatitis.

F Reduce the amount of phenylalanine and increase the amount of tyrosine.

G Less messy, more convenient to carry around.

H (a) Endoscopy. (b) X-ray.

CHAPTER 29

A Infection with *Salmonella typhimurium* does not always cause the disease – some people are carriers.

B (a) Gloves, masks, sterile instruments. (b) Gloves, mask.

C There are now effective vaccines for measles, mumps and diphtheria, but none for the common cold.

D If you don't have the gene to produce the cell receptor for a particular virus, it cannot get into your cells.

E Antibiotic to kill the bacteria, antitoxin to neutralise the toxin they release when they die.

F Bacteria do not divide as quickly at lower temperatures so curve will be lower and flatter.

G Salmonellosis develops when the bacteria that cause it multiply inside the cells that line the intestine. Other types of food poisoning are caused by bacteria that do not multiply in the body. They simply release toxins, and so cause symptoms more quickly.

H Attack from antibodies and cells of the immune system.

I More people are in the water to get infected in the day.

J No problems from osmosis.

CHAPTER 30

A Warm, moist, lots of food.

B Mucous secretions containing dust and dead cells, which should be swept by cilia to the exterior, remain trapped in the airways to cause congestion and infection.

C Anticoagulants stop blood clotting so that leech can feed freely. Uses: microsurgery, plastic surgery, dispersal of blood clots, treatment of chronic skin ulcers.

D Two basophils, a lymphocyte and a monocyte.

E Not likely to cause an allergic reaction.

F Some of the donors would have been blood group O, some of the recipients would have been blood group AB, and so no adverse reactions would have occurred.

G AB blood has A and B antigens on the surface of the red blood cells. These react with antibodies present in the blood of people from the other three groups.

H Agglutination occurs when red cells stick together due to antibodies making them sticky. Blood clotting is the complex reaction that turns fibrinogen into a fibrin mesh, in which red cells get stuck.

I Fungi are eukaryotes while antibiotics work by interfering with prokaryote metabolism, and don't affect eukaryotes.

J If the course is not finished, not all the bacteria will be killed. The most susceptible die first, leaving the more resistant ones to breed back. The behaviour therefore speeds up the evolution of superbugs.

CHAPTER 31

A They live longer.

B Ozone hole, people holiday in more exotic locations.

C Need for stress relief, peer pressure, addictive nature of nicotine.

D It uses more energy than a muscle at rest.

E (a) More oxygen and fuel to muscles. (b) Don't have to go to the loo as often.

F The interplay between control genes and genes that code for the LDL receptor might result in a lower risk of developing high blood cholesterol.

G (a) They were told to resume their normal lifestyle. (b) Hypertension and smoking.

H Enzymes digest adhesion proteins that hold normal cells together, making it easier for tumour cells to grow.

I Doing things that are bad for you is fun. Eating lettuce, drinking mineral water and doing lots of strenuous exercise, generally, isn't fun.

J Cost:benefit ratio is poor.

CHAPTER 33

A Temperate deciduous forest.

B Abiotic: (a), (c), (d). Biotic: (b), (e), (f).

C Habitat is where an organism lives, the term ecological niche also describes what it does and how it relates to its physical environment and to other organisms.

CHAPTER 34

A We still need to maintain body temperature, breathe, keep our hearts beating and keep cellular processes going. All this requires energy.

B False. Some heterotrophs eat other heterotrophs (eg lions eat gazelles).

C The transfer of energy from one organism to another.

E Provided – water, carbon dioxide; taken away – oxygen, glucose/sugar.

F Oxidation is the loss of electrons, reduction is gain (OILRIG).

G When a molecule is reduced it gains energy.

H More gas can be exchanged per unit time. Thin-walled, moist, good vascular network.

I Organic 'fuel' is oxidised to release energy and do useful work, and it produces heat as waste.

J It would take too long to break glucose down. ATP releases energy when one phosphate group is removed.

K The first uses a substrate as phosphate source, the second uses free phosphate and is powered by energy gained from redox reactions in the electron transport chain.

L An extra electron.

M It goes into the blood, is taken to the lungs and is breathed out.

N Because each glucose provides two pyruvates and therefore two acetyl-CoAs.

O These more metabolically active cells need more ATP to power their processes and therefore need more mitochondria (the sites of ATP synthesis in the cell).

P Active transport requires energy.

Q (a) Each one of the 20 amino acids has a different number of carbon atoms. In respiration, the reactions which break each one down produce different volumes of CO_2 and use different volumes of O_2. (b) CO_2 is produced but no O_2 is used: any value divided by 0 is infinity.

R It prevents them being eaten.

S 50 units.

T The end-products are different (though the processes are similar).

CHAPTER 35

A Nitrogen fixing bacteria in the root nodules of clover would increase the amount of nitrogen compounds in the soil. When the clover was ploughed in at the end of the year, these compounds would decompose into nitrates and would increase crop fertility. The advantage of this from an environmental point of view is that the farmer would need to use fewer artificial nitrogen fertilisers.

CHAPTER 36

A Intra: within a species. Inter: between different species.

B The oak wood.

CHAPTER 37

A Sustainable: used but continually replaced, eg solar power, hydroelectric power. Unsustainable: when used it is gone forever, eg coal deposits, oil, gas.

B Take two water samples A and B. Record the oxygen concentration of A. Incubate B in the dark at 20 °C for five days. Then measure the oxygen concentration of B. The BOD is B – A.

ACKNOWLEDGEMENTS

The publisher thanks examination boards (OCR, AQA, IBO, Edexcel, WJEC, AEB, UCLES, NEAB) for their permission to reproduce examination questions.

Every effort has been made to contact the holders of copyright material, but if any have been inadvertently overlooked the publishers will be pleased to make the necessary arrangements at the first opportunity.

Photographs

The publishers would like to thank the following for permission to reproduce photographs (T = Top, B = Bottom, C = Centre, L= Left, R = Right):

A-Z Botanical Collection/M Nimmo, 26.6R; Allsport/T Duffy, 3.24, M Powell, 7.12, A Bello, 10.A1, M Steele, 14.1, S Forster, 14.4, 34.35, B Radford, 14.7,G Mortimore, 14.A1, 14.A2; Animal Photography/Sally Anne Thompson, 19.1; Heather Angel, 33.6d, 33.11C, 34.17, 34.47, 36.23, Natural Visions/B Rogers, 33.11T, I Tait, 33.14; Aquarius 27.1; Ardea/S Gooders, 33.2L, D & K Urry, 36.6; Associated Press AP, p.322; From: the F G Banting Papers, Thomas Fisher Rare Book Library, University of Toronto, 11.4; BBC Natural History Unit/L M Stone, 18.9, D Weschler, 36.24; Biophoto Associates, 1.15, 1.16, 1.18, 1.20, 1.28, 1.30, 1.33, 1.34, 1.A2, 2.4, 4.18b&c, 4.22, 4.A1, 7.1a, 7.1b, 7.8, 8.2L, 8.3L, 8.5L, 8.6L, 11.5d, 11.7, 15.13b, 17.3b, 17.10T, 18.7a, 19.2, 21.16a, 23.1a, 23.6, 23.9, 23.10L, 23.11T, 23.13, 23.16b, 28.3, 29.18e, 30.5, 34.3CR, 34.7, 34.18, 34.42a, 34.52, 36.15L, 36.18R; www.JohnBirdsall.co.uk, 12.12a, 12.A2, 15.A3, p.362, 21.1R, p.484, 25.1; © 1992-2002 Warren Bishop, M.D, and The University of Iowa. All rights reserved, 8.2R, 8.5R; Anthony Blake Photo Library/T Hill, 6.A1a, Maximilian, p.150; From: The Discovery of Insulin by Michael Bliss, published by Paul Harris Publishing, Edinburgh, 1983, p.202; Chris Bonington Picture Library/D Scott, 9.9; Miko Boyle, 30 A2; The Three Graces by Reubens, Prado, Madrid/Bridgeman Art Library, London 3.20; Adam and Eve by Titian, Prado, Madrid/Index/Bridgeman Art Library, London, p.442; Bruce Coleman Ltd, 19.13R, R Williams, 19.13L, W Layer, 21.1L, A Purcell, 24.9CR, J Grayson, 24.9BL, F Labhardt, 24.9BR, K Taylor, 24.9CL, 24.14b, 26.14b, 26.19, 36.11, 36.13, A Bacchella, 26.9, S Krasemann, 26.17L, P Davey, 26.22, J Shaw, 34.3d, Dr F Sauer, 34.8L, D Davies, 34.8CL, J Burton, 36.16, J & D Bartlett, p.658; Bubbles Photo Library, 18.A2, I West, 7.13; Martyn Chillmaid, 9.6, 16.11; Cephas, 34.49, M Schilder, 6.11R, TOP/C Adam, 6.11B, Stockford, 34.49, A Proust, 34.49; Cleveland Museum of Natural History, Ohio, 22.3T, 26.23; Collections/A Sieveking, p.340, G Howard, 21.11a; Corbis/S Chenn, 14.5; Corbis-Bettmann/UPI, 29.2; Courtesy of Centers for Disease Control and Prevention, 29.8; Brian Culcheth, 33.3; Environmental Images/R Brook, 37.4b&c, 37.10, 37.11R, P Glendell, 37.11L, G Burns, 37.13; Vivien Fifield, 26.3; Format Photographers/J Harrison, 7.11, M Murray, 20.1; Gettyimages, 14.10, 18.1, 22.5, 31.2, p.638; GlucoWatch® Biographer (registered trademark of Cygnus Inc, Redwood City, California, USA), 6.20; Ronald Grant Archive, 24.14a; Green Moon Ltd, 12.3; Sally & Richard Greenhill Photo Library, 12.4R, 21.2, 21.3; Professor D Grierson, Nottingham University, 6.10; GSF Photo Library, 26.6L; © David Hoffman, 12.4L; Holt Studios International/P Peacock, 6.4, N Cattlin, 34.26, 34.27, 34.A1, 35.8, 37.3C&B, G Roberts, 37.1, R Anthony, 37.3T; Hulton Getty Picture Collection, 28.10; Courtesy Kobal Collection/20th Century Fox/Paramount, 12.5; Courtesy Kobal Collection/Michaels, Darren/Columbia, 22.1; Andrew Lambert, 3.4, 3.10, 3.13, 3.16, p.86T, 5.A1, 7.5, 13.2, 14.8, 31.25; Frank Lane Picture Agency/M Ranjit, 15.A1, M J Thomas, 25.7, R Wilmshurst, 26.4, 36.4, S Hosking, 26.5, E & D Hosking, 26.16, T Whittaker, 26.17R, Foto Natura Stock, 34.15, P Perry, 36.10, M Gore, 36.14BR; Lommond Mountain Rescue Team/G Baird, 12.A1; MPL International/R Adshead, 34.14; The Muscular Dystrophy Group, 25.15; NASA, p.296; Nature & Science AG/F-L Vaduz, 13.7a; NHM Photo Library, 16.8, 26.1; NHPA/M Stelzner, 6.3, Agence Nature, 6.7, B Wood, 12.6; From: Damasio H, Grabowski T, Frank R, Galaburda AM, Damasio AR: The return of Phineas Gage: Clues about the brain from a famous patient. Science, 264: 1102-1105, 1994. Dept of Neurology & Image Analysis Facility, University of Iowa, p.280; Novo Nordisk A/S, 11.A1, 11.A2; Oxford Scientific Films Ltd/D Broomhall, 10.23B, H Pooley, 12.1, M Hill, 33.1, Tui de Roy, 33.6a, S Osolinski, 33.6b, R Redfern, 33.9, D H Thompson, 33.12, D Macdonald, 33.13 G I Bernard, 34.1, W Shattil & B Rozinski, 36.12; Panos Pictures/T Page, 7.6, P Harrison, 28.1, H J Davies, 29.20, J Hammond, 29.23; Dr Colin Paterson, Ninewells Hospital, Dundee, 3.39; Planet Earth/J Scott, 14.A3, E Darack Photography, 26.14a, M Snyderman, 26.14cL, J Jackson, 26.14cR, Purdy & Matthews, 34.3T, A Bartschi, 36.14T N Garbutt, 36.14BL, M Welby, 36.5; Popperfoto, p.308, Reuter, 7.A2; Professional Sport/T Hindley, 18.A1; Punch Ltd, 28.2; Purdue News Service. Photo by David Umberger, p.28; Redferns/Michael Och's Archive, p.422; Rex Features Ltd, 3.21, 7.4, p.216, 13.1, 21.7, p.404; Roselin

Institute, Edinburgh, 23.1b&c, 27.12; Royal Holloway College, University of London, EM Unit, 1.23, 1.31; Science Photo Library, 1.A4, 1.25, 10.12, 23.5, 25.2, 29.12, 29.18b, p.518, 31.13, Moredun Animal Health Ltd, 1.7, 36.18L, D Phillips, 1.A3, 2.1, D Fawcett, 1.13, A B Dowsett, p.2, p.496, 34.43, D Scharf, 1.3, 1.4, 1.6, 9.A1, 23.3, 34.50, Dr G Murti, 1.9, 24.A1T, 27.11, Omikron, 1.11, 24.A1B, CNRI, 1.14, 4.6, 6.24, 8.4R, 10.3c, 15.10b, 16.4b, 20.6, 21.16b, 22.2, p.388B, 23.7, 28.9, S Sinclair, 2.A1T, M Devlin, 2.A1C, 16.2a, John Radcliffe Hospital, 2.A1B, BSIP Leca, 2.A3, Detlev van Ravensway, p.38, C D Winters, 3.6, Quest, 1.A1, 3.18, C Pouderas, Eurelios, 3.19, K Eward, 3.32, Professor K Seddon & Dr T Evans, Queen's University, Belfast, 3.35, B Nelson/Custom Medical Stock, p.68, GJLP, 4.6, 16.2b, O Burriel, 4.17, 7.3a&b, 7.14, J C Revy, 5.2, 14.2, 27.5, 30.11, 32.6, 34.33, W Baumeister, 5.8, A Bartel, 5.16, 22.3CR, J Mason, p.102, D Leah, 6.2, F Sauze, 6.14, Saturn Stills, 6.19, 11.2, 28.14, 30.13, H Young, 6.25, 23.16a, 28.5, 31.6, 32.1R, P Plailly, 6.26, 27.4, 31.28, Rosenfeld Images Ltd, 6.A1c, SCIMAT, 6.A1d, J King-Holmes & D Mercer, p.130, M Clarke, 7.7, 8.13, 20.16, C Pedrazzini, 7.10, Dr C Liguory/CNRI, 8.3R, J Burbridge, 8.4L, D M Martin MD, 8.6R, D Parker, 8.10, 12.10, 36.22, A Tsiaras, p.166, 20.A3, M Dohrn, Royal College of Surgeons, 9.1h, 34.30, A & H Frieder Michler, 9.5a, BSIP, Edwige, 9.7, Professor P Motta, Dept of Anatomy, University, 'La Sapienza', Rome, 9.8, 13.7b, 17.2c, 19.8, 20.4b, J Durham, 9.A2, Biophoto Associates, 10.3b, 15.3R, 31.19, G Tompkinson, 10.10, 19.7, A Crump, TDR, WHO, 10.15, 29.A4, E Reschke, Peter Arnold Inc, 10.16, E Grave, 10.20a, 29.18a, 29.18c, 36.15R, J L Martra, Publiophoto, Diffusion, 11.5a, Dr R Clark & M R Goff, 12.2, Dr P Marazzi, 12.15, 21.16c, 26.20, 30.1, 31.26, 37.9, H Morgan, p.230, 16.12, 16.15a, 20.A1, 27.6, M Chillmaid, 13.15, J King-Holmes, p.244, 22.3BL, 27.7, 27.10, p.536, M Kage, 15.3L, 19.3a, D Lovegrove, 15.9, Eye of Science, 15.13a, 29.5, 29.18d, Department of Clinical Radiology, Salisbury Hospital, 15.18, S Ford, 15.A2, 29.17, 29.22, 34.51, G Watson, 16.4a, A Syred, 17.2b, 34.16, P Plailly, Eurelios, 17.8, p.378, P Menzel, 17.12, 22.3BR, 32.7b, H C Robinson, 18.2, P Jude, 19.5, A McClenaghan, 19.15b, P M Motta, G Macchiavelli, S A Nottola, 20.8b, S Terry, 20.11, Dr Y Nikas, 20.A2, 27.14, A Pasieka, 21.14, Petit Format, 22.3CL, S Fraser, Royal Victoria Infirmary, Newcastle upon Tyne, p.388T, P M Motta & J van Blerkom, 23.2, M Sklar, 23.4, A Hart-Davis, 25.16, © 1995 Worldstat International and J Knighton, 26.A1, National Library of Medicine, 26.2, T Craddock, 26.10, 37.5, Dept of Clinical Cytogenetics, Addenbrookes Hospital, 27.2, J Berger, Max Planck Institute, 27.13, 32.1L, Deep Light Productions, 28.4, L Georgia, 28.8, A Dex, Publiphoto Diffusion, 28.11, Dr L Stannard, 28.15, P Saada, Eurelios, 28.16, H Pincis, 29.3, Peter Arnold, 29.4, M Dohrn, 29.6, S Stammers, 29.23, 29.A2, NIBSC, 29.7, 30.4, Dr V Bradbury, 30.6, Dr A Brody, 30.7, J Varney, 31.17, Dr K Lounatmaa, 30.18, J Ashton, 31.21, J Stevenson, 31.24, P Jerrican, 31.27a, Breast Screening Unit, Kings College Hospital, London, 31.27b, A Pol, CNRI, 31.A2, R Sutherland, 32.3, St Bartholomew's Hospital, 32.4, K Guldbrandsen, 32.7a, M Kulyuk, 32.9, Tom van Sant/Geosphere Project, Santa Monica, p.584, Professors P M Motta, K R Porter, P M Andrews, 34.2, J Howard, 34.3CL, © Health & Safety Laboratory, 34.31,Dr K Porter, 34.42b, S Fraser, 36.7, 37.4a, 37.7, Y Hamel, 37.4d, NOAA, 37.8; Science and Society Picture Library, 9.11; www.shoutpictures.com, p.180, 31.5; Still Pictures/N Cobbing, 6.6, E Parker, 6.17, J Schytte, 6.18, C Pye-Smith, 6.A1b, M Edwards, 13.3, p.260, F Bavendam, 33.6c, N Dickinson, 33.4, J Isaac, 34.29, H Schwarzbach, 36.A1; Photo courtesy of the College of Veterinary Medicine, Texas A&M University, p.466; C&S Thompson, 13.4; Illustration provided courtesy of UC Davies Health System, Sacramento, CA, USA, 2.3; John Walmsley Photography, 33.2R; The Wellcome Trust, 7.A1, 12.12b, 24.12, 28.6, 30.8, 31.8, p.562, 32.2; Robin Westlake, US National Marine Fisheries Service, Southwest Fisheries Science Center, p.628; Woodfall Wild Images/B Gibbons, 36.25; Photo courtesy of Eyam Parish Church/Mike Williams, 29.1; Zeneca Agrochemicals & Seeds, 6.9.

Front cover: DNA strands
Images supplied by: Tony Stone Imaging/Gettyimages

INDEX